S0-AUZ-932

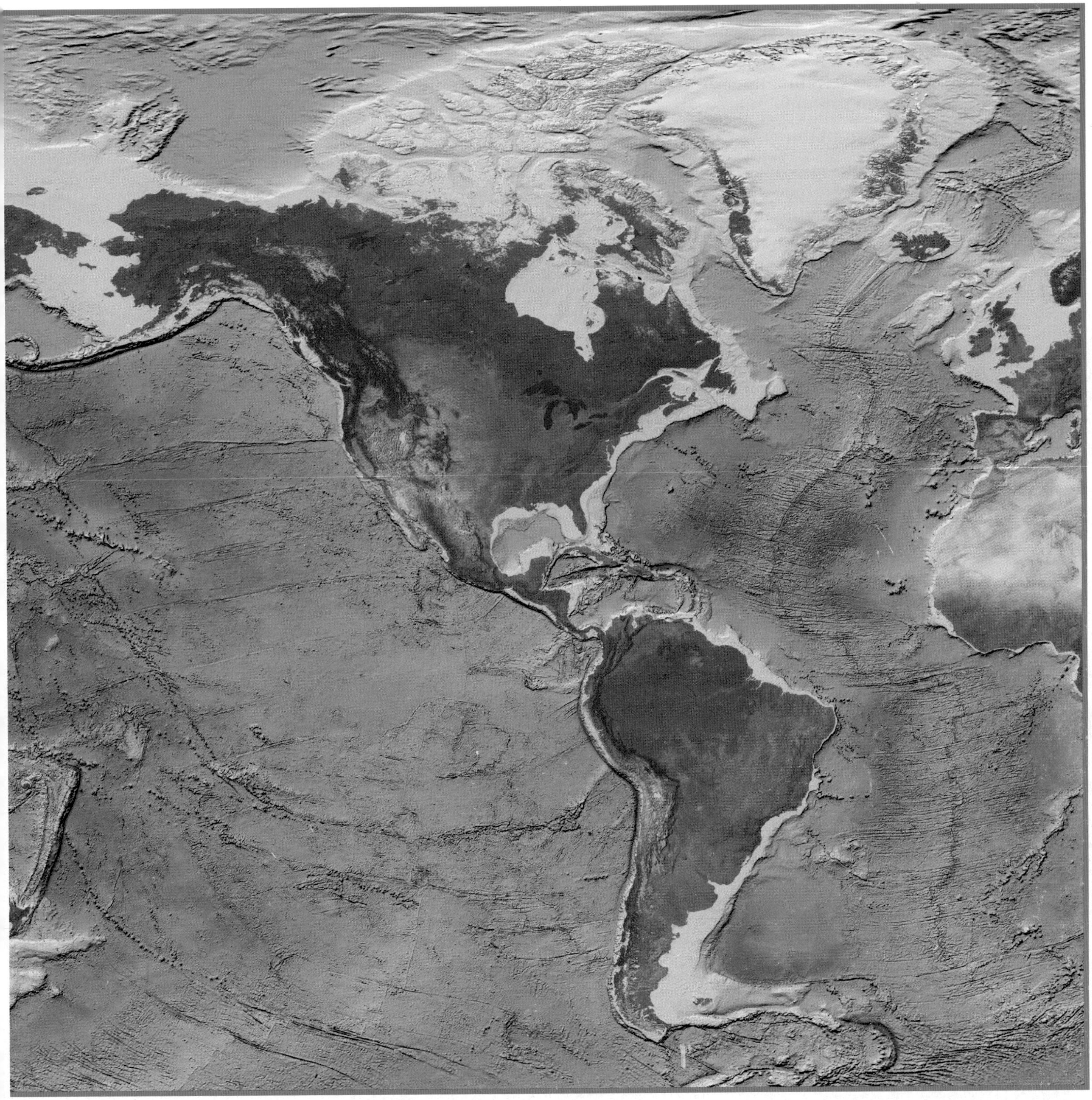

About the Cover

This photograph by *Michael Collier* shows an aerial view looking south along Comb Ridge in southwestern Utah. The San Juan River is flowing away from the viewer. The high peak in the top center of the image is the Mule Ear. We chose this photograph because it illustrates many aspects that are fundamental to geology. Hard rocks of the mountains provide an elegant example of how rocks and geologic structures control the landscape, displaying striking patterns that reflect the interaction of topography with folded orange and gray layers. Nearby, the folded layers trap petroleum in several important oil fields. Dark rounded hills to the right of the Mule Ear expose volcanic material containing blocks from different levels of Earth's crust and the upper mantle, providing us with a glimpse into Earth's subsurface. The river and its associated floodplain illustrate the importance of water and sediment to life, creating a green corridor that cuts across the dry landscape and provides habitats for plants, animals, and people.

EXPLORING GEOLOGY

Stephen J. Reynolds
Arizona State University

Julia K. Johnson
Arizona State University

Michael M. Kelly
Michael M. Kelly and Associates

Paul J. Morin
University of Minnesota
National Center for Earth-surface Dynamics

Charles M. Carter

Boston Burr Ridge, IL Dubuque, IA New York San Francisco St. Louis
Bangkok Bogotá Caracas Kuala Lumpur Lisbon London Madrid Mexico City
Milan Montreal New Delhi Santiago Seoul Singapore Sydney Taipei Toronto

McGraw-Hill Higher Education

EXPLORING GEOLOGY

Published by McGraw-Hill, a business unit of The McGraw-Hill Companies, Inc., 1221 Avenue of the Americas, New York, NY 10020.

This book is printed on acid-free paper.

1 2 3 4 5 6 7 8 9 0 DOW/DOW 0 9 8 7

ISBN 978–0–07–313515–1
MHID 0–07–313515–1

Publisher: *Thomas D. Timp*
Executive Editor: *Margaret J. Kemp*
Developmental Editor: *Liz Recker*
Senior Marketing Manager: *Lisa Nicks*
Senior Project Manager: *Gloria G. Schiesl*
Senior Production Supervisor: *Sherry L. Kane*
Media Producer: *Daniel M. Wallace*
Senior Designer: *David W. Hash*
(USE) Cover Image: *San Juan River, ©Michael Collier*
Interior Design Layout: *S. Reynolds, J. Johnson, and M. Kelly*
Senior Photo Research Coordinator: *Lori Hancock*
Photo Researcher: *Pronk and Associates*
Compositor: *Aptara*
Typeface: *9/11pt Avenir 35 Light*
Printer: *R. R. Donnelley Willard, OH*
Unless otherwise credited photographs: © *Stephen J. Reynolds*

The credits section for this book begins on page C-1 and is considered an extension of the copyright page.

Library of Congress Cataloging-in-Publication Data

Exploring geology / Stephen J. Reynolds ... [et al.]. -- 1st ed.
p. cm.
Includes index.
ISBN 978–0–07–313515–1 --- ISBN 0–07–313515–1 (hard copy : alk. paper)
1. Geology--Textbooks. I. Reynolds, Stephen J.
QE26.3.E97 2008
550--dc22

2007014243

www.mhhe.com

Brief Contents

Contents

CHAPTER 3 Plate Tectonics 52

CHAPTER 4 Earth Materials 74

CHAPTER 5

Igneous Environments 106

CHAPTER 6

Volcanoes and Volcanic Hazards 136

CHAPTER 7

CHAPTER 8

Deformation and Metamorphism 202

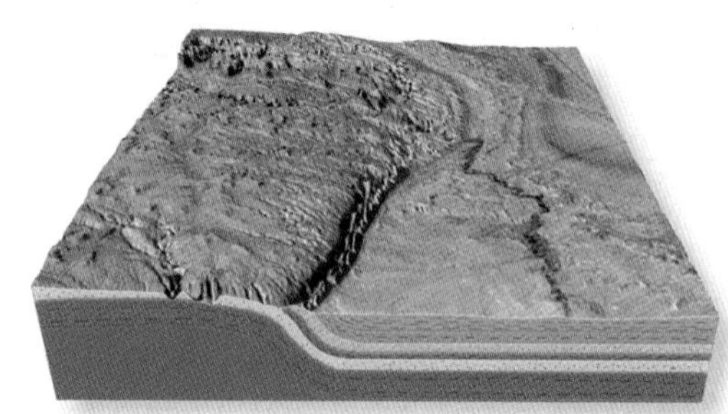

CHAPTER 9

Geologic Time 238

CHAPTER 11

Mountains, Basins, and Continents 300

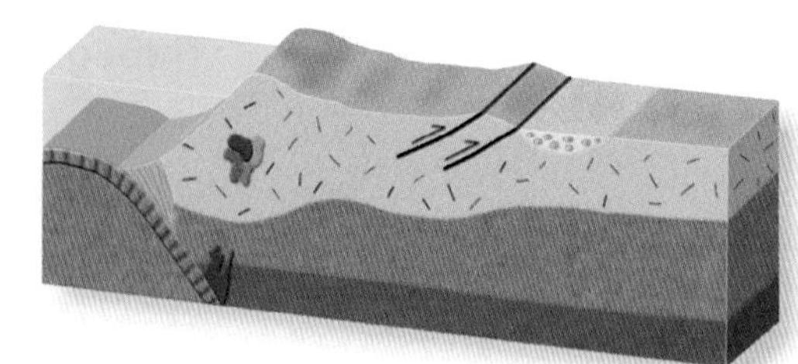

CHAPTER 12

Earthquakes and Earth's Interior 330

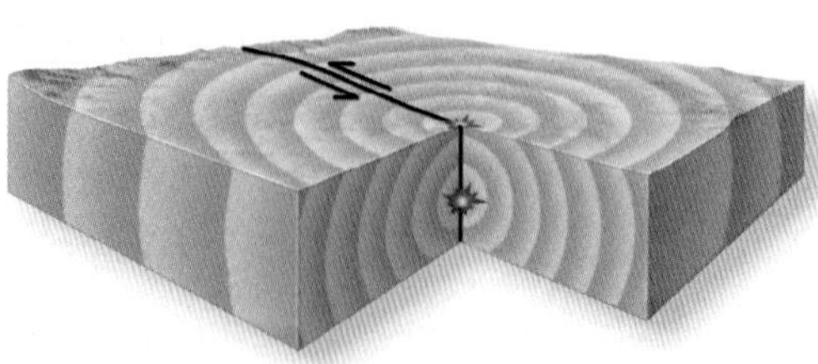

CHAPTER 13 Climate, Weather, and Their Influences on Geology 366

CHAPTER 14

Shorelines, Glaciers, and Changing Sea Levels 398

CHAPTER 15

Weathering, Soil, and Unstable Slopes 434

CHAPTER 17

Water Resources 498

CHAPTER 18

Energy and Mineral Resources 524

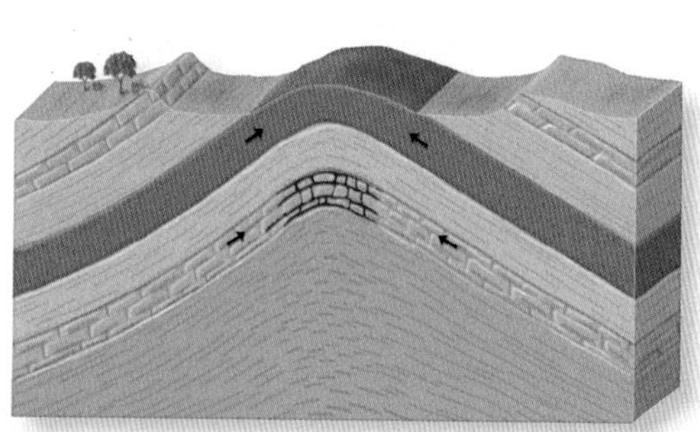

How Is This Textbook Different Than Other Physical Geology Textbooks?

AS YOU EXAMINE *EXPLORING GEOLOGY* you will notice that it is different from other textbooks. You might ask: Why do most pages have few large blocks of text? Why are there nearly 2,700 illustrations when most introductory geology textbooks have less than 1,000 illustrations? Why do the authors focus on illustrations? In addition to answering these questions, the Preface will explain the text's approach, how a typical chapter is organized, and how you can most efficiently use this textbook for learning geologic concepts and scientific inquiry.

A Why Is the Book Designed Around Figures?

How do we learn new things? Scientists who study thinking and learning speak about two types of memory, working memory and long-term memory. *Working memory*, also called *short-term memory*, holds information that our minds are actively processing, whereas *long-term memory* is like a storehouse in which we file information until we need it. The amount of knowledge we retain depends on transferring information from *working memory* to *long-term memory* and on linking the new information with our existing mental framework.

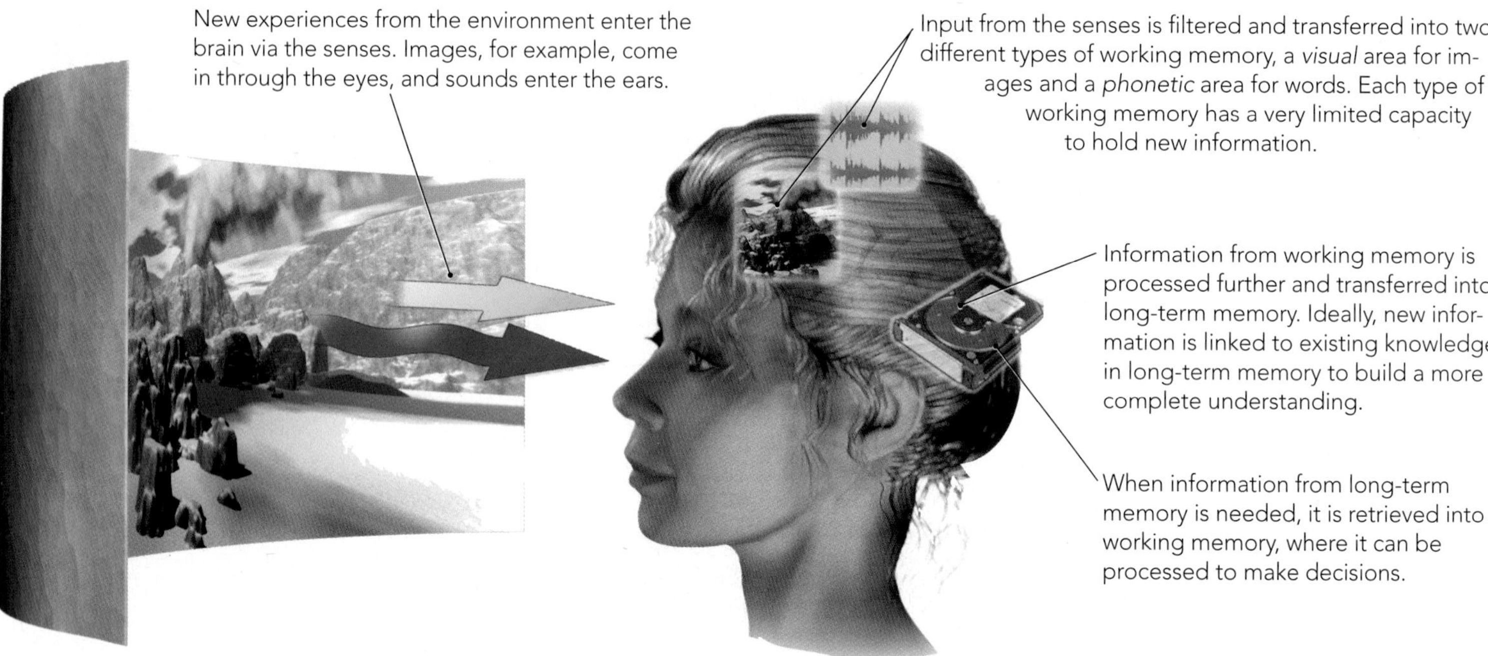

To utilize both types of working memory and to help link together visual and text information, most information in this book is presented as central figures that are tightly integrated with small blocks of text. Examine this figure and the associated text to see how easily your mind makes sense of the information. ▷

Rocky hills and mountains produce rocks that are large and angular.

As gravity and streams transport the rocks down and away from the mountains, the corners of the rocks are worn away and so the rocks become rounder and smaller.

With enough transport, the rocks become small and well rounded. This process explains why stones in a riverbed commonly have rounded shapes.

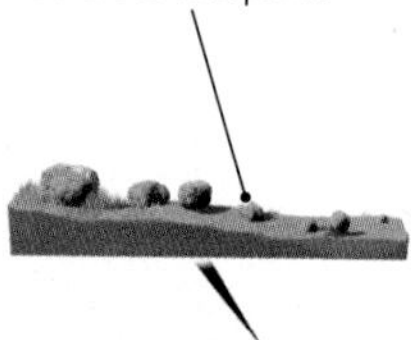

B What Is the Style of the Book?

This book is a new type of textbook intended for an introductory college geology course, such as Physical Geology. The book consists entirely of *two-page spreads* organized in chapters. Each two-page spread, like the one shown below, focuses on one or more important geologic concepts or approaches to geologic problems. These spreads help students learn and organize geologic knowledge in a new and exciting way.

Each two-page spread is a self-contained block of information about a specific topic. Each spread has a unique number, such as 5.6 for the 6th topical two-page spread in chapter 5.

The title of each spread is a *question* about an important aspect of geology, a way we study geologic problems, or the geology of an interesting place.

Most two-page spreads are subdivided into sections, each of which is a coherent piece of the larger topic. The sections guide students through the pages in an easy-to-follow way and provide a clear break for students to consolidate their knowledge before moving on.

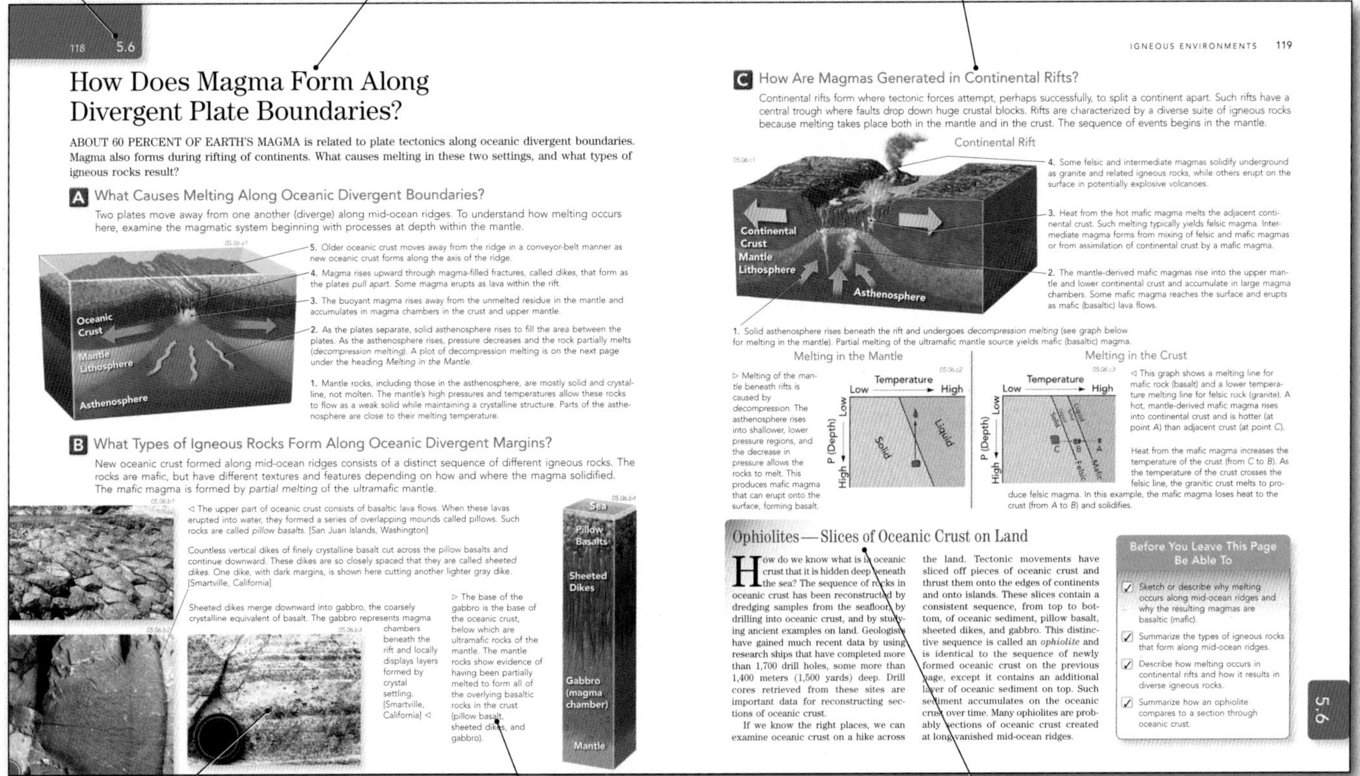

118 5.6

How Does Magma Form Along Divergent Plate Boundaries?

ABOUT 60 PERCENT OF EARTH'S MAGMA is related to plate tectonics along oceanic divergent boundaries. Magma also forms during rifting of continents. What causes melting in these two settings, and what types of igneous rocks result?

A What Causes Melting Along Oceanic Divergent Boundaries?

Two plates move away from one another (diverge) along mid-ocean ridges. To understand how melting occurs here, examine the magmatic system beginning with processes at depth within the mantle.

5. Older oceanic crust moves away from the ridge in a conveyor-belt manner as new oceanic crust forms along the axis of the ridge.

4. Magma rises upward through magma-filled fractures, called *dikes*, that form as the plates *pull apart*. Some magma erupts as lava within the rift.

3. The buoyant magma rises away from the unmelted residue in the mantle and accumulates in magma chambers in the crust and upper mantle.

2. As the plates separate, solid asthenosphere rises to fill the area between the plates. As the asthenosphere rises, pressure decreases and the rock partially melts (*decompression melting*). A plot of decompression melting is on the next page under the heading *Melting in the Mantle*.

1. Mantle rocks, including those in the asthenosphere, are mostly solid and crystalline, not molten. The mantle's high pressures and temperatures allow these rocks to flow as a weak solid while maintaining a crystalline structure. Parts of the asthenosphere are close to their melting temperature.

B What Types of Igneous Rocks Form Along Oceanic Divergent Margins?

New oceanic crust formed along mid-ocean ridges consists of a distinct sequence of different igneous rocks. The rocks are mafic, but have different textures and features depending on how and where the magma solidified. The *mafic* magma is formed by *partial melting* of the *ultramafic* mantle.

◁ The upper part of oceanic crust consists of basaltic lava flows. When these lavas erupted into water, they formed a series of overlapping mounds called pillows. Such rocks are called *pillow basalts*. [San Juan Islands, Washington]

Countless vertical dikes of finely crystalline basalt cut across the pillow basalts and continue downward. These dikes are so closely spaced that they are called *sheeted dikes*. One dike, with dark margins, is shown here cutting another lighter gray dike. [Smartville, California]

Sheeted dikes merge downward into gabbro, the coarsely crystalline equivalent of basalt. The gabbro represents magma chambers beneath the rift and locally displays layers formed by crystal settling. [Smartville, California] ◁

▷ The base of the gabbro is the base of the oceanic crust, below which are ultramafic rocks of the mantle. The mantle rocks show evidence of having been partially melted to form all of the overlying basaltic rocks in the crust (pillow basalt, sheeted dikes, and gabbro).

IGNEOUS ENVIRONMENTS 119

C How Are Magmas Generated in Continental Rifts?

Continental rifts form where tectonic forces attempt, perhaps successfully, to split a continent apart. Such rifts have a central trough where faults drop down huge crustal blocks. Rifts are characterized by a diverse suite of igneous rocks because melting takes place both in the mantle and in the crust. The sequence of events begins in the mantle.

Continental Rift

4. Some felsic and intermediate magmas solidify underground as granite and related igneous rocks, while others erupt on the surface in potentially explosive volcanoes.

3. Heat from the hot mafic magma melts the adjacent continental crust. Such melting typically yields felsic magma. Intermediate magma forms from mixing of felsic and mafic magmas or from assimilation of continental crust by a mafic magma.

2. The mantle-derived mafic magmas rise into the upper mantle and lower continental crust and accumulate in large magma chambers. Some mafic magma reaches the surface and erupts as mafic (basaltic) lava flows.

1. Solid asthenosphere rises beneath the rift and undergoes *decompression melting* (see graph below for melting in the mantle). Partial melting of the ultramafic mantle source yields mafic (basaltic) magma.

Melting in the Mantle

▷ Melting of the *mantle* beneath rifts is caused by *decompression*. The asthenosphere rises into shallower, lower pressure regions, and the decrease in pressure allows the rocks to melt. This produces mafic magma that can erupt onto the surface, forming basalt.

Melting in the Crust

◁ This graph shows a melting line for mafic rock (basalt) and a lower temperature melting line for felsic rock (granite). A hot, mantle-derived mafic magma rises into continental crust and is hotter (at point *A*) than adjacent crust (at point *C*).

Heat from the mafic magma increases the temperature of the crust (from *C* to *B*). As the temperature of the crust crosses the felsic line, the granitic crust melts to produce felsic magma. In this example, the mafic magma loses heat to the crust (from *A* to *B*) and solidifies.

Ophiolites—Slices of Oceanic Crust on Land

How do we know what is in oceanic crust that it is hidden deep beneath the sea? The sequence of rocks in oceanic crust has been reconstructed by dredging samples from the seafloor, by drilling into oceanic crust, and by studying ancient examples on land. Geologists have gained much recent data by using research ships that have completed more than 1,700 drill holes, some more than 1,400 meters (1,500 yards) deep. Drill cores retrieved from these sites are important data for reconstructing sections of oceanic crust.

If we know the right places, we can examine oceanic crust on a hike across the land. Tectonic movements have sliced off pieces of oceanic crust and thrust them onto the edges of continents and onto islands. These slices contain a consistent sequence, from top to bottom, of oceanic sediment, pillow basalt, sheeted dikes, and gabbro. This distinctive sequence is called an *ophiolite* and is identical to the sequence of newly formed oceanic crust on the previous page, except it contains an additional layer of oceanic sediment on top. Such sediment accumulates on the oceanic crust over time. Many ophiolites are probably sections of oceanic crust created at long vanished mid-ocean ridges.

Before You Leave This Page Be Able To

- ✓ Sketch or describe why melting occurs along mid-ocean ridges and why the resulting magmas are basaltic (mafic).
- ✓ Summarize the types of igneous rocks that form along mid-ocean ridges.
- ✓ Describe how melting occurs in continental rifts and how it results in diverse igneous rocks.
- ✓ Summarize how an ophiolite compares to a section through oceanic crust.

5.6

The textbook is built around large and small illustrations that convey knowledge in a visual way.

Short blocks of text, which label and explain key features one at a time and in complete sentences, accompany each figure. This approach allows our minds to better integrate the various types of information and to envision how the different parts and processes are related. The figures are designed so that key points are described in a logical order.

Most two-page spreads contain a major block of text at the end of the spread. This text block provides a more in-depth discussion of some important and interesting aspect of the topic.

Our Goals for This Book

Geology may be the only science course students take in college, or it may be one of a number of science classes that they take. Whatever the circumstance, we designed, wrote, and illustrated this book with several clear goals in mind. One major goal is to help students become better at observing the world around them and at understanding how geology controls so much of what they see in landscapes. Another goal is to help students think logically about the natural world and to help them know how to study natural phenomena using a scientific approach. We also want to help students appreciate how geology is relevant to their life, on a local and a global scale, and how a knowledge of geology can help them avoid natural hazards, such as flooding and landslides. Finally, we, like most geologists, love traveling across and studying different parts of our planet. We want to share our excitement for how geology shapes our world and how geologists contribute to modern society.

How Is Each Chapter Organized?

EACH CHAPTER INCLUDES FOUR TYPES of two-page spreads: a chapter *opening* spread, a number of *topical* spreads, a chapter *application* spread, and a chapter-closing *investigation* spread. Each type of spread has its own distinguishing characteristics and was designed to be approached in a specific way.

How Does Each Chapter Begin?

Chapter-*opening* two-page spreads focus on an interesting place that illustrates the main aspects of the chapter and why the information covered is relevant to society.

The opening paragraph introduces some central ideas from the chapter and previews some main issues related to this topic.

Chapter-opening spreads are generally organized around one large figure, such as a map or computer-generated satellite image.

A list of topics helps students navigate the chapter and anticipate which topics will be covered.

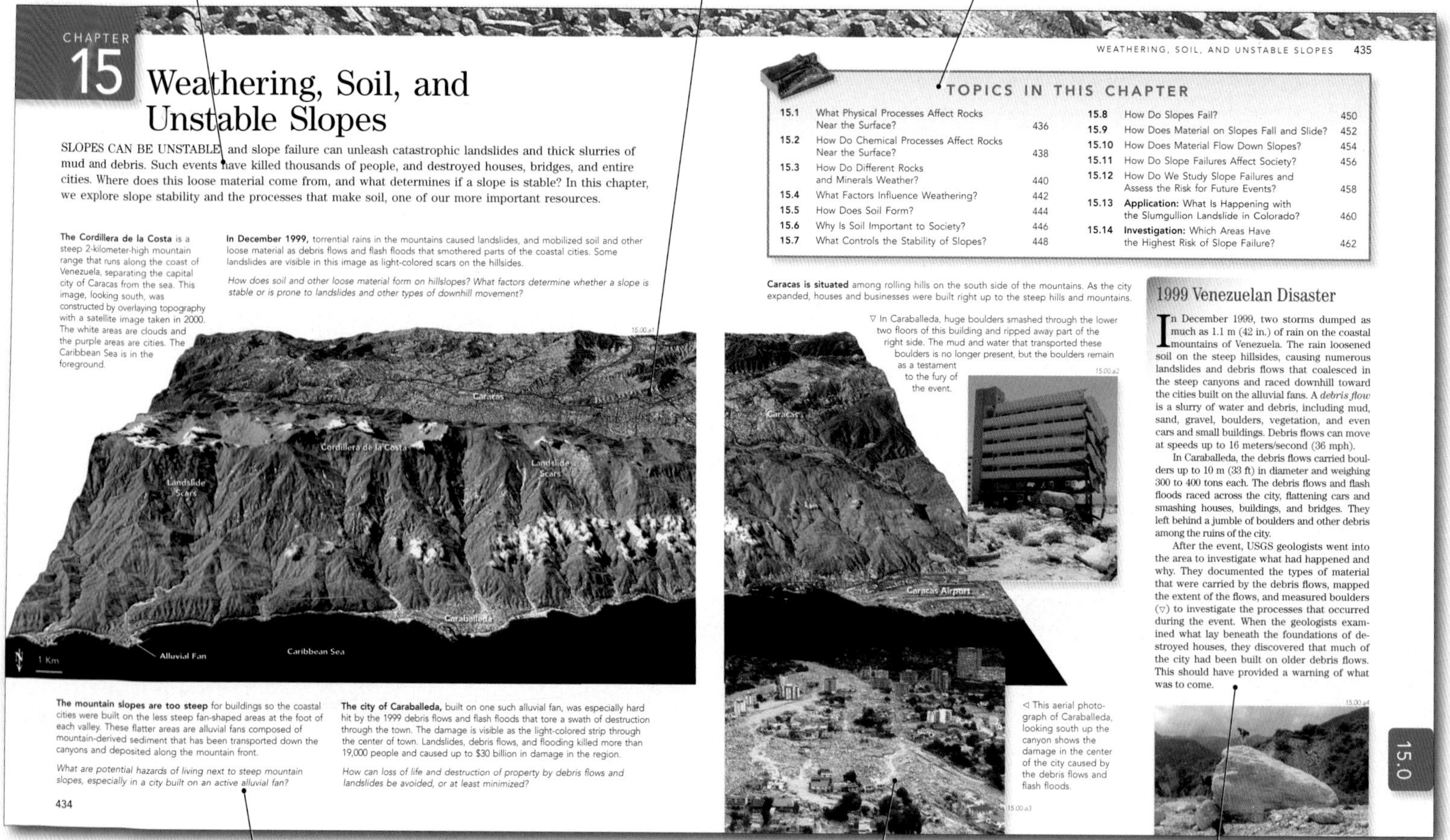

CHAPTER 15

Weathering, Soil, and Unstable Slopes

SLOPES CAN BE UNSTABLE and slope failure can unleash catastrophic landslides and thick slurries of mud and debris. Such events have killed thousands of people, and destroyed houses, bridges, and entire cities. Where does this loose material come from, and what determines if a slope is stable? In this chapter, we explore slope stability and the processes that make soil, one of our more important resources.

The Cordillera de la Costa is a steep 2-kilometer-high mountain range that runs along the coast of Venezuela, separating the capital city of Caracas from the sea. This image, looking south, was constructed by overlaying topography with a satellite image taken in 2000. The white areas are clouds and the purple areas are cities. The Caribbean Sea is in the foreground.

In December 1999, torrential rains in the mountains caused landslides, and mobilized soil and other loose material as debris flows and flash floods that smothered parts of the coastal cities. Some landslides are visible in this image as light-colored scars on the hillsides.

How does soil and other loose material form on hillslopes? What factors determine whether a slope is stable or is prone to landslides and other types of downhill movement?

15.00.a1

Caracas
Cordillera de la Costa
Landslide Scars
Landslide Scars
Caraballeda
Alluvial Fan
Caribbean Sea
1 Km

The mountain slopes are too steep for buildings so the coastal cities were built on the less steep fan-shaped areas at the foot of each valley. These flatter areas are alluvial fans composed of mountain-derived sediment that has been transported down the canyons and deposited along the mountain front.

What are potential hazards of living next to steep mountain slopes, especially in a city built on an active alluvial fan?

The city of Caraballeda, built on one such alluvial fan, was especially hard hit by the 1999 debris flows and flash floods that tore a swath of destruction through the town. The damage is visible as the light-colored strip through the center of town. Landslides, debris flows, and flooding killed more than 19,000 people and caused up to $30 billion in damage in the region.

How can loss of life and destruction of property by debris flows and landslides be avoided, or at least minimized?

434

WEATHERING, SOIL, AND UNSTABLE SLOPES 435

TOPICS IN THIS CHAPTER

Caracas is situated among rolling hills on the south side of the mountains. As the city expanded, houses and businesses were built right up to the steep hills and mountains.

▽ In Caraballeda, huge boulders smashed through the lower two floors of this building and ripped away part of the right side. The mud and water that transported these boulders is no longer present, but the boulders remain as a testament to the fury of the event.

15.00.a2

Caracas
Caracas Airport

◁ This aerial photograph of Caraballeda, looking south up the canyon shows the damage in the center of the city caused by the debris flows and flash floods.

15.00.a3

1999 Venezuelan Disaster

In December 1999, two storms dumped as much as 1.1 m (42 in.) of rain on the coastal mountains of Venezuela. The rain loosened soil on the steep hillsides, causing numerous landslides and debris flows that coalesced in the steep canyons and raced downhill toward the cities built on the alluvial fans. A *debris flow* is a slurry of water and debris, including mud, sand, gravel, boulders, vegetation, and even cars and small buildings. Debris flows can move at speeds up to 16 meters/second (36 mph).

In Caraballeda, the debris flows carried boulders up to 10 m (33 ft) in diameter and weighing 300 to 400 tons each. The debris flows and flash floods raced across the city, flattening cars and smashing houses, buildings, and bridges. They left behind a jumble of boulders and other debris among the ruins of the city.

After the event, USGS geologists went into the area to investigate what had happened and why. They documented the types of material that were carried by the debris flows, mapped the extent of the flows, and measured boulders (▽) to investigate the processes that occurred during the event. When the geologists examined what lay beneath the foundations of destroyed houses, they discovered that much of the city had been built on older debris flows. This should have provided a warning of what was to come.

15.00.a4

15.0

The text blocks highlight important attributes of this locality or of the geologic phenomena featured in the chapter. These text blocks include questions that a student should be able to answer by the end of the chapter.

Smaller photographs and figures address different features or manifestations of the geologic processes that are explored further within the chapter.

A major text block provides a more detailed narrative about some interesting aspect of the region, such as the important geologic events that affected the region, its cities, and its people.

B What Do Topical Two-Page Spreads Include?

Topical two-page spreads are the foundation of the book. Each spread is about one or more closely related topics. Topical spreads convey the geologic content and help organize knowledge. Each chapter contains at least one two-page spread illustrating how geology impacts society and another two-page spread that specifically describes how we study geologic problems.

Topical two-page spreads begin with a question and then explore that question within the rest of the two-page spread.

Information is conveyed by the illustrations, photographs, and associated text blocks. The blocks are numbered if they should be read in a specific order.

This book contains many unique illustrations designed to help convey the content, to explain different types of geologic figures, and to provide a visual representation for calculations.

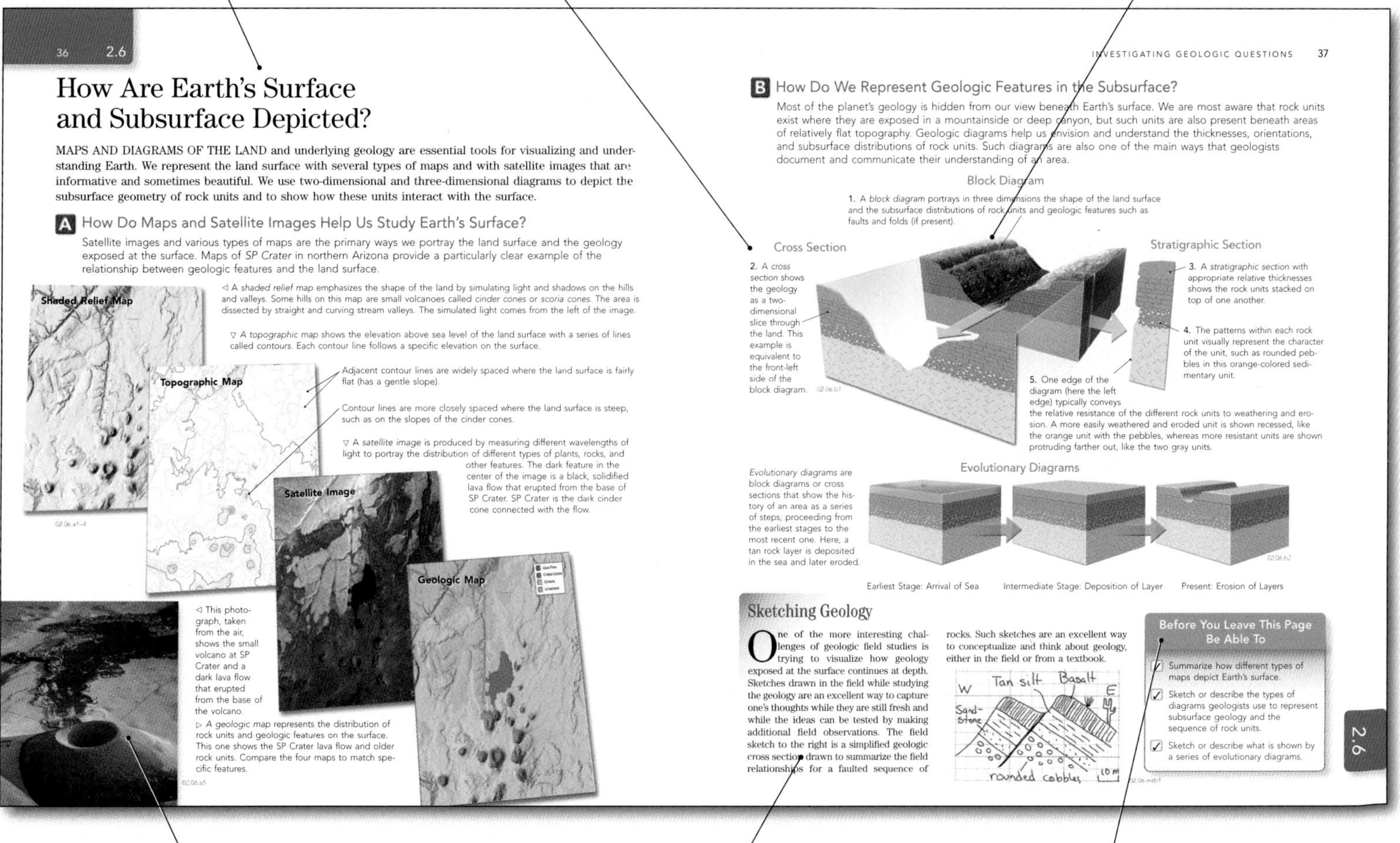

36 2.6

How Are Earth's Surface and Subsurface Depicted?

MAPS AND DIAGRAMS OF THE LAND and underlying geology are essential tools for visualizing and understanding Earth. We represent the land surface with several types of maps and with satellite images that are informative and sometimes beautiful. We use two-dimensional and three-dimensional diagrams to depict the subsurface geometry of rock units and to show how these units interact with the surface.

A How Do Maps and Satellite Images Help Us Study Earth's Surface?

Satellite images and various types of maps are the primary ways we portray the land surface and the geology exposed at the surface. Maps of *SP Crater* in northern Arizona provide a particularly clear example of the relationship between geologic features and the land surface.

◁ A *shaded relief map* emphasizes the shape of the land by simulating light and shadows on the hills and valleys. Some hills on this map are small volcanoes called *cinder cones or scoria cones.* The area is dissected by straight and curving stream valleys. The simulated light comes from the left of the image.

▽ *A topographic map* shows the elevation above sea level of the land surface with a series of lines called *contours.* Each contour line follows a specific elevation on the surface.

Adjacent contour lines are widely spaced where the land surface is fairly flat (has a gentle slope).

Contour lines are more closely spaced where the land surface is steep, such as on the slopes of the cinder cones.

▽ A *satellite image* is produced by measuring different wavelengths of light to portray the distribution of different types of plants, rocks, and other features. The dark feature in the center of the image is a black, solidified lava flow that erupted from the base of SP Crater. SP Crater is the dark cinder cone connected with the flow.

02.06.a1–4

◁ This photograph, taken from the air, shows the small volcano at SP Crater and a dark lava flow that erupted from the base of the volcano.

▷ *A geologic map* represents the distribution of rock units and geologic features on the surface. This one shows the SP Crater lava flow and older rock units. Compare the four maps to match specific features.

02.06.a5

INVESTIGATING GEOLOGIC QUESTIONS 37

B How Do We Represent Geologic Features in the Subsurface?

Most of the planet's geology is hidden from our view beneath Earth's surface. We are most aware that rock units exist where they are exposed in a mountainside or deep canyon, but such units are also present beneath areas of relatively flat topography. Geologic diagrams help us envision and understand the thicknesses, orientations, and subsurface distributions of rock units. Such diagrams are also one of the main ways that geologists document and communicate their understanding of an area.

Block Diagram

1. A *block diagram* portrays in three dimensions the shape of the land surface and the subsurface distributions of rock units and geologic features such as faults and folds (if present).

Cross Section

2. A *cross section* shows the geology as a two-dimensional slice through the land. This example is equivalent to the front-left side of the block diagram. 02.06.b1

Stratigraphic Section

3. A *stratigraphic section* with appropriate relative thicknesses shows the rock units stacked on top of one another.

4. The patterns within each rock unit visually represent the character of the unit, such as rounded pebbles in this orange-colored sedimentary unit.

5. One edge of the diagram (here the left edge) typically conveys the relative resistance of the different rock units to weathering and erosion. A more easily weathered and eroded unit is shown recessed, like the orange unit with the pebbles, whereas more resistant units are shown protruding farther out, like the two gray units.

Evolutionary Diagrams

Evolutionary diagrams are block diagrams or cross sections that show the history of an area as a series of steps, proceeding from the earliest stages to the most recent one. Here, a tan rock layer is deposited in the sea and later eroded.

Earliest Stage: Arrival of Sea — Intermediate Stage: Deposition of Layer — Present: Erosion of Layers

02.06.b2

Sketching Geology

One of the more interesting challenges of geologic field studies is trying to visualize how geology exposed at the surface continues at depth. Sketches drawn in the field while studying the geology are an excellent way to capture one's thoughts while they are still fresh and while the ideas can be tested by making additional field observations. The field sketch to the right is a simplified geologic cross section drawn to summarize the field relationships for a faulted sequence of rocks. Such sketches are an excellent way to conceptualize and think about geology, either in the field or from a textbook.

02.06.mtb1

Before You Leave This Page Be Able To

- ✓ Summarize how different types of maps depict Earth's surface.
- ✓ Sketch or describe the types of diagrams geologists use to represent subsurface geology and the sequence of rock units.
- ✓ Sketch or describe what is shown by a series of evolutionary diagrams.

2.6

Photographs show examples of the features and processes being examined.

Most topical two-page spreads include a major text block that highlights important details or examples of the geologic processes.

A *Before You Leave This Page* list indicates what is important and what students are expected to understand or be able to do before they move on. This list contains learning objectives for the spread. Test questions are tightly articulated with this list.

C What Is in the Chapter Application Two-Page Spread?

The next-to-last two-page spread in each chapter is an *application spread*, which is designed to help students integrate the various concepts from the chapter and to show how these concepts can be applied to an actual location. The application spread also serves as preparation for the following *investigation* two-page spread.

Application spreads are about real places that nicely illustrate the geologic concepts and features covered in the chapter.

Many application spreads explicitly illustrate how a type of geologic problem is investigated and how geologic processes have relevance to society.

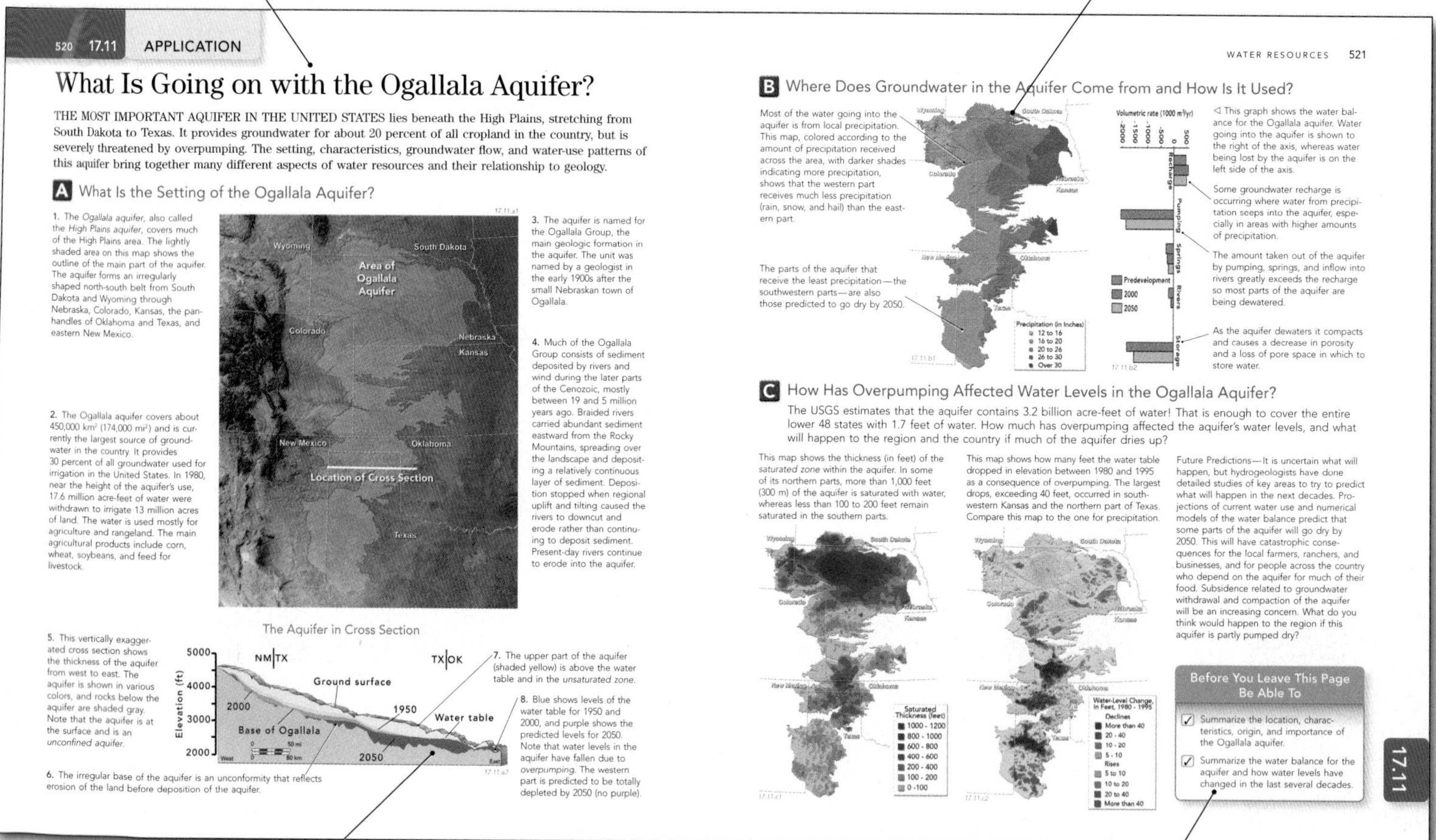

520 17.11 APPLICATION

What Is Going on with the Ogallala Aquifer?

THE MOST IMPORTANT AQUIFER IN THE UNITED STATES lies beneath the High Plains, stretching from South Dakota to Texas. It provides groundwater for about 20 percent of all cropland in the country, but is severely threatened by overpumping. The setting, characteristics, groundwater flow, and water-use patterns of this aquifer bring together many different aspects of water resources and their relationship to geology.

A What Is the Setting of the Ogallala Aquifer?

1. The *Ogallala aquifer*, also called the *High Plains aquifer*, covers much of the High Plains area. The lightly shaded area on this map shows the outline of the main part of the aquifer. The aquifer forms an irregularly shaped north-south belt from South Dakota and Wyoming through Nebraska, Colorado, Kansas, the panhandles of Oklahoma and Texas, and eastern New Mexico.

2. The Ogallala aquifer covers about 450,000 km^2 (174,000 mi^2) and is currently the largest source of groundwater in the country. It provides 30 percent of all groundwater used for irrigation in the United States. In 1980, near the height of the aquifer's use, 17.6 million acre-feet of water were withdrawn to irrigate 13 million acres of land. The water is used mostly for agriculture and rangeland. The main agricultural products include corn, wheat, soybeans, and feed for livestock.

3. The aquifer is named for the Ogallala Group, the main geologic formation in the aquifer. The unit was named by a geologist in the early 1900s after the small Nebraskan town of Ogallala.

4. Much of the Ogallala Group consists of sediment deposited by rivers and wind during the later parts of the Cenozoic, mostly between 19 and 5 million years ago. Braided rivers carried abundant sediment eastward from the Rocky Mountains, spreading over the landscape and depositing a relatively continuous layer of sediment. Deposition stopped when regional uplift and tilting caused the rivers to downcut and erode rather than continuing to deposit sediment. Present-day rivers continue to erode into the aquifer.

17.11.a1

The Aquifer in Cross Section

5. This vertically exaggerated cross section shows the thickness of the aquifer from west to east. The aquifer is shown in various colors, and rocks below the aquifer are shaded gray. Note that the aquifer is at the surface and is an *unconfined aquifer*.

6. The irregular base of the aquifer is an unconformity that reflects erosion of the land before deposition of the aquifer.

7. The upper part of the aquifer (shaded yellow) is above the water table and in the *unsaturated zone*.

8. Blue shows levels of the water table for 1950 and 2000, and purple shows the predicted levels for 2050. Note that water levels in the aquifer have fallen due to *overpumping*. The western part is predicted to be totally depleted by 2050 (no purple).

17.11.a2

WATER RESOURCES 521

B Where Does Groundwater in the Aquifer Come from and How Is It Used?

Most of the water going into the aquifer is from local precipitation. This map, colored according to the amount of precipitation received across the area, with darker shades indicating more precipitation, shows that the western part receives much less precipitation (rain, snow, and hail) than the eastern part.

The parts of the aquifer that receive the least precipitation—the southwestern parts—are also those predicted to go dry by 2050.

17.11.b1

17.11.b2

◁ This graph shows the water balance for the Ogallala aquifer. Water going into the aquifer is shown to the right of the axis, whereas water being lost by the aquifer is on the left side of the axis.

Some groundwater recharge is occurring where water from precipitation seeps into the aquifer, especially in areas with higher amounts of precipitation.

The amount taken out of the aquifer by pumping, springs, and inflow into rivers greatly exceeds the recharge so most parts of the aquifer are being dewatered.

As the aquifer dewaters it compacts and causes a decrease in porosity and a loss of pore space in which to store water.

C How Has Overpumping Affected Water Levels in the Ogallala Aquifer?

The USGS estimates that the aquifer contains 3.2 billion acre-feet of water! That is enough to cover the entire lower 48 states with 1.7 feet of water. How much has overpumping affected the aquifer's water levels, and what will happen to the region and the country if much of the aquifer dries up?

This map shows the thickness (in feet) of the *saturated zone* within the aquifer. In some of its northern parts, more than 1,000 feet (300 m) of the aquifer is saturated with water, whereas less than 100 to 200 feet remain saturated in the southern parts.

17.11.c1

This map shows how many feet the water table dropped in elevation between 1980 and 1995 as a consequence of overpumping. The largest drops, exceeding 40 feet, occurred in southwestern Kansas and the northern part of Texas. Compare this map to the one for precipitation.

17.11.c2

Future Predictions—It is uncertain what will happen, but hydrogeologists have done detailed studies of key areas to try to predict what will happen in the next decades. Projections of current water use and numerical models of the water balance predict that some parts of the aquifer will go dry by 2050. This will have catastrophic consequences for the local farmers, ranchers, and businesses, and for people across the country who depend on the aquifer for much of their food. Subsidence related to groundwater withdrawal and compaction of the aquifer will be an increasing concern. What do you think would happen to the region if this aquifer is partly pumped dry?

Before You Leave This Page Be Able To

- ✓ Summarize the location, characteristics, origin, and importance of the Ogallala aquifer.
- ✓ Summarize the water balance for the aquifer and how water levels have changed in the last several decades.

17.11

The locality is explored from various perspectives, using maps, photographs, and other figures to show how different topics in the chapter relate to one another and to the larger picture.

Some application spreads do not have a major text block, but they do have a *Before You Leave This Page* list, indicating what a student should be able to do before beginning the investigation.

D How Does Each Chapter End?

Each chapter ends with an *investigation* two-page spread that is an exercise in which students apply the knowledge, skills, and approaches learned in the chapter. These exercises commonly involve virtual places that students will explore and investigate to answer a series of geologic questions. They are modeled after the types of problems geologists investigate and use the same kinds of data and illustrations encountered in the chapter.

A list of goals for the exercise indicates what students will need to do to complete the investigation exercise. The exercise assumes that students have mastered the content and approaches described in previous parts of the chapter.

An instructor might do the investigation as an in-class exercise, use it to assess what students have learned, assign it as homework, or integrate it into the laboratory.

Many investigation exercises involve (a) making observations from photographs, data, or rock samples, (b) completing well-explained calculations, or (c) constructing graphs, maps, or cross sections.

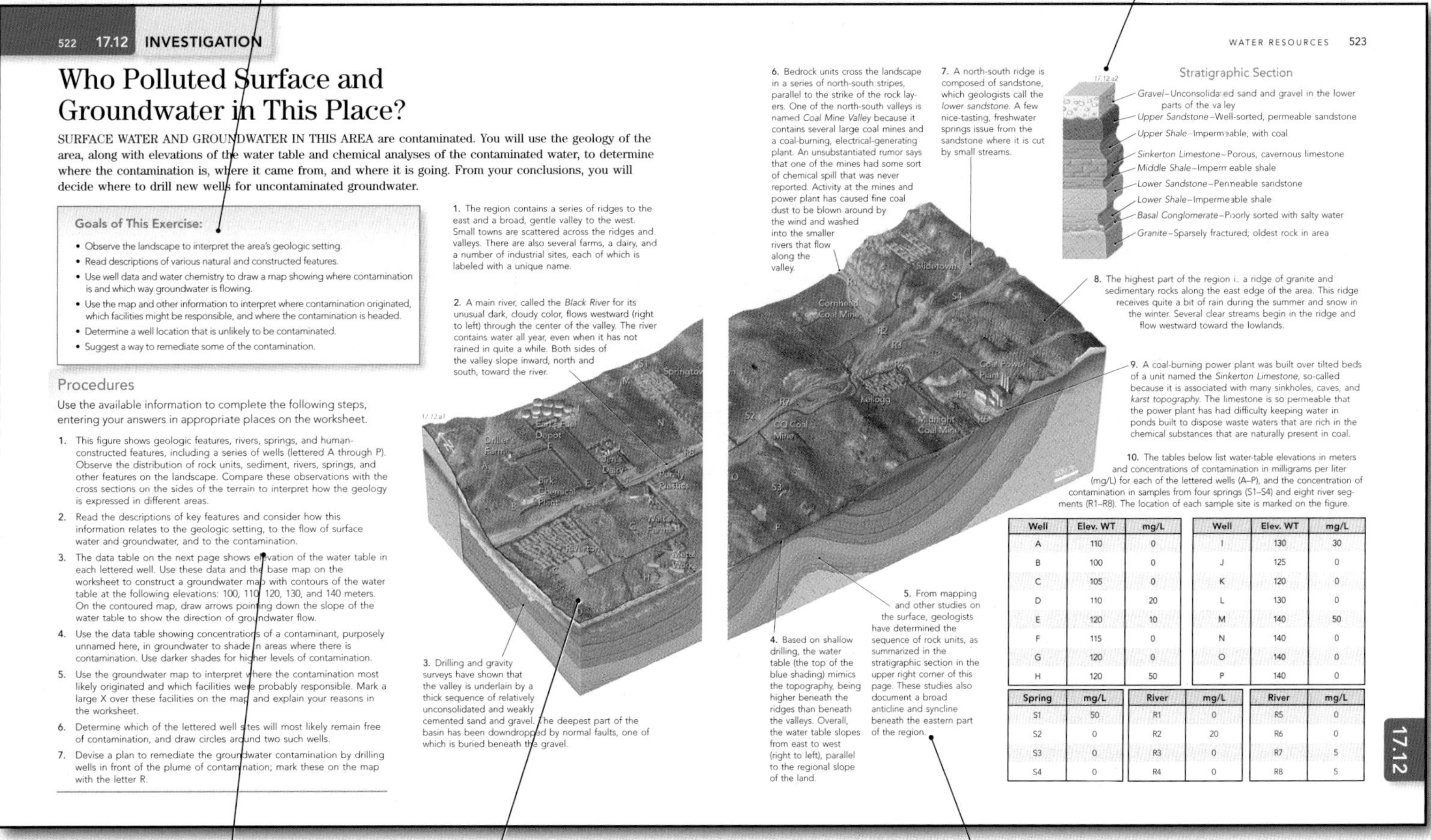

522 17.12 INVESTIGATION

Who Polluted Surface and Groundwater in This Place?

SURFACE WATER AND GROUNDWATER IN THIS AREA are contaminated. You will use the geology of the area, along with elevations of the water table and chemical analyses of the contaminated water, to determine where the contamination is, where it came from, and where it is going. From your conclusions, you will decide where to drill new wells for uncontaminated groundwater.

Goals of This Exercise:

- Observe the landscape to interpret the area's geologic setting.
- Read descriptions of various natural and constructed features.
- Use well data and water chemistry to draw a map showing where contamination is and which way groundwater is flowing.
- Use the map and other information to interpret where contamination originated, which facilities might be responsible, and where the contamination is headed.
- Determine a well location that is unlikely to be contaminated.
- Suggest a way to remediate some of the contamination.

Procedures

Use the available information to complete the following steps, entering your answers in appropriate places on the worksheet.

1. This figure shows geologic features, rivers, springs, and human-constructed features, including a series of wells (lettered A through P). Observe the distribution of rock units, sediment, rivers, springs, and other features on the landscape. Compare these observations with the cross sections on the sides of the terrain to interpret how the geology is expressed in different areas.
2. Read the descriptions of key features and consider how this information relates to the geologic setting, to the flow of surface water and groundwater, and to the contamination.
3. The data table on the next page shows elevation of the water table in each lettered well. Use these data and the base map on the worksheet to construct a groundwater map with contours of the water table at the following elevations: 100, 110, 120, 130, and 140 meters. On the contoured map, draw arrows pointing down the slope of the water table to show the direction of groundwater flow.
4. Use the data table showing concentrations of a contaminant, purposely unnamed here, in groundwater to shade in areas where there is contamination. Use darker shades for higher levels of contamination.
5. Use the groundwater map to interpret where the contamination most likely originated and which facilities were probably responsible. Mark a large X over these facilities on the map and explain your reasons in the worksheet.
6. Determine which of the lettered well sites will most likely remain free of contamination, and draw circles around two such wells.
7. Devise a plan to remediate the groundwater contamination by drilling wells in front of the plume of contamination; mark these on the map with the letter R.

1. The region contains a series of ridges to the east and a broad, gentle valley to the west. Small towns are scattered across the ridges and valleys. There are also several farms, a dairy, and a number of industrial sites, each of which is labeled with a unique name.

2. A main river, called the *Black River* for its unusual dark, cloudy color, flows westward (right to left) through the center of the valley. The river contains water all year, even when it has not rained in quite a while. Both sides of the valley slope inward, north and south, toward the river.

3. Drilling and gravity surveys have shown that the valley is underlain by a thick sequence of relatively unconsolidated and weakly cemented sand and gravel. The deepest part of the basin has been downdropped by normal faults, one of which is buried beneath the gravel.

4. Based on shallow drilling, the water table (the top of the blue shading) mimics the topography, being higher beneath the ridges than beneath the valleys. Overall, the water table slopes from east to west (right to left), parallel to the regional slope of the land.

5. From mapping and other studies on the surface, geologists have determined the sequence of rock units, as summarized in the stratigraphic section in the upper right corner of this page. These studies also document a broad anticline and syncline beneath the eastern part of the region.

6. Bedrock units cross the landscape in a series of north-south stripes, parallel to the strike of the rock layers. One of the north-south valleys is named *Coal Mine Valley* because it contains several large coal mines and a coal-burning, electrical-generating plant. An unsubstantiated rumor says that one of the mines had some sort of chemical spill that was never reported. Activity at the mines and power plant has caused fine coal dust to be blown around by the wind and washed into the smaller rivers that flow along the valley.

7. A north-south ridge is composed of sandstone, which geologists call the *lower sandstone*. A few nice-tasting, freshwater springs issue from the sandstone where it is cut by small streams.

WATER RESOURCES 523

Stratigraphic Section

Gravel–Unconsolidated sand and gravel in the lower parts of the valley
Upper Sandstone–Well-sorted, permeable sandstone
Upper Shale–Impermeable, with coal
Sinkerton Limestone–Porous, cavernous limestone
Middle Shale–Impermeable shale
Lower Sandstone–Permeable sandstone
Lower Shale–Impermeable shale
Basal Conglomerate–Poorly sorted with salty water
Granite–Sparsely fractured; oldest rock in area

8. The highest part of the region is a ridge of granite and sedimentary rocks along the east edge of the area. This ridge receives quite a bit of rain during the summer and snow in the winter. Several clear streams begin in the ridge and flow westward toward the lowlands.

9. A coal-burning power plant was built over tilted beds of a unit named the *Sinkerton Limestone*, so-called because it is associated with many sinkholes, caves, and *karst topography*. The limestone is so permeable that the power plant has had difficulty keeping water in ponds built to dispose waste waters that are rich in the chemical substances that are naturally present in coal.

10. The tables below list water-table elevations in meters and concentrations of contamination in milligrams per liter (mg/L) for each of the lettered wells (A–P), and the concentration of contamination in samples from four springs (S1–S4) and eight river segments (R1–R8). The location of each sample site is marked on the figure.

Well	Elev. WT	mg/L
A	110	0
B	100	0
C	105	0
D	110	20
E	120	10
F	115	0
G	120	0
H	120	50

Well	Elev. WT	mg/L
I	130	30
J	125	0
K	120	0
L	130	0
M	140	50
N	140	0
O	140	0
P	140	0

Spring	mg/L
S1	50
S2	0
S3	0
S4	0

River	mg/L
R1	0
R2	20
R3	0
R4	0

River	mg/L
R5	0
R6	0
R7	5
R8	5

17.12

Step-by-step instructions guide students while they record observations, calculations, and results. Each investigation is accompanied by a worksheet.

Most investigation exercises are built around three-dimensional perspectives of a place that provides the context for the problems. These places are chosen to have realistic and challenging, but solvable, issues.

The figure is accompanied by a series of small text blocks that provide descriptions of different geologic features that you could observe if you were actually traversing across the region. These text blocks provide clues to help investigate various alternatives, answer the geologic questions, or determine the best course of action.

How Should Students Use This Book?

THE DESIGN OF THIS BOOK aims to make reading and studying as efficient and enjoyable as possible. In normal textbooks, two major challenges are trying to decide *what is most important* and *how the text relates to the figures*. This book simplifies that process because the text and figures are tightly integrated. The book also showcases the key concepts and ideas of geology because less important aspects are deliberately excluded. The authors think that everything in this book is important!

A How Do We Recommend You Read and Study This Book?

This book poses questions at all levels, from titles of each two-page spread to questions embedded within text blocks and lists. You should be able to answer any question posed in the opening two-page spread, in the title of a two-page spread, in a section heading, and especially in the *Before You Leave This Page* list. Finally, know how to complete the investigation at the end of the chapter.

Read the title of a two-page spread first, and then think about the question. This should activate *prior knowledge* and help link new information into an existing mental framework, making the new information easier to learn, retain, and use.

For each section, read and think about its question. Begin the section by closely observing the main figure. After mentally absorbing the whole scene, read each small text block associated with the figure and integrate the text and visual images into memory.

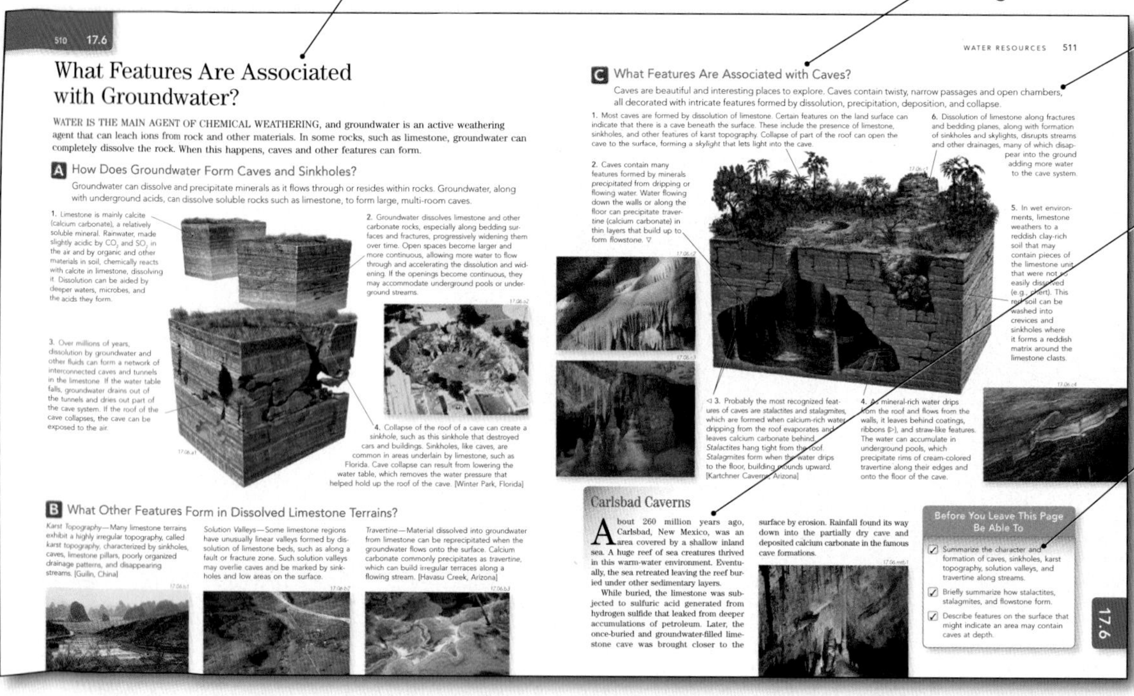

510 17.6

What Features Are Associated with Groundwater?

WATER IS THE MAIN AGENT OF CHEMICAL WEATHERING, and groundwater is an active weathering agent that can leach ions from rock and other materials. In some rocks, such as limestone, groundwater can completely dissolve the rock. When this happens, caves and other features can form.

A How Does Groundwater Form Caves and Sinkholes?

Groundwater can dissolve and precipitate minerals as it flows through or resides within rocks. Groundwater, along with underground acids, can dissolve soluble rocks such as limestone, to form large, multi-room caves.

1. Limestone is mainly calcite (calcium carbonate), a relatively soluble mineral. Rainwater, made slightly acidic by CO_2 and SO_2 in the air and by organic and other materials in soil, chemically reacts with calcite in limestone, dissolving it. Dissolution can be aided by deeper waters, microbes, and the acids they form.

2. Groundwater dissolves limestone and other carbonate rocks, especially along bedding surfaces and fractures, progressively widening them over time. Open spaces become larger and more continuous, allowing more water to flow through and accelerating the dissolution and widening. If the openings become continuous, they may accommodate underground pools or underground streams.

3. Over millions of years, dissolution by groundwater and other fluids can form a network of interconnected caves and tunnels in the limestone. If the water table falls, groundwater drains out of the tunnels and dries out part of the cave system. If the roof of the cave collapses, the cave can be exposed to the air.

4. Collapse of the roof of a cave can create a *sinkhole*, such as this sinkhole that destroyed cars and buildings. Sinkholes, like caves, are common in areas underlain by limestone, such as Florida. Cave collapse can result from lowering the water table, which removes the water pressure that helped hold up the roof of the cave. [Winter Park, Florida]

B What Other Features Form in Dissolved Limestone Terrains?

Karst Topography—Many limestone terrains exhibit a highly irregular topography, called *karst topography*, characterized by sinkholes, caves, limestone pillars, poorly organized drainage patterns, and disappearing streams. [Guilin, China]

Solution Valleys—Some limestone regions have unusually linear valleys formed by dissolution of limestone beds, such as along a fault or fracture zone. Such solution valleys may overlie caves and be marked by sinkholes and low areas on the surface.

Travertine—Material dissolved into groundwater from limestone can be reprecipitated when the groundwater flows onto the surface. Calcium carbonate commonly precipitates as *travertine*, which can build irregular terraces along a flowing stream. [Havasu Creek, Arizona]

WATER RESOURCES 511

C What Features Are Associated with Caves?

Caves are beautiful and interesting places to explore. Caves contain twisty, narrow passages and open chambers, all decorated with intricate features formed by dissolution, precipitation, deposition, and collapse.

1. Most caves are formed by dissolution of limestone. Certain features on the land surface can indicate that there is a cave beneath the surface. These include the presence of limestone, sinkholes, and other features of karst topography. Collapse of part of the roof can open the cave to the surface, forming a *skylight* that lets light into the cave.

2. Caves contain many features formed by minerals *precipitated* from dripping or flowing water. Water flowing down the walls or along the floor can precipitate *travertine* (calcium carbonate) in thin layers that build up to form *flowstone*. ▽

◁ 3. Probably the most recognized features of caves are *stalactites* and *stalagmites*, which are formed when calcium-rich water dripping from the roof evaporates and leaves calcium carbonate behind. *Stalactites* hang tight from the roof. *Stalagmites* form when the water drips to the floor, building mounds upward. [Kartchner Caverns, Arizona]

4. As mineral-rich water drips from the roof and flows from the walls, it leaves behind coatings, ribbons (▷), and straw-like features. The water can accumulate in underground pools, which precipitate rims of cream-colored travertine along their edges and onto the floor of the cave.

5. In wet environments, limestone weathers to a reddish clay-rich soil that may contain pieces of the limestone unit that were not so easily dissolved (e.g., chert). This red soil can be washed into crevices and sinkholes where it forms a reddish matrix around the limestone clasts.

6. Dissolution of limestone along fractures and bedding planes, along with formation of sinkholes and skylights, disrupts streams and other drainages, many of which disappear into the ground adding more water to the cave system.

Carlsbad Caverns

About 260 million years ago, Carlsbad, New Mexico, was an area covered by a shallow inland sea. A huge reef of sea creatures thrived in this warm-water environment. Eventually, the sea retreated leaving the reef buried under other sedimentary layers.

While buried, the limestone was subjected to sulfuric acid generated from hydrogen sulfide that leaked from deeper accumulations of petroleum. Later, the once-buried and groundwater-filled limestone cave was brought closer to the surface by erosion. Rainfall found its way down into the partially dry cave and deposited calcium carbonate in the famous cave formations.

Before You Leave This Page Be Able To

- ✓ Summarize the character and formation of caves, sinkholes, karst topography, solution valleys, and travertine along streams.
- ✓ Briefly summarize how stalactites, stalagmites, and flowstone form.
- ✓ Describe features on the surface that might indicate an area may contain caves at depth.

17.6

Before moving to the next section, take a moment to reflect on what you just learned in order to transfer this knowledge into long-term memory.

Read the major text block after completing everything else on the two-page spread. This text will help consolidate the ideas presented and provide an example of how the geologic concepts apply to a real place or a societal issue.

Lastly, read the *Before You Leave This Page* list and make sure you can answer all the questions. Record answers in your notes to solidify the knowledge while it is still fresh and to save time when studying later. Writing or sketching thoughts is a great way to help retain the main points because it makes us examine and clarify our ideas and, thus, transfer them into long-term memory.

B How Are Geologic Terms Introduced in This Book?

This book includes only those terms that we consider to be essential to thinking, writing, and talking about geology. We generally have readers observe a feature or process before providing its name, as shown below.

▷ Examine this photograph from the American Southwest. What do you observe, and what are the characteristics of the main feature shown?

00.03.b1

This feature is called a *mesa*, which is a flat-topped hill or mountain bounded by cliffs or steep slopes on at least one side. Key terms in this book are in *italics*.

Observing a feature before learning its name makes it easier to remember what the term means. Then, this compact word can be used to refer to these kinds of features.

C How Can Annotated Sketches Help Students Learn Geology?

The following illustrations demonstrate a way that text and figures can be integrated to produce a coherent understanding of geologic concepts and processes. To begin to understand this approach, observe the photograph of the Moon and identify some of its main features. Then examine the accompanying sketch and the words that summarize what we know about these features.

00.03.c1

00.03.c2

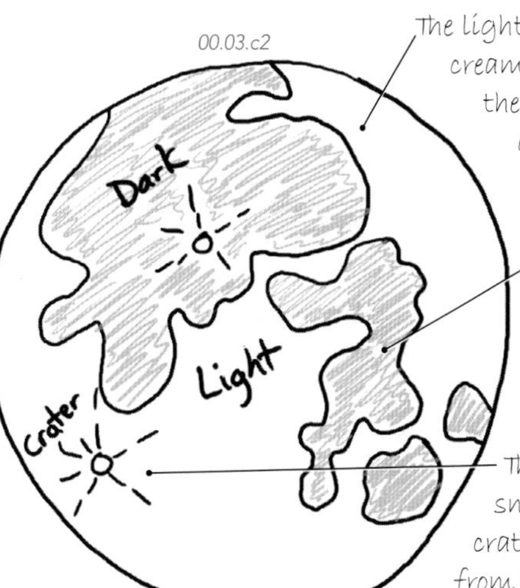

The light-colored regions of the Moon are composed of cream-colored crystals that rose to the surface when the Moon was molten during its early history. These areas are very old and are covered with craters of all sizes.

The dark, somewhat circular regions are huge craters that were filled with dark lava flows erupted onto the surface. They formed after the light-colored areas.

The Moon has many craters formed when meteorites smashed onto the surface. Some of the younger craters have light-colored rays radiating out from them.

D How Do You Construct a Concept Sketch?

The annotated sketch of the Moon is an example of what we call a *concept sketch*. Here's how to make one.

Begin by observing a natural phenomenon, landscape, animation, photograph, or textbook illustration. Make a list of the main features or processes that you think should be depicted.

Draw a sketch that is simple but clearly shows the essential information.

Annotate your sketch with complete sentences to identify features, describe how they form, and summarize the main geologic processes.

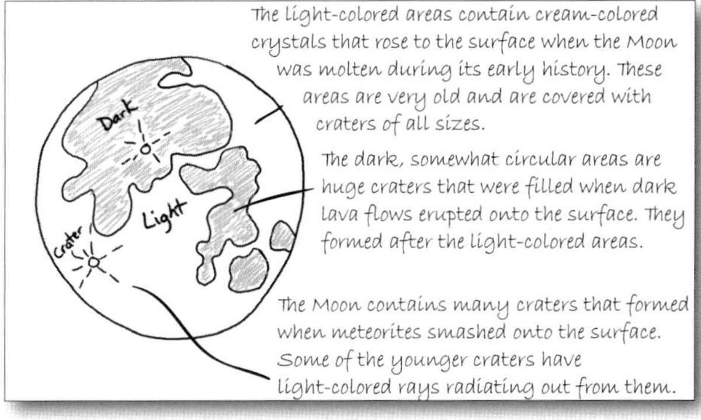

Draw lines connecting your text to the appropriate features on the sketch.

You can increase your understanding by explaining the concept sketch to yourself and to a classmate, describing the features and processes shown.

If you can draw, label, and accurately explain a concept sketch, you probably understand that concept well. Such understanding will help with whatever types of questions you are asked to answer.

Learning with Concept Sketches

We have used concept sketches in our own classes for years, and we regard them as one of the best ways to learn and teach geology. They are an excellent way to construct knowledge from readings, lectures, and observations of landscapes and other rock exposures. Constructing a concept sketch enables students to visualize the different parts of a geologic system in their proper spatial context. They require students to identify the important aspects of the system and to de-emphasize the less important complexities present in all natural systems. You will note that this book consists largely of illustrations that resemble concept sketches.

We recommend that for each section you construct your own concept sketches from our more detailed illustrations and text. In essence, you are striving to put these illustrations and descriptions, and the processes and concepts they represent, into your own words and mind. This book is ideally suited for learning with concept sketches, if you choose to do so.

Making concept sketches also links students with the way geologists approach geologic questions. Geology has a long heritage of using pictures to convey observations and interpretations. Geologists draw sketches in the field or any time they are trying to share ideas with others. Drawing sketches is a natural way to conceptualize hidden places and distant times.

For more information about concept sketches, see Johnson, J.K., and Reynolds, S.J., 2005, *Journal of Geoscience Education*, v. 53, pp. 85–95.

What Resources Support This Textbook?

McGraw-Hill offers various tools and technology products to support *Exploring Geology*. Instructors can obtain teaching aids by calling the Customer Service Department at 800-338-3987 or contacting their local McGraw-Hill sales representative.

Course Management

ARIS

McGraw-Hill's ARIS—Assessment, Review, and Instruction System for *Exploring Geology* (www.mhhe.com/reynolds) is a complete, online tutorial, electronic homework, and course management system, designed for greater ease of use than any other system available. Instructors can create and share course materials and assignments with colleagues with a few clicks of the mouse. All PowerPoint® lectures, assignments, quizzes, animations, and Internet activities are directly tied to text-specific materials in *Exploring Geology*. Instructors can also edit questions, import their own content, and create announcements and due dates for assignments. ARIS has automatic grading and reporting of easy-to-assign homework, quizzing, and testing. All student activity within McGraw-Hill's ARIS is automatically recorded and available to the instructor through a fully integrated grade book that can be downloaded to Excel®. Contact your local McGraw-Hill Publisher's representative for more information on getting started with ARIS.

Digital Assets

ARIS Presentation Center

Build instructional materials where ever, whenever, and however you want!

ARIS Presentation Center is an online digital library containing assets such as photos, artwork, animations, PowerPoints, and other media types that can be used to create customized lectures, visually enhanced tests and quizzes, compelling course websites, or attractive printed support materials.

Access to your book, access to all books!

Nearly all 2,693 illustrations from *Exploring Geology* will be available as digital files. In addition, the Presentation Center library includes thousands of assets from many McGraw-Hill titles. This ever-growing resource gives instructors the power to utilize assets specific to an adopted textbook as well as content from all other books in the library.

Nothing could be easier!

Accessed from the instructor side of your textbook's ARIS website, Presentation Center's dynamic search engine allows you to explore by discipline, course, textbook chapter, asset type, or keyword. Simply browse, select, and download the files you need to build engaging course materials. All assets are copyrighted by McGraw-Hill Higher Education but can be used by instructors for classroom purposes.

Author Resources

Instructor's Testing and Resource CD-ROM

This cross-platform CD, prepared by the authors, provides a wealth of resources for the instructor. Among the supplements featured on this CD is a computerized test bank that uses testing software to quickly create customized exams. The user-friendly program allows instructors to search for questions by topic, format, or difficulty level; edit existing questions or add new ones; and scramble questions for multiple versions of the same test. Word files of the test bank questions are provided for those instructors who prefer to work outside the test-generator software.

The Instructor's Guide is also included on the Instructor's Testing and Resource CD. This manual contains valuable information on how to teach using *Exploring Geology*. A quick guide to teaching each chapter is provided. Chapter sections include suggestions for using PowerPoint presentations and interactive media in the classroom. Suggestions are also given for using concept sketches for assessment and classroom presentation. The Instructor's Guide is keyed to the PowerPoint presentations.

Bonus Media

Media DVD for Instructors

This DVD, prepared by the authors, will bring the geology experience right into the classroom! It provides lecture PowerPoint files with key points built around the spectacular art from the book. Certain slides in the presentations contain built-in links to launch media files, most of which consist of interactive versions of figures from the book. The media includes images that can be rotated, images that can be used with a GeoWall, 3-D maps, and animations.

Discovery Channel DVD

This exciting DVD offers 50 short (3–5 minute) videos on topics ranging from conservation to volcanoes. Begin your class with a quick peek at science in action.

Classroom Performance System and Questions

McGraw-Hill has partnered with eInstruction to provide the revolutionary Classroom Performance System (CPS) and to bring interactivity into the classroom. CPS is a wireless response system that gives the instructor and students immediate feedback from the entire class. The wireless response pads are essentially remotes that are easy to use and engage students. CPS allows you to motivate student preparation, interactivity, and active learning so you can receive immediate feedback and know what students understand. A text-specific set of questions, formatted for both CPS and PowerPoint, is available via download from the Instructor area of *Exploring Geology* ARIS site.

Custom Options

Custom Publishing

Did you know that you can design your own text or lab manual using any McGraw-Hill text and your personal materials to create a custom product that correlates specifically to your syllabus and course goals? Because of the two-page spread format, *Exploring Geology* is the perfect candidate for this. Contact your McGraw-Hill sales representative to learn more about this option.

Acknowledgments

Writing a totally new type of introductory geology textbook would not be possible without the suggestions and encouragement we received from instructors who reviewed various drafts of this book and its artwork. We are especially grateful to people who contributed entire days either reviewing or attending symposia to openly discuss the vision, challenges, and refinements of this kind of new approach. Parts of the manuscript received special attention from reviewers Scott Linneman, Richard Sedlock, Bill Dupre, and Grenville Draper. The accuracy and presentation of information was improved greatly through these thoughtful and valuable comments. Many of our colleagues enthusiastically encouraged us onward, including Bruce Herbert, Scott Linneman, Steve Semken, Diane Clemens-Knott, Jeff Knott, and Barbara Tewksbury. Over the years, we have also received many ideas from our colleagues, mentors, and students in the geology, science-education, and cognitive fields. For all of this we are very grateful.

This book contains nearly 2,700 figures, two to three times more than a typical introductory geology textbook. This massive art program required great effort and artistic abilities by the artists who turned our vision and sketches into what truly are pieces of art. In addition to our coauthor Chuck Carter, we greatly appreciate the dedication and artistic touches of illustrators Susie Gillatt, Cindy Shaw, Daniel Miller, David Fierstein, Karen Carter, and Ren Olsen. We also benefited from interactions with designers Chris Willis and David Hash, who helped translate our ideas about pedagogy into a workable and aesthetically sound design. Many people went out of their way to provide us with photographs, illustrations, and advice, in some cases going out into the field to take the photographs we needed. These helpful people included Vince Matthews, Ron Blakey, Michael Collier, Matthew Larsen, Allen Glazner, Karen Carr, Ed Garnero, Ramón Arrowsmith, Don Burt, Phil Christensen, Tom Sharp, Steve Semken, Doug Bartlett, Spencer Lucas, Michael Ort, Nancy Riggs, Peg Owens, Tom McGuire, and Barbara Tewksbury.

We used a number of data sources to create many illustrations. Reto Stöckli of the Department of Environmental Sciences at ETH Zürich and NASA Goddard produced the Blue Marble and Blue Marble Next Generation global satellite composites. We used data from the ZULU server at of the NASA Earth Science Enterprise Scientific Data Purchase Program for hundreds of figures in this book. Brian Davis of the USGS EROS Data Center was quick to find elusive data, and Collin Bode of the National Center for Earth-surface Dynamics was indispensable in helping us process GIS data. Debbie Leedy provided mineralogy and chemistry 3D files, and Melanie Busch and Joshua Coyan provided other 3D files.

We have treasured our interactions with the wonderful Iowans at McGraw-Hill Higher Education, who enthusiastically supported our vision, needs, and progress. We especially thank Marge Kemp, our publishing editor and most steadfast champion, for encouraging our nontraditional approach and for providing timely reality checks. Liz Recker skillfully and cheerfully guided the development of the book during the entire publication process, making it all happen. We also appreciate the support, cooperation, guidance, and enthusiasm from Kent Peterson, Thomas Timp, Todd Turner, Lisa Nicks, Dan Wallace, Daryl Bruflodt, Joyce Berendes, Gloria Schiesl, Judi David, Lori Hancock, Sherry Kane, Margaret Horn, Tammy Ben, Traci Andre, Steven Hoffman, Fran Aitkens, Mary Ellen Tuccillo, Ellen Osterhaus, Annie Langel, and many others who worked hard to make this book a reality.

Finally, a project like this is truly life consuming, especially when the author team is doing the writing, illustration, photography, near-final page layout, media development, and development of assessments and teaching ancillaries. We are extremely appreciative of the support, patience, and friendship we received from family members, friends, colleagues, and students who shared our sacrifices and successes during the creation of this new vision of a textbook. We thank Susie Gillatt, Annabelle Louise, Sarah Kelly, Lisa Logan, August Morin, Oliver Morin, and Karen Carter. We thank you all so much!

Reviewers

Special thanks and appreciation go out to all reviewers. This first edition (through several stages of manuscript development) has enjoyed many beneficial suggestions, new ideas, and invaluable advice provided by these reviewers. We appreciate all the time they devoted to reviewing manuscript chapters, attending focus groups, reviewing art samples, surveying students, and promoting this text to their colleagues:

Martin Appold *University of Missouri, Columbia*
Suzanne L. Baldwin *Syracuse University*
Julie K. Bartley *University of West Georgia*
J. Bret Bennington *Hofstra University*
Elisa Bergslein *Buffalo State College*
David M. Best *Northern Arizona University*
Theodore Bornhorst *Michigan Technical University*
Steve Boss *University of Arkansas*
Douglas Britton *Long Beach City College*
Pamela C. Burnley *Georgia State University*

John H. Burris *San Juan College*
T.J. Callahan *College of Charleston*
James L. Carew *College of Charleston*
Cinzia Cervato *Iowa State University*
Renee M. Clary *Mississippi State University*
Diane Clemens-Knott *California State University, Fullerton*
Chuck Connor *University of South Florida*
Peter Copeland *University of Houston*
Ellen A. Cowan *Appalachian State University*
Randel Tom Cox *University of Memphis*
Jim Criswell *Cape Fear Community College*
Margaret E. Crowder *Western Kentucky University*
Kathleen Devaney *El Paso Community College*
Craig Dietsch *University of Cincinnati*
Rebecca L. Dodge *University of West Georgia*
Grenville Draper *Florida International University*
Bill R. Dupre *University of Houston*
David E. Fastovsky *University of Oklahoma*
Mark D. Feigenson *Rutgers University, New Brunswick, NJ*
Stan Finney *California State University, Long Beach*
Mark Fischer *Northern Illinois University*
Mark Frank *Northern Illinois University*
Kyle C. Fredrick *Buffalo State College*
Yongli Gao *East Tennessee State University*
Ed Garnero *Arizona State University*
Dennis Geist *University of Idaho*
Francisco Gomez *University of Missouri, Columbia*
G. Michael Grammer *Western Michigan University*
Ronald Greeley *Arizona State University*
Roy Haggerty *Oregon State University*
Duane Hampton *Western Michigan University*
Thor A. Hansen *Western Washington University*
Timothy Heaton *University of South Dakota*
Bruce Herbert *Texas A&M University*
Curtis L. Hollabaugh *University of West Georgia*
Mary Hubbard *Kansas State University*
Paul F. Hudak *University of North Texas*
Marilyn C. Huff *New Mexico State University*
John R. Huntsman *University of North Carolina, Wilmington*
Jason Janke *Metropolitan State College of Denver*
Steven C. Jaume *College of Charleston*
Steve Kadel *Glendale Community College*
Jeffrey A. Karson *Syracuse University*
G. Randy Keller *University of Oklahoma*
David T. King, Jr. *Auburn University, Auburn*
Kent C. Kirkby *University of Minnesota*
Jeffrey Knott *California State University, Fullerton*
Mark A. Kulp *University of New Orleans*
Ming-Kuo Lee *Auburn University*
Robert A. Leighty *Mesa Community College*
Adrianne A. Leinbach *Wake Technical Community College*
Stephen D. Lewis *California State University–Fresno*
Scott R. Linneman *Western Washington University*
Brian E. Lock *University of Louisiana, Lafayette*
James Martin-Hayden *University of Toledo*
Stephen Mattox *Grand Valley State University*
Kyle Mayborn *Western Illinois University*
Joseph Meert *University of Florida*
Gretchen Miller *Wake Technical Community College*
Kula C. Misra *University of Tennessee, Knoxville*
Jared R. Morrow *University of Northern Colorado*
Michael A. Murphy *University of Houston*
Thomas Naehr *Texas A&M Corpus Christi*
Pamela Nelson *Glendale Community College*
Steven R. Newkirk *University of Memphis*
Peter A. Nielsen *Keene State College*
Clair Russell Ossian *Tarrant County College, Hurst, TX*
Kate Pound *Saint Cloud State University*
Steven Ralser *University of Wisconsin, Madison*
John Renton *West Virginia University, Morgantown*
Carl Richter *University of Louisiana at Lafayette*
Nancy Riggs *Northern Arizona University*
Bethany D. Rinard *Tarleton State University*
Delores Robinson *University of Alabama, Tuscaloosa*
Scott Rowland *University of Hawaii*
Cassandra J. Runyon *College of Charleston*
Randye L. Rutberg *Hunter College*
Dewey D. Sanderson *Marshall University*
Richard Sedlock *San Jose State University*
Yuch-Ning Shieh *Purdue University, West Lafayette*
Eric Small *University of Colorado, Boulder*
Abe Springer *Northern Arizona University*
Neptune Srimal *Florida International University, Miami*
Mark J. Sutherland *College of Dupage*
Michael Taber *University of Northern Colorado*
Jan Tullis *Brown University*
Lensyl Urbano *University of Memphis*
Stacey Verardo *George Mason University*
Mari Vice *University of Wisconsin, Platteville*
Adil M. Wadia *The University of Akron, Wayne College*
Stephen Wareham *California State University, Fullerton*
Richard Warner *Clemson University*
Barry Weaver *University of Oklahoma*
John Weber *Grand Valley State University*
David A. Williams *Arizona State University*
Wendi J. W. Williams *University of Arkansas, Little Rock*
Kenneth Windom *Iowa State University*
Lorraine W. Wolf *Auburn University*
Aaron Yohsinobu *Texas Tech University*

2006 Geology Survey Participants

We would also like to thank those who participated in our geology survey. Your feedback will continue to be used by McGraw-Hill to expand our geology resources:

Mead A. Allison *Tulane University*
Jeffrey Amato *New Mexico State University*
John Anderson *Georgia Perimeter College*
Martin Appold *University of Missouri, Columbia*
Mark Baskaran *Wayne State University*
Timothy Bralower *Pennsylvania State University*
Nathalie Nicole Brandes *Montgomery College*
Pamela C. Burnley *Georgia State University*
Sean Chamberlin *Fullerton College*
Kevin Cole *Grand Valley State University*
Margaret E. Crowder *Western Kentucky University*
Stewart S. Farrar *Eastern Kentucky University*
Heather Gallacher *Cleveland State University*
Danny Glenn *Wharton County Junior College*
Andrew M. Goodliffe *University of Alabama*
Nathan L. Green *University of Alabama*
Willis Hames *Auburn University*
Thor A. Hansen *Western Washington University*
Michael J. Harrison *Tennessee Tech University*
David Hirsch *Western Washington University*
Curtis L. Hollabaugh *University of West Georgia*
Bernard A. Housen *Western Washington University*
John R. Huntsman *University of North Carolina at Wilmington*
Amanda Palmer Julson *Blinn College*
Steve Kadel *Glendale Community College*
Michael Katuna *College of Charleston*
Ming-Kuo Lee *Auburn University*
William W. Little *Brigham Young University*
Richard Lozinsky *Fullerton College*
Neil Lundberg *Florida State University*
Dan Moore *Brigham Young University*
John E. Mylroie *Mississippi State University*
Ravi Nandigam *University of Texas at Brownsville*
Pamela J. Nelson *Glendale Community College*
Philip M. Novack-Gottshall *University of West Georgia*
Clair Russell Ossian *Tarrant County College, Hurst, TX*
William C. Parker *Florida State University*
Roy E. Plotnick *University of Illinois at Chicago*
Kenneth Rasmussen *Northern Virginia Community College*
Bethany D. Rinard *Tarleton State University*
Randye L. Rutberg *Hunter College*
Roy Schlische *Rutgers University*
Neptune Srimal *Florida International University*
Paula J. Steinker *Bowling Green State University*
J. Robert Thompson *Glendale Community College*
Johnny Waters *Appalachian State University*
Barry Weaver *University of Oklahoma*
Harry Williams *University of North Texas*
Lorraine W. Wolf *Auburn University*
Aaron Yoshinobu *Texas Tech University*

Art Review Team

We extend a special thanks to the Art Review Team whose members provided constructive criticism to the art work as it was rendered by our author team. The team members included:

Mark D. Feigenson *Rutgers University*
G. Michael Grammer *Western Michigan University*
Scott R. Linneman *Western Washington University*
Brian E. Lock *University of Louisiana at Lafayette*
Stephen Wareham *California State University, Fullerton*

Sales Champions

Thank you also goes to our sales champions. This team of McGraw-Hill sales representatives has provided valuable feedback throughout the development of *Exploring Geology*. They have supported *Exploring Geology* from its early stages in development and are now ready to promote this text to the market.

Nicole Prevanas
Catherine Riley
Scott Rubin
Oliver Tillman
Kathyrn Ziesel

About the Authors

Stephen Reynolds

Stephen Reynolds received an undergraduate geology degree from University of Texas at El Paso, and M.S. and Ph.D. degrees in structure/tectonics and regional geology from the University of Arizona. He then spent ten years directing the geologic framework and mapping program of the Arizona Geological Survey, where he completed the 1988 Geologic Map of Arizona. Steve currently is a professor in the School of Earth and Space Exploration at Arizona State University, where he has taught Physical Geology, Structural Geology, Field Geology, Orogenic Systems, Cordilleran Regional Geology, Teaching Methods in the Geosciences, and others. He helped establish the ASU Center for Research on Education in Science, Mathematics, Engineering, and Technology (CRESMET), and was President of the Arizona Geological Society. He has authored or edited nearly 200 geologic maps, articles, and reports, including the 866-page *Geologic Evolution of Arizona*. He also coauthored *Structural Geology of Rocks and Regions*, a widely used Structural Geology textbook, and *Observing and Interpreting Geology*, a laboratory manual for Physical Geology. For the last ten years, he has done science-education research on student learning in college geology courses, especially the role of visualization. Steve is known for innovative teaching methods, has received numerous teaching awards, and has an award-winning website. As a National Association of Geoscience Teachers (NAGT) distinguished speaker, he traveled across the country presenting talks and workshops on how to infuse active learning and inquiry into large introductory geology classes. He also has been a long-time industry consultant in mineral, energy, and water resources, and has received outstanding alumni awards from UTEP and the University of Arizona.

Julia K. Johnson

Julia K. Johnson received B.S. and M.S. degrees from the Department of Geological Sciences at Arizona State University, and she currently is a full-time faculty member in the School of Earth and Space Exploration at ASU. Her M.S. and Ph.D. research involved structural geology and geoscience education research. The main focus of her geoscience education research is on student- and instructor-generated sketches for learning, teaching, and assessment in college geology classes. Prior to coming to ASU, she did groundwater studies of copper deposits and then taught full time in the Maricopa County Community College District, teaching Physical Geology, Environmental Geology, and their labs. At ASU, she teaches Introduction to Geology to nearly 1,000 students per year and supervises the associated introductory geology labs. She also coordinates the introductory geology teaching efforts of the School of Earth and Space Exploration, helping other instructors incorporate active learning and inquiry into large lecture classes. Julia is recognized as one of the best science teachers at ASU and has received student-nominated teaching awards and very high teaching evaluations in spite of her challenging classes. In recognition of her teaching, she was a Featured Faculty of the Month on ASU's website in 2005. She has authored publications on geology and science-education research, including an article in the *Journal of Geoscience Education* on concept sketches. She coauthored *Observing and Interpreting Geology* and also developed a number of websites used by geology students around the world, such as the *Visualizing Topography* and *Biosphere 3D* website.

Michael Kelly

Michael Kelly received an undergraduate geology degreee from the University of California, Santa Cruz and an M.S. degree in geology from Northern Arizona University. His graduate research defined ductile structures and strain in Mojave Desert mountain ranges. As a USGS geologist, he mapped in the western U.S., coauthoring several geologic maps and performing paleomagnetic research and laboratory studies on Columbia River Basalts. As Senior Geologist at EMCON Associates, he led environmental investigations into geologically complex groundwater industrial contamination sites across the Pacific Northwest. He returned to the southwest as the director of the Center for Research and Evaluation of Advanced Technologies in Education (CREATE) at NAU and was adjunct faculty in Environmental Sciences. Here, Kelly's research activities centered around the use of virtual reality to enhance undergraduate science education. He was Co-Investigator on numerous NSF science education projects and is author or coauthor on numerous publications resulting from the CREATE's efforts. Kelly is lead author on a virtual reality geology laboratory curriculum that has been in use by undergraduates since 1999. Today, Kelly designs and assesses media-enhanced science curriculum. His recent research focuses on spatial learning, particularly how landscapes and terrain are interpreted and how they can be used as frameworks for understanding connected science domains. Kelly recently designed virtual reality exhibits installed in two national parks, and his 3-D terrain software *ROMA* is used nationally for geoscience education in secondary and undergraduate institutions. He recently published in the *Journal of Geoscience Education* a study on the effectiveness of the GeoWall in undergraduate geology.

Paul Morin

Paul Morin specializes in earth-science visualization at the University of Minnesota, Department of Geology and Geophysics and the National Center of Earth-surface Dynamics. His interests have to do with the effect of artistic technique and technology on the efficacy of visualizations in the hands of students. Paul co-founded the GeoWall Consortium, a group of over 200 teaching institutions using visualization and stereo projection in the classroom. Currently, over 25% of all students taking introductory Physical Geology Lab in the U.S. have access to a GeoWall. Over the past five years, Paul has been instrumental in bringing earth science visualization to science museums around the world. He is currently co-investigator and co-developer of *Water Planet*, a 5,000 to 10,000 square foot traveling exhibit about water's role in shaping Earth. He was a major contributor of interactive visualizations to the NAGT-sponsored laboratory manual for Physical Geology. For several years, he has been an NAGT distinguished speaker, visiting universities and colleges to present talks on the role of visualization in geology courses. He is regarded by many people as one of the top visualization developers in the geosciences. Paul's current research has been on the replacement of standard geologic and topographic maps in the classroom with GeoWall and anaglyph versions. Other professional interests include the visualization of data sources that are traditionally viewed as being too complex for students to understand, such as three-dimensional spherical convection, seismic tomography, and paleontology.

Chuck Carter

Chuck Carter has been working in the artistic end of the science and entertainment industries for more than 20 years. His illustration and animation work has been used extensively by *National Geographic*, and his illustrations and layouts are featured in books published by *National Geographic* to feature the best of their artwork. He was the first freelance artist hired by *National Geographic* to create a 3-page digital illustration (on dinosaur evolution) and in 1994 was instrumental in helping launch *National Geographic Online.* He also has worked with Harcourt Education, McGraw-Hill Higher Education, *Knight-Ridder News in Motion*, and other clients for more than 18 years. He has produced illustrations and animations for the U.S. Navy, U.S. Department of Defense, and various defense contractors. His entertainment projects include being lead illustrator on the computer game *Myst* and he has worked on more than 20 other video games, including the popular *Command and Conquer* series, as a digital artist, animator, writer, art director, and computer-graphics supervisor. While working with Threshold Entertainment, he worked as a digital matte painter for shows like *Babylon 5.* He is lead illustrator and a coauthor of this book.

EXPLORING GEOLOGY

CHAPTER 1

The Nature of Geology

GEOLOGY HAS MANY EXPRESSIONS in our world. Geologic processes reshape Earth's interior and sculpt its surface. They determine the distribution of metals and petroleum and control which places are most susceptible to natural disasters, such as volcanoes and floods. Geology impacts factors that are critical to ecosystems, such as climate and the availability of water. In this book, we explore geology, *the science of Earth*, and examine why an understanding of geology is important in our modern world.

North America and the surrounding ocean floor have a wealth of interesting features. The large image below (▽) is a computer-generated image that combines different types of data to show features on the land and on the bottom of the ocean. The shading and colors on land are from space-based satellite images, whereas colors and shading on the seafloor indicate depths below the surface of the sea. Can you find the region where you live? What is there?

◁ **The dramatic scenery of Glacier National Park** in Montana was carved by glaciers that formed during the Ice Ages, during the last 2 million years.

What controls Earth's climate, and what evidence is there for past climatic changes?

The 1980 eruption of Mount St. Helens in southwestern Washington (▽) ejected huge amounts of volcanic ash into the air, toppled millions of trees, and unleashed floods and mudflows down nearby valleys. Geologists study volcanic phenomena to determine how and when volcanoes may erupt and what hazards volcanoes may cause.

How do geologic studies help us determine where it is safe to live?

TOPICS IN THIS CHAPTER

Rocks of New England and easternmost Canada record a fascinating history, which includes an ancient ocean that was destroyed by the collision of two landmasses. Many of these rocks, such as those in Nova Scotia (▽), have contorted layers, and some rocks provide evidence of forming at a depth of 30 kilometers below the surface.

How do layers in rocks get squeezed, and how do such deep rocks get to the surface where we now find them?

01.00.a4

Everglades National Park in southern Florida (▽) is one of the most environmentally threatened regions on the planet because the water needs of humans conflict with those of the ecosystem.

How can geologists help study and protect this and other natural treasures?

A View of North America

North America is a diverse continent, ranging from the low, tropical rain forests of Mexico to the high Rocky Mountains of northern Canada. The relatively high standard of living of people in the United States and Canada is largely due to an abundance of natural resources, such as coal, petroleum, minerals, and soils.

In the large image of North America on the left, the colors on land are from satellite images that show the distribution of rock, soil, plants, lakes, and cities. Green colors represent dense vegetation, including forests shown in darker green and fields and grassy plains shown in lighter green. Brown colors represent deserts and other regions that have less vegetation, including regions where rock and sand are present on the land surface. Lakes are shown with a solid blue color. Note that there are no clouds in this artificial picture.

The ocean floor is shaded according to depth below sea level. Light blue colors represent shallow areas, whereas dark blue represents places where the seafloor is deep. Take a minute or two to observe the larger features in this image, including those on land and at sea. Ask yourself the following questions: What is this feature? Why is it located here, and how did it form? In short, what is its story?

Notice that the two sides of North America are very different from each other and from the middle of the continent. The western part of North America has a more complex appearance because it has numerous mountains and valleys. The mountains in the eastern United States are more subdued, and the East Coast is surrounded by a broad shelf (shown in light blue) that continues out beneath the Atlantic Ocean. The center of the continent has no large mountains but is instead covered with broad plains, hills, and river valleys.

All of the features that you observe on this globe are part of geology. Geologic history explains why the mountains on the two sides of the continent are so different and when and how each mountain range formed. Geology explains how features on the seafloor came to be. It also explains why the central United States and Canada are the agricultural heartland of the continent and why some regions are deserts.

01.00.a5

How Does Geology Influence Where and How We Live?

GEOLOGY INFLUENCES OUR LIVES IN MANY WAYS. Geologic features and processes constrain where people can live by determining whether a site is safe from landslides, floods, or other natural hazards. Some areas are suitable building sites, but other areas are underlain by unstable geologic materials that could cause damage to any structure built there. Geology also controls the distribution of energy resources and the materials required to build houses, cars, and factories.

A Where Is It Safe to Live?

The landscape around us contains many clues about whether a place is relatively safe or whether it is a natural disaster waiting to happen. What are the important clues that should guide our choice of a safe place to live?

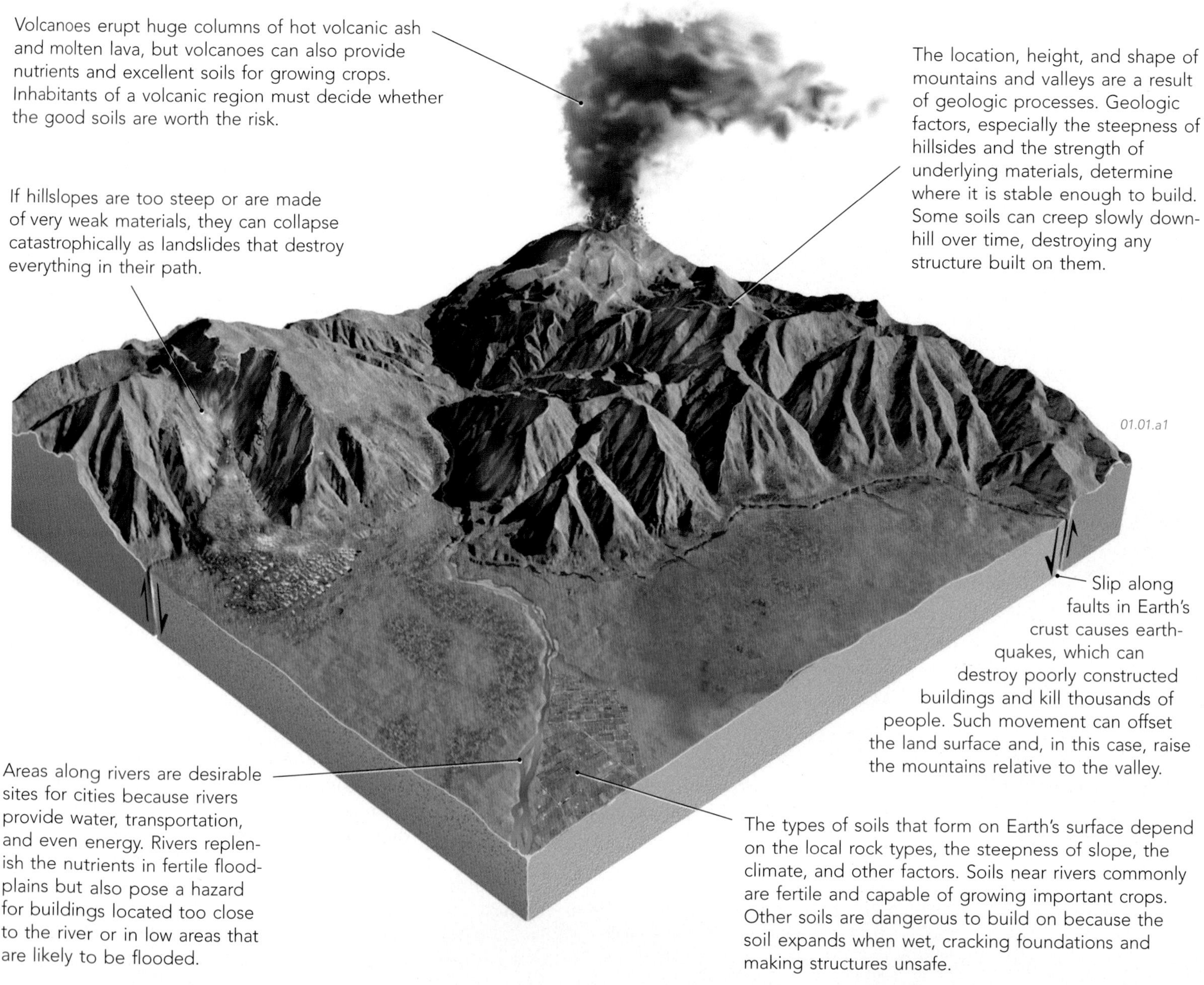

B What Controls the Distribution of Natural Resources?

This map of North America shows the locations of large, currently or recently active copper mines (orange dots) and iron mines (blue dots). What do you notice about the distribution of each type of mine?

1. After observing this pattern, we might ask several questions: What determines where such mines are located? Why are there so few of these mines in the southeastern United States? Why do the mines occur in regions of the continent that look quite different? The answer is that these mines reflect differences in the ages and types of rocks.

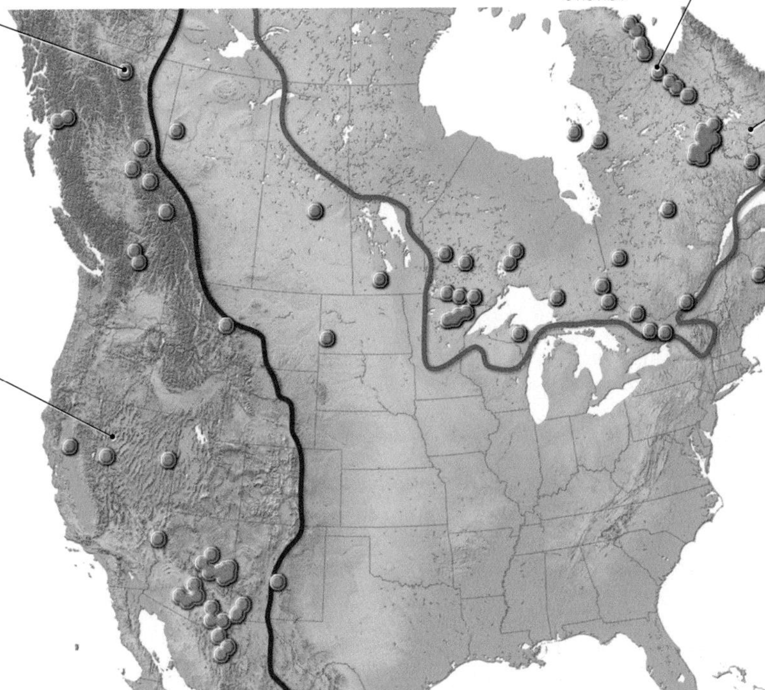

01.01.b1

2. The large *copper* mines (orange dots) are restricted to the western part of the continent, west of the purple line that separates the mountains to the west from the plains to the east.

3. The copper mines are in the mountainous west because this part of the continent was invaded by magma (molten rock), especially around 70 million years ago. These magmas formed distinctive kinds of copper-rich granites and were related to geologic events that occurred only along the western side of the continent.

4. Large *iron* mines (blue dots) are most common in the Great Lakes region and in eastern Canada, within an area called the *Canadian Shield* (inside the red line).

5. Rocks of the Canadian Shield are much older than the western copper deposits. The iron-rich rocks accumulated early in Earth's history, when oxygen became more abundant in Earth's atmosphere and caused iron to precipitate out of the seas and into vast iron-rich layers. Iron-rich rocks of this age (about 2 billion years) did not form in most other parts of the continent. In fact, some parts of North America did not exist at this time.

6. Both types of mines are in rocks of a specific type, age, and origin. Such geologic factors control where mineral resources, such as copper and iron, are located. Resources often are not located where humans would prefer them to be for logistical, political, or environmental reasons.

Ways That Geologic Studies Help Us Understand Ecosystems

Some geologists study ecosystems to better protect them. The types of research that these geologists do are as diverse as the ecosystems they study, which range from deserts to rain forests to icy glaciers. In such studies, geologists commonly work with biologists, climate specialists, and the government agencies that decide how the land will be managed.

The most important factor in the viability of many ecosystems is availability of clean water. In most geologic settings, water moves between the surface and subsurface, and geologic studies are the primary basis for determining how much water is available, how water flows, and the purity of water. Geologists study materials on the surface and beneath the surface to understand how these materials influence the flow of water. Geologists guide cleanup efforts by sampling water (△), pinpointing sources of contamination, and predicting which way the contamination is likely to move in the future.

01.01.mtb1

In addition to water, the viability of an ecosystem generally depends on the health of its soils. Geologic studies document how soils develop, how they are changing, and the best ways to prevent erosion or other types of soil deterioration. Such studies also determine the natural fluxes of organic material and other chemical constituents among the soil, plants, and water. Geologists investigate these factors in order to understand how a region is responding to environmental changes, which may be a result of natural or human causes.

Before You Leave This Page Be Able To

- ✓ Sketch or list some ways that geology determines where it is safe to live.
- ✓ Summarize some ways in which geology influences the distribution of natural resources.
- ✓ Summarize how geologic studies help us understand ecosystems.

How Does Geology Help Explain Our World?

THE WORLD HAS INTERESTING FEATURES at all scales. From space, one sees oceans, continents, and mountains. Traveling through the countryside, we notice smaller things—a beautiful rock formation or soft, green hills. Upon closer inspection, the rocks may include fossils that provide evidence of ancient life and past climates. Here, we give examples of how geology explains big and small features of our world.

A Why Do Continents Have Different Regions?

Examine the shaded-relief map of part of the northwestern United States. The map is centered on Yellowstone National Park in northwestern Wyoming. Note the differences among regions in this part of the continent.

01.02.a1

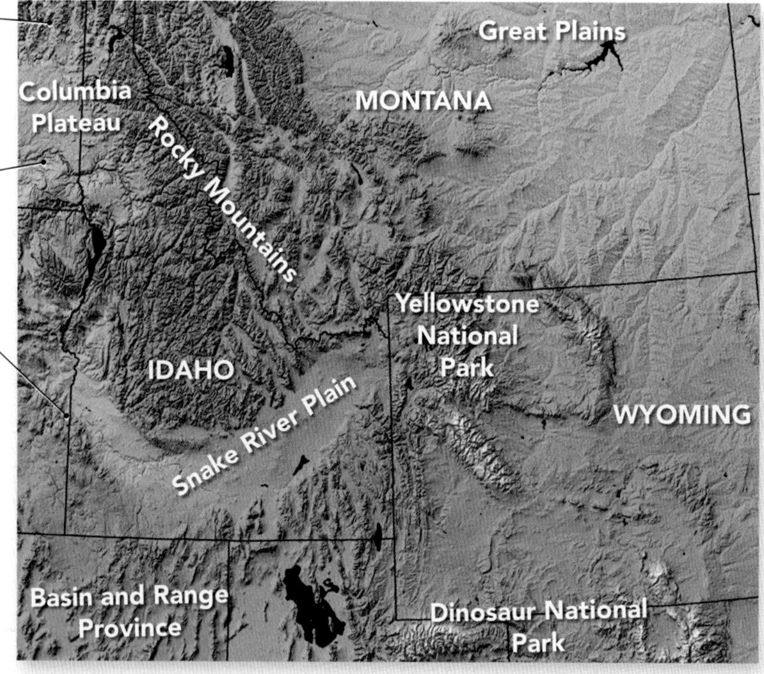

Some regions on this map have rough landscapes with mountains and steep valleys.

Some regions consist of broad plains or low hills.

Other areas contain large but isolated mountain ranges, or alternating mountains and valleys.

From your observations, you might ask "What controls whether a region is mountainous, hilly, or flat, and why are mountains located where they are?"

Different regions exist because each has a different geologic history; that is, certain geologic events affected some areas but not others. The mountains in the western half of this area formed as a result of activity that occurred along the western edge of the continent, but the effects did not reach inland to the interior of the continent. As a result, the rough land surface to the west reflects its complex geologic history, but the plains to the east express their simpler history. Other features, such as the Snake River Plain and the Columbia Plateau, are relatively smooth because these regions were covered by massive outpourings of lava (molten rock) that buried the landscape.

B What Story Do Landscapes Tell?

Observe this photograph taken along the edge of a cliff and note any questions that you have.

The land below the cliff has sunlit plateaus (flat-topped areas) and shadowy canyons.

Rocks in the cliff are reddish brown. These include a large, angular block of rock that is perched on the edge of the cliff.

01.02.b1

Several questions about the landscape come to mind. What are the reddish-brown rocks? Why did a cliff form here? How long will it take for the block of rock to tumble off the cliff?

The answer to each question helps explain part of the scene. The first question is about the *present*, the second is about the *past*, and the third is about the *future*. The easiest questions to answer are usually about the present, and the hardest ones are about the past or the future.

The reddish-brown rocks consist of sand grains that have been packed and stuck together to form a rock called *sandstone*.

The cliff, like most other cliffs, formed because its rocks (the reddish-brown sandstone) are harder to wear away than the rocks above and below.

It is difficult to say when the block will fall, but in geologic terms it will be soon, probably in less than 1,000 years.

C What Is the Evidence That Life in the Past Was Different from Life Today?

Museums and action movies contain scenes, like the one below, of dinosaurs lumbering or scampering across a land covered by exotic plants. Where does the evidence for these strange creatures come from?

△ **1.** This mural shows what types of life were on our planet during the Jurassic Period, approximately 160 million years ago. It was a world dominated by dinosaurs and non-flowering plants.

◁ **2.** Fossil bones of Jurassic dinosaurs are common in Dinosaur National Park, Utah. From such bones and other information, geologists infer what these creatures looked like, how big they were, how they lived, and why they died. Studying the rocks in which the bones are found provides clues to the local and global environment at the time of the dinosaurs.

D How Has the Global Climate Changed Since the Ice Ages?

These computer-generated images show where glaciers and large ice sheets were during the last ice age and where they are today. Note how the extent of these features changed in this relatively short period of time. What caused this change, and what might happen in the future because of global warming or cooling?

28,000 Years Ago

Twenty-eight thousand years ago, Earth's climate was slightly cooler than it is today. Cool climates permitted continental ice sheets to extend across most of Canada and into the upper Midwest of the United States. Ice sheets also covered parts of northern Asia and Europe.

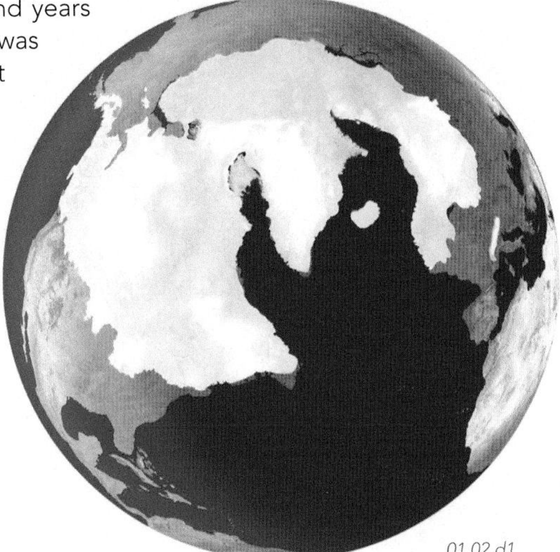

Today

During the past 15,000 to 20,000 years, Earth's climate warmed enough to melt back the ice sheets to where they are today. Our knowledge of the past extent of ice sheets comes from geologists, who examine the landscape for clues, such as glacial features and deposits that remained after the glaciers retreated.

Geology at Various Scales

Geology shapes our world at all scales of observation. Geologic history governs the location and character of mountains, plains, and coastlines. Geologic processes form various types of rocks and sculpt these rocks into the landscapes we see today. Within landscapes is evidence for the history of our planet, including past environments, past events, and ancient life, such as dinosaurs.

The study of geology helps us better understand our surroundings, from the smallest grain of sand to an entire continent. Knowing some geology will enable you to see the world in a new and different way. It will guide what you observe and how you interpret what you see. Geology helps us recognize what is happening today, reconstruct past events, and appreciate how these events are expressed in the landscapes around our homes and in the rest of our world.

Before You Leave This Page Be Able To

- ✓ Summarize some ways in which geology is used to help explain the world around us.
- ✓ Describe some things we can learn about Earth's past by observing its landscapes, rocks, and fossils.

What Is Inside Earth?

HAVE YOU WONDERED WHAT IS INSIDE EARTH? You can directly observe the uppermost parts of Earth, but what else is down there? Earth consists of concentric layers that have different compositions. The outermost layer is the *crust*, which includes *continental crust* and *oceanic crust*. Beneath the crust is the *mantle*, Earth's most voluminous layer. The molten *outer core* and the solid *inner core* are at Earth's center.

A How Does Earth Change with Depth?

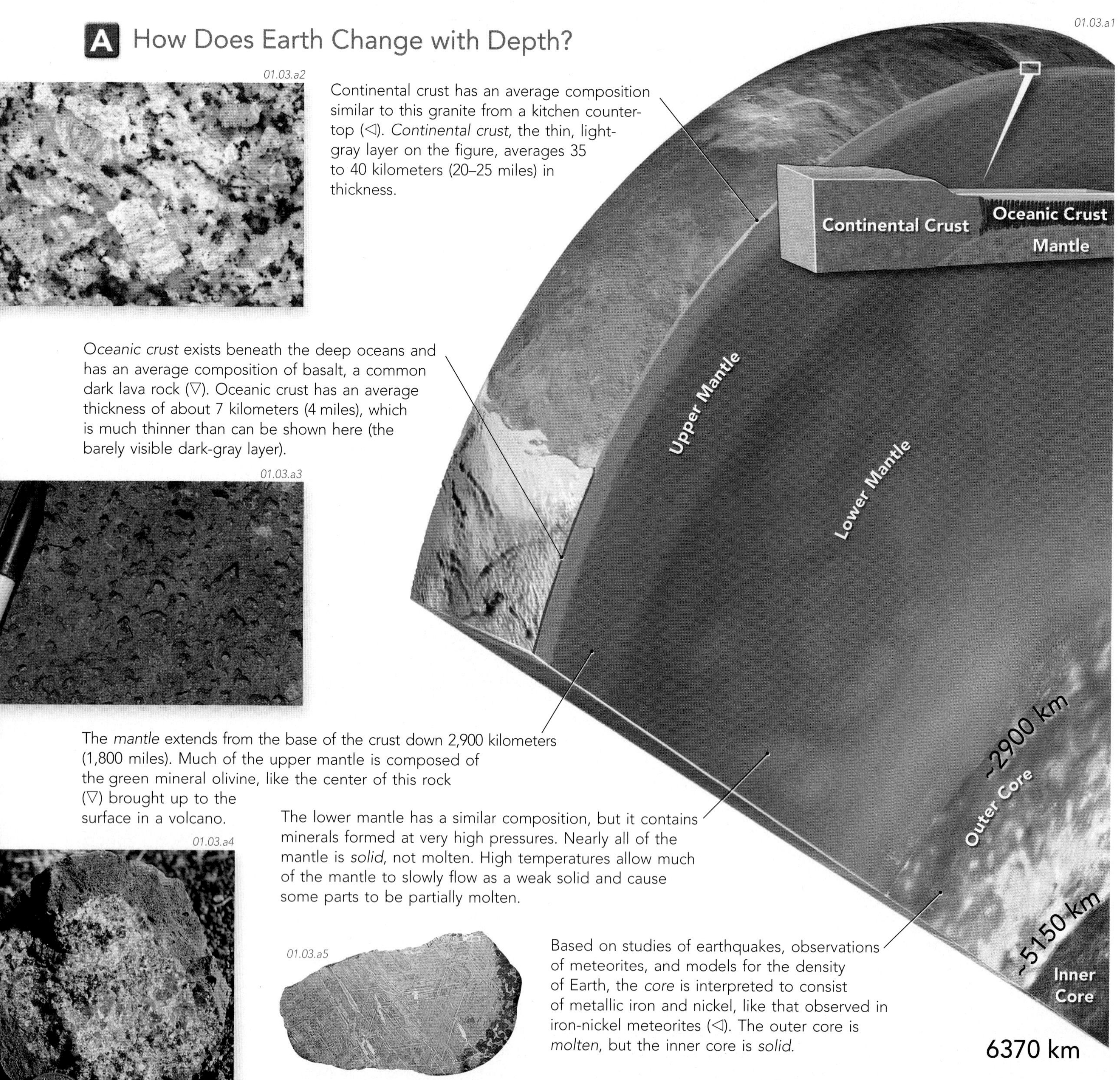

Continental crust has an average composition similar to this granite from a kitchen countertop (◁). *Continental crust*, the thin, light-gray layer on the figure, averages 35 to 40 kilometers (20–25 miles) in thickness.

Oceanic crust exists beneath the deep oceans and has an average composition of basalt, a common dark lava rock (▽). Oceanic crust has an average thickness of about 7 kilometers (4 miles), which is much thinner than can be shown here (the barely visible dark-gray layer).

The *mantle* extends from the base of the crust down 2,900 kilometers (1,800 miles). Much of the upper mantle is composed of the green mineral olivine, like the center of this rock (▽) brought up to the surface in a volcano.

The lower mantle has a similar composition, but it contains minerals formed at very high pressures. Nearly all of the mantle is *solid*, not molten. High temperatures allow much of the mantle to slowly flow as a weak solid and cause some parts to be partially molten.

Based on studies of earthquakes, observations of meteorites, and models for the density of Earth, the *core* is interpreted to consist of metallic iron and nickel, like that observed in iron-nickel meteorites (◁). The outer core is *molten*, but the inner core is *solid*.

B Are Some Layers Stronger Than Others?

In addition to layers with different compositions, Earth has layers that are defined by strength and how easily the material in the layers fractures or flows when subjected to forces.

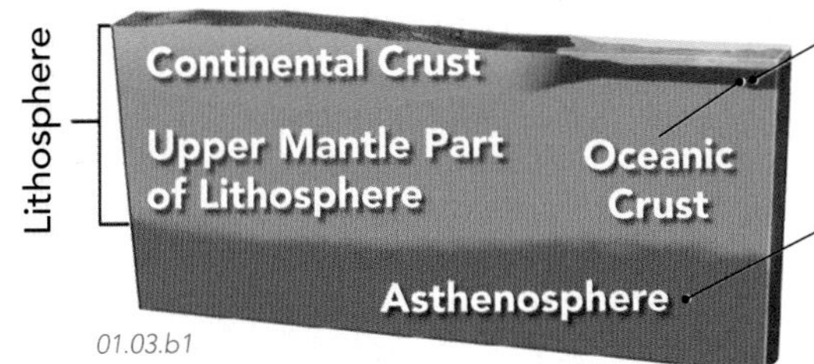

The uppermost part of the mantle is relatively strong and is joined together with the crust into an outer, rigid layer called the *lithosphere* (*lithos* means *"stone"* in Greek).

The mantle beneath the lithosphere is solid, but it is hotter than the rock above and can flow under pressure. This part of the mantle is called the *asthenosphere*. It functions as a soft, weak zone over which the lithosphere may move. The word *asthenosphere* is from a Greek term for *not strong*.

C Why Do Some Regions Have High Elevations?

Why is the Gulf Coast of Texas near sea level, and why are the Colorado mountains 3 to 5 km (2 to 3 mi) above sea level? Why are the continents mostly above sea level, but the ocean floor is typically below sea level? The primary factor controlling the elevation of a region is the thickness of the underlying crust.

The thickness of continental crust ranges from less than 25 km (16 mi) to more than 60 km (37 mi). Regions that have high elevation generally have thick crust. The crust beneath Colorado is commonly more than 45 km (28 mi) thick.

The crust, beneath regions of low elevation, such as Texas, is thinner. If the crust is thinner than 30 to 35 km (18 to 20 mi), the area will probably be below sea level, but it can still be part of the continent.

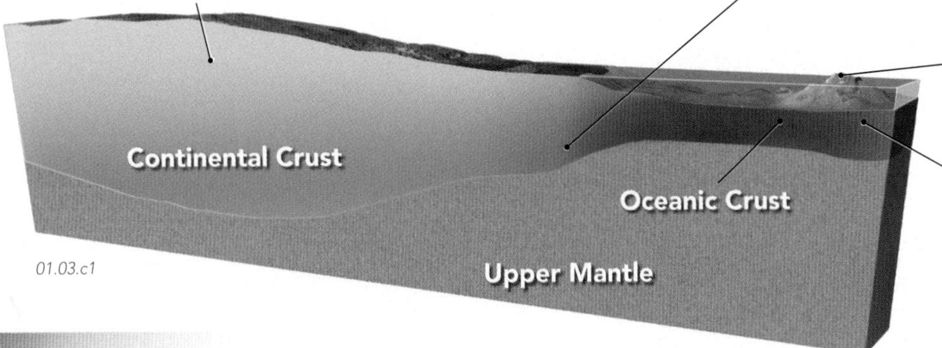

Most islands are volcanic mountains built on oceanic crust, but some are small pieces of continental crust.

Oceanic crust is thinner and consists of denser rock than continental crust, so regions underlain by oceanic crust generally are well below sea level.

Isostasy

The relationship between regional elevation and crustal thickness is similar to that of wooden blocks of different thicknesses floating in water. Thicker blocks of wood, like thicker parts of the crust, rise to higher elevations than do thinner blocks of wood.

For Earth, we envision the crust being supported by the surrounding mantle, which, unlike the liquid used in the wooden-block example, is solid. This concept of different thicknesses of crust riding on the mantle is called *isostasy*. Isostasy explains most of the variations in elevation from one region to another, and commonly is paraphrased by saying *mountain belts have thick crustal roots*.

As in the case of the floating wooden blocks, most of the change in crustal thickness occurs at depth and less occurs near the surface. Smaller, individual mountains do not necessarily have thick crustal roots. They can be supported by the strength of the crust, like a small lump of clay riding on one of the wooden blocks.

The density of the rocks also influences regional elevations. The fourth block shown below is the same thickness as the third block, but consists of a dense type of wood. It therefore floats lower in the water. Likewise, a region of Earth underlain by especially dense crust or mantle is lower in elevation than a region with less dense crust or mantle, even if the two regions have similar thicknesses of crust. If the lithosphere in some region has been abnormally heated, it expands, becoming less dense, and so the region rises in elevation. The temperature controls the *thickness* of the lithosphere, which also affects a region's elevation. We'll revisit this subject later.

Before You Leave This Page Be Able To

- ✓ Draw a sketch of the major layers of Earth.
- ✓ Summarize the difference between continental crust and oceanic crust and between lithosphere and asthenosphere.
- ✓ Discuss and sketch how the principle of isostasy can explain differences in regional elevation.

What Processes Affect Our Planet?

EARTH IS SUBJECT TO VARIOUS FORCES. Some forces are generated within Earth, and others are imposed by Earth's celestial neighbors. What happens on Earth is a consequence of these forces and how these forces interact with Earth's materials—its land, water, air, and inhabitants.

The weight of the air in the atmosphere presses down on Earth's surface and on its inhabitants. *Atmospheric pressure* is greater at sea level than at the top of mountains because there is more air above. ▽

01.04.a2

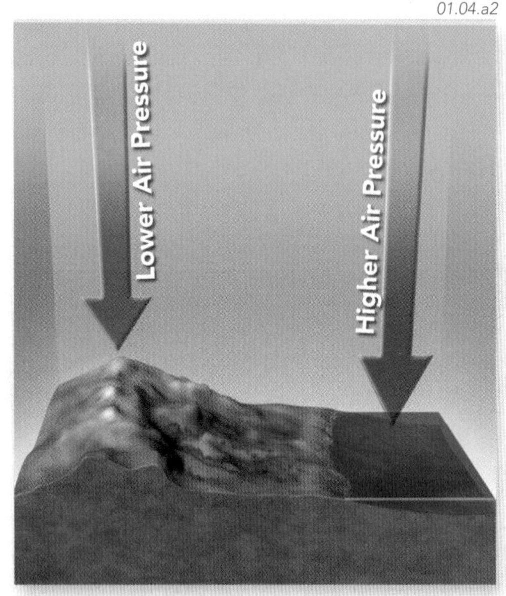

Water, in either liquid or frozen forms, moves downhill in rivers and glaciers, transporting rocks and other debris and carving downward into the landscape.

01.04.a1

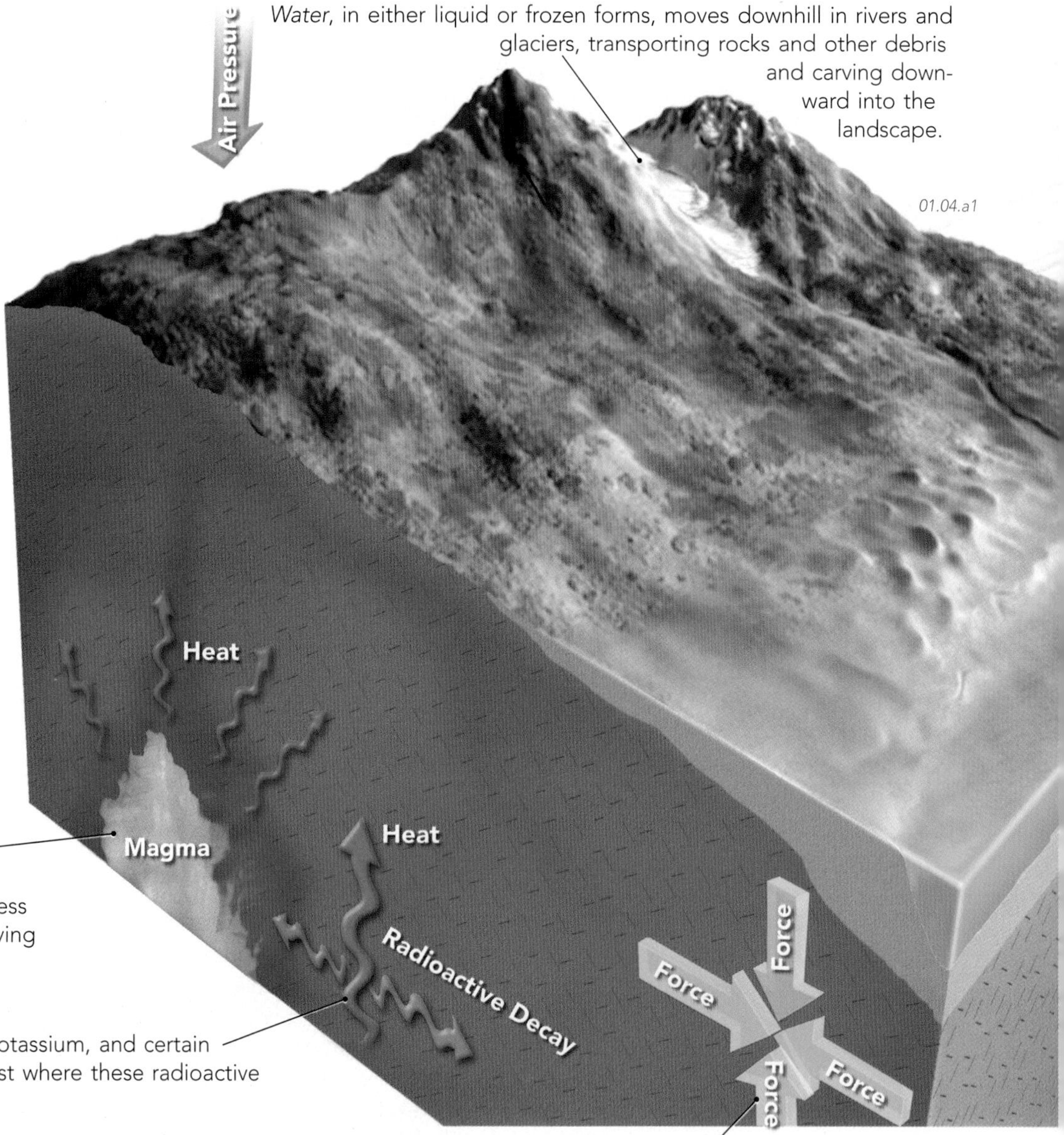

Air Pressure

Heat

Magma

Heat

Radioactive Decay

Force

Force

Force

Force

Heat from deeper in Earth rises upward toward the cooler surface. Some heat is transferred from hotter rock to cooler rock by direct contact, a process called *conduction*. Other heat is transferred by moving material, such as rising molten rock (magma).

Radioactive decay of naturally occurring uranium, potassium, and certain other elements produces heat, especially in the crust where these radioactive elements are concentrated.

The *weight of rocks* exerts a downward force on all underlying rocks, and this force increases deeper in Earth as more and more rocks are on top (▷). The downward force is balanced by forces pushing back in all directions (not shown). These forces squeeze rocks at depth from all directions.

01.04.a3

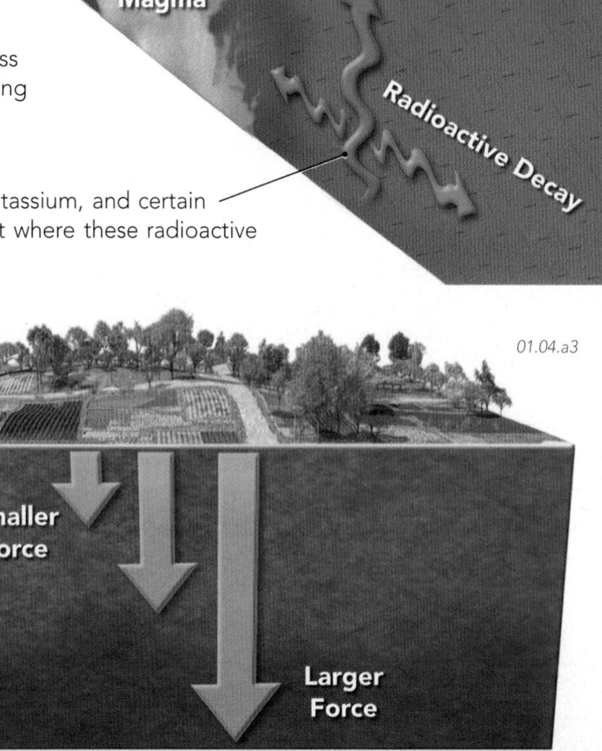

Forces produced by the weight of overlying rocks push down, and these forces are countered by forces pushing up and in from the sides. Additional forces arise by processes deep within Earth, such as from the subsurface movement of rocks and magma. Such forces generated in one area can be transferred to an adjacent area, causing sideways pushing or pulling on the rocks.

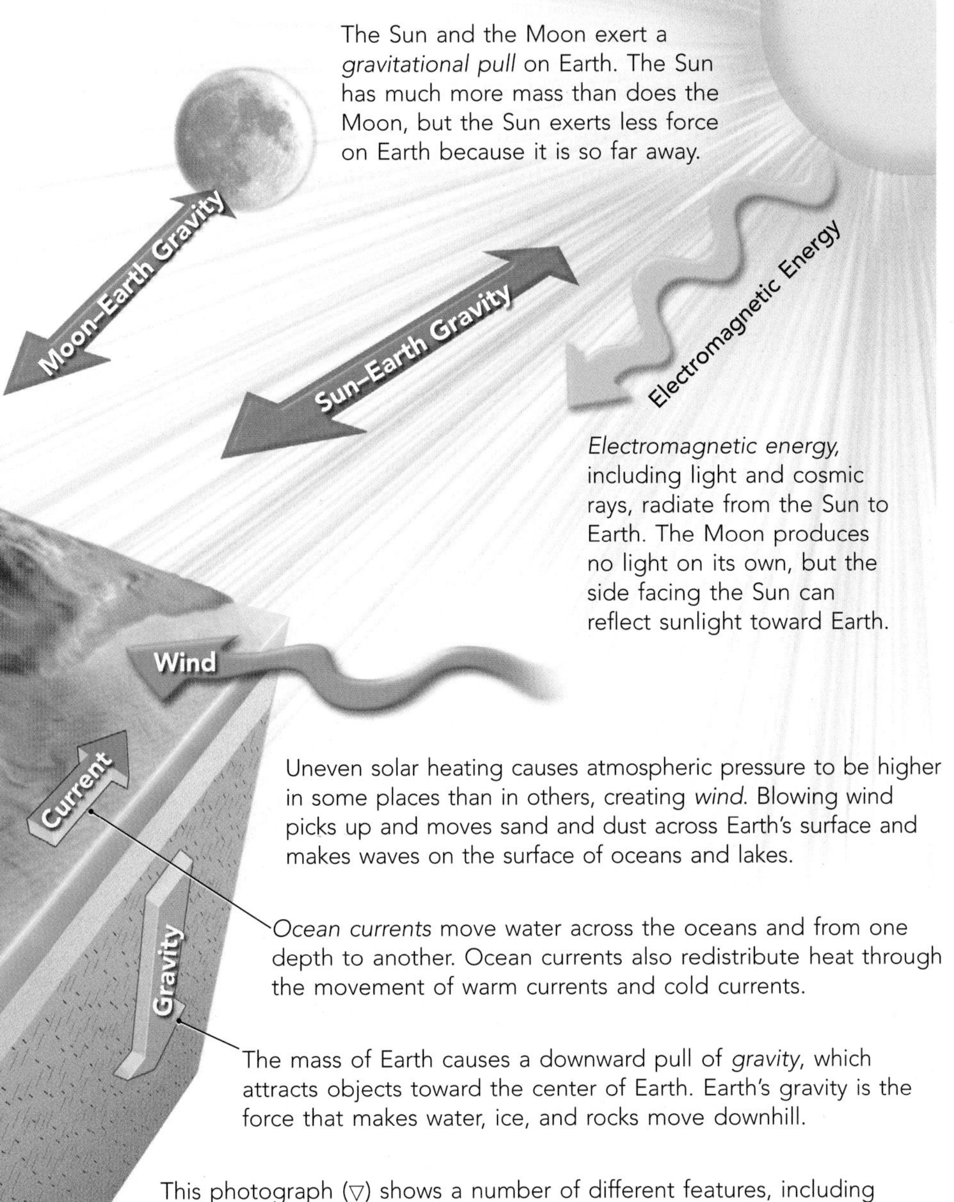

The Sun and the Moon exert a *gravitational pull* on Earth. The Sun has much more mass than does the Moon, but the Sun exerts less force on Earth because it is so far away.

Electromagnetic energy, including light and cosmic rays, radiate from the Sun to Earth. The Moon produces no light on its own, but the side facing the Sun can reflect sunlight toward Earth.

Uneven solar heating causes atmospheric pressure to be higher in some places than in others, creating *wind*. Blowing wind picks up and moves sand and dust across Earth's surface and makes waves on the surface of oceans and lakes.

Ocean currents move water across the oceans and from one depth to another. Ocean currents also redistribute heat through the movement of warm currents and cold currents.

The mass of Earth causes a downward pull of *gravity*, which attracts objects toward the center of Earth. Earth's gravity is the force that makes water, ice, and rocks move downhill.

This photograph (▽) shows a number of different features, including clouds, snow, slopes, and a grassy field with cows (the small dark spots). Observe the scene and for each feature you recognize, think about how the processes described above influence what is there and what is occurring. Finally, think about how geology influences the life of the cows. [Henry Mountains, Utah]

01.04.a4

Energy and Forces

Earth's energy supply originates from internal and external sources. *Internal energy* comes from within Earth and includes heat energy trapped when the planet formed and heat produced by radioactive decay. This heat drives many internally generated processes, such as melting of rocks at depth to produce magma.

The most significant source of *external energy* is the Sun, which bathes Earth in light, thermal energy, and other electromagnetic energy waves. Thermal energy and light from the Sun impact the equatorial areas of Earth more than the polar areas, causing temperature differences in the atmosphere and oceans. These temperature differences help drive wind and ocean currents. The Sun's light is also the primary energy source for plants, through the process of photosynthesis.

Early in Earth's history, the planet was bombarded by meteoroids and other objects left over from the formation of the solar system. Such impacts brought a tremendous amount of energy to the early Earth. Some of this energy remains stored in Earth's hot interior.

Internal forces also affect Earth. All objects that have mass exert a gravitational attraction on other masses. If a mass is large and close, the pull of gravity is relatively strong. Earth's gravity acts to pull objects toward the center of Earth. Gravity is probably the most important agent on Earth for moving material from one place to another. It causes loose rocks, flowing glaciers, and running water to move downhill from high elevations to lower ones, and drives ocean currents and wind.

Objects on Earth also feel an *external* pull of gravity from the Sun and the Moon. The Sun's gravity maintains Earth's orbit around the Sun. The Moon's pull of gravity on Earth is stronger than that of the Sun and causes more observable effects, such as the rise and fall of ocean tides.

Before You Leave This Page Be Able To

- ✓ Sketch and summarize the major processes that affect our planet.
- ✓ Describe or sketch internal forces, external forces, and sources of energy, and explain how they affect Earth.
- ✓ Summarize the importance of gravity for moving material on Earth's surface.

How Do Rocks Form?

THE VARIOUS PROCESSES THAT OPERATE on and within Earth produce the variety of rocks we see. Many common rock types form in familiar environments on Earth's surface, such as in rivers or beaches. Other rocks form in less familiar environments under high pressures deep within Earth or at high temperatures, such as those near a volcano. To understand the different types of rock that are possible, we explore the types of materials that are present in different environments.

A What Types of Rocks Form in Familiar Surface Environments?

Much of the surface of Earth is dominated by environments that many people have seen, such as mountains, rivers, and lakes. Think back to what you have observed on the ground in these types of places—probably sand, mud, and boulders. These loose pieces of rock are called *sediment* and formed by the breaking and wearing away of other rocks in the landscape. Although more hidden, sediment also occurs beneath the sea.

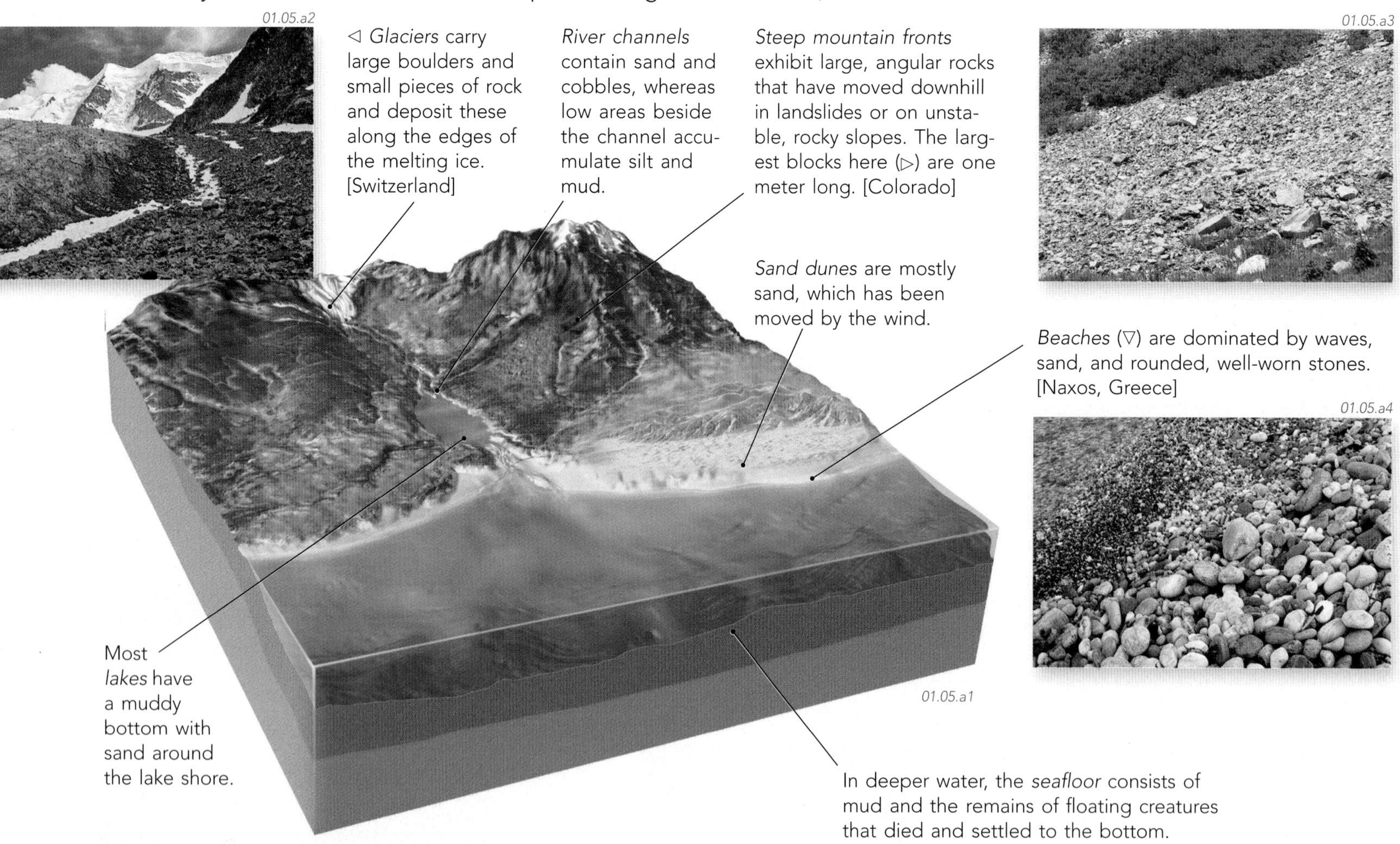

Sediments and Sedimentary Rocks

When studying sediments, we begin by observing a *modern environment* and the types of sediment that are there. Then, we infer that these same types of sediment would have been produced in older, *prehistoric* versions of that environment. By doing this, we use modern examples to understand ancient environments.

Over time, if the loose sediment is buried, it can become consolidated into hard rock. We use the term *sedimentary rocks* for sediment that originally accumulated in these familiar environments and later hardened into rock. Most sedimentary rocks form from pieces of sediment in this way, but some are produced by coral and other organisms that extract material directly from water.

B What Types of Rocks Form in Hot or Deep Environments?

Some rocks form in environments that are foreign to us. These environments may include molten rocks or may be hidden from our view, deep within Earth. A different suite of rock types forms in such environments because the rocks, instead of being constructed from sediment, form from molten rock, from hot water, or under high temperatures and pressures that can transform one type of rock into another type of rock.

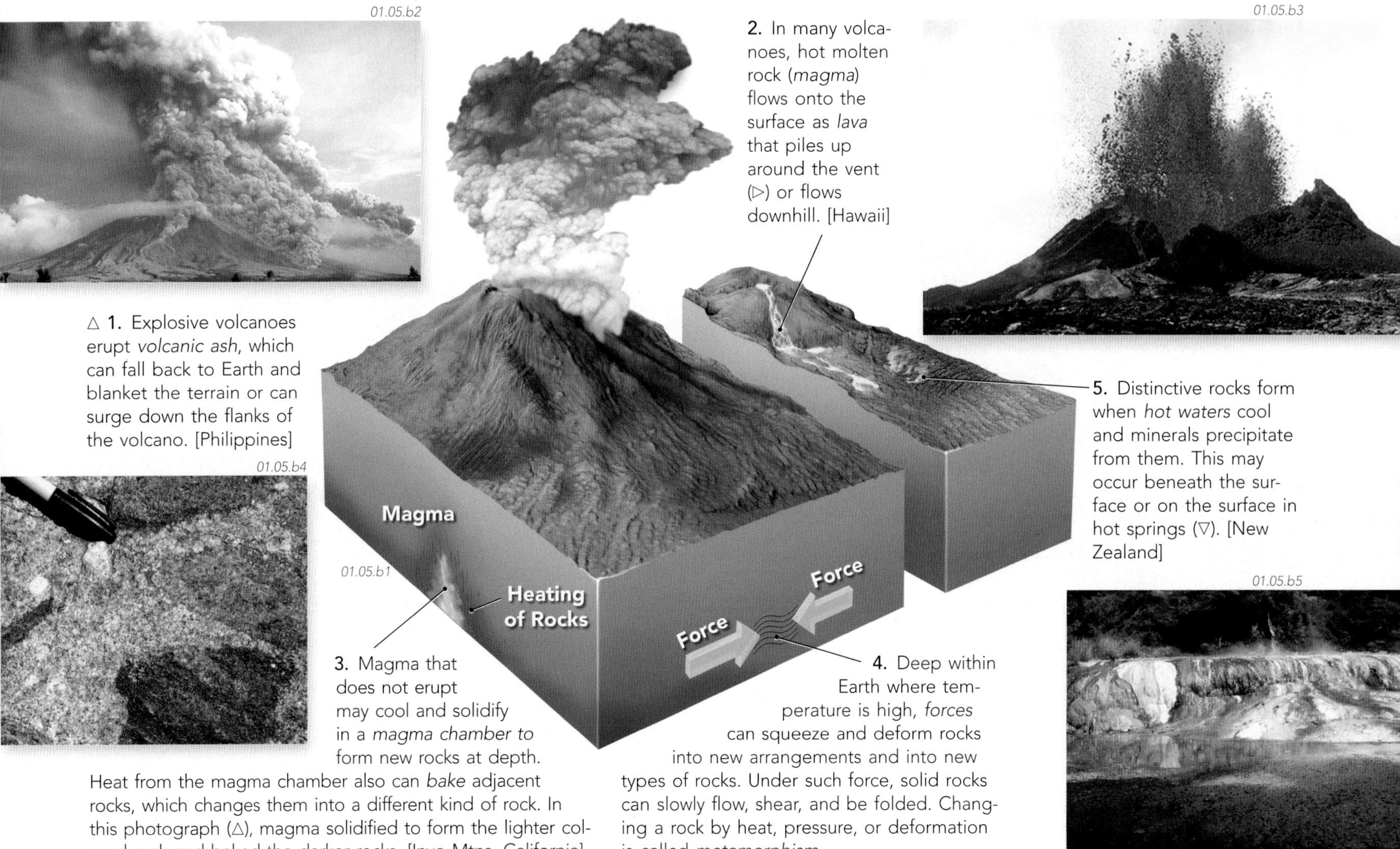

△ **1.** Explosive volcanoes erupt *volcanic ash*, which can fall back to Earth and blanket the terrain or can surge down the flanks of the volcano. [Philippines]

2. In many volcanoes, hot molten rock (*magma*) flows onto the surface as *lava* that piles up around the vent (▷) or flows downhill. [Hawaii]

3. Magma that does not erupt may cool and solidify in a *magma chamber* to form new rocks at depth. Heat from the magma chamber also can *bake* adjacent rocks, which changes them into a different kind of rock. In this photograph (△), magma solidified to form the lighter colored rock and baked the darker rocks. [Inyo Mtns, California]

4. Deep within Earth where temperature is high, *forces* can squeeze and deform rocks into new arrangements and into new types of rocks. Under such force, solid rocks can slowly flow, shear, and be folded. Changing a rock by heat, pressure, or deformation is called *metamorphism*.

5. Distinctive rocks form when *hot waters* cool and minerals precipitate from them. This may occur beneath the surface or on the surface in hot springs (▽). [New Zealand]

Families of Rocks

The diverse environments shown on these pages produce many different types of rocks that, depending on the classification scheme, are grouped into three or four families. The families of rocks compose the world around us and are important for understanding Earth's landscapes, history, and resources.

Sedimentary rocks form in normal, low-temperature environments and are the most common rocks in our landscapes. Rocks formed from magma that has cooled and solidified are called *igneous rocks*. These form as volcanoes erupt ash and lava or as magma crystallizes in magma chambers at depth.

Rocks that have been changed because they are exposed to different temperatures, pressures, or deformation are called *metamorphic rocks*. Such rocks may start as sedimentary or igneous rocks before being transformed into metamorphic rock. Finally, rocks that precipitate directly from hot water are called *hydrothermal rocks*. Some geologists classify these rocks with metamorphic rocks and so discuss three families, but in this book we separate them as a fourth family.

Before You Leave This Page Be Able To

- ✓ Distinguish the origins of the four families of rocks.
- ✓ Sketch or summarize at least five settings where sedimentary rocks form.
- ✓ Sketch or summarize one setting each for where igneous, metamorphic, and hydrothermal rocks form.

What Can Happen to a Rock?

AFTER A ROCK FORMS, many things can happen to it. It can break apart into sediment or be buried deeply and metamorphosed. If temperatures are high enough, a rock can melt and then solidify to form an igneous rock. Metamorphic and igneous rocks may be uplifted to the surface and then be broken down into sediment. If you were a rock, what could happen during your lifetime?

1. Weathering

A rock on the surface is exposed to sunlight, rain, wind, plants, and animals. As a result, it may be mechanically broken apart or altered by chemical reactions via the process of *weathering*. This can create loose pieces of rock (▽), which are called *sediment*. [Tasmania, Australia]

01.06.a2

2. Erosion and Transport

Rock pieces that have been loosened or dissolved by weathering can be stripped away by *erosion*, where sediment is removed from its source. The materials can be *transported* by gravity, glaciers, flowing water, or the wind.

01.06.a1

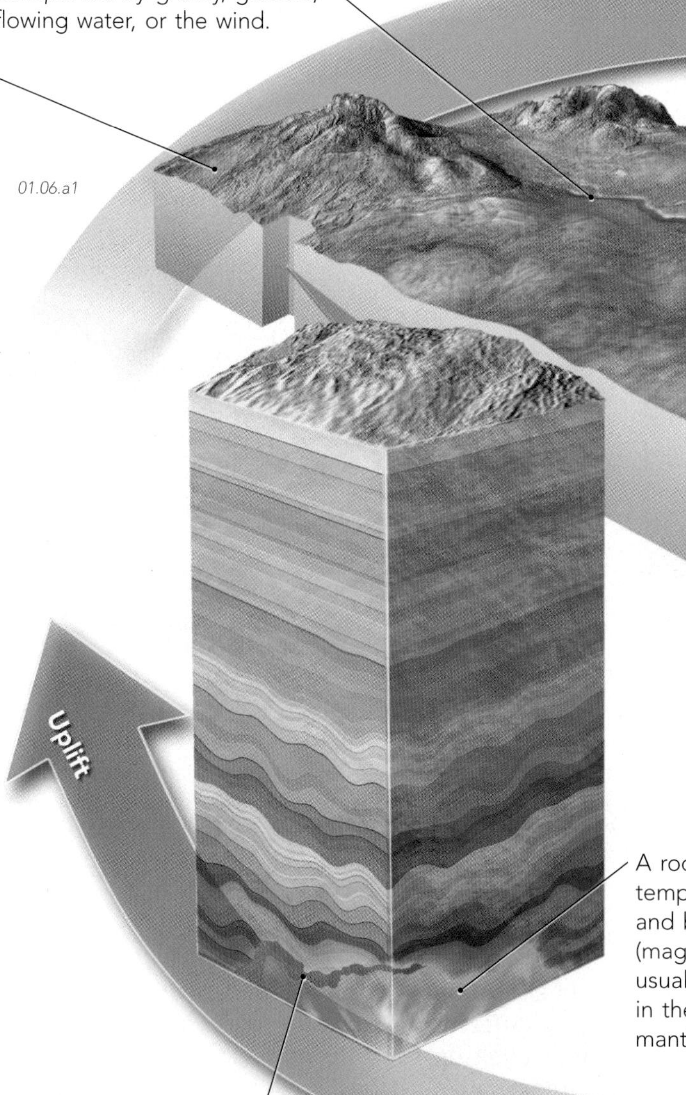

8. Uplift

At any point during its history, a rock may be uplifted back to the surface where it is again exposed to weathering. Uplift commonly occurs in mountains, but it can also occur over broad regions.

6. Melting

A rock exposed to high temperatures may melt and become molten (magma). Such melting usually occurs at depth in the lower crust or the mantle.

01.06.a3

7. Solidification

As magma cools, either at depth or after being erupted onto the surface, it begins to crystallize and harden back into solid rock, a process called *solidification*. This rock (◁) with large, well-formed crystals formed from magma that solidified at depth. The rock was later uplifted to the surface. [Capetown, South Africa]

The Life and Times of a Rock — The Rock Cycle

This process, in which a rock may be moved from one place to another or even converted into a new type of rock, is called the *rock cycle*. Many rocks do not go through the entire cycle, but instead move through only part of the cycle. Importantly, the different steps in the rock cycle can happen in almost any order. The steps are numbered here only to guide your reading. Let's follow *possible* sequences of events for a single rock.

Suppose that a rock has been recently uplifted and exposed at Earth's surface. The rock will be weathered, and broken into smaller pieces. Eventually the fragments will be eroded and transported at least a short distance before being deposited. Under the right conditions, the rock fragments will be buried beneath other sediments or perhaps beneath volcanic rocks that are erupted onto the surface. Many times, however, sediment is not buried, but only weathered, eroded, transported, and deposited again. As an example of this circumstance, imagine a rounded rock in a river. When the river currents are strong enough, the rock is picked up and carried downstream, where it might be deposited beside the channel and remain for years, centuries, or even millions of years. At a later time, a flood that is larger than the last one may pick up the rock and transport it farther downstream.

If the rock does get buried, it has two possible paths. It can be buried to some depth and then be uplifted back to the surface to be weathered, eroded, and transported again, or the rock may be buried so deeply that it is metamorphosed under high temperatures and pressures. The metamorphic rock then may be uplifted toward the surface.

If the rock, however, remains at depth and is heated to even higher temperatures, it can melt. The magma that forms may remain at depth or may be erupted onto the surface. In either case, the magma eventually will cool and solidify into an igneous rock. Igneous rocks formed at depth may be uplifted to the surface or they may remain at depth, where the rocks can be metamorphosed or even remelted.

A key point to remember is that the rock cycle illustrates the possible things that can happen to a rock. Most rocks do not complete the cycle because of the many paths, interruptions, and shortcuts a rock can take. The path a rock takes through the cycle depends on the geologic events that happen and the order in which they occur.

3. Deposition

After transport, the sediment is laid down, or *deposited*. Deposition can occur when the sediment reaches the sea or at any point along the way, such as beside the river. The river gravels below are at rest for now but could be picked up and moved by the next large flood. [Tibet]

01.06.a4

4. Burial and Lithification

Once deposited, sediment can be buried and compacted by the weight of overlying sediment. The pieces of sediment can also be cemented together by chemicals in the water. The process of sediment turning into rock is called *lithification*.

Uplift

5. Deformation and Metamorphism

After it is formed, a rock can be subjected to strong forces that squeeze the rock and fold its layers, a process called *deformation*. If buried deeply enough, a rock can be heated, deformed, and converted into a new kind of rock by the process of *metamorphism*. The rock to the left has been strongly deformed and metamorphosed. [Kettle Falls, Washington]

01.06.a5

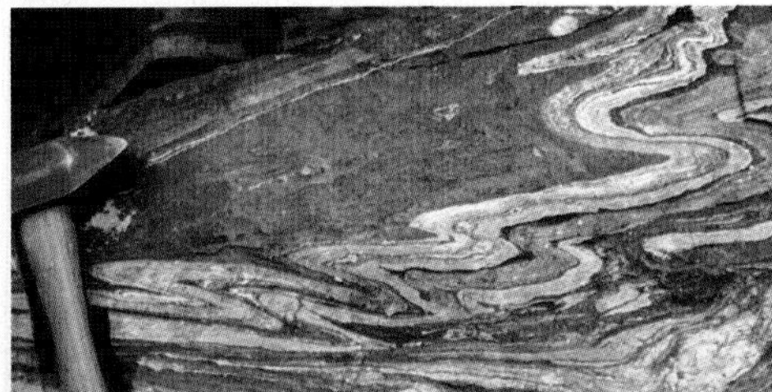

Before You Leave This Page Be Able To

- ✓ Sketch a simple version of the rock cycle, labeling and explaining, in your own words, the key processes.
- ✓ Describe why a rock might not experience the entire rock cycle.

How Do the Atmosphere, Water, and Life Interact with Earth's Surface?

WHAT SHAPES THE SURFACE OF EARTH? The elevation of Earth's surface is a reflection of crustal thickness, mountain building, and other geologic processes, but various processes give the landscape its detailed shape. Three important factors that affect landscapes are water and its movement on the surface, the atmosphere and its movement around Earth, and the impact of all forms of life.

 How Does Water Move on Our Planet?

Water on Earth resides in oceans, lakes, rivers, glaciers, and in soil and rock. Much of this water is in constant motion as it flows under the influence of gravity.

1. Rainfall impacts Earth's surface and imperceptibly erodes it. As raindrops coalesce on the surface, they form a thin film of liquid water that coats rocks and soils, and causes them to weather. As more drops accumulate, water begins to flow downhill as *runoff*.

2. When winter snows don't melt completely away, as is common at higher elevations and at polar latitudes, ice accumulates in *glaciers*, which are huge flowing fields, or tongues, of ice. As glaciers move, they transport sediment and carve the underlying landscape.

01.07.a1

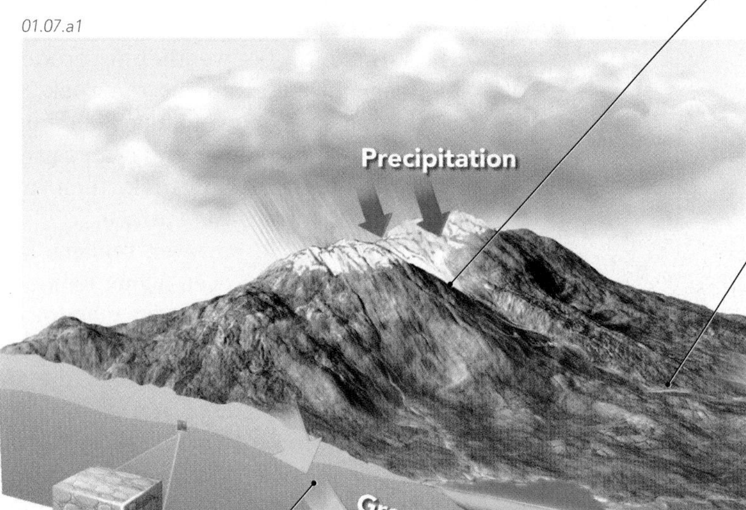

3. As moving water and the sediment it carries encounter obstructions, like solid rocks and loose debris, the water breaks them apart and picks up and transports the pieces (▷). Flowing water is the most important agent for sculpting Earth. [Mt. Cook, New Zealand]

01.07.a2

4. Water in lakes, rivers, and runoff can sink into the ground and travel through cracks and other empty spaces in rocks and soils. Such *groundwater* can react chemically with rocks through which it flows. It typically flows toward lower areas, where it may emerge back on Earth's surface as springs. [Grand Canyon, Arizona] ▽

01.07.a3

5. The uppermost part of the oceans is constantly in motion, partly due to friction between winds in the atmosphere and the surface of the oceans. Winds in the oceans cause waves (▷) that erode and shape shorelines. [Australia]

01.07.a4

B How Does the Atmosphere Affect Earth's Surface?

The atmosphere is a mix of mostly nitrogen and oxygen gas that surrounds Earth's surface. It shields Earth from cosmic radiation, transfers water from one place to another, and permits life as we know it to exist. Like the oceans, the atmosphere is constantly moving. It produces winds and storms that impact the surface.

1. The atmosphere includes a low percentage of water vapor, most of which *evaporated* from Earth's oceans. Under certain conditions, the water vapor condenses to produce clouds, which are made of water droplets. Rain, snow, and hail may fall from clouds back to the surface as *precipitation*.

2. Uneven heating of the atmosphere causes winds that blow across Earth's surface. When strong enough, wind can erode soils and rock, helping to shape the landscape. Wind can move particles as large as sand grains, which can form sand dunes and can sandblast and polish rock faces.

3. Liquid water on the surface can *evaporate*, becoming water vapor in the atmosphere. Most water vapor comes from evaporation in the oceans, but some also comes from evaporation of water bodies on land and from water vapor released by plants.

4. In the upper levels of the atmosphere, oxygen absorbs most of the Sun's harmful *ultraviolet radiation* and prevents it from reaching Earth's surface, where it would have a detrimental effect on many forms of life.

5. Light from the Sun passes through the atmosphere to be absorbed by Earth.

6. Some of the light that strikes Earth is converted to *infrared energy*. Some of this energy radiates upward and is trapped by the atmosphere, causing it to warm. This process regulates temperatures on the planet. The moderate temperatures on Earth allow water to exist as liquid water, gaseous water vapor, and solid ice.

Precipitation

Ultraviolet Energy

Infrared Energy

Wind

Evaporation

01.07.b1

The Role of the Biosphere

The *biosphere* includes life and all of the places it can exist on, below, and above Earth's surface. How does life interact with the surface of Earth? How much does life affect the landscape?

Life on Earth is currently the main source of the oxygen and carbon dioxide in the atmosphere. Why does this matter? We need oxygen to survive. Also, both gases play a critical role in breaking down material on Earth's surface. Oxygen combines readily with many earth materials and causes them to weather so that they can be broken apart and carried away by water or wind. Carbon dioxide, when combined with water, makes a weak acid that also can chemically weaken rocks. Vegetation changes the way water moves across the surface of Earth, and allows water to remain in contact with rocks and soils longer. This increases the rate at which weathering breaks down earth materials.

Humans have affected the surface of Earth by removing vegetation that would compete with crops, villages, and cities. This fact is clearly shown in satellite data that show, for example, clear-cutting of forests near Mt Rainier, Washington. Whenever we clear forests or build cities, we change the chemical balance of the atmosphere and usually increase the power of water to erode into the land.

Before You Leave This Page Be Able To

- ✓ Draw a sketch that includes the major ways that water moves over, on, and under Earth's surface.
- ✓ Summarize how moving water, ice, and wind can sculpt Earth's surface.
- ✓ Explain how sunlight heats the atmosphere.
- ✓ Explain the connections between life, the atmosphere, and the land.

1.7

What Is Earth's Place in the Solar System?

EARTH IS NOT ALONE IN SPACE. It is part of a system of planets and moons associated with the Sun, which together comprise a *solar system*. The Sun is the most important object for Earth because it provides light and heat, without which life would be difficult if not impossible. Earth has a number of neighbors, including the Moon, in our solar system.

A What Are Earth's Nearest Neighbors?

Earth has five nearby neighbors—three other planets, one moon, and the Sun. The Sun and the Moon have the greatest effect on Earth, but the three other planets provide a glimpse of how Earth might have turned out. Earth and three other inner planets are rocky and are called *terrestrial planets*.

1. The Sun is the center of our solar system. It is by far the largest object in our solar system, but it is only a medium-sized star compared with other suns in other solar systems in our galaxy. The Sun's gravity is strong enough to keep all the planets orbiting around it. On Earth, a year is defined as the time it takes Earth to complete one orbit around the Sun.

2. The Sun creates light, heat, and other types of energy by fusing together hydrogen atoms in a process called *nuclear fusion*. This process is different than the process of *nuclear fission*, which causes atoms to break apart and is how Earth generates much of its internal heat. The Sun is the only object in our solar system that generates its own light—all the planets and moons, including our own, are bright because they reflect the Sun's light.

SUN

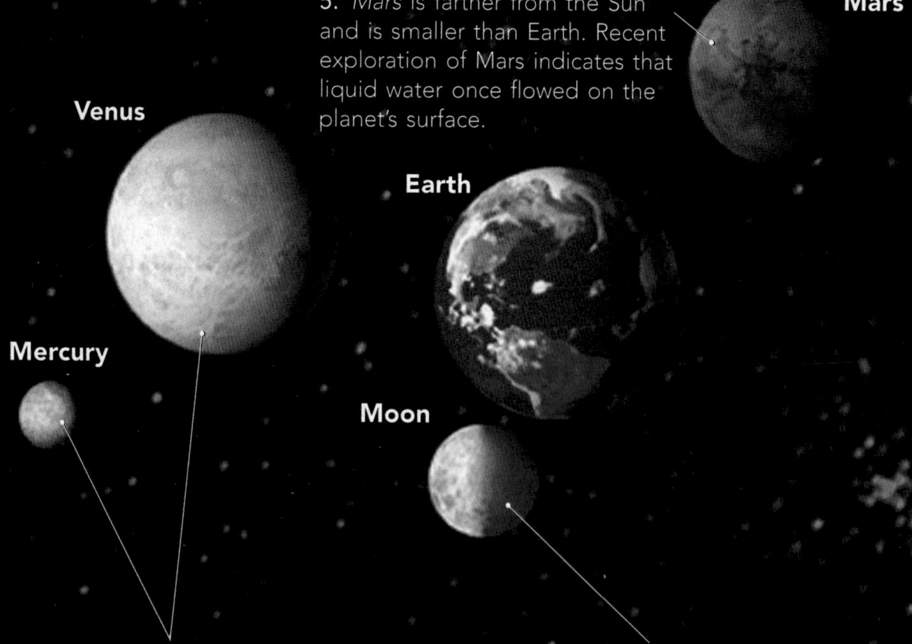

5. *Mars* is farther from the Sun and is smaller than Earth. Recent exploration of Mars indicates that liquid water once flowed on the planet's surface.

3. *Mercury* and *Venus* are planets that are closer to the Sun than Earth. Both planets are much warmer than Earth but for different reasons. Mercury is close to the Sun and has virtually no atmosphere. Venus has a thick atmosphere that traps heat like a greenhouse. Venus is shrouded in clouds, but it is shown here with no clouds.

4. The closest object to Earth is the *Moon*, which reflects sunlight back toward Earth, providing at least some light on most nights. The Moon's surface is covered with craters produced by meteoroid impacts. Many craters are large enough to be seen from Earth with binoculars. The Moon's gravity causes the rise and fall of tides along ocean shorelines.

01.08.a1

B What Are Some Characteristics of the Outer Planets?

The outer planets and many asteroids are beyond Mars. Four outer planets are called gas giants because of their large size and gas-rich character. Pluto is a small, distant object that is no longer considered to be a planet.

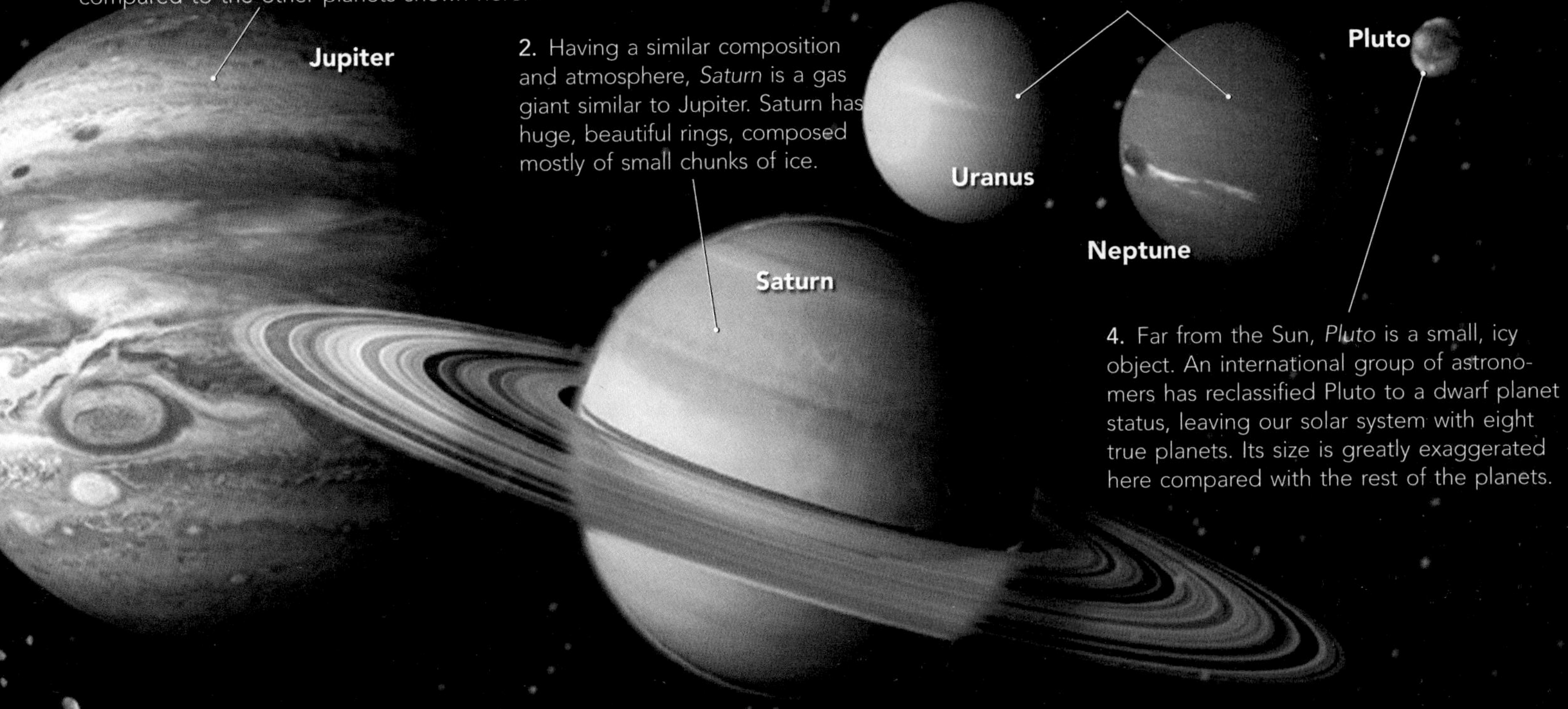

C What Is the Shape and Spacing of the Orbits of the Planets?

This is what the orbits of the inner planets and Jupiter would look like if viewed from above the plane within which the planets orbit the Sun. In other words, if you traveled to Earth's North Pole and straight up, this is the view you would have.

Observe that all of the planet orbits, including Earth's orbit, are almost circular. In other words, Earth is at about the same distance from the Sun during all times of the year.

Note how far from the Sun Jupiter is compared with the inner planets. The distance from Jupiter to Saturn is greater than the distance from the Sun to Mars, and the distances to Uranus and Neptune are even larger. Jupiter is much larger, relative to the other planets, than shown here.

The sizes of the planets are greatly exaggerated here relative to the size of the Sun.

Because Earth's orbit is essentially circular, it receives nearly the same amount of light and heat at all times of the year. Earth's seasons (summer and winter), therefore, are not caused by changes in the distance between Earth and the Sun. The seasons have another explanation, which we will explore later.

Before You Leave This Page Be Able To

- ✓ Sketch a view of the Solar System, from the Sun outward to Jupiter.
- ✓ Explain why the Sun and the Moon are the most important objects to Earth.
- ✓ Summarize how the outer planets are different from the inner planets.

How Is Geology Expressed in the Black Hills and in Rapid City?

THE BLACK HILLS OF SOUTH DAKOTA AND WYOMING are a geologic wonder. The area is famous for its gold and for the presidents' faces carved into granite cliffs at Mt. Rushmore. Rapid City, at the foot of the mountains, was devastated by a flash flood in 1972. In this area, the impacts of geology are dramatic and provide an opportunity to examine how geologic concepts presented in this chapter apply to a real place.

A What Is the Setting of the Black Hills?

01.09.a1

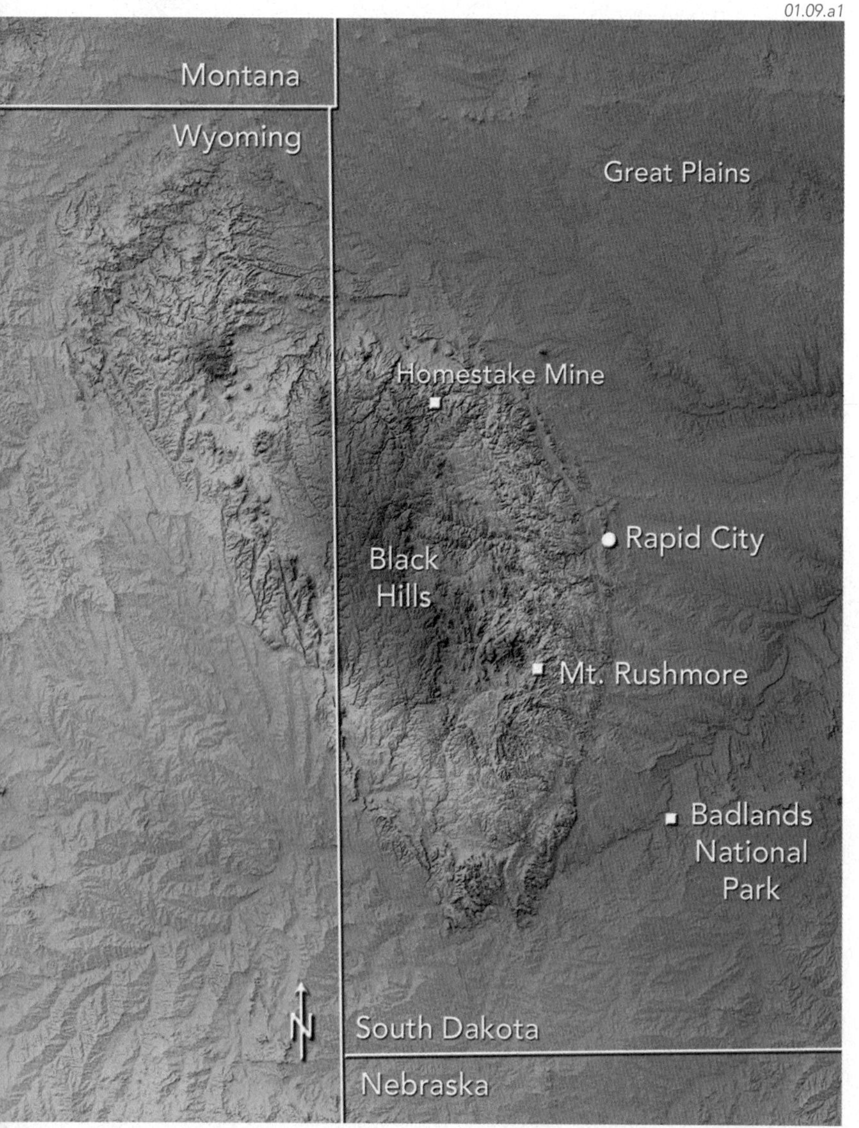

◁ As seen in this shaded relief map, the Black Hills are an isolated mountainous area that rises above the surrounding Great Plains. The region has moderately high elevation, more than 1,000 m (3,000 ft) above sea level, because the continental crust beneath the area is thick (about 45 km, or 28 mi).

Famous gold deposits of the Homestake Mine formed at submarine hot springs nearly 1.8 billion years ago. The rocks were then buried deep within the crust, where they were heated, strongly deformed, and metamorphosed. Much later, uplift of the Black Hills brought the rocks closer to the surface.

Rapid City is on the eastern flank of the Black Hills. To the south, Badlands National Park, known for its intricately eroded landscapes, is carved into soft sedimentary rocks. The region around Rapid City is also famous for its caves.

The presidents' faces at Mt. Rushmore (▽) were chiseled into a granite that solidified in a magma chamber 1.7 billion years ago. The granite and surrounding metamorphic rocks were uplifted to the surface when the Black Hills formed 60 million years ago.

01.09.a2

▽ This figure shows the geometry of rock units beneath the surface of the Black Hills. The Black Hills rose when horizontal forces squeezed the region and warped the rock layers upward. As the mountains were uplifted, weathering and erosion stripped off the upper layers of rock exposing an underlying core of ancient igneous and metamorphic rocks (shown here in brown). Rapid City is near the boundary between the hard, ancient bedrock of the mountains (shown in brown) and sedimentary rocks of the plains, (shown in purples, blues, and greens).

01.09.a3

B What Geologic Processes Affect the Rapid City Area?

This view of Rapid City is a satellite image superimposed over topography. Examine this scene and think about the geologic processes that might be occurring in each part of the area.

Rapid City is located along the mountain front, partly in the foothills and partly on the plains. Some parts of the city are on low areas next to Rapid Creek, which begins in the Black Hills and flows eastward through a gap in a ridge and then through the center of the city.

Upturned rock layers form a ridge that divides the city into two halves. Some of the homes are right along the creek, whereas others are on the steep hillslopes.

The plains contain sedimentary rocks, some of which were deposited in a great inland sea and then buried by other rocks. With uplift of the mountains and erosion, the rocks came back to the surface where they are weathered and eroded today.

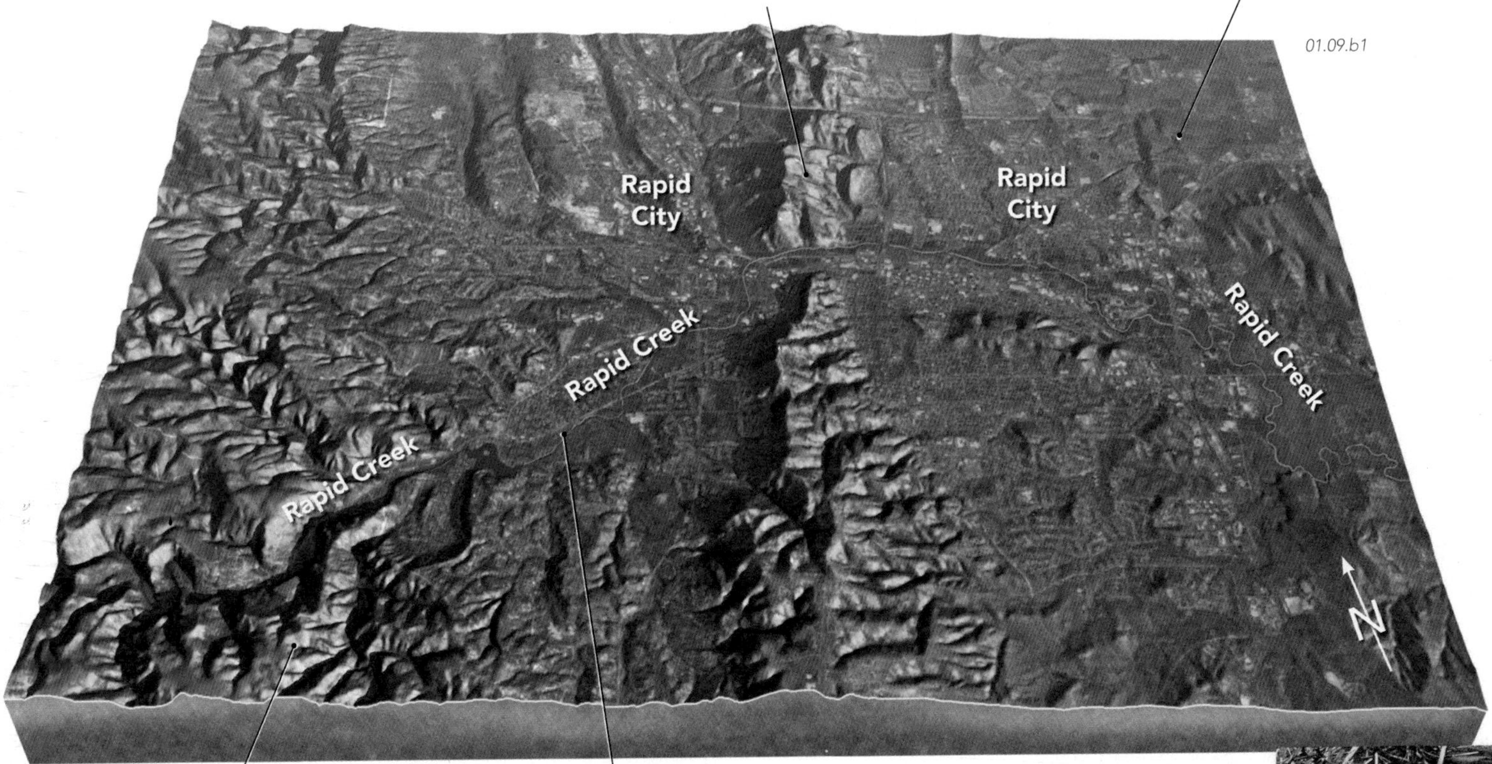

This part of the Black Hills consists of hard igneous and metamorphic rocks that form steep mountains and canyons. Farther north (not in this view), the world-famous *Homestake Mine* produced 39 million ounces of gold, more than any other mine in the Western Hemisphere. The underground mine reached depths of more than 2.5 km (8,000 ft)!

Rapid Creek drains a large area of the Black Hills and flows through the middle of Rapid City. A small dam forms Canyon Lake just above the city.

01.09.b2

The area along Rapid Creek (▷) was littered with shattered lumber and other debris from homes that were destroyed by a flash flood that occurred in 1972, an example of the hazards of living too near flowing water.

The Rapid City Flash Flood of 1972

In June of 1972, winds pushed moist air westward up the flanks of the Black Hills, forming severe thunderstorms. The huge thunderstorms remained over the mountains, where they dumped as much as 15 inches of rain in one afternoon and evening. This downpour unleashed a flash flood down Rapid Creek that was ten times larger than any previously recorded flood on the creek. The swirling floodwaters breached the dam at Canyon Lake, which increased the volume of the flood downstream through Rapid City. The floodwaters raced toward the center of the city. These floodwaters killed 238 people and destroyed more than 1,300 homes. They caused 160 million dollars in damage. Most of the damage occurred along the river channel, where many homes had been built too close to the creek, and in areas low enough to be flooded by this large volume of water.

Before You Leave This Page Be Able To

- ✓ Briefly explain or sketch the landscape around Rapid City and how geology affects this landscape.
- ✓ Identify and explain ways that geology affects the people of Rapid City.
- ✓ Summarize the events that led to the Rapid City Flood of 1972 and why there was so much damage.

How Is Geology Affecting This Place?

GEOLOGY HAS A MAJOR ROLE, from global to local scales, in the well-being of our society. The image below shows satellite data superimposed on topography for an area near St. George, Utah. In this investigation you will identify some important geologic processes operating in this region and think about how geology affects the people who live here.

Goals of This Exercise:

- Determine where important geologic processes are occurring.
- Interpret how geology is affecting the people who live here.
- Identify a place that is away from geologic hazards and where it is safe to live.

Begin by reading the *procedures* list on the next page. Then examine the figure and read the descriptions flanking the figure.

1. Most of this region receives only a small amount of rain and is fairly dry. The low areas are part of a desert that has little vegetation and is hot during the summer. The dry climate, coupled with erosion, provides dramatic exposures of the various rock types.

2. A high, pine-covered mountain range flanked by steep cliffs is next to the valley (▽). The mountains receive abundant winter snow and torrential summer rains, which cause flash flooding down canyons that lead into the valley. This photograph shows the valley and mountains, viewed toward the northwest.

01.10.a2

3. The topography (height) of the area has been exaggerated in this view to better show the features. The mountains are shown twice as high and twice as steep as they really are. Exaggerating the topography in this way is called *vertical exaggeration*.

4. The Virgin River receives water from precipitation in mountains around Zion National Park. It enters the valley through a narrow gorge. There are hot springs at the end of the gorge, where the river flows through the cliffs.

Reservoir

2 miles

Procedures

Use the figure and descriptions to complete the following steps. Record your answers in the worksheet.

1. Using the image below, explore this landscape. Make observations about the land and the geologic processes implied in the landscape. Next, mark on the provided worksheet at least one location where the following geologic processes would likely occur: weathering, erosion, transport of sediment, deposition, formation of igneous rock, flooding, and landsliding.
2. Using your observations and interpretations, list on the worksheet all the ways that geology might influence the lives of the people who live here. Think about each landscape feature and each geologic process, and then decide whether it has an important influence on the people.
3. Using all your information, select a location away from geologic hazards that would be a relatively safe place to live. Mark this location on your worksheet with the word *Here*.

5. Several of these dark lumpy hills are volcanoes that have erupted in the recent geologic past (last million years). When the volcanoes erupted, they poured molten rock (lava) onto the surface and launched hot volcanic projectiles into the air.

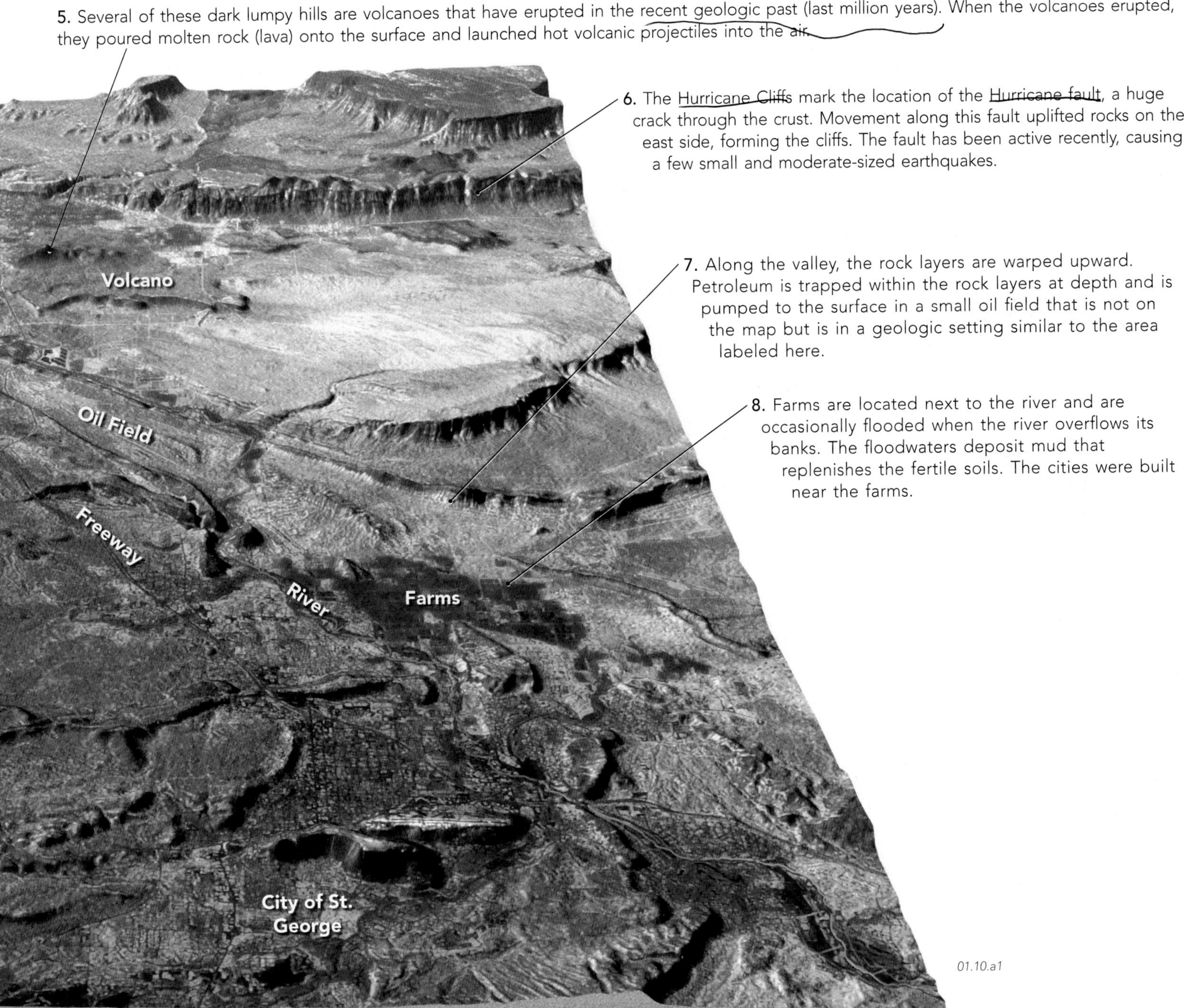

6. The Hurricane Cliffs mark the location of the Hurricane fault, a huge crack through the crust. Movement along this fault uplifted rocks on the east side, forming the cliffs. The fault has been active recently, causing a few small and moderate-sized earthquakes.

7. Along the valley, the rock layers are warped upward. Petroleum is trapped within the rock layers at depth and is pumped to the surface in a small oil field that is not on the map but is in a geologic setting similar to the area labeled here.

8. Farms are located next to the river and are occasionally flooded when the river overflows its banks. The floodwaters deposit mud that replenishes the fertile soils. The cities were built near the farms.

01.10.a1

1.10

CHAPTER

2

Investigating Geologic Questions

OUR WORLD IS FULL OF GEOLOGIC MYSTERIES. Investigating these mysteries requires asking appropriate questions and then knowing what to observe, how to interpret what we see, and how to analyze the problem from different viewpoints. The investigation of geologic questions leads to new ideas and theories about Earth. This chapter explores ways to investigate geologic questions, beginning with a mystery about the Mediterranean Sea.

This image of the Mediterranean region shows the seafloor colored in shades of blue according to depth, with darker blue representing deeper water. On land, satellite data show rock and sand in shades of brown or tan and areas with forests and grasslands in shades of green.

Trace with your finger the entire coast of the Mediterranean Sea. Where does the Mediterranean Sea connect with an ocean?

Drilling from a research ship encountered thick layers of salt within sediments at the bottom of the Mediterranean Sea. Such salt layers usually form when large volumes of water evaporate, such as in hot, dry climates.

How do you get salt deposits at the bottom of a sea?

02.00.a1

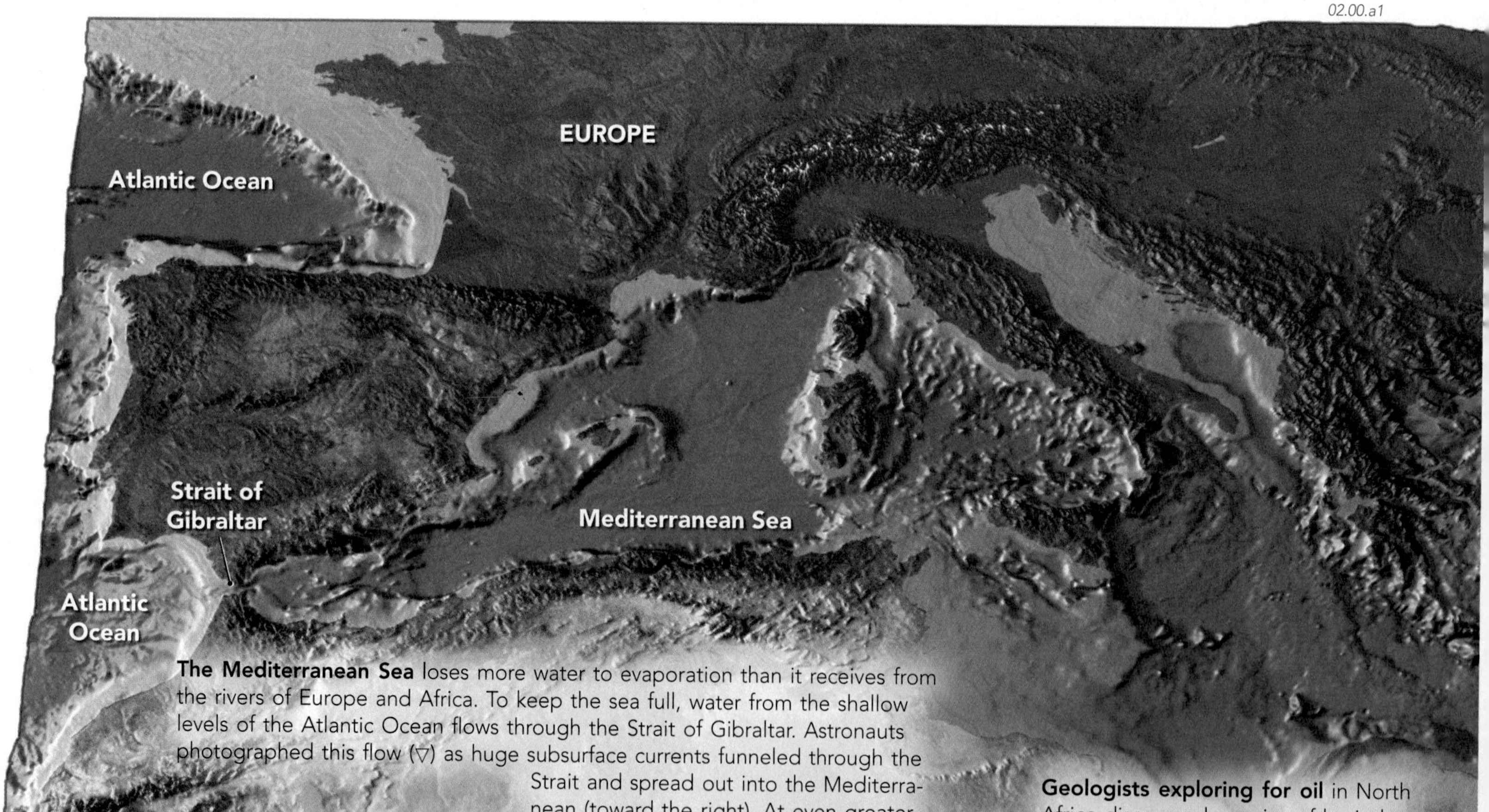

The Mediterranean Sea loses more water to evaporation than it receives from the rivers of Europe and Africa. To keep the sea full, water from the shallow levels of the Atlantic Ocean flows through the Strait of Gibraltar. Astronauts photographed this flow (▽) as huge subsurface currents funneled through the Strait and spread out into the Mediterranean (toward the right). At even greater depths, some water also flows from the Mediterranean out to the Atlantic.

What would happen if this flow to and from the Atlantic Ocean was blocked?

Geologists exploring for oil in North Africa discovered a series of buried canyons beneath the Nile River Valley and the sands of the Saharan Desert.

When and how did these buried canyons form?

02.00.a2

TOPICS IN THIS CHAPTER

About six million years ago, many species of animals around the Mediterranean Sea and Black Sea suddenly became extinct. Marine organisms that lived in the deeper parts of the Mediterranean Sea were most affected.

What dramatic changes in the environment caused these species to perish?

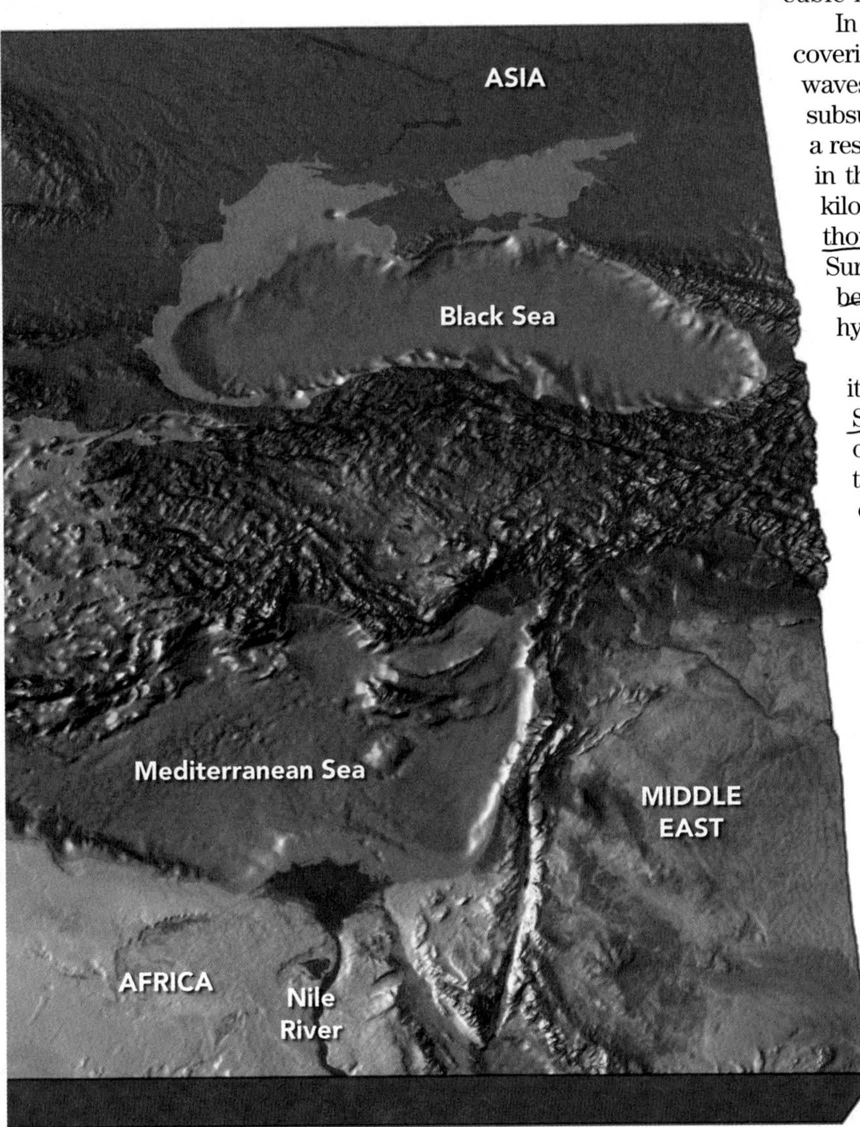

When the Mediterranean Sea Was a Desert

The picturesque Mediterranean Sea is 4,000 km (8,700 mi) long and except for its connection to the Atlantic Ocean through the Strait of Gibraltar is surrounded by land. The Mediterranean is, on average, 1,500 m (4,900 ft) deep and holds 1.6 million cubic kilometers (1.0 million cubic miles) of water.

In the 1960s and early 1970s, geologists made a series of puzzling discoveries in the Mediterranean Sea. Surveys of the seafloor made using sound waves revealed some unusual features, similar to those observed for large subsurface layers of salt. To investigate these features, scientists brought in a research ship, the *Glomar Challenger*, which was capable of drilling holes in the seafloor and retrieving samples, even from water depths of several kilometers. This and later drilling revealed that layers of salt hundreds to thousands of meters thick existed within the sediments on the seafloor. Surprisingly, the drilling also encountered sands that show evidence of being deposited by wind. From these and other studies arose an amazing hypothesis—that in the past the Mediterranean Sea had totally dried up.

According to this hypothesis, the flow of Atlantic water into the Mediterranean Sea was blocked six million years ago by bedrock near the Strait of Gibraltar. The blockage occurred because of volcanism, uplift of bedrock by mountain building, or a worldwide drop in sea level. As the water in the Mediterranean evaporated, it deposited layer upon layer of salt. The large thickness of salt requires that seawater spilled into the Mediterranean basin from the Atlantic Ocean many times and then evaporated. After several hundred thousand years, the Mediterranean Sea evaporated totally and became a dry, hot, salt flat, similar to parts of Death Valley, but 1,500 to 3,000 m (5,000 to 10,000 ft) below sea level. Rivers draining into this new deep basin eroded down through the land and cut canyons hundred of meters deep. The drying of the Mediterranean Sea caused profound climate changes in the region, and perhaps across much of the planet, leading to extinction of local organisms on a massive scale.

By 5.3 million years ago, a global rise of sea level caused Atlantic Ocean water to spill over the bedrock and cascade into the Mediterranean Sea. A gigantic waterfall at the Strait of Gibraltar was 800 m (2,600 ft) high and is estimated to have carried 1,000 times more water than Niagara Falls. Even with this torrent of water, geologists calculate that it took more than 100 years to refill the Mediterranean basin. As the Mediterranean Sea rose, adjacent rivers deposited sediment, which filled and buried the recently cut canyons. This hypothesis for the Mediterranean Sea illustrates how an observation can lead to questions and to a possible hypothesis, which can become an accepted scientific explanation after being tested thoroughly.

What Can We Observe in Landscapes?

EARTH'S HISTORY IS RECORDED in its rocks and landscapes. To understand this history, we often begin by observing a landscape to determine what is there. Most geologic landscapes display a variety of features, such as different rock layers and numerous fractures. These pages provide a guide for observing a landscape in order to read its story.

A What Features Do Landscapes Display?

Observe the top photograph, trying to identify distinct parts of the scene and then focusing on one part at a time. After examining the photograph, read the accompanying text.

02.01.a1

Color commonly catches our attention. These rocks are various shades of red, tan, and gray. Close examination of these rocks by geologists reveals that the rocks consist of consolidated sand and mud, and therefore are *sedimentary rocks*. [Canyonlands National Park, Utah]

Another thing to notice is that this hill has different parts. At the very top there is a small knob of light-colored rocks.

Below the knob is a reddish and tan slope and a small reddish cliff.

There is a main, light-colored cliff, the upper part of which has a tan color and is fairly smooth and rounded.

Some parts of the cliff have horizontal lines that can be followed around corners of the cliff. These lines are the outward expression of layers within the rock. Such layers in sedimentary rocks are called *bedding*.

Lower parts of the cliff have a darker reddish-brown color and display many sharp angles and corners. Some of these corners coincide with vertical cracks, or *fractures*, that extend back into the rock. The red color on the lower cliff is a natural stain on the *outside* of the rocks.

Below the cliff is a slope that has pinkish-red areas that are locally covered by loose pieces of light-colored rock. A reasonable interpretation is that the loose pieces have fallen off the main cliff.

02.01.a2

In this figure, color overlays accentuate different features and parts of the hill. Compare these features with the photograph above.

The uppermost three rock units (numbered 1, 2, and 3) are shaded tan and orange. Rocks of the main cliff (4) are shaded a light purplish color. On the lower slope, the reddish rocks are shaded orange (5), whereas the covering of loose rocks is shaded gray (6).

Brown lines highlight obvious and subtle layers (bedding) in the rock units.

Black lines mark *fractures* cutting the rocks.

Simplifying this scene into a few types of features makes it easier to observe, describe, and understand the landscape. In this landscape, we observe a few rock layers and numerous fractures. Some layers are more resistant to weathering and form cliffs, whereas less resistant ones form slopes. Weathering has rounded off corners on the top of the cliff, removed the reddish stain, and loosened pieces that fell off the cliff, covering a slope of underlying, reddish rocks.

Reexamine the top photograph. Do you look at the scene differently? Try this strategy when observing features where you live.

B Should We Pay Equal Attention to Everything in a Landscape?

To recognize key geologic information in a landscape, it helps to determine which parts of the scene are distractions. Both photographs below have aspects that distract our view from the exposed rocks.

If we are focusing on the sequence of rock layers in this scene, distractions include the loose brown soil in the foreground and the shape of the hill, especially the small ridges jutting out from the hill. [Petrified Forest, Arizona]

02.01.b1

02.01.b2

These rocks form cliffs separated by narrow, snow-covered slopes and ledges, the expression of nearly horizontal layers. A scenic distraction is provided by trees and fall colors. [Durango, Colorado]

C What Are Some Aspects to Observe in a Landscape?

Observe the photograph below and try to recognize some of the same features, such as layers and fractures, that were present in the earlier photographs. After you have made your observations, read the accompanying list that identifies some aspects to observe in any landscape.

02.01.c1

- Shapes into which the rocks have eroded
- Colors of the different rocks, including whether some colors are a natural stain on only the outside surfaces of the rock
- Cliffs and ledges, which generally represent harder rocks, versus slopes or soil-covered areas, which commonly mark materials that are more easily weathered
- Rock that is still a part of the *bedrock* (solid parts of the land) versus loose pieces of weathered or eroded sediment
- The presence of rock layers and the orientation and geometry of the layers, including whether the layers are horizontal, tilted, or folded
- Fractures in the rock, including their orientation and variation in spacing from place to place
- The different kinds of rocks, such as sandstone, that appear to be present

Strategies for Observing Landscapes

Certain strategies are useful for observing natural landscapes or rock that is exposed in a construction cut along a road (called a *roadcut*).

- Begin by carefully observing the entire landscape or roadcut, and then focus on smaller parts, one part at a time.
- Identify the different *types of features*, such as layers, fractures, or loose pieces of rock.
- Focus on one type of feature at a time and try to determine where the feature is present, where it is not, and how it is expressed in the scene.
- Examine relationships between different features, such as whether loose pieces of rock are mostly below a certain cliff.
- Finally, in this book you will learn how to use the characteristics of each rock type to infer the environment in which that rock formed, and how the environment might have changed through time. That is, you will observe a scene and begin to interpret its geologic history.

Before You Leave This Page Be Able To

- ✓ Summarize or sketch the main components of a relatively simple landscape, like those shown here.
- ✓ Summarize the different features you can observe in a landscape.
- ✓ Summarize the strategies for observing a landscape, roadcut, or photograph of a geologic scene.

How Do We Interpret Geologic Clues?

LANDSCAPES AND ROCKS CONTAIN MANY CLUES about their geologic history. From the characteristics of a rock, we can infer the environment in which it formed. We can also apply some simple principles to determine the age of one rock unit or geologic feature relative to another. Changes in landscapes through time provide additional clues about how a place has changed and why it has its present appearance.

A How Can We Infer the Environment in Which a Rock Formed?

To infer how a rock formed, compare the characteristics of the rock, such as the size and roundness of its clasts, to those of deposits in modern environments and decide which environment is the best match. Observe the characteristics of the rock in the large photograph and compare the rock with the small photos of two modern environments. Which of these environments is most similar to the one in which the rock formed?

▷ Many river channels contain large, rounded stones surrounded by a matrix of sand. Note the marking pen for scale.

◁ Steep mountain fronts typically contain large, angular rocks of many sizes in a matrix of mud, sand, or small rock fragments. Which environment looks most like the rock in the photograph?

B How Can We Envision the Slow Change of Landscapes Through Time?

Most landscapes evolve so slowly that we rarely notice any large changes in our lifetime. To get around this limitation, geologists use a strategy called *trading location for time*. This strategy means observing different parts of a landscape and mentally arranging the parts into a logical progression of how the landscape is interpreted to have changed, or will change, through time. The approach is illustrated below using three models, each of which might represent a real place or an interpreted stage in the evolution of a landscape.

Erosion attacks a sequence of rock layers, carving a mountain with steep sides and a broad top. Such a flat-topped mountain is commonly called a *mesa*.

With time, erosion wears away the edges of the mesa, forming a smaller, steep-sided mountain, which geologists commonly call a *butte*.

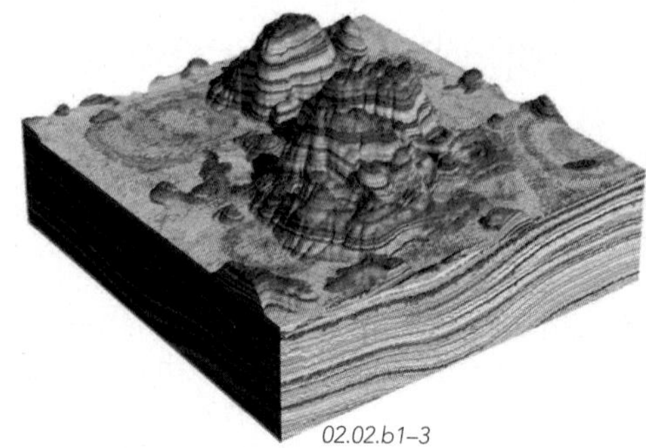

Erosion continues to strip away the terrain, producing a series of low, rounded hills and isolated knobs. This is how a mesa or butte might look like in the future.

C How Do We Determine the Sequence of Past Geologic Events?

When exploring the geology of an area, we like to know the sequence of events that formed different rocks and geologic features. We determine the *relative ages* of rocks and geologic features by using common-sense principles, including the four shown below.

The youngest rock layer is on top. Any sedimentary or volcanic layer must be younger than any rock unit on which it is deposited. Here, the reddish layer on top is the youngest.

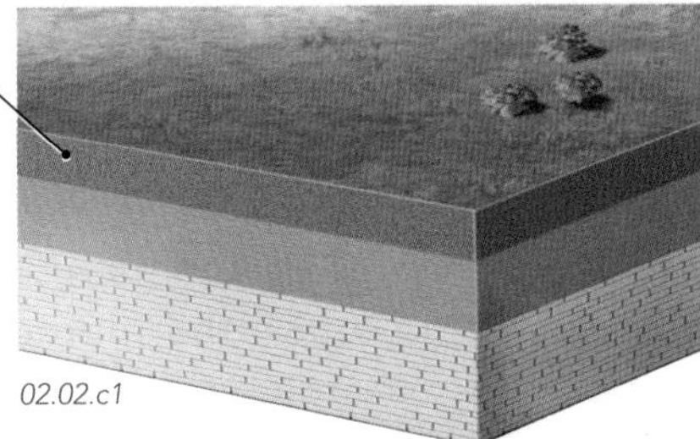

02.02.c1

02.02.c2

A geologic feature is younger than any rock or feature it crosscuts. For example, the fault shown must be younger than rock layers it crosses and offsets. This fault is also younger than the formation of the land surface.

02.02.c3

A small cliff exposes a tan sandstone that overlies darker colored metamorphic rocks. The metamorphic rocks had to already be there to have the sand deposited on top. The cliff formed later.

02.02.c4

A small fault cuts across volcanic layers in a roadcut, so the cross-cutting fault must be younger than the layers. Scratch marks on the rocks were made during construction of the highway.

A younger rock can include pieces of an older rock. For such pieces to have been incorporated into the younger rock, the pieces had to already exist.

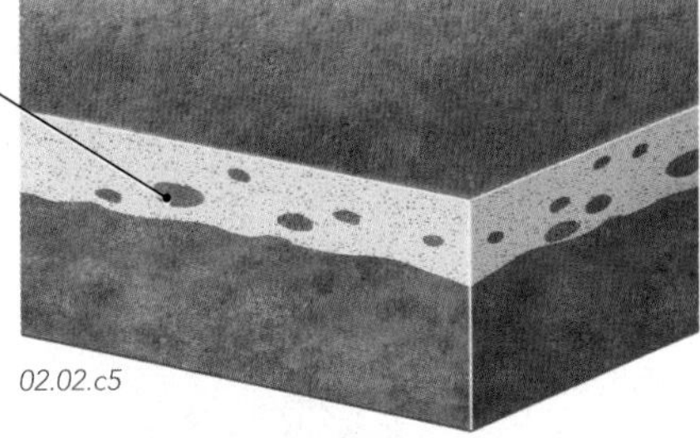

02.02.c5

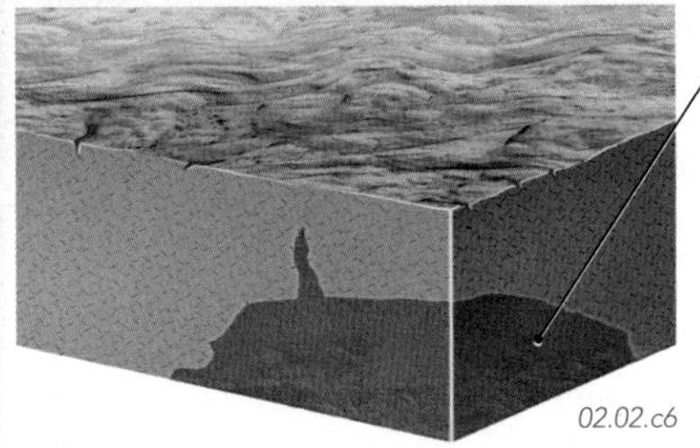

02.02.c6

A younger magma can bake or metamorphose older rocks. Hot magmas bake or otherwise change pre-existing rocks with which the magma comes in direct contact, either underground or on the surface.

02.02.c7

Weathering and erosion broke off pieces of the gray bedrock on the lower right, and the pieces became incorporated into a younger reddish sediment that was deposited on top.

02.02.c8

Tan sedimentary rocks, in the bottom half of the photograph, show a reddish zone caused by baking next to a magma that solidified into the dark igneous rock in the top of the photograph.

Figure It Out!

A really fun part of geology is visiting a new place and trying to figure out what events happened and in what order. This outcrop shows several meters of an upper, tan and gray unit that overlies a dark volcanic rock unit. Use the principles described above to interpret which rock layer is older: the upper gray one or the lower black one. Not all principles may apply.

02.02.mtb1

Before You Leave This Page Be Able To

- ✓ Describe the overall philosophy used to infer the environment in which a rock formed.
- ✓ Summarize or sketch what is meant by trading location for time.
- ✓ Sketch or summarize four principles used to determine the relative ages of rocks and geologic features.

How Do We Investigate Geologic Questions?

EVERY REGION CONTAINS A WEALTH of interesting geologic questions. Geologists observe Earth and its processes through their senses and by using scientific instruments. Geologists use these observations to ask questions and then employ a series of logical steps that build from observations to explanations.

A What Are Observations?

Everyone learns about our world by gathering *observations* with our senses. Scientific instruments can provide additional information about aspects of the world that we cannot sense, or discriminate finely enough. For example, we might sense that the temperature outside is below freezing, but a thermometer can be used to measure the precise value. Every day we make judgments about our observations to determine whether they are worth remembering or are unambiguous and reliable enough to choose a course of action.

1. Geologists, like other scientists, are trained to constantly judge the validity of their *observations*, such as when examining these layers of volcanic ash. An observation that is judged to be valid becomes a piece of *data* that can be used to develop possible explanations.

02.03.a1

2. Compasses and other scientific instruments are checked and calibrated to make sure that they function properly and ensure that the measurements become valid and trustworthy data. All measurements are recorded in a field notebook or in a portable computer and then analyzed and archived.

3. Evaluating the validity of observations is critical, so geologists commonly repeat measurements to compare values and may bring other geologists out to the field to check and discuss their observations and ideas. ▽

02.03.a2

 B How Are Interpretations Different from Data?

Data, by themselves, are not very useful until we begin analyzing them in the context of existing ideas, perhaps identifying the need for a new interpretation of some problem. The recent history of volcanic eruptions near Yellowstone National Park is used below to contrast data and interpretations.

DATA: This map shows a belt of relatively smooth, lower elevation terrain, outlined in red. The belt trends northeast across the mountains of southern Idaho and northern Nevada.

INTERPRETATION: Some process formed a belt of low topography after most of the mountains had formed.

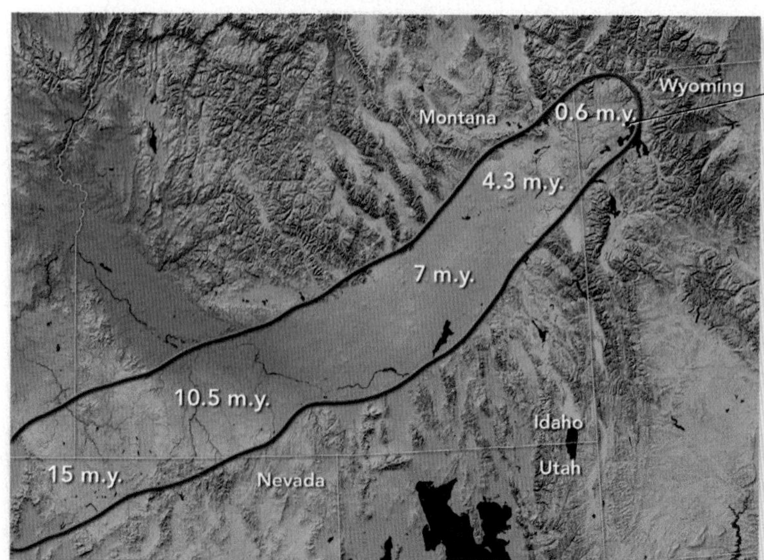

02.03.b1

DATA: The belt of smooth topography ends near Yellowstone, an active volcanic area in the corner of Wyoming.

INTERPRETATION: Recent volcanism at Yellowstone may be related to the process that smoothed the belt's topography.

DATA: The ages (shown in white) of volcanic rocks, as measured in the laboratory, get younger toward the northeast, from 15 million years in Nevada to less than one million years near Yellowstone.

INTERPRETATION: The smoothed belt did not form all at once, but is the result of a continental mass moving over a deep, stationery source of magma.

C How Did Continental Drift Explain Glacial Deposits in Unusual Places?

Geologists working on continents in the Southern Hemisphere were puzzled by evidence that ancient glaciers had once covered places that today are close to the equator. It was a real geologic mystery.

02.05.c1

02.05.c2

1. This rounded outcrop in South Africa has a polished and scratched surface that is identical to those observed at the bases of modern glaciers.

2. Sedimentary rocks above the polished surface contain an unsorted collection of clasts of various sizes. Some of the clasts have scratch marks, like those seen near glaciers.

3. The scratch marks on the polished surface are used to interpret the direction that glaciers moved across the land as they gouged the bedrock. The various observations are most easily interpreted as evidence that glaciers moved across the area about 280 million years ago. [Kimberly, South Africa]

4. The overall directions of glacial movement inferred from the scratch marks made it seem as if the glaciers had come from the oceans, something that is not seen today. Wegener discovered that these data made more sense when the continents were pieced back together into a larger, ancient continent, as shown in this illustration. According to this model, a polar ice cap was centered over South Africa and Antarctica, and the directions of glacial ice movement were those shown by the blue arrows.

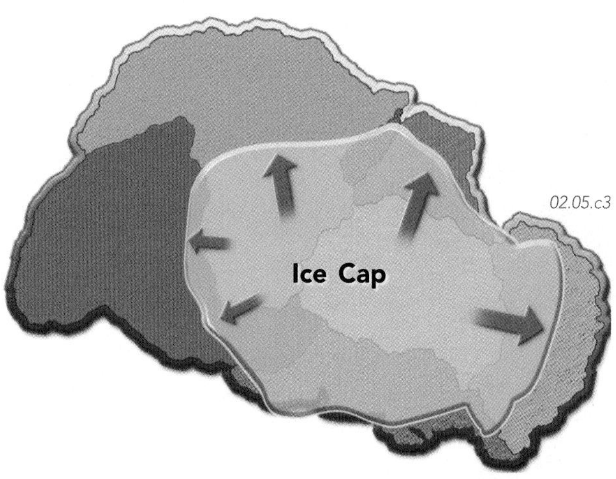

02.05.c3

Old and New Ideas About Continental Drift

The hypothesis of *continental drift* received mixed reviews from geologists and other scientists. Geologists working in the Southern Hemisphere were intrigued by the idea because it explained the observed similarities in rocks, fossils, and geologic structures on opposite sides of the Atlantic Ocean. Geologists working in the Northern Hemisphere were more skeptical, in part because many had not seen the Southern Hemisphere data for themselves.

We now know that Wegener (with the evidence he gathered) was on the right track. A crucial weakness of his hypothesis, however, was the need for a mechanism that allowed continents to move. Wegener imagined that continents plowed through or over oceanic crust in the same way that a ship plows through the ocean. Scientists of his day, however, could demonstrate that this mechanism was not feasible. Continental crust is not strong enough to survive the forces needed to move a large mass across such a great distance while pushing aside oceanic crust. Because scientists of Wegener's time could show with experiments and calculations that this mechanism was unlikely, they practically abandoned the hypothesis, in spite of its other appeals. The hypothesis probably would have been more widely accepted if Wegener or another scientist of that time had been able to conceive a mechanism by which continents could move.

In the late 1950s, the idea of drifting continents again surfaced as new information about the seafloor's topography, age, and magnetism became available. These data showed, for the first time, that the ocean floor had long submarine mountain belts, such as the Mid-Atlantic Ridge (▽) in the middle of the Atlantic Ocean. Harry Hess and Robert Dietz, two geologists familiar with Wegener's work, examined the new data on ocean depths, as well as new data on magnetism of the seafloor. The magnetic data had largely been acquired in the search for enemy submarines during World War II. Hess and Dietz both proposed that oceanic crust was spreading apart along these mountain belts, carrying the continents apart. This process of *seafloor spreading* rekindled interest in Alfred Wegener's idea of continental drift.

02.05.mtb1

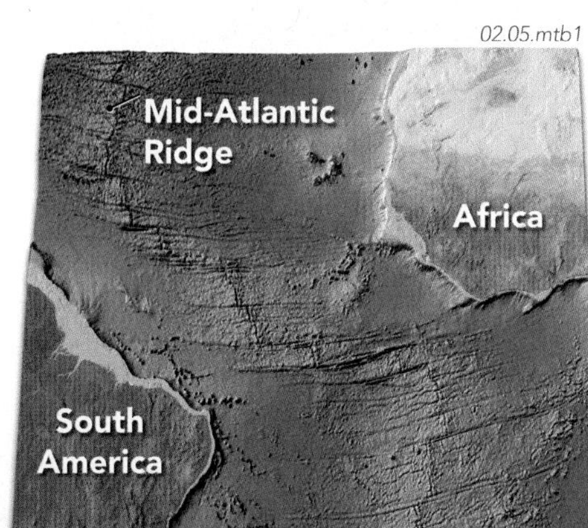

Before You Leave This Page Be Able To

- ✓ Summarize the observations that Wegener used to support the hypothesis of continental drift.
- ✓ Summarize why the hypothesis was not widely accepted.
- ✓ List some discoveries that brought a renewed interest in the idea of continental drift.

How Are Earth's Surface and Subsurface Depicted?

MAPS AND DIAGRAMS OF THE LAND and underlying geology are essential tools for visualizing and understanding Earth. We represent the land surface with several types of maps and with satellite images that are informative and sometimes beautiful. We use two-dimensional and three-dimensional diagrams to depict the subsurface geometry of rock units and to show how these units interact with the surface.

A How Do Maps and Satellite Images Help Us Study Earth's Surface?

Satellite images and various types of maps are the primary ways we portray the land surface and the geology exposed at the surface. Maps of *SP Crater* in northern Arizona provide a particularly clear example of the relationship between geologic features and the land surface.

◁ A *shaded relief map* emphasizes the shape of the land by simulating light and shadows on the hills and valleys. Some hills on this map are small volcanoes called *cinder cones* or *scoria cones*. The area is dissected by straight and curving stream valleys. The simulated light comes from the left of the image.

▽ A *topographic map* shows the elevation above sea level of the land surface with a series of lines called *contours*. Each contour line follows a specific elevation on the surface.

Adjacent contour lines are widely spaced where the land surface is fairly flat (has a gentle slope).

Contour lines are more closely spaced where the land surface is steep, such as on the slopes of the cinder cones.

▽ A *satellite image* is produced by measuring different wavelengths of light to portray the distribution of different types of plants, rocks, and other features. The dark feature in the center of the image is a black, solidified lava flow that erupted from the base of SP Crater. SP Crater is the dark cinder cone connected with the flow.

02.06.a1–4

◁ This photograph, taken from the air, shows the small volcano at SP Crater and a dark lava flow that erupted from the base of the volcano.

▷ *A geologic map* represents the distribution of rock units and geologic features on the surface. This one shows the SP Crater lava flow and older rock units. Compare the four maps to match specific features.

02.06.a5

B How Do We Represent Geologic Features in the Subsurface?

Most of the planet's geology is hidden from our view beneath Earth's surface. We are most aware that rock units exist where they are exposed in a mountainside or deep canyon, but such units are also present beneath areas of relatively flat topography. Geologic diagrams help us envision and understand the thicknesses, orientations, and subsurface distributions of rock units. Such diagrams are also one of the main ways that geologists document and communicate their understanding of an area.

Block Diagram

1. A *block diagram* portrays in three dimensions the shape of the land surface and the subsurface distributions of rock units and geologic features such as faults and folds (if present).

Cross Section

2. A *cross section* shows the geology as a two-dimensional slice through the land. This example is equivalent to the front-left side of the block diagram. 02.06.b1

Stratigraphic Section

3. A *stratigraphic section* with appropriate relative thicknesses shows the rock units stacked on top of one another.

4. The patterns within each rock unit visually represent the character of the unit, such as rounded pebbles in this orange-colored sedimentary unit.

5. One edge of the diagram (here the left edge) typically conveys the relative resistance of the different rock units to weathering and erosion. A more easily weathered and eroded unit is shown recessed, like the orange unit with the pebbles, whereas more resistant units are shown protruding farther out, like the two gray units.

Evolutionary Diagrams

Evolutionary diagrams are block diagrams or cross sections that show the history of an area as a series of steps, proceeding from the earliest stages to the most recent one. Here, a tan rock layer is deposited in the sea and later eroded.

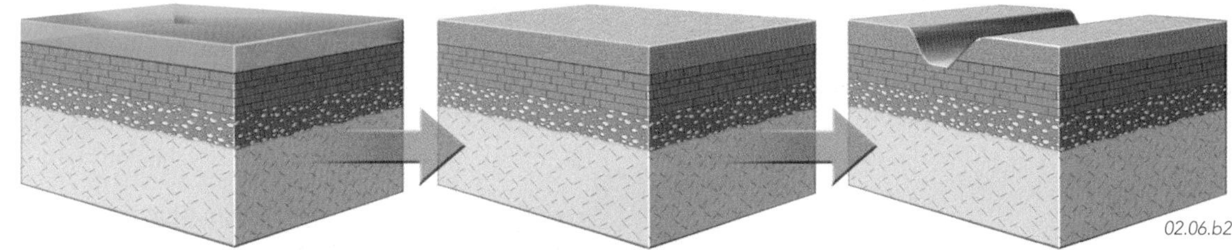

Earliest Stage: Arrival of Sea — Intermediate Stage: Deposition of Layer — Present: Erosion of Layers

Sketching Geology

One of the more interesting challenges of geologic field studies is trying to visualize how geology exposed at the surface continues at depth. Sketches drawn in the field while studying the geology are an excellent way to capture one's thoughts while they are still fresh and while the ideas can be tested by making additional field observations. The field sketch to the right is a simplified geologic cross section drawn to summarize the field relationships for a faulted sequence of rocks. Such sketches are an excellent way to conceptualize and think about geology, either in the field or from a textbook.

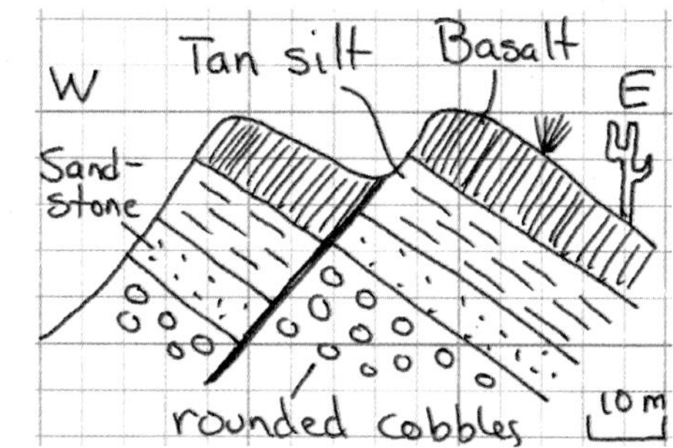

02.06.mtb1

Before You Leave This Page Be Able To

- ✓ Summarize how different types of maps depict Earth's surface.
- ✓ Sketch or describe the types of diagrams geologists use to represent subsurface geology and the sequence of rock units.
- ✓ Sketch or describe what is shown by a series of evolutionary diagrams.

How Are Geologic Problems Quantified?

GEOLOGISTS APPROACH PROBLEMS IN MANY WAYS. Geologists ask questions about Earth processes and then try to collect the data that can help answer these questions. Some questions require *quantitative* data, which can be analyzed and visualized using graphs and numerical calculations.

A What Is the Difference Between Qualitative and Quantitative Data?

02.07.a1

When Augustine Volcano in Alaska erupts, geologists collect various types of observations and measurements. Some observations are *qualitative*, like simple descriptions, and others are measurements that are more *quantitative*.

02.07.a2

Qualitative data are conveyed with words or labels. We can describe this recently formed volcanic unit on Augustine Volcano with phrases such as "contains large, angular fragments," "releases steam," or "represents a dangerous situation."

02.07.a3

Quantitative data are numbers that represent measurements. They are collected with scientific instruments, such as the thermal camera above, or with simple measuring devices like a compass. Geologists collect such data in the field and in the lab.

B What Kinds of Quantitative Data Do Geologists Use?

Geologists often describe features qualitatively, but many of the data they collect are quantitative, consisting of numbers measured with scientific tools or instruments. The cube of rock below illustrates some of the quantitative data with which geologists work.

COMPOSITION: The abundances of different types of grains in a rock provide information about what materials are present and the conditions under which the materials formed or were modified.

PHYSICAL PROPERTIES: Density, strength, and other physical properties of a rock, as measured in the laboratory, form the basis for modeling how rocks behave when subjected to forces.

AGE: Certain rocks, including a volcanic layer shown here, can be dated using precise analytical instruments that measure the ratios between different types of radioactive elements.

SPECTRA: The way that light is reflected off the surface of the rock tells us about the composition and the surface texture or roughness.

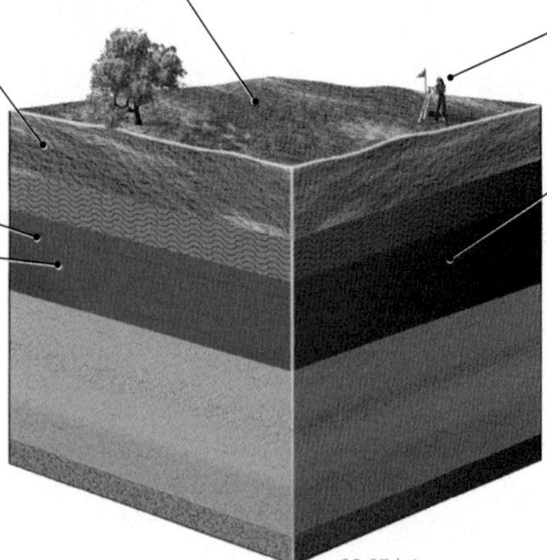

02.07.b1

VELOCITY: All rocks are in motion because they are part of a moving plate. Some rocks are also being uplifted or lowered by deformation. The velocity of this motion can be measured using satellite data and other methods.

MAGNETISM: When a rock forms, it may record Earth's magnetic field in its internal structure. If the rock is later rotated or tilted, we can use the original magnetism to estimate how much its orientation and position have changed.

WATER FLOW AND QUALITY: Measurements can record the velocity and volumes of water flowing in rivers and in groundwater, and chemical analyses measure what the water contains. Below, the USGS measures flow in the Rappahannock River of Virginia. ▽

02.07.b3

02.07.b2

◁ *ORIENTATION:* Geologists observe and measure the orientation of geologic features, such as layers, fractures, and folds. This geologist is using a level on a handheld compass to determine how much these sedimentary layers have been tilted during faulting.

C How Are Graphs Useful for Answering Geologic Questions?

Numeric data are often easier to understand if they are presented graphically. Graphs allow scientists to visually interpret data sets and see relationships that might not be apparent by examining a long list of numbers. Graphs are powerful tools for evaluating correlations, extrapolating trends, and making predictions.

In a graph, the horizontal axis generally is called the *X-axis* and the vertical axis is the *Y-axis*. In most graphs, both axes start with low values near the origin (lower left corner of the graph) and have increasing values away from the origin.

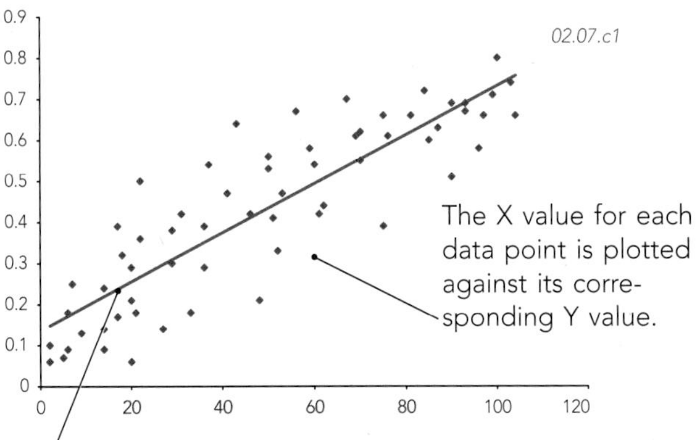

A line drawn through the data shows the average relationship between X and Y. In some cases, the line can be used to predict the value of X for any value of Y.

Each of the graphs below provides an answer to a question a geologist might ask.

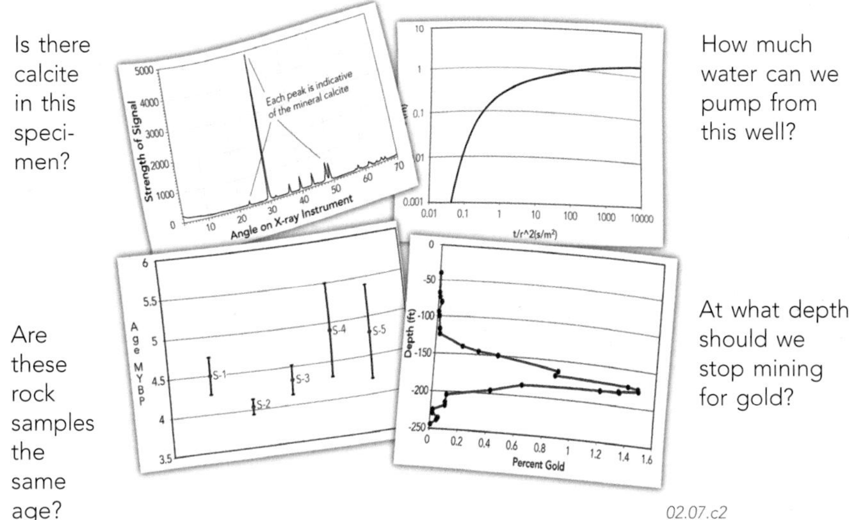

D What Types of Calculations Can Be Applied to Earth?

Many geologic questions involve quantitative data and calculations. How fast will contamination reach a water well? How much damage could an earthquake along the New Madrid fault cause in St. Louis? How long will Yucca Mountain protect us from the nuclear waste we store under it? The answers to such questions require us to collect data and then do calculations, as in this example from the 2004 Indian Ocean tsunami.

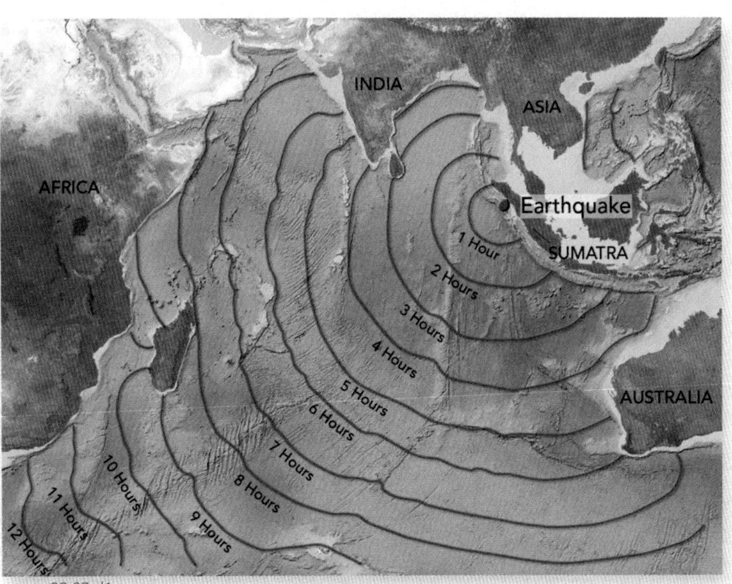

The December 26, 2004, Indian Ocean tsunami was caused by a large earthquake and resulting upward movement of the seafloor. A huge area of the seafloor was displaced forming both a mound and a trough on the surface of the overlying ocean. This movement started the tsunami wave, which spread across the Indian Ocean.

The blue lines indicate the amount of time in hours that the tsunami took to get to each region that it impacted. The velocity of the tsunami is related to the depth of the ocean—the wave goes faster in deeper water.

How much time did people have before disaster struck? A simple calculation provides an estimate for how fast the Indian Ocean tsunami wave was traveling:

$$Velocity\ (m/s) = \sqrt{(9.8\ m/s^2 \times ocean\ depth\ (m))}$$

In the case of the Indian Ocean tsunami, which originated in about 5,000 meters of water, scientists estimate that the top wave velocity was 748 kilometers per hour (468 miles per hour)! If you lived along the Indian Ocean within 100 kilometers of the origin of the tsunami, you may have had less than 10 minutes to reach safety.

Before You Leave This Page Be Able To

- ✓ Explain how qualitative data differ from quantitative data.
- ✓ Describe several types of quantitative data that geologists use.
- ✓ Explain why graphs are useful.
- ✓ Provide an example of a geologic question that can be answered using numeric data and calculations.

How Do We Measure Geologic Features?

MEASUREMENTS ARE AN IMPORTANT PART OF SCIENCE, including geology. The physical properties of an object include length, area, volume, weight, and density. For topographic features, we measure elevation, relief, and slope. On these pages, we summarize some important units of measurements used in geology.

A How Do We Refer to the Linear Dimensions of Geologic Features?

The standard unit for length, width, thickness, or diameter of geologic objects is the *meter*. Objects smaller than a meter are measured in *centimeters*, *millimeters*, and even smaller units. Features much larger than a meter are measured in *kilometers*.

Meters

▽1. Although this figure is not shown in its actual size, it illustrates that a *meter* is equivalent to 39 inches, a little more than three feet. Three feet is also called a *yard*.

02.08.a1–2

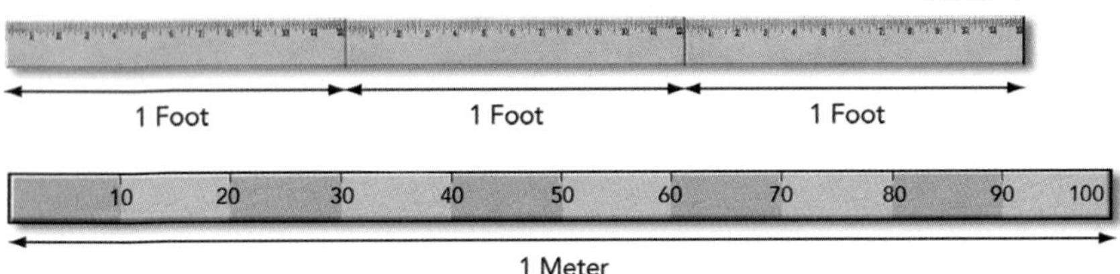

△ 2. Ten centimeters is 10/100 of a meter (or 1/10 of a meter) and is equivalent to 4 inches.

Centimeters and Millimeters

02.08.a3

3. An *inch*, a nonmetric unit, is just a little larger than the diameter of a U.S. quarter.

02.08.a4

02.08.a5

4. A *centimeter* (1/100 of a meter) is equivalent to half the diameter of a U.S. nickle. One inch equals 2.54 centimeters.

5. A *millimeter* (1/1000 of a meter) is approximately equivalent to the thickness of a U.S. dime.

Hundreds of Meters and Kilometers

02.08.a6

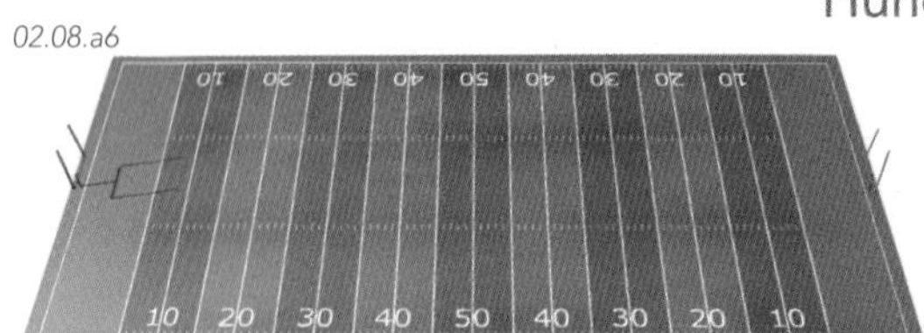

6. On a larger scale, 100 meters is a little longer than a 100-yard-long, American-style football field, minus the end zones.

02.08.a7

7. For longer distances, we use the *kilometer*, which is 1,000 meters or a little less than 11 football fields (one km = 1,093 yds). One kilometer is 0.6 miles, and one mile is 1.6 kilometers.

B How Do We Refer to Differences in Topography?

The height of a feature above sea level is called *elevation*. Elevation can be described in *meters* or *kilometers* above sea level, but it is also commonly given in *feet* on many older maps and signs.

We can also refer to the height of a feature above an adjacent valley. The difference in elevation of one feature relative to another is called *topographic relief*. Like elevation, relief is measured in meters or feet.

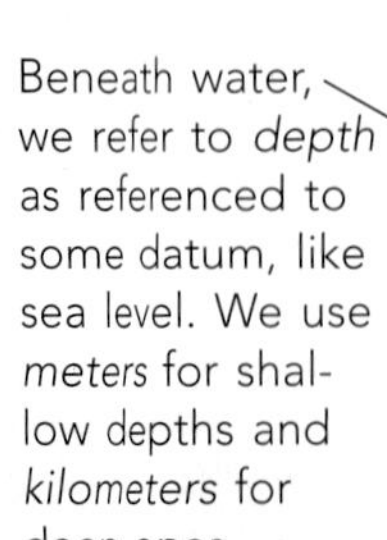

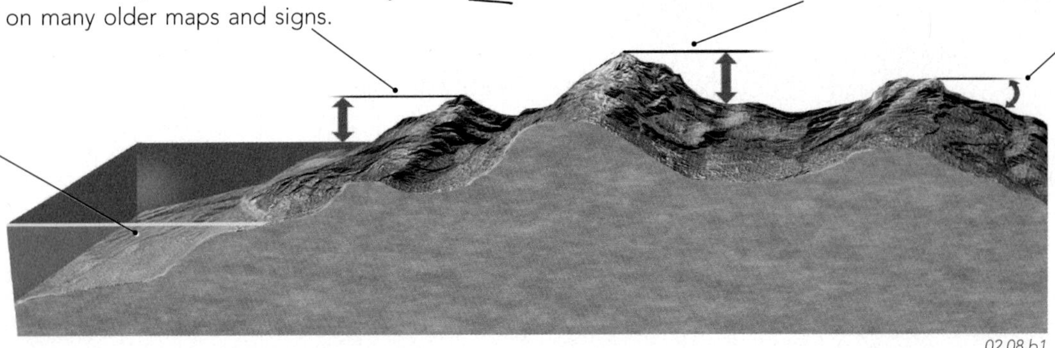

Beneath water, we refer to *depth* as referenced to some datum, like sea level. We use *meters* for shallow depths and *kilometers* for deep ones.

The steepness of a slope can be described in *degrees* from horizontal, such as a 30-degree slope. We also refer to *gradient*, such as a 70-meter drop in elevation over one kilometer (also expressed as 0.070, or 70 m/1000 m).

02.08.b1

C How Do We Refer to Areas and Volumes of Geologic Features?

In addition to linear measurements, geologists refer to *areas* of the land surface and to *volumes* of geologic materials, such as rock, water, and magma. The meter is the standard unit, and such measurements are *metric*.

Area

02.08.c1

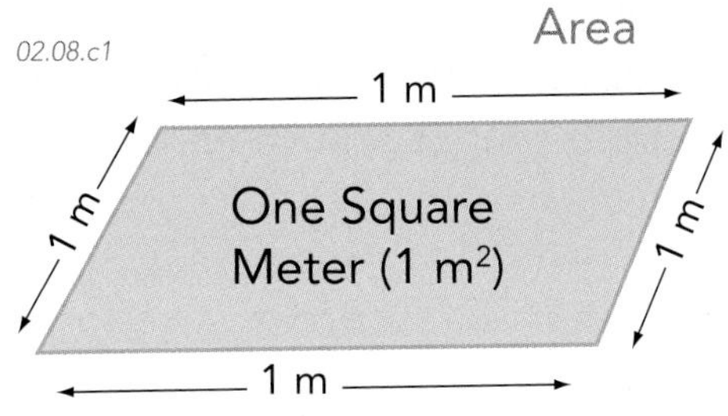

The *area* of moderate-sized geologic objects is measured in the *square meter*. One square meter is equivalent to a surface that is one meter long and one meter wide. It is written as 1 m^2. Larger areas are measured in *square kilometers* (1 km by 1 km, written as km^2).

Calculation of Area

02.08.c2

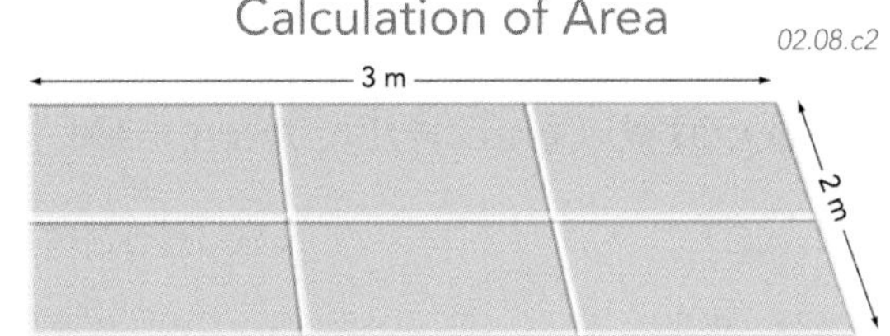

To calculate the area of a rectangular surface, you multiply the length of the two sides. For the example above, the area is

$$2\ m \times 3\ m = 6\ m^2.$$

Volume

02.08.c3

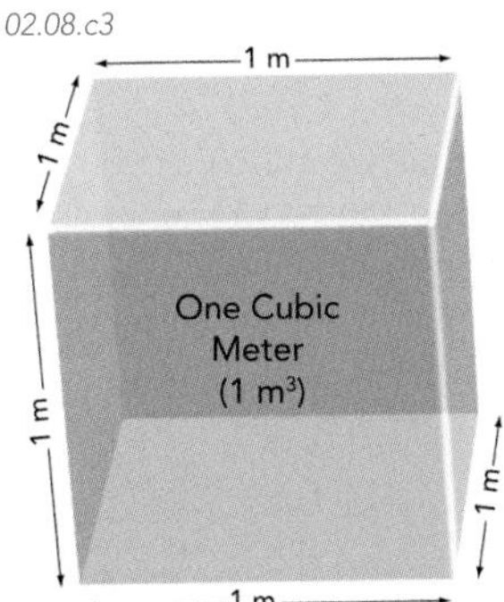

The *volume* of geologic objects is measured in units of a *cubic meter*, which is equivalent to the volume of a cube 1 m long, 1 m wide, and 1 m high. One cubic meter is written as 1 m^3.

Larger volumes are measured in the cubic *kilometer*. A volume of one cubic kilometer is equal to a cube 1 km by 1 km by 1 km, and is written as 1 km^3.

Calculation of Volume

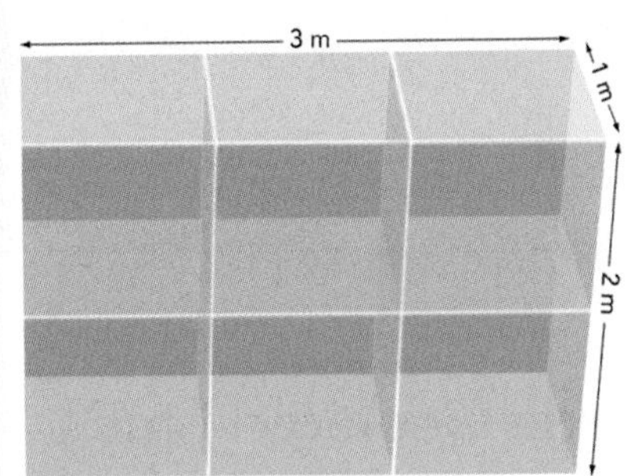

02.08.c4

This object is 3 m long, 1 m wide, and 2 m high. To calculate the *volume* of the object, multiply its length times its width times its height. In this case, the volume is

$$3\ m \times 1\ m \times 2\ m = 6\ m^3.$$

Count the number of boxes here to verify that this is true.

D What Is the Difference Between Weight and Density?

Density

Density refers to how much mass (substance) is present in a given volume. Here, a wooden block, cube of water, and stone block all have the same volume but different amounts of mass. The wood is less dense than water and floats, but the stone is more dense and sinks. The cube of water has the same density as the surrounding water and so does not sink to the bottom or float on the surface.

02.08.d1

Density is calculated using this formula:

$$density = mass/volume$$

Mass is typically measured in grams or kilograms, and volumes are measured in cubic centimeters, cubic meters, or liters. Densities are written in units such as gm/cm^3. By definition, water has a density of 1 gm/cm^3.

Weight

The *weight* of an object is how much downward force it exerts under the pull of gravity. Weight depends on how much *mass* the object contains and where we measure it. A person weighing 180 pounds on Earth will weigh only 30 pounds on the Moon.

02.08.d3

02.08.d2

Before You Leave This Page Be Able To

- ✓ Explain what units we use to measure length, area, and volume.
- ✓ Calculate the area and volume of simple objects.
- ✓ Explain elevation, depth, and topographic relief.
- ✓ Describe what weight and density are and how to calculate density.

How Do Geologists Refer to Rates and Time?

TIME IS ONE OF THE MOST IMPORTANT ASPECTS of geology and, in part, makes geology different than most other sciences. Geologists commonly investigate events that happened thousands, millions, or billions of years in the past. Many geologic events and processes occur through very long durations of time and may occur at rates that are so slow as to be nearly imperceptible. Geologic time needs a special language and calendar.

A How Do We Refer to Rates of Geologic Events and Processes?

We calculate rates of geologic processes in the same way that we calculate rates such as the speed of a car or the speed of a runner. The major difference is that some geologic rates are measured in millimeters per year or centimeters per year instead of miles per hour or kilometers per hour.

A runner provides a good reminder of how to calculate rates. A rate is how much something changed divided by the amount of time required for the change to occur.

If this runner sprinted 40 meters in 5 seconds, the runner's average speed is calculated as follows:

40 m/5 s = 8 m/s

02.09.a1

Geologic processes that are much faster than this runner include the motion of the ground during earthquakes or the speed of an explosive volcanic eruption.

Geologic processes that are much slower than the runner include the movement of groundwater, the motion of continents, and the uplift and erosion of the land surface.

B How Do We Subdivide Geologic Time?

The geologic history of Earth is long, so geologists commonly refer to time spans in millions of years (m.y.) or billions of years (b.y.). If we are referring to times before the present, we use the abbreviation *Ma* (mega-annum) for millions of years before present and *Ga* (giga-annum) for billions of years before present. We also use a special calendar to refer to the four main chapters of Earth history.

The most recent chapter in Earth history is the *Cenozoic Era*, which began at 65 Ma (65 m.y. ago) and continues to the present.

The next oldest chapter is the *Mesozoic Era*, which refers to the time interval from 251 Ma to 65 Ma. It largely coincides with the time when dinosaurs roamed the planet.

The *Paleozoic Era* started at 542 Ma, a date that marks the first record of creatures with shells and other hard body parts. It ended at 251 Ma when many Paleozoic organisms became extinct.

The oldest chapter in Earth's history is the *Precambrian*. It represents most of Earth's history and extends from 4,500 Ma (that is 4.5 billion years before present) to 542 Ma, the start of the Paleozoic Era.

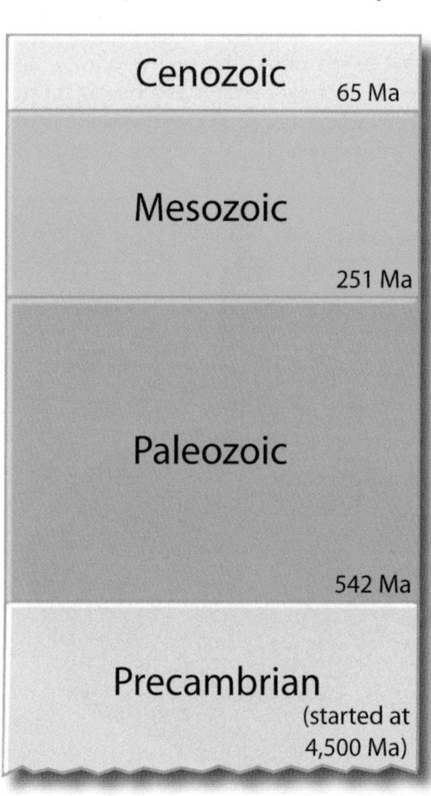

02.09.b1

Each of the four chapters represents a different amount of geologic time. This figure (▷) shows the entire Precambrian, which began with the origin of Earth at 4.5 Ga (b.y. ago). The Precambrian, which has early and late parts, represents approximately 90 percent of geologic time, whereas the Cenozoic Era represents only 1.4 percent of geologic time.

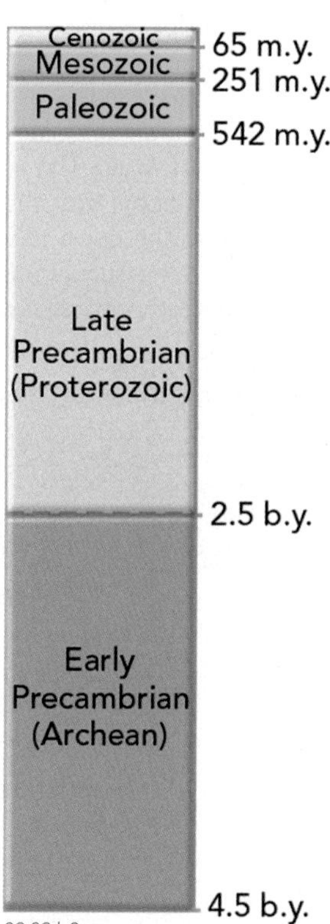

02.09.b2

C What Are Some Important Times in Earth's History?

If the entire 4.5-billion-year-long history of Earth is scaled to a single calendar year, the Precambrian takes up the first ten months and part of November. On this calendar, Earth formed on January 1st.

On Earth's calendar, the oldest dated rocks (about 3.9 to 4.0 b.y. old) would fall in early March. The oldest known fossils are only a little younger (in late March).

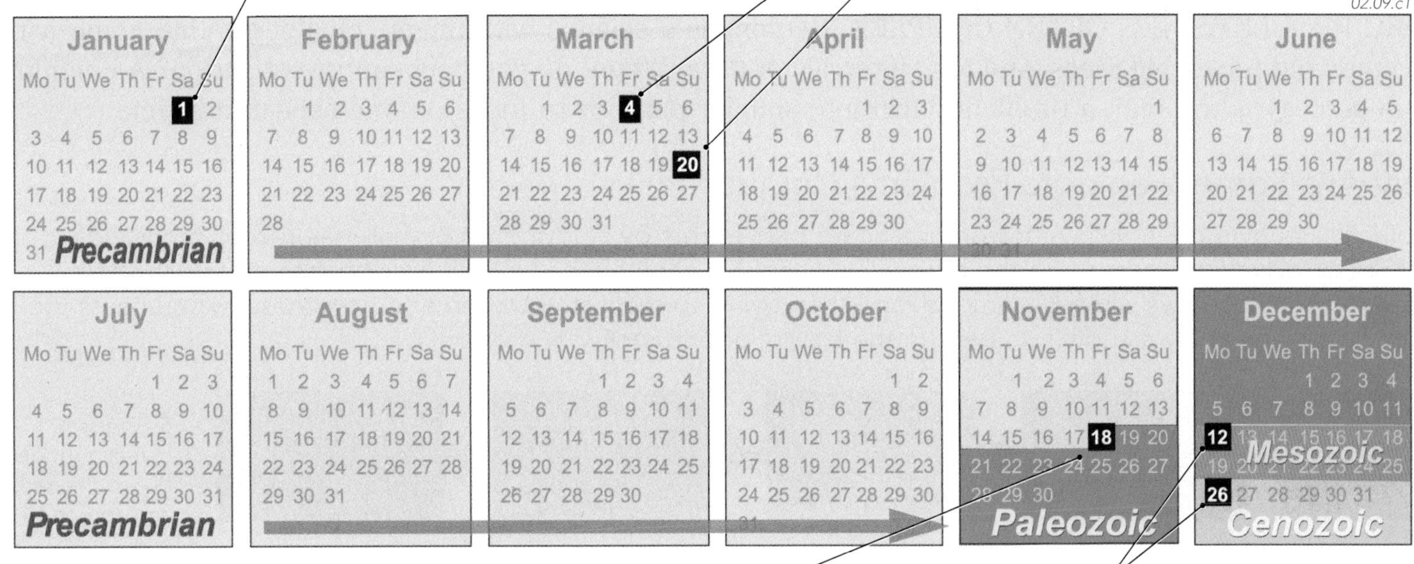

Animals having hard shells arrived at about 542 Ma. On our calendar, this is the middle of November. This event is used to define the beginning of the *Paleozoic Era*, the second of Earth's main chapters. The long time before the Paleozoic Era is called the *Precambrian* (shown in brown).

Earth's final two chapters began in December. The Paleozoic Era ended and the *Mesozoic Era* started at 251 Ma in mid-December. The Mesozoic Era ended and the Cenozoic Era began at 65 Ma, equivalent to December 26.

Fossils, Absolute Ages, and the Geologic Timescale

Many rocks include *fossils*, which are shells, bones, leaf impressions, and other evidence of prehistoric animals and plants. Geologists have discovered that the types of fossils change from one layer to another up or down in a sequence of sedimentary rocks. By comparing one location to another, geologists defined certain age periods on the basis of their fossils. These age periods are named from the place where they were first recognized, such as the Pennsylvanian Period for Pennsylvania. When arranged in their proper order, these periods and their subdivisions constitute the *geologic timescale*. ▷

The generalized time scale presented here has four main divisions. The *Precambrian* is the time before life developed shells and other hard body parts. The Precambrian is followed by the *Paleozoic*, *Mesozoic*, and *Cenozoic*, from oldest to youngest. These four divisions of geologic time appear throughout this book and are an essential part of the vocabulary of geology.

Geologists calculate the actual, or *numeric age* of rocks in thousands, millions, or billions of years before present. Many rocks contain atoms that change by natural radioactive decay into a different type of atom, such as potassium (K) to argon (Ar). Calculating the numeric age of a rock involves precisely measuring in the laboratory the rock's abundance of both types of atoms. These calculations, when combined with a knowledge of fossils, provide ages for the boundaries between the divisions of the geologic timescale. When referring to a geologic event, we may use the name of the time period, such as the Mississippian, or the age in millions of years before present. The timescale is not set in stone, as evidenced by a recent change (◁) in the subdivisions of the Cenozoic.

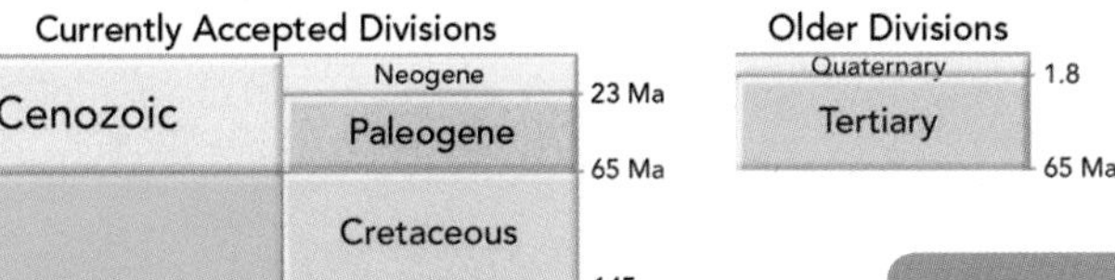

Era	Period	Boundary age
Mesozoic	Cretaceous	145
	Jurassic	200
	Triassic	251 Ma
Paleozoic	Permian	300
	Pennsylvanian	320
	Mississippian	359
	Devonian	416
	Silurian	444
	Ordovician	488
	Cambrian	542 Ma
Precambrian		

02.09.mtb1

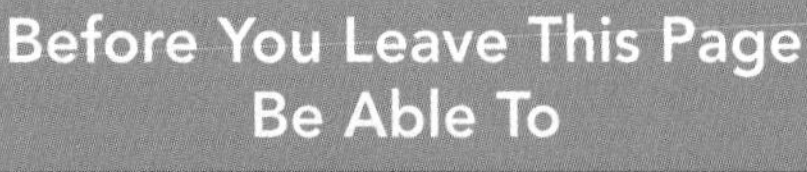

- ✓ Give an example of a rate and how a rate is calculated.
- ✓ Summarize or sketch the four main chapters of Earth history, showing which chapter is longest and which one is shortest.
- ✓ Discuss what the geologic timescale is and the kinds of data that were used to construct it.

What Are Some Strategies for Approaching Geologic Problems?

GEOLOGIC PROBLEMS ARE OFTEN COMPLEX. Geology is a science with unique challenges that come with trying to study old rocks, big areas, and long intervals of time. Where do you begin an investigation? A common strategy in geology is to break a problem down into simpler parts or to focus on one aspect at a time.

 Why Are Some Geologic Problems Unique?

Geologists study modern processes or investigate past events by studying rocks that may record thousands, millions, or billions of years of Earth's history. Studying the geologic past is like walking into a vast laboratory where experiments have already occurred, and then trying to reconstruct when and how these experiments took place. The figures below show five steps in the history of one rock.

02.10.a1–5

1. Deposition on a sandy beach forms layers of sandstone.
2. The sandstone is eroded, which produces clasts that are transported away.
3. The sandstone clasts are deposited as part of another sediment.
4. The sediment with the clasts is buried and lithified into sedimentary rock.
5. The sedimentary rock is buried, heated, and converted into a metamorphic rock. With the rock in this condition, it might be difficult to reconstruct the entire history.

B How Does the Scale of a Problem Affect Our Approach?

Geologic problems come in all sizes. Some can be answered on a single hill or with one rock sample. Others require examining the geology of an entire state or even the entire Earth. The types of data we collect and how we think about a problem depend on the scale. Most geologic problems require using different scales of observation. Data and interpretations gained at one scale often guide our approach to a problem at a different scale.

▽ Built at the foot of steep, unstable slopes, La Conchita is a small oceanside town in southern California. In 1988, 1995, and 2005, parts of the slopes collapsed, and landslides plowed through a subdivision. The medium-size scale of this problem allows a few geologists to map the landslide and suggest ways to save lives. Geologists have recommended abandoning the town.

02.10.b1

▽ Landslides can also be viewed at a broader scale, as in this map of landslide hazards for the conterminous United States. Areas shown in red, yellow, and green have significant risk for landslides, with red areas showing the highest risk. This map required the efforts of hundreds of geologists, each mapping and studying individual landslides. Viewing landslides at this scale is useful for allocating funds in anticipation of future landslide disasters.

02.10.b2

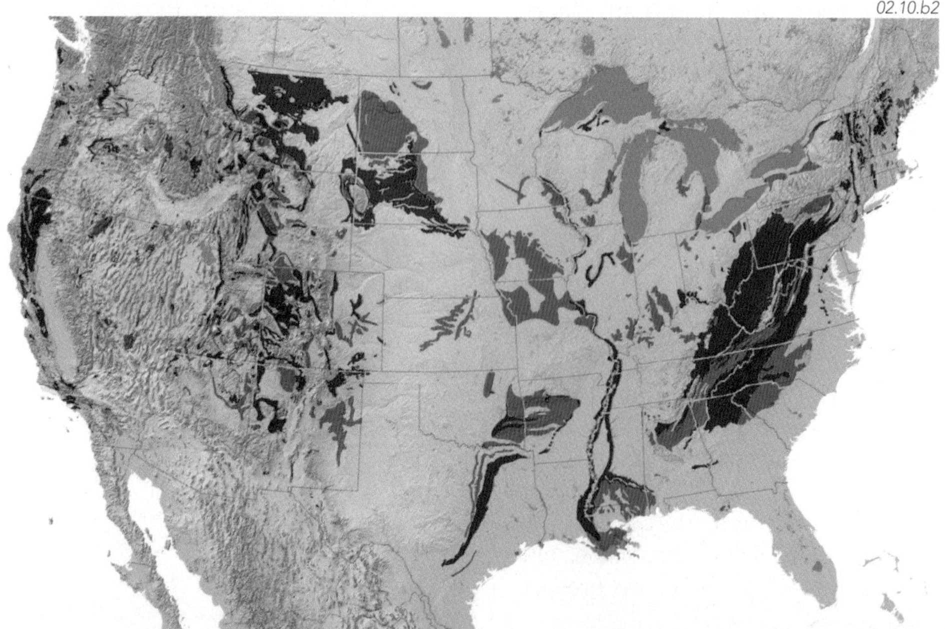

C Are There Ways to Simplify and Conceptualize Complex Geologic Problems?

Many geologic problems have complexities at various scales that can be somewhat overwhelming. One strategy for addressing a complex problem is to simplify it by appropriately combining or excluding some aspects of the problem based on the goals and purposes of the study.

1. This terrain has a simple landscape but moderately complex geology in the subsurface.

2. There are four types of materials: a thin upper layer of reddish-brown sediment, a tan-colored sedimentary unit with cobbles, a gray metamorphic unit, and some pinkish igneous rocks.

3. A fault has offset all of the rock units except the top layer of river gravels.

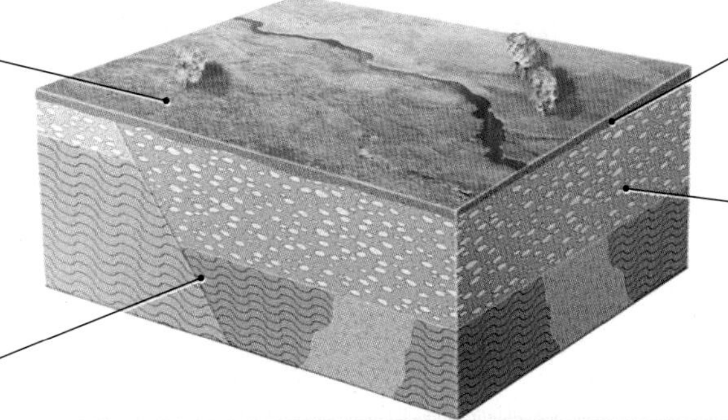

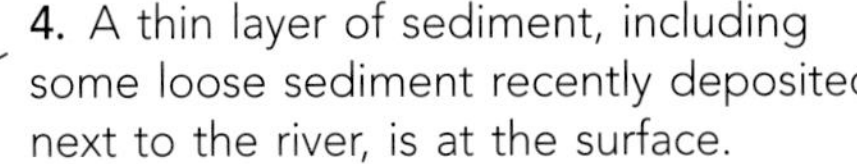

4. A thin layer of sediment, including some loose sediment recently deposited next to the river, is at the surface.

5. The sedimentary rocks overlie the older metamorphic and igneous rocks and are different thicknesses on the two sides of the fault.

6. Different questions can be addressed by considering only the aspects that pertain to that question.

7. Where would we explore for petroleum?

9. How does groundwater flow?

11. Where are potential earthquake hazards?

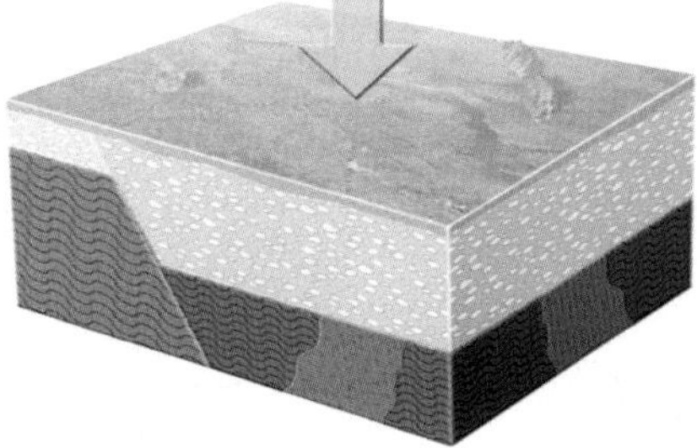

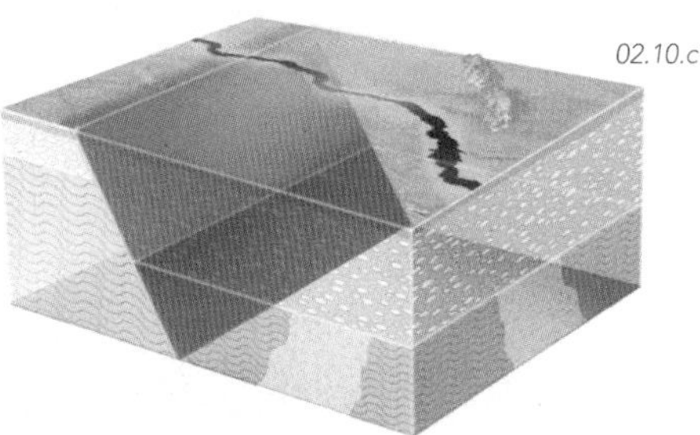

02.10.c1–4

8. If exploring for petroleum, we may only need to consider the sedimentary rocks, the most common rocks in which petroleum is found. So we would focus our attention on the top two units.

10. If evaluating the flow of groundwater, the terrain can be simplified into three units each having certain properties that control the movement of groundwater. The fault could also affect groundwater flow.

12. If anticipating the severity of ground shaking during an earthquake caused by movement on the fault, the most hazardous areas would be near and above the fault or on loose, easily shaken sediment along the river.

The Problem of Incomplete Data

An important problem that plagues many geologic investigations is working with incomplete data. When reconstructing the geology of a mountain range, it is impossible to walk over and examine every exposure of rock and every square meter of ground. A field geologist must make many generalizations, such as *this rock type is the same as the one observed 20 meters away*. We rarely have as much data as we would like or have time to collect.

The problem of incomplete data is illustrated in the figure to the right, which shows two water wells drilled near each other. One well is in a sandstone that contains abundant water between the sand grains, but the other well intercepted an unseen granite having water only in its fractures. Both wells encountered water at the same depth, but the well drilled into granite went dry when pumped because there was so little water in the fractures. Although from the surface and the initial water levels the two wells seemed comparable, the well in granite was later abandoned—another victim of incomplete data. For this situation, we would need to know the geometry of the rocks in the subsurface and those characteristics of the rocks that influence the flow of groundwater. The number of fractures in the granite would be relevant information, but the age of the granite is probably not.

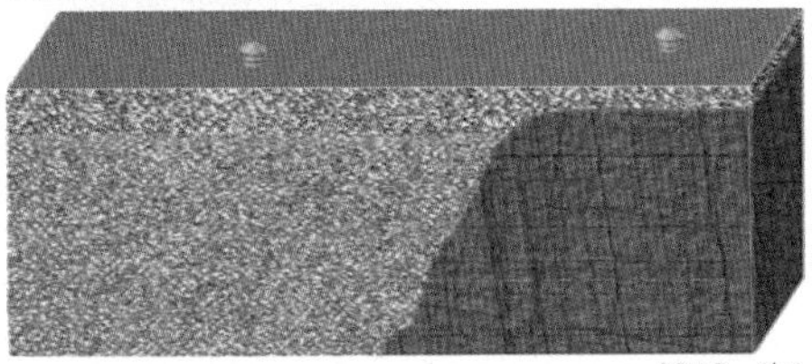

Successful well   Poor well

02.10.mtb1

Before You Leave This Page Be Able To

- ✓ Explain why some geologic problems are unique.
- ✓ Summarize an example of viewing a problem from different scales.
- ✓ Explain how you could simplify a complex problem.
- ✓ Summarize an example of the problem of incomplete data.

What Does a Geologist Do?

GEOLOGISTS ADDRESS A DIVERSE RANGE OF PROBLEMS relevant to our society. There are many types of geologists, each with interesting questions to explore. Some questions can be explored directly in the field, whereas others require sophisticated computers or other technology to study places that are inaccessible, such as other planets or the interior of Earth. Some questions address active processes, and others address the interpretation of ancient processes from rocks, structures, and fossils.

A How Do Geologists Investigate Questions in the Field?

The traditional view of a geologist is a person outdoors with backpack, hiking boots, and rock hammer traversing a scenic mountain ridge. From field studies, geologists produce descriptions, maps, and other data needed to reconstruct a piece of Earth's history and determine how this history affects how and where we live.

▷ During their field studies, geologists observe various aspects of the natural environment, record these observations, and propose explanations for what they discover. A common goal is to understand the area's geologic processes and history. Sometimes, just getting to the field site is an outdoor adventure. [Antarctica]

02.11.a1

02.11.a2

△ Many field geologists construct a geologic map by identifying rock types in the field, finding the boundaries between adjacent rock units, and drawing these on a map. The geologic map becomes the main basis for making interpretations about subsurface geology and the geologic history of the area. This geologic map shows rock layers and river deposits near St. Paul, Minnesota.

02.11.a3

△ Geologists help find most of the energy and mineral resources on which we depend. They conduct field studies to find areas favorable for a certain resource and to determine through drilling the location and size of the resource. Energy and mineral companies have traditionally employed many geologists and have taken their employees to many parts of the world, such as 3 km (10,000 ft) below the ground in a South African gold mine.

02.11.a4

△ Some field geologists study volcanoes, earthquakes, landslides, floods, or other natural hazards. To help us avoid these hazards, geologists study the processes that are operating, determine how often hazardous events occur, and map areas that are most likely to be affected. This geologist is extracting samples of lava on Hawaii.

B How Do Geologists Study Places That Are Inaccessible?

Many geologic questions involve places where we cannot do field studies, such as a magma chamber beneath a volcano, rock layers that are kilometers beneath the seafloor, or the surfaces of other planets. To address these questions, scientific instruments and remote probes are used to collect pertinent data.

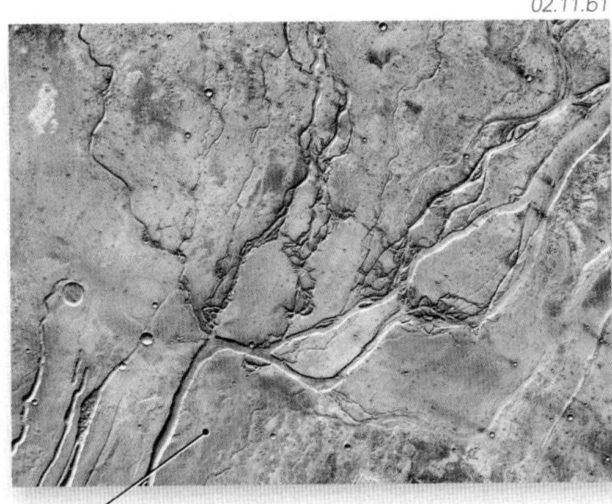

An exciting field of geology is exploring other planets and their moons, principally by sending spacecraft that orbit a planet or that land on its surface. From such observations, planetary geologists try to understand what processes are reshaping the surface of the moon or planet and whether there is a possibility of water and life. This recent image (△) shows channels on Mars that are interpreted to have been carved by water flowing on the surface.

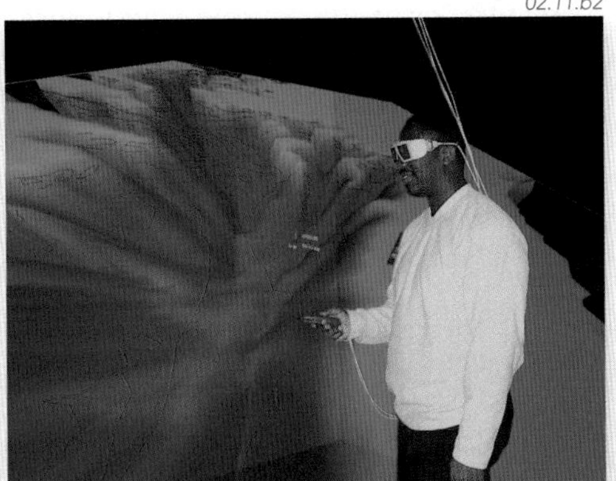

◁ Many geologic problems require visualizing data in three dimensions, by using virtual-reality displays and computer simulations.

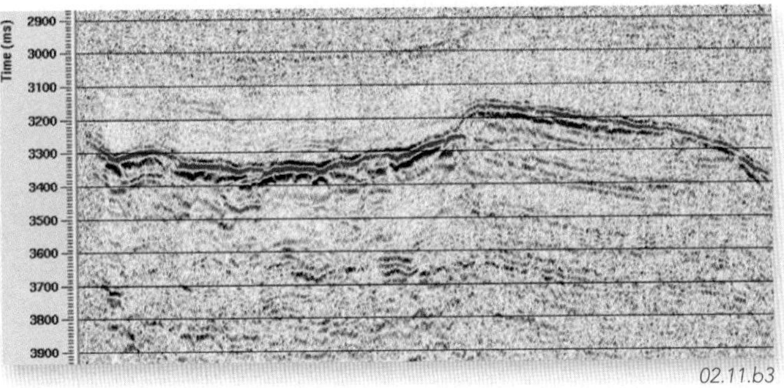

△ Most oil, gas, and coal reside within sedimentary rocks, so exploration for these resources focuses on understanding the sequence and geometry of rock layers and the conditions in which each layer was deposited. This computer-generated cross section of subsurface rock layers, called a *seismic profile*, was produced by sending powerful vibrations into the ground and recording their arrivals back on the surface. This seismic line shows folded sedimentary layers along the southern coast of the United States.

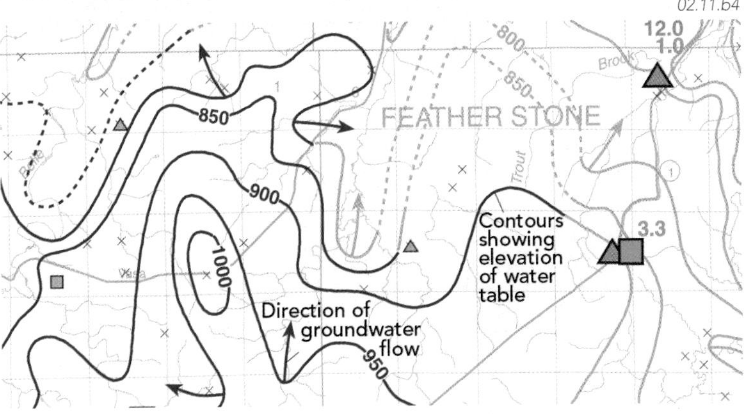

△ Clean water is our most important natural resource, and many geologists study water supplies and water contamination. By drilling holes into the subsurface, geologists construct groundwater maps that show where groundwater is present and which way it flows. Groundwater maps, such as this one of part of Minnesota, are the primary tool for understanding groundwater contamination and for designing cleanup efforts. The blue lines show the elevation of the top of groundwater, and the arrows show which way groundwater flows in different places.

The Education and Training of Geologists

Professional geologists have various levels of education, ranging from bachelor's degrees to master's degrees and doctorate degrees. A university education emphasizes the major processes of geology, such as those that form rocks, minerals, fossils, sequences of layers, geologic structures, resources, and landscapes. Some classes cover exploring the subsurface of Earth using sound and electromagnetic waves and how water and petroleum move through the subsurface. Some geology students are interested in the exploration of other planets, an active field with many recent discoveries. Most fields of geology involve precise analytical instruments, sophisticated computer processing, and high-tech displays of data. Many students direct their attention to learning how to teach geology and to becoming employed as one of the many geology and earth science teachers in K-12 schools, colleges, and universities. In the United States, more than 120,000 geologists are currently employed in private industry, government agencies, and schools.

Before You Leave This Page Be Able To

- ✓ Summarize the kinds of questions geologists investigate with field studies.
- ✓ Summarize how geologists find energy and mineral resources.
- ✓ Summarize how geologists help us avoid geologic hazards.
- ✓ Summarize some places or questions that geologists investigate using techniques other than field studies.

How Did This Crater Form?

A CRATER ON THE PLATEAU OF NORTHERN ARIZONA is more like those on the Moon than any nearby feature. The crater is a huge pit, more than 1,250 m (4,100 ft) across and 170 m (560 ft) deep, with a raised rim of broken rocks. Some dramatic geologic event must have occurred here. What could it be? We explore this mystery to see how we evaluate competing explanations for the origin of a geologic feature.

A What Would You Observe at the Crater?

Observe the photographs below and think about what your observations might imply about how the crater formed. What other types of information would you like to have to understand the crater's origin?

02.12.a1

▷ The raised rim of the crater consists of huge, angular blocks of limestone and sandstone, many of which are fractured and shattered. Because they are pieces of the rock layers exposed in the crater walls, these blocks are interpreted to have been thrown out of the crater.

02.12.a2

02.12.a3

◁ Rocks in the walls of the crater contain unusual minerals that can be seen using a microscope. Laboratory experiments show that these minerals can form only under extremely high pressures.

B What Are Some Ways to Show the Geologic Setting of This Crater?

This geologic block diagram shows the geometry of rock layers through the crater. A geologist constructed this cross section on the front side by observing the rocks exposed in the crater and in the surrounding plains, by examining results from drilling in the floor of the crater, and by extrapolating these observations into the subsurface.

Numerous meteorites are scattered on the surface around the crater. Geologists have collected many small and large pieces to understand what information the meteorites provide about the origin of the crater and about the solar system.

The layers of sandstone and limestone are nearly horizontal away from the crater, but the layers along the rim of the crater have been peeled back and are now locally completely upside down.

02.12.b1

C What Are Some Possible Explanations for the Origin of the Crater?

Since the crater was discovered, geologists have proposed and tested several explanations for its origin. Among these explanations are a volcanic explosion, warping by a rising mass of salt, and excavation by a meteoroid impact.

	Volcanic Explosion	Warping by a Rising Mass of Salt	Meteoroid Impact
Possible Explanation	A volcanic explosion blasts open the crater during a violent eruption of gas and magma.	The rock layers are upwarped by a rising mass of salt. Rainwater later dissolves the salt to form the crater.	A large meteoroid streaks through Earth's atmosphere and blasts the crater when it collides with the surface.
Predictions from Each Explanation	Volcanic materials, such as ash, should be present around the crater, and solidified magma might underlie the crater floor.	A mass of salt should exist directly beneath the floor of the crater.	Fragments of a meteoroid might remain beneath the crater or on the surrounding plains. Much of the meteoroid could have been vaporized during the blast.
Results of Testing Each Explanation	No volcanic material was found around the crater or by drilling beneath the crater, so this explanation seems improbable.	No salt was found beneath the crater, despite extensive drilling into the crater floor.	Drilling did not encounter a large mass of meteorite beneath the crater, but numerous small meteorite fragments are scattered across the surrounding plains.

An Incident 50,000 Years Ago

Of the three explanations, only the meteoroid-impact explanation remains viable because the crater does not contain the rock types predicted by the other two explanations. Other types of data also support an origin by meteoroid impact. Laboratory experiments indicate that the unusual minerals in the walls of the crater can form only at very high pressures, such as when a shock wave from an impact passes through rocks. These experiments also show that rock layers near the surface can be peeled back and shattered by the collision. Computer models of meteoroid impacts predict that large meteoroids, such as the one required to form this crater, are mostly vaporized by the impact. Therefore, little of the meteoroid would remain in the crater, but scattered pieces might survive around the crater. It is estimated that 30 tons of meteorite specimens have been collected around the site. The accepted origin for this crater is reflected in its famous name, *Meteor Crater*.

From the presently available information, geologists conclude that the crater formed about 50,000 years ago when a meteoroid 30 to 50 meters in diameter smashed onto the surface at a speed of about 11 km/s (25,000 miles/hr). A recent study suggests that the meteoroid broke apart and slowed before it hit. Humans were not yet living in this area to observe the fireball blazing across the sky, but animals, such as the woolly mammoth, were around to witness this unexpected catastrophe.

Before You Leave This Page Be Able To

- ✓ Summarize or sketch the three explanations for the origin of the crater and which observations support or do not support each possible explanation.
- ✓ Summarize how geologists interpret Meteor Crater to have formed.

2.12

What Is the Geologic History of Upheaval Dome?

A PERPLEXING GEOLOGIC FEATURE CALLED UPHEAVAL DOME is located in Canyonlands National Park of Utah. In most of Canyonlands, the colorful sedimentary rocks are nearly horizontal, but around Upheaval Dome they are abruptly warped upward and eroded into a very unusual circular feature.

Goals of This Exercise:

- Make some observations and develop some questions about Upheaval Dome.
- Determine the sequence of geologic events that formed the rock layers in the dome.
- Suggest some ways to test possible explanations for the origin of the dome.

A What Are Some Observations and Questions About the Dome?

The three-dimensional perspective below shows an unusual circular feature called *Upheaval Dome*. Make some observations about this landscape, and record your observations on a sheet of paper or on the worksheet. Also record any questions that you have as you study this landscape.

The rocks shown in this image are all sedimentary and of Mesozoic age. The rock layers away from the dome are nearly horizontal and form various benches and flat-topped mesas.

The rock layers in the dome are tilted outward in all directions.

B What Sequence of Geologic Events Formed the Rocks and the Dome?

Shown below are (1) a stratigraphic section with the sequence of rock layers and (2) a geologic cross section across the dome. Using these two figures and the strategies in section 2.2, determine the order in which the layers were formed, and write your answers on the worksheet.

Stratigraphic Section

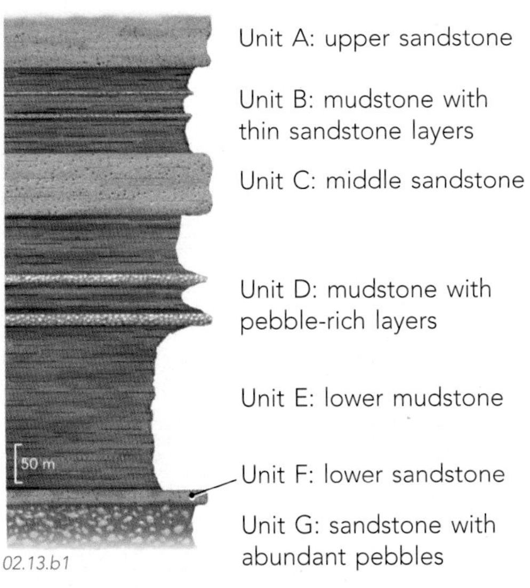

Unit A: upper sandstone

Unit B: mudstone with thin sandstone layers

Unit C: middle sandstone

Unit D: mudstone with pebble-rich layers

Unit E: lower mudstone

Unit F: lower sandstone

Unit G: sandstone with abundant pebbles

02.13.b1

Units A, B, and C have a Jurassic age, whereas units D and E are Triassic. Units F and G are Permian (see timescale in Section 2.9).

Cross Section

The letters A–G mark the units shown in the stratigraphic section to the left. The letters are assigned in order from top to bottom, not in the order in which the units formed. Some units, such as B and D, contain a series of related sedimentary rock layers rather than just a single type of rock.

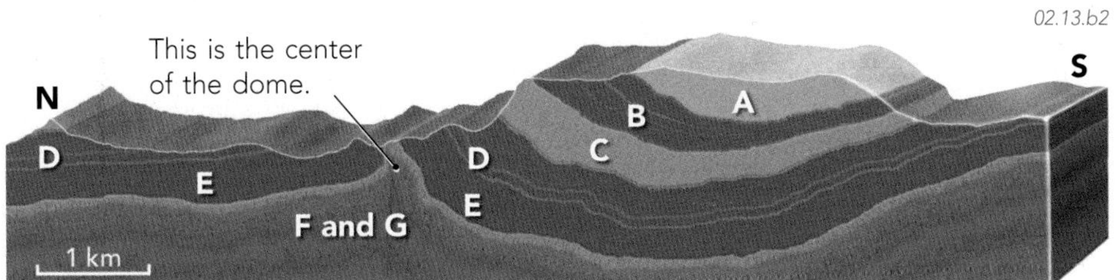

02.13.b2

It is uncertain what types of rocks lie at depth below the dome (i.e., below units F and G).

All of the rock units have been folded into the dome. Any layers deposited after the dome was formed have been eroded away.

Many faults and folds (not shown here) have thickened the rock layers in some places and thinned them in others.

C How Would You Test Possible Explanations for the Origin of the Dome?

The origin of Upheaval Dome is controversial, and three competing explanations currently are being debated. There is no single right answer.

1. The rock layers were warped upward by a rising mass of salt. A thick salt layer is known to be present beneath much of the Canyonlands region, and the salt would have risen upward because it is less dense than the surrounding rocks.
2. The dome formed as a result of rising magma. Igneous rocks formed from such magmas are common elsewhere in the region, where they have bowed up and baked the surrounding rock layers.
3. The dome is part of a larger, circular crater formed by a meteoroid impact. Many of the larger meteoroid-impact craters on the Moon and elsewhere have a central peak, or dome, which is interpreted to form by converging shock waves.

The age of the dome is poorly constrained. The dome is younger than all of the rock layers in the vicinity. When the dome formed, the presently exposed rocks were several kilometers deep, buried beneath overlying rock layers that have since eroded away. This erosion, therefore, removed some key evidence for the origin of the dome.

Procedures for Possible Explanations of the Dome

1. For each of the three explanations, draw a simple sketch on the worksheet illustrating which types of rocks you predict to find at depth.
2. List a prediction that follows from each explanation. Then, explain how that prediction could be tested.
3. List the types of information you would like to know about this location to further constrain the origin of the dome.

CHAPTER 3

Plate Tectonics

THE SURFACE OF EARTH IS NOTABLE for its dramatic mountains, beautiful plains, and intricate coastlines. Beneath the sea are equally interesting features, such as undersea mountain ranges, deep oceanic trenches, and thousands of submarine mountains. In this chapter, we examine the distribution of these features, along with the locations of earthquakes and volcanoes, to explore the theory of plate tectonics.

On these images of the world, the land is colored using satellite data overlain on topography, whereas the seafloor is shaded blue according to water depth (lighter blue colors represent shallower water).

Offshore of western North America is a long, fairly straight step in the seafloor that trends east-west and ends abruptly at the coastline. North of this step, a curious ridge called the Juan de Fuca Ridge, zigzags across the seafloor.

What are these features on the seafloor and how did they form?

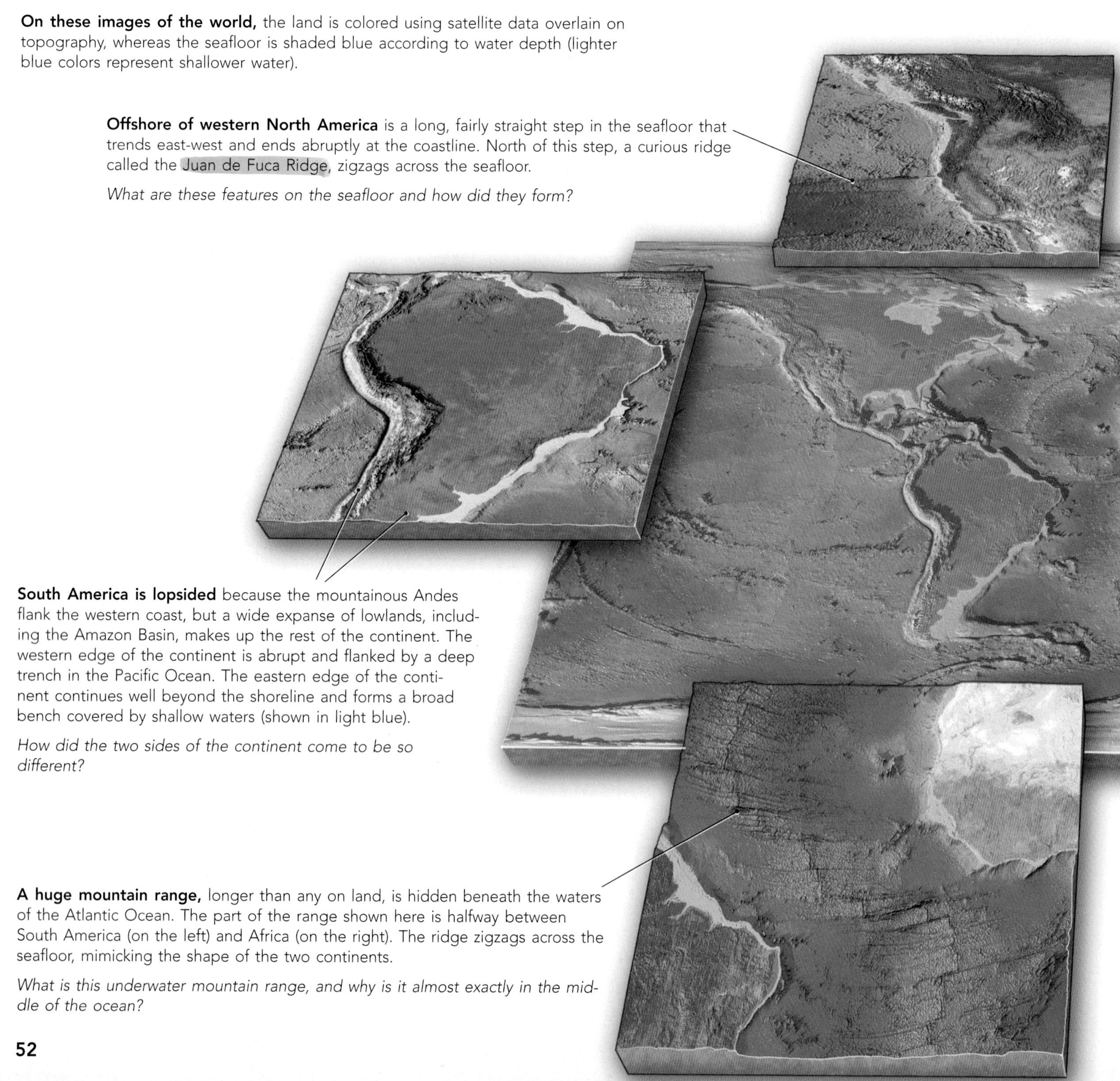

South America is lopsided because the mountainous Andes flank the western coast, but a wide expanse of lowlands, including the Amazon Basin, makes up the rest of the continent. The western edge of the continent is abrupt and flanked by a deep trench in the Pacific Ocean. The eastern edge of the continent continues well beyond the shoreline and forms a broad bench covered by shallow waters (shown in light blue).

How did the two sides of the continent come to be so different?

A huge mountain range, longer than any on land, is hidden beneath the waters of the Atlantic Ocean. The part of the range shown here is halfway between South America (on the left) and Africa (on the right). The ridge zigzags across the seafloor, mimicking the shape of the two continents.

What is this underwater mountain range, and why is it almost exactly in the middle of the ocean?

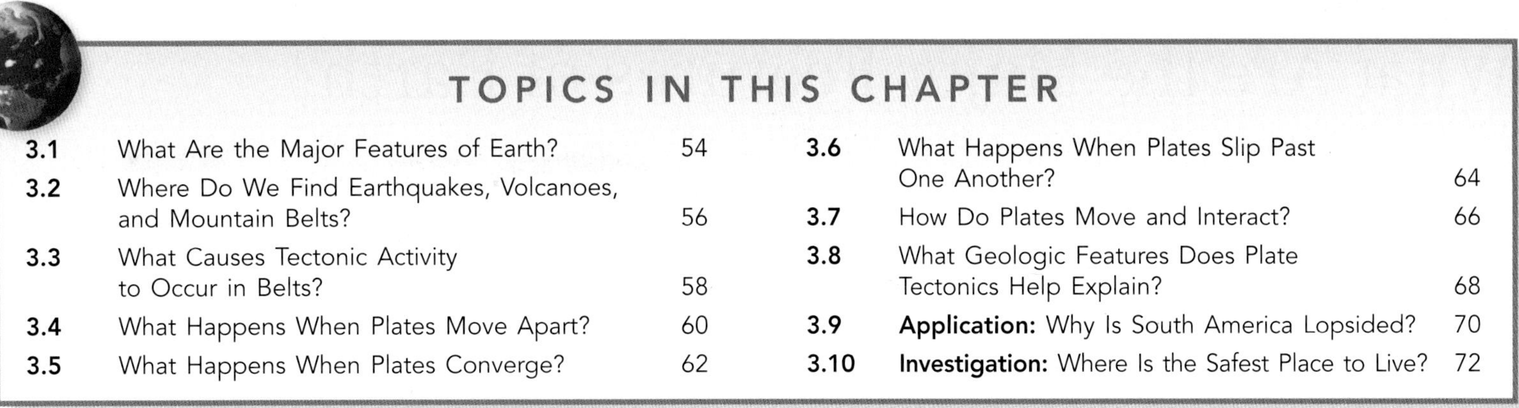

TOPICS IN THIS CHAPTER

03.00.a1

The Tibetan Plateau of southern Asia rises many kilometers above the lowlands of India and Bangladesh. The Himalaya mountain range with Mount Everest, the highest mountain on Earth, is perched on the southern edge of this plateau.

Why does this region have such a high elevation?

Japan is along the intersection of large, curving ridges that are mostly submerged beneath the ocean. Each ridge is flanked to the east by a deep trench in the seafloor. This area is well known for its destructive earthquakes and for Japan's picturesque volcano, Mount Fuji.

Do submarine ridges and trenches play a role in earthquake and volcanic activity?

The Arabian Peninsula is important for geology and world economics. East of the peninsula, the Persian Gulf has a shallow and smooth seafloor and is flanked by the world's largest oil fields. West of the peninsula, the Red Sea has a well-defined trough or fissure down its center.

How did the Red Sea form, and what processes are causing its seafloor to be disrupted?

What Are the Major Features of Earth?

THE SURFACE OF EARTH has three major types of features: continents, ocean basins, and islands. The continents display a remarkable diversity of landforms, from broad coastal plains to steep, snow-capped mountains. Features of the ocean floor include deep trenches and a submarine mountain range that encircles much of the globe. Some islands are large and isolated, but other islands define arcs, ragged lines, or irregular clusters. What are the characteristics of each type of feature?

This map shows large features on land and on the seafloor. The colors on land are from images taken by satellites orbiting Earth and show vegetated areas (green), rocky areas (brown), and sandy areas (tan). Greenland and Antarctica are white and light gray because they are mostly covered with ice and snow. Ocean colors show the depth of the seafloor and range from light blue where the seafloor is shallow, to darker blue where it is deep.

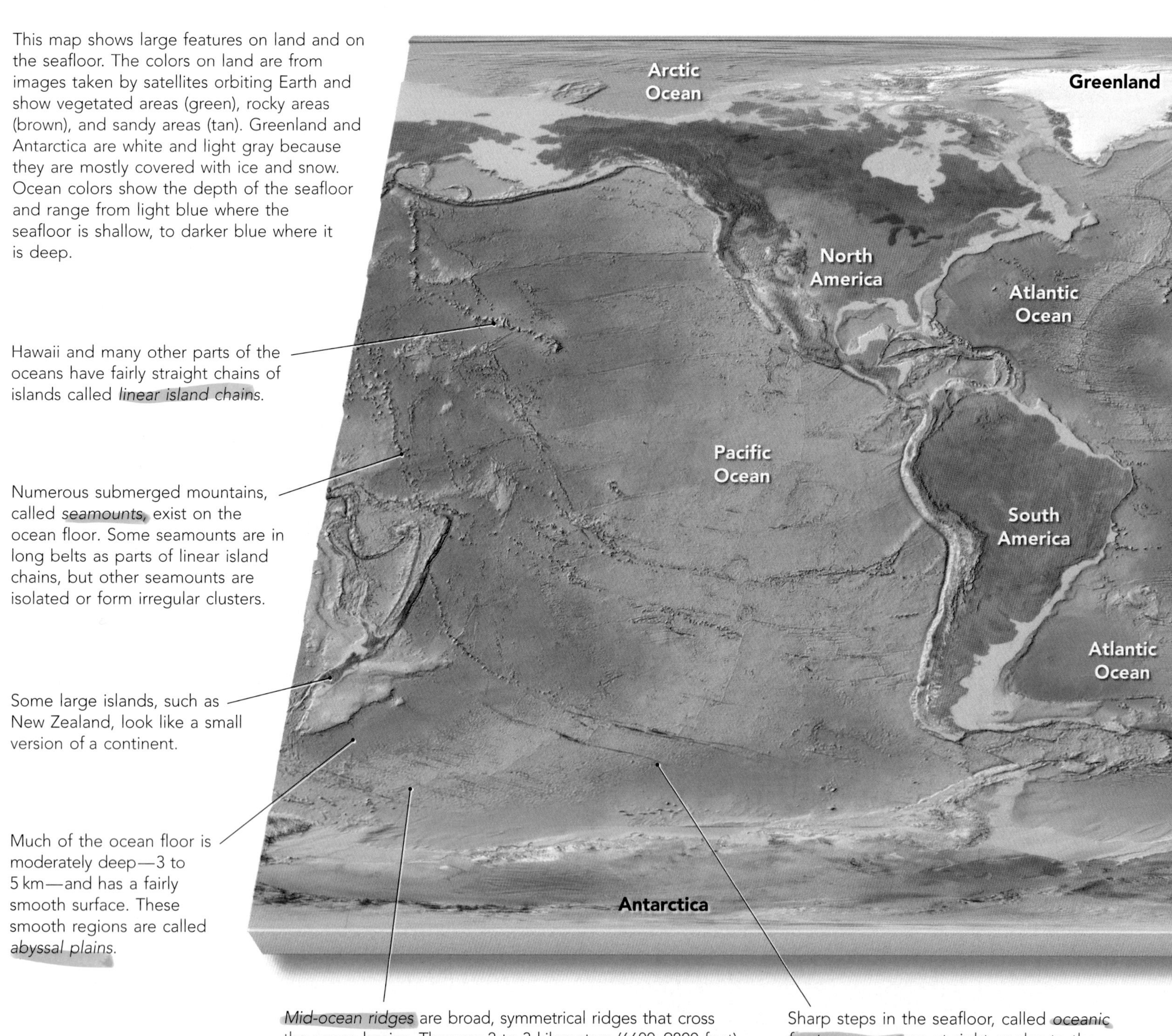

Hawaii and many other parts of the oceans have fairly straight chains of islands called *linear island chains*.

Numerous submerged mountains, called *seamounts*, exist on the ocean floor. Some seamounts are in long belts as parts of linear island chains, but other seamounts are isolated or form irregular clusters.

Some large islands, such as New Zealand, look like a small version of a continent.

Much of the ocean floor is moderately deep—3 to 5 km—and has a fairly smooth surface. These smooth regions are called *abyssal plains*.

Mid-ocean ridges are broad, symmetrical ridges that cross the ocean basins. They are 2 to 3 kilometers (6600–9800 feet) higher than the average depth of the seafloor.

Sharp steps in the seafloor, called *oceanic fracture zones*, are at right angles to the mid-ocean ridges.

Before You Leave This Page Be Able To

- ✓ Identify on a world map the named continents and oceans.
- ✓ Identify on a world map the main types of features on the continents and in the oceans.
- ✓ Summarize the main characteristics for each type of feature, including whether it is found in the oceans, on continents, or as islands.

Some continents continue outward from the shoreline under shallow seawater for hundreds of kilometers, forming submerged benches (light blue in this image), known as *continental shelves*. Such shelves surround Great Britain and Ireland.

All of the continents have large regions that are broad plains with gentle topography. Some continents have mountains along their edges or within their interiors.

Most continental areas have elevations of less than 1 to 2 kilometers (3,300 to 6,600 feet). Broad, high regions, called *plateaus*, reach high elevations such as on the Tibetan Plateau of southern Asia.

03.01.a1

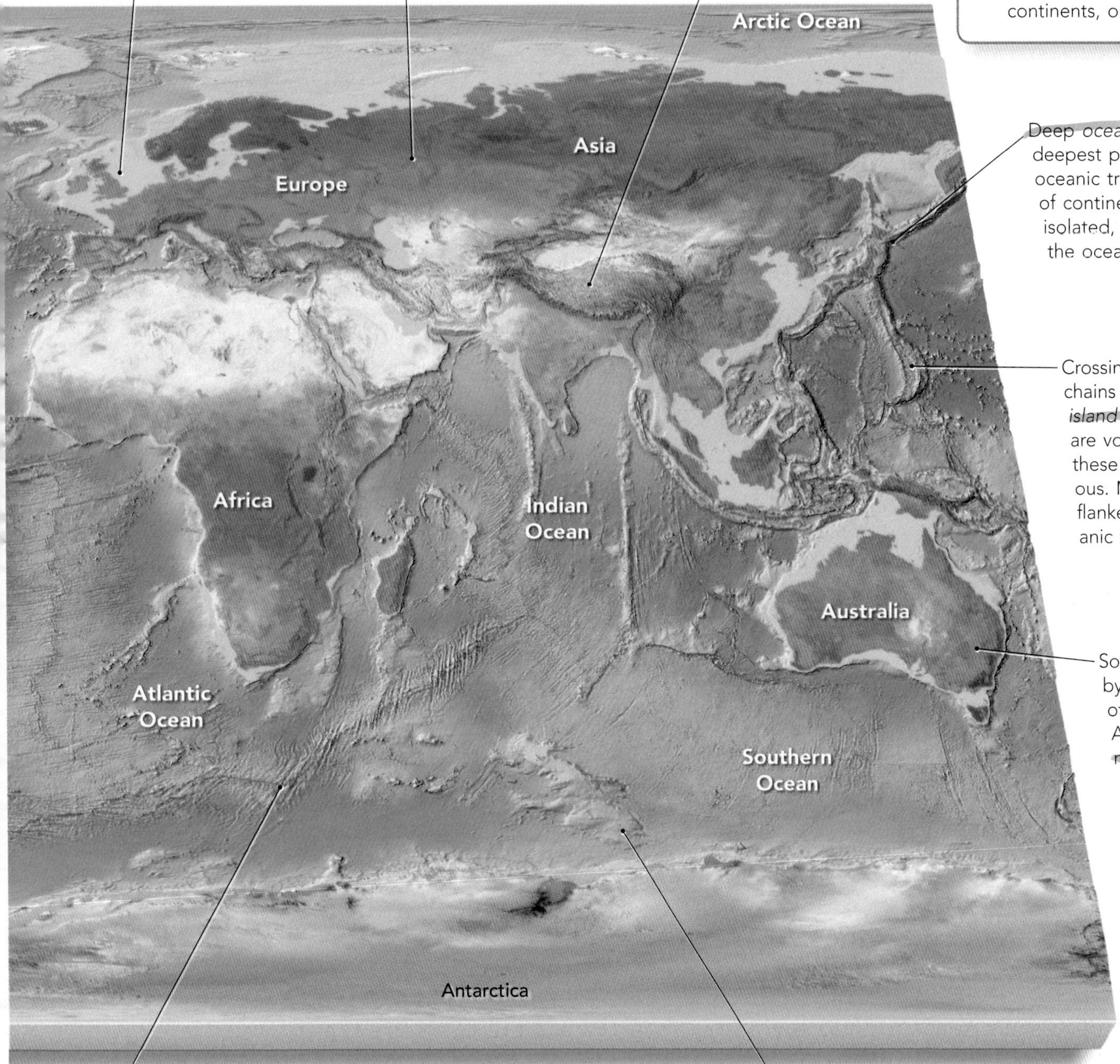

Deep *oceanic trenches* make up the deepest parts of the ocean. Some oceanic trenches follow the edges of continents, whereas others form isolated, curving troughs out in the ocean.

Crossing the seafloor are curving chains of islands, known as *island arcs*. Most of the islands are volcanoes, and many of these are active and dangerous. Most island arcs are flanked on one side by an oceanic trench.

Some continents are flanked by oceanic trenches, but other continents, such as Australia and Africa, have no nearby trenches.

Mid-ocean ridges and their associated fracture zones encircle much of the globe and, in the Atlantic and Southern Oceans, occupy a position halfway between the continents on either side.

Within the oceans are several broad, elevated regions called *oceanic plateaus*. The Kerguelen Plateau near Antarctica is one example.

3.1

Where Do We Find Earthquakes, Volcanoes, and Mountain Belts?

EARTHQUAKES, VOLCANOES, AND MOUNTAIN BELTS are spectacular expressions of geology. Many of these features are in remote, far-off places, but some are close to where we live. The distributions of earthquakes, volcanoes, and mountain belts are not random. They instead define clear patterns that reflect important, large-scale Earth processes.

A Where Do Most Earthquakes Occur?

On this map, yellow circles show the locations of moderate to strong earthquakes that occurred between 1973 and 2000. Observe the distribution of earthquakes on this map before reading on. What patterns do you notice? Which regions have many earthquakes? Which regions have few or no earthquakes?

Earthquakes are not distributed uniformly across the planet. Instead, they are mostly concentrated in discrete belts, such as the one that runs along the western coast of North America.

Most earthquakes in the oceans occur along the winding crests of mid-ocean ridges. Where the ridges curve or zigzag, so does the pattern of earthquakes.

In several regions, earthquakes are abundant in the middle of a continent. Such earthquake activity happens in parts of the Middle East and especially in China and Tibet.

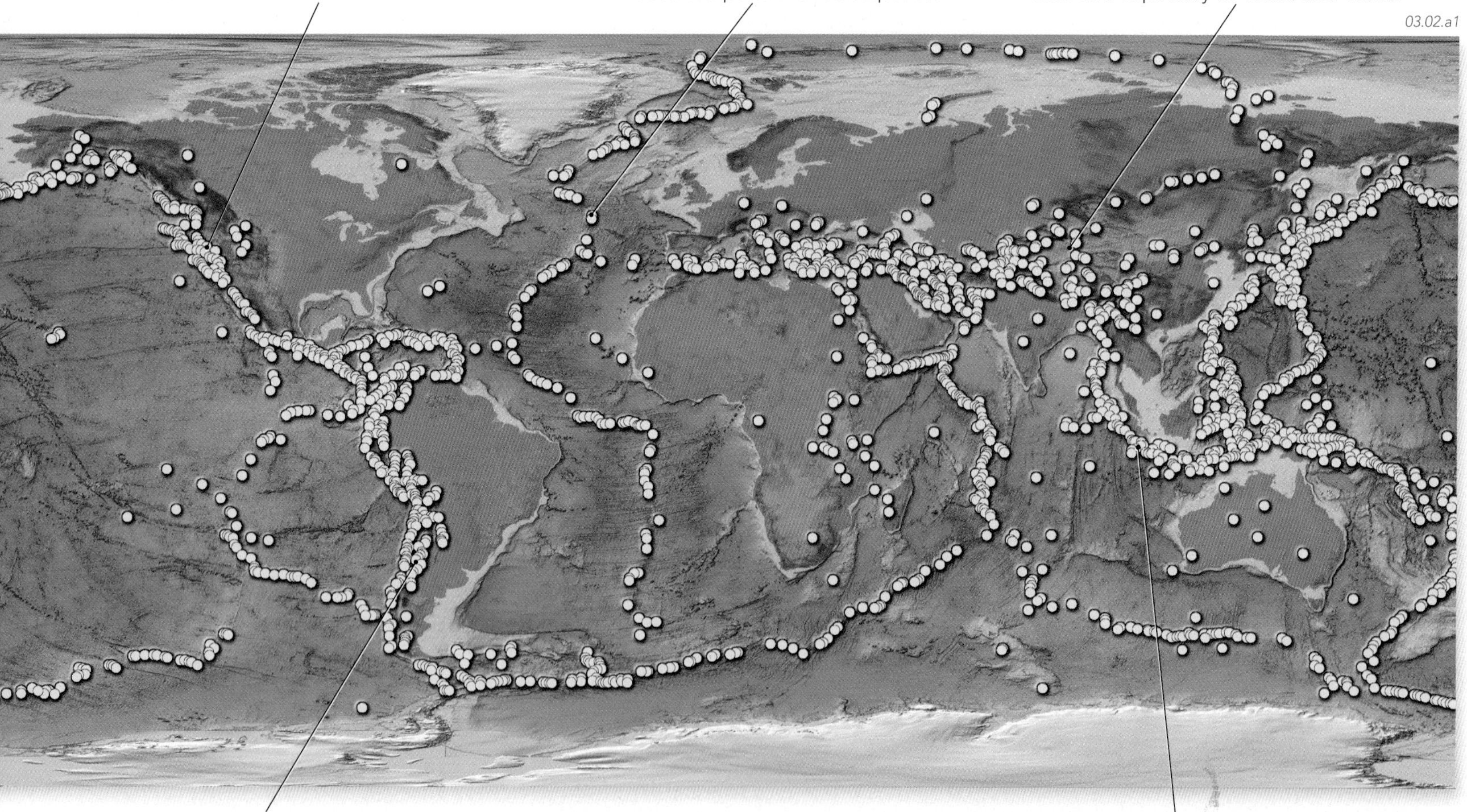

Some continental edges have many earthquakes, but others have few. In South America and North America, earthquakes are common along the western coasts, but not along the eastern coasts.

Earthquakes are not common in the interiors of some continents, such as in eastern North America, eastern South America, western Africa, and the northern parts of Asia and Europe.

Earthquakes are common along the curving oceanic trenches and the associated island arcs. In fact, many of the world's largest and most deadly earthquakes occur in regions near an oceanic trench. A recent example was the large earthquake that unleashed deadly waves in the Indian Ocean in December of 2004.

B Which Areas Have Volcanoes?

On the map below, orange triangles show the locations of recently active volcanoes. Observe the distribution of volcanoes and note any patterns before reading on. Which areas have many volcanoes? Which areas have none?

Volcanoes, like earthquakes, are not distributed everywhere, but commonly occur in belts. One belt extends along the western coasts of North and South America.

There are clusters of volcanoes in the oceans such as near Iceland. Iceland is a large volcanic island along the mid-ocean ridge in the North Atlantic Ocean.

Volcanoes occur along the western edge of the Pacific Ocean, extending from north of Australia through the Philippines and Japan.

03.02.b1

Beneath the oceans are many volcanic mountains, only the largest of which are shown here. Volcanic activity also occurs along most of the mid-ocean ridge.

Some volcanoes form in the middle of continents, such as in China and the eastern part of Africa.

Volcanoes, many of which are active, define island arcs, such as the curving island chain of Java.

The map below is the base map on which the earthquakes and volcanoes are plotted. The colors show elevation, with high elevations in brown and low elevations in green. The blue colors of the oceans darken with depth. Using the three maps, compare the distributions of earthquakes, volcanoes, and elevations. Try to identify areas where there are (1) mountains but no earthquakes, (2) mountains but no volcanoes, and (3) earthquakes but no volcanoes. Make a list of these areas, or mark the areas on a map.

03.02.b2

Before You Leave This Page Be Able To

- ✓ Show on a relief map of the world the major belts of earthquakes and volcanoes.
- ✓ Describe how the distribution of volcanoes corresponds to that of earthquakes.
- ✓ Compare the distributions of earthquakes, volcanoes, and elevations.

What Causes Tectonic Activity to Occur in Belts?

WHY DO EARTHQUAKES AND VOLCANOES occur in belts around Earth's surface? Why are there vast regions that have comparatively little of this activity? What underlying processes cause this observed pattern? The best explanation is the *theory of plate tectonics*.

A What Do Earthquake and Volcanic Activity Tell Us About Earth's Lithosphere?

1. Examine the map below, which shows earthquakes (yellow circles) and volcanoes (orange triangles). After noting the patterns, compare this map with the lower map and then read the associated text.

2. On the upper map, there are large regions that have few earthquakes and volcanoes. These regions are relatively stable and intact blocks. There are a dozen or so of these blocks, whose edges are defined by belts of earthquakes and volcanoes.

03.03.a1

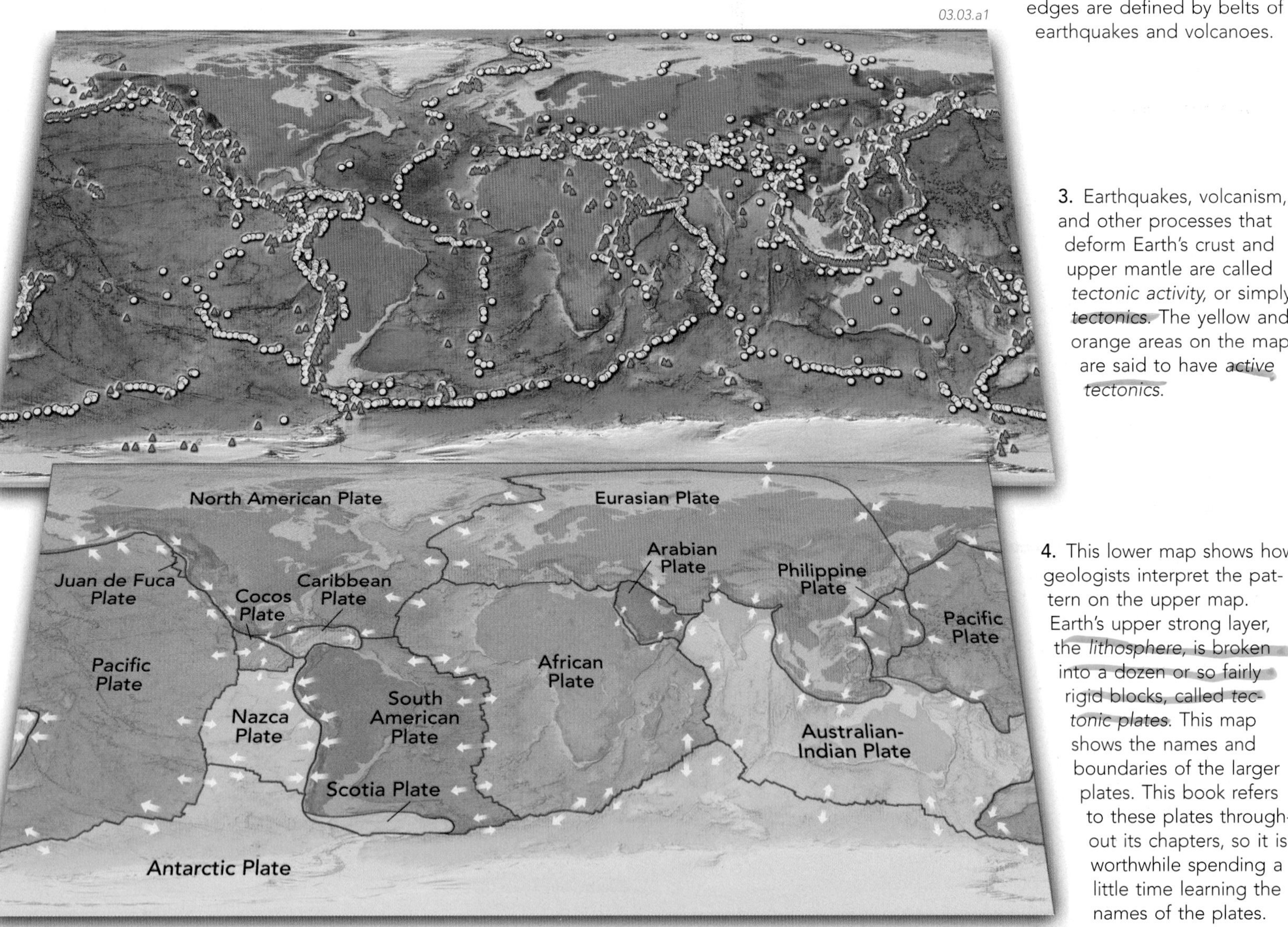

3. Earthquakes, volcanism, and other processes that deform Earth's crust and upper mantle are called *tectonic activity*, or simply *tectonics*. The yellow and orange areas on the map are said to have *active tectonics*.

4. This lower map shows how geologists interpret the pattern on the upper map. Earth's upper strong layer, the *lithosphere*, is broken into a dozen or so fairly rigid blocks, called *tectonic plates*. This map shows the names and boundaries of the larger plates. This book refers to these plates throughout its chapters, so it is worthwhile spending a little time learning the names of the plates.

5. Compare the two maps and note how the distribution of tectonic activity, especially earthquakes, outlines the shapes of the plates. Earthquakes are a better guide to plate locations than are volcanoes. Most volcanoes do lie near plate boundaries, but many plate boundaries have no volcanoes. Some volcanoes and earthquakes occur in the middle of plates.

B How Do Plates Move Relative to One Another?

The boundary between two plates has earthquakes and other tectonic activity because the plates are moving *relative to one another*. For this reason, we talk about *relative motion* of plates across a plate boundary. Two plates can *move away*, *move toward*, or *move sideways* relative to one another. Based on these types of relative motion, three types of plate boundaries are defined: *divergent*, *convergent*, and *transform*.

Divergent Boundary

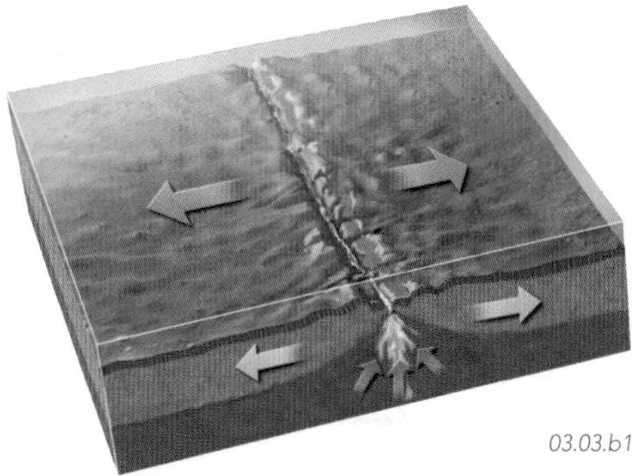

At a *divergent boundary*, two plates move apart relative to one another.

Convergent Boundary

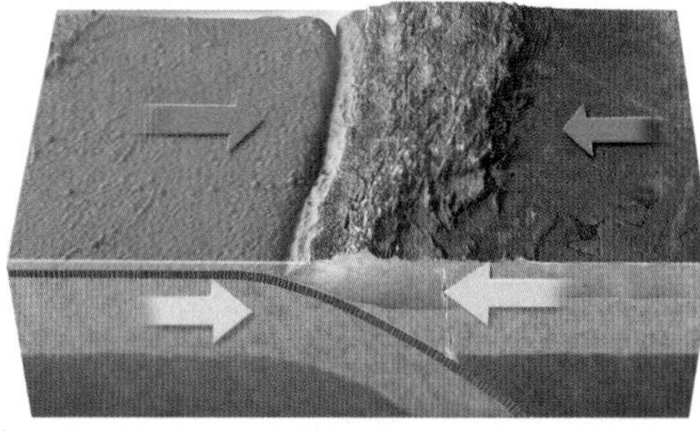

At a *convergent boundary*, two plates move toward one another.

Transform Boundary

At a *transform boundary*, two plates move horizontally past one another, as shown by the yellow arrows on the top surface.

C Where Are the Three Types of Plate Boundaries?

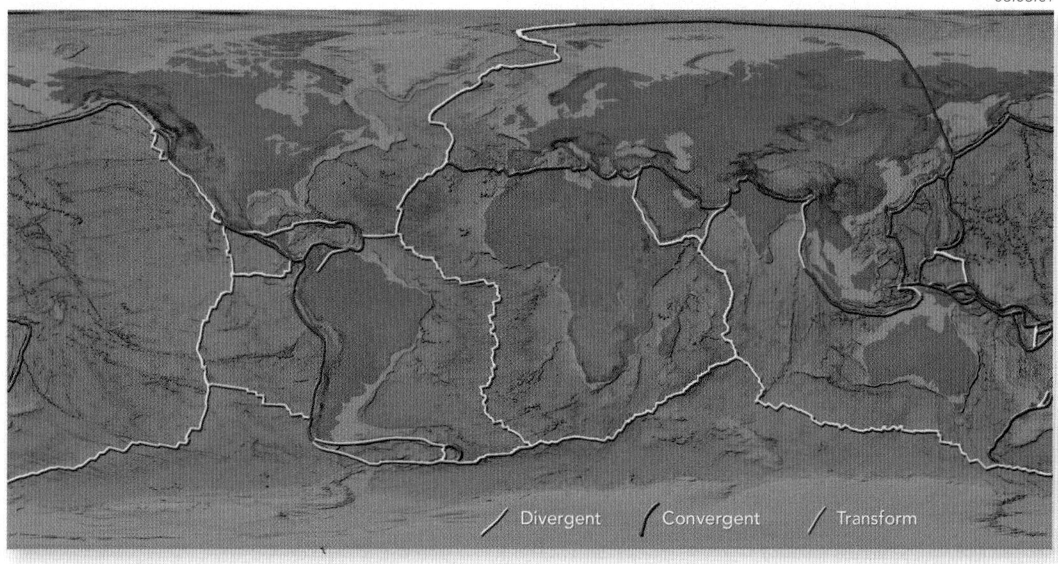

On this map, plate boundaries are colored according to type. Compare this map with the top map in *Part A* and with those shown earlier in the chapter. Determine whether each type of plate boundary is *generally* associated with each of the following features:

- Earthquakes
- Volcanoes
- Mountain Belts
- Mid-Ocean Ridges
- Oceanic Trenches

Rigid and Not-So-Rigid Plates

Plate tectonics, in its most strict sense, would have purely rigid plates with tectonic activity happening only at the boundaries. Based on the maps of earthquakes and volcanoes, this is not precisely true. Most tectonic activity occurs at or near inferred plate boundaries, but some tectonic activity happens well away from plate boundaries. Clearly, Earth is more complicated than the plate-tectonic ideal. This is largely because some parts of the lithosphere are weaker than other parts. Forces can be transmitted through the strong parts, causing the weaker parts to break and slip. In a few cases, tectonic activity within a plate indicates that a new plate boundary may be forming.

Before You Leave This Page Be Able To

- ✓ Explain what plate tectonics is and the three types of plate boundaries.
- ✓ Compare the three types of plate boundaries with the distributions of earthquakes, volcanoes, mountain belts, mid-ocean ridges, and trenches.

What Happens When Plates Move Apart?

AT MID-OCEAN RIDGES, Earth's tectonic plates are moving apart (diverging), forming new oceanic lithosphere. Such boundaries are the sites of numerous small earthquakes and submarine volcanism. Divergent motion can split a continent into two pieces and form a new ocean basin as the rifted pieces move apart.

 How Do Mid-Ocean Ridges Form?

Mid-ocean ridges mark divergent boundaries in the ocean where two oceanic plates move apart relative to one another. These boundaries are also called *spreading centers* because of the way the plates spread apart.

1. A narrow trough, called a *rift*, runs along the axis of most mid-ocean ridges. The rift forms as large blocks of crust are down-faulted to accommodate spreading. The faulting is accompanied by earthquakes.

2. As the plates move apart, solid mantle in the asthenosphere rises toward the surface and partially melts in response to the decrease in pressure. The molten rock, or magma, rises along narrow conduits and accumulates in magma chambers beneath the rift.

3. Along the rift, magma erupts onto the seafloor as submarine lava flows, while other magma solidifies at depth. These magmatic additions create new oceanic crust along the spreading center and add to the oceanic plates as they move away. This entire process is called *seafloor spreading.*

4. Mid-ocean ridges are higher than the surrounding seafloor because they have hotter, less dense rocks and a thinner lithosphere. The elevation of the seafloor decreases away from the ridge as the rock cools and contracts, and as the less dense asthenosphere cools into more dense lithosphere.

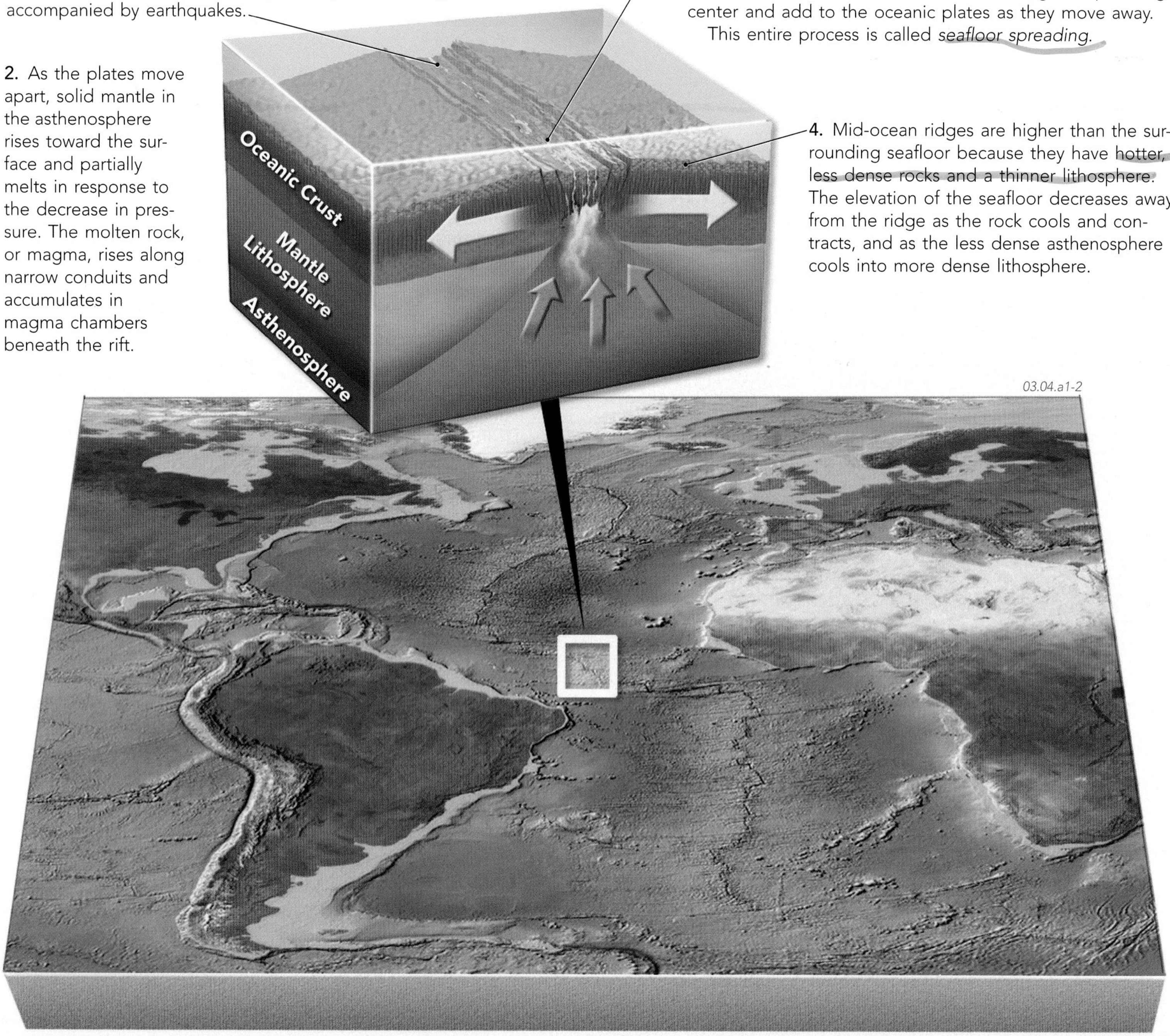

Oceanic Crust
Mantle Lithosphere
Asthenosphere

03.04.a1-2

B What Happens When Divergence Splits a Continent Apart?

A divergent boundary can form within a continent, causing a *continental rift* such as the *Great Rift Valley* in East Africa. Such rifting, if it continues, leads to seafloor spreading and the formation of a mid-ocean ridge and new ocean basin, following the progression shown here.

The initial stage of continental rifting commonly includes broad uplift of the land surface as mantle-derived magma ascends into and heats the crust. The magma can melt parts of the continental crust, producing additional magma. Heating of the crust causes it to expand, which results in further uplift.

Stretching of the crust causes large crustal blocks to drop down along faults, forming a *continental rift*. The down-dropped blocks are *basins* that can trap sediment and water. At depth, rifting causes solid mantle material in the asthenosphere to continue flowing upward and partially melt. The resulting magma erupts from volcanoes and long fissures on the surface, or can solidify in the subsurface. The entire crust thins as it is pulled apart, so the rifted region will drop in elevation as it cools. A modern example of this stage is the *East African Rift*.

If rifting continues, the continent splits into two pieces and a narrow ocean basin forms by the onset of seafloor spreading. A modern example of this stage is the *Red Sea*, which formed when the Arabian Peninsula rifted away from Africa.

With continuing seafloor spreading, the ocean basin becomes progressively wider, eventually becoming a broad ocean like the modern-day *Atlantic Ocean*. The Atlantic Ocean basin formed when North and South America rifted away from Europe and Africa, following the sequence shown here. Seafloor spreading continues today along the ridge in the middle of the Atlantic Ocean, transporting the Americas farther away from Europe and Africa.

Before You Leave This Page Be Able To

- ✓ Sketch, label, and explain an oceanic divergent boundary.
- ✓ Sketch, label, and explain a divergent boundary within a continent (i.e., a continental rift).
- ✓ Sketch, label, and explain how continental rifting can lead to the formation of a new ocean basin.

What Happens When Plates Converge?

CONVERGENT BOUNDARIES FORM where two plates move together. Convergence of plates can occur between two oceanic plates, between an oceanic plate and a continental plate, or between two continental plates. Oceanic trenches, island arcs, and many of Earth's largest mountain belts form at convergent boundaries. Many of Earth's most dangerous volcanoes and largest earthquakes also occur along these boundaries.

A What Happens When Two Oceanic Plates Converge?

1. Convergence of two oceanic plates forms an *ocean-ocean convergent boundary.* One plate is pulled down beneath the other plate along an inclined zone marked by earthquakes. This process of one plate sliding beneath another plate is called *subduction*.

2. An oceanic trench forms where the subducting plate starts to bend down. Sediment in the trench and slices of oceanic crust are scraped off, forming a wedge of highly sheared rocks called an *accretionary prism*. This name signifies that material is being added (accreted) over time to the wedge-shaped region.

3. Chemical reactions in the subducting plate release water that causes melting in the overlying asthenosphere. The magmas are buoyant and rise into the overlying plate.

4. Some magma erupts onto the surface from dangerous, explosive volcanoes. Eventually, the erupted lava and volcanic ash construct a curving belt of islands that rise above sea level as an *island arc*. An example is the arc-shaped Aleutian Islands chain of Alaska.

5. Some magma solidifies at depth, also adding to the volume of the crust. Over time, the crust gets thicker and volcanoes build together to form a more continuous strip of land, such as along the Java region of Indonesia.

B What Happens When an Oceanic Plate and a Continental Plate Converge?

1. Convergence between an oceanic and a continental plate forms an *ocean-continent convergent boundary.* Along this boundary, the denser oceanic plate is subducted beneath the more buoyant continental plate.

2. An oceanic trench marks the plate boundary and receives sediment from the adjacent continent. This sediment and material scraped off the oceanic plate form an *accretionary prism.*

3. Volcanoes form on the surface of the overriding continental plate. These volcanoes erupt volcanic ash and lava onto the landscape and pose a hazard for people who live nearby. Examples include the large volcanoes atop the Cascade Range of the Pacific Northwest and those within the Andes Mountains of South America.

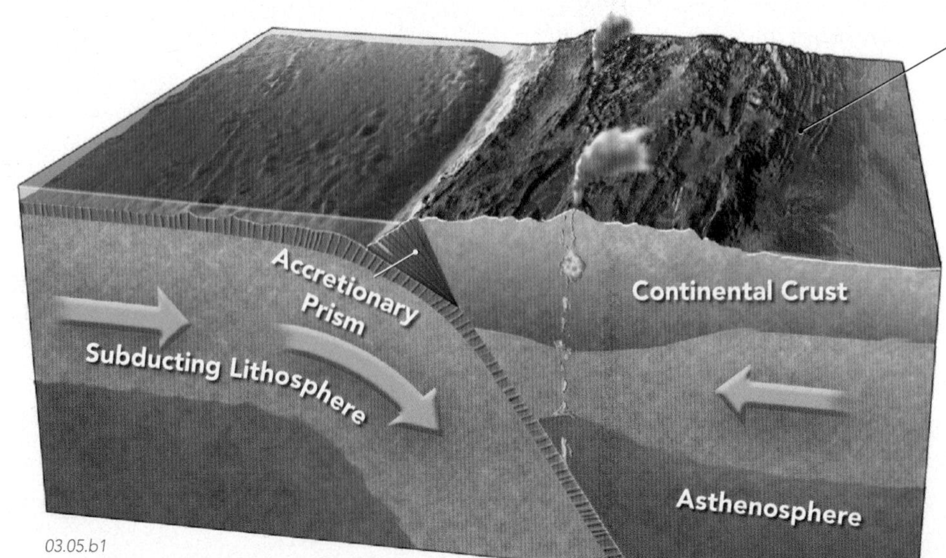

4. Compression associated with the convergent boundary squeezes the crust for hundreds of kilometers into the continent. The crust deforms and thickens, forming a high mountain range, such as the Andes.

5. Magma forms by melting of the asthenosphere above the subduction zone. Some magma rises into the overlying continental crust, while other magma solidifies at depth.

C What Causes the Pacific Ring of Fire?

Volcanoes surround the Pacific Ocean, forming the *Pacific Ring of Fire*, as shown in the map below. The volcanoes extend from the southwestern Pacific, through the Philippine Islands, Japan, and Alaska, and down the western coasts of the Americas. The Ring of Fire results from subduction on both sides of the Pacific Ocean.

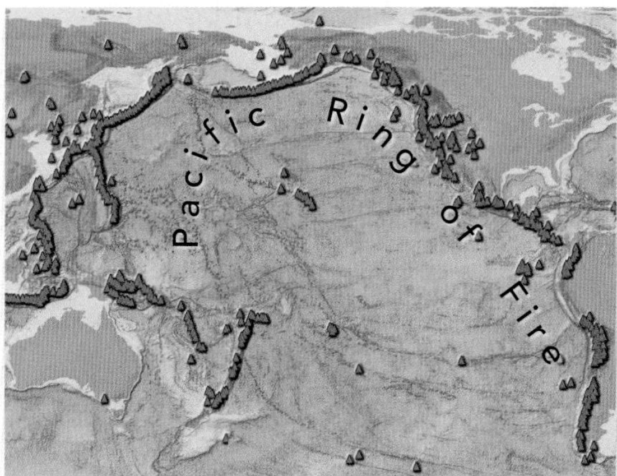

03.05.c1

New oceanic lithosphere forms along a mid-ocean ridge, called the *East Pacific Rise*. Once formed, the new lithosphere moves away from the ridge as seafloor spreading continues.

Subduction of this oceanic lithosphere occurs along both sides of the Pacific Ocean, producing oceanic trenches on the seafloor and volcanoes on the overriding plates. The Pacific Ocean is wider than shown here and the spreading center is much closer to the Americas (to the right).

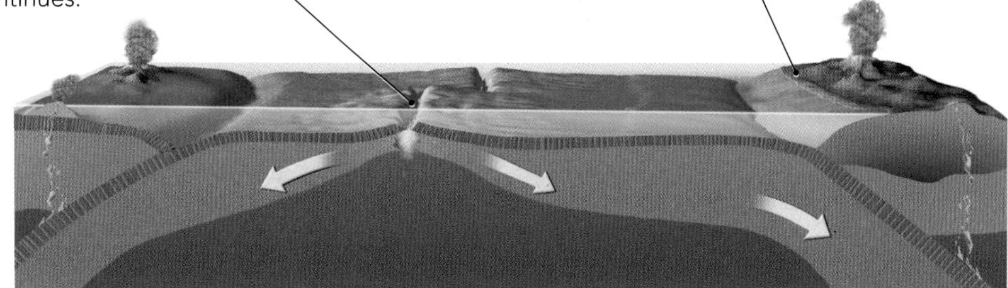

03.05.c2

D What Happens When Two Continents Collide?

Two continental masses may converge along a *continent-continent convergent boundary*. This type of boundary is commonly called a *continental collision*.

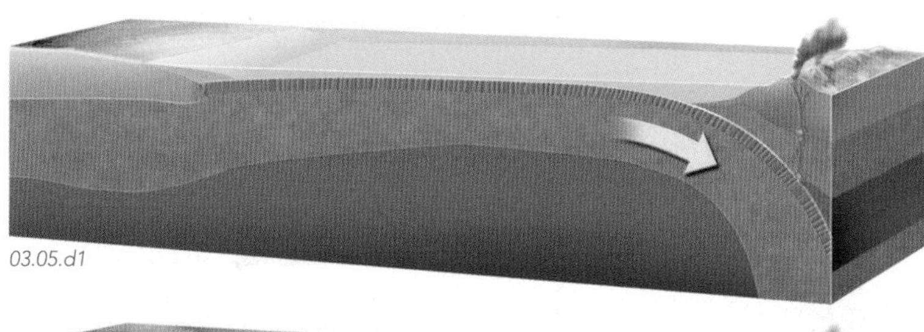

03.05.d1

The plate on the left is partly oceanic and partly continental, and the oceanic part can be subducted at a convergent boundary. For a continental collision to occur, the overriding plate (on the right) must be a continental plate.

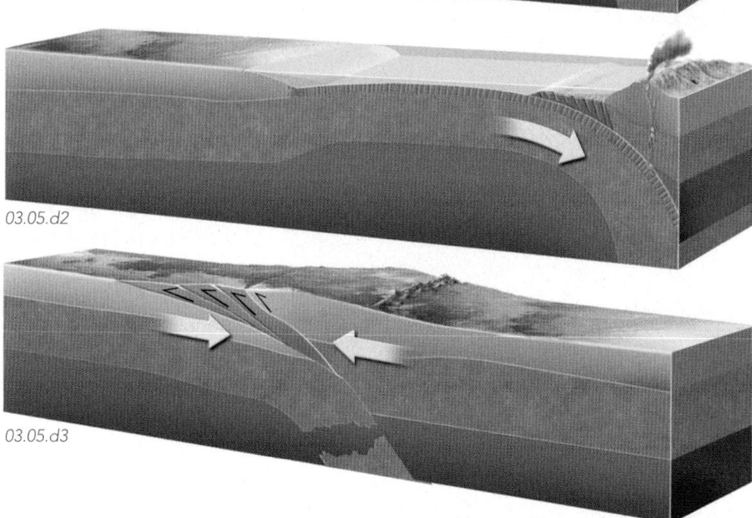

03.05.d2

Subduction of the oceanic part of the downgoing plate may bring the continent progressively closer to the trench and to the convergent boundary. Note that all magmatic activity occurs in the overriding plate, not on the approaching continent.

03.05.d3

When the continent arrives at the boundary, it is dragged partially under the other continent or simply clogs the subduction zone as the two continents collide. The continent cannot be subducted deeply into the asthenosphere because it is too buoyant. Collision can form enormous mountain belts and high plateaus, such as the Himalaya and Tibetan Plateau of southern Asia. In this region, the continental crust of India collided with and was shoved beneath the southern edge of Asia.

Before You Leave This Page Be Able To

- ✓ Sketch, label, and explain the three types of convergent boundaries, including the features and processes associated with each.
- ✓ Sketch, label, and explain the steps leading to a continental collision (continent-continent convergent boundary).

What Happens When Plates Slip Past One Another?

AT TRANSFORM BOUNDARIES, PLATES SLIP HORIZONTALLY past each other. In the oceans, such boundaries are *transform faults* and generally are associated with mid-ocean ridges. Transform faults alternate with spreading centers to form a zigzag pattern on the seafloor. Some transform faults link other types of plate boundaries, such as a mid-ocean ridge and a trench. Transform boundaries can also cut across continents, sliding one large crustal block past another, as occurs along the San Andreas fault.

A Why Do Mid-Ocean Ridges Have a Zigzag Pattern?

To understand why mid-ocean ridges have the shape they do, examine how the two parts of this pizza have pulled apart, just like two *diverging* plates.

The break in the pizza did not follow a straight line. It took jogs to the left and the right, following cuts where the pizza was the weakest.

The gaps created where the pizza pulled apart represent the segments of a mid-ocean ridge that are spreading apart. In a mid-ocean ridge, there are no open gaps because new material derived from the underlying mantle fills the space as fast as it opens.

The spreading segments are linked by breaks along which the two parts of the pizza simply slid by one another. There are no gaps here, only horizontal movement of one side past the other. The arrows show the direction of relative motion. In geology, a break along which movement has occurred is called a *fault*. A fault that accommodates the horizontal movement of one plate past another is a *transform fault*.

03.06.a1

Transform Faults Along the Mid-Ocean Ridge

1. Mid-ocean ridges, such as this one in the South Atlantic Ocean, have a zigzag pattern similar to the broken pizza.

2. In this region, there are north-south segments along which spreading is occurring.

3. East-west segments are *transform* faults along which the two diverging plates simply slide past one another, like the breaks in the pizza. These transform faults link the spreading segments and have the relative motion shown by the arrows.

4. Transform faults along mid-ocean ridges are generally *perpendicular* to the axis of the ridge. As in the pizza example, transform faults are *parallel* to the direction in which the two plates are spreading apart (diverging).

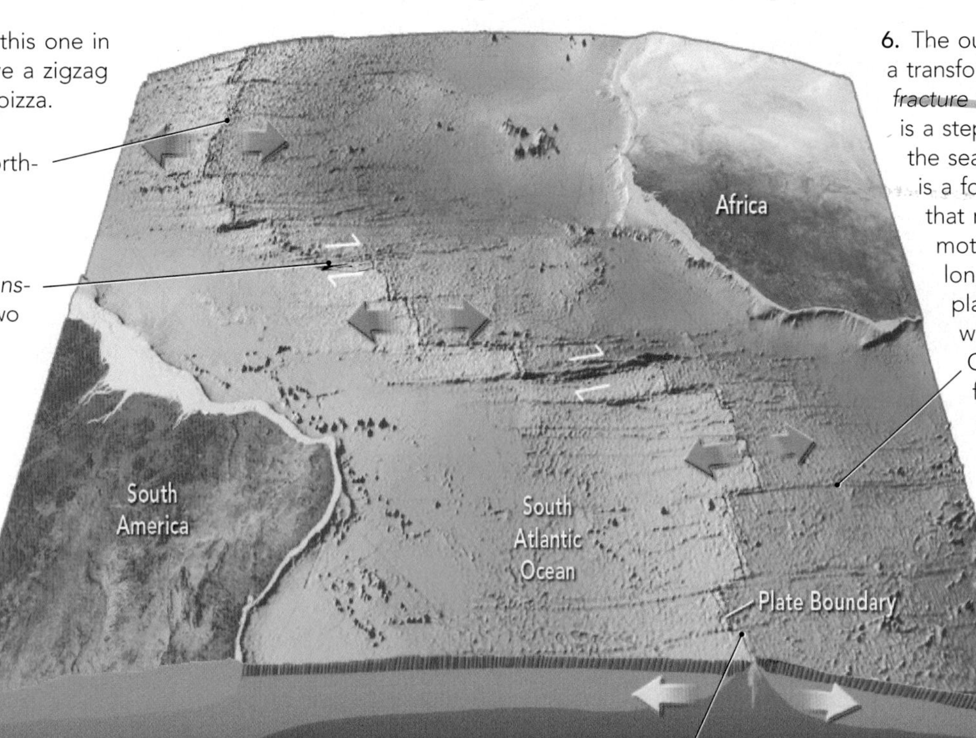

03.06.a2

5. Transform faults exist where seafloor spreading along the ridge is moving the two plates in opposite directions relative to each other. This movement requires horizontal slip along the transform faults to link adjacent segments that are spreading.

6. The outward continuation of a transform fault is an oceanic *fracture zone*, which generally is a step in the elevation of the seafloor. A fracture zone is a former transform fault, that now has no relative motion across it. It no longer separates two plates and instead is within a single plate. Opposite sides of the fracture zone have different elevations because they formed by seafloor spreading at different times in the past and so had different amounts of time to cool and subside after forming at the spreading center. Younger parts of the plate are higher than older parts.

B What Are Some Other Types of Transform Boundaries?

The Pacific seafloor and western North America contain several different transform boundaries. The boundary between the Pacific plate and the North American plate is mostly a transform boundary, with the Pacific plate moving northwest relative to the rest of North America.

1. The Queen Charlotte transform fault lies along the edge of the continent, from north of Vancouver Island to southeastern Alaska.

2. The zigzag boundary between the Pacific plate and the small Juan de Fuca plate has two transform faults (shown in green) that link different segments of an oceanic spreading center.

3. The San Andreas transform fault extends from north of San Francisco to southeast of Los Angeles. The part of California west of the fault is on the Pacific plate and is moving approximately 5 cm/year to the northwest relative to the rest of North America. South of this map the transform boundary continues into the Gulf of California.

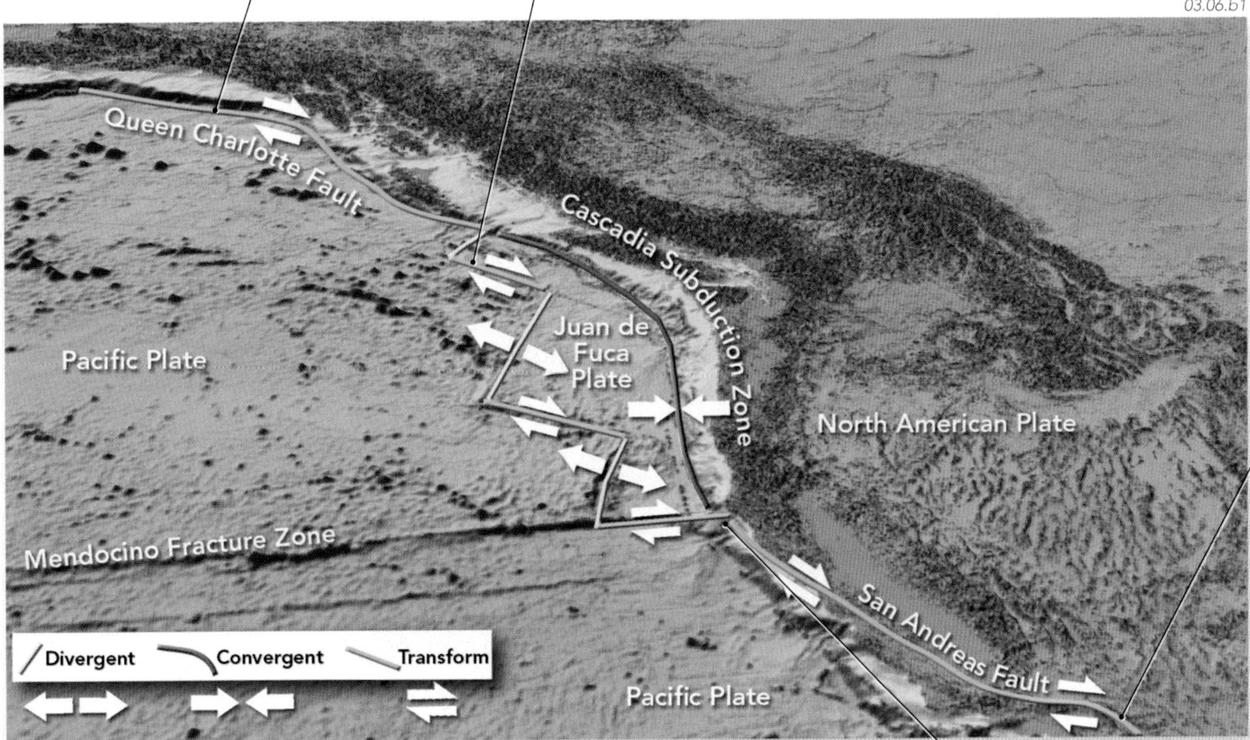

4. A transform fault links a spreading center on the Juan de Fuca plate with the Cascadia subduction zone and the San Andreas fault. The place where the three plate boundaries meet is called a *triple junction*.

5. To the west, the Mendocino fracture zone formed as a transform fault, but is now entirely within the Pacific plate and has no relative displacement. Oceanic crust to the north is higher because it is younger than oceanic crust to the south.

Field Trip to a Transform Fault Zone

Californians have a transform fault in their backyard. The fault poses severe dangers from a major earthquake, and it is also responsible for uplifting the steep mountains around Los Angeles and the beautiful rolling hills around San Francisco. California is one of the best places in the world to visit an active transform fault.

In central and northern California, the San Andreas fault forms linear valleys, abrupt mountain fronts, and lines of lakes. Some stream valleys jog to the right where they cross the fault, a reflection of the relative movement of the two sides. This can be seen in this photograph of the Carrizo Plain, where a linear gash in the topography marks the fault. A large stream follows part of the fault. The North American plate is to the left, and the Pacific plate is to the right and being displaced toward the viewer.

03.06.mtb1

Before You Leave This Page Be Able To

- ✓ Sketch, label, and explain an oceanic transform boundary related to seafloor spreading at a mid-ocean ridge.
- ✓ Locate on a map modern examples of a transform fault associated with a mid-ocean ridge.
- ✓ Sketch, label, and explain the motion of transform faults along the West Coast.

How Do Plates Move and Interact?

THE MOTION OF PLATES transports material back and forth between the asthenosphere and the lithosphere. Some asthenosphere becomes lithosphere at mid-ocean spreading centers and then takes a slow trip across the surface of Earth and back down into the asthenosphere at a subduction zone. This process is the major way that Earth transports heat to the surface.

A What Moves the Plates?

How exactly do plates move? To move, an object must be *subjected to a driving force* (a force that drives the motion). Second, the driving force must exceed the *resisting forces*—those forces that resist the movement, such as friction and any resistance from other material that is in the way. What forces drive the plates?

Slab Pull—The subducting oceanic lithosphere is denser than the surrounding asthenosphere, so gravity pulls the plate downward into the asthenosphere. Slab pull is considered to be the strongest force driving the plates, and plates that are being subducted generally move faster than plates that are not being subducted.

Ridge Push—The mid-ocean ridge is higher than the ocean floor away from the ridge because lithosphere near the ridge is thinner and hotter. Gravity causes the plate to slide away from the topographically high ridge and push the plate outward.

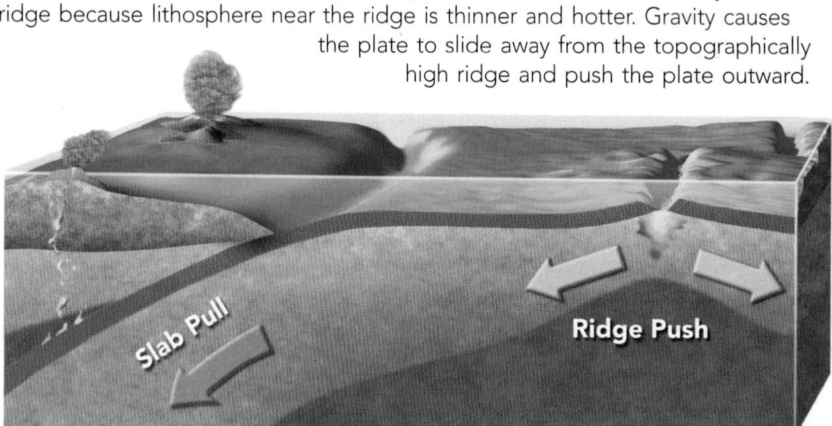

03.07.a1

Other Forces—Plates also are affected by *convection currents* in the mantle, by centers of upwelling mantle material called *hot spots*, and by the tendency of the slab to sag backward away from the trench. Another important source of forces is the motion of a plate with respect to the underlying mantle.

B How Fast Do Plates Move Relative to One Another?

The motion of plates relative to one another ranges from 1 to 15 cm/yr, about as fast as your fingernails grow. This map shows the relative motion along the major plate boundaries. Arrows indicate whether the plate boundary has divergent (outward pointing), convergent (inward pointing), or transform (side by side) motion.

03.07.b1

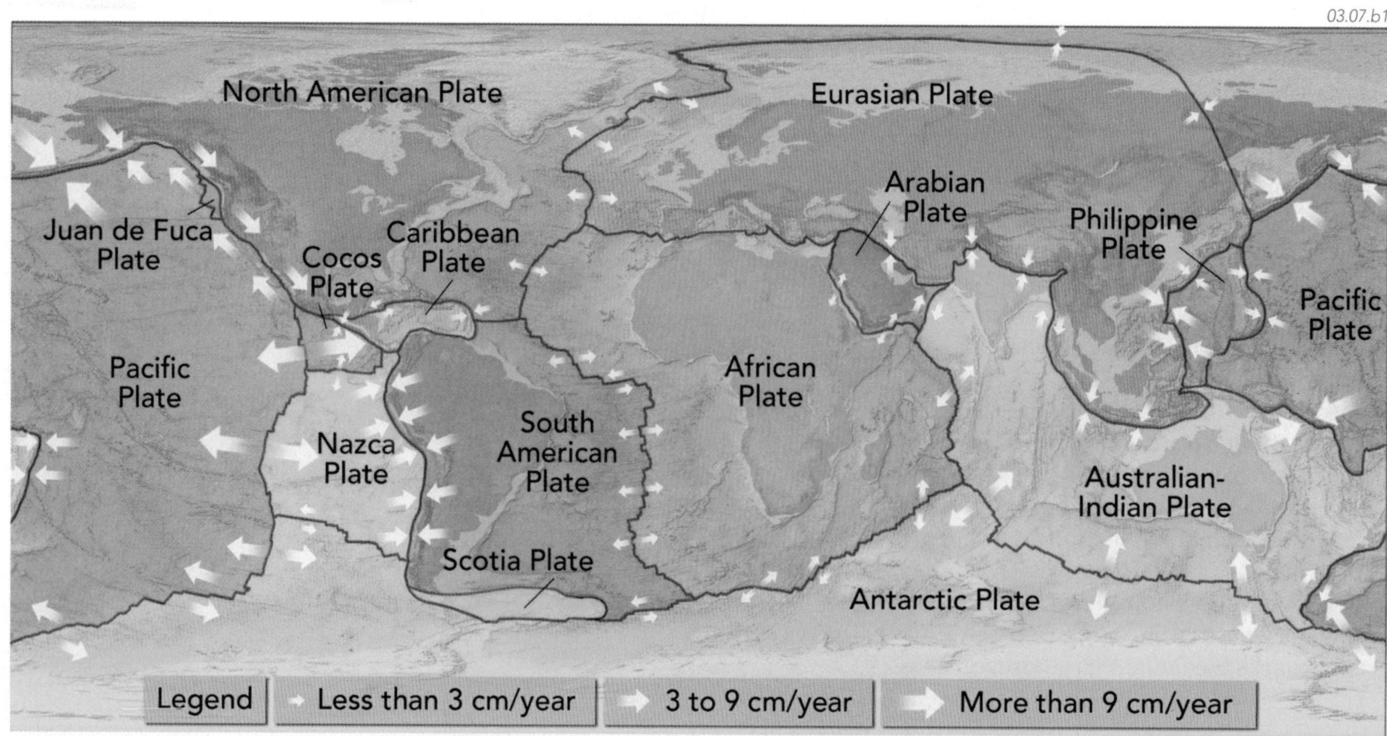

C Is There a Way to Directly Measure Plate Motions?

Modern technology allows direct measurement of modern-day plate motions by using satellites, lasers, and other methods. Motions determined by these data are consistent with our current understanding of plates.

03.07.c1

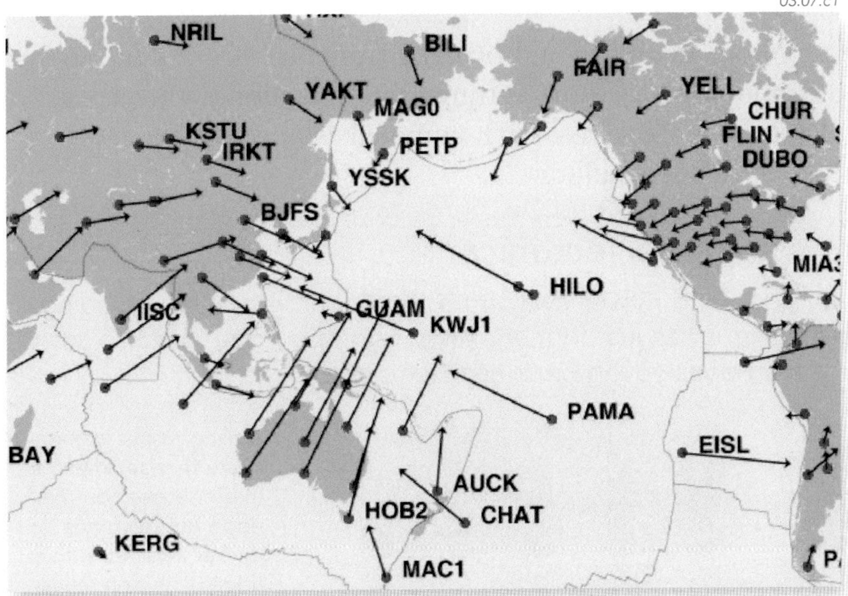

Global Positioning System (GPS) is a very accurate location technique that uses small radio receivers to record signals from several dozen Earth-orbiting satellites.

By attaching GPS receivers to a number of sites on land and monitoring any changes in position over time, geologists have produced this map showing the velocity of each site. The direction and length of the arrows depict the direction and rate of motion of each site. The name of each site is abbreviated.

Note that Hawaii (labeled HILO on the map), on the Pacific plate, is moving northwest, while western North America is moving southwest. This direction matches predictions from the theory of plate tectonics.

Although not shown here, current plate motions can also be measured by bouncing laser light off satellites or by measuring the slight differences in arrival time of natural radio signals from space.

D What Happens Where Plate Boundaries Change Their Orientation?

A boundary between two plates can change to a different type of boundary, such as from divergent to transform, as it changes orientation compared to the direction of relative plate movement. Nearly all plate boundaries contain curves or abrupt bends, so most boundaries change type as they curve across Earth's surface.

▽ 1. As these two interlocking blocks pull apart, two gaps (spreading centers) are linked by a transform boundary where the blocks slip horizontally by one another.

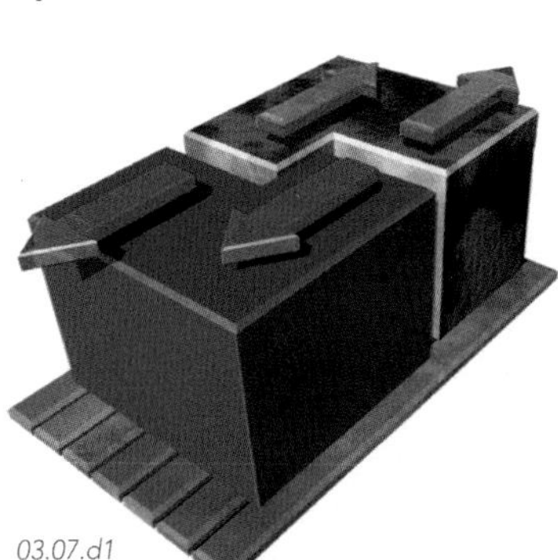

03.07.d1

03.07.d2

South American Plate
Scotia Plate
Antarctic Plate
South America
Antarctica
Divergent Convergent Transform

3. The boundary between the South American plate and Antarctic plate is mostly convergent along the west coast of South America, but becomes a transform boundary as it wraps around the southern tip of South America.

4. The boundary of the small Scotia plate changes in character as its orientation varies. Its eastern side is convergent beneath the small Scotia island arc, where the South American plate subducts beneath the Scotia plate. It turns into a transform boundary where it trends east-west, parallel to relative motion between the two plates.

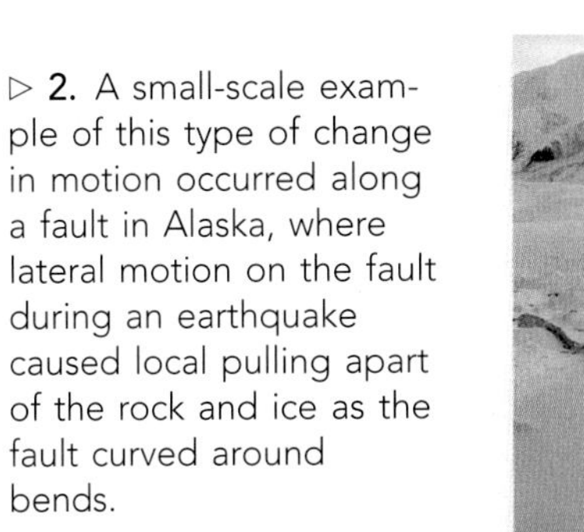

▷ 2. A small-scale example of this type of change in motion occurred along a fault in Alaska, where lateral motion on the fault during an earthquake caused local pulling apart of the rock and ice as the fault curved around bends.

03.07.d3

Before You Leave This Page Be Able To

- ✓ Summarize the driving forces of plate tectonics.
- ✓ Describe the typical rates of relative motion between plates.
- ✓ Describe some ways to directly measure plate motion.
- ✓ Sketch, label, and explain how a plate boundary can change its type as its orientation changes.

What Geologic Features Does Plate Tectonics Help Explain?

MANY OF EARTH'S LARGE FEATURES, including mid-ocean ridges and oceanic trenches, are related to plate tectonics. Many mountain ranges are in predictable plate-tectonic settings, such as along convergent boundaries. The theory of plate tectonics also explains other features, such as linear island chains, continents that look like they would fit together, and the age of the seafloor.

A Is the Age of the Seafloor Consistent with Plate Tectonics?

If oceanic crust forms by spreading at a mid-ocean ridge and moves away from the ridge by further spreading, then the crust should be youngest near the ridge, where it has just formed. Away from the ridge, oceanic crust will be older and have a thicker cover of sediment than does young crust near the ridge.

Since 1968, ocean-drilling ships have drilled hundreds of deep holes into the world's seafloor, measured the thickness of sediment, and removed samples of rock, sediment, and fossils for analysis.

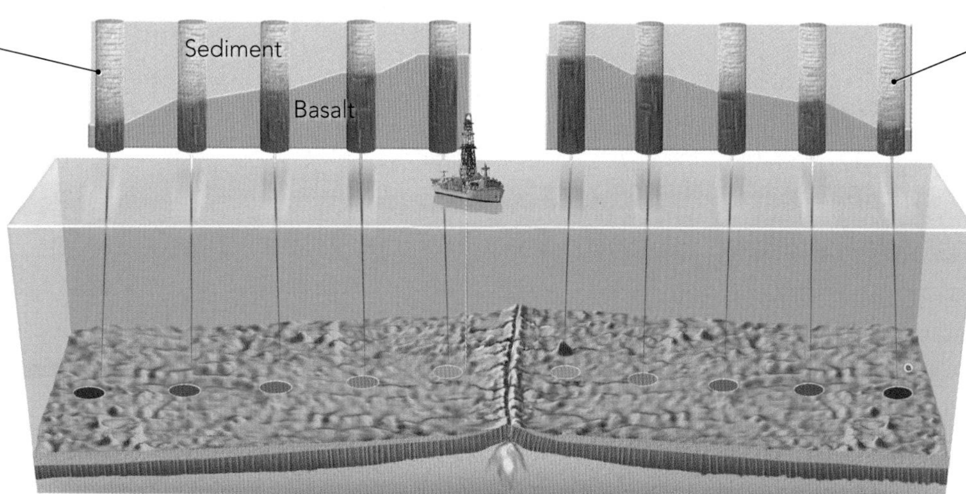

Drilling results demonstrate that sediment is thin or absent on the ridge but becomes thicker away from the ridge. Age determinations from fossils in the sediment and from underlying lava flows show that oceanic crust gets systematically older away from mid-ocean ridges. The drilling results strongly support the theory of plate tectonics.

03.08.a1

B If Continents Have Rifted Apart, Do Their Outlines and Geology Match?

According to the theory of plate tectonics, the South Atlantic Ocean formed when South America rifted away from Africa. If the continents are moved back together, their outlines should match. In this figure, the continents have been moved most of the way back together to allow their outlines to be compared. Observe where one continent juts out, the other curves in.

There are a few places where the continents would appear to overlap if moved back together, such as where the upper right corner of Brazil would overlap Africa. In this case, the yellow-colored material along the coast of Africa is sediment deposited since the continents rifted apart. This material was added after rifting.

In examining such matches, we include the continental shelf, which is part of the continent, although hidden beneath shallow seas.

03.08.b1

Africa

Central Africa

Brazil

South America

If the continents were once joined, they likely shared many geologic features, such as the same ages and types of rocks. If restored back to their joined, pre-rifting position, these geologic features should match. The colors on this map show the ages of geologic units in brown, blue, green, and yellow from oldest to youngest. The yellow units and some of the green units were deposited after the continents rifted apart.

A good example of this matching geology occurs between Brazil and central Africa. The dark brown colors represent very old Precambrian rocks. Detailed geologic comparisons of the age and character of these rocks in Brazil with those in central Africa indicate that these two areas are closely related, consistent with the theory of plate tectonics.

C How Does Plate Tectonics Help Explain the Formation of Linear Island Chains?

Fairly straight lines of oceanic islands and submarine mountains (seamounts) cross some parts of the ocean floor. They are called *linear island chains* to distinguish them from the curved shapes of subduction-related island arcs. How do linear chains of islands and seamounts form, and are they consistent with plate tectonics?

03.08.c1

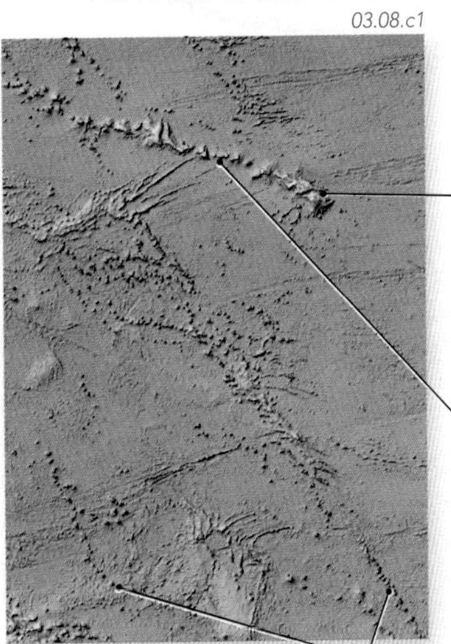

1. Most of the examples of linear island chains are in the Pacific Ocean. This large region around Hawaii has several lines of islands, including the Hawaiian Island chain. Like Hawaii, all of these islands have a volcanic origin.

2. A line of seamounts continues northwest from the Hawaiian Islands. One mountain in this chain is high enough to form Midway Island, scene of a pivotal air and sea battle during World War II.

3. Southwest of Hawaii, other lines of islands and seamounts also trend in a northwest direction.

03.08.c2

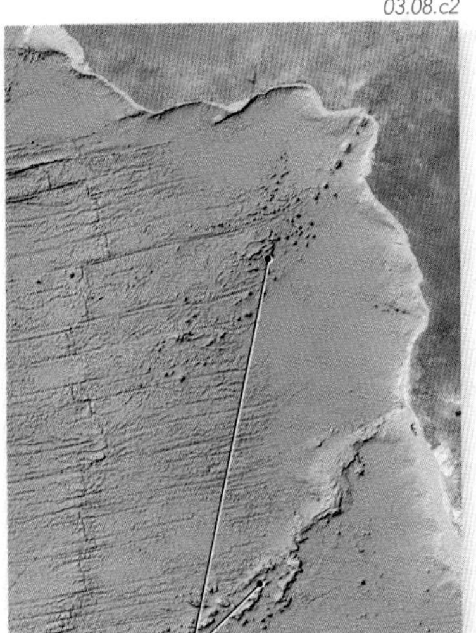

4. Linear island chains are located west of Africa and line up with active volcanoes on the continent or on the seafloor.

03.08.c3

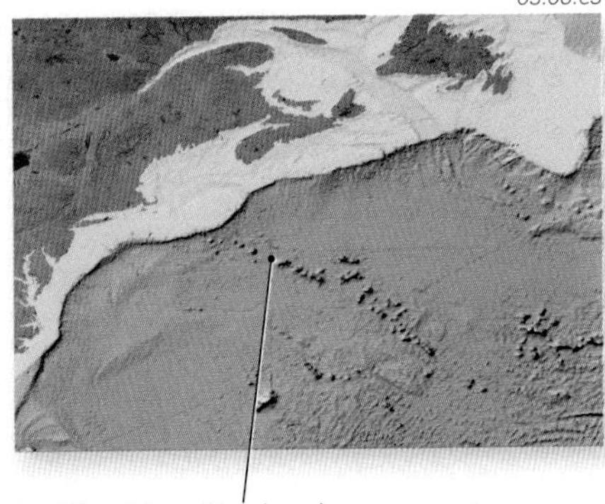

5. The New England seamounts form a line beneath the Atlantic Ocean off the eastern coast of North America.

A Plate-Tectonic Model for the Formation of Linear Island Chains

Linear island chains and most clusters of islands in the oceans have several things in common. They have a volcanic origin and commonly are near sites that geologists interpret to be above unusually high-temperature regions in the deep crust and upper mantle. Such anomalously hot regions are called *hot spots*.

03.08.c4

1. This figure shows how linear island chains are interpreted to be related to a plate moving over a hot spot. A hot spot is interpreted to be a place where anomalously hot mantle rises and melts, forming magma that ascends into the overlying plate. In this example, the plate above the hot spot is moving relative to the hot spot.

Hot Spot

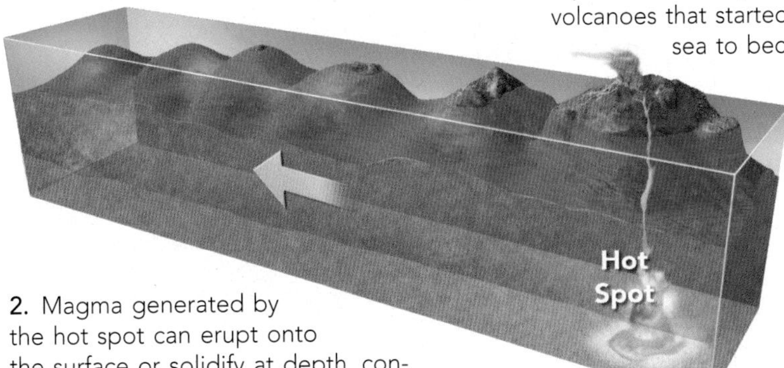

2. Magma generated by the hot spot can erupt onto the surface or solidify at depth, constructing a volcanic mountain that can grow high enough above the seafloor to become an island.

3. As the plate moves beyond the hot spot, it cools and subsides, so volcanoes that started out as islands may subside beneath the sea to become *seamounts*. In this way, a hot spot will make a track of volcanic islands and seamounts, each constructed when it was over the hot spot. Volcanoes closest to the hot spot are the youngest, and the volcanoes and seamounts farthest from the hot spot are the oldest. If the plate is not moving, the hot spot forms a cluster of volcanic islands and seamounts.

Ages determined on volcanic rocks along the Hawaiian line of islands systematically increase to the northwest. Such ages are consistent with the hot spot model and the calculated motion of the Pacific plate on which Hawaii rides.

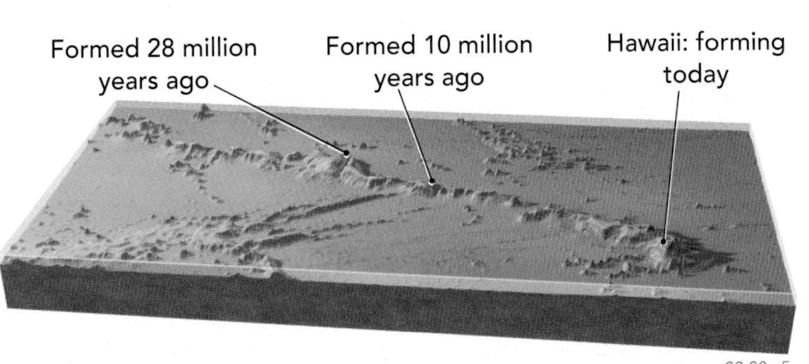

03.08.c5

Before You Leave This Page Be Able To

- ✓ Predict the relative ages of seafloor from place to place using a map of an ocean with a mid-ocean ridge.
- ✓ Summarize how plate tectonics can explain similar continental outlines and geology on opposite sides of an ocean.
- ✓ Describe the characteristics of a linear island chain and summarize how they are interpreted to be related to hot spots.

Why Is South America Lopsided?

THE TWO SIDES OF SOUTH AMERICA are very different. The western side is mountainous whereas the eastern side has much less relief. The differences are a reflection of the present plate boundaries and the continent's history over the last 200 million years. South America is a perfect place to bring together the various aspects of plate tectonics and to show how to analyze a large region of Earth.

A What Is the Present Setting of South America?

The perspective below shows the region around South America. Observe the topography of the continent, its margins, and the adjacent oceans. From these features, infer the locations of plate boundaries, including those in oceans, and predict what type of motion (divergent, convergent, or transform) is likely along each boundary. Make your observations and predictions before reading the accompanying text.

The Galapagos Islands are located in the Pacific Ocean, west of South America. They consist of a cluster of volcanic islands, flanked by seamounts. Some of the islands are volcanically active and are interpreted to be over a hot spot.

The center of the continent has low, subdued topography because it is away from any plate boundaries. It is a relatively stable region that has no large volcanoes and few significant earthquakes. It is not tectonically active.

This part of the Mid-Atlantic Ridge is a divergent boundary between the South American and African plates. Seafloor spreading creates new oceanic lithosphere and moves the continents farther apart at a rate of 3 centimeters per year.

The Andes mountain range follows the west coast of the continent and has dangerous earthquakes and volcanoes. Offshore along this edge of the continent is a deep ocean trench that marks where oceanic plates of the Pacific are subducted eastward under South America.

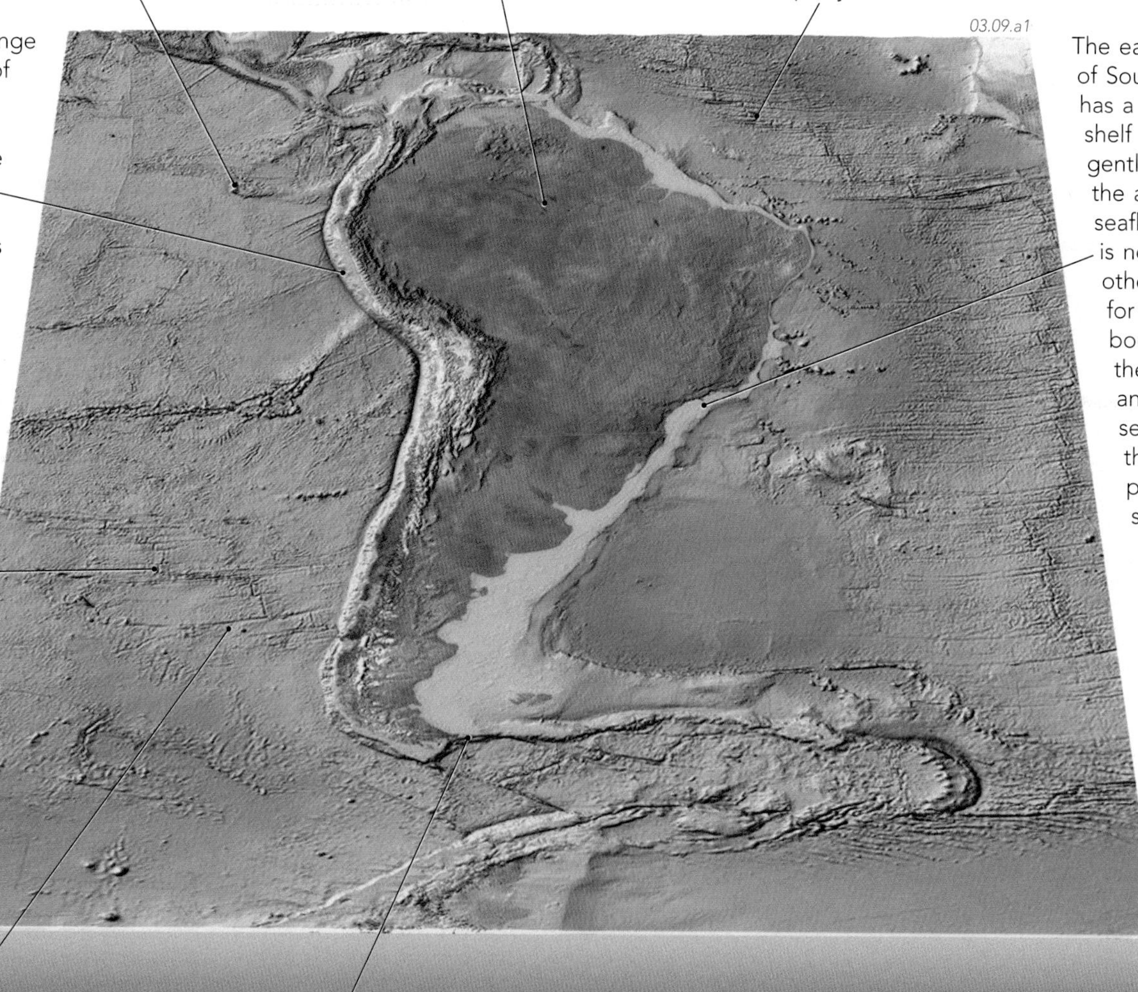

The eastern side of South America has a continental shelf that slopes gently toward the adjacent seafloor. There is no trench or other evidence for a plate boundary, so the continent and adjacent seafloor to the east are part of the same plate.

The Pacific seafloor contains mid-ocean ridges with the characteristic zig-zag pattern of a divergent boundary with transform faults.

Oceanic fracture zones cross the seafloor, but they are not plate boundaries.

The southern edge of the continent is very abrupt and has a curving "tail" extending to the east. This edge of the South American plate is a transform boundary, along which South America is moving west relative to the Antarctic and Scotia plates to the south.

B What Is the Geometry of the South American Plate and Its Neighbors?

This cross section shows how geologists interpret the configuration of plates beneath South America and the adjacent oceans. Compare this cross section with the plate boundaries you inferred in part A.

Along the Mid-Atlantic Ridge, new oceanic lithosphere is added to the trailing edges of the African and South American plates as they move apart. As this occurs, the oceanic part of the South American plate gets wider over time.

A subduction zone dips under western South America and takes oceanic lithosphere beneath the continent. This subduction zone causes large earthquakes near the west coast and dangerous volcanoes in the Andes.

Along the eastern edge of South America, continental and oceanic parts of the plate are simply joined together. There is no subduction, no seafloor spreading, or any type of plate boundary. As a result, large volcanoes and earthquakes are absent.

C How Did South America Get into Its Present Plate Tectonic Situation?

If South America is on a moving plate, where was it in the past? When did it become a separate continent, and when did its current plate boundaries develop? Here is one interpretation.

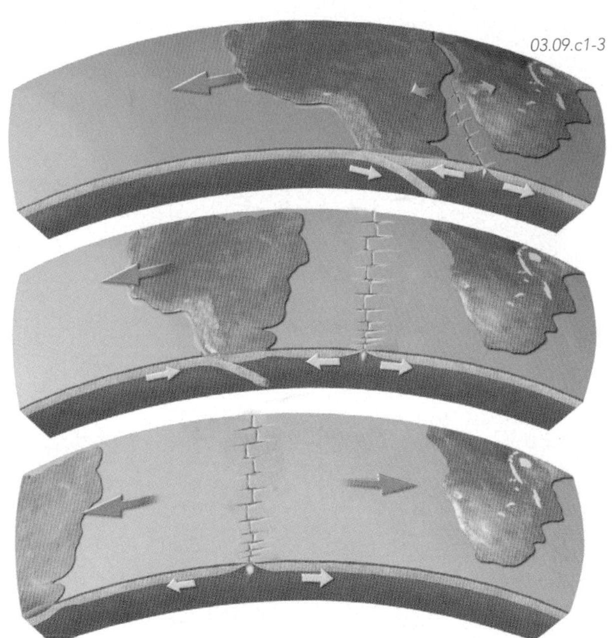

Before 140 million years ago, Africa and South America were part of a single large supercontinent called *Gondwana*. At about 140 million years, a *continental rift* began splitting South America away from the rest of Gondwana and carving it into a separate continent.

By 100 million years ago, Africa and South America had rifted completely apart, forming the South Atlantic Ocean. Spreading along the Mid-Atlantic Ridge moved the two continents farther apart with time, as depicted in this figure. While the Atlantic Ocean was opening, oceanic plates in the Pacific were subducting beneath western South America at various times over the last 140 million years. This subduction thickened the crust by compressing it horizontally and by adding magmas, resulting in the formation and rise of the Andes mountain range.

Today, Africa and South America are still moving apart at a rate of several centimeters per year. As spreading along the mid-ocean ridge continues, the Atlantic Ocean gets wider. Earth is not getting larger through time, so this widening must be accommodated by subduction elsewhere on the planet, such as by present-day subduction along the western coast of South America, out of this view.

These photographs contrast the rugged Patagonian Andes of western South America with landscapes further east that have more gentle relief and are not tectonically active.

Before You Leave This Page Be Able To

- ✓ Summarize or sketch the present plate-tectonic setting of South America and describe how it explains the large features on the continent and adjacent oceans.
- ✓ Summarize the plate tectonic evolution of South America over the last 140 million years.

Where Is the Safest Place to Live?

AN UNDERSTANDING OF PLATE TECTONICS is important for assessing potential risks for earthquakes and volcanoes. In this regard, knowing the locations and types of any plate boundaries is especially important. In this exercise, you will examine an unknown ocean between two continents and identify the locations of plate boundaries and other features. Using this information, you then will predict the risk for earthquakes and volcanoes and determine the safest sites to live.

Goals of This Exercise:

- Use the features of an ocean and two continental margins to identify possible plate boundaries and their types.
- Use the types of plate boundaries to predict the likelihood of earthquakes and volcanoes.
- Determine the safest site for two cities, considering the earthquake and volcanic hazards.
- Draw a cross section of your plate boundaries, to show the geometry of the plates at depth.

Procedures for the Map

This perspective view shows two continents, labeled A and B, and an intervening ocean. Use the topography to identify possible plate boundaries and complete the following steps. Mark your answers on the map on the worksheet.

- Use the topographic features to identify possible plate boundaries and mark the location of each plate boundary on the map in the worksheet. Propose whether each boundary is divergent, convergent, or transform, and mark this on the map. You can use colored pencils to better highlight the different types of boundaries.
- Draw circles [O] at any place, on land or in the ocean, where you think earthquakes are likely.
- Draw triangles [△] at any place, on land or in the ocean, where you think volcanoes are likely. Remember that not all volcanoes form *directly on* the plate boundary; some form off to one side. Also, a line of islands and seamounts could mark the track of a hot spot, and may not be on a plate boundary.
- Determine a safe place to build one city on each continent. Show each location with a large plus sign [+] on the map. On the worksheet, explain your reasons for choosing these as the safest sites.

Procedures for the Cross Section

The figure below shows a cross section across the area. On the worksheet version of this figure, draw a simple cross section of the geometry of the plates in the subsurface. Use other figures in this chapter as a guide to the thicknesses of the lithosphere and to the geometries typical for each type of boundary. Some features are not located along the front edge of the figure and so cannot be shown on the cross section.

- Draw the geometries of the plates at depth for any spreading center or subduction zone.
- Show the variations in thickness of the crust and variations in thickness of the lithosphere.
- Draw arrows to indicate which way the plates are moving relative to each other.
- Show where melting is occurring at depth to form volcanoes on the surface.

CHAPTER 4

Earth Materials

EARTH'S SURFACE IS COMPOSED of diverse materials: black lava flows, white sandy beaches, red cliffs, and hills of gray granite. Also, Earth is a treasury of gemstones and other mineral resources that are useful to society. What kinds of materials are common on Earth, and how did the less common ones, such as gemstones, form? Here, we explore Earth materials from landscapes to atoms.

This large perspective shows satellite data superimposed over topography for southernmost California and adjacent Baja California, Mexico. The Peninsular Ranges, a forested mountainous area east of San Diego, are in the center of the image. The white line across the image, added for reference, marks the border between the United States and Mexico.

What are the gray rocks that make up the hills and mountains of the Peninsular Ranges? ▽

The Peninsular Ranges consist mostly of various types of grayish-colored rocks like granite. When viewed up close (▽), the granite displays different kinds of crystals that are cream colored, pink, transparent gray, and black. The largest crystals below are 1 to 2 cm (0.4 to 0.8 in.) in diameter.

What are rocks made from, and what controls the color and other properties of a rock?

San Diego County is a famous source of beautiful crystals of tourmaline (◁), which can be pink, purple, green, or all three colors.

What are crystals, how do they form, and where do we find them?

TOPICS IN THIS CHAPTER

East of the Peninsular Ranges, the land drops down into the lowlands of the Salton Trough, which contains sandy deserts, farmlands of the Imperial Valley, and several large salty lakes, including the blue Salton Sea. The sand in the Salton Trough was eroded from the adjacent mountains or brought into the area by rivers or strong winds.

What are most sand grains composed of?

The Peninsular Ranges

The Peninsular Ranges are a broad, uplifted region that crosses southernmost California and the northern Baja area of Mexico. In this image, the mountains are shaded green because they are mostly covered by forests and other types of vegetation. The Salton Trough to the east (right) of the mountains receives much less rain and has a lighter color in this image because vegetation is sparse and sand and rocks cover the surface.

The mountains and lowlands expose a variety of rocks and other Earth materials. The Peninsular Ranges are topographically high because the mountain block was uplifted by faulting and because the uplift brought up deep, relatively hard rocks that resist erosion. Most of the range consists of gray granite formed from magma that solidified deep within the crust. The range also includes rocks that were buried and then heated, deformed, and converted into metamorphic rocks.

The mountains were uplifted mostly during the last 10 million years, long after the granite and metamorphic rocks formed. Once at the surface, these rocks were weathered and eroded to produce sand, gravel, and other sediment that were transported, mostly by streams, into the lowlands to the east and west. These sediments are relatively unconsolidated and therefore easily eroded. The contrast of landscapes between the soft-appearing lowlands and the more rugged, granitic mountains reflects differences in geologic history and the differences in geologic materials.

What Is the Difference Between a Rock and a Mineral?

WHAT MATERIALS MAKE UP THE WORLD around us? What do we see if we walk close to a rock exposure? How does the rock look when viewed with a magnifying glass? Let's investigate these questions using the beautiful scenery of Yosemite National Park in California.

A What Materials Make Up a Landscape?

04.01.a1

1. Observe this photograph of Yosemite Valley, the heart of Yosemite National Park. What do you notice about the landscape?

2. This landscape is dominated by the green valley and the dramatic cliffs and steep slopes of massive gray rock. The valley is famous for waterfalls and for huge rock faces, such as those on the appropriately named Half Dome at the right side of the photograph. What would we see if we got closer to this landscape?

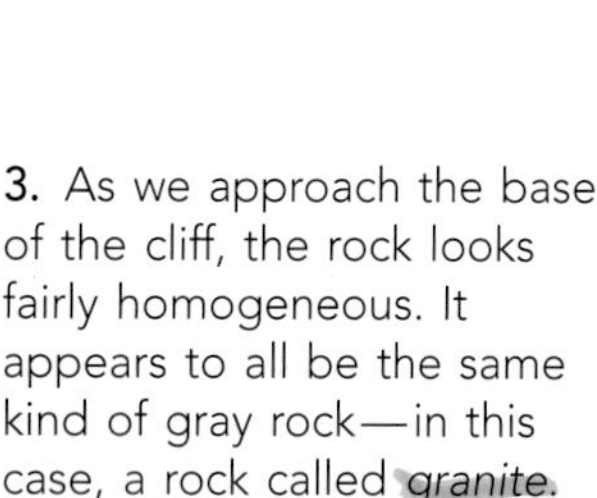

3. As we approach the base of the cliff, the rock looks fairly homogeneous. It appears to all be the same kind of gray rock—in this case, a rock called *granite*.

04.01.a2

04.01.a3

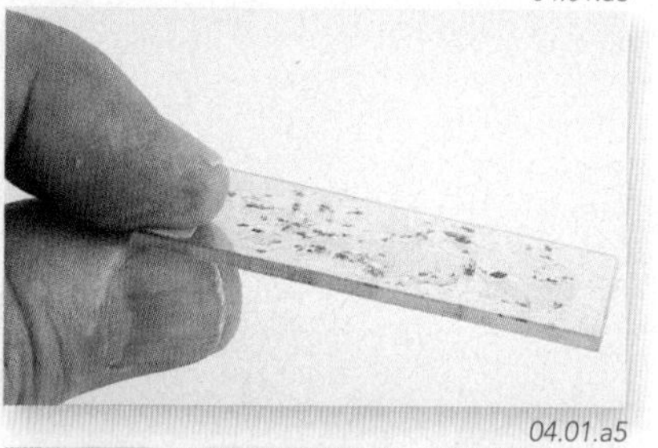

5. To examine the rock in more detail, a geologist cuts a very thin slice from the rock and mounts the slice on a glass slide. This mounted slice, called a *thin section*, is then examined using a microscope.

4. Up close, the granite displays several different colors: cream, clear gray, and black. To better observe a rock at this scale, a geologist commonly collects a hand-sized piece of the rock, called a *hand specimen*. The geologist then examines the hand specimen using a magnifying glass or *hand lens*.

04.01.a4

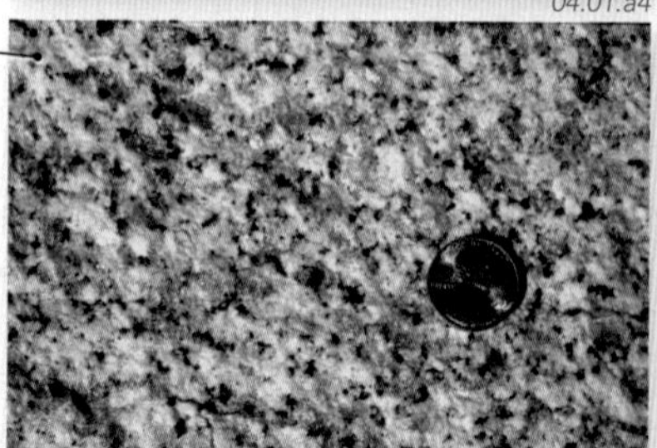

04.01.a5

6. When polarized light shines through a thin section, the internal structure of crystals in the thin section changes the light in a systematic way that causes each type of crystal to have certain colors and characteristics.

04.01.a6

7. By examining this rock at these various scales, we find that it consists mostly of several different types of crystals that correspond to the different colors in the rock. All of the clear gray crystals share similar properties, such as chemical composition, and represent a single kind of material called a *mineral*. The next page explores what criteria a material must meet to be a mineral.

B What Is and What Is Not a Mineral?

What characteristics define a mineral? To be considered a mineral by geologists, a substance must fulfill all of the criteria listed below.

Solid

04.01.b1

04.01.b2

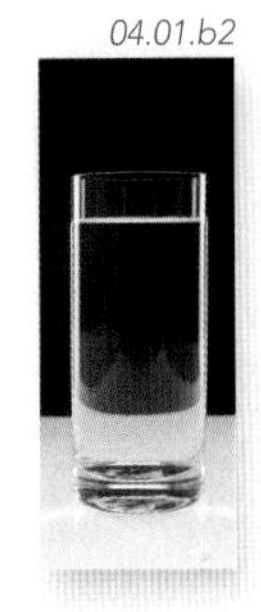

All minerals are solid, not liquid or gaseous. Ice is solid and a mineral, but liquid water is not. Ice and water have the same chemistry, a compound of hydrogen and oxygen (H_2O).

Natural

04.01.b3

04.01.b4

A mineral must be natural. Crystals on the left grew naturally from hot waters flowing through rocks, but the artificial crystals on the right were grown in a laboratory.

Inorganic

04.01.b5

04.01.b6

A mineral generally is *inorganic*, such as the crystal on the left, which grew without any influence from plants or animals. In contrast, the wooden log came from a tree.

Ordered Internal Structure

04.01.b7

04.01.b8

A mineral has an internal structure that is ordered, which means that atoms are arranged in a regular framework. The mineral on the left has an orderly internal structure that is expressed by the shape of the crystals. The volcanic glass on the right does not have an organized structure. Its atoms are arranged in a haphazard way, so the glass is not a mineral.

Specific Chemical Composition

04.01.b9

04.01.b10

A mineral has a fairly consistent chemical composition. Table salt, which is the mineral *halite*, contains atoms of the chemical elements *sodium* and *chlorine* in approximately equal amounts. The rock on the right is not a mineral because different parts of the rock have very different compositions.

Rocks, Minerals, and "Minerals"

We often hear the word *mineral* used in the context of *vitamins and minerals*. Are these minerals the same as the minerals described above? The answer is *no*. In the pharmacy, a mineral is a chemical element, such as potassium or zinc. This type of mineral is different from the orderly solids that geologists call minerals.

In geology, most minerals consist of at least two different chemical elements, such as *sodium* and *chlorine* in the mineral *halite* (salt). Many minerals have three, four, or even more chemical elements. A few minerals, however, include only one chemical element. The mineral *diamond* consists entirely of the element *carbon*.

It is the geological types of minerals that are the components of rocks. Most rocks, like the granite from Yosemite National Park, include several different minerals. Some rocks consist of a single mineral. The rock *limestone*, for example, consists of the mineral *calcite*. Calcite, in turn, consists of the chemical elements *calcium*, *carbon*, and *oxygen*. Rocks are made of minerals, *and* minerals are made of elements. The vitamin pill you take with breakfast may not contain any geologic minerals, but most of the chemical elements in the pill were chemically extracted from geologic minerals.

Before You Leave This Page Be Able To

- ✓ Explain the difference between a rock and a mineral.
- ✓ Explain each characteristic that a material must have to be called a mineral, and provide for each characteristic an example that is a mineral and an example that is not a mineral.
- ✓ Explain the relationship between rocks, minerals, and chemical elements.
- ✓ Explain the difference between a mineral in a vitamin pill and a geologic mineral.

How Are Minerals Put Together in Rocks?

ROCKS DISPLAY A WONDERFUL VARIETY of appearances. Geologists use the term *texture* to refer to the visual appearance of a rock, especially the way in which the minerals are arranged. What controls the texture of a rock? How are minerals in a rock connected to one another? What can we determine about a rock from its texture and from the types of minerals it contains?

A How Are Minerals Put Together in Rocks?

The beautiful and geologically interesting Engineer Mountain in the San Juan Mountains of southwestern Colorado contains two very different parts that illustrate the two main ways in which minerals occur in rocks.

04.02.a1

◁ The main mountain has two different parts, an upper gray part and lower reddish-brown part. Loose pieces of the upper gray part tumble down the hillside forming gray slopes that cover some of the red rocks.

04.02.a2

△ A close-up of the side of the mountain displays the different character of the two parts. The upper gray part has vertical fractures but no obvious layers, whereas the lower reddish-brown part has well-defined, nearly horizontal layers.

Two Types of Rocks

04.02.a3

The upper gray rock displays light-colored *crystals* surrounded by a gray matrix of microscopic crystals (too small to see in this photograph). A rock composed of crystals is called a *crystalline rock.* Most crystalline rocks contain interlocking crystals that grew together. Crystalline rocks typically form in high-temperature environments from crystallization of magma, from metamorphism, or from precipitation from hot water. Some rocks contain crystals formed from precipitation of minerals in cooler waters.

04.02.a4

The lower reddish-brown layers include distinct *pieces of rock,* ranging from pebbles to sand grains. These pieces are called *clasts,* and a rock consisting of pieces derived from other rocks is called a *clastic rock.* Most clastic rocks form on Earth's surface in low-temperature environments such as sand dunes, rivers, and beaches. Clasts also make up some volcanic rocks, including those that form when a volcanic explosion blasts apart pieces of a volcano and its magma.

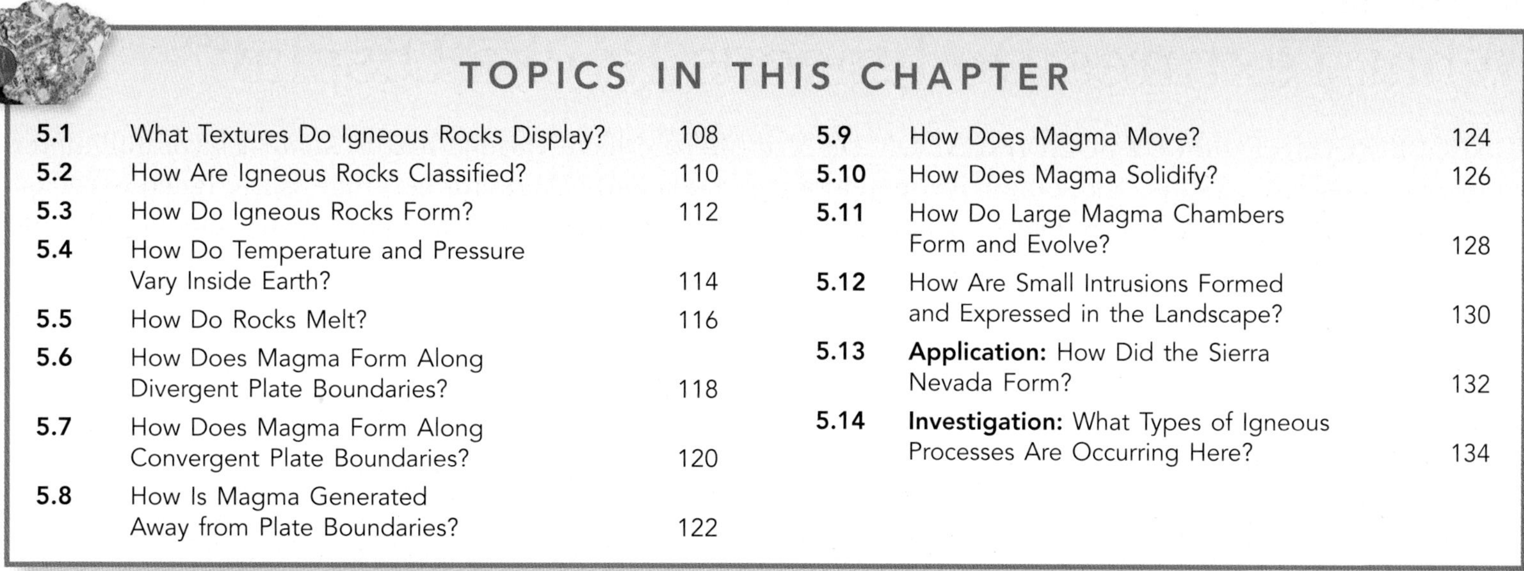

TOPICS IN THIS CHAPTER

The Harding Pegmatite Mine east of the Rio Grande has large igneous crystals (▽), some as long as 2 meters (about 6 feet). Compare the size of the light-purple crystals with the rock hammer in the photograph below. These unusual rocks must have formed in a very different igneous environment than the volcanic rocks to the west.

What factors control whether crystals in igneous rocks are microscopic or meters long?

05.00.a4

Many small volcanoes and dark lava flows form the small Cerros del Rio volcanic field east of the Rio Grande, the large river that cuts diagonally across the area. Unlike the explosive eruptions of the Valles Caldera, magma from the smaller volcanoes flowed onto the surface in a less violent manner and constructed dark volcanic layers, like these exposed elsewhere along the Rio Grande. ▽

Where and how does magma form, and what factors determine whether magma erupts as an explosion of hot ash or an outpouring of less explosive lava?

05.00.a5

Valles Caldera and Bandelier National Monument

The Valles Caldera of the Jemez Mountains is one of the most studied volcanic features in the world. It was here that geologists first figured out how the collapse of a caldera is related to explosive eruptions of volcanic ash. The caldera has since been explored using deep drill holes to study its subsurface geometry, to investigate the potential for geothermal energy, and to better understand these large volcanic features.

At 1.2 million years ago, a huge volume of magma rose from deep in the crust and accumulated in a magma chamber several kilometers below the surface. Some of the magma erupted explosively, forming a turbulent cloud of pumice, volcanic ash, rock fragments, and hot, toxic gases that raced outward at speeds of hundreds of kilometers per hour. As magma escaped from the underground chamber, the roof of the chamber collapsed downward forming the roughly circular depression visible today. After the main explosive eruption, smaller volumes of magma reached the surface as slow-moving lava that piled up into dome-shaped mounds within the caldera.

Volcanic ash that was erupted from the caldera blanketed most of the mountains and became compacted by the weight of additional ash. Streams later eroded steep canyons, within which the ancient puebloan peoples of the Southwest built cliff dwellings and other structures preserved within Bandelier National Monument.

What Textures Do Igneous Rocks Display?

IGNEOUS ROCKS FORM BY SOLIDIFICATION OF MAGMA and have various textures. Most have millimeter- to centimeter-sized crystals, but some have meter-long crystals and others are very fine glass. Igneous rocks vary from nearly white to nearly black, or can be various colors. They can have holes or fragments. What do these various textures tell us about where and how the magma solidified?

A What Textures Are Common in Igneous Rocks?

The *texture* of a rock refers to the sizes, shapes, and arrangement of different components. The texture of igneous rocks mostly reflects crystal size and the presence of other features, such as holes and rock fragments.

05.01.a1

Some igneous rocks contain very large crystals, which are typically centimeters long. Such rocks are called *pegmatite*.

05.01.a2

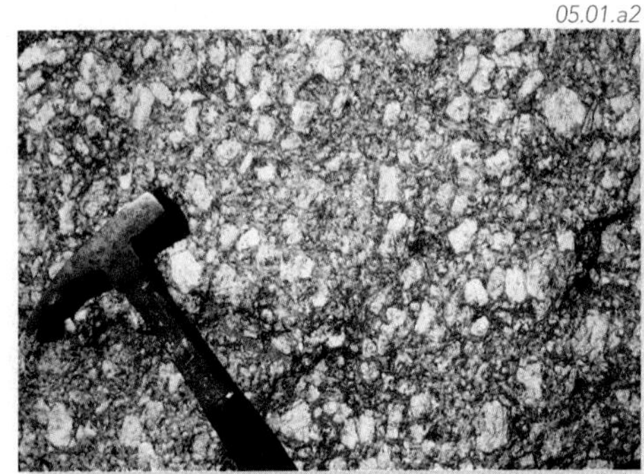

This rock is *coarsely crystalline* (or *coarse grained*). It has crystals larger than several millimeters; many are several centimeters.

05.01.a3

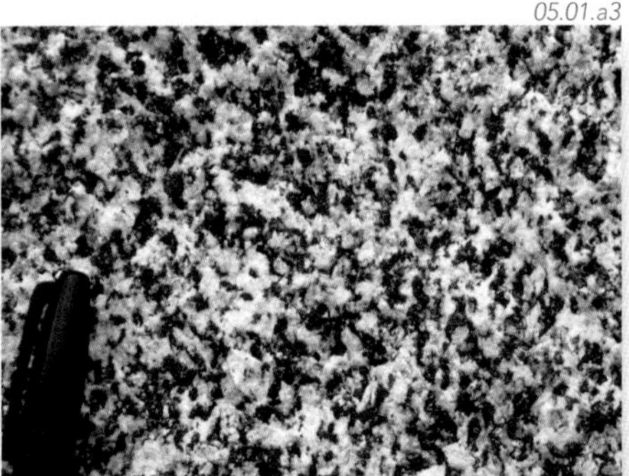

Medium-grained rocks have crystals that are easily visible to the unaided eye.

05.01.a4

The crystals in *fine-grained* igneous rocks can be too small to see without a magnifying glass.

05.01.a5

Some igneous rocks consist of glass rather than minerals. These are called *volcanic glass* or are referred to as being *glassy*.

05.01.a6

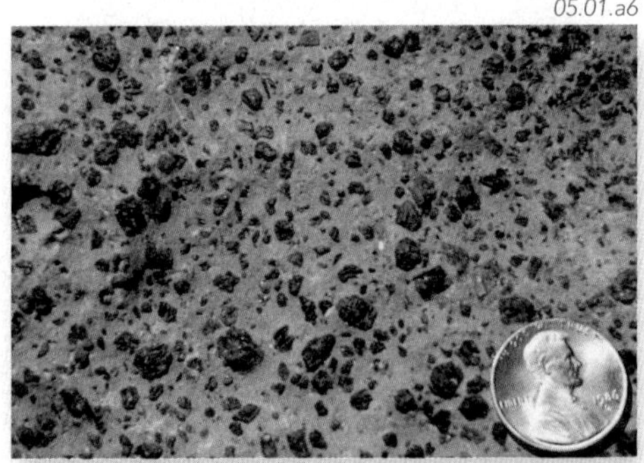

Igneous rocks can include larger crystals in a finer grained matrix. The crystals are *phenocrysts* and the texture is *porphyritic*.

05.01.a7

Many volcanic rocks contain small holes known as *vesicles*. The holes represent bubbles of gas released by the magma.

05.01.a8

Hot volcanic ash and pumice can become compacted by overlying materials, in which case the rock is said to be *welded*.

05.01.a9

Some volcanic rocks contain angular fragments in a finer matrix, and are called a *volcanic breccia*.

B In What Settings Does Each Igneous Texture Form?

The different textures of igneous rocks largely indicate the environment in which the magma solidified, such as whether it solidified at depth or was erupted onto the surface. Examine below where each texture shown on the previous page would form near a volcano and its underground magma chambers.

Vesicles form when gases dissolved in the magma accumulate as bubbles. They can form only under low pressures on or very near the surface. Many lavas have vesicles, and much of the material in volcanic ash forms when the thin walls between vesicles burst, shattering the rocks or magma into sharp particles. Most volcanic ash is broken vesicles.

Volcanic *breccia* can form in many ways, including explosive eruptions of ash and rock fragments, a lava flow that breaks apart as it solidifies and flows, or from volcano-triggered mudflows and landslides on the steep and unstable slopes of the volcano.

Volcanic *glass* forms when magma erupts on the surface and cools so quickly that crystals do not have time to form. This can happen in volcanic ash or in a lava flow solidifying at the surface.

For a *porphyritic* texture to form, some crystals must have time to grow in magma chambers at depth. Later, the magma rises closer to the surface where the remaining magma solidifies into the fine-grained matrix around the larger crystals.

05.01.b1

Some *volcanic ash* erupts vertically in a column and settles back to Earth. This ash cools significantly before accumulating on the surface. Because it is relatively cool and strong, the ash does not become welded.

Other volcanic ash erupts in thick clouds of hot gas, ash, and rock fragments, called *pyroclastic flows,* that flow rapidly downhill. The ash deposited by pyroclastic flows is very hot, and so most parts are *welded* to some extent.

Fine-grained igneous rocks form where the magma has only enough time to grow small crystals. This commonly occurs where magma solidifies at shallow depths beneath the surface. Medium-grained rocks would form deeper.

Coarse-grained igneous rocks form at depth, where the magma cools at a rate that is slow enough to allow large crystals to grow.

Pegmatite forms when water dissolved in the magma helps large crystals grow. This occurs near the sides and top of a magma chamber and in local pockets within the magma. Most pegmatite forms at moderate to deep levels within Earth's crust.

Igneous Textures Under the Microscope

Many aspects of igneous textures can be observed from a hand specimen of the rock or by using a magnifying glass (hand lens). Other aspects are best observed by examining a thinly sliced section of the rock, called a *thin section,* under a microscope. This thin section, placed between two polarizing filters, shows the interlocking crystals in a granite. The gray minerals are quartz and feldspar and the brown one is biotite.

05.01.mtb1

Before You Leave This Page Be Able To

- ✓ Summarize the various textures displayed by igneous rocks.
- ✓ Sketch an igneous system and show where the main igneous textures form.

How Are Igneous Rocks Classified?

IGNEOUS ROCKS HAVE VARIED COMPOSITIONS. Some are composed entirely of dark minerals, whereas others contain only light-colored minerals. How do we subdivide the family of igneous rocks so that we can use a few key words to identify rocks that have a similar composition and that form in similar ways?

A How Do the Characteristics of Igneous Rocks Vary?

Compare the colors and sizes of crystals in these rock samples, each of which is approximately 5 to 10 cm (2 to 4 in.) across. Consider how you would arrange these rocks if you wanted to classify them.

05.02.a1

B How Does the Composition of Igneous Rocks Vary?

Geologists typically organize igneous rocks according to two criteria. One criterion is the size of crystals in the rock. The top row of images below includes rocks that have coarse crystals that can be seen easily. The second criterion is the composition of the rock's minerals. Rocks that have a light color and contain abundant quartz and feldspar are *felsic rocks,* whereas rocks that are dark and contain minerals rich in magnesium and iron are *mafic* rocks.

Felsic | Intermediate | Mafic | Ultramafic

Coarsely Crystalline

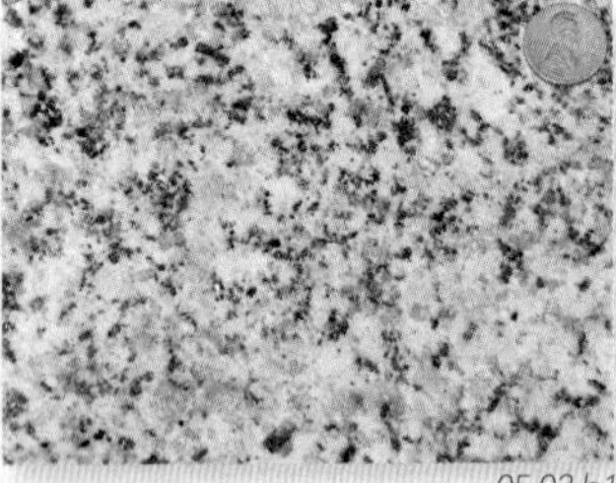

05.02.b1

Granite is a coarsely crystalline, light-colored rock consisting mostly of feldspar and quartz, with some mica.

05.02.b2

Diorite contains more mafic minerals than does granite. It is *intermediate* between felsic and mafic compositions.

05.02.b3

Gabbro consists of mafic minerals, such as dark-colored pyroxene, and calcium-rich feldspar.

05.02.b4

Peridotite has more magnesium and iron minerals, such as green olivine and dark pyroxene, than do mafic rocks.

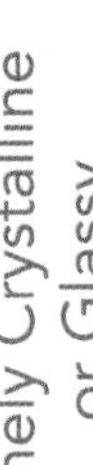

05.02.b5

Rhyolite is the fine-grained equivalent of granite. It has fine crystals, glass, or, as here, pieces of pumice and ash.

05.02.b6

Andesite is the fine-grained equivalent of diorite. It is commonly gray or greenish and may have phenocrysts.

05.02.b7

Basalt is a common and familiar dark lava rock. Most samples are dark gray to nearly black and have some vesicles.

05.02.b8

Ultramafic lavas erupted early in Earth's history. They were very hot and grew crystals that are unusually long for a lava flow.

C How Do We Observe the Percentages of Different Minerals?

To better identify the minerals in a rock and to estimate their percentages, geologists observe coarse-grained rocks by cutting a slab or by examining them with a *hand lens*. Fine-grained rocks require a microscope.

These two photographs are of granite slabs approximately 20 cm (8 in.) across. The rocks are different-varieties of very coarse-grained granite. How many types of minerals do you recognize in each rock, and in what percentage is each mineral present?

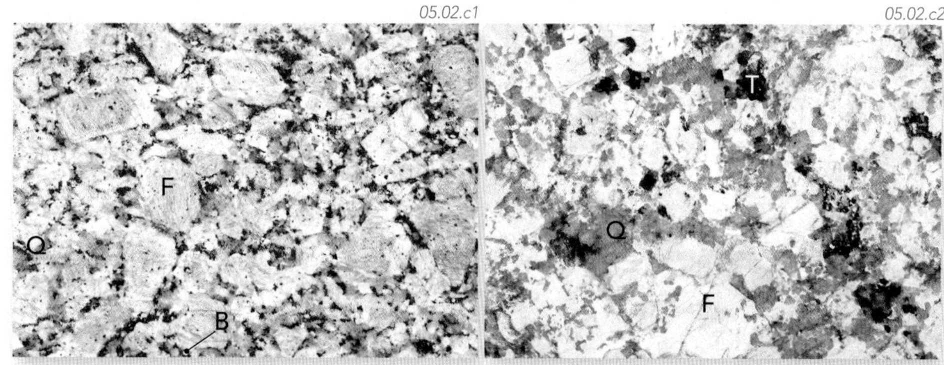

Minerals:

F: Feldspar (pink or cream colored)

Q: Quartz (partially transparent gray)

B: Biotite mica (black flakes)

T: Tourmaline (black)

D What Are Some Other Common Igneous Rocks?

Some common igneous rocks do not fit into the classification in part B.

Obsidian is normally a medium gray to black, shiny volcanic glass. Most obsidian has a composition equivalent to that of rhyolite and forms from rapidly cooled lava flows.

Tuff is a volcanic rock composed of various percentages of volcanic glass, pumice, mineral crystals, and rock fragments. Some tuff consists only of fine volcanic ash that has been compacted.

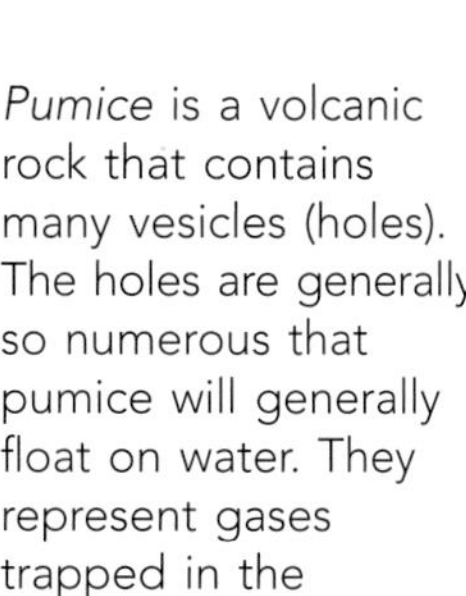

Pumice is a volcanic rock that contains many vesicles (holes). The holes are generally so numerous that pumice will generally float on water. They represent gases trapped in the magma.

Scoria is a dark gray, black, or red volcanic rock that is very vesicular. It consists of fragments that are centimeters to meters across and it usually has the composition of basalt or andesite.

The Chemical Composition of Igneous Rocks

The chemical composition of a rock helps control the percentages of different minerals in the rock. Granite and related *felsic* igneous rocks consist mostly of quartz and feldspar and typically contain high amounts of silica (SiO_2), commonly 70% to 76% SiO_2. Rhyolite and some obsidian have the same chemical composition as granite but largely consist of material that is too fine to reveal the percentages of minerals without using a microscope.

Silica is also the major chemical constituent in *mafic* rocks, such as basalt and gabbro, but has lower concentrations (44% to 50% SiO_2) in these rocks. Compared to felsic rocks, mafic rocks contain more magnesium, iron, and calcium, and these elements cause the darker, mafic minerals to be more abundant. Intermediate rocks, such as andesite and diorite, contain intermediate amounts of silica (about 60% SiO_2) compared to felsic and mafic rocks.

Before You Leave This Page Be Able To

- ✓ Summarize or sketch how igneous rocks are classified.
- ✓ List some common igneous rocks and a few characteristics of each.
- ✓ Summarize the main differences between felsic and mafic rocks.

How Do Igneous Rocks Form?

THE DIFFERENT IGNEOUS COMPOSITIONS AND TEXTURES reflect the type of material that was melted, the way the magma solidified, and whether the magma solidified at depth or was erupted onto the surface. How do these processes create so many different types of igneous rocks, and can we use the composition and texture of an igneous rock to infer something about the rock's origin?

A What Processes Are Involved in the Formation of Igneous Rocks?

Igneous rocks form by melting of rocks at depth, usually followed by movement of magma toward the surface, and then solidification of the magma into solid rock. With such a history, igneous systems are best described from the bottom up, so begin with number one at the bottom of this figure.

05.03.a1

6. Magma that reaches the surface erupts as lava (molten rock that flows on the surface) or as volcanic ash. Volcanic ash forms when dissolved gases in the magma expand and blow the magma apart into small fragments of volcanic glass. Any igneous rock that forms on the surface is called an *extrusive* rock because it forms from magma *extruded* onto the surface. More commonly, we simply call it a *volcanic* rock.

5. Many magma chambers are only several kilometers below the surface, such as inside a volcano. Magma may inject a little at a time into the chamber, and some magma may solidify before the next batch comes in. Any batch of magma may crystallize in the chamber or rise to the surface.

4. As magma rises through the crust, it may stop in, or pass through, a series of magma chambers. A body of molten rock in the subsurface is referred to as an *intrusion* because of the way the magma intrudes (invades) the surrounding rocks. Any igneous rock that solidifies *below the surface* is called an *intrusive* rock.

3. Magma can accumulate to form a *magma chamber*. The magma may solidify in this chamber and never reach the surface, or it may reside in the chamber only temporarily before continuing its journey upward. An igneous rock that solidified at a considerable depth (more than several kilometers) is referred to as a *plutonic* rock, and the body of rock is called a *pluton*. Granite is a common plutonic rock and forms within plutons.

2. Once magma has formed in the source area, separate pockets of magma may accumulate to form a larger volume of magma. The magma rises because it is less dense than rocks around it.

1. The first stage in the formation of an igneous rock is melting at depth (40–150 km) in the deeper parts of the crust or in the mantle. The place where melting occurs is called the *source area*. Most of the source area remains solid rather than melting into a single large region of magma.

B How Does Melting Affect the Composition of a Magma?

The initial composition of a magma indicates what type of rock was melted in the source area and whether rocks in the source area were completely melted or only partially melted.

Partial and Nearly Complete Melting

1. If a magma could be generated from nearly *complete melting* of the source region, it would have a similar composition to the source. Complete melting is probably not common.

2. Most rocks melt by *partial melting* where some minerals melt before others. Felsic minerals generally melt at lower temperatures than mafic minerals, so *partial melting* will produce a magma that is *more felsic* than the source. For example, partial melting of a mafic source can yield an intermediate magma.

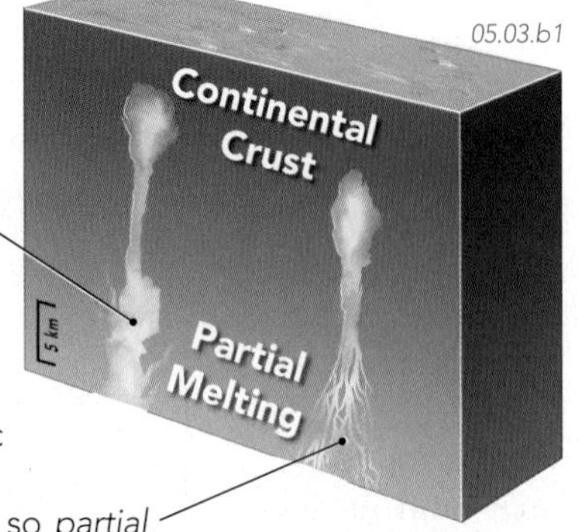

Source Area

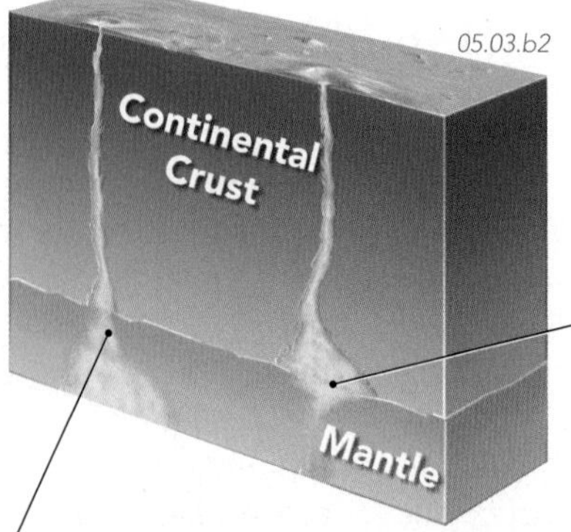

4. If a more felsic source area, such as continental crust, is melted, the magma will be felsic. If an intermediate source is nearly completely melted, the magma will have an intermediate composition.

3. If an ultramafic source such as the mantle is melted, the magma generally has a mafic (basaltic) composition not a felsic composition. Most mafic magma is derived from partial melting of the mantle.

C How Are Magmas Generated in Continental Rifts?

Continental rifts form where tectonic forces attempt, perhaps successfully, to split a continent apart. Such rifts have a central trough where faults drop down huge crustal blocks. Rifts are characterized by a diverse suite of igneous rocks because melting takes place both in the mantle and in the crust. The sequence of events begins in the mantle.

Continental Rift

05.06.c1

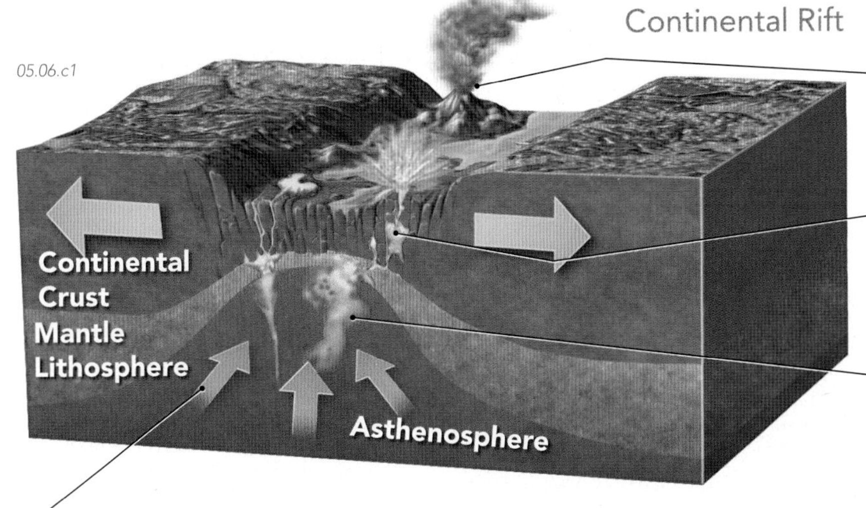

4. Some felsic and intermediate magmas solidify underground as granite and related igneous rocks, while others erupt on the surface in potentially explosive volcanoes.

3. Heat from the hot mafic magma melts the adjacent continental crust. Such melting typically yields felsic magma. Intermediate magma forms from mixing of felsic and mafic magmas or from assimilation of continental crust by a mafic magma.

2. The mantle-derived mafic magmas rise into the upper mantle and lower continental crust and accumulate in large magma chambers. Some mafic magma reaches the surface and erupts as mafic (basaltic) lava flows.

1. Solid asthenosphere rises beneath the rift and undergoes *decompression melting* (see graph below for melting in the mantle). Partial melting of the ultramafic mantle source yields mafic (basaltic) magma.

Melting in the Mantle

▷ Melting of the *mantle* beneath rifts is caused by *decompression*. The asthenosphere rises into shallower, lower pressure regions, and the decrease in pressure allows the rocks to melt. This produces mafic magma that can erupt onto the surface, forming basalt.

05.06.c2

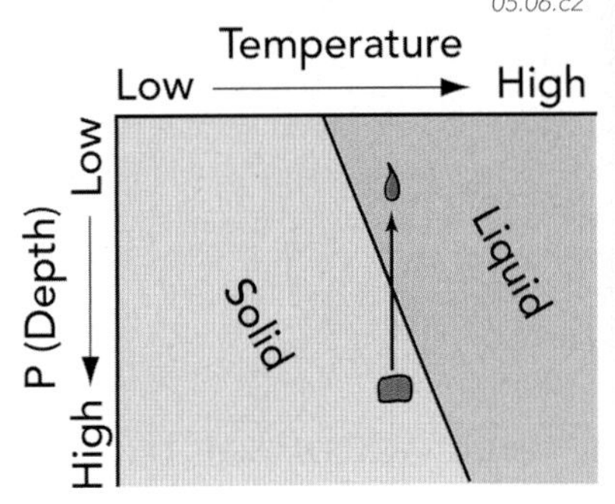

Melting in the Crust

05.06.c3

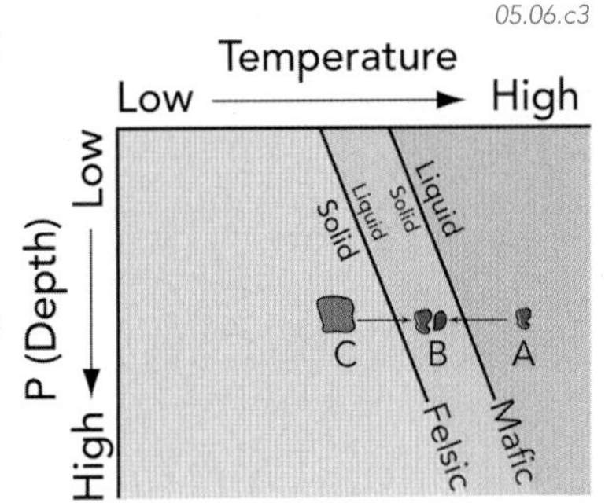

◁ This graph shows a melting line for mafic rock (basalt) and a lower temperature melting line for felsic rock (granite). A hot, mantle-derived mafic magma rises into continental crust and is hotter (at point *A*) than adjacent crust (at point *C*).

Heat from the mafic magma increases the temperature of the crust (from *C* to *B*). As the temperature of the crust crosses the felsic line, the granitic crust melts to produce felsic magma. In this example, the mafic magma loses heat to the crust (from *A* to *B*) and solidifies.

Ophiolites—Slices of Oceanic Crust on Land

How do we know what is in oceanic crust that it is hidden deep beneath the sea? The sequence of rocks in oceanic crust has been reconstructed by dredging samples from the seafloor, by drilling into oceanic crust, and by studying ancient examples on land. Geologists have gained much recent data by using research ships that have completed more than 1,700 drill holes, some more than 1,400 meters (1,500 yards) deep. Drill cores retrieved from these sites are important data for reconstructing sections of oceanic crust.

If we know the right places, we can examine oceanic crust on a hike across the land. Tectonic movements have sliced off pieces of oceanic crust and thrust them onto the edges of continents and onto islands. These slices contain a consistent sequence, from top to bottom, of oceanic sediment, pillow basalt, sheeted dikes, and gabbro. This distinctive sequence is called an *ophiolite* and is identical to the sequence of newly formed oceanic crust on the previous page, except it contains an additional layer of oceanic sediment on top. Such sediment accumulates on the oceanic crust over time. Many ophiolites are probably sections of oceanic crust created at long-vanished mid-ocean ridges.

Before You Leave This Page Be Able To

- ✓ Sketch or describe why melting occurs along mid-ocean ridges and why the resulting magmas are basaltic (mafic).
- ✓ Summarize the types of igneous rocks that form along mid-ocean ridges.
- ✓ Describe how melting occurs in continental rifts and how it results in diverse igneous rocks.
- ✓ Summarize how an ophiolite compares to a section through oceanic crust.

How Does Magma Form Along Convergent Plate Boundaries?

MANY MAGMAS ARE GENERATED ALONG CONVERGENT BOUNDARIES where two plates move toward one another. What type of melting produces this magma? Are there differences in magmas generated at the three types of convergent boundaries: ocean-ocean, ocean-continent, and continent-continent?

A How Is Magma Generated Along Subduction Zones?

Approximately 20 percent of Earth's magma forms where an oceanic plate subducts down into the mantle along ocean-ocean and ocean-continent convergent boundaries.

05.07.a1

1. In *ocean-ocean* and *ocean-continent* convergence, an oceanic plate consisting of oceanic crust and lithospheric mantle subducts into the mantle. As the plate descends, pressure increases and temperature gradually increases.

2. As a result of these changing conditions, existing minerals convert into new ones through the process of *metamorphism*. Water-bearing minerals, such as mica, break down, forcing water out of the crystalline structures. Water is also supplied by wet sediment. Some water (shown in blue) is released from minerals deep within the subducting plate and then rises into the overlying asthenosphere.

3. The added water lowers the melting temperature of the mantle above the subducting plate by changing its conditions from dry to wet as shown in the graph below. Mantle-derived magmas rise into the overriding plate, where they may erupt onto the surface or be trapped at depth.

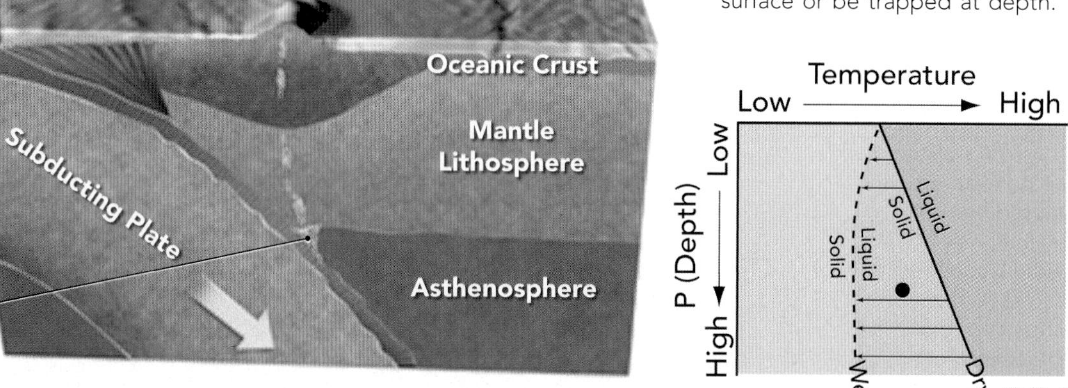

05.07.a2

B What Happens When Subduction-Derived Magmas Encounter the Crust?

Subduction-derived magmas rise into the overridding plate, which may be an oceanic plate or a continental plate. The magmas interact with and modify the crust that they encounter and may themselves be modified. Begin at the bottom.

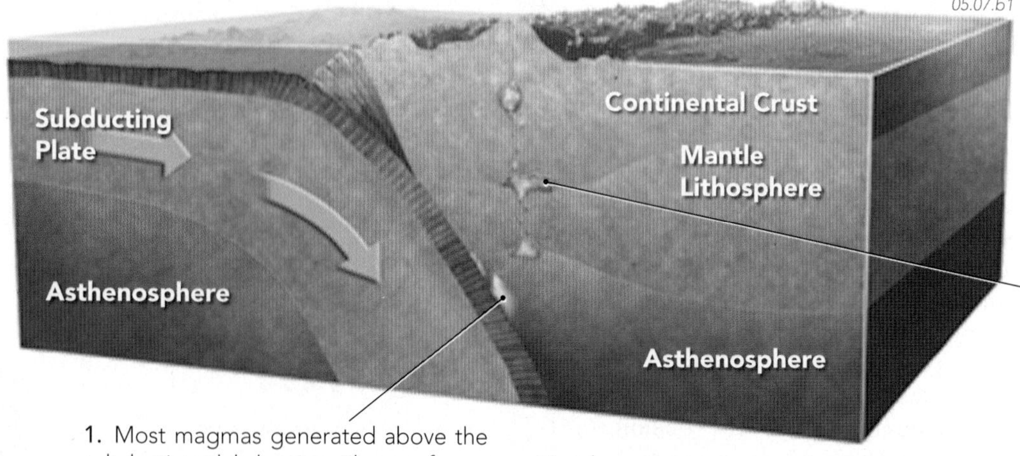

05.07.b1

1. Most magmas generated above the subducting slab begin with a *mafic* composition because the ultramafic mantle undergoes partial melting. In some cases, partial melting may also generate magma of *intermediate* composition.

2. If the overridding plate is a continental plate, the rising magma encounters thick continental crust that traps some magma. The magma heats the surrounding rocks causing localized *partial melting* that produces *felsic* or *intermediate* magma.

3. Most subduction-related magma probably never reaches the surface, but some erupts forming clusters or belts of volcanoes. Where the overlying crust is continental, most volcanoes are part of a mountain belt. Where the overlying crust is oceanic (as shown above in *Part A*), subduction-generated magma constructs individual volcanoes along an *island arc*. In both cases, such magma generally has an *intermediate* composition and forms andesite.

C What Controls How Easily Magma Moves?

Viscosity is a measure of a material's resistance to flow. A viscous magma does not flow easily, whereas a less viscous magma flows more easily. Magmas are considerably more viscous than other hot liquids with which you are familiar. A magma's viscosity is controlled by its temperature, composition, and crystal content.

1. *Viscous magma* strongly *resists* flowing. When viscous magma erupts on the surface, it does not spread out but piles up, forming mounds or domes of lava.

2. *Less viscous magma* flows more easily and may spread out in thin layers on the surface. This magma can travel longer distances from its source and cover large areas with lava.

Temperature

3. *Low Temperature*—Temperature is the most important control of viscosity. Magma at relatively low temperature, such as one barely hot enough to be molten, is *viscous* and flows only with difficulty.

4. *High Temperature*—Magma that is very hot, such as basalt, has low viscosity and so flows very easily. Mafic magma is hotter and less viscous than felsic magma.

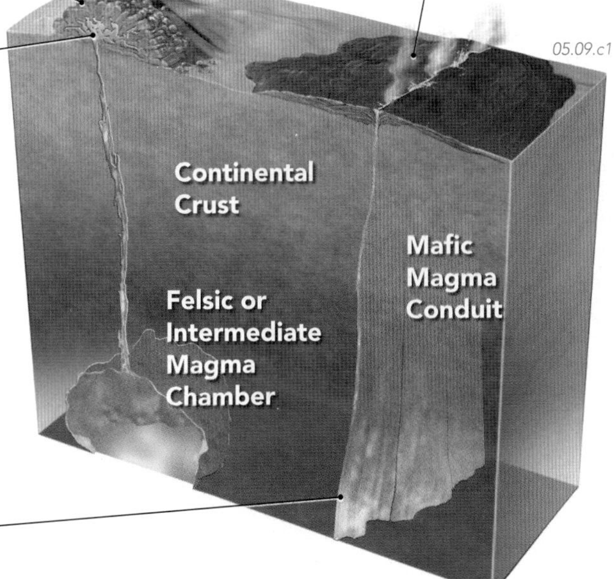

Silicate Chains

5. *Abundant Chains*—Silicate molecules in magma can link into long chains that do not bend or move easily out of the way of one another. A magma that has abundant chains, such as most felsic and intermediate magma, will be very viscous.

6. *Few Chains*—The silicate molecules in more mafic magma are less connected or are in short chains. This allows the magma to move more easily (be *less* viscous). Water dissolved in magma disrupts long chains, decreasing the viscosity.

Percentage of Crystals

Abundant Crystals—As a magma cools, crystals begin to form within the melt. The crystals in the flowing magma get in each other's way and cause the magma to flow more slowly. A magma with abundant crystals is *more* viscous.

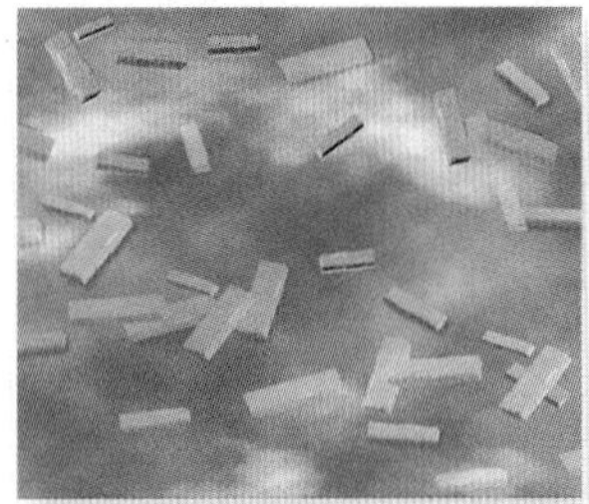

Few Crystals—A magma that has few crystals has few internal obstructions and flows more easily (is *less* viscous). Such magma flows more smoothly and thus can flow faster and farther.

How Viscous Is Your Breakfast?

One way to think about viscosity is to examine a typical Sunday breakfast that might be eaten by a student while visiting a relative's well-stocked home. The fluids you encounter during breakfast are much less viscous than magma but illustrate some important aspects of viscosity.

The orange juice that begins the feast has *low viscosity* and so pours easily, like a very hot basalt without many crystals. Next on the menu is oatmeal, which is *more viscous*, like a crystal-rich, felsic magma with long silicate chains. When thick, it piles up in a dome-shaped mound that spreads out over time. Adding milk separates the oatmeal flakes allowing them to move past one another, like scattered crystals in a magma. If a stick of butter for your toast is out on the counter too long, it softens, becoming *less viscous*, and starts to flow. It was firmer and *more viscous* when cold. Temperature clearly has an effect on viscosity. Now, what to put on the hash brown potatoes—low-viscosity catsup, low-viscosity salsa, or high-viscosity chunky salsa? Who knew that thinking about viscosity could be such an important part of breakfast?

Before You Leave This Page Be Able To

- ✓ Describe three ways that magmas rise through the crust.
- ✓ Summarize factors that influence how far a magma rises toward the surface.
- ✓ Explain the factors that control viscosity of a magma.
- ✓ Describe what factors might be combined to form very high viscosity magma or very low viscosity magma.

How Does Magma Solidify?

MAGMAS EVENTUALLY COOL AND SOLIDIFY. The general term *solidify* is used here instead of the more specific term *crystallize* because a magma can cool so rapidly that crystals do not have time to form. The magma instead cools quickly to volcanic glass. Magma solidifies when it loses thermal energy to its surroundings. Its rate of cooling affects the size and shape of any resulting crystals.

A Under What Conditions Does Magma Solidify?

For a magma to solidify, it must lose enough thermal energy to its surroundings to pass from the liquid to the solid state. This generally happens when a magma has risen to a place that is cooler than it is.

When magma reaches the surface, it transfers thermal energy to air and possibly water through conduction and radiation.

Magma loses thermal energy to surrounding rocks by *conduction*. As the wall rocks are heated, their temperature increases, possibly causing metamorphism or even melting.

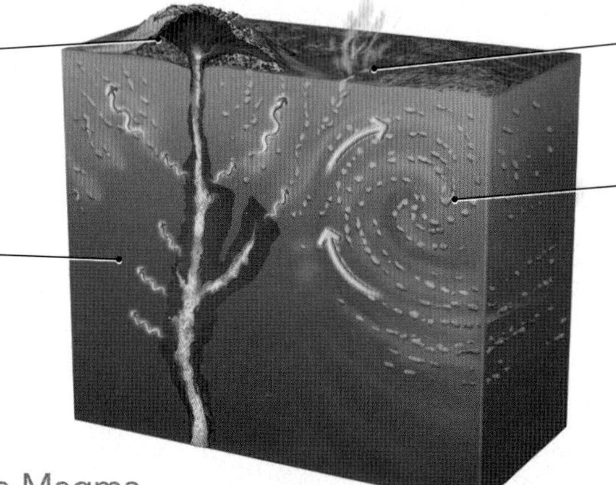

Magma also loses heat when it releases gases, including water, into the wall rocks or onto the surface.

Water in rocks near the magma receives heat by conduction from the magma or the hot wall rocks. As the water gets hotter, its density decreases and the water rises. The upward flowing water is replaced by an inflow of cooler water, causing *convection*. Such convection of water may be the primary way some magma cools.

05.10.a1

Cooling History of a Magma

This graph plots conditions of temperature versus depth. We can track the history of an idealized magma by plotting points representing the way it cools and reaches the surface. Follow the numbered changes by starting at the bottom.

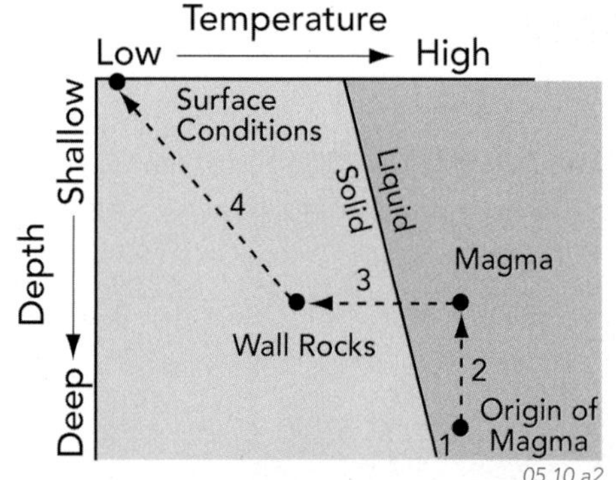

05.10.a2

4. At some later time, the now-solidified magma and its wall rocks are uplifted to the surface, where they cool to low temperature.

3. The magma cools by losing thermal energy to the surrounding wall rocks. It crosses the solid-liquid line and crystallizes.

2. Most of the magma rises some distance in the mantle or crust and so is in a place where it is surrounded by cooler rocks.

1. A magma forms at depth, where temperature is high enough to overcome pressure and to cause melting.

B What Happens If Magma Cools Slowly or Rapidly?

A magma solidifies as minerals crystallize or as other types of chemical bonding form glass. The size of crystals in a magma largely reflects the rate at which the magma cools. Magma cools slowly when it is in hot surroundings or is thermally insulated by wall rocks. Water dissolved in magma can help form very large crystals (pegmatite). The rocks below are felsic but differ mainly by how fast the magma cooled.

Slow Cooling

05.10.b1

Coarse Granite Pegmatite

Medium Cooling

05.10.b2

Finely Crystalline Granite

Very Fast Cooling

05.10.b3

Obsidian (felsic glass)

Slow Then Fast Cooling

05.10.b4

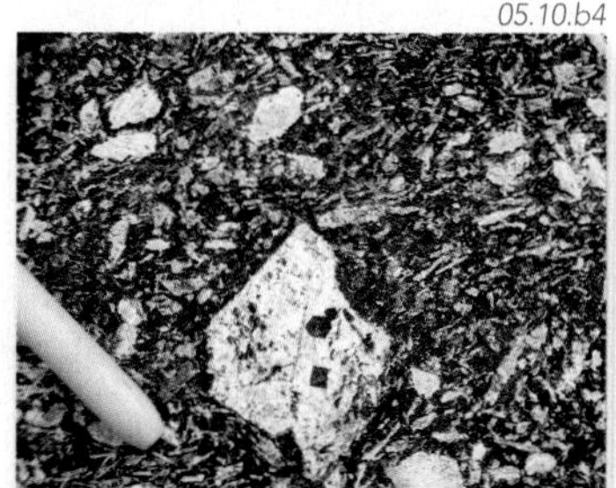

Porphyritic Intrusive Rock

C In What Order Do Minerals Crystallize?

Minerals melt at different temperatures—felsic minerals melt before mafic ones. Minerals crystallize in the opposite order that they melt—mafic minerals crystallize before felsic minerals. The figure below shows a general sequence of mineral crystallization called *Bowen's Reaction Series*.

1. Mafic minerals, like olivine and pyroxene (▷), are the first to crystallize from a mafic magma. They typically do not crystallize from a felsic magma, which lacks the high contents of iron and magnesium required by these minerals. The shape of the colored area for each mineral indicates whether the amount of crystallization increases or decreases with temperature.

05.10.c2

3. Plagioclase feldspar (▷) displays a continuum from calcium-rich to sodium-rich end members. Calcium (Ca)-rich plagioclase crystallizes first from a magma, followed by more sodium (Na)-rich varieties.

05.10.c3

2. As the magma cools, other mafic minerals including amphibole (▽) and biotite begin to crystallize, provided that the magma has a composition that is not too mafic.

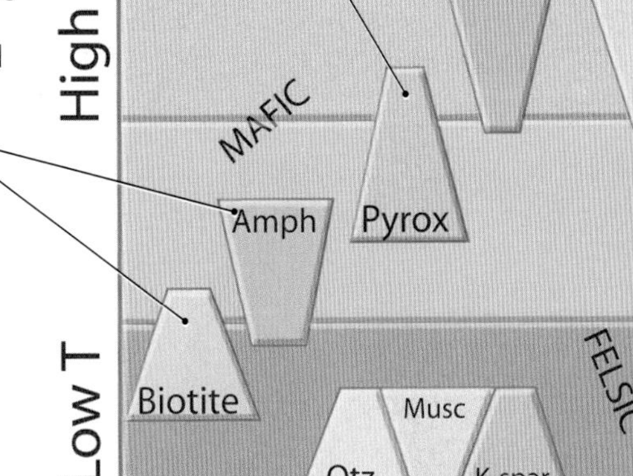

05.10.c5

4. Light-colored felsic minerals, such as quartz (▷), K-feldspar (▽), and muscovite, crystallize at the lowest temperatures. These minerals, along with Na-rich plagioclase, may be the only minerals formed from *felsic* magmas, which lack the chemical components required to grow *mafic* minerals. They typically do not grow from mafic magmas, which lack the required constituents.

05.10.c4

05.10.c6

5. Minerals that crystallize early in the crystallization sequence can grow unimpeded in the magma, and so commonly have well-defined crystal shapes, like these well-formed crystals of light-colored feldspar. ▷

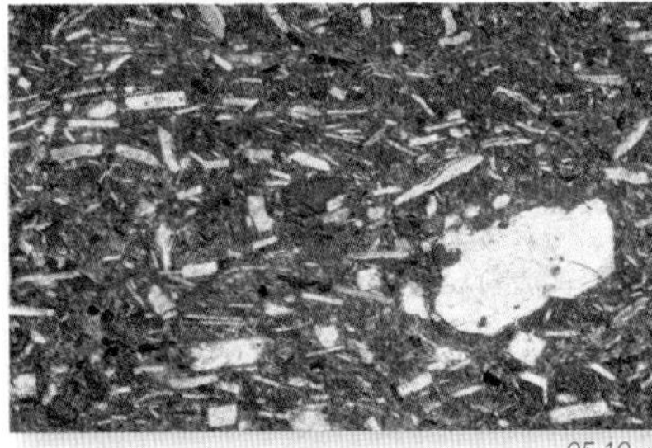

05.10.c7

05.10.c8

6. Minerals that crystallize late in the sequence must grow around preexisting crystals so may grow in irregular, poorly defined crystal shapes. The white crystals (◁) grew late and so had to fill around dark crystals that formed early.

How Crystallization Changes the Composition of a Magma

As minerals crystallize from a cooling magma, they remove the chemical constituents that are incorporated into the crystals. Once within a crystal, atoms and molecules are somewhat sheltered from interaction with the remaining magma. Therefore, the chemical composition of the remaining magma changes as minerals crystallize.

If *mafic* minerals crystallize first from a magma, they extract the mafic constituents, such as magnesium, iron, and calcium. As these crystals are removed from the magma, the remaining magma contains less of these constituents over time. That is, the magma becomes *less mafic* and *more intermediate or felsic*.

This graph illustrates the effects of such crystallization on a magma. It shows contents of magnesium (expressed as magnesium oxide, MgO) and silicon (expressed as SiO_2) for a series of rocks produced by a single crystallizing magma. In this example, the magma is crystallizing the Mg-rich mineral olivine, causing the Mg content of the remaining magma to decrease over time. In other words, crystallization of mafic minerals is making the remaining magma less mafic.

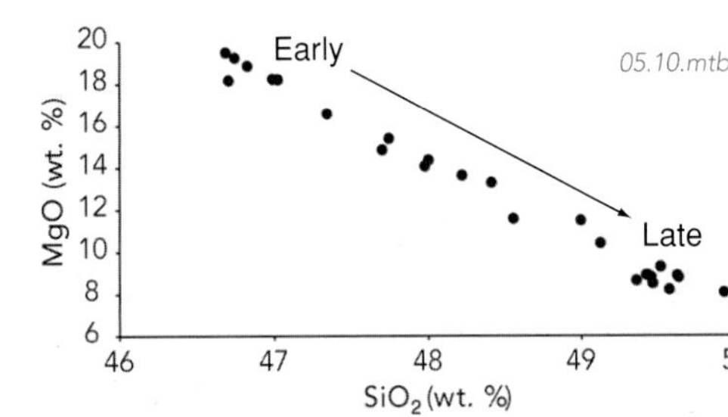

Before You Leave This Page Be Able To

- ✓ Explain or sketch the processes by which a magma cools.
- ✓ Describe or sketch the cooling history of a magma as it rises in the crust.
- ✓ Explain the order in which minerals crystallize from a magma (Bowen's Reaction Series), and compare it to the order in which they melt.
- ✓ Describe how the rate of magma cooling affects the size and shape of crystals that form.
- ✓ Explain how the crystallization of minerals can change the composition of remaining magma.

How Do Large Magma Chambers Form and Evolve?

MAGMA CAN ACCUMULATE IN UNDERGROUND CHAMBERS, some of which contain thousands of cubic kilometers of molten rock. How do these chambers form, what are their shapes, and what processes occur within them? What do they look like after they have solidified and are uplifted to the surface?

A What Is a Magma Chamber and What Processes Occur in Large Chambers?

A *magma chamber* is an underground body of molten rock. Think of it as an always-full reservoir or holding tank that allows magma to enter from below and perhaps exit out the top. Magma chambers are interpreted to be very dynamic, with magmas evolving, crystallizing, and being replenished by additions of new magma.

Large magma chambers can consist of a single magma but generally involve more than one influx of magma.

During crystallization, early formed minerals remove chemical constituents from the magma and may rise or sink (*crystal settling*) within the chamber.

As new pulses of magma are injected into the chamber, they add thermal energy and perhaps *mix* with existing magma.

A magma undergoing *crystallization* could be intruded by a new, hotter pulse of magma into the chamber. The new magma can transfer heat to the old magma causing newly formed minerals to melt back into the magma.

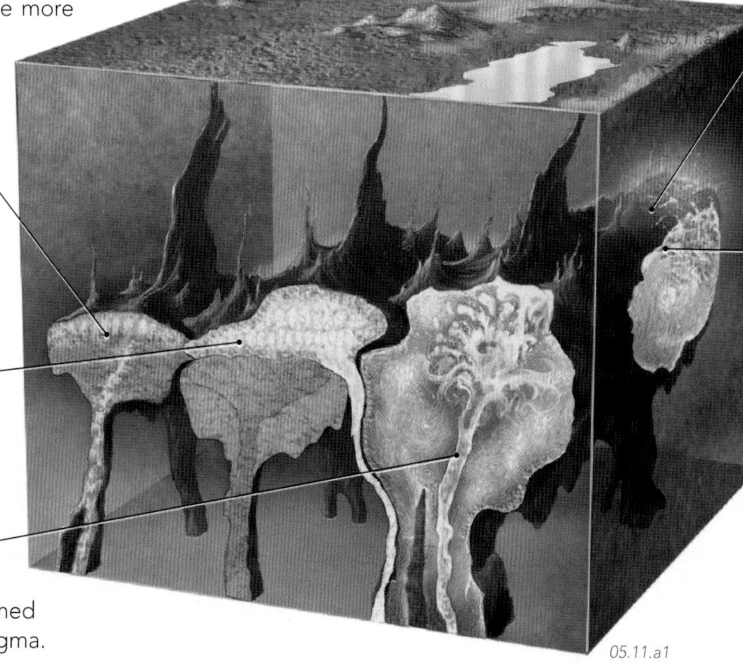

05.11.a1

The magma can heat and *partially melt* the wall rocks, forming a new magma with a different composition. Such melting is aided by heat brought into the chamber by new batches of magma from below.

Magma produced by partial melting of the wall rocks can be *assimilated* into the existing magma or can rise out of the chamber without interacting chemically with other magma.

How two magmas mix depends on their relative densities and the temperatures at which they crystallize. Magmas of similar density may form well-mixed magma, whereas magmas of different densities may form a patchwork of magma types.

B In What Settings Do Large Magma Chambers Form?

A large influx of magma is required to form a large magma chamber. This, in turn, requires melting on a large scale that is possible only in certain tectonic settings.

In oceanic lithosphere, large magma chambers form above hot spots and within mid-ocean ridges. In both cases, the mantle-derived magmas are mafic.

Large dikes and other chambers of intermediate and felsic magma form above subduction zones, either within magmatic arcs on continents or within oceanic island arcs.

Hot spots and rifts within continents produce large amounts of mantle-derived magma that can melt continental crust to form large felsic magmas.

Continental collisions cause crustal thickening, which can lead to melting of continental crust. Large amounts of felsic magma may be trapped at depth.

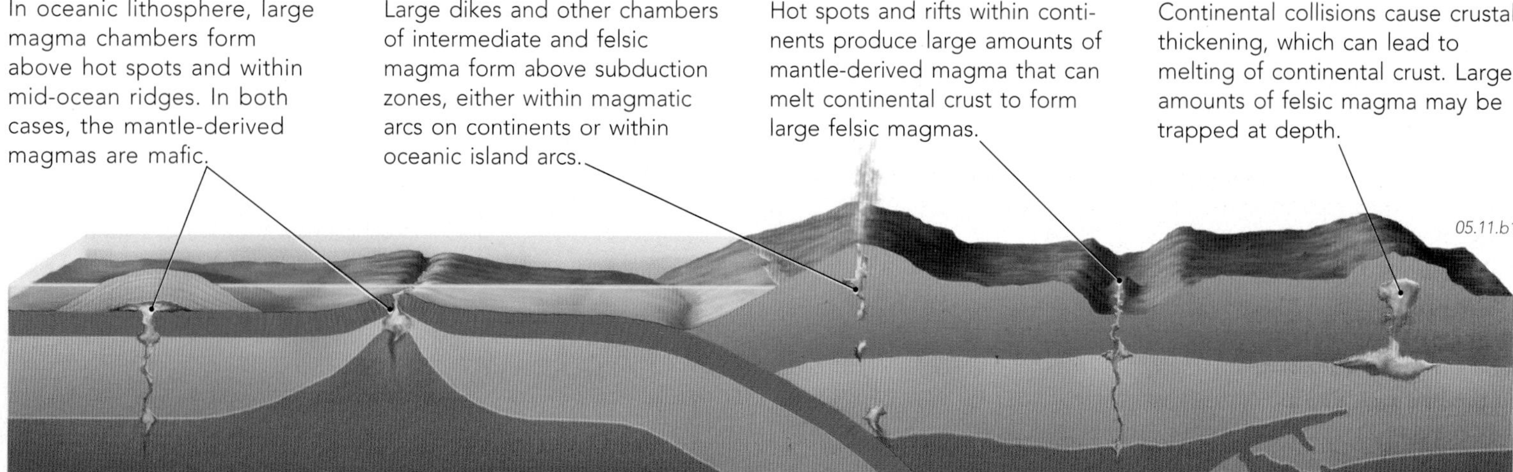

05.11.b1

C How Are Large Solidified Magma Chambers Expressed in the Landscape?

A solidified magma chamber is called a *pluton*. A pluton can be cylindrical, sheetlike, or have a very irregular shape. Several generations of magma may intrude the same region, forming a complex mass of plutons with various compositions, textures, and geometries. Plutons are classified according to their size and geometry.

Irregular Plutons

A pluton with an exposed area of less than 100 km² is called a *stock*.

05.11.c1–2

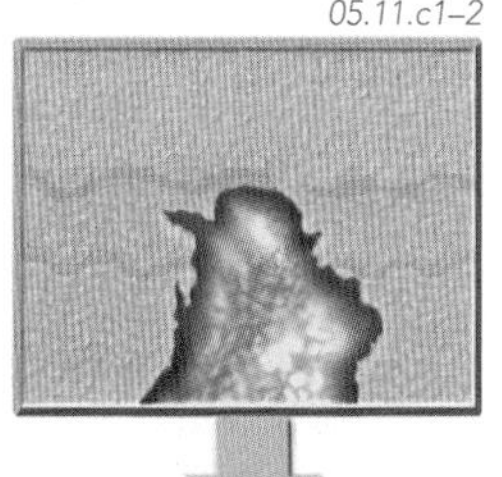

Most stocks are irregularly shaped. Many have a shape like a steeply oriented pipe or cylinder.

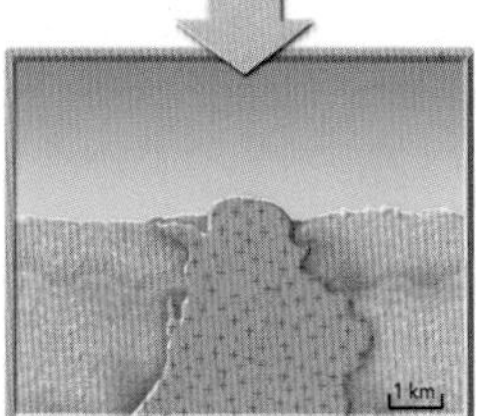

On the surface, most stocks have steep boundaries and may resist erosion more than surrounding rocks.

05.11.c7

A stock of bold, gray rocks represents a magma that solidified at depth and later was uplifted. [Toyabe Range, Nevada]

Sheetlike Plutons

Some plutons have the shape of a thick sheet.

05.11.c3–4

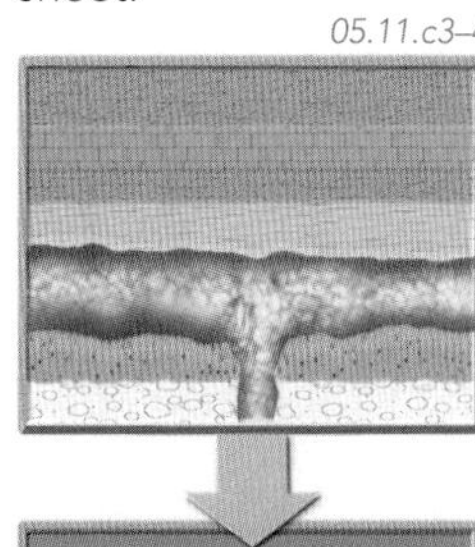

Plutonic sheets can be horizontal, vertical, or inclined, and may be parallel to or cutting across layers in the wall rocks.

When horizontal sheets are exposed at the surface, their tops and bottoms may be visible.

05.11.c8

The gray granitic rocks were a horizontal sheet of magma that squeezed into dark metamorphic rocks. [Cuernos del Paine, Chile]

Batholiths

A batholith is one or more contiguous plutons that cover more than 100 km².

05.11.c5–6

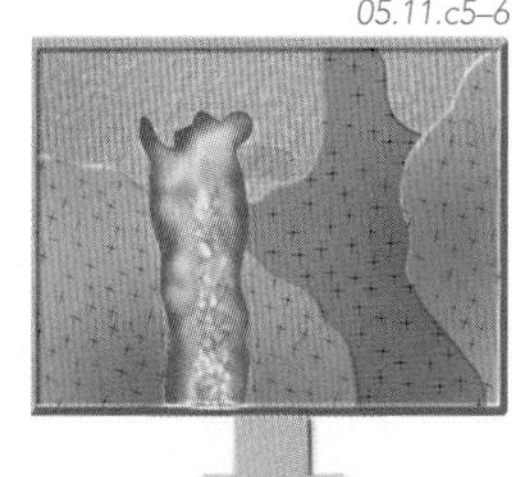

Most batholiths form as multiple magmas are emplaced into the same part of the crust through a long period of time.

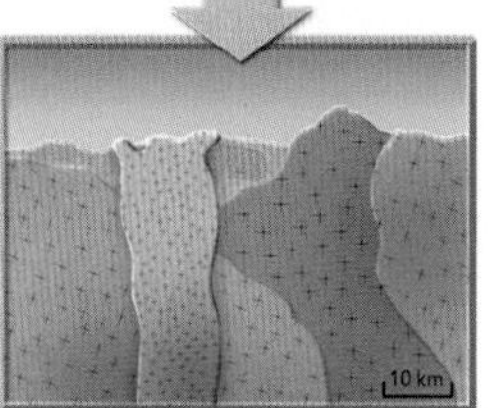

Exposed batholiths are characterized by plutonic rocks that cover a huge region.

05.11.c9

A huge expanse of gray granite characterizes the Sierra Nevada batholith of California, as in this view from the east.

The White Mountain Batholith of New England

The White Mountain batholith is centered in the middle of New Hampshire. Granitic rocks of the batholith form high peaks of the White Mountains and many of the area's scenic landmarks.

The batholith consists of several dozen individual plutons (shown in red and yellow) that were emplaced between 200 and 155 million years ago. The plutons represent separate injections of magma emplaced at somewhat different times. Some plutons are cylindrical, whereas others are like curved dikes.

Geologists interpret the White Mountain batholith as being related to a hot spot that melted its way into continental crust. The age of the batholith coincides with the rifting of North America from Africa that led to the opening of the Atlantic Ocean. A line of submerged volcanic mountains in the Atlantic Ocean, called the *New England Seamount Chain*, is interpreted to mark the path of the North American plate over the hot spot.

05.11.mtb1

Before You Leave This Page Be Able To

- ✓ Summarize what a magma chamber is and the processes that occur in one.
- ✓ Sketch or summarize the tectonic settings in which large magma chambers form.
- ✓ Sketch the different geometries of large magma chambers and summarize how these are expressed in the landscape.
- ✓ Summarize the character of the White Mountain batholith and how it is interpreted to have formed.

How Are Small Intrusions Formed and Expressed in the Landscape?

MANY INTRUSIONS ARE SMALL OR THIN FEATURES that are commonly exposed on a single small hill or in a roadcut. Small intrusions can have a sheetlike, pipelike, or even blisterlike geometry. Where exposed at the surface, small intrusions form distinctive landscape features.

A Why Is Some Magma Injected as Sheets?

Many small intrusions have the shape of thin or thick sheets, typically ranging in thickness from several centimeters to several tens of meters. These form when underground forces allow magma to generate fractures or to open up and inject into existing fractures. In some cases, magma squeezes between preexisting layers in the wall rocks.

Dike

05.12.a1

A dike is a sheetlike intrusion that cuts across layers or is steep. Dikes form because magma can most easily push apart the rocks in a horizontal direction, perpendicular to the dike. Dikes are also common within many larger magma chambers.

05.12.a2

The Greek island of Santorini erupted catastrophically probably around 1650 B.C. Steep dikes in the walls of the volcanic crater cut across the volcanic layers. Some of the dikes are along faults.

Sill

05.12.a3

An intrusion that is parallel to layers or is subhorizontal is called a *sill*. A sill forms by pushing adjacent rocks upwards rather than sideways. The Palisades along the Hudson River of New York is a large sill. Sills commonly have many steep joints.

05.12.a4

These dark, mafic sills intruded parallel to layers of light-colored, sedimentary wall rocks. The sills contain steep fractures formed by cooling of the sills after they solidified. [Salt River Canyon, Arizona]

Laccolith

05.12.a5

In some areas, ascending magma encounters gently inclined layers and begins squeezing parallel to them as a sill. The magma then begins inflating a blister-shaped magma body called a *laccolith*. As the magma chamber grows, the layers over the laccolith are tilted outward and eventually define a dome.

05.12.a6

The Four Corners region of the American Southwest contains some of the world's most famous stocks and laccoliths, including these in the Henry Mountains of southern Utah. The laccoliths formed at a depth of several kilometers about 25 million years ago and were later uncovered by erosion. Igneous rocks of the laccolith are medium grained, porphyritic, and an intermediate composition.

B What Kind of Magma Chambers Form Within and Beneath Volcanoes?

Magma that erupts from volcanoes is fed through conduits that may be circular, dike shaped, or both. After the volcano erodes away, the solidified conduit can form a steep topographic feature called a *volcanic neck*.

1. A small volcano has been partially eroded revealing a cross section through the volcano. A resistant and jointed volcanic conduit is within the center of the volcano. [Mount Taylor, New Mexico]

3. Shiprock is a famous volcanic neck that rises above the landscape of New Mexico. It consists of fragmented mafic rocks and connects to dikes that radiate out from the central conduit.

2. Many volcanic necks, like the one above, form as erosion wears down a volcano, exposing the harder, more resistant rocks that solidified inside the magmatic conduit of the volcano. ▷

◁ 4. Some volcanic necks, including Shiprock, were not originally *inside* a volcano but instead were magmatic conduits that formed well *beneath* the surface. The volcano above Shiprock was not a mountain, but a crater (pit) excavated by a violent explosion. The explosion occurred when magma ascending up a conduit encountered groundwater and generated huge amounts of steam. Erosion after the event removed the crater and hundreds of meters of rock that once overlay the area around the conduit.

Columnar Joints

Many igneous rock bodies display distinctive fracture-bounded columns of rocks, like the ones in Devil's Postpile National Monument in California (▽). Similar fractures outline columns hundreds of meters high in Wyoming's Devils Tower (▷), which is a famous landmark used in the filming of the movie *Close Encounters of the Third Kind*. These fractures, known as *columnar joints*, form when a hot but solid igneous rock contracts as it cools. The fractures carve out columns that commonly have five or six sides. Columnar joints are common in basaltic lava flows, felsic ash flows, sills, dikes, and some laccoliths.

05.12.mtb1

Before You Leave This Page Be Able To

- ✓ Sketch the difference between a dike and a sill, and explain why each has the orientation that it does.
- ✓ Sketch or discuss the geometry of a laccolith.
- ✓ Sketch and explain two ways that a volcanic neck can form.
- ✓ Describe what columnar joints are.

How Did the Sierra Nevada Form?

ONE OF THE WORLD'S MOST STUDIED BATHOLITHS forms the scenic granite peaks of the Sierra Nevada of central California. The batholith contains a diverse suite of plutonic rocks that cover an area of 40,000 km^2 (16,000 mi^2). The batholith was constructed by separate pulses of magma, largely between 140 and 80 million years ago when oceanic plates subducted beneath western North America.

A What Is the Nature of the Sierra Nevada Batholith?

The Sierra Nevada batholith includes hundreds of individual plutons, some of which cover more than 1,000 km^2 (380 mi^2). The batholith also includes small stocks that are only hundreds of meters across as well as countless dikes and sills of various compositions. Rocks within and around the batholith tell its geologic story.

05.13.a2

◁ **1.** The scenery of the Sierra Nevada is dominated by peaks, cliffs, and rounded domes of massive gray granite. [Stately Dome, Sierra Nevada]

2. The figure below shows the landscape of the region colored according to rock type. The Sierra Nevada is the broad, high mountain range and is mostly granitic rocks (colored gray) with smaller areas of metamorphic rocks (colored green). Patches of volcanic rocks much younger than the batholith are shown in red and pink. The valley east of the Sierra Nevada is the Owens Valley, which is underlain by recent sediments (colored yellow).

05.13.a3

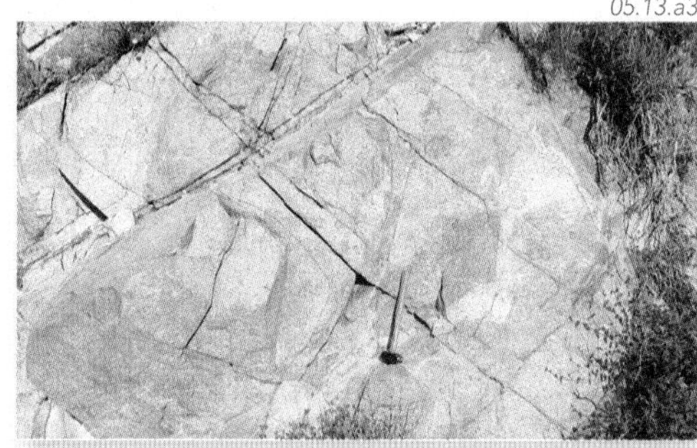

△ **3.** The most common rocks in the batholith are light- to medium-gray granite and other plutonic rocks. The plutons solidified slowly so have medium-grained to coarse-grained crystals.

05.13.a1

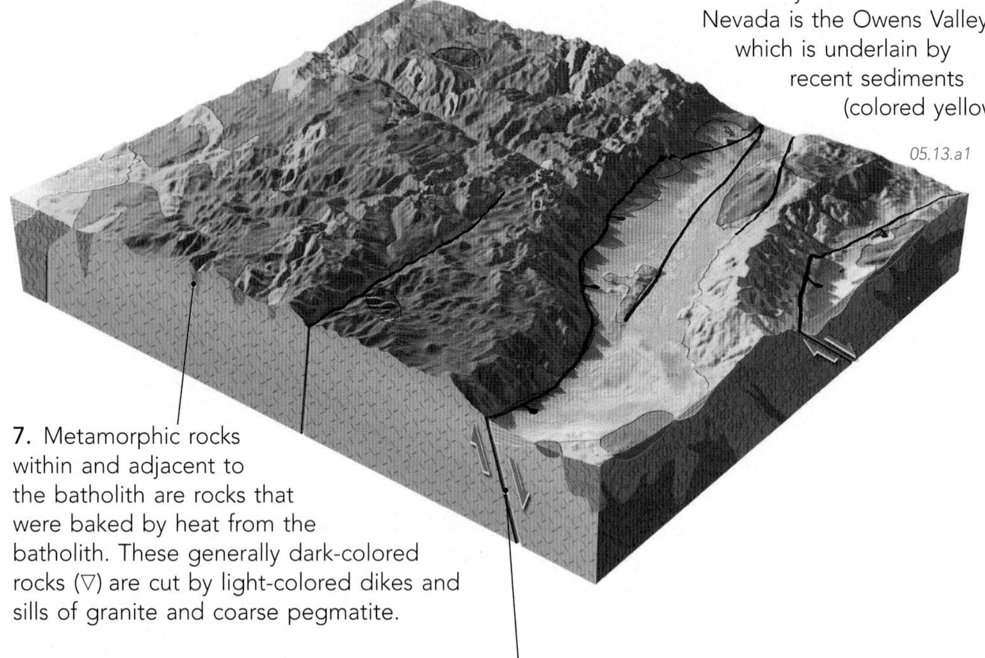

7. Metamorphic rocks within and adjacent to the batholith are rocks that were baked by heat from the batholith. These generally dark-colored rocks (▽) are cut by light-colored dikes and sills of granite and coarse pegmatite.

05.13.a4

△ **4.** Some outcrops, including this one, show great diversity in composition, including light-gray granite, dark intermediate and mafic rocks, and thin cream-colored felsic dikes. These represent different magmas, commonly emplaced at different times.

05.13.a5

6. The steep east side of the Sierra Nevada is along a fault that downdropped Owens Valley relative to the mountains. During faulting, the entire Sierra was tilted, raising the eastern side of the range so that it is now higher and steeper than the western side.

▷ **5.** Some plutons display compositional variations that record *crystallization* and *settling* of early-formed crystals. In this photograph a light-colored, more felsic part is to the left and a darker mafic one is to the right.

05.13.a6

B What Is the Tectonic History of the Batholith and Surrounding Areas?

The Sierra Nevada batholith is a product of plate tectonics—it formed by subduction-related partial melting of mantle and lower continental crust. Its origin illustrates how different magmas are generated.

Plate-Tectonic History of the Batholith

1. This figure shows the interpreted setting of the batholith 100 million years ago, when North America was converging with oceanic plates in the Pacific Ocean. Most of the batholith formed between 140 and 80 million years ago when oceanic plates in the Pacific Ocean were being subducted eastward beneath North America.

05.13.b1

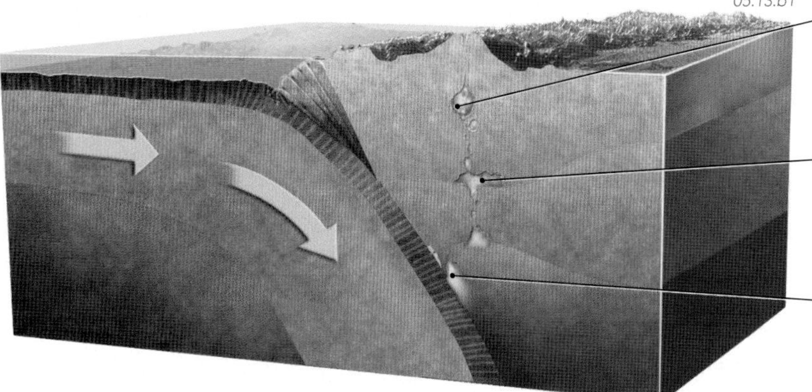

2. Water driven from minerals in the subducting slab rose into the overlying mantle, causing partial melting because the water lowered the melting temperature. Melting generated mafic magma that rose toward the crust.

3. Heat transfer from the mantle-derived mafic magma caused partial melting of continental crust, which generated felsic magma. Mixing of felsic and mafic magmas, along with partial crystallization, produced a wide range of igneous compositions.

4. Magma rose in the crust and large volumes solidified at depth as plutons and dikes. Sixty million years of sustained magmatism, with numerous discrete magmas, constructed the regionally extensive batholith.

5. While the batholith formed underground, large volumes of magma reached the surface and erupted in explosive volcanoes. The volcanoes were eroded away, but their record is preserved in sedimentary rocks deposited west of the batholith (between the batholith and the offshore trench).

6. At about 80 million years ago, a change in the plate-tectonic setting shut off magmatism in the Sierra Nevada and shifted magmatism eastward into Nevada and Arizona.

7. Between 80 and 30 million years ago, the Sierra Nevada batholith was slowly uplifted until the plutonic rocks, which formed at depths of 10 to 20 kilometers, and their metamorphic wall rocks were exposed at the surface.

05.13.b2

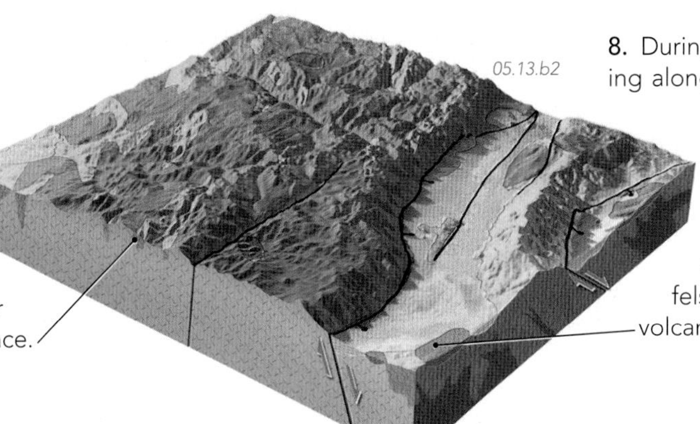

8. During the last 5 million years, faulting along the east side of the batholith uplifted the Sierra Nevada to the majestic mountain range it is today. This faulting, part of an episode of intraplate rifting, was accompanied by eruption of felsic domes and other volcanic units shown in red and pink.

How Do Geologists Study the Sierra Nevada?

Many geologists study the Sierra Nevada to reconstruct the geologic history of this special place and to study the processes of magma chambers.

To study the batholith and its magmatic processes, geologists first do geologic field work by hiking up and down the ridges examining the rocks, identifying boundaries between different plutons, and collecting samples for later analysis. From the field studies, geologists construct a geologic map and geologic sections that represent the distribution of the plutons. Geologists cut thin sections from the rock samples to determine what minerals are present and in what order the minerals crystallized from the magma. Chemical analyses of the samples for potassium, silicon, and other elements document how the magma evolved over time. Analysis of isotopes helps determine the age of the rocks and the types of source rocks that were melted to form the magmas. It is an excellent and scenic place to study igneous processes.

05.13.mtb1

Before You Leave This Page Be Able To

- ✓ Describe the Sierra Nevada batholith and what rocks it contains.
- ✓ Sketch the plate-tectonic setting that formed the Sierra Nevada batholith.
- ✓ Sketch or describe how the magmas of the batholith formed.
- ✓ Summarize how the deep batholithic rocks ended up on Earth's surface.
- ✓ Briefly summarize the kinds of data geologists collect in studying the batholith.

What Types of Igneous Processes Are Occurring Here?

IGNEOUS ACTIVITY IS NOT DISTRIBUTED UNIFORMLY ON EARTH. As a result, different regions have different potential for volcanic eruptions and other igneous activity. In this exercise, you will investigate five sites to interpret the types of igneous rocks likely to be present, the style of eruption, and the probable cause of melting.

Goals of This Exercise:

- Use the regional features of an ocean and two continents to infer the tectonic setting and cause of melting at five sites.
- Observe and identify nine rock types and infer the cooling history of each rock based on its texture.
- For each site, predict the viscosity of the magma and probable style of eruptions.

A Tectonic Settings of Igneous Activity

The perspective view below shows two continents and an intervening ocean basin. The area has five sites, labeled A, B, C, D, and E, where igneous activity has been observed. For each site, investigate the igneous processes responsible for the activity and enter your results in the worksheet using the steps listed below.

1. Use the features on this map to infer whether the tectonic setting of each site is associated with a plate boundary and, if so, which type of plate boundary is present. The possible tectonic settings for this region are as follows: (1) an oceanic or continental divergent boundary, (2) one of the three types of convergent boundaries, or (3) a hot spot in a continent, ocean, or both. All of these settings are not present in this area.
2. For each site, determine the likely cause of melting. The options are (1) decompression melting, (2) melting by adding water, and (3) melting of continental crust caused by an influx of mantle-derived magmas. More than one of these causes might apply to each site. Think about the kinds of igneous rocks you would predict to find at each site, including those that solidify at depth (plutonic) and those erupted onto the surface (volcanic). Your instructor may ask you to list these.

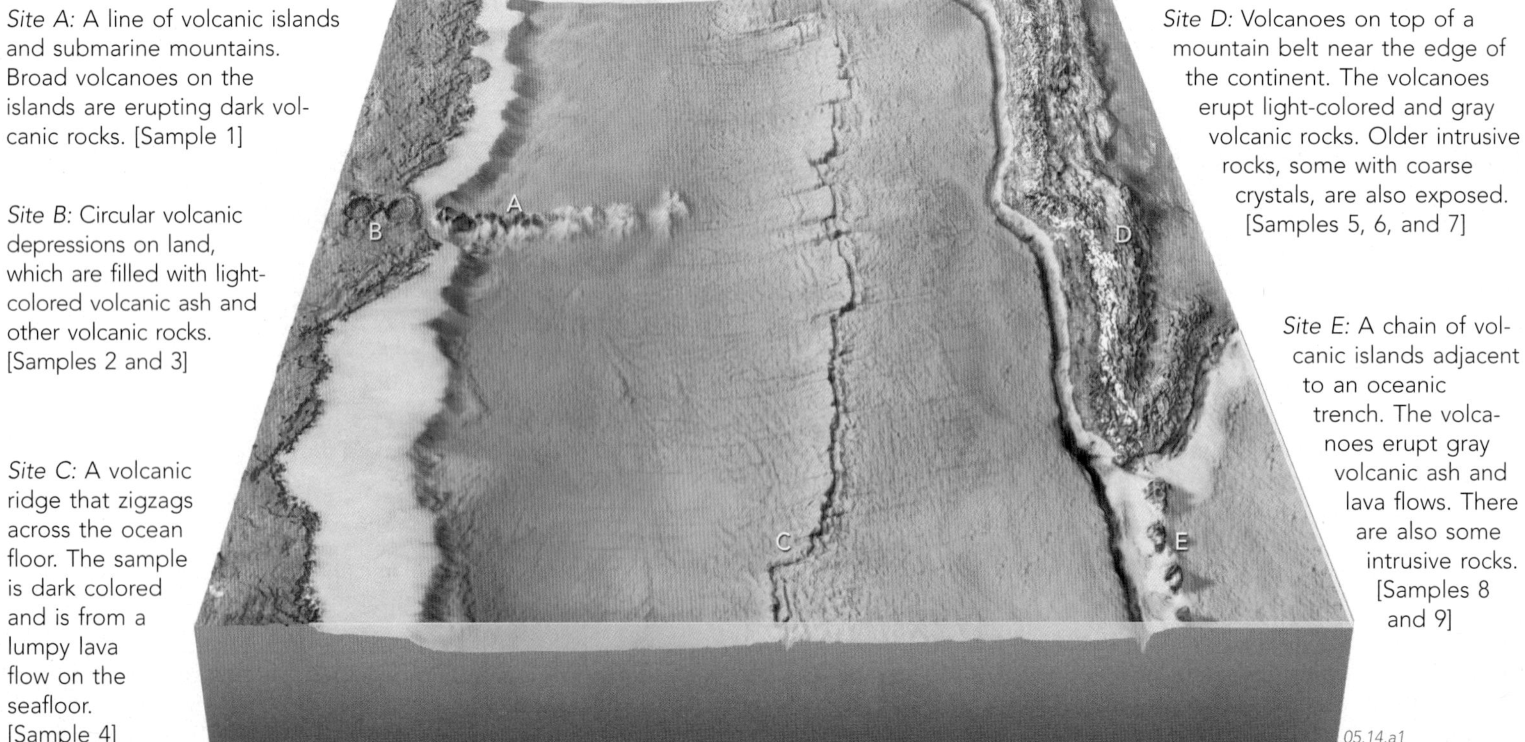

Site A: A line of volcanic islands and submarine mountains. Broad volcanoes on the islands are erupting dark volcanic rocks. [Sample 1]

Site B: Circular volcanic depressions on land, which are filled with light-colored volcanic ash and other volcanic rocks. [Samples 2 and 3]

Site C: A volcanic ridge that zigzags across the ocean floor. The sample is dark colored and is from a lumpy lava flow on the seafloor. [Sample 4]

Site D: Volcanoes on top of a mountain belt near the edge of the continent. The volcanoes erupt light-colored and gray volcanic rocks. Older intrusive rocks, some with coarse crystals, are also exposed. [Samples 5, 6, and 7]

Site E: A chain of volcanic islands adjacent to an oceanic trench. The volcanoes erupt gray volcanic ash and lava flows. There are also some intrusive rocks. [Samples 8 and 9]

B Predicting the Types of Igneous Rocks and Eruptions at Each Site

Photographs below show nine different rock types. Your instructor may provide you with samples of each rock or may substitute a different suite of rocks. Observe each of these rocks and complete the steps below.

1. Your instructor may have you write a short description of each photograph or of actual samples.
2. On the worksheet, indicate (1) whether each rock shown is coarsely crystalline, finely crystalline, or has other distinctive igneous textures, (2) whether it is probably mafic, intermediate, or felsic, and (3) the name you would apply to such a rock.
3. Predict the cooling and solidification history for each rock sample based on its texture (slow, moderate, fast, slow then fast, or slow cooling in the presence of water).
4. For each site, use the rock samples that you interpret to be volcanic to predict whether the magma for that site has a high or low viscosity, and what type of volcanic eruption probably formed the rock sample.
5. Your instructor may have you use the various types of information to explain how the samples are consistent with the tectonic setting of each site. Alternatively, your instructor may have you infer the entire sequence of events including (1) what caused the initial melting event, (2) what processes might have occurred in the magma chamber, (3) where and how the rock cooled and solidified, and (4) whether uplift and erosion are required to expose the rock.

05.14.b1

CHAPTER 6

Volcanoes and Volcanic Hazards

THE ERUPTION OF A VOLCANO is one of nature's most spectacular events. Clouds of scalding volcanic ash are blasted into the air, or orange streams of molten rock pour down the volcano's flank. Volcanoes represent an obvious geologic hazard, and past eruptions have claimed the lives of tens of thousands of people at a time. In this chapter, we explore volcanoes, their landforms, and their associated hazards.

Mount St. Helens in southwestern Washington was one of the most beautiful and symmetrical high peaks (▽) in the Cascade Range of the Pacific Northwest. Its shape changed forever in May of 1980 when the sleeping volcano erupted violently. The volcano's north flank was blown apart, and a huge crater was excavated where the mountain peak used to be. Within the newly formed crater, continuing eruptions have built the steaming lava dome shown in the larger photograph.

What is a volcano, and how do we recognize one?

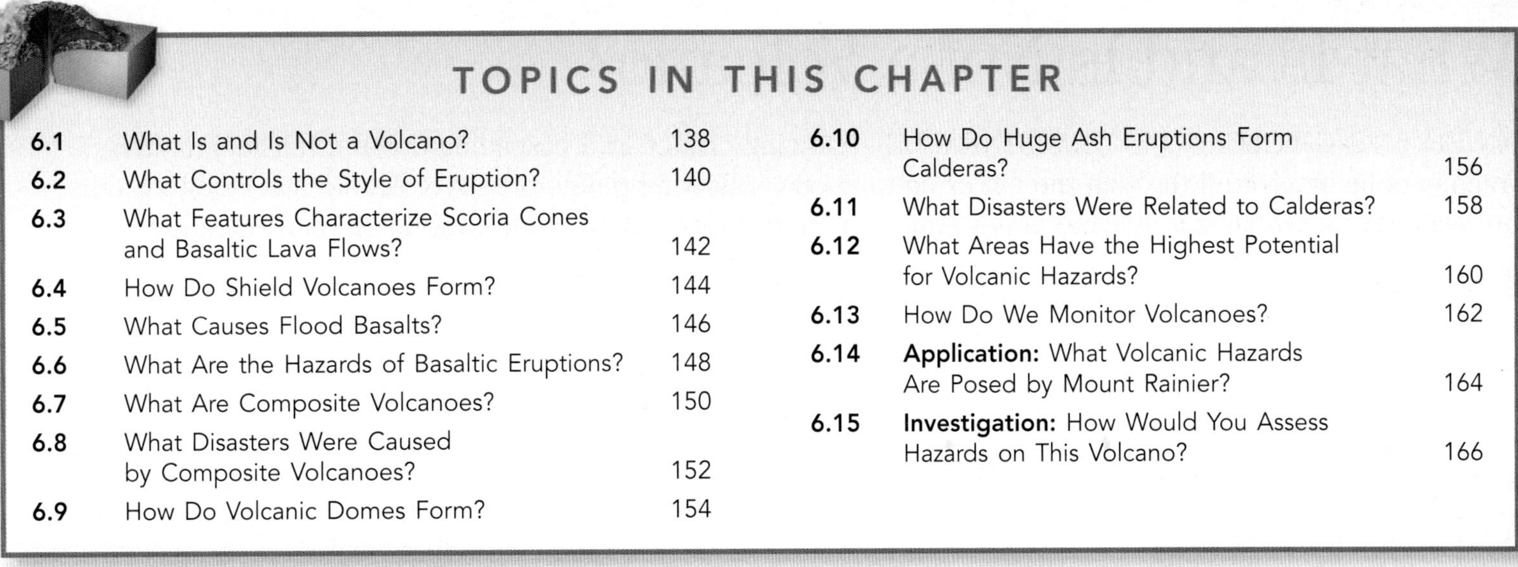

TOPICS IN THIS CHAPTER

06.00.a4

The May 1980 eruption started with a northward-directed blast that knocked over millions of trees and unleashed a *pyroclastic flow,* a swirling, hot cloud of dangerous gases, volcanic ash, and angular rock fragments. The pyroclastic flow swept downhill and across the landscape, burying and killing most living things in its path. This was followed immediately by a huge column of volcanic ash that rose 25 kilometers into the atmosphere (◁). The ash was carried eastward by the wind and blocked the Sun as it settled back to Earth across a large area of Washington, Idaho, and Montana.

What are the different ways that volcanoes erupt, and what hazards are associated with each type of eruption?

06.00.a5

Since the main eruption, magma rising through the throat of the volcano has piled on top of the vent as a lava dome (△). Collapse of part of the unstable dome unleashes explosions or avalanches of hot volcanic ash and rocky fragments of the dome.

What factors determine whether magma erupts as an explosion of hot ash or a slow outpouring of lava?

06.00.a6

Geologists monitor the volcano by observing the processes and the resulting rocks and features. In this photograph, geologists are using a helicopter to sample rocks in an angular part of the volcanic dome. Geologists also use instruments to monitor temperatures, gas emmisions, and tilting of the land surface.

What do different types of information tell us about whether a volcano is likely to erupt soon?

The May 1980 Eruption

With eruptions continuing into 2007, Mount St. Helens is the most active of 15 large volcanoes that crown the Cascade Range of the Pacific Northwest. The mountain was constructed entirely within the last 40,000 years and is the youngest volcano in the range. Prior to 1980, the volcano last erupted in the mid-1800s. Before 1980, a team of geologists from the U.S. Geological Survey studied the geology of the mountain and recognized that past eruptions had unleashed vast amounts of volcanic ash, lava, and volcanic mudflows.

The volcano reawakened in March of 1980 when it vented steam, shook the area with numerous earthquakes, and pushed out an ominous bulge of rock on its north flank. One morning in May 1980, an earthquake caused the oversteepened north flank to collapse downhill in a huge landslide that carried rock pieces as large as buildings. This catastrophic removal of rock released pressure on the magma inside the volcano, which exploded northward in a cloud of scalding and suffocating volcanic ash. The pyroclastic flow raced across the landscape at speeds of up to 1,000 kilometers (620 miles) per hour. The eruption blasted away most of the north flank of the mountain and forever changed the peak's appearance. It turned the surrounding countryside into a barren wasteland smothered by a thick blanket of volcanic ash. Evacuations helped limit the loss of life to 57 people, but damage estimates for the eruption exceed one billion dollars, making it the most expensive and deadly volcanic eruption in U.S. history.

What Is and Is Not a Volcano?

AN ERUPTING VOLCANO IS UNMISTAKABLE—glowing orange lava cascading down a hillside, molten fragments being ejected through the air, or an ominous, billowing plume of gray volcanic ash rising into the atmosphere. But what if a volcano is not erupting? How do we tell whether a mountain is a volcano?

A What Are the Characteristics of a Volcano?

How would you describe a volcano to someone who had never seen one? Examine the two volcanoes photographed below and look for common characteristics.

06.01.a1

A volcano is a *vent* where magma and other volcanic products have erupted onto the surface. The volcano on the left (Hawaii) erupts molten lava, whereas the one below (Alaska) erupts volcanic ash. Most volcanoes are hills or mountains that have been *constructed* by volcanic eruptions, and some geologists require such a hill or mountain before using the term *volcano*.

Most volcanoes have a *crater*, which is a roughly circular topographic depression that is usually located near the top of the volcano. Some volcanoes have no obvious crater or are nothing but a crater. Over time, volcanoes and craters are degraded and disguised by erosion.

Volcanoes consist mostly of *volcanic rocks,* which formed from lava, pumice, volcanic ash, and other products of volcanism.

Not all magma comes from volcanoes having the classic shape of a cone. Magma erupts from fairly linear cracks called *fissures* and from huge circular depressions called *calderas.*

06.01.a2

Many volcanoes display evidence of having been active during the last several hundred to several million years or during the last several days. Such evidence can include a layer of volcanic ash on hillslopes (left side of volcano shown above) or lava and ash flows that are relatively unweathered and lack a well-developed soil.

B Is Every Hill Composed of Volcanic Rocks a Volcano?

If a landscape lacks most of the diagnostic features described above, it is probably not a volcano. Many mountains and hills are not volcanoes, and some volcanoes are not mountains or hills.

▽ The flat-topped hill below, called a mesa, has a cap of volcanic rocks, but it is not a volcano. It did not form over the volcanic vent. Instead, it is simply an eroded remnant of a once more extensive lava flow, as illustrated to the right. [Hopi Buttes, Arizona]

06.01.b1

1. Lava erupts from a central volcanic vent or from a linear vent called a *fissure*. Once erupted, the lava cools.

2. Erosion begins to remove the edges of the lava flow and works inward toward a central remnant.

3. The past location of the fissure is marked by a dike that cuts through the rocks and across the landscape.

4. The lava flow is more resistant to erosion than underlying rocks and so forms a steep-sided, flat-topped mesa. It is a hill, but it is not over the vent and is not a volcano.

06.01.b2–4

C What Are Some Different Types of Volcanoes?

Volcanoes have different sizes and shapes and contain different types of rocks. These variations reflect differences in the composition of the magmas and the style of the eruptions. There are four common types of volcanoes that construct hills and mountains: scoria cones, volcanic domes, composite volcanoes, and shield volcanoes. Later in this chapter, we describe other types of volcanoes that are not hills or mountains.

Scoria Cone

△ Scoria cone in northern Arizona

Scoria cones are cone-shaped hills several hundred meters high, or higher, usually with a small crater at their summit. They also are called *cinder cones* because they contain loose black or red, pebble-sized volcanic *cinders* (scoria), with larger volcanic *bombs.* They are basaltic or, less commonly, andesitic in composition. Some are next to or on the flanks of composite and shield volcanoes.

Shield Volcano

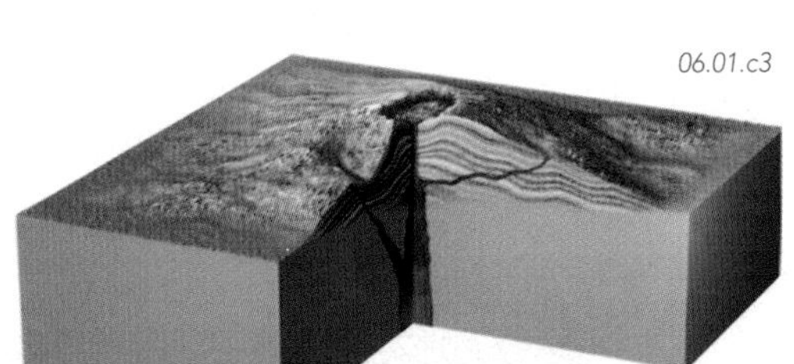

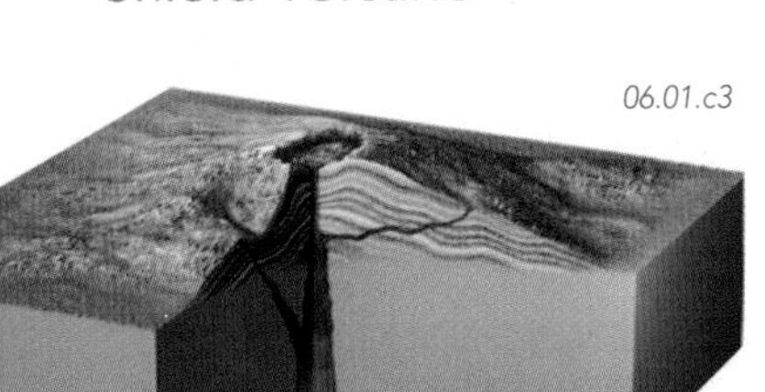

△ Shield volcano in central Iceland

Shield volcanoes have broad, gently curved slopes and can be less than a kilometer across or form huge mountains tens of kilometers wide and thousands of meters high. They commonly contain a crater or line of craters and have fissures along their summit. Shield volcanoes consist mostly of basaltic lava flows with smaller amounts of scoria and volcanic ash.

Composite Volcano

△ Composite volcano in Ecuador, South America

Composite volcanoes commonly are fairly symmetrical mountains thousands of meters high and with moderately steep slopes. Their name derives from the interlayering of lava flows, pyroclastic deposits, and volcanic mudflows. They consist mostly of intermediate-composition rocks, such as andesite, but can also contain felsic and mafic rocks.

Volcanic Dome

△ Dome in the crater of Mount St. Helens, Washington

Volcanic domes are dome-shaped volcanoes that are hundreds of meters high. They consist of solidified lava, which can be highly fractured or intact and dense. Domes include some volcanic ash intermixed with rock fragments. They form where felsic or intermediate magma erupts and is so viscous that it piles up around a vent. Many domes are within craters of composite volcanoes.

The Relative Sizes of Different Types of Volcanoes

Volcanoes vary from small hills less than a hundred meters high to broad mountains tens of kilometers across. There is also much variation of size within each type of volcano (e.g., shield volcanoes), but we can make a few generalizations about the relative sizes of the different volcano types.

The figure below illustrates that some types of volcanoes are larger than others, but the relative sizes of the volcanoes are not to scale in this figure because of the huge difference between the smallest scoria cones and the largest shield volcanoes.

Scoria cones and domes, which typically form during a single eruptive episode, are the smallest volcanoes. Shield volcanoes and composite volcanoes are much larger because they were constructed, layer by layer, by multiple eruptions. Shield volcanoes have gentler slopes than scoria cones, domes, or composite volcanoes.

Scoria Cone
Small Dome
Shield
Composite Volcano
06.01.mtb1
Large Shield

Before You Leave This Page Be Able To

- ✓ Summarize or sketch the diagnostic characteristics of a volcano.
- ✓ Describe or sketch why every hill composed of volcanic rocks is not a volcano.
- ✓ Sketch and describe the four main types of volcanoes that construct hills and mountains.
- ✓ Sketch or describe the relative sizes of different types of volcanoes.

What Controls the Style of Eruption?

THE DIFFERENT TYPES OF VOLCANOES reflect differences in the style of eruption. Some eruptions are explosive, whereas others are comparatively calm. What causes these differences? The answer involves magma chemistry and gas content, which both control how magma behaves near the surface.

A What Are Ways That Magma Erupts?

Magma has various behaviors once it reaches Earth's surface. During explosive *pyroclastic eruptions*, bits of lava, volcanic ash, and other particles are thrown into the atmosphere. During nonexplosive eruptions, lava issues from a vent onto the surface. Both types of eruptions can occur from the same volcano.

Lava Flows and Domes

A *lava flow* forms when magma erupts onto the surface and flows away from a vent. Erupted lava can be fairly fluid, flowing downhill like a river of molten rock. Some lava flows are not so fluid and travel a limited distance before solidifying. [Kilauea, Hawaii]

06.02.a1

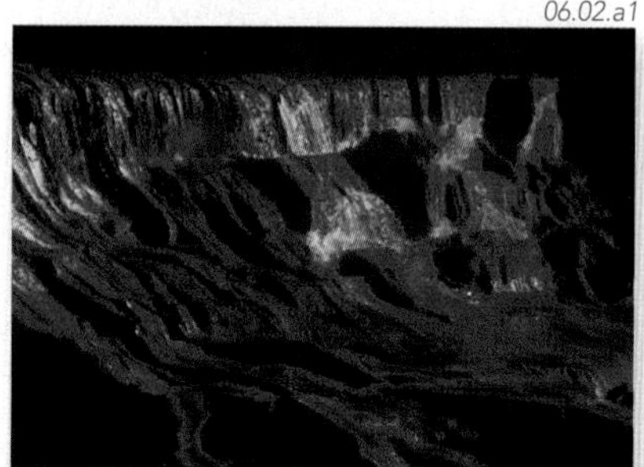

A lava dome is constructed from lava that does not flow easily (is *viscous*). The high viscosity of the lava generally is due to a high silica content and causes the lava to pile up around the vent instead of flowing away. Domes are accompanied by several types of explosive eruptions. [Mount St. Helens]

06.02.a3

Pyroclastic Eruptions

06.02.a2

Some explosive eruptions propel molten pieces of lava into the air. A *lava fountain*, such as shown here, can accompany basaltic volcanism and results from a high initial gas content in a less viscous lava. The gas propels the lava and separates it into discrete pieces. [Kilauea, Hawaii]

06.02.a4

Other explosive eruptions eject a mixture of volcanic ash, pumice, and rock fragments into the air. Such airborne material is called *tephra*, and tephra particles that are sand sized or smaller are *volcanic ash*. Tephra is derived from fragmented volcanic glass, pumice, and shattered rocks. [Augustine volcano, Alaska]

Two Different Eruptions of Tephra from the Same Volcano

The Augustine volcano in Alaska erupts tephra in two eruptive styles—an eruption column and a pyroclastic flow.

06.02.a5

△ *Eruption Column*—Tephra, which forms when magma is ripped apart by volcanic gases, erupts high into the atmosphere, forming an *eruption column*. The tephra in the column falls back to Earth as solidified and cooled pieces of rock. Finer particles of ash drift kilometers away from the volcano and slowly settle down to the ground.

▽ *Pyroclastic Flow*—Some ash does not jet straight up but collapses down the side of the volcano as a dense, hot cloud of ash particles and gas. This eruption style is called a *pyroclastic flow* or simply an *ash flow*. A pyroclastic flow is an extremely devastating eruption because of its high speed (more than 100 km/hr) and high temperature (exceeding 500°C).

06.02.a6

The two eruptions differed primarily because of the gas content of the magma.

The *eruption column* formed when large volumes of volcanic gas came out of the magma and made the cloud of tephra and gas buoyant in the atmosphere.

The *pyroclastic flow* formed when the gas flux was lower and could not support the eruption column, which then rapidly collapsed downhill under the force of gravity.

06.02.a7

B How Do Gases Affect Magma?

06.02.b1

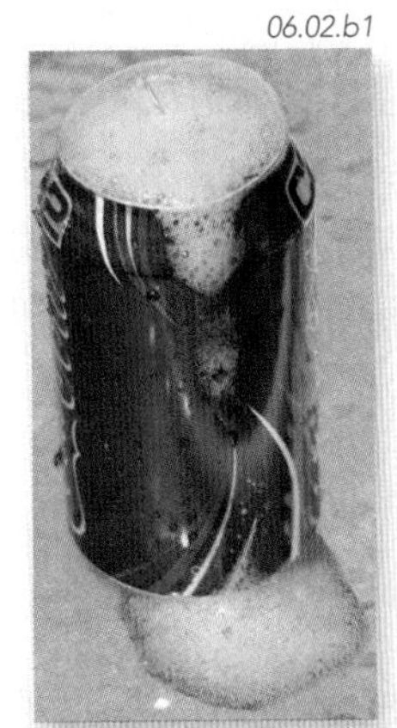

1. To envision gas in magma, think of what happens when you open a can of soda. The liquid may have no bubbles until it is opened, at which time bubbles appear in the liquid, rise to the top, and perhaps cause the soda to spill out. The dissolved gas was always there in the liquid, but only became visible bubbles when you opened the top and released the pressure that held the gas in solution.

2. Magma, like the soda, contains some dissolved gases, such as H_2O (water vapor), CO_2 (carbon dioxide) and SO_2 (sulfur dioxide). These gases have a critical effect on eruption style and help the magma rise toward the surface.

3. At depth, confining pressure keeps most of the gases in solution and keeps bubbles from forming.

06.02.b2

4. As the magma approaches the surface, confining pressure diminishes and the gases cannot remain in solution. Bubbles of gas form in the magma. If enough bubbles form fast enough, the bubbles cause the magma to be more buoyant and help it rise toward the surface and erupt out of the volcano.

C How Does Viscosity Affect Gases in Magma?

Viscosity, the resistance to flow, dictates how fast a magma can flow and how fast crystals and gas can move through the magma. When gas in a magma comes out of solution, movement of the resulting bubbles is resisted by the magma's viscosity. If the bubbles become trapped, the magma is potentially more explosive.

More Viscous

06.02.c1

▷ More viscous magma, with higher amounts of silica chains, prevents gas from escaping easily. Gas builds up in the magma and greatly increases the pressure on the surrounding rock. This can cause explosive eruptions. [Mount St. Helens]

Less Viscous

06.02.c2

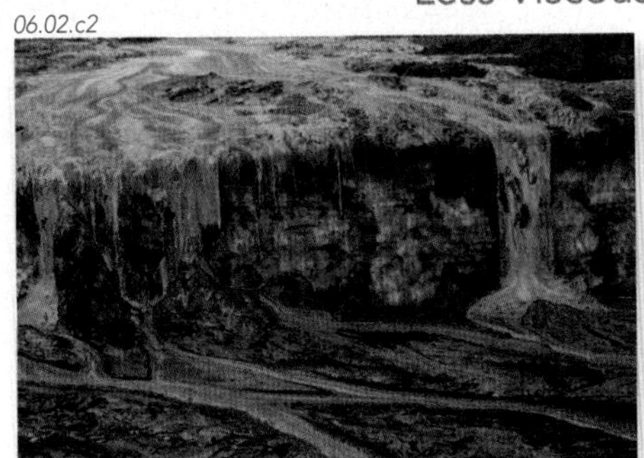

◁ Less viscous magma, such as those with a basaltic composition, allows gas bubbles to escape relatively easily. This can lead to a fairly nonexplosive eruption, such as this basaltic lava flow. [Kilauea, Hawaii]

How Water and Gases Exist in Magma

Visualizing water and gases in magma is challenging. Red, glowing lava seems totally inhospitable to, and incompatible with, water.

Gases such as water (H_2O) can be dissolved in magma because of the chemical bonding of the gas molecules to silica chains. Water in magma is present as molecular water and in the form of the OH^- molecule. Carbon dioxide (CO_2) is in the form of the $CO_3^=$ molecule when dissolved in magma, but it reverts back to CO_2 gas as it comes out of solution. At higher pressures, the chemical bonding is very effective at keeping molecules of gas isolated from one another.

The amount of dissolved gas that a magma can hold is related to temperature and pressure. A magma can hold more dissolved gas at high temperatures than it can at low temperatures, but rocks cool relatively slowly, especially if they are in the subsurface where they are insulated by surrounding rocks.

Pressure is the most important control of gas content in magma because confining pressures can change rapidly as a magma rises through the crust. When a magma nears Earth's surface, a drop in pressure permits gas molecules to come out of solution and form gas bubbles in the magma. The expanding bubbles cause the volume of the magma to increase, driving the magma toward the surface and to an explosive eruption.

Before You Leave This Page Be Able To

- ✓ Describe four ways that magma erupts.
- ✓ Summarize the difference between an eruption column and pyroclastic flow, and the role that gas plays in eruptive style.
- ✓ Explain how gas is dissolved in magma and how it behaves at different depths in a magma.
- ✓ Describe how viscosity influences how explosive an eruption is.

What Features Characterize Scoria Cones and Basaltic Lava Flows?

ERUPTIONS OF BASALTIC MAGMA form various rock types and landforms. This variety is largely a result of how gas in the magma affects the style of eruption and the solidification of lava. A single eruption of basaltic magma can produce a wide range of volcanic features and rock textures.

A What Are Scoria Cones and Basalt Flows?

Basaltic magma has a relatively low viscosity compared to other magmas and erupts in characteristic ways.

Basaltic Eruptions—At the beginning of a basaltic eruption, gases explode bits of lava into the air forming a *lava fountain*. The airborne bits of lava cool and then fall around the vent as loose pieces called *scoria*. The lava fountain may be followed or accompanied by the eruption of a basaltic lava flow.

06.03.a1

Scoria Cones—Pieces of scoria from the lava fountain gradually build up a cone-shaped hill called a *scoria cone* (also known as a *cinder cone*). Ejected fragments can be as small as sand grains to as large as huge boulders. Scoria cones typically form in a short amount of time, from a few months to a few years, and generally are no more than 300 meters (980 feet) high.

Basaltic Lava Flows—Fluid basaltic lava pours from the vent and flows downhill. Sometimes, as shown here, the lava fills up and overtops the crater in the scoria cone, but other lava flows issue from the base of the scoria cone after most of the cone has been constructed.

Rock Types

06.03.a2

Vesicular basalt contains abundant gas pockets called *vesicles*. The vesicles were gas bubbles that expanded in the magma (as pressure decreased) and were trapped when the lava solidified. Vesicles occur in lava flows and in ejected material such as scoria.

06.03.a4

Basaltic magma may not contain enough gas to form bubbles, so the lava solidifies into *nonvesicular basalt*. A magma may have a low content of gas because it started out with a low content of dissolved gas or because it lost gas (degassed) somewhere along the way.

06.03.a6

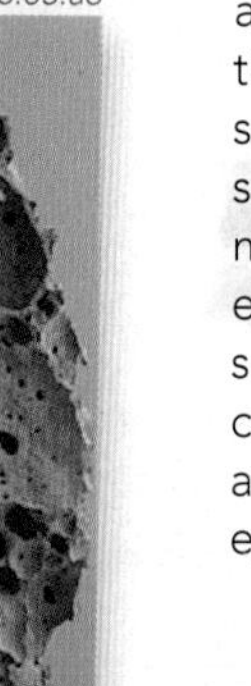

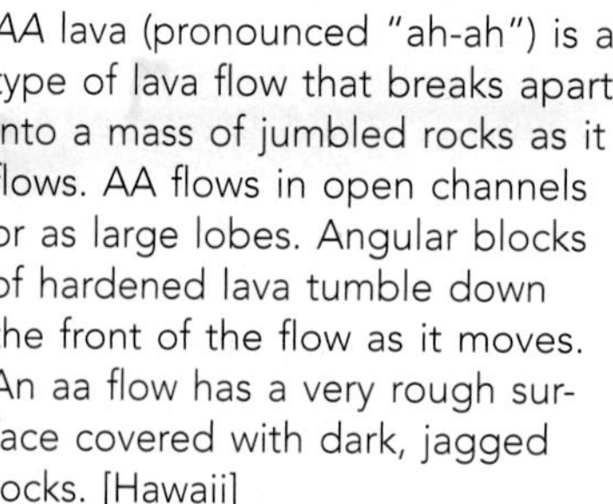

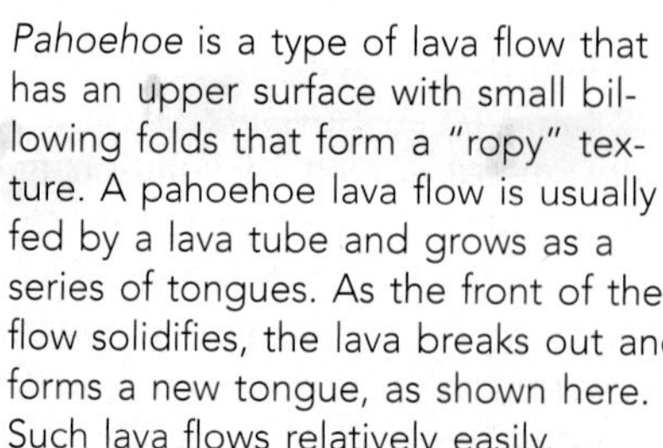

Scoria is a vesicular rock that has a basaltic or andesitic composition and is ejected from a vent, such as in a *lava fountain*. It starts as frothy blobs of lava and may be liquid or solid when ejected. Small blobs cool and solidify in the air to form solid cinders. Large blobs of magma and solid angular blocks are ejected as *volcanic bombs*.

Features of Lava Flows

06.03.a3

Lava tubes form when the surface of a lava flow solidifies to form an insulating roof over the hot, still-moving interior of the flow. Lava flows insulated by lava tubes can flow farther than lava flows on the surface. If the tube drains, it becomes a curving, tube-shaped cave. [Hawaii]

06.03.a5

AA lava (pronounced "ah-ah") is a type of lava flow that breaks apart into a mass of jumbled rocks as it flows. AA flows in open channels or as large lobes. Angular blocks of hardened lava tumble down the front of the flow as it moves. An aa flow has a very rough surface covered with dark, jagged rocks. [Hawaii]

06.03.a7

Pahoehoe is a type of lava flow that has an upper surface with small billowing folds that form a "ropy" texture. A pahoehoe lava flow is usually fed by a lava tube and grows as a series of tongues. As the front of the flow solidifies, the lava breaks out and forms a new tongue, as shown here. Such lava flows relatively easily.

B How Do Scoria Cones and Basalt Flows Form Around the Same Vent?

Early Formation of a Scoria Cone—If basaltic magma contains enough dissolved gas, the gas comes out of solution as the magma approaches the surface. The gas expands dramatically and propels clots of frothy lava out of the vent, forming a scoria cone. This generally occurs early in a basaltic eruption because the magma has not had time to degas in the magma chamber.

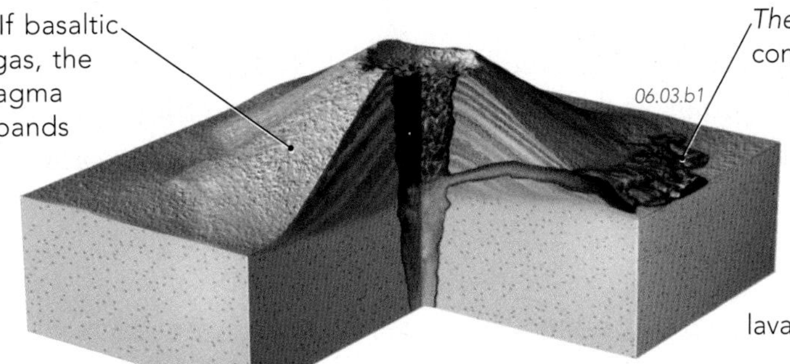

The Switch to Lava Flows—After most of the cone is built, magma that contains less gas reaches the surface and erupts nonexplosively as a lava flow. Taking the easiest way out of the vent, the magma squeezes out near the base of the scoria cone rather than rising to erupt from the summit crater. Some scoria cones are not accompanied by a lava flow, and vice versa.

C What Do Scoria Cones and Basalt Flows Look Like in the Landscape?

Scoria cones are loose piles of material that erode rapidly. Basalt flows, on the other hand, are considerably more resistant to weathering and erosion, and are more likely to survive in the geologic record. Scoria cones and lava flows typically occur in clusters called *volcanic fields*.

Scoria Cones

Most scoria cones begin with a conical shape and a central crater at the top of the cone. Young scoria cones lack well-developed soil or extensive vegetation, and commonly are associated with dark lava flows. [Galápagos Islands]

Over time, erosion wears away the summit of a scoria cone, making the cone into a rounded hill without a central crater. Erosion cuts into the slopes, and the slopes have more soil and plants. [Northern Arizona]

Lava Flows

Young lava flows have steep flow fronts commonly with discrete, protruding lobes and embayments. The top of the flow is typically rough and displays flow features such as aa and pahoehoe. [SP Flow, northern Arizona]

Older lava flows are more subdued because they are eroded and sediment has accumulated in low spots on the surface. Flow tops lose their small features, such as pahoehoe, and become covered with soil and plants. [Reunion, Indian Ocean]

Battling Lava Flows with Seawater in Iceland

In 1973, the volcano Eldfell ("Fire Mountain" in Icelandic) erupted next to a fishing village on the island of Heimaey in Iceland. Scoria from the basaltic eruption accumulated on roofs and caused houses to collapse and burn. Blocky lava flows (aa) issued from the base of the crater, buried buildings, and encroached on the harbor, threatening to destroy the fishing economy of the island. Local fishermen and others began pumping cold seawater on the advancing flow hoping to solidify it and save their harbor. By the end of the eruption, 1.5 billion gallons of seawater was pumped onto the flow. There is some debate about how effective the pumped water was in slowing down the flow, but the lava flow did stop before it totally closed the harbor. The town is somewhat back to normal and a tourist destination.

Before You Leave This Page Be Able To

- ✓ Explain the characteristics of scoria cones and basalt flows, including how you might distinguish between young and old examples of each.
- ✓ Describe how vesicular and nonvesicular basalt rocks differ in appearance and formation.
- ✓ Describe how basaltic magma may form a scoria cone, lava flow, or both.

How Do Shield Volcanoes Form?

SHIELD VOLCANOES ARE SOME OF THE LARGEST VOLCANOES ON EARTH. Many of the world's volcanic islands, including the Hawaiian and Galápagos Islands, are shield volcanoes. How do we recognize a shield volcano, how is one constructed, and where does all the magma come from?

A What Is a Shield Volcano?

Shield volcanoes have a broad, *shield-shaped* form and fairly gentle slopes when compared to other volcanoes. They are constructed by a succession of basaltic lava flows and lesser amounts of scoria and ash. Shield volcanoes can form in any tectonic setting that produces basaltic magma, but the largest ones, such as those in Hawaii, form on oceanic plates in association with hot spots.

1. This image shows satellite data superimposed on topography of the Big Island of Hawaii. The island consists of three large volcanoes and two smaller ones. Green areas are heavily vegetated, and recent lava flows are brown or dark gray.

2. Mauna Loa, the central mountain, is the world's largest volcano. It rises 9,000 meters (29,520 feet) above the seafloor and is 4,170 meters (13,680 feet) above sea level. From seafloor to peak, Mauna Loa is Earth's largest mountain.

3. Kilauea volcano, probably the most active volcano in the world, is on the southeast side of the island. Recent lava flows (shown in dark gray) flowed eastward, destroying roads and housing subdivisions.

06.04.a1

06.04.a2

4. A roughly circular depression, called a *caldera*, is at Kilauea's summit. A caldera forms when magma is removed from an underground chamber, which causes the land above to collapse downward.

7. Mauna Kea is an inactive shield volcano and the site of astronomical observatories.

5. Eruptions from Kilauea issue from linear fissures that are interpreted to form over the top of vertical, sheetlike magma chambers (dikes). The dikes are fed from below by larger and deeper magma chambers.

06.04.a3

6. The spine of Mauna Loa is also a fissure from which mafic (basaltic) lava flows erupt. The fissure is the surface expression of one or more dikes.

◁ 8. At the start of the 1983 Kilauea eruption, a new volcanic vent formed, and early, gas-rich magma shot into the air in a lava fountain. The cooled and hardened pieces fell back around the vent to form a scoria cone. Such scoria cones are common companions to shield volcanoes.

9. The basaltic lava flows have low viscosity and flow fluidly downhill, in some cases all the way to the ocean. When the molten rock reaches the ocean, it causes seawater to boil in rising clouds of steam. These flows add new land to the island. ▷

06.04.a4

B How Do Shield Volcanoes Erupt?

Shield volcanoes erupt mostly low-viscosity basaltic lava and so are dominated by relatively nonexplosive outpourings of lava from fissures and vents. Early phases of eruptions are commonly marked by spectacular fountains in which molten rock is ejected hundreds of meters into the air from fissures or central vents.

06.04.b1

Fissure Eruption—A fissure eruption occurs when magma rises through a dike and erupts onto the surface from a long fissure. Large volumes of lava can flow out of the fissure, and smaller amounts are thrown into the air as a fiery curtain. [Mauna Loa, Hawaii]

06.04.b2

Lava Flow—Fluid basaltic lava typically flows downhill as a river of molten rock. Flows can divide, rejoin, spread out, or constrict as they encounter variations in topography. They can even have waves and rapids, like those in a river of water. [Kileuea, Hawaii]

06.04.b3

Pillow Basalt—When fluid lava, such as basalt, erupts into water, the lava grows forward as small, individual tongues that form rounded shapes called *pillows*. Pillows are reliable evidence that lava erupted into water, and most commonly formed in the sea. [San Juan Islands]

C How Would You Recognize a Shield Volcano in the Landscape?

Although they vary greatly in size, shield volcanoes are recognized by their fairly consistent shape. They also are constructed of dark basalt flows with some red or black scoria.

06.04.c1

This photograph of Mauna Loa shows the gentle shape of the shield volcano. The height of the mountain is hard to appreciate because the mountain's flanks are so gentle.

06.04.c2

This small shield volcano is less than a kilometer across, has gentle flanks, and barely rises above the surrounding basaltic plain. [central Arizona]

Mauna Loa and Kilauea—Two Hawaiian Volcanoes

Two volcanoes on the Big Island are extremely active, building and reshaping the island before our eyes. Mauna Loa, which in Hawaiian means "long mountain," is the larger of the two and has erupted 33 times since 1843. The most recent eruption was in 1984, and the U.S. Geological Survey reports that the volcano is "certain to erupt again." USGS geologists closely monitor the volcano to anticipate, or perhaps predict, the next eruption.

Kilauea volcano to the east is regarded as the home of *Pele*, the Hawaiian volcano goddess. Kilauea has been even more active than Mauna Loa, erupting nearly continuously since the 1800s, with 61 historic eruptions. It erupted 34 times since 1952 and has been erupting nearly nonstop since 1983. The 1983 eruption began with the construction of a new scoria cone during an initial lava-fountaining event. Since that time, lava from the volcano has flowed down toward the sea. When the red hot lava flows enter the ocean, they cool and solidify, and new land is added to the island. All of the Hawaiian Islands formed by this process—the eruption of basaltic lava flows that make new land where there once was sea.

Before You Leave This Page Be Able To

- ✓ Describe the type of magma and other general characteristics of a shield volcano.
- ✓ Explain how shield volcanoes erupt.
- ✓ Sketch or summarize how you would recognize a shield volcano in the landscape.
- ✓ Summarize how active the two main volcanoes on Hawaii, Mauna Loa and Kilauea, have been.

What Causes Flood Basalts?

FLOOD BASALTS INVOLVE HUGE VOLUMES OF MAGMA and represent the largest igneous eruptions on Earth. Flood basalts cover tens of thousands of square kilometers in Siberia, South Africa, Brazil, India, and the Pacific Northwest. They also form large oceanic plateaus at or below sea level. How do such huge eruptions occur, and do they affect climate and life on Earth?

A What Are Flood Basalts?

Flood basalts are sequences of basaltic lava flows that cover vast areas and commonly have thicknesses of several kilometers. Individual lava flows can cover thousands of square kilometers and contain more than 1,000 cubic kilometers of magma—equal to emptying a cube-shaped magma chamber that is 10 kilometers (over 6 miles) on a side. Flood basalts are fed by long fissures that are the upward continuation of subsurface dikes.

The Columbia Plateau of the Pacific Northwest consists of a thick sequence of basalt flows, which cover parts of Washington, Oregon, and western Idaho. Several individual lava flows within the basalt cover a surprisingly large area, including one that covers most of the area shown in pink.

06.05.a1

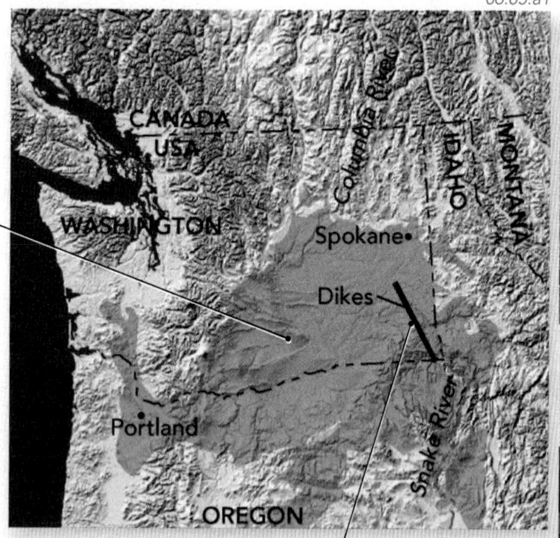

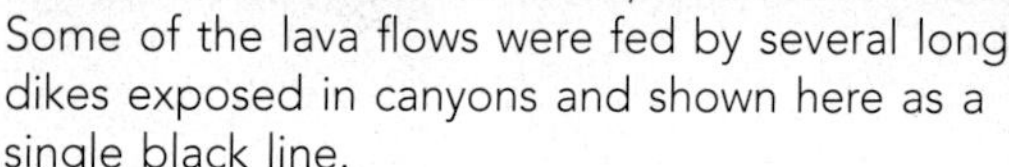

Some of the lava flows were fed by several long dikes exposed in canyons and shown here as a single black line.

06.05.a2

◁ The canyons of the Columbia Plateau expose a sequence of basalt flows, each of which forms a ledge. Each basalt flow represents a single eruption, and most of the flows formed about 15 million years ago.

06.05.a3

◁ This fissure eruption is similar to those that formed the basalts of the Columbia Plateau, but the volume of lava that poured from the fissures on the Columbia Plateau was much greater.

B How Do Flood Basalts Erupt onto the Surface?

Instead of erupting from a single, central volcanic vent, flood basalts erupt from one or more long, nearly continuous fissures or from a linear string of vents. This permits large volumes of magma to erupt onto the surface in a geologically short period of time. Examine the figure below, starting on the lower left.

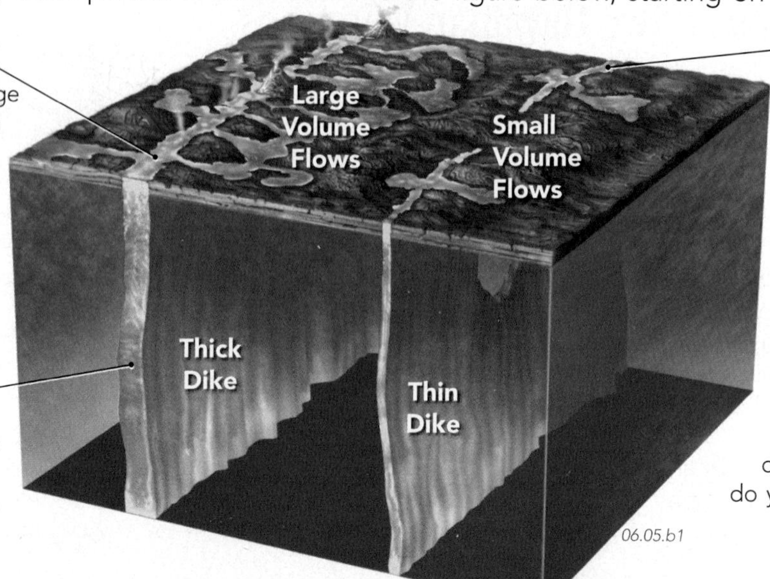

1. Mantle-derived basaltic magma rises through the crust along vertical dikes. Pressure from the magma pushes outward against the wall rocks of the dike, holding them apart and allowing magma to pass.

2. The combination of a thick dike and low-viscosity magma allows large volumes of magma to pass through the dike and erupt along a linear fissure. A high flow of magma results in individual lava flows that remain hot and molten for a longer time, flow long distances, and cover large areas.

3. If a dike is thin, less magma can flow through the conduit and resulting lava flows have smaller volumes and travel shorter distances away from the vent.

4. One basalt flow on the Columbia Plateau represents more than 1,000 km^3 of magma that probably erupted very quickly, perhaps in only several decades. Through what type of dike do you think it was erupted?

C Where Does the Magma for Flood Basalts Originate?

Most flood basalts probably form at hot spots as rising mantle plumes encounter the lithosphere. It is uncertain whether the magmas come from melted lithosphere, melted asthenosphere, or directly from the plume.

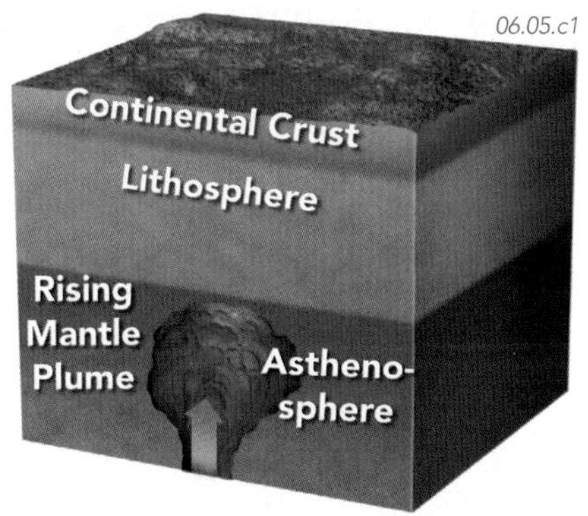

A mantle plume rising through the mantle is probably mostly solid and acquires an inverted teardrop shape as it flows.

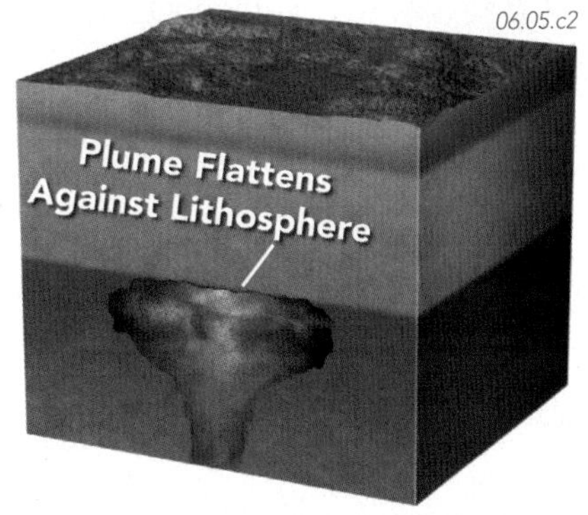

When the rising plume encounters the base of the lithosphere, it meets increased resistance and spreads laterally.

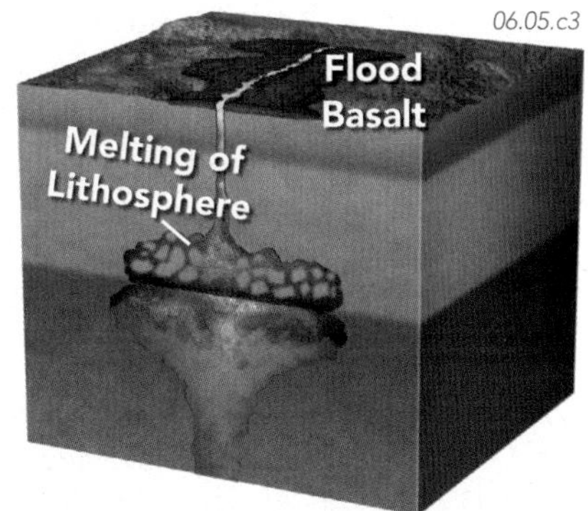

The rising plume can melt because of decompression, or the plume can melt the adjacent asthenosphere or lithosphere.

Some Important Flood Basalts of the World

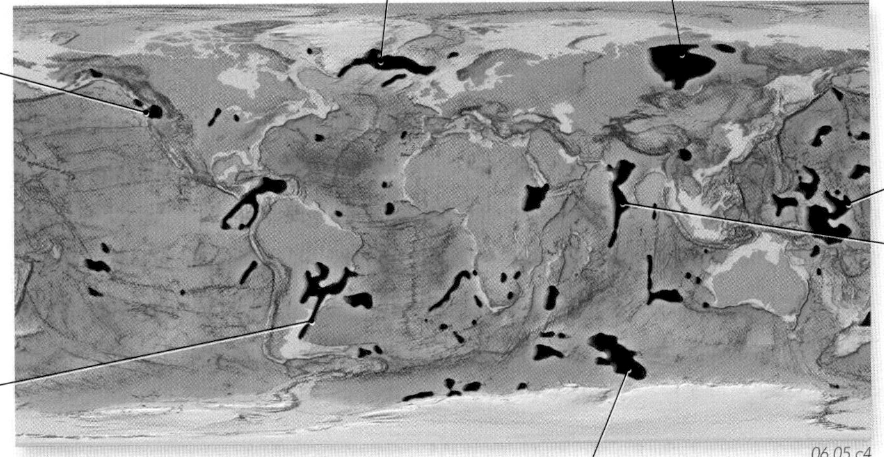

Flood basalts on the Columbia Plateau erupted 15 Ma and were probably related to the hot spot now beneath Yellowstone.

Thick sequences of basalt along the margins of Greenland and Scotland formed above the hot spot that is presently below Iceland.

Flood basalts in Siberia erupted at about 250 Ma, the same time as a major extinction event that affected life in the seas and on land. Were these events related?

Vast oceanic plateaus of basalt rise above the ocean floor in the western Pacific Ocean but are mostly below sea level. They are interpreted to have formed over oceanic hot spots.

Large volumes of basalts erupted about 133 Ma along the southeastern flank of South America. The magma probably formed when the continent was over a hot spot in the South Atlantic. Some flood basalts are on land and others are offshore.

Flood basalts cover large areas of India, where they are called the *Deccan Traps*. The basalts erupted 65 Ma, at the same time as the dinosaurs became extinct.

Basaltic eruptions related to a hot spot constructed the huge *Kerguelen Plateau* near Antarctica.

Flood Basalts, Fissure Eruptions, Climate, and Life

Large volumes of flood basalt erupted during a short time period may change Earth's climate and negatively impact life. Such eruptions release large amounts of sulfur dioxide (SO_2) gas that can reflect sunlight and cause atmospheric cooling. Alternatively, the eruption of flood basalts may release carbon dioxide (CO_2) gas, which can act as a greenhouse gas and cause global warming.

Some of Earth's most voluminous eruptions of flood basalts coincide with major extinction episodes of marine and land animals. The Deccan Traps of India coincide with the extinction of the dinosaurs at the Mesozoic-Cenozoic boundary (65 Ma). Flood basalts in Siberia erupted at 250 Ma, a time when more than 90 percent of marine species became extinct during one of Earth's most devastating extinction events. This massive extinction defines the boundary between the Paleozoic and Mesozoic Eras.

Geologists are currently investigating whether the flood basalts had a role in the two major extinctions or whether the extinctions have other causes, such as meteorite impacts.

Before You Leave This Page Be Able To

- ✓ Describe the characteristics of flood basalts and how they erupt.
- ✓ Identify at least three areas on Earth that contain flood basalts.
- ✓ Sketch and describe the interpreted relationship between flood basalts and mantle plumes.
- ✓ Summarize how flood basalts could affect climate and life on Earth.

What Are the Hazards of Basaltic Eruptions?

HAZARDS ASSOCIATED WITH BASALTIC ERUPTIONS usually affect only nearby areas. Basaltic scoria destroys only a limited vicinity, but basaltic flows are relatively fluid and can destroy buildings and crops tens of kilometers away from the volcanic vent. Surprisingly, they can also cause huge floods.

A What Is Meant by a Hazard and a Risk?

The terms *hazard* and *risk* may seem more appropriate for a lesson about insurance, but geologists frequently apply these terms to the effects geologic events can have on humans and society. What is the difference between a hazard and a risk?

A *hazard* is the existence of a potentially dangerous situation or event, such as a landslide off a steep slope or a lava flow erupting from a volcano. The *hazard* in this photograph was a basaltic lava flow. [Kilauea, Hawaii]

06.06.a1

Risk is an assessment of whether the hazard might have some *societal impact*, such as loss of life, damage to property, loss of employment, destruction of fields and forests, or implications for local or global climates. Remnants of destroyed houses, cars, and roads demonstrate that this area had a high *risk* for volcanic *hazards*.

06.06.a2

The risk was extreme for people living on the flanks of an active volcano in Zaire, Africa. In 1977, a fast-moving (50 km/hr) lava flow killed as many as 300 people living in villages near the volcano. If no people were living near this volcano, a hazard would still exist but there would be essentially no risk.

B What Are the Hazards Associated with Scoria Cones?

Scoria cones can be deadly and destructive, especially to nearby areas, as they hurl lava and solid rock and spew out dangerous gases. Fine ash ejected high into the air can cause widespread damage.

Falling Objects

Hazards that exist close to an erupting scoria cone include being impacted and burned by cinders and being struck by blobs of magma and other projectiles ejected from the cone. Most scoria falls back to Earth near the vent and piles up on the scoria cone. [Hawaii]

06.06.b1

Larger projectiles, called *volcanic bombs*, pose a severe hazard close to the erupting cone.

06.06.b2

Gases

Gases associated with basaltic eruptions can be a significant hazard. Gases such as carbon dioxide (CO_2) can asphyxiate if concentrated. Other gases, such as hydrogen sulfide (H_2S), cause death by paralysis. Gaseous sulfur dioxide (SO_2), hydrochloric acid (HCl), sulfuric acid (H_2SO_4), and fluorine compounds expelled during eruptions can destroy crops, kill livestock, and poison drinking water for people and animals. [Krafla, Iceland]

06.06.b3

Volcanic Ash

Sand-sized cinders and finer particles of ash can be carried by wind, bury nearby structures, and cause breathing problems. This photograph of the Paricutin volcano in Mexico shows scoria and ash settling out of the column erupted from the volcano. Eruption columns from scoria cones typically reach heights of several kilometers and can impact areas downwind of the volcano.

06.06.b4

C What Are the Hazards Associated with Lava Flows?

Lava flows move slowly enough that people can usually get out of their way, but lava flows can completely destroy any structures in their path. Destruction occurs by fire or by being buried by lava.

When Lava Comes to Town

06.06.c1

A lava flow will cause wooden structures and vegetation to catch fire. This house in Hawaii burst into flames when touched by a basaltic lava flow from Kilauea. Any structure in the path of a moving lava flow will likely be engulfed by fire and then crushed or bulldozed by the weight of the lava.

Lava flows, such as these encroaching into a subdivision in Hawaii in 1983, change the landscape and produce an uninhabitable environment. Communities rarely can be rebuilt after such an event.

06.06.c2

D How Do Basaltic Eruptions Cause Floods of Water, Ice, and Debris?

A special type of hazard is associated with erupting volcanoes beneath ice sheets. The heat from such eruptions can melt large quantities of ice and produce huge floods.

06.06.d1

Because of its location just south of the Arctic Circle and its position on top of a mid-ocean ridge and a hot spot, Iceland is a land of ice and fire. Iceland has many glaciers (the Icelandic term for glacier is *jökull*), including a large ice sheet that covers nearly 25 percent of the country. Beneath the ice sheet are a half dozen basaltic volcanoes, similar to this one, which is only partially buried.

In 1996, a volcano beneath the ice sheet erupted, melted the surrounding ice, and released a catastrophic flood of meltwater (jökulhlaup in Icelandic) carrying blocks of ice, rock, and other debris. The flood carried an enormous volume of water and sediment, destroying everything in its path. The flood deposited so much sediment that it formed new land and moved the coastline out several kilometers.

06.06.d2

Laki and the Summer That Wasn't

In 1783, an Icelandic fissure at a place called Laki unleashed Earth's largest known historical eruption (16 cubic kilometers of magma). The eruption caused the climate to cool in most of Europe because it released a large amount of ash and sulfur dioxide (SO_2) gas. Sulfur dioxide gas combines with water in the atmosphere to form sulfuric acid (H_2SO_4) in very small drops called *aerosols*. These drops and the volcanic ash drifted over northern Europe for eight months and were thick enough to dim the sunlight. The summer of the eruption was dismal, and the winter was unusually cold. The following summer was defined by crop failure and famine, which continued for the next three years during which the climate remained cooler than normal. The cooling effects of sulfur dioxide in the upper atmosphere are interpreted to last only a few years after a big eruption.

06.06.mtb1

Before You Leave This Page Be Able To

- ✓ Explain how risk is different than hazard, and provide an example of each.
- ✓ Summarize the difference between hazards associated with scoria cones and hazards associated with basaltic flows.
- ✓ Explain how a volcanic eruption can cause a flood.
- ✓ Summarize the effects of the Laki eruption of 1783.

What Are Composite Volcanoes?

COMPOSITE VOLCANOES FORM STEEP, CONICAL MOUNTAINS that look like a volcano. They are common above subduction zones, such as along the Pacific Ring of Fire. They contain diverse volcanic rock types that reflect different compositions of magma and several styles of eruption.

A What Are Some Characteristics of a Composite Volcano?

Composite volcanoes are so named because they are a *composite* of interlayered lava flows, pyroclastic flows, tephra falls, and volcano-related mudflows and other debris. They are long lived, which explains their large size and complex internal structure.

06.07.a1

◁ 1. *Eruption Column*—Composite volcanoes produce a distinctive column of tephra and gas that rises upward many tens of kilometers into the atmosphere. Coarser pieces of tephra settle around the volcano, but the finer particles (volcanic ash) can drift long distances in the prevailing winds. [Mount St. Helens, Washington]

▷ 2. *Pyroclastic Flows*—These are the most violent eruptions from the volcano. They form when the eruption column collapses downward as a dense, swirling cloud of hot gases, volcanic ash, and angular rocks. Pyroclastic flows are one of the primary mechanisms by which these volcanoes are constructed. [Mount Mayon, Philippines]

06.07.a3

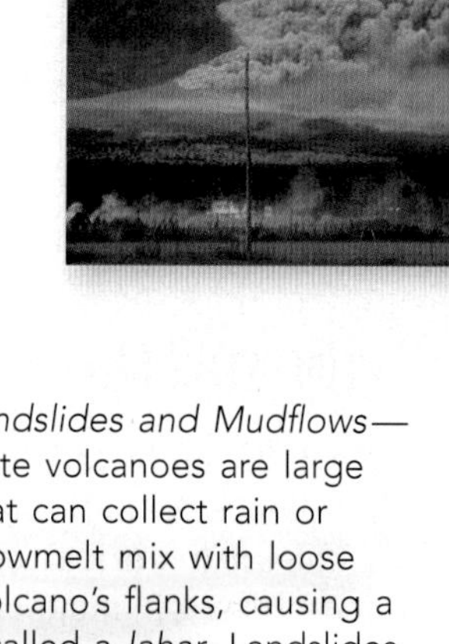

6. *Shape*—Composite volcanoes display the classic volcano shape because most material erupts out of a central vent. They have steep slopes because the volcanoes are built from small eruptions of viscous lava flows and domes that pile up on the flanks of the volcano. The shape represents one snapshot in a series of stacked volcanic mountains that have been built over time.

06.07.a2

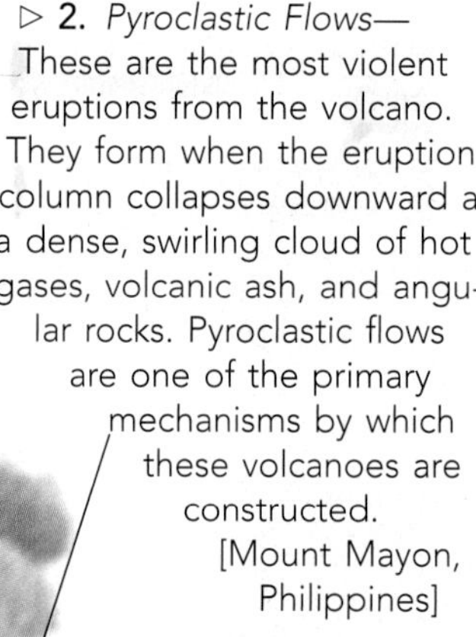

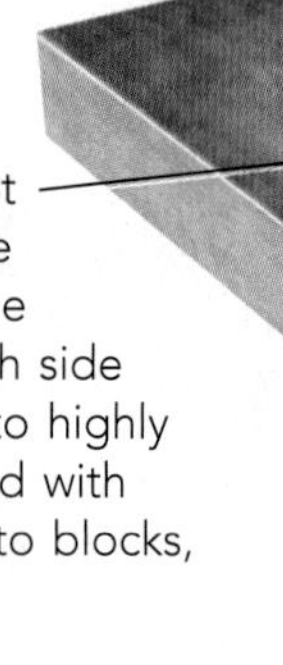

5. *Lava Flows and Domes*—Lava flows and domes can erupt from most levels of a composite volcano. Lava can erupt from the summit crater or escape through side vents. Most lava is moderately to highly viscous and so moves slowly and with difficulty. The lava may break into blocks, as did the dark lava below. [Augustine, Alaska]

3. *Landslides and Mudflows*—Composite volcanoes are large mountains that can collect rain or snow. Rain and snowmelt mix with loose ash and rocks on the volcano's flanks, causing a volcano-related mudflow called a *lahar*. Landslides and debris flows (▽) are also a hazard because of the steep slopes, loose rocks, and abundant clay minerals produced when hot water interacts with the volcanic rocks. [Augustine, Alaska]

4. *Rocks*—The volcano consists of alternating layers of pyroclastic flows, lava flows, and landslide and mudflow deposits. The modern mountain, formed during eruptions from the long-lived vent, is built on and around earlier versions of the volcano. The present peak hides a complex interior that was constructed by eruptions on earlier versions of the mountain.

06.07.a4

06.07.a5

B What Types of Rocks and Deposits Form on Composite Volcanoes?

Composite volcanoes consist mostly of intermediate lava and ash, but they can also include felsic and mafic materials. The combination of diverse magma compositions and different eruptive styles produce a variety of rock types and volcanic features. Both modern and ancient examples reflect these complexities.

06.07.b1

◁ This exposure contains a dark lava flow above a sequence with different kinds of volcanic rock fragments that range in composition from basalt to andesite to rhyolite. [Mount Shasta, California]

06.07.b2

◁ The most common rock is andesite, a gray, greenish-gray, or purplish volcanic rock that may contain dark- or light-colored phenocrysts.

06.07.b5

▷ Andesite from ancient composite volcanoes commonly has a greenish-gray or purplish-gray color, forms outcrops that are somewhat layered, and may include angular fragments from volcanic breccia and mudflows. [southwestern Colorado]

06.07.b3

◁ An eruption column deposits *tephra*, which contains fragments of pumice, crystals, and rock in a matrix of volcanic ash. Consolidated tephra is called *tuff*, and has distinct layers that record variations in the eruption over time. [northern Arizona]

06.07.b4

◁ A pyroclastic flow forms *ash-flow tuff*, which contains similar materials to tephra falls, but is less sorted. Ash-flow tuff is also harder and denser because the ash is hotter when deposited and is compressed by overlying ash. [Mount Redoubt, Alaska]

06.07.b6

◁ Volcano-derived mudflows (lahars) and landslide deposits consist of angular fragments (clasts) of rock usually in a matrix of finer materials, such as mud, derived from ash that has been weathered or hydrothermally altered. [Mount Redoubt, Alaska]

Some Famous Composite Volcanoes of the World

Composite volcanoes are not distributed uniformly on Earth. Most are above subduction zones at ocean-ocean or ocean-continent convergent boundaries. Many composite volcanoes have names and appearances that you recognize from newscasts, nature shows, or history and geography courses. Here, we describe a few of the more famous or interesting ones.

The beautiful and symmetrical *Mount Fuji* (▽) is the landmark composite volcano of Japan. It last erupted in 1708 and is above a subduction zone where the Pacific plate subducts beneath Japan.

06.07.mtb1

Mount Etna, on the island of Sicily, is often shown on newscasts as erupting rivers of lava or ejecting glowing volcanic bombs from the crater (▷). Eruptions are visible in this photograph from the *International Space Station* (▽).

06.07.mtb2

Mount Kilimanjaro, the highest mountain in Africa (▽), is along the East African Rift of Tanzania and Kenya.

06.07.mtb3

06.07.mtb4

Before You Leave This Page Be Able To

- ✓ Describe or sketch the characteristics of a composite volcano, including its internal structure.
- ✓ Describe the processes on composite volcanoes and the rocks they form.
- ✓ Describe the tectonic setting of most composite volcanoes.
- ✓ Identify some examples of composite volcanoes from around the world.

What Disasters Were Caused by Composite Volcanoes?

COMPOSITE VOLCANOES ARE DANGEROUS because they can be very explosive and unleash pyroclastic flows, toxic gases, and other deadly materials. They are responsible for numerous human disasters, including the destruction of Pompeii in Italy, St. Pierre in Martinique, and the area around Mount St. Helens.

A How Did Vesuvius Destroy Pompeii?

Vesuvius is an active composite volcano near the city of Naples in southwestern Italy. In A.D. 79, a series of pyroclastic flows erupted down the flank of the volcano, destroyed the coastal towns of Pompeii and Herculaneum, and killed the cities' inhabitants, estimated at 20,000 people and 5,000 people, respectively.

1. This image is an artist's conception of an explosive volcanic eruption striking the city of Naples and its surroundings, where three million people currently live.

2. Archeologic and geologic evidence from Pompeii indicate that the catastrophe began with earthquakes and the formation of an eruption column that deposited a layer of loose tephra over Pompeii.

3. The tephra fall was immediately followed by six pyroclastic flows that raced down the mountainside. Three of these flows hit Pompeii. The first probably burned most of the remaining survivors, and the last was strong enough to complete the destruction of standing buildings. People were smothered or suffocated, died from thermal shock, or were crushed by collapsing buildings. The bodies of victims in the ash decomposed, leaving mostly hollow molds, which archeologists filled with plaster to reveal the victims' last moments. ▷

4. The dashed red line marks the outward limit of pyroclastic flows, but tephra from the eruption column covered a wider area. Note how much of the present city of Naples is within the devastation of 79 A.D.

B What Happened at St. Pierre, Martinique?

Mount Pelée, a composite volcano on the Caribbean island of Martinique, is part of an island arc over a subduction zone. On May 8, 1902, the volcano erupted and sent a pyroclastic flow into the town of St. Pierre.

1. This view shows the island of Martinique, which consists of several distinct volcanoes, including Mt. Pelée, the northernmost peak.

2. The coastal town of St. Pierre is at the foot of the volcano.

3. Before the main eruption, the volcano emitted obvious warning signs, including noisy explosions, earthquakes, sulfurous gases, and small eruptions that dusted nearby areas with ash. People from the surrounding countryside sought shelter in the town of St. Pierre where they witnessed minor eruptions of ash, the formation of a lava dome in the crater, and a few small pyroclastic flows.

4. During the main eruption, a massive pyroclastic flow, estimated to have traveled at 500 km/hr, entered the town. Every building was mostly or completely destroyed (◁). Almost all of the 30,000 residents died within minutes. Most deaths were probably caused by asphyxiation as people breathed hot gas and ash.

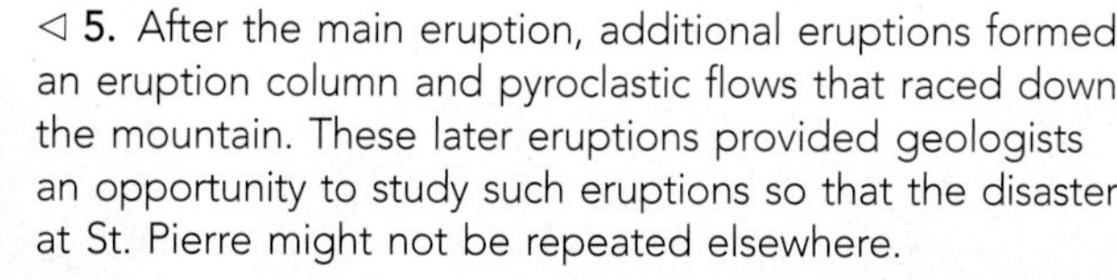

◁ **5.** After the main eruption, additional eruptions formed an eruption column and pyroclastic flows that raced down the mountain. These later eruptions provided geologists an opportunity to study such eruptions so that the disaster at St. Pierre might not be repeated elsewhere.

C What Events Preceded and Accompanied the Mount St. Helens Eruption?

The Cascade Range of the Pacific Northwest has produced some historic eruptions, such as the one at Mount Lassen in 1915. Geologists consider these volcanoes to be dangerous, and native people remember, through oral traditions, other cataclysmic eruptions at places such as Crater Lake in the Cascades. The eruption of Mount St. Helens in Washington state was the first major composite volcano eruption to occur in the age of television, and the world watched the destruction.

Geologic Studies Before the Eruption

1. Geologists studied Mount St. Helens before the eruption. They mapped the volcano and its surroundings, and constructed a geologic cross section through the mountain. This cross section shows that, before the eruption, the volcano consisted of interlayered lava flows, pyroclastic rocks, and mudflows, and had a domelike central conduit. In other words, it was a typical composite volcano. ▷

WSW Mount St. Helens ENE

06.08.c1

Dome and Conduit

Lava Flows and Dikes

Pyroclastic Rocks and Mudflows

0 1 km

2. The geologists determined that the volcano's eruptive history through the last 40,000 years included pyroclastic flows, tephra falls, mudflows, lava flows, and dome building. The geologists also recognized evidence for *horizontal* blasts of ash, one of the first cases in which this process was recognized.

A Volcano Awakens–Precursors to the Eruption

3. In March 1980, Mount St. Helens began to shake from earthquakes that geologists interpreted as being caused by magma moving beneath the mountain. Moderate quakes, such as the one recorded on this seismogram (▷), were the signal to geologists that something was going to happen at Mount St. Helens.

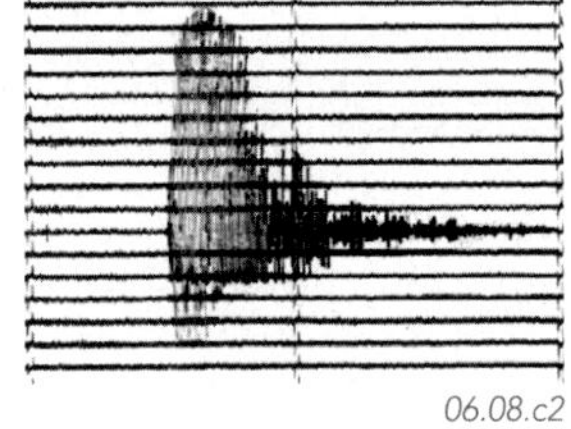

06.08.c2

▷ 4. In April 1980, a bulge formed on the north side of the mountain and then continued to grow. Geologists recognized that the bulge had unstable slopes, so they monitored its growth carefully.

06.08.c3

"Vancouver! Vancouver! This is it!"

5. David Johnston, the 30-year-old USGS geologist who spoke these words, was collecting data on a ridge near Mount St. Helens at the moment of the eruption. His observation post was considered to have low risk provided that the volcano erupted out the top as expected. Also, available data suggested that an eruption was not imminent. His last recorded scientific observation was undeniably correct. This was it!

6. On May 18, 1980, at 8:32 a.m., an earthquake triggered a massive avalanche and the bulge slid off the north side of the mountain. As this sequence of images show (▷), the lowering of pressure on the magma caused a lateral blast and an upward growth of an eruption column that spread ash over several states. Pyroclastic flows ravaged the landscape near the volcano. In addition to David Johnston, 56 other people died in the eruption, mainly from asphyxiation by hot gas and ash.

06.08.c4

06.08.c5

06.08.c6

06.08.c7

7. Pyroclastic flows, such as this one in 1980, and other eruptions continued on Mount St. Helens until 1986. After a lull of several years, the volcano built, destroyed, and rebuilt several new domes.

Before You Leave This Page Be Able To

- ✓ Summarize the type of eruption that occurred at Vesuvius and the cause of most of the deaths.
- ✓ Summarize the eruption at Mt. Pelée and events that preceded it.
- ✓ Summarize the eruption of Mount St. Helens, and the data that warned of its dangers.

How Do Volcanic Domes Form?

MANY VOLCANIC AREAS CONTAIN DOME-SHAPED HILLS constructed as volcanic domes. These represent eruptions of viscous lava that mounds up around the vent. When domes collapse, they can release small, but still deadly, pyroclastic flows. Volcanic domes form distinctive rocks and features in the landscape.

A What Are Some Characteristics of a Volcanic Dome?

Some volcanic domes have a nearly perfect domal shape, but most have a more irregular shape because some parts of the dome have grown more than other parts or because one side of the dome has collapsed downhill. Domes are typically hundreds of meters high and approximately a kilometer across, but they can be much smaller or much larger.

06.09.a1

This rubble-covered dome formed near the end of the 1912 eruption in the Valley of Ten Thousand Smokes in Alaska. Volcanic domes commonly have this type of rubbly appearance because their outer surface consists of angular blocks as small as several centimeters or as large as houses. The blocks form when solidified lava fractures by movement of the dome or when steep slopes collapse downhill.

06.09.a2

Most domes do not form in isolation but occur in clusters or in association with another type of volcano. Domes form within the craters of composite volcanoes, such as these at Mount St. Helens, and also develop within large calderas. In composite volcanoes and calderas, they commonly represent viscous magma erupted after a major eruptive event.

B How Are Volcanic Domes Formed and Destroyed?

Domes form as viscous lava flows to the surface and solidifies near the vent. Domes can grow in two different ways. They can grow from the inside or can grow from the outside. Domes can also be destroyed in two different ways—collapse or explosion.

Growth of a Dome

Domes mostly grow from the inside as magma is injected into the interior of the dome. This new material causes the dome to expand upward and outward, carrying the blocks of rubbly, solidified lava that coat the outside of the dome.

Domes can also grow as magma breaks through to the surface and flows outward as thick, slow-moving lava. As the dome advances, the front of the flow cools and can collapse into angular blocks and ash.

06.09.b1

Collapse or Destruction of a Dome

▷ Domes can be partially destroyed when steep flanks of the dome collapse and break into a jumble of blocks and ash that flow downhill as small-scale pyroclastic flows.

06.09.b2

◁ Domes can also be destroyed by explosions from within the dome. These typically occur when magma solidifies in the conduit and traps gases that build up until the pressure can no longer be held.

06.09.b3

C What Type of Rocks and Landscapes Characterize Domes?

Most volcanic domes consist of andesite or rhyolite, or a rock composed of the two. They are distinctive features when they form and after they have been partially eroded. They consist of solidified lava that has several different textures, and they typically are associated with pyroclastic rocks and other debris that formed when the dome partially collapsed or was blown apart.

Rock Types

06.09.c1

Some parts of domes cool rapidly into volcanic glass, such as *obsidian* which, although dark, has a felsic composition. This example has layers, called *flow bands,* formed by shearing and other processes during flow. [Alaska]

06.09.c2

The outer parts of domes cool, solidify, and fracture into angular blocks, which can become consolidated or incorporated into the magma to produce *volcanic breccia.* [central Arizona]

06.09.c3

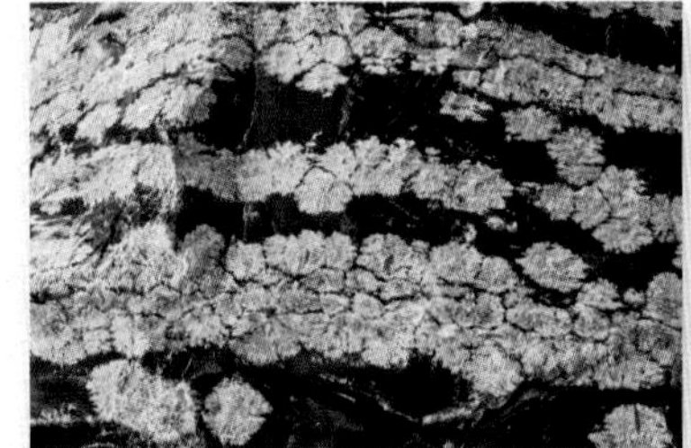

Volcanic glass is unstable. It eventually changes from unordered glass into rhyolite consisting of very small crystals. In some cases, the conversion to rhyolite produces spherical patches of distinct crystals. [California]

06.09.c4

When a volcanic dome collapses, the release of pressure causes pyroclastic flows and avalanches of rock and other debris. These collapses form *tuff* or *volcanic breccia* consisting of pieces of the dome in an ash-rich matrix. [Flagstaff, Arizona]

Expression in the Landscape

06.09.c5

These domes were formed only 550 to 600 years ago and so have relatively uneroded shapes and unaltered volcanic glass. [Inyo Mountains, eastern California]

06.09.c6

This peak is the remnant of a dome that has been extensively eroded. Over time, the glass has converted to finely crystalline rhyolite. [Castle Mountains, Nevada]

Deadly Collapse of a Dome at Mount Unzen, Japan

Mount Unzen towers above a small city in southern Japan. The top of the mountain is a steep volcanic dome that collapsed repeatedly between 1990 and 1995. The collapsing flanks unleashed more than 10,000 small pyroclastic flows (top photograph) toward the city below. The opportunity to observe and film these small pyroclastic flows attracted volcanologists and other onlookers to the mountain. In 1991, partial collapse of the dome caused a pyroclastic flow larger than had occurred previously. This larger flow killed 43 journalists and volcanologists and left a path of destruction through the valley (lower photograph). Note that damage was concentrated along the lowest part of the valley.

06.09.mtb1

06.09.mtb2

Before You Leave This Page Be Able To

- ✓ Describe the characteristics of a volcanic dome.
- ✓ Explain or sketch the two ways by which a volcanic dome can grow.
- ✓ Explain or sketch how a volcanic dome can collapse or be destroyed by an explosion.
- ✓ Describe how you might recognize a volcanic dome in the landscape.
- ✓ Summarize the types of rocks associated with volcanic domes.

How Do Huge Ash Eruptions Form Calderas?

CALDERA ERUPTIONS ARE NATURE'S MOST VIOLENT SPECTACLE. They can spread volcanic ash over areas larger than one thousand square kilometers and erupt more than one thousand cubic kilometers of magma. As the magma withdraws from the magma chamber, the roof of the chamber collapses to form a depression tens of kilometers across. The caldera can fill with ash, lava flows, and sediment.

A What Is a Caldera?

The term *caldera* describes a volcanic-related topographic depression and the underlying geologic structure. The low central part of a caldera is surrounded by a topographic escarpment, or *wall*, of the caldera. The Valles Caldera of New Mexico nicely illustrates the important features of a caldera because it formed only two million years ago and its subsurface has been explored by drilling and other geologic studies.

1. This image shows satellite data superimposed on topography. The circular Valles Caldera is expressed in the topography as a central depression surrounded by steep walls.

2. The caldera formed when a huge volume of magma was erupted from a shallow magma chamber, producing an eruption column and pyroclastic flows.

3. As shown in this cross section, the caldera contains a series of faulted blocks that have been downdropped relative to rocks outside the caldera. Faulting helped accommodate caldera subsidence and occurred at the same time as the main eruption of tephra. As a result, thicker amounts of tephra (now consolidated into volcanic ash and shown in light maroon) were trapped within the caldera than accumulated outside the caldera.

06.10.a1

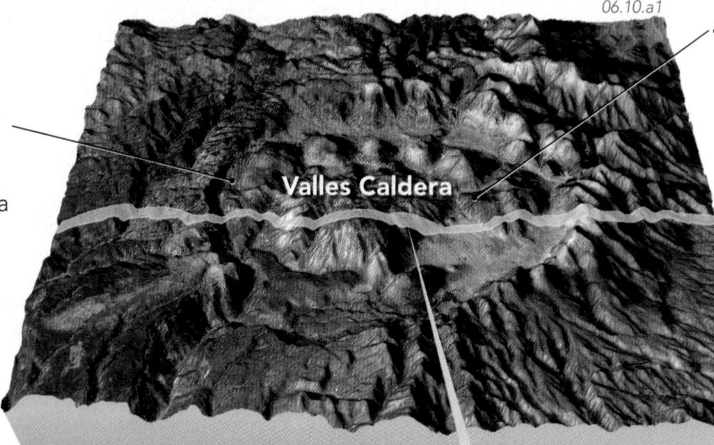

4. Small, rounded mountains inside the caldera are *rhyolite domes* constructed within the caldera after the main eruption. The domes were fed by dikes that tapped magma that remained after the main eruption. Some of this magma solidified at depth into granite, but some may still be molten.

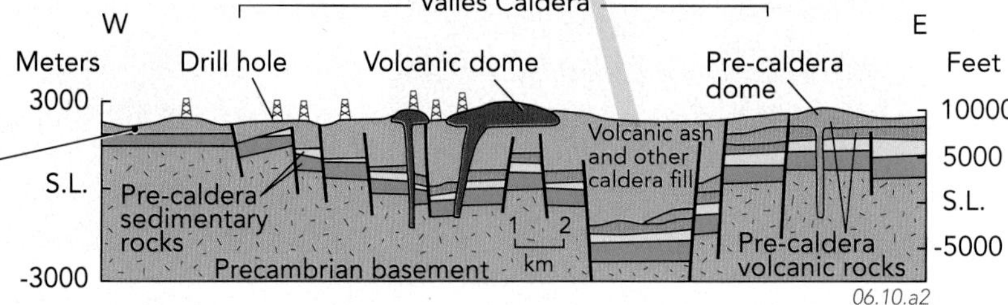

06.10.a2

B What Types of Rocks and Landscapes Typify Calderas?

Calderas that formed recently and that still have a clear expression in the landscape are moderately easy to identify. Ancient examples are more difficult to recognize because they may have been eroded, covered by other rocks, or disrupted by faulting or other events. A caldera should be suspected, however, at any location where there is an extremely thick (more than 100 meters) single layer of strongly welded ash-flow tuff.

06.10.b1

◁ Long Valley caldera of eastern California has a central depression with hills that are domes. Part of the caldera wall is expressed as the mountain on the left.

The caldera formed 760,000 years ago when pyroclastic eruptions emptied an underlying magma chamber, causing the caldera to subside. Pyroclastic deposits (▷), called the *Bishop Tuff*, accumulated inside the caldera and over much of the western United States.

06.10.b2

C How Does a Caldera Form?

The formation of a caldera and the associated eruption occur simultaneously—the caldera subsides in response to the removal of magma from the underlying chamber. The largest calderas erupted volcanic ash more than 1,000 meters thick. The formation of a caldera is illustrated below.

Formation of a Caldera

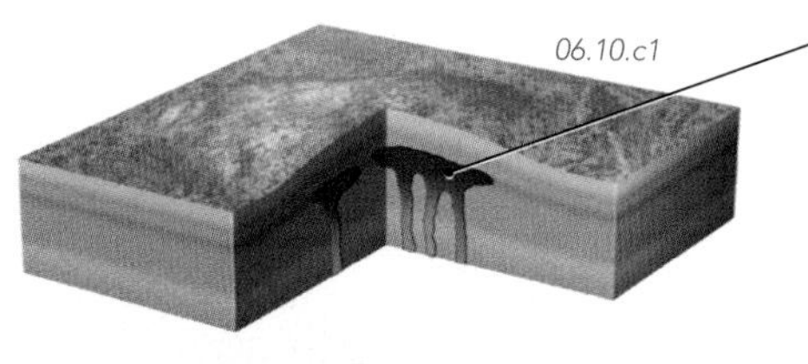

1. The first stage in the formation of a caldera is the generation, mostly by crustal melting, of felsic magma. The magma rises and accumulates in one or more chambers that can be kilometers thick and tens of kilometers across. The chamber or chambers may be within several kilometers of the surface.

2. Next, some magma reaches the surface and erupts. As the magma chamber loses material, the roof of the chamber subsides to occupy the space that is being vacated. Circular fractures form, allowing the block to drop and providing numerous new conduits to the surface for magma.

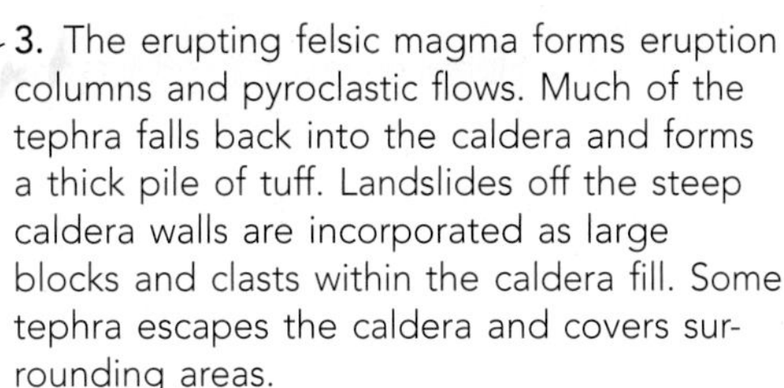

3. The erupting felsic magma forms eruption columns and pyroclastic flows. Much of the tephra falls back into the caldera and forms a thick pile of tuff. Landslides off the steep caldera walls are incorporated as large blocks and clasts within the caldera fill. Some tephra escapes the caldera and covers surrounding areas.

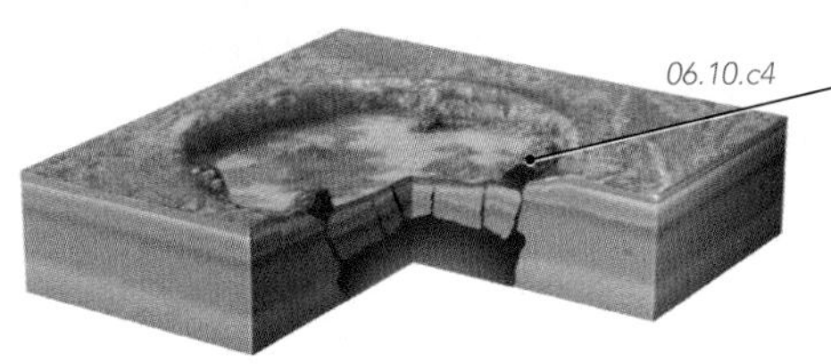

4. As the eruption diminishes, magma rises through fractures (dikes) along the edge in the interior of the caldera, erupting on the surface as volcanic domes. If the caldera remains a closed depression, it may become a lake or may trap later sedimentary and volcanic deposits.

A Section Through a Caldera

The rocks below are arranged from top to bottom in the way they occur in a caldera.

The top of a caldera may contain lake beds and ash-rich sediments trapped within the caldera. [Creede, Colorado]

Ash and larger tephra within the caldera may be thick and strongly welded into a hard, compact rock with or without crystals. [Sierra Madre, Mexico]

Deep in the caldera, finely crystalline granite represents the crystallized magma chamber. [Silverbell, Arizona]

Crater Lake and the Eruption of Mount Mazama

Crater Lake National Park contains a spectacular caldera, which is now filled by a beautiful deep-blue lake. The caldera formed about 7,700 years ago when the top of an ancient, large composite volcano called *Mount Mazama* was destroyed. The eruption was more than 50 times larger than the 1980 eruption of Mount St. Helens. As the magma erupted in a huge eruption column, the roof of the magma chamber subsided downward forming the main crater (caldera). Ash from this eruption can be traced all the way to southern Canada, 1,200 kilometers (750 miles) from its source. Wizard Island, shown here in the center of the caldera, is a small scoria cone that grew on the floor of the caldera after it formed. The beauty and serenity of Crater Lake today seem incompatible with its fiery, cataclysmic origin 7,700 years ago.

Before You Leave This Page Be Able To

- ✓ Describe or sketch the characteristics of a caldera, including its geometry in the subsurface.
- ✓ Describe how you could recognize a recent caldera and an ancient caldera.
- ✓ Sketch and explain the stages in the formation of a caldera.
- ✓ List the kinds of rocks you might find in a caldera from top to bottom.
- ✓ Explain the formation of Crater Lake.

What Disasters Were Related to Calderas?

CALDERA ERUPTIONS ARE AMONG THE MOST LETHAL natural disasters. Evidence of their past destruction is recorded by geology and by historical accounts in many parts of the world. These pages explore two such ancient disasters, Thera and Krakatau, and one possible future disaster.

A Did the Eruption of a Caldera Destroy a Civilization Near Greece?

Santorini, east of the Greek mainland, is a group of volcanic islands that records a major caldera collapse 3,500 years ago, about the time of the collapse of the Minoan civilization on the island of Crete.

06.11.a1

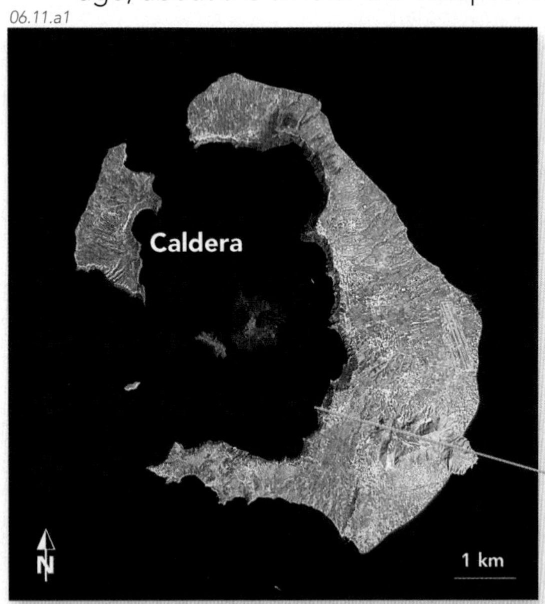

◁ **1.** This view from space shows the islands of Santorini, including *Thera*, the largest island. Thera has curving cliffs (▷) that expose pumice, volcanic ash, and rock fragments, which record explosive volcanism that began one to two million years ago.

2. The islands encircle a submerged caldera that formed when the center of a larger volcanic island collapsed, leaving the modern islands as remnants. The caldera collapse occurred as an eruption of magma of intermediate composition emptied a large magma chamber.

3. The steep cliffs encircling the caldera are eroded remnants of the original wall of the caldera. Islands in the middle of the caldera represent more recent eruptions.

06.11.a2 06.11.a3

4. The main caldera-forming eruption produced an ash column estimated to have been almost 40 kilometers (25 miles) high, followed by pyroclastic flows. The collapse of the caldera evidently unleashed a large destructive wave that traveled southward across the sea, probably helping lead to the downfall of the Minoan civilization on the island of Crete.

5. The eruption buried the remaining parts of Thera, including the city excavated above (△), with up to 50 meters of pumice and ash. This destruction of the civilization on Santorini and collapse of the volcanic island into the sea may have started legends about the sinking of a land mass into the sea.

B What Happened During the Deadly Eruption of Krakatau?

One of the largest historic volcanic eruptions struck the Indonesian Island of Krakatau in 1883. Eruption of a caldera killed about 40,000 people, mostly by large waves and pyroclastic flows formed during the collapse.

06.11.b1

1. Before 1883, the area contained three islands, the largest of which had three volcanoes and was called *Krakatau*. At this time, the region was densely populated and many people lived along the coast. Both factors contributed to the heavy death toll from the 1883 eruption.

06.11.b2

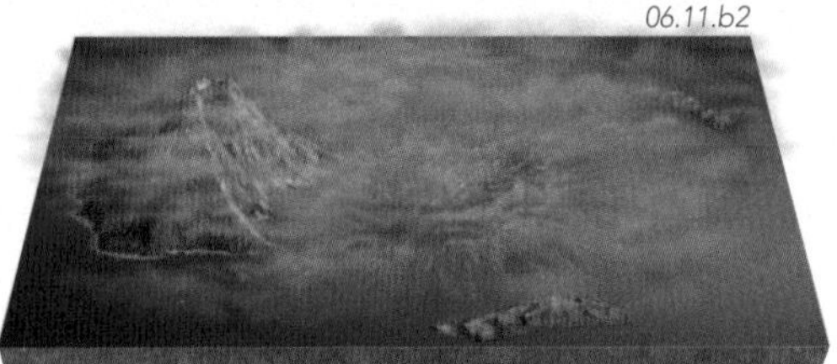

2. In 1883, massive amounts of tephra erupted from a magma chamber beneath the islands, forming a huge eruption column and pyroclastic flows. The eruption was accompanied by huge explosions, landslides, caldera collapse, and destruction of two of the volcanoes and nearly half of another. Large waves struck ships and adjacent coasts. One explosion was heard thousands of kilometers away and is the loudest sound in recorded history!

06.11.b3

3. After the eruption, only part of Krakatau remained. Since the eruption, a small volcano grew within the caldera, forming a new island called *Anak Krakatau* (child of Krakatau), which is shown in this satellite image. ▷

06.11.b4

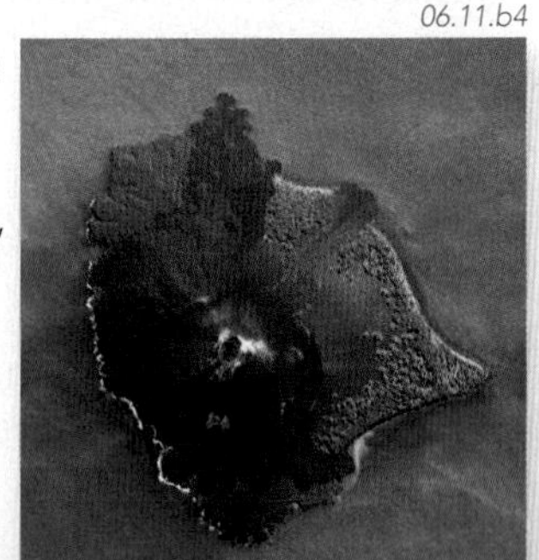

C Could the Yellowstone Caldera Cause a Future Disaster?

Yellowstone is one of the world's largest active volcanic areas. Abundant geysers, hot springs, and other hydrothermal activity are evidence of its recent volcanic history. During the last two million years, the Yellowstone region experienced three huge, caldera-forming eruptions. What is the possibility that Yellowstone could erupt again and rain destructive ash over the Rocky Mountains and onto the Great Plains?

06.11.c1

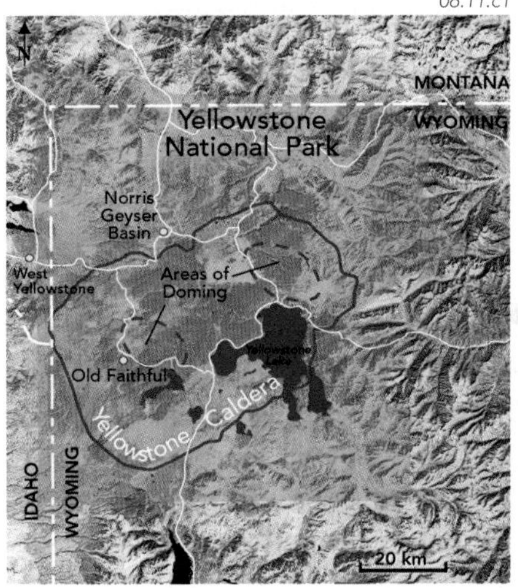

◁ 1. This image shows the outline of the youngest Yellowstone caldera, which formed 640,000 years ago. The boundaries of the caldera have been partially obscured by erosion, deposition of sediment, and lava flows that erupted after the caldera formed.

06.11.c2

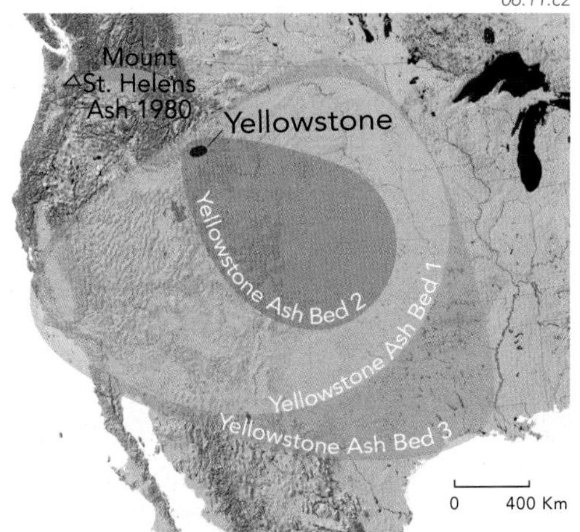

◁ 2. Ash from the three Yellowstone eruptions was carried by the wind and deposited over a huge area that extends from northern Mexico to southern Canada and as far east as the Mississippi River. A repeat of such an eruption could devastate the region around Yellowstone and cause extensive crop loss in the farmlands of the Great Plains and Midwest.

06.11.c3

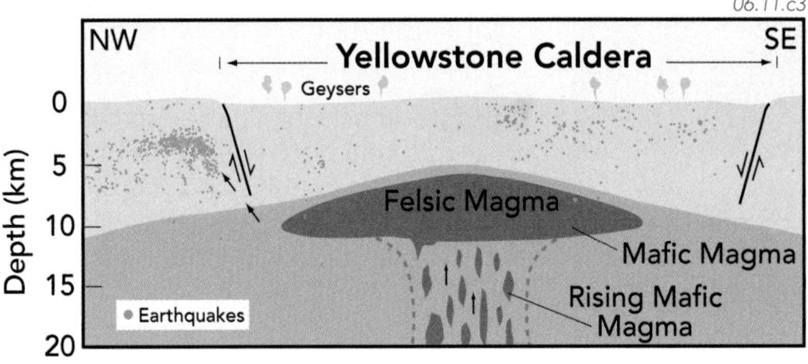

△ 3. A cross section through Yellowstone shows the youngest caldera, and associated geysers, earthquakes (dots), and a magma chamber interpreted to be present at depth.

▷ 4. To study hazards posed by the caldera, geologists use radar on satellites to precisely measure ground movements over time. Colors on this computer-generated image show a bull's-eye of recent uplift within the caldera, perhaps indicating that magma is rising beneath the surface and could erupt in the future.

06.11.c4

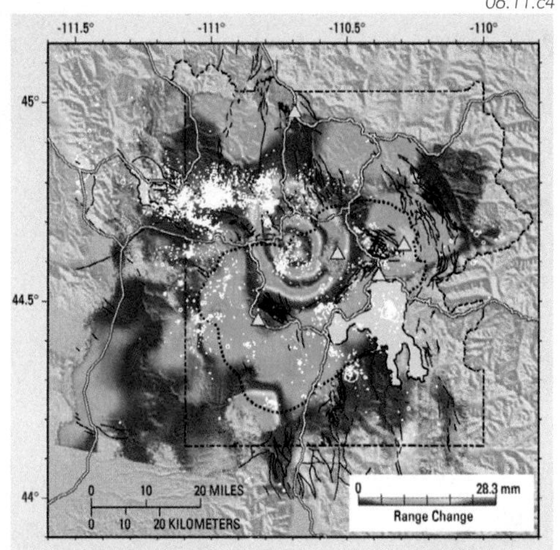

The Yellowstone Hot Spot and Related Volcanic Features

Large caldera-forming eruptions in the Yellowstone region occurred three times during the last 2.1 million years: 2.1, 1.3, and 0.64 million years ago. The average time between eruptions is about 700,000 years. Because 640,000 years have passed since the last eruption, Yellowstone could perhaps erupt again.

Where is all the magma coming from, and is melting still occurring beneath the region? The underlying cause of volcanism is interpreted to be a *hot spot* currently under Yellowstone. According to this model, as North America moved southwestward over the hot spot, the hot spot burned a path across southern Idaho forming the mostly basaltic Snake River Plain. Ages of basaltic and felsic volcanic rocks on the Snake River Plain are youngest near Yellowstone and become older to the southwest, which is consistent with the movement of North America over the hot spot. The Columbia Plateau flood basalts of Washington and Oregon could have formed when the same mantle plume reached the lithosphere and spread to the north.

06.11.mtb1

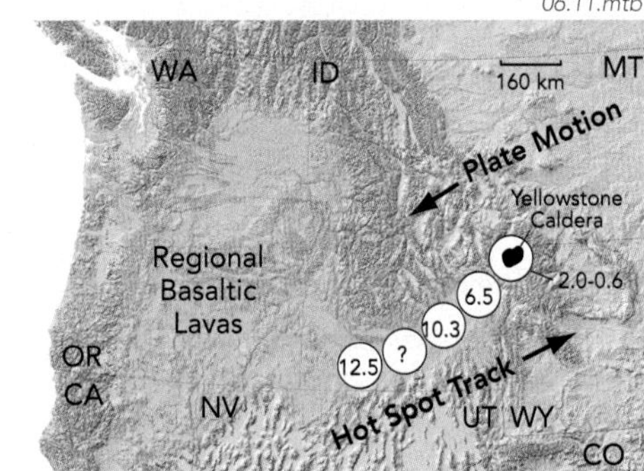

Before You Leave This Page Be Able To

- ✓ Explain how a volcanic eruption destroyed Santorini.
- ✓ Describe what happened during the eruption of Krakatau.
- ✓ Summarize the volcanic history of Yellowstone, including the distribution of volcanic ash.
- ✓ Summarize or sketch how volcanism at Yellowstone is related to a hot spot.

What Areas Have the Highest Potential for Volcanic Hazards?

SOME AREAS HAVE HIGHER POTENTIAL for volcanic hazards than others. Volcanic eruptions are more likely in Indonesia than in Nebraska. Also, there are different types of volcanoes that have different eruptive styles, so some volcanoes are more dangerous than others. What factors should we consider when determining which areas are the most dangerous and which are the safest?

A How Do We Assess the Danger Posed by a Volcano?

The potential hazards of a volcano depend on the type of volcano, its age, and its recent history.

Shape—The shape of a volcano provides important clues about how dangerous it might be. Volcanoes that have steep slopes, such as composite volcanoes, are more dangerous because they form from potentially explosive, viscous magma. Volcanoes that have relatively gentle slopes typically form from less explosive basaltic eruptions.

Rock Type—The types of rocks on a volcano are excellent indicators of the potential hazards. If a volcano contains welded tuffs, it has erupted pyroclastic flows. If it consists of rhyolite or andesite, it is more dangerous than a volcano composed of basalt. Chemical analyses of the amount of silica in the rocks can be used to help classify the rocks and thereby assess their danger.

06.12.a1

Age—The age of a volcano is essential information. The form of a volcano, especially whether it still has a volcano shape or has been degraded by erosion is one indicator of a volcano's age. Isotopic measurements on a rock, which can determine how long ago the rock formed in a volcanic event, are more accurate indicators of a volcano's age.

History—Important clues are provided by a volcano's history, which can be reconstructed from historical records, including oral histories of nearby people. Geologic studies of the sequence and ages of volcanic layers also provide insight into how often eruptions recur.

B What Areas Around a Volcano Have the Highest Risk?

Once we have determined the type of volcano that is present, we identify other factors that might influence which areas near the volcano have the highest potential for damage caused by a volcanic eruption.

1. *Proximity*—The biggest factor determining potential hazard is proximity, or nearness to the volcano. The most hazardous place is inside a summit crater. The potential hazard decreases with increasing distance away from the volcano.

2. *Valleys*—Lava flows, small pyroclastic flows, and mudflows can be channeled within valleys carved into the volcano and surrounding areas. Such valleys are more dangerous than nearby ridges.

06.12.b1

3. *Wind Direction*—Volcanic ash and pumice that are thrown from the volcano are carried farthest from the volcano in the direction that the wind is blowing at the time of the eruption. Because most regions have a *prevailing wind direction*, a greater hazard exists downwind from a volcano.

4. *Particulars*—Each volcano has its own peculiarities, and these may influence which part of the volcano is most dangerous. For example, steeper parts of a volcano are riskier. One side of a volcano may contain a dome that may be prone to collapse and pyroclastic flows.

5. This image shows three small villages around a volcano. Is one village located in a more hazardous site than the others? Which one is in the least hazardous area, and why do you think this site is less hazardous?

C What Regions Have the Highest Risk for Volcanic Eruptions?

We can think on a broader scale about which regions are most dangerous. In North America, volcanoes are relatively common along the west coast and virtually absent along the east coast. *Tectonic setting,* especially proximity to certain types of plate boundaries, is the major factor controlling regional variations of volcanic hazards. The map below shows the distribution of active or recently active volcanoes.

06.12.c2

The largest concentration of composite volcanoes is along the Pacific Ring of Fire. The volcanoes form above subduction zones on island arcs and continental arcs. Some subduction-zone volcanoes erupt so vigorously that calderas form.

Divergent margins are sites of copious eruptions from fissures and vents on the seafloor of fluid basaltic lava. Such eruptions pose little risk to humans because almost all of these occur at the bottom of the ocean. The island of Iceland, where the divergent margin coincides with a hot spot, is an exception.

06.12.c3

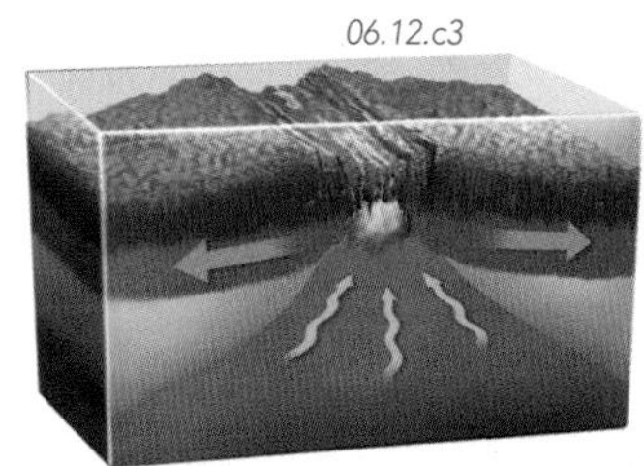

06.12.c1

06.12.c4

Many shield volcanoes occur along lines of islands and submarine mountains in the Pacific and other oceans. Most of these linear island chains and some clusters of islands formed above hot spots, such as those beneath Hawaii and the Galápagos Islands. Shield volcanoes also occur in other settings.

Some volcanic features, including flood basalts, calderas, and composite volcanoes, are in the middle of continents. Most of these form over hot spots or in continental rifts, such as the East African Rift.

06.12.c5

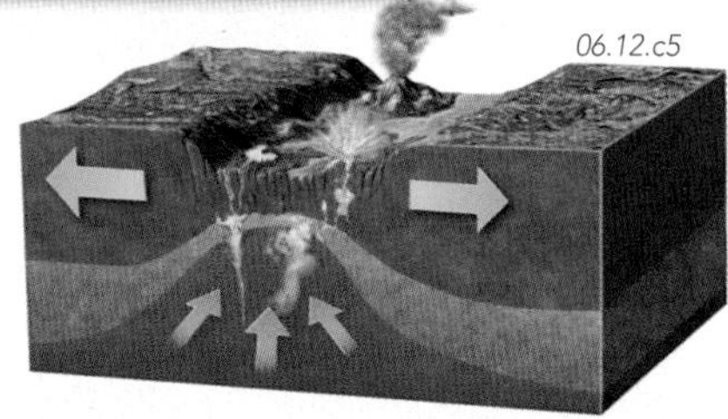

Forecasts, Policy, and Publicity

Predicting volcanic eruptions is currently an imprecise science. There have been some fabulous successes and some disappointing failures. Volcanologists have successfully predicted some major eruptions by studying clusters of small earthquakes generated as magma rises through the crust, by measuring changes in the amount of gas released by volcanoes, and through other types of investigations. Some predictions have saved lives because government officials (policymakers) acted on the scientific evidence. Some predictions have been unsuccessful because an eruption that was considered possible or even likely did not occur. In these cases, predictions, policy, and publicity can interact in detrimental ways. Negative publicity about a failed prediction can cause policymakers, or the public, to discount and not act on predictions. The consequences will be disastrous if a prediction is not persuasive, people are not evacuated, and an eruption does occur.

Before You Leave This Page Be Able To

- ✓ Summarize ways to assess the potential danger of a volcano and to identify which areas around a volcano have the highest potential hazard.
- ✓ Summarize which parts of the world have the highest potential for volcanic eruptions, and explain why.
- ✓ Summarize how the plate tectonic setting of a region influences its potential for volcanic hazards.

How Do We Monitor Volcanoes?

GEOLOGISTS MONITOR VOLCANOES using instruments that measure changes in ground deformation, ground shaking, heat flow, gas output, and water chemistry. Any of these changes may indicate that an eruption is imminent. Some monitoring is done using devices that remotely transmit data from a volcano.

A Can Seismic Activity Signal an Eruption?

As magma moves, expands, or contracts, it exerts force on the surrounding rocks and causes them to break, crack, or bend. This deformation causes *ground shaking*, which can be recorded with seismic instruments.

06.13.a1

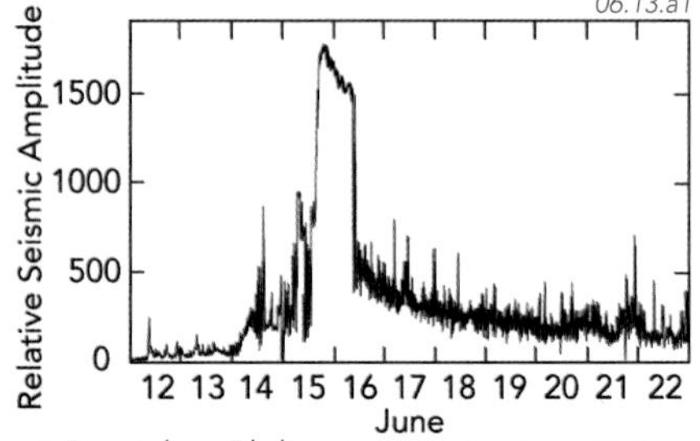

Seismometers are instruments that measure shaking of the ground. Increased shaking, called *seismic activity*, accompanies movement of magma. Seismic energy travels past the seismic station and is recorded as a function of time. In Mount Pinatuba, Philippines, seismic activity increased before a major eruption on June 15.

06.13.a2

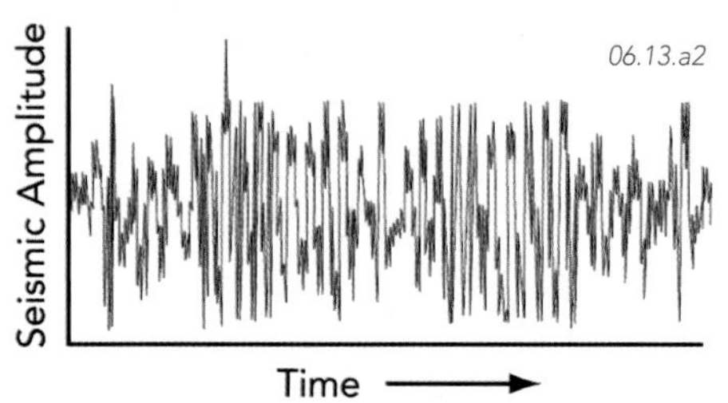

Magma flowing through conduits produces several distinctive types of seismic activity, including a fairly rhythmic, repeating pattern on seismic plots. Such patterns often precede and accompany a volcanic eruption and so have been used to predict that an eruption is imminent or already occurring.

B How Does Gas Output Change Before an Eruption?

Gases dissolved in a rising magma may come out of solution, expand, and provide the driving force for an eruption. The flow of gases from a volcano, therefore, may indicate that magma is rising and losing its gas.

06.13.b1

Sulfur dioxide (SO_2) gas emissions have increased just before some volcanic eruptions. Such increases, when integrated with other information, may indicate that a volcanic eruption is likely. [Volcano Island, Italy]

06.13.b2

Various ground and airplane-mounted instruments can measure the amount of SO_2 coming from a volcano. These instruments allow measurements of the gas flow from a volcano to be made from a safe distance.

06.13.b3

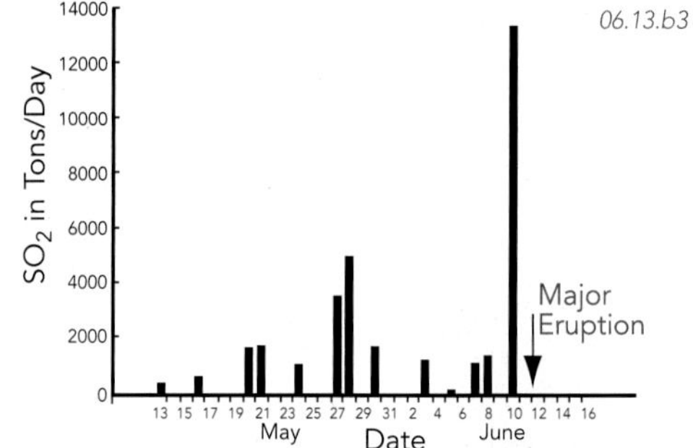

This graph shows measured SO_2 emissions just before the 1991 eruption of Mount Pinatubo in the Philippines. The amount of gas coming from the volcano increased dramatically before the eruption.

C How Are Changes in Heat Flow Measured?

Eruptions of lava or tephra or the release of gases can warm parts of a volcano before a major eruption. Such changes in thermal output can be detected by geologists and instruments on the volcano or by satellites, which are especially useful for monitoring volcanoes in remote locations.

Steam eruptions on volcanoes happen when water or ice come in contact with hot magma. Heat from the steam and magma can be detected by instruments that measure thermal energy. [Mount Spurr, Alaska]

06.13.c1

06.13.c2

This satellite image was specially processed to emphasize hot areas. Yellow and red colors mark magma and hot tephra. Bright blue colors represent cooler ash and volcanic rocks on the ground. [Augustine, Alaska]

D How Are Changes in a Volcano's Topography Monitored?

Before an eruption, the surface of a volcano may bulge or otherwise change shape by centimeters to hundreds of meters as magma inflates the mountain. Such changes in Earth's surface, called *ground deformation*, alert geologists to volcanic activity and can be measured by several methods.

▷ 1. *Global Positioning Satellite (GPS)* units are relatively small devices that use satellites to precisely determine the unit's position. As a volcano's surface changes slightly, a GPS station can track its changing position.

06.13.d1

06.13.d2

◁ 3. Special surveying instruments can accurately and precisely measure the distance between the instrument and a distant target. As a volcano inflates, the target moves toward the instrument.

▷ 2. *Tiltmeters* are instruments used to determine whether a measuring station is being tilted in one way or another. This is one of the oldest methods for monitoring the inflation or deflation of a volcano and tilting near a magmatic bulge.

06.13.d3

▷ 4. Another method for studying the surface of a volcano uses satellite radar to map the topography of the volcano at two different times, perhaps months apart. Colorful maps show differences in topography resulting from ground deformation.

06.13.d4

E Can Volcano Related Mudflows Be Detected Remotely?

Composite volcanoes commonly are covered with snow and ice that will melt during an eruption and mix with any loose volcanic material on the slopes of the volcano. This thick, muddy mixture, called a *mudflow* or a *lahar*, travels down river channels, destroys houses and bridges, and poses a threat to people downstream.

06.13.e1

◁ This eruption-related mudflow destroyed a bridge and highway. To detect such mudflows, geologists use a network of remote monitoring stations (▷) to detect the characteristic rumble made by the mudflows as they rush down valleys. This monitoring allows for rapid warning and evacuation of downstream communities.

06.13.e2

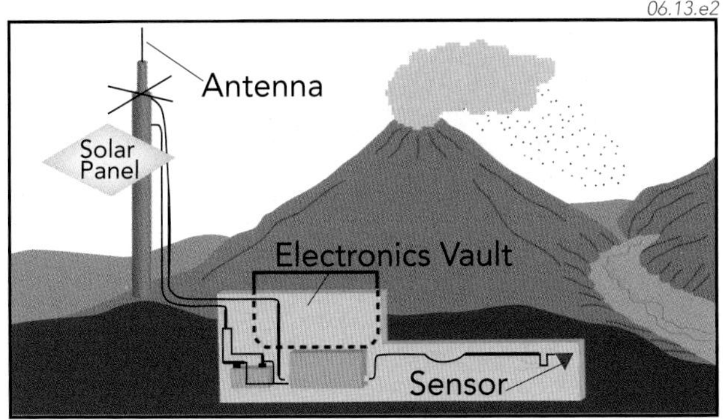

Monitoring a Volcano Using Various Approaches

Geologists use as many of these methods as possible to monitor a volcano and better understand its behavior and potential dangers. Such monitoring, therefore, requires a team of geologists, each with a different field of expertise (seismic records, volcanic gases, and others). Team members compare and discuss results from the various methods to develop a coherent interpretation that is consistent with all of the data. Remote monitoring and observations, however, only go so far.

For a geologist, there are few professional thrills more invigorating than observing, in person, a volcanic eruption. When an eruption does occur, a geologist knows what to observe and what kinds of additional data to collect.

06.13.mtb1

Before You Leave This Page Be Able To

- ✓ Discuss why seismic measurements are helpful for predicting an eruption.
- ✓ Explain why and how volcanic gases and thermal energy are measured.
- ✓ Summarize how topographic changes on a volcano could precede an eruption and how they are measured.
- ✓ Briefly discuss how a volcano-related mudflow can be detected remotely.

What Volcanic Hazards Are Posed by Mount Rainier?

MOUNT RAINIER IS ONE OF A CHAIN OF VOLCANOES above the Cascadia subduction zone of the Pacific Northwest. What kind of volcano is Mount Rainier, how did it form, when and how did it last erupt, and what risks does it pose to people living in the valleys below? Mount Rainier provides an opportunity to examine the various aspects of volcanoes, including styles of eruption, setting, and hazards.

A What Kind of Volcano Is Mount Rainier?

Mount Rainier rises ominously above the city of Tacoma (▷). The steep, symmetrical shape of the mountain identifies it as a dangerous composite volcano. A composite volcano plus a city equals high risk.

06.14.a1

△ This image shows the position of Mount Rainier and the suburbs of Tacoma. The top of the volcano is covered by glacial ice and snow. River valleys, only one of which is labeled, begin on the flanks of the volcano and continue into the suburbs. These provide a pathway from the volcano to the people.

06.14.a2

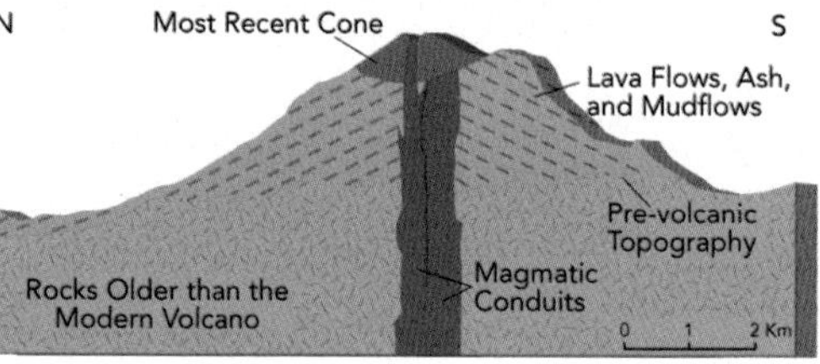

06.14.a3

△ A geologic cross section of Mount Rainier shows that the andesitic composite volcano was built on an eroded surface of granitic rocks and was fed by a pipelike magmatic conduit. The top of the mountain, largely covered by ice, is a younger volcanic cone that was constructed within an older crater.

B What Is the Plate-Tectonic Setting of Mount Rainier?

Mount Rainier is one of the volcanoes that cap the Cascade Range. The Cascade volcanoes exist because of melting associated with the Cascadia subduction zone, which is an ocean-continent convergent boundary.

The large composite volcanoes of the Cascades are related to a plate boundary between the North American plate to the east and the small Juan de Fuca plate to the west. The Juan de Fuca plate is moving eastward with respect to North America and subducting into the mantle.

06.14.b1

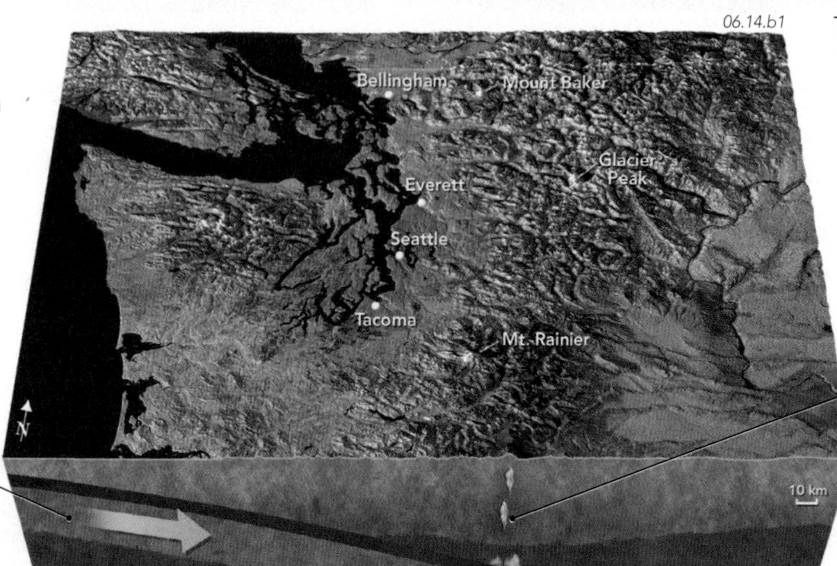

The Cascade Range is a north-south belt of mountains and is capped by snow-covered composite volcanoes, including the ones labeled. Mount Rainier and the other large volcanoes erupt mostly viscous, intermediate (andesitic) magmas that form thick, slow-moving lava flows and domes and explosive pyroclastic materials. Additional Cascade volcanoes are in Canada to the north and Oregon and California to the south.

The magmas form by melting of mantle above the subduction zone and rise to interact with overlying continental crust. The result is intermediate-composition magma and dangerous composite volcanoes.

C What Hazards Does Mount Rainier Pose to the Surrounding Area?

Mount Rainier is considered to be a very dangerous volcano. It has had at least eleven significant pyroclastic eruptions in the last 10,000 years. The most recent occurred in 1820. According to the U.S. Geological Survey, mudflows from Mount Rainier constitute the greatest volcanic hazard in the Cascade Range.

06.14.c1

1. The figure below shows hazard zones for lava flows, pyroclastic flows, and mudflows. The green zone around Mount Rainier has the highest hazard for lavas and pyroclastic flows. The yellow and orange colors show the potential for mudflows of different size and recurrence intervals (how often, on average, they occur). Yellow is used for least frequent but large mudflows, whereas red is used for more frequent and small mudflows.

06.14.c2

2. Mount Rainier erupts viscous lava flows, pyroclastic flows, and columns of ash. The hazards from lava flows are mostly near the volcano. Small explosive eruptions also have the most impact close to the summit, but pyroclastic flows can travel tens of kilometers away from the summit, in part following valleys. During a major eruption, a large eruption column could spread ash and pumice across the region. Prevailing winds would probably spread the ash to the east, but could blow in any direction depending on the weather conditions.

06.14.c3

Explanation
- Small mudflow event, not necessarily associated with volcanism
- Moderate-size mudflow event
- Large mudflow event
- Lava flows and pyroclastic flows

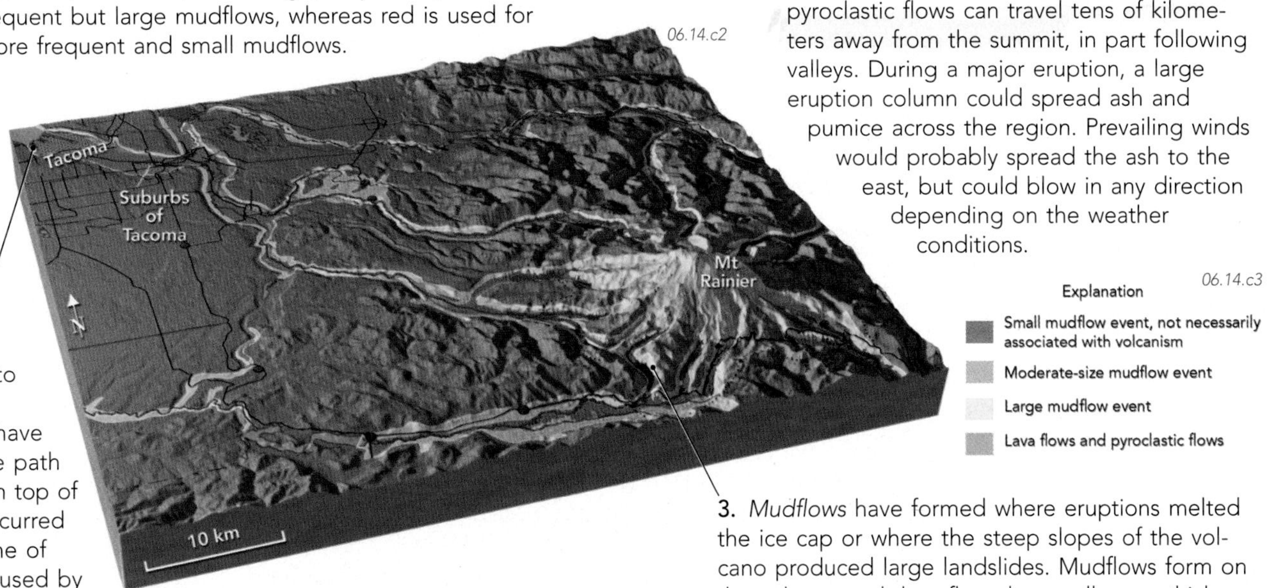

3. *Mudflows* have formed where eruptions melted the ice cap or where the steep slopes of the volcano produced large landslides. Mudflows form on the volcano and then flow down valleys as thick slurries of mud, pyroclastic material, rocks, and almost anything else that gets in the way.

4. Mudflows could flow northwest all the way into Tacoma and its suburbs (shown above). Houses have been built directly in the path of mudflows and even on top of a huge mudflow that occurred only 600 years ago. Some of these mudflows were caused by avalanches of rock and ice, not by a volcanic eruption. The risk posed by mudflows to these houses is very great! Would you buy one of these houses? How would you know if a house was at risk? Hazard maps like the one above are a good place to start.

Recent Eruptions of Volcanoes in the Cascade Range

Mount Rainier and Mount St. Helens are only two of the dangerous volcanoes in the Cascades. Ten other volcanoes in the U.S. part of the Cascades have erupted during the last 4,000 years.

This figure (▷) shows the locations of the large Cascade volcanoes and the number of times each has erupted during the last 4,000 years. This figure shows that Mount St. Helens is the most active of the Cascade volcanoes, but that Glacier Peak, Medicine Lake, and Mount Shasta have each erupted six or more times during the last 4,000 years. Seven of the volcanoes, including Mount Rainier, have erupted during the last 200 years. Living on the flanks of these volcanoes, or in the valleys below, carries the risk of mudflows, ash falls, and even pyroclastic flows. All of these volcanoes and the associated hazards exist because of subduction beneath North America.

06.14.mtb1

Cascade Eruptions During the Past 4,000 Years

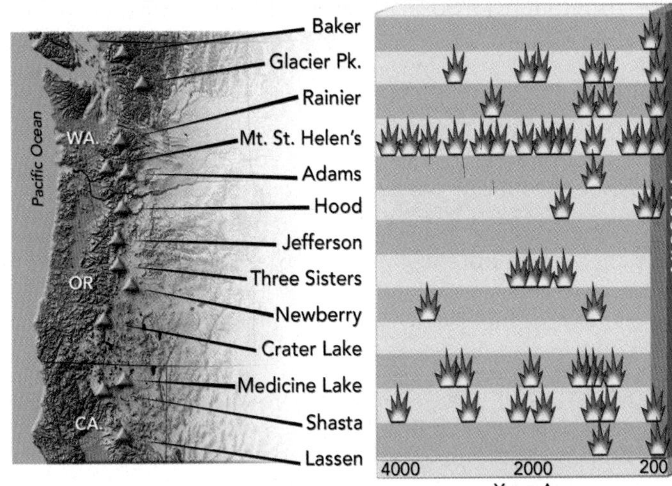

Before You Leave This Page Be Able To

- ✓ Describe the type of volcano that Mount Rainier is and the types of rocks it contains.
- ✓ Sketch and explain a cross section showing the plate-tectonic setting of Mount Rainier.
- ✓ Summarize the volcanic hazards near Mount Rainier.
- ✓ Briefly summarize how active the Cascade volcanoes have been during the last 4,000 years.

How Would You Assess Hazards on This Volcano?

DECIDING WHERE TO LIVE requires careful consideration. An overriding factor is whether a place is safe. In this exercise, you will investigate a volcanic island to determine what types of eruptions have occurred in the past, assess the volcanic hazards, and find the least dangerous place on the island to live.

Goals of This Exercise:

- Observe the physical characteristics of the volcano and the rock types it contains.
- Use your observations to determine what type of volcano is present and how it would likely erupt.
- Assess the potential for volcanic hazards in different parts of the island and determine the least dangerous place to live.

06.15.a1

A Observing the Characteristics of the Volcano

The study of a volcano begins by observing its physical characteristics, such as its size, shape, steepness of slopes, locations of ridges and valleys, and any unusual topographic features. The next step is to observe the types of rocks that are present, determine the aerial distribution of each rock type, and interpret how and in what order each rock formed. Follow the steps below and record your answers from each step in the accompanying worksheet.

1. Observe the image to the right. Record any important characteristics of the volcano on the worksheet.
2. Observe the photographs of rock samples and describe each rock's key attributes, such as whether it contains fragments. Use these attributes to identify the rock types (basalt, rhyolite, tuff, for example) by comparing the photographs to those in chapters 5 and 6. Alternatively, your instructor may provide hand specimens of the rocks for you to observe and identify.
3. Use your rock identifications to infer the *style of eruption* by which each rock formed.

B Assessing the Volcanic Hazards of the Island

Assess the general volcanic hazards of the volcano, and then assess the relative hazards of each part of the island compared to the others. Using your hazard assessments, determine the most dangerous places and the relatively least dangerous place to live.

1. Consider your rock identifications in the context of the topography and geologic features in the areas where the samples were collected. From these combined data, interpret what types of volcanic features, such as craters or scoria cones, are present in different parts of the island. A newspaper account of previous eruptions (bottom of next page) provides some useful clues.
2. Assess how each volcanic feature contributes to the hazard potential in different parts of the island. On the map in the worksheet, draw boundaries around and label those areas that have high, medium, or low hazard potential compared to the rest of the island. The differences between the three hazard zones will be fairly *subjective*. Use your best judgement and be consistent.
3. From your investigations, identify the areas you interpret to be the most dangerous and the least dangerous places to live. When choosing between two sites that are equally safe, you may consider other factors, such as the scenery, whether the sites are level enough to build on safely, and whether they are subject to storms, landslides, floods, and other natural hazards.

07.01.a5

◁ In high mountains or at high latitudes (close to the North or South Pole), snow can accumulate faster than it is removed by melting or other processes. Over time, the snow becomes compacted into ice, which may flow downhill as a *glacier*. As glaciers move, they erode underlying materials and carry the sediment away. The sediment and water are released upon melting of the ice. [French Alps]

▷ Rivers that flow over gentle terrain commonly *meander* gracefully from one side to the other. Most rivers are flanked by relatively flat land that may be covered when the river floods (a *floodplain*). Floodplains of meandering rivers are built from layer upon layer of mud carried by the floodwaters.

07.01.a6

07.01.a7

◁ Where a stream or river enters a standing body of water, such as an ocean or lake, its current slows, which causes most of its sediment to spread out and be deposited along the waterfront. The sediment piles up and forms a *delta* that builds out into the ocean or lake. This photograph shows a small delta formed where a desert wash deposits sediment into a very slow-moving river. [Canyonlands, Utah]

07.01.a8

▷ *Lakes* contain a range of environments, from quiet, deep water in the center, to more active water with waves caused by wind along the shoreline. Some lakes remain filled with water, but others dry completely as the water evaporates or seeps into underlying materials. The lake shown is Lake Superior, one of the Great Lakes.

In very wet environments, such as those adjacent to lakes and on deltas, the soil may become saturated with water, allowing swamps and ponds to form (▽). Such *wetlands* typically have abundant vegetation, which may become an important component of the sediment. [Florida]

07.01.a9

Before You Leave This Page Be Able To

✓ Summarize or sketch the main sedimentary environments on land, and describe some characteristics of each.

What Sedimentary Environments Are Near Shorelines and in Oceans?

OCEANS AND THEIR SHORELINES are dominated by wind, waves, ocean currents, and sediment that may be eroded along the coastline or brought in from elsewhere. The characteristics of each environment, especially the types of sediment, depend mostly on the proximity to shore, the availability of sediment, and the depth and temperature of the water.

07.02.a2

▷ *Beaches* are stretches of coastline along which sediment has accumulated. Some shorelines have bedrock all the way to the ocean and so have little or no beach. Most beaches consist of sand, pieces of shell, and rounded gravel, cobbles, or boulders. The setting determines which of these components is most abundant.

07.02.a1

07.02.a3

△ The water near the shoreline may be sheltered by a reef or islands. The sheltered water, called a *lagoon,* is commonly shallow, quiet, and perhaps warm. The near-shore parts of lagoons contain sand and mud derived from land, whereas the outer parts may have calcite sand and pieces of coral eroded from a reef.

07.02.a4

△ Where ocean water is shallow, warm, and clear, coral and other marine creatures construct *reefs,* which can parallel the coast, encircle islands, or form irregular mounds and platforms. Reefs typically buffer the shoreline from the energetic, big waves of the deeper ocean. [Red Sea, Egypt]

The seafloor in deep parts of the ocean is a dark, cold environment that commonly is several kilometers beneath the surface. It generally receives less sediment than areas closer to land, and its sediment is dominated by fine, wind-blown dust and by the remains of mostly single-celled organisms.

07.02.a5

◁ Sandy dunes that are inland from beaches are called *coastal dunes*. These dunes commonly form where sand and finer sediment from the beach are blown or washed inland and reshaped by the wind.

▷ Some shorelines include low areas, called *tidal flats*, that are flooded by the seas during high tide but exposed to the air during low tide. Most tidal flats are covered by mud and sand or are rocky. Some low parts of the land adjacent to tidal flats can accumulate salt and other evaporite minerals as seawater and terrestrial waters evaporate under arid (dry) conditions.

07.02.a6

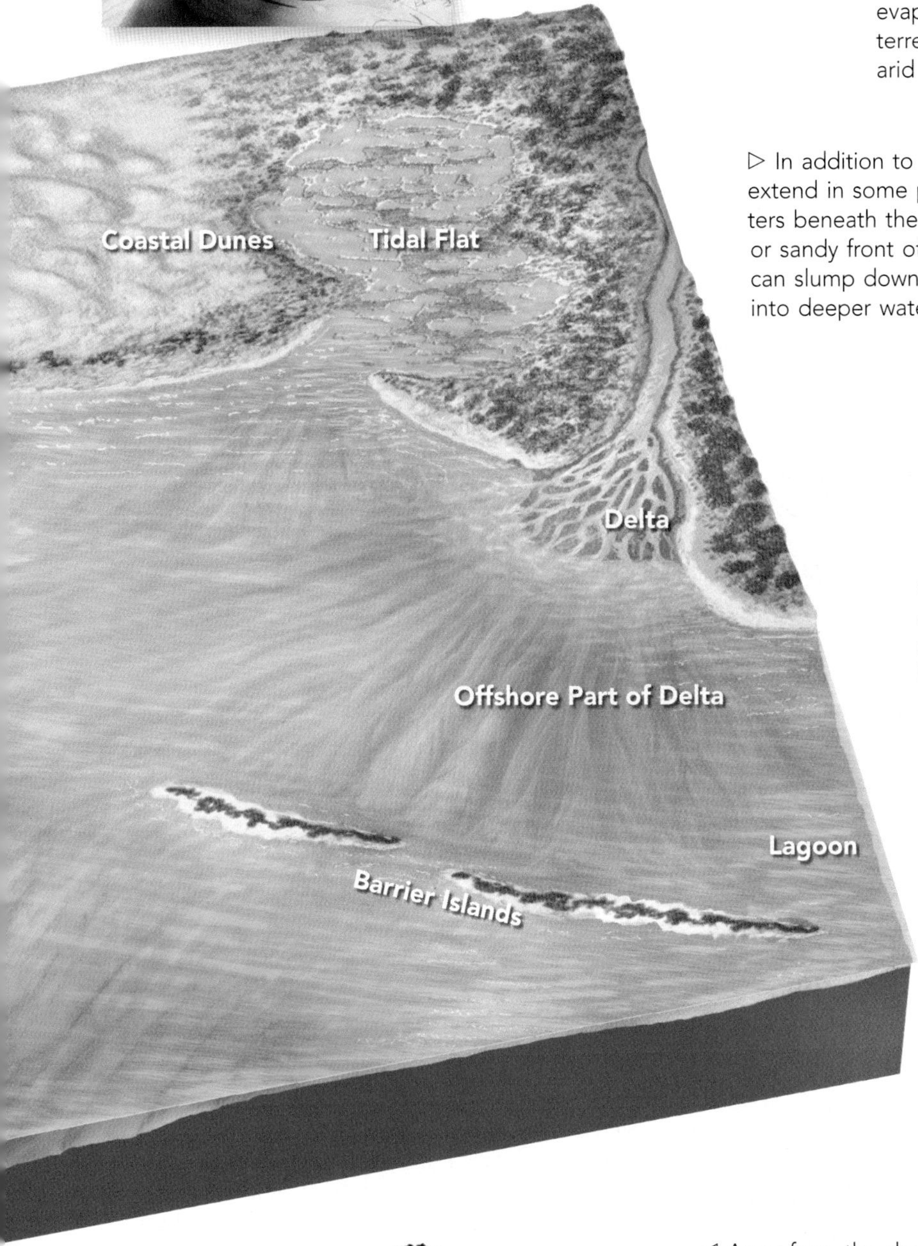

Coastal Dunes

Tidal Flat

Delta

Offshore Part of Delta

Lagoon

Barrier Islands

07.02.a7

▷ In addition to the part of deltas on land, *deltas* extend in some places for many tens of kilometers beneath the waters of the ocean. The muddy or sandy front of the delta may be unstable and can slump down the slope, sending sediment into deeper water. [Mississippi Delta]

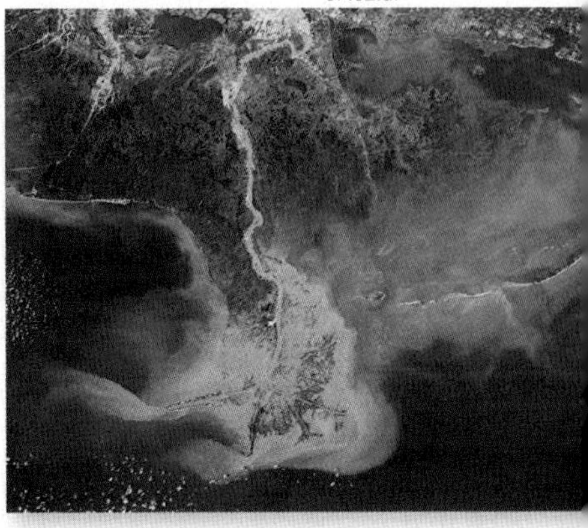

Other accumulations of sand rise above the shallow coastal waters as long, narrow islands, called *barrier islands*. Most barrier islands, such as the one below, are only hundreds of meters wide. The areas between barrier islands and the shoreline are commonly shallow lagoons or saltwater marshes. [Maryland-Virginia] ▽

07.02.a8

07.02.a9

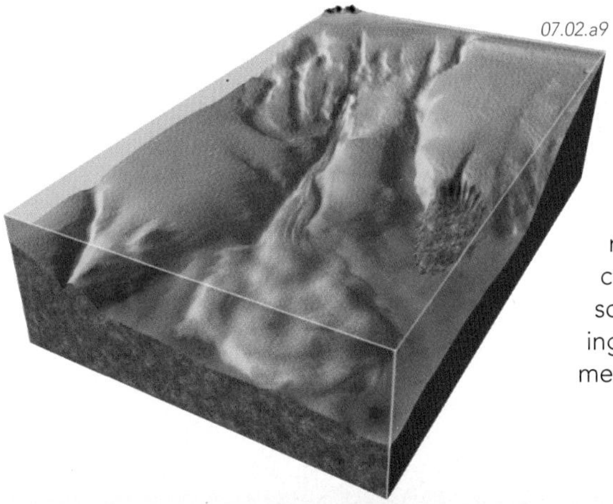

◁ Away from the shoreline, many landmasses are flanked by continental shelves and slopes consisting of layers of mud, sand, and carbonate minerals. Material from these sites can collapse down the slope in landslides or in turbulent, flowing masses of sand, mud, and water, called *turbidity currents*. The slopes of some continents are incised by branching *submarine canyons* that funnel sediment toward deeper waters.

Before You Leave This Page Be Able To

✓ Summarize or sketch the main sedimentary environments in oceanic and near-shore environments.

Where Do Clasts Come From?

SEDIMENTARY ROCKS CONSIST OF MATERIALS that came mostly from other locations. Most sediment is pieces, or *clasts*, of other rocks formed by weathering and transport. Other sediment is extracted from water by chemical reactions or by coral and other aquatic creatures.

 A How Do Physical and Chemical Weathering Produce Sediment?

Most sediment forms by the process of *weathering,* where physical and chemical processes on or near Earth's surface attack rocks, loosen pieces, and dissolve some parts. Different sizes, shapes, and types of sediment form depending on the material that is weathered and the conditions during weathering. The processes of weathering are summarized below and are discussed in detail in a later chapter.

Physical Weathering

Physical weathering is the physical breaking apart of rocks that are exposed to the environment. There are four major causes of physical weathering.

Near-Surface Fracturing—Many processes on or near the surface break rock into smaller pieces. These include fracturing caused by rocks pulling away from a steep cliff. Fractures also result when rocks expand as they are uplifted toward the surface and are progressively exposed to less pressure.

Thermal Expansion—Different sides of a stone are heated by the Sun from different directions during the course of a day. The stresses that arise from thermal expansion of the different sides may be enough to crack some rocks. Intense heating and thermal expansion during wildfires may also cause some fractures.

Frost and Mineral Wedging—Rocks can be pried apart as water freezes and expands in fractures. Crystals of salt and other minerals that grow in thin fractures can also push open cracks.

Biological Activity—Roots can grow downward into fractures and pry rocks apart as the root diameter increases. Burrowing animals can transport rock and soil from depth and move it to the surface where it is exposed to the elements.

07.03.a1

Chemical Weathering

Chemical weathering includes four types of chemical reactions that affect a rock by breaking down some minerals, causing new minerals to form, or removing soluble material from the rock. It affects loose rock fragments and bedrock.

Dissolution—Some minerals are soluble in water, especially the weakly acidic waters that are common in the environment. These minerals and the rocks they compose can dissolve in water, and the released ions can be carried away in rivers, streams, or groundwater.

Hydrolysis—When silicate minerals are exposed to water, especially water that is somewhat acidic, the hydrogen ions from the water react chemically with the minerals. This process commonly converts the original materials to clay minerals and removes positive ions. It is involved in the formation of many soils.

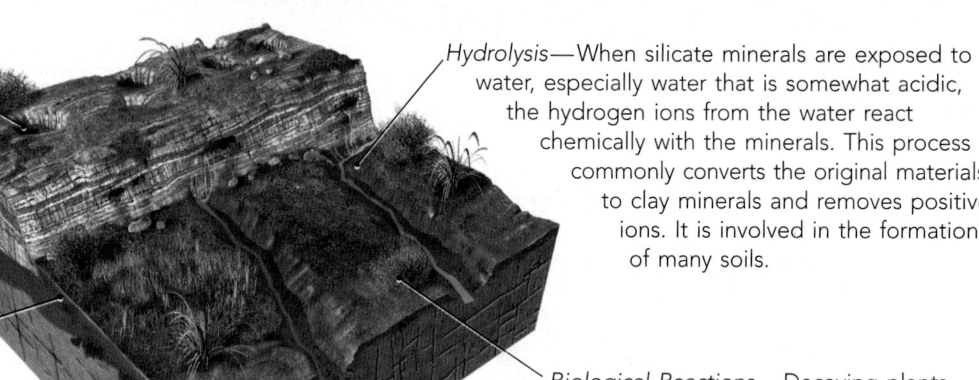

Oxidation—Some minerals are unstable when exposed to oxygen in Earth's atmosphere. These minerals can combine with oxygen to form oxide minerals, such as iron oxides, the reddish and yellowish material that forms when metal rusts.

Biological Reactions—Decaying plants produce acids, and some bacteria consume rocks. These processes break the minerals into their constituent elements.

07.03.a2

B What Are the Starting Materials and the Products of Weathering?

Physical and chemical weathering affect various starting materials, including rocks of different composition, grain size, strength, and solubility. These variations result in many different sizes and types of sediment.

Rock Type

07.03.b1

Rocks exposed on Earth's surface have varying composition, grain size, and texture and so weather in different ways. Some rocks and minerals weather chemically into clay and iron oxides, some dissolve completely, but others resist chemical weathering and so break physically into boulder- to clay-sized pieces (*clasts*). Quartz crystals in granite (◁) become grains of sand, but feldspar crystals can weather to clay minerals or to sand grains. Quartz and feldspar are the sand grains in this sedimentary rock (▷), which also contains quartz-rich pebbles that resist weathering. Soluble material dissolved to make some of the holes.

07.03.b2

Intensity of Fracturing

07.03.b3

Rocks that are fractured are weaker and more easily weathered than rocks that are intact. Fracturing increases the surface area of the rock that is exposed to the environment. Fractures permit water, air, and organisms to invade the rock, which causes more chemical and physical weathering. In the left photograph, fractures in the sandstone cliff allow blocks to detach from the cliff and fall onto the slope below. In the right photograph, lines of trees and soil mark a set of parallel fractures in sandstone. [Capitol Reef National Park, Utah (◁) and Zion National Park, Utah (▷)]

07.03.b4

C How Do Erosion and Transport Produce Sediment?

Once weathering has loosened pieces of the bedrock, the pieces can be transported by rivers, glaciers, waves, wind, and other currents. During transport, larger clasts can be broken to produce smaller ones.

07.03.c1

As large clasts, such as boulders and cobbles, are transported, they can break apart as they collide with other large clasts. Through this process, boulders can become a higher number of cobbles, and cobbles can break down into smaller pebbles. Some clasts end up as sand.

07.03.c2

Silt, sand, and larger clasts carried in water can cause *abrasion* of other clasts in a channel or of bedrock along the channel. This process is akin to sandpapering and can smooth rough edges, remove small pieces, and scour pits and recesses into hard bedrock, as shown here. [Grand Canyon, Arizona]

Before You Leave This Page Be Able To

- ✓ Summarize the main processes of physical and chemical weathering.
- ✓ Summarize how the rock type, degree of fracturing, and solubility of starting materials influence the type of sediment that is produced.
- ✓ Summarize how rocks can be broken during transport.

What Are the Characteristics of Clastic Sediments?

SEDIMENT CONSISTS OF LOOSE FRAGMENTS, or *clasts*, of rocks and minerals. When clastic sediment becomes sedimentary rock, the name assigned to the rock depends on the size and shape of the clasts. Other characteristics of the sediment, such as rounding, are used to further describe the resulting rock.

A How Are Clastic Sediments Classified?

Clastic sediment is derived from various environments and rock types. These varied conditions impart distinctive characteristics to the sediment, including the sizes of the clasts, how the sizes of the clasts in the sediment vary, and the shape of clasts. Together these attributes reveal aspects of the sediment's origin and history.

Size of Clasts

07.04.a1

Clasts larger than sand include, from largest to smallest, *boulders*, *cobbles*, and *pebbles*. A softball has a size that is in the middle of the size range of cobbles; boulders are much larger, and pebbles are much smaller.

07.04.a2–4

Sand ranges from coarse to medium to fine sizes. *Medium sand* is between 1/4 and 1/2 millimeters, and fine sand can have diameters smaller than 1/10 millimeters.

07.04.a5

Particles finer than sand are *clay* and *silt*. Clay is so fine that it commonly feels slippery between fingertips, and *silt*, which is coarser than clay, feels gritty.

Amount of Sorting

07.04.a6

Sorting describes the size range of clasts in sediment. *Poorly sorted* sediment has a wide range of clast sizes, such as in this rock.

07.04.a7

Many sediments have moderate sorting of clasts, such as from sand to pebbles, or from silt to clay.

07.04.a8

Well-sorted sediment consists of clasts that all have the same size. Sand dunes (shown here) are composed of well-sorted sand.

Shape of Clasts

07.04.a9

Angular clasts have sharp corners and edges. They can be blocky or shaped like chips or plates with angular edges.

07.04.a10

Many clasts have an angular shape with edges and corners that have been partially rounded.

07.04.a11

Rounded clasts have smooth, curving surfaces and shapes like eggs, flattened balloons, or objects that are nearly spheres.

B What Are the Characteristics of Conglomerate, and How Does Conglomerate Form?

Conglomerate differs from breccia because it has rounded, instead of angular, clasts. Rounded clasts indicate longer distances of transport down a stream or river or continued pounding by waves along a shoreline.

Clast Size

07.08.b1

Conglomerate contains large, rounded clasts usually in a matrix of sand and silt. This conglomerate contains well-rounded clasts, many of which are more than 30 centimeters (12 inches) across.

07.08.b3

Some conglomerate is finer grained, having few clasts larger than a centimeter or two. Such conglomerate represents less turbulent conditions or a source region that lacked large clasts.

Sorting

07.08.b2

Some conglomerate is well sorted, which means that most clasts have nearly the same size. In this rock, the clasts rest directly on one another instead of being completely separated by matrix.

07.08.b4

Other conglomerate is less sorted, having scattered large clasts dispersed in a fine-grained matrix. The pebbles shown here are surrounded by a matrix of sand and silt.

Environments of Formation

07.08.b5

Conglomerate can form as rivers and streams deposit sediment in or adjacent to the channel. The clasts can be rounded by even a moderate amount of transport in a river.

07.08.b6

Conglomerate is deposited by braided rivers, which move back and forth across a broad alluvial plain and deposit a widespread layer of coarse debris. The debris can harden into conglomerate. [Tibet]

07.08.b7

Waves incessantly pound stones on many beaches, churns them, and rounds their corners and edges. Smaller pebbles and sand can be moved much of the time, but larger cobbles and boulders may only move during extremely high tides or during storms. [Naxos, Greece]

07.08.b8

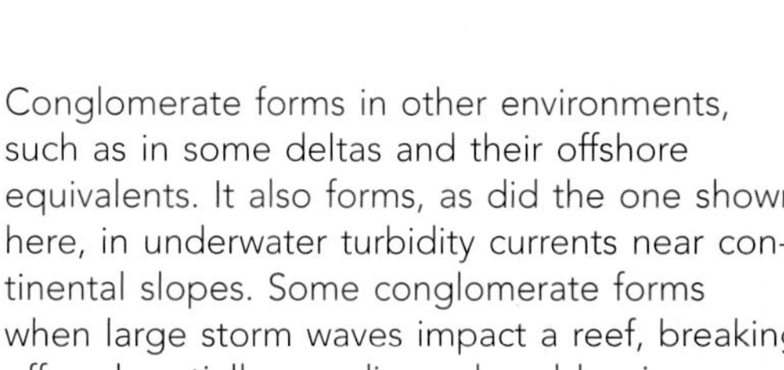

Conglomerate forms in other environments, such as in some deltas and their offshore equivalents. It also forms, as did the one shown here, in underwater turbidity currents near continental slopes. Some conglomerate forms when large storm waves impact a reef, breaking off and partially rounding vulnerable pieces.

Before You Leave This Page Be Able To

- ✓ Summarize or sketch the characteristics of a breccia, and identify some environments in which this rock forms.
- ✓ Summarize or sketch the characteristics of a conglomerate, and identify some environments in which this rock forms.
- ✓ Contrast breccia and conglomerate, and explain reasons why one rock type might form instead of the other.

Where Does Sandstone Form?

SANDSTONE IS A COMMON SEDIMENTARY ROCK because sand occurs in many environments, ranging from sand dunes on land to submarine canyons beneath the oceans. Each environment produces a different variety of sandstone with characteristics diagnostic of that environment. In what environments does sandstone form, and how do the resulting sandstones differ?

A What Are the Characteristics of Sandstone?

Sandstones, even those formed in different environments, share many characteristics. Sandstone is mostly or wholly composed of sand-sized grains and commonly is at least moderately sorted. Quartz is the dominant mineral in most sandstone, but feldspar, iron oxides, calcite, and other minerals are locally common.

07.09.a1

Sandstone contains mostly sand grains. Some sandstone, such as the one shown here, consists entirely of sand-sized grains. The individual sand grains can be well rounded or angular.

07.09.a2

Most sandstone has layers defined by variations of color, grain size, or composition of the grains. Such layers can be parallel beds, as shown here, or cross beds centimeters to tens of meters high.

07.09.a3

Sandstone can contain other clast sizes, such as silt or scattered pebbles. Such rock is still called sandstone as long as sand is the dominant clast size.

07.09.a4

Some sandstone layers appear massive from a distance because they have little variation of grain size, as in the massive sandstone cliff in the center of this photograph. [Yampa River, northern Utah]

B In What Land Environments Does Sandstone Form?

Sandstone forms in rivers, migrating sand dunes, and in other land environments. Such units commonly have a tan or red color because of oxidation of iron-bearing minerals during exposure to air and groundwater.

07.09.b1

Sand dunes form sandstone that is almost all sand grains because wind blows away finer sediment. Wind cannot pick up large clasts, so sand dunes usually do not include grains larger than sand. [Morocco]

07.09.b2

Rivers deposit sand in channels and on the adjacent floodplain. In this photograph, the floodplain is the relatively flat area with grass and trees. It is only covered with water during large floods. [Yampa River, Utah]

07.09.b3

This sandstone, deposited by dunes, is mostly sand grains and has a reddish color because of iron-oxide minerals that cement the sand grains. The unit has large cross beds that reflect the shapes of original dunes. [Snow Canyon, Utah]

07.09.b4

Sandstone deposited by a river usually has discrete layers reflecting floods and shifting positions of the channel. It can be interlayered with siltstone and other rock types. These sandstone ledges alternate with slope-forming siltstone. [Grand Canyon, Arizona]

C How Does Sandstone Form Along Shorelines?

Most beaches are dominated by sand, so many beach deposits become sandstone. Sand also dominates many parts of deltas, especially the channels and shallow-water parts of the delta.

07.09.c1

A typical ocean beach includes sand with shells, pebbles, and locally some larger blocks derived from nearby rock exposures. The resulting sandstone includes these same items. The shells indicate a possible beach origin.

07.09.c2

This sandstone formed in a delta that built out into a shallow sea. The lower beds include shale deposited in the sea, whereas the upper sandstone-dominated part records the approach of the delta.

07.09.c3

The sandstone in the right side of this photograph formed along a beach as seas moved across the land. The beach sands banked up against an island formed from the darker rocks in the lower left of the image. [Yampa River, Utah]

07.09.c4

Sandstones can form in other nearshore environments. The cross-bedded sandstone beds shown here formed near a barrier island.

D How Does Sandstone Form in Offshore Environments?

Sand also accumulates at sites farther from shore or in deeper water. The sand is derived from erosion of continents, islands, deltas, reefs, and barrier islands.

07.09.d1

Continental shelves and other offshore areas can accumulate sand derived from erosion of land. The sands are moved out onto the shelf by waves and currents and are buried by later sediment. [northern California]

07.09.d2

Sands deposited near volcanic islands contain less quartz than do most sands, but contain more grains that are crystals or pieces of volcanic rock. Such settings can produce black or even green sandstone. [Hawaii]

07.09.d3

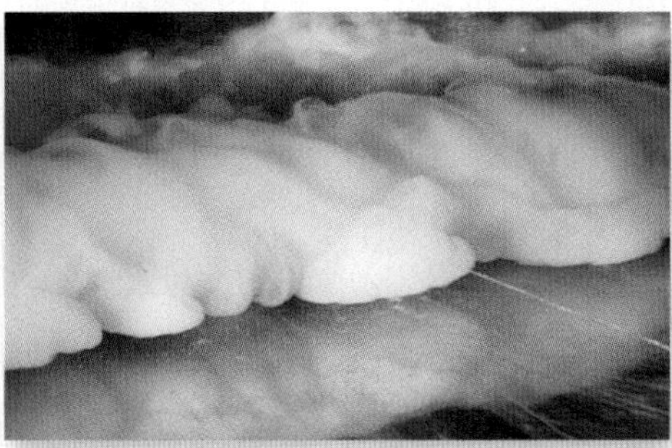

If loose sand on the continental shelf or slope becomes unstable, it flows down the sloping seafloor as a thick slurry of sediment and water as a *turbidity current*. The beds deposited by these currents contain sand, mud, and larger clasts and commonly display graded bedding that reflects a gradual slowing of the current's velocity. This example was created in a laboratory tank.

07.09.d4

These tan sandstone beds are interpreted to represent a turbidity current because they exhibit graded bedding and are interbedded with dark, deep marine shales.

Before You Leave This Page Be Able To

- ✓ Summarize the characteristics of sandstones, including their expression in landscapes.
- ✓ Describe the land environments in which sandstone forms, and how you might distinguish sandstone formed by sand dunes from those formed by rivers.
- ✓ Describe how sandstone forms along shorelines and in offshore environments.

How Do Fine-Grained Clastic Rocks Form?

MANY SEDIMENTS ARE FINE GRAINED. These sediments consist of grains that are smaller than sand. Compared to coarser sediment, fine sediment is easily transported by water or wind, even by slow water or wind, and remains in transit until it reaches fairly quiet conditions, such as a lake or sea. The resulting sedimentary rocks include shale, one of the most common rocks exposed on land.

 What Are the Characteristics of Fine-Grained Clastic Rocks?

Fine-grained clastic rocks consist mostly of clasts of *silt* and *clay*. Silt is slightly finer than sand and becomes the rock *siltstone*. *Clay* represents the smallest sedimentary particles and becomes *shale* when lithified. *Mud* includes both silt and clay, and the resulting rock is *mudstone* or simply *mudrock*.

Siltstone consists of small grains that are not visible with the unaided eye but are visible through a hand lens. Siltstone can contain some fine sand grains, as in this example. It commonly occurs in centimeter-thick beds.

07.10.a1

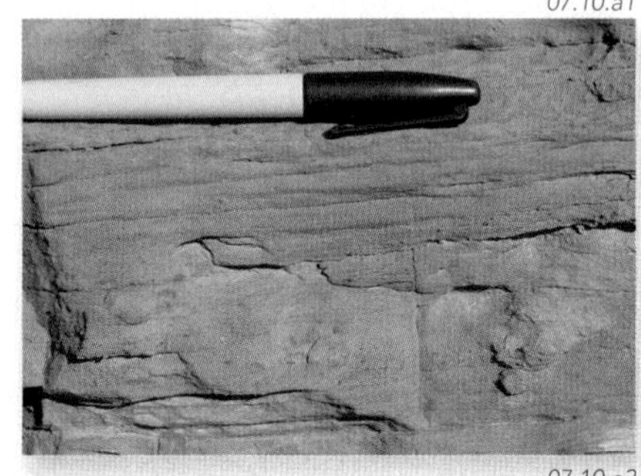

07.10.a2

Mudstone is a fine-grained rock that is similar in many ways to shale, so some people do not distinguish the two rock types. Mudstone breaks into pieces that tend to be more rounded than the thin chips into which shale weathers.

Shale has particles that are too small to see even with a hand lens. It has very thin beds and characteristically splits into thin flakes or chips, as in this photograph. It commonly is poorly exposed.

07.10.a3

07.10.a4

Shale and mudstone can form thick sequences that are almost entirely fine grained or are interbedded with sandstone or limestone. Depending on the environment, shale and mudstone can have various colors. [northern New Mexico]

B In What Land Environments Do Siltstone and Shale Form?

Silt and clay are deposited in several terrestrial (on-land) settings. In most of these settings, deposition occurs in slow-moving or even stagnant water. These fine-grained sediments can also be deposited by wind.

Floodplains of meandering rivers are dominated by silt and fine-grained sand. During floods, silt is carried farther from the channel than are coarser sediments like sand and gravel. [southern Utah]

07.10.b1

07.10.b2

Wind transports and deposits silt over large areas. Wind-blown silt was especially abundant during periods of glaciation, when moving ice sheets ground rock into powdery silt-sized particles. [China]

The bottoms of lakes are covered with mud carried to the lakes by streams, rivers, wind, and erosion of adjacent hillslopes. Lakes produce soft, thin-bedded rocks that can be gray from organic material. [northern New Mexico]

07.10.b3

07.10.b4

Chemical weathering converts many minerals, such as feldspar, into clay, which then accumulates as a layer of *soil* on the surface. If such soils are preserved, they usually form fine-grained rocks. [Nepal]

C Where Along Shorelines and Farther Offshore Do Silt, Clay, and Mud Accumulate?

Silt, clay, and mud form in several settings near shorelines and farther out to sea. Clay covers more of the seafloor than any other type of sediment, so shale is a common sedimentary rock.

Mud flats occur along some shorelines and are flooded by high tides and during storms. These flats show the influence of saltwater by commonly containing salt, gypsum, and other evaporites. [Drakes Estero, California]

07.10.c1

07.10.c2

Mud can also accumulate in shallow continental seas, on continental shelves, and on adjacent continental slopes. The multicolored shales shown here formed in a shallow sea within the North American continent. [southwestern Colorado]

The relatively calm water of a lagoon is an efficient trap of mud and clay carried from the land by streams, rivers, and wind. The green and gray shales shown here accumulated slightly offshore. [Grand Canyon, Arizona]

07.10.c3

07.10.c4

Many shales form in deep water in an ocean basin, where fine particles are carried by wind and ocean currents. Deep-water shales usually are dark gray, which reflects a high organic content and lack of oxidation. [Hudson Valley, New York]

D How Are Fine-Grained Clastic Rocks Expressed in the Landscape?

Shale, mudstone, and siltstone are relatively erodible rocks. Where exposed, these rocks form soft slopes covered by small, loose chips derived from weathering of the thinly bedded rocks. [Grand Junction, Colorado]

07.10.d1

07.10.d2

Fine-grained rocks commonly are partially or entirely covered by soil or loose debris (talus) from overlying, more-resistant rocks. Here, outcrops of red mudstone project through a surficial cover of light-colored debris. [Wilson Cliffs, Nevada]

During erosion of landscapes, rivers preferentially carve their channels into shale and siltstone because these rocks are so easily eroded. Many rivers follow shale-rich units across the land surface, and may even follow the shales around folds. This desert wash follows a layer of fine-grained rocks around the bend of a broad fold. [Comb Ridge, Utah]

07.10.d3

Shale and associated fine-grained rocks form another distinctive type of landscape—*badlands*. Badlands have a soft, rounded appearance that reflects the softness of the rocks. Badlands also have an intricate network of small drainages and eroded ridges because erosion is not restrained by strong beds in the rocks. [Petrified Forest, Arizona]

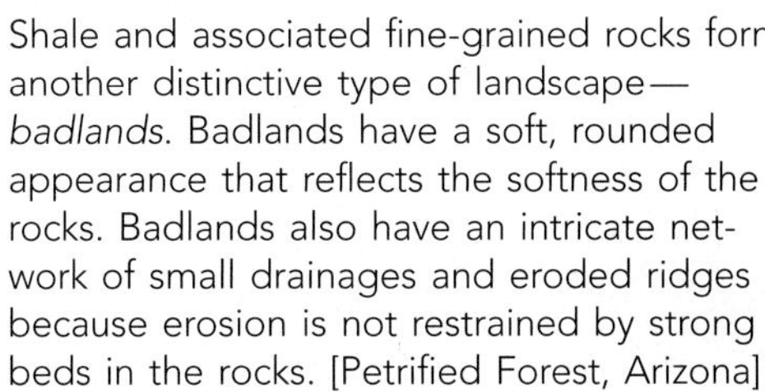

07.10.d4

Before You Leave This Page Be Able To

- ✓ Summarize the main characteristics of shale and siltstone, including which rock has the finest particles.
- ✓ Describe the land environments in which shale, mudstone, and siltstone form.
- ✓ Describe the environments near shorelines and farther offshore in which shale and siltstone form.
- ✓ Describe how some shale and siltstone are expressed in the landscape, including some of the landscape features they form.

How Do Carbonate Rocks Form?

LIMESTONE AND CERTAIN OTHER SEDIMENTARY ROCKS are called *carbonate rocks* because they consist of a carbonate ion combined with calcium, magnesium, or other elements. Most carbonate rocks form directly from water through chemical or biological processes, but some are clastic rocks consisting of pieces derived from shells, coral, or the erosion of carbonate bedrock.

A What Are the Characteristics of Carbonate Rocks?

Limestone is a common rock and exists in many varieties, all of which consist mostly of the mineral *calcite* ($CaCO_3$). Calcite can convert to the mineral *dolomite* by the addition of magnesium (Mg), which produces the carbonate rock *dolostone*. Limestone and dolostone commonly occur together.

Limestone typically is a gray rock. Its color ranges from almost white to dark gray, but it can also have shades of yellow, tan, or brown. It is soluble, so it frequently has a "dissolved" appearance.

07.11.a1

07.11.a2

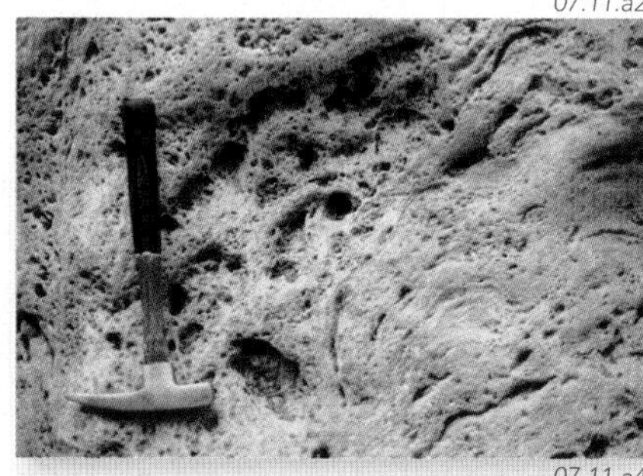

Travertine is a variety of limestone that usually is cream-colored and porous (has open spaces). It can precipitate in springs, lakes, and caves. Most travertine is banded because one coating precipitated after another.

Limestone frequently includes fossils, such as shells, corals, and other marine organisms, as shown here. Limestones that form in lakes may have fossils of nonmarine organisms, such as freshwater fish.

07.11.a3

07.11.a4

Dolostone, consisting of the mineral dolomite, resembles limestone, but it has a "less-dissolved" appearance because it is less soluble. Dolostone can be gray, but it commonly is tan, light brown, or even slightly orange.

Erosion and reworking of coral, shells, or carbonate bedrock produces carbonate clasts. The resulting rock has a clastic texture, but a carbonate composition, such as this coarse sandstone composed of calcite grains.

07.11.a5

07.11.a6

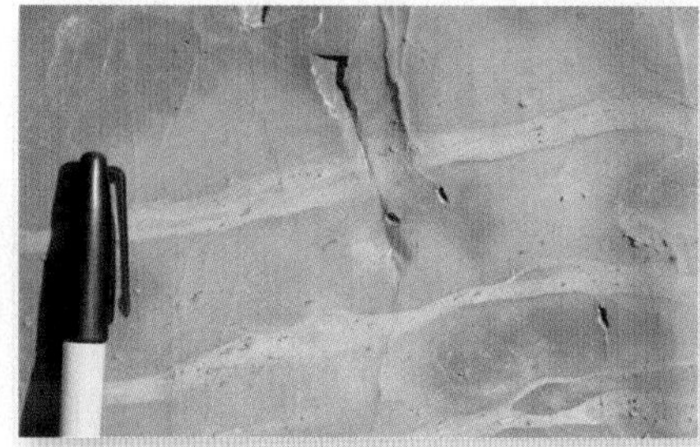

Some sedimentary rocks contain a mixture of carbonate and noncarbonate clastic material, such as clay particles and grains of quartz. Intermixed clay usually gives the rock a greenish, tan, dark gray, or pinkish tint. It can also give the rock a mottled appearance.

B In What Nonmarine Environments Do Carbonate Rocks Form?

Most limestones form in marine environments, but limestone can also be deposited around springs, in certain lakes, and as coatings and various features in caves. Carbonate forms during soil development in dry climates.

Limestone forms in lakes that experience large amounts of evaporation, such as lakes in hot, dry climates. These limestones usually have light color, as do these cream-colored limestones and shales. [Green River, Utah]

07.11.b1

07.11.b2

Limestone in some lakes occurs as coatings and irregular masses of white carbonate material. In some cases, this carbonate forms pillars, such as these at Mono Lake in eastern California.

C How Do Carbonate Rocks Form in Marine and Nearshore Environments?

Most carbonates form in marine settings, such as reefs and as carbonate accumulations on the continental shelf. Carbonates also form on low-lying mud flats and along shorelines dominated by carbonate sand.

Reefs are important carbonate environments. Coral and other reef organisms extract calcium carbonate from the water to build their skeletons, shells, and stems. Reef-formed limestones normally preserve these fossils.

07.11.c1

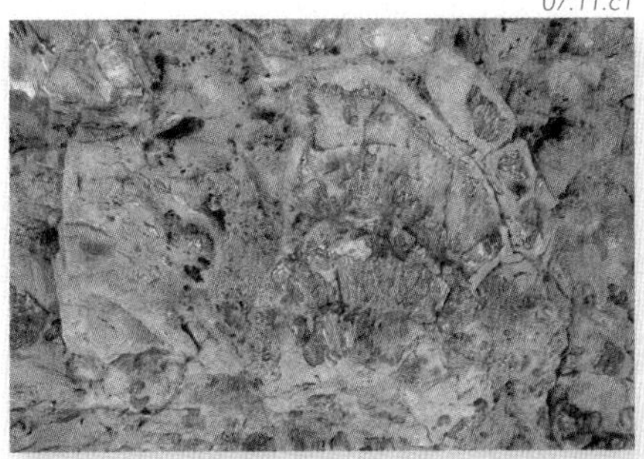

07.11.c2

Storms and waves break off pieces of reef and grind the pieces, along with pieces of shells, into calcite sand, which forms white, sandy beaches. If buried and lithified, the calcite beach sand becomes a type of clastic limestone.

Limestones formed from the remains of carbonate-secreting organisms or through chemical processes accumulate on continental shelves. Such deposits can form thick limestone sequences. [Provo, Utah]

07.11.c3

07.11.c4

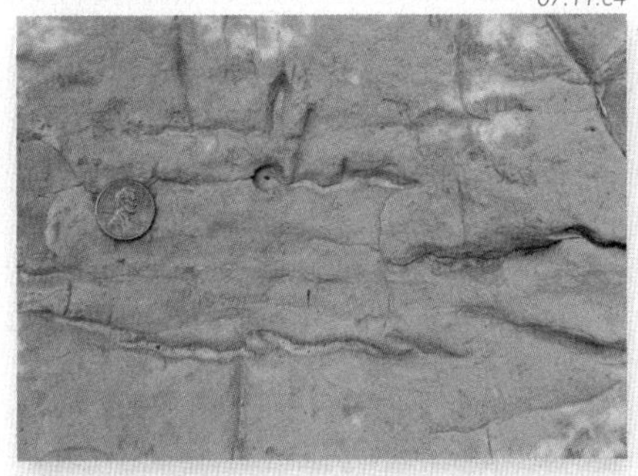

Dolostone forms when magnesium-carrying fluids interact with limestone. The fluids cause the calcite to be replaced by dolomite. This replacement is usually not complete, so the rocks are part limestone and part dolostone, each weathering in its distinctive way.

D How Are Carbonate Rocks Expressed in the Landscape?

Limestone is a very common and distinctive sedimentary rock that covers large areas of North America and other continents. It is generally recognized from a distance by its gray color and well-bedded character.

In some climates, limestone and dolostone are fairly resistant rocks. The rocks form gray cliffs and steep slopes and have individual beds of slightly different thickness and color. [Chamonoix, French Alps]

07.11.d1

07.11.d2

In wet climates, weathering, erosion, and dissolution of limestone may not affect all areas equally but instead may leave behind pillars of gray limestone, such as these pillars in China.

In very wet climates, limestone, and to a lesser extent dolostone, are very soluble and so are not very resistant to weathering and erosion. They commonly contain numerous caves and small openings formed where the carbonate rocks dissolved away. Dissolving of limestone is usually most pronounced along fractures, as is shown here. [Austrian Alps]

07.11.d3

Many caves form where groundwater dissolves limestone in the subsurface. The groundwater carries away the soluble carbonate and leaves behind a cave. When the roof of such a cave collapses, a closed depression, or *sinkhole* forms on the surface. Sinkholes can damage buildings and roads. [Florida]

07.11.d4

Before You Leave This Page Be Able To

- ✓ Summarize the characteristics of limestone and dolostone.
- ✓ Summarize the environments in which limestone and other deposits of calcium carbonate form.
- ✓ Describe how dolostone forms.
- ✓ Describe how carbonate rocks are expressed in the landscape, including sinkholes and limestone pillars.

How Do Changing Environments Deposit a Sequence of Sediments?

MOST SEDIMENTARY ROCKS ARE IN A SEQUENCE, where one rock layer overlies another. The layers in a sequence may all be the same rock type, such as limestone, but more commonly include a variety of sedimentary rock types, such as sandstone, shale, and limestone. What changes in environment occur in order to deposit a sequence of different rock types?

A What Happens When Environments Shift Through Time?

Environments that move across Earth's surface over time can result in a sequence of different sedimentary rocks. What would happen, for example, if seas advanced across a region? The figures below illustrate one possible outcome.

Time 1

1. At the earliest time, a shoreline separates marine environments to the left from beach and land environments to the right.

2. Sand with pieces of shell is deposited along the beach and outward into nearby shallow water.

3. The land is being eroded instead of receiving sediment. There may be no sedimentary record on land of the time when the beach sand was forming. The time is instead represented only by an erosion surface.

07.12.a1

Time 2

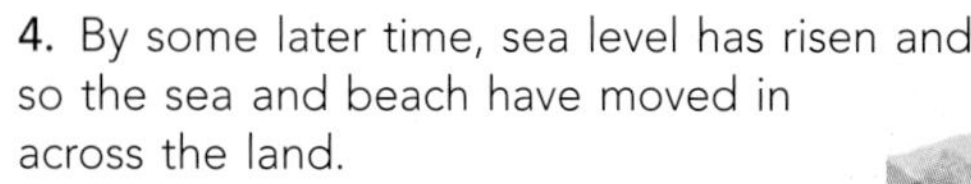

4. By some later time, sea level has risen and so the sea and beach have moved in across the land.

5. The area that used to be covered by beach sand is now far enough from the shoreline to have clear water in which a coral reef flourishes.

6. A lagoon is between the reef and shoreline. The relatively tranquil water of the lagoon traps mud that accumulates on the seafloor.

7. As the sea advances, beach sand is deposited over areas that used to be land. The base of the sand layer is a buried erosion surface.

8. The lagoon is located where a sandy bottom used to exist, so lagoon mud is deposited over the older layer of beach sand.

07.12.a2

Time 3

9. As the sea moves farther inland, the center of the area becomes a reef. Comparing this figure to previous ones reveals that the area of the reef was first near a beach and then was a lagoon.

10. The offshore progression results in a sequence of beach sand overlain by lagoon mud, which is overlain by limestone.

11. Therefore, when the sea advances across the land, shoreline deposits are progressively overlain by sediments that represent areas farther and farther offshore. An advance of the sea across the land is called a *transgression*.

12. Note that different environments can exist at the same time, such as a lagoon and beach. A different type of sediment is deposited in each environment, and each type of sediment is called a *sedimentary facies*. We have here a beach facies, a lagoon facies, and a reef facies. The term *facies* is used for the sediment and the resulting rock.

07.12.a3

B What Happens When the Sea Moves Out?

The opposite of a transgression is where the sea retreats and more land is exposed. A retreat of the sea is a *regression* and is illustrated below.

Time 1

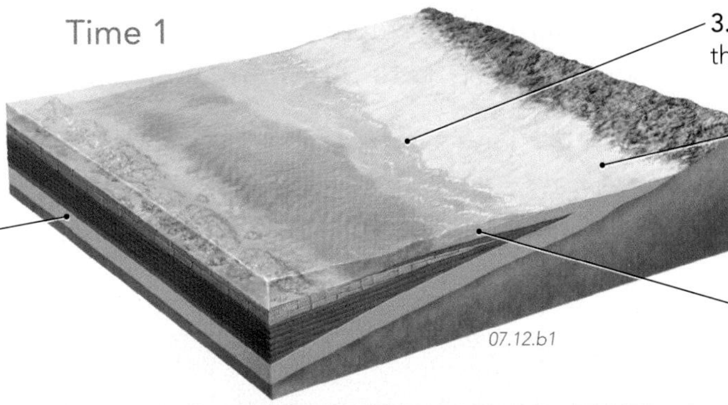

1. As the sea moves out during a regression, such as when sea level becomes lower, the sedimentary facies shift toward the sea (to the left in this series of figures).

2. The sedimentary sequence deposited during the regression will be built on the previous three layers that formed during the transgression.

3. As the shoreline retreats toward the sea, the beach and lagoon follow.

4. Sand that previously was close to the beach is now being eroded and is available for reworking by the wind.

5. Farther from shore, beach sand is deposited over lagoon mud, and lagoon mud is deposited over reef limestone.

Time 2

07.12.b2

6. As the regression continues, the sedimentary facies shift farther toward the sea. The reef is now out of view.

7. Lagoon mud builds out over the limestone (all the way to the left edge of the model).

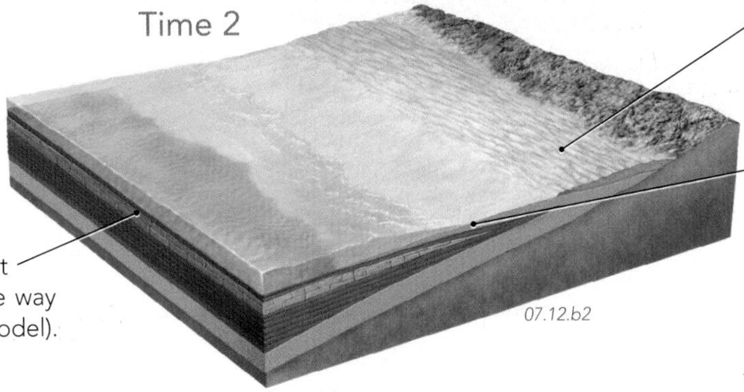

8. Wind remobilizes beach sand into a series of coastal sand dunes that follow the shoreline toward the sea.

9. The beach sand and dune sand build toward the sea, partially covering the lagoon mud, which in turn overlies a limestone.

Time 3

07.12.b3

10. During a regression, the sea retreats, and deeper marine sediment is successively overlain by shallower marine sediment, shoreline deposits, and, if the sea retreats far enough, land facies.

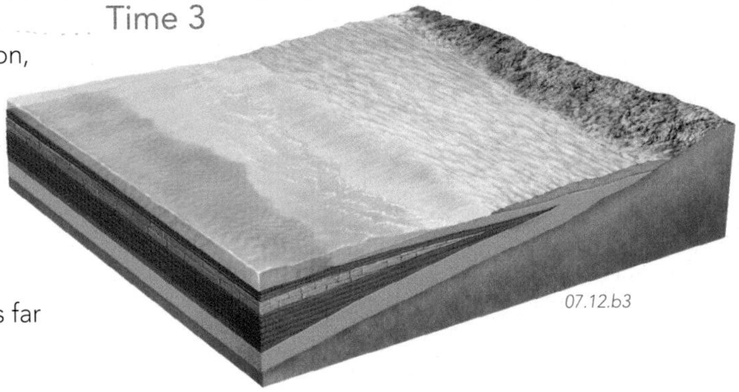

Stratigraphic Sections

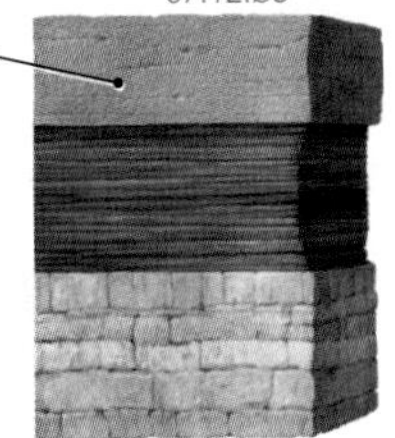

11. This stratigraphic section shows the sequence deposited during the *regression* (sea moving out). Limestone is overlain by mud, which is overlain by beach sandstone.

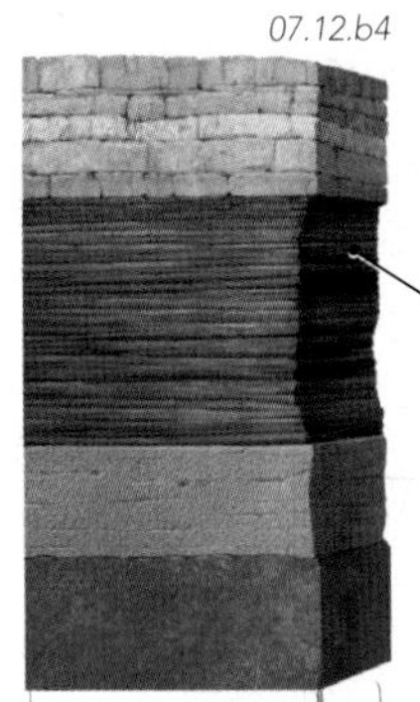

12. This section shows the sequence of sediments deposited during the *transgression* (sea moving in). An erosion surface is successively overlain by beach sand, mud, and limestone.

Why Sedimentary Layers End

All sedimentary layers eventually end. One reason layers end is because a facies ends, as shown by the thinning out and disappearance of the lagoon facies toward the right side of the illustrations above.

A layer also can end because it is deposited only within a channel, such as the river channel shown to the right. Coarse gravel accumulates inside the channel, but does not extend outside the channel.

Before You Leave This Page Be Able To

- ✓ Summarize or sketch what happens during a transgression and during a regression, including which way sedimentary facies shift.
- ✓ Sketch an example of a sequence of rocks formed during a transgression and contrast it with a sequence formed during a regression.
- ✓ Summarize or sketch two reasons why sedimentary layers end.

How Do We Study Sedimentary Sequences?

An important goal of geology is to reconstruct past events and environments. We do this by observing a sequence of rocks and noting characteristics that provide clues to the environment in which each rock unit formed. We then interpret the past environments by comparing these characteristics to sediment in modern environments. Studying a sequence of rocks allows us to infer how conditions changed over time.

A What Attributes of Sedimentary Rocks Are Indicators of Environment?

Sedimentary rocks contain many clues about the environment in which they formed. Nearly every attribute, such as the size or shape of clasts, provides some information with which to infer the past environment.

Color of Rocks

07.13.a1

Red sedimentary rocks generally form on land where they can be oxidized by the atmosphere, whereas *dark gray* sedimentary rocks usually form under water and in low-oxygen conditions.

Clast Size, Shape, and Sorting

07.13.a2

Large, angular, poorly sorted clasts reflect strong currents and limited transport. Small clasts indicate weak currents, and rounded, well-sorted clasts reflect more transport or reworking by waves.

Thickness of Bedding

07.13.a3

Thick bedding implies bigger events, faster rates of deposition, or longer times between environmental changes. Thin bedding implies smaller events or more rapidly changing conditions.

Types of Bedding

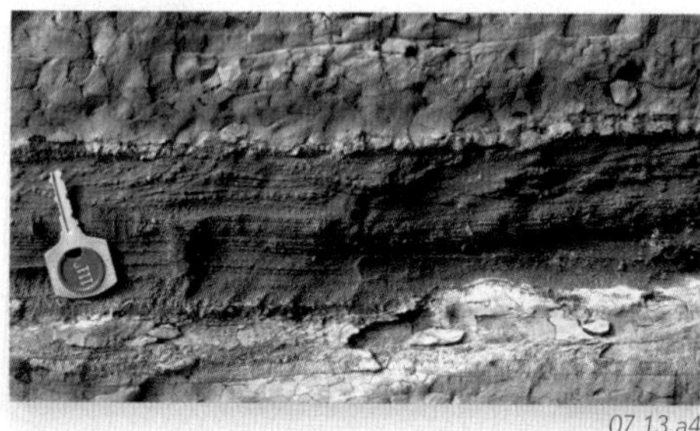

07.13.a4

Certain types of bedding reflect specific conditions of formation. Graded beds, shown above, indicate that the strength of the current decreased through time, either during a turbidity current or a flood.

07.13.a7

Large-scale cross beds in a well-sorted sandstone usually indicate that the sand was deposited by wind as a series of large sand dunes.

Mudcracks

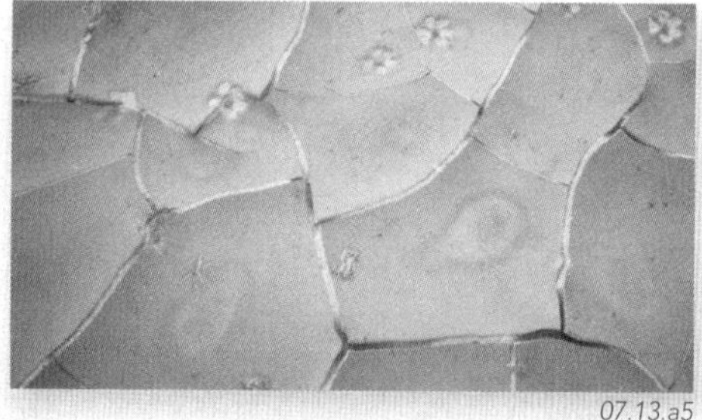

07.13.a5

When wet mud and clay dry, the sediment contracts and produces polygon-shaped pieces surrounded by cracks that can fill with sand, as in the modern example (with animal tracks) above.

07.13.a8

Sedimentary rocks having *mudcracks* must have been deposited on land and in environments where wet sediment could periodically dry.

Fossils

07.13.a6

Fossils of land plants indicate that the sediment was deposited on land and can provide information about temperature, elevation, amount of rainfall, and other environmental conditions.

07.13.a9

Marine fossils are diagnostic of deposition in seawater and can provide information about temperature, salinity, and clarity of the water.

Indicators of the Direction that Water or Wind Currents Flowed

07.13.a10

Cross beds, whether they form in rivers (as these did), sand dunes, or beneath the sea, are inclined toward the direction in which the current flowed, in this case toward the left.

07.13.a11

Large clasts in a river normally lean toward the direction that the current flows (left in this example). This property, called *clast imbrication*, can be used to infer past flow directions from a conglomerate.

07.13.a12

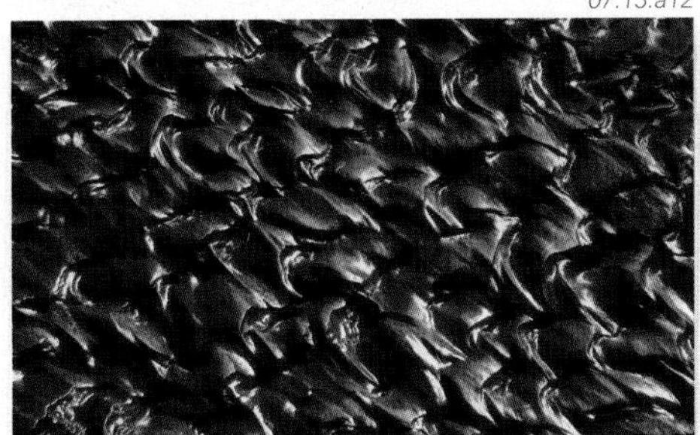

Ripple marks are small ridges and troughs formed by moving current. In some ripples, the ridges have a steeper side toward the direction of current flow. In this example, current flowed toward the lower right.

B What Can We Observe and Interpret in a Sequence of Sedimentary Rocks?

We can often infer past events, environments, and changes in the environment by carefully observing a sequence of rock layers. For each rock layer, we make observations about the attributes of the sedimentary rock types. These observations are the basis for interpreting the environment that each rock layer represents. Try this reasoning by reading each observation below, thinking about any clues, and then reading the interpretation.

Sequence of Sedimentary Rocks

07.13.b1

Observations	Interpretations
A reddish-gray breccia is at the top of the sequence. It is poorly sorted with angular boulders of granite in a mud-rich matrix.	The large, angular clasts indicate only minor transport of the clasts, like in a debris flow, and a nearby steep terrain, such as a mountain. The terrain consisted of granite.
The underlying layer is a tan sandstone. Lower parts of the unit include broken shells, but the upper part has fossils of land plants and coal.	The change of the type of fossils in the sandstone is consistent with the unit having formed along a shoreline, perhaps in a beach or delta.
The middle of the sequence is thick, gray shale with an intervening layer of limestone. The shale contains shallow-water marine fossils, such as clams, and the limestone contains fossil coral.	The shale and limestone accumulated in the shallow part of a sea. The shale may represent offshore muds or a lagoon; the limestone represents a reef.
Another layer of tan sandstone is beneath the gray shale and includes marine shell fragments.	This tan sandstone is interpreted as a beach sand or a sand that formed in shallow marine water.
A red sandstone at the base of the sequence is well sorted and contains large cross beds.	The well-sorted sand and large cross beds are consistent with the sandstone having been deposited on land as a series of large sand dunes.

Interpretation of the Change in Sediment Type over Time

In addition to specific interpretations about each unit, we can infer *changes in the environment* by comparing each unit with the unit above it. Begin at the bottom with the oldest unit, and work upward toward younger units.

1. The change from the sand dunes to overlying beach sands and marine shales with limestone is evidence of a transgression, an advance of the sea toward the land.

2. The sea probably reached its maximum advance during deposition of the limestone. The shoreline at this time was far enough away to allow coral to grow in clear water.

3. Sometime after the limestone formed, the sea retreated during a *regression*. Delta sands with land plants were deposited over the marine shales.

4. Finally, steep, granite mountains formed during some tectonic event. The mountains shed large granite clasts onto an adjacent area, perhaps in a series of debris flows.

Before You Leave This Page Be Able To

- ✓ Summarize the attributes that we observe in sedimentary rocks and how each indicates something about the rock's origin and environment.
- ✓ Given a sequence of rocks and a list of key attributes, interpret the environment of each rock and how the environment changed.

Why Are Sediments and Sedimentary Rocks Important to Our Society?

SEDIMENTARY ROCKS AND THEIR RESOURCES are essential to our society. Sediments and sedimentary rocks are the main hosts for groundwater, oil and natural gas, coal, salt, and material for making cement and construction aggregate. The study of sedimentary rocks also helps us understand the geologic history of Earth, including climate change and how life originated and changed through time.

A How Do Sedimentary Rocks Control the Distribution of Resources?

Sedimentary rocks host many of our most important resources. Some resources, such as coal and salt, originated as sedimentary deposits, whereas other resources are most common in sedimentary rocks because these rocks permit the flow and entrapment of fluids, such as water and oil.

Groundwater

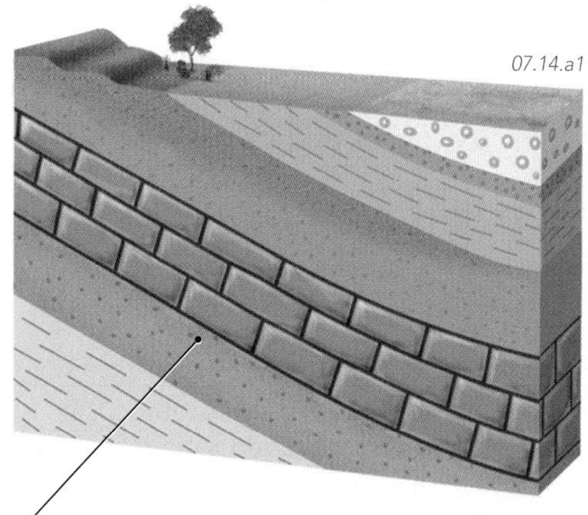

Groundwater, shaded here in blue, occurs predominately in sediment and sedimentary rocks. Most groundwater resides in the pore spaces between sedimentary grains and in fractures.

Petroleum

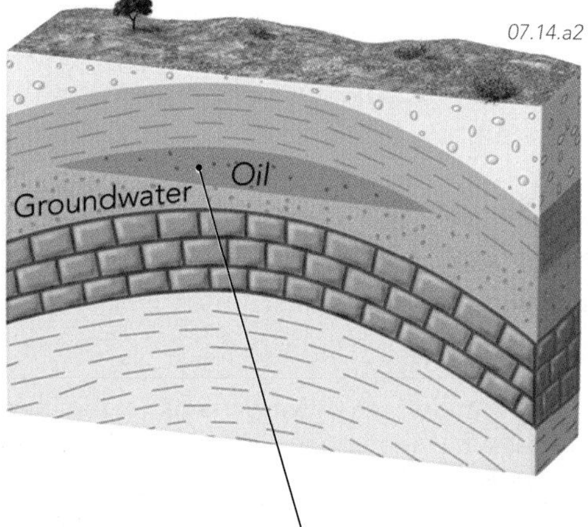

Oil and natural gas occur mostly in sedimentary rocks. Oil and gas form in organic-rich sedimentary rocks and then migrate upward until they reach the surface or are trapped at depth.

Coal

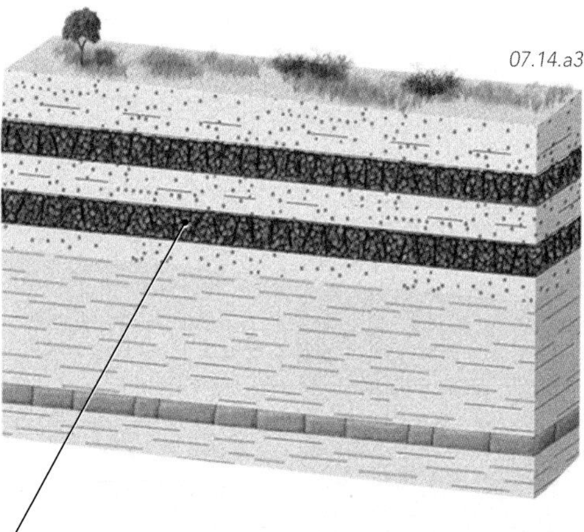

Coal forms as a sedimentary rock through the consolidation of plant remains that accumulate in swamps, deltas, and other wetland environments. Most coal is used to generate electrical energy.

Cement from Limestone

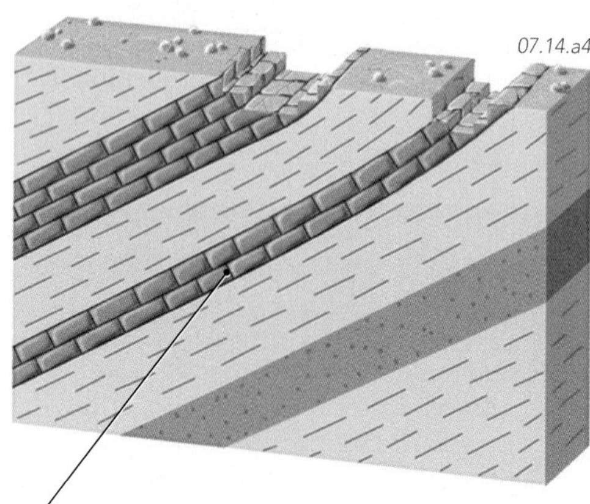

Cement is produced from limestone that is relatively free of sand, silt, chert, and other impurities. Cement is used to make concrete for highways, bridges, building foundations, and other constructions.

Salt

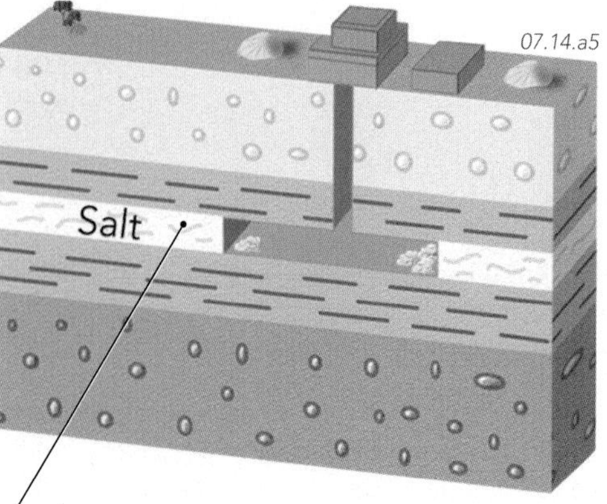

Salt is either mined from ancient sedimentary salt layers or is harvested by evaporating salty water. It is used in the preparation of food, medicines, and various industrial products.

Uranium

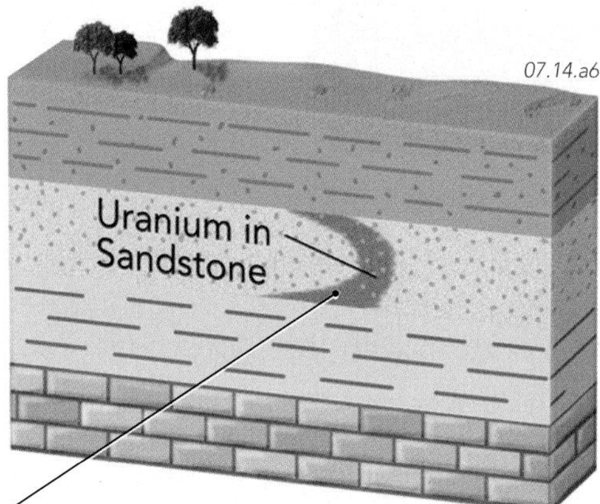

Uranium deposits commonly occur in sandstone and other sedimentary rocks, but the uranium often was derived from elsewhere and brought into the area by migrating groundwater.

B How Do Sedimentary Rocks Help Us Understand Earth's Geologic History?

Sedimentary rocks are the primary source of information about ancient environments, climate change, and past events. Fossils in sedimentary rocks are the main record of how life originated and evolved.

07.14.b1

Geologists *study modern environments* to understand the processes that are occurring, the types of sediment produced, and how these environments may impact where we live and what we do. Such studies are also the foundation for understanding an area's ecology. [Indonesia]

07.14.b2

By studying ancient sedimentary rocks, we observe the record of *past environments.* From these observations, we can interpret the character and distribution of environments, how environments changed through time, and how resources formed in the environments. [Grand Canyon, Arizona]

07.14.b3

Sedimentary rocks provide important data for *investigating climate change*. By understanding the severity and possible causes of past climate changes, we can better understand possible consequences of future climate changes, such as global warming or cooling.

07.14.b4

Sedimentary deposits and rocks allow us to *examine the record of past events,* such as landslides (shown here), storms, and earthquakes. Studying these deposits and rocks enables us to infer the processes that occurred and consider how the events affect the landscape and life around them.

07.14.b5

Sedimentary rocks are the main way we *study the sequence of past events*, such as advances of the seas, migration of ancient deserts, and erosion of mountains. The succession of rock layers above, from beach sandstone at the base to upper limestone cliffs, records a transgression. [Yampa River, Utah]

07.14.b6

Fossils allow us to *study ancient life*, including the types of organisms that lived at different times and the environments in which they lived. By studying the succession of fossils from one layer to the next, we observe how life on Earth evolved and we may infer the causes of the observed changes.

Sand and Gravel—the Most Used Sedimentary Resource

Resources such as gold, oil, and diamonds easily capture our interest because they are so precious or because we depend so highly on them. However, sand and gravel, based on the amount of material extracted, is our most widely used resource.

The phrase *sand and gravel*, defined as a resource, refers to sediment that commonly is excavated from pits and used in various types of construction. It includes clasts of various sizes, from clay, silt, and sand, to pebbles, cobbles, and boulders. The material sometimes is used as it is, but it more commonly is poured through large screens to sort out certain sizes of clasts, such as those that are too large. In some cases, clasts are crushed to achieve a desired size.

After the material is sorted into the correct sizes, it can be added to cement to make concrete and concrete blocks, added to clay to make tile, or used as fill beneath buildings and roads. In 2006, more than 900,000 million tons of sand and gravel were used in the United States, approximately 3 tons per person.

Before You Leave This Page Be Able To

- ✓ Summarize or sketch some of the main resources that occur in sedimentary rocks.
- ✓ Summarize how sedimentary rocks help us understand modern and ancient environments, events, and life.
- ✓ Describe what the phrase *sand and gravel* means and how we use these important materials.

How Did Sedimentary Layers West of Denver Form?

THE FOOTHILLS OF THE FRONT RANGE west of Denver, Colorado contain spectacular exposures of sedimentary rocks. The layers have been folded and tilted along the mountain front and form dramatic landscapes. The area provides an example of how to integrate various aspects of sedimentary rocks to interpret the geologic history of a region.

A How Are the Sedimentary Layers Exposed?

The figure below shows a geologic map superimposed on topography of an area west of Denver. The colors show the distribution of different sedimentary layers and other rock types. The front of the figure is a geologic cross section that shows the interpreted geometry of rock layers at depth. Begin on the lower left.

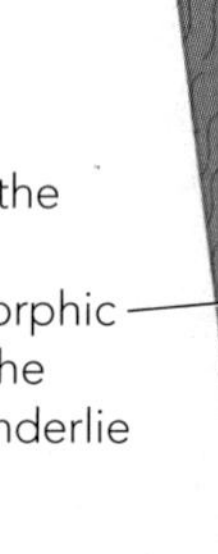

07.15.a1

1. The Front Range is part of the Rocky Mountains and consists mostly of Precambrian metamorphic and granitic rocks. These are the oldest rocks in the area and underlie all other rock units.

2. Red sandstone and conglomerate (colored blue on this map because they are Paleozoic in age) were deposited on top of the metamorphic and igneous basement. The layers were later tilted and now dip eastward off the mountain front. These rocks form the dramatic exposures at Red Rocks Amphitheater, a famous venue for music concerts.

3. A sequence of diverse sedimentary rocks, including marine shale, beach and delta sandstone, river deposits, and wind-blown sandstone, overlies the red rocks.

4. Green Mountain, a round hill, is east of the mountains and contains the highest and youngest sedimentary rocks in the map area.

5. A long, gently curving ridge called the Dakota Hogback is a dominant feature of the landscape. This ridge is held up by relatively resistant sandstone of the Dakota Formation, which is tilted eastward away from the mountains. This sandstone is colored green on this figure.

6. Low areas on either side of the hogback are underlain by more easily eroded sedimentary layers, such as shale.

7. The rock sequence continues into the subsurface where it is folded and cut by a series of faults. An oil field was found by studying the sedimentary layers and predicting where to drill to find oil several kilometers beneath the surface.

B How Are the Sedimentary Layers Expressed in the Landscape?

07.15.b1

Red Rocks Amphitheater nestles within the lowest sedimentary unit, a series of reddish conglomerate and sandstone layers that dip away from the mountain.

07.15.b2

The Dakota Hogback is a ridge held up by a tilted, resistant layer called the *Dakota Sandstone*. The sandstone is tilted to the right (east) and Red Rocks is to the left.

07.15.b3

A spectacular roadcut along I-70 exposes tilted Mesozoic rocks. A trail through the sequence is accompanied by descriptions of each rock formation.

C What Is the Exposed Sequence and History of Sedimentary Rocks?

The history is depicted below in a stratigraphic section, which is accompanied by a summary of key characteristics, photographs, and maps depicting the interpreted environments. Begin at the bottom of the section.

07.15.c1

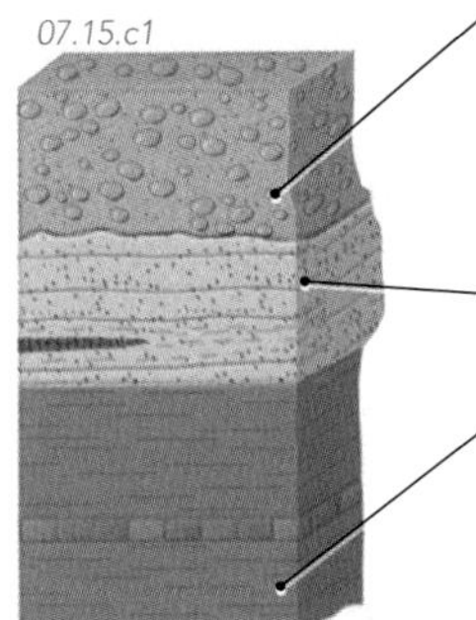

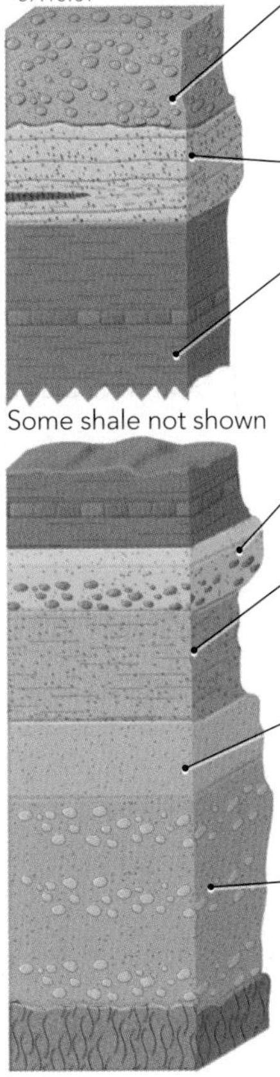

7. Cenozoic boulder conglomerate and sandstone at the top of the section include clasts derived from the Precambrian crystalline rocks of the Front Range and from some local volcanic terrains. These rocks record the formation and erosion of the mountains 70 m.y. to 40 m.y. ago.

6. The shale is overlain by late Mesozoic sandstone, shale, and coal with plant fossils. These layers were deposited on land and along shorelines.

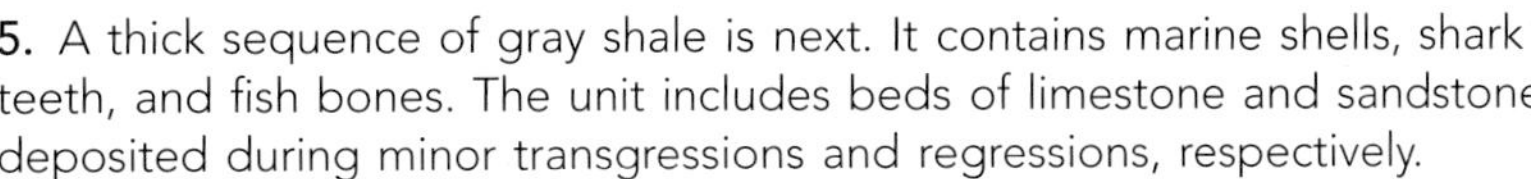

5. A thick sequence of gray shale is next. It contains marine shells, shark teeth, and fish bones. The unit includes beds of limestone and sandstone deposited during minor transgressions and regressions, respectively.

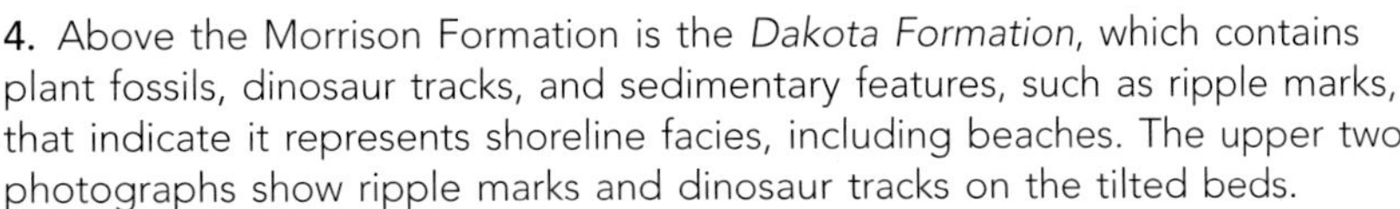

4. Above the Morrison Formation is the *Dakota Formation*, which contains plant fossils, dinosaur tracks, and sedimentary features, such as ripple marks, that indicate it represents shoreline facies, including beaches. The upper two photographs show ripple marks and dinosaur tracks on the tilted beds.

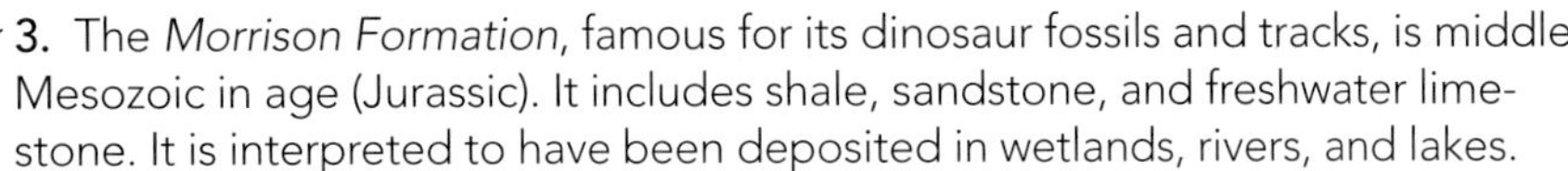

3. The *Morrison Formation*, famous for its dinosaur fossils and tracks, is middle Mesozoic in age (Jurassic). It includes shale, sandstone, and freshwater limestone. It is interpreted to have been deposited in wetlands, rivers, and lakes.

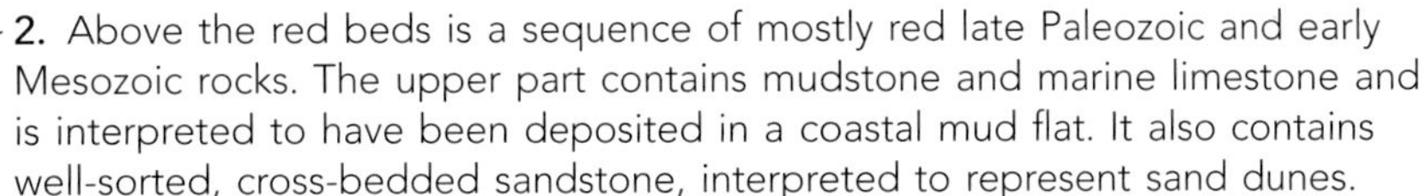

2. Above the red beds is a sequence of mostly red late Paleozoic and early Mesozoic rocks. The upper part contains mudstone and marine limestone and is interpreted to have been deposited in a coastal mud flat. It also contains well-sorted, cross-bedded sandstone, interpreted to represent sand dunes.

1. The lowest sedimentary unit is a reddish sequence of sandstone and poorly sorted, coarse conglomerate and breccia (▷). This unit is interpreted to have been deposited by rivers and debris flows that drained an ancient mountain range called the *Ancestral Rockies*. The unit is late Paleozoic in age and rests on Precambrian rocks.

07.15.c2

07.15.c3

07.15.c4

Paleogeographic Maps for Three Chapters in the Geologic History of Colorado

07.15.c5

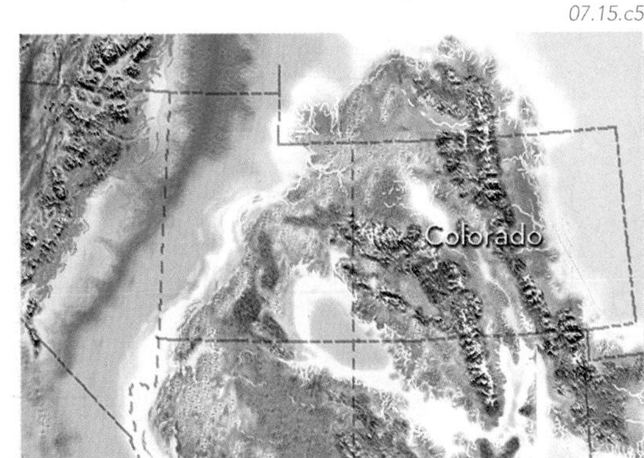

1. *Late Paleozoic*—Sandstone and coarse, reddish conglomerate formed from sediment shed off the Ancestral Rockies. Salt and other evaporites formed in inlets of seawater that evaporated.

07.15.c6

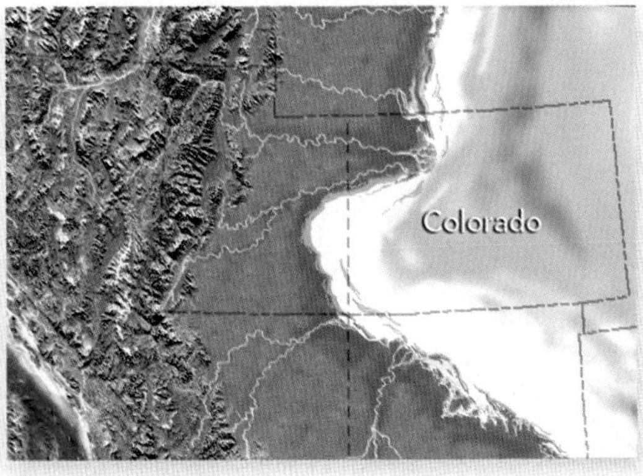

3. *Late Mesozoic*—A shallow sea stretches from the Arctic to the Gulf of Mexico and is overrun by deltas from the west. The Dakota Formation accumulated during the transgression, and marine shales accumulated in the shallow sea.

07.15.c7

2. *Early and Middle Mesozoic*—The region is a continental environment dominated by mud flats, sand dunes, lakes, and river systems. These environments change their distribution over time and so deposit a sequence of different sedimentary layers. These include mudstone, sandstone, and conglomerate. Dinosaurs roamed the landscape, leaving tracks and bones in the Morrison Formation of Jurassic age.

Before You Leave This Page Be Able To

✓ Summarize how the characteristics and sequence of sedimentary rocks can be used to reconstruct the geologic history of an area. Use examples from sedimentary rocks west of Denver.

What Is the Sedimentary History of This Plateau?

A plateau in southern Utah exposes a sequence of different kinds of sedimentary rocks. Some sedimentary units were deposited on land, and others were deposited by shallow seas. Using key observations about each rock unit, you will reconstruct the history of these sedimentary rocks.

Goals of This Exercise:

- Use photographs or samples to make observations about the sedimentary layers.
- Interpret a possible environment for each sedimentary layer.
- Use a stratigraphic section to infer how the environment changed through time as the layers were deposited.

A Observe the Sequence and Characteristics of Sedimentary Layers

Two photographs show the upper and lower parts of a sequence of layers. Observe the photographs and try to identify boundaries between different sedimentary units. Next, compare your observations with the observations and interpretations next to each photograph and with the information on the next page. Your instructor may also provide you with rock samples. Record your observations in the worksheet.

07.16.a1

Locality Number 1: Top Part of the Sequence

1. The highest rock unit exposed in this area forms a series of tan, brown, and gray ledges. There are obvious layers represented by the alternation of ledges and slopes. The ledges are similar in appearance to those formed by sandstone.

2. Below the ledges is a gray slope that is locally covered by pieces eroded from the overlying ledges. The rock forming the gray slope has layers, but all of the layers look similar. It looks fairly soft and nonresistant to erosion, as is common for fine-grained rocks like shale. The gray color of the rock implies that the unit was deposited in conditions that were not rich in oxygen.

07.16.a2

Locality Number 2: Lower Part of the Sequence

3. In a nearby area, the soft gray rocks shown in the upper photograph directly overlie a tan and cream-colored cliff. The cliff-forming unit is resistant to erosion and weathers like many sandstones do.

4. The cliff overlies a series of soft, thinly layered rocks that are maroon, reddish brown, gray, and cream colored. The soft rocks are probably composed of fine-grained, easily eroded material. Most rocks that have this reddish color were deposited on land.

B Interpret the Sedimentary History of the Sequence of Layers

The stratigraphic section below shows the relative thicknesses of the units. The oldest unit is on the bottom and the youngest is at the top. Photographs and brief descriptions of each rock unit accompany the section. Your instructor may provide you with samples of similar rocks. Follow the steps below to propose a plausible interpretation for the environment of deposition for each sedimentary unit and for how the environment changed from one rock unit to the next. Write your answers to the following questions on the worksheet or on a sheet of paper.

1. What is your interpretation of the environment for each of the four rock units? List two key attributes of each unit that support your interpretation.
2. What is the oldest environment represented by this rock sequence?
3. Does the change of environment from the base of the section up to the thick gray shale indicate an advance (transgression) or retreat (regression) of the sea? Explain the reasons for your answer.
4. Does the change from the thick gray shale to the overlying sandstone indicate a transgression or a regression? Explain the observations that support your answer.
5. Which of the following phrases summarizes the history of the entire sequence: (a) a transgression, (b) a regression, (c) a transgression followed by a regression, or (d) a regression followed by a transgression?
6. Compare this sedimentary sequence to the one exposed west of Denver (in the previous two-page spread). What names from the Denver area would you apply to the lowest unit and to the overlying yellowish-tan sandstone in this plateau?

Stratigraphic Section

07.16.b1

This unit includes sandstone, mudstone, and layers of coal (shown in black). The upper part of the unit contains sandstone beds with small cross beds. The mudstone has mudcracks and plant fossils. The lowest part of the unit contains tan sandstone with broken marine shells. The photograph to the right shows layers of black coal in this unit. ▷

07.16.b2

This shale is medium to dark gray because it has a high amount of organic matter. It contains fossils of clams and other marine organisms. Thin limestone beds are locally present in the middle of the unit, but are not shown in the section. The shale and limestone contain abundant marine fossils. The photograph to the right shows a close-up of the transition from the shale to the overlying sandstone. ▷

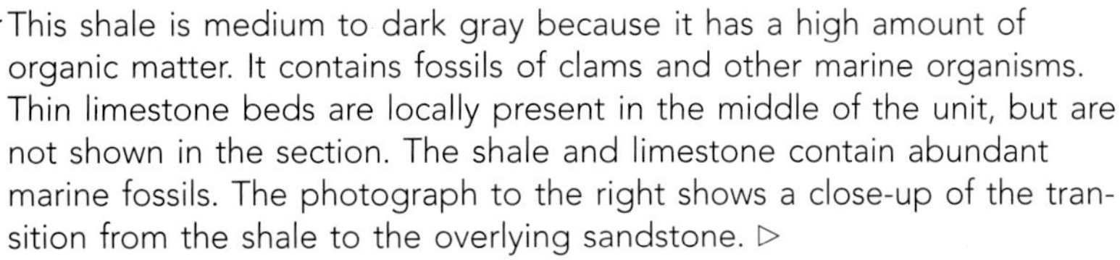

07.16.b3

This unit is mostly a yellowish-tan sandstone containing quartz sand with small pieces of marine shells. As shown in the photograph to the right (▷), the very base of the unit is a thin conglomerate that overlies an erosion surface. This lower part locally contains fossils of wood and leaves.

07.16.b4

The lowest unit includes a basal conglomerate with moderately rounded pebbles and coarse sand containing scattered rounded pebbles and pieces of fossilized wood. The conglomerate is overlain by reddish, maroon, and gray shale and mudstone with plant fossils. The photograph to the right (▷) shows a nearly circular dinosaur track where a large, plant-eating dinosaur with huge round feet stepped into and pressed down the then-soft sediment.

07.16.b5

CHAPTER 8

Deformation and Metamorphism

ROCKS CAN BE DEFORMED, heated, and subjected to higher pressure as they are buried. The new conditions change the rocks by causing existing minerals to grow, new minerals to form, and new structural and metamorphic fabrics to develop—the process of *metamorphism*.

The Appalachian Mountains and adjacent parts of the eastern United States display a wide range of landscapes. This image shows satellite data superimposed on topography for part of southeastern Pennsylvania. The image shows curving mountains and ridges (green) alternating with lowlands (pinkish brown). The large river is the Susquehanna River, which flows south and cuts across the ridges and lowlands.

Choose a ridge and follow it across the region. What does it do? What other features do you observe as you examine this image?

This distinctive region has alternating ridges and valleys. Some of the ridges and valleys are straight, but others curve back and forth across the landscape. This region is called the *Ridge and Valley Province* by some people and the *Valley and Ridge Province* by other people.

How did these unusual landscapes form, and what do they tell us about the architecture of the underlying rocks?

08.00.a1

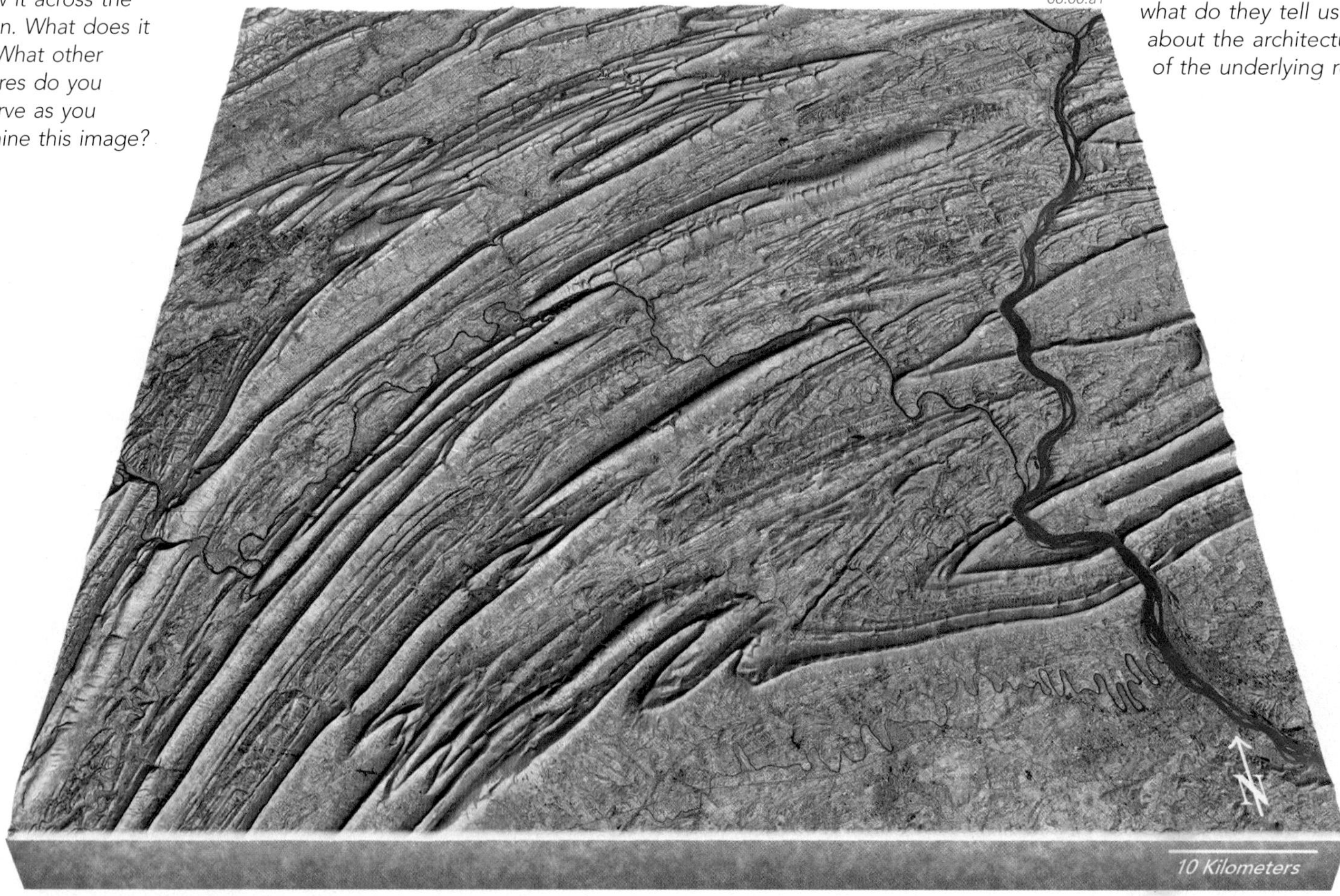

10 Kilometers

The landscapes of the Ridge and Valley Province, as shown in this cross section for an area south of the map, reflect large folds in the Paleozoic sedimentary rocks. The folds may be tens of kilometers across and more than 100 kilometers (62 miles) long.

How do we determine that large folds are present in an area?

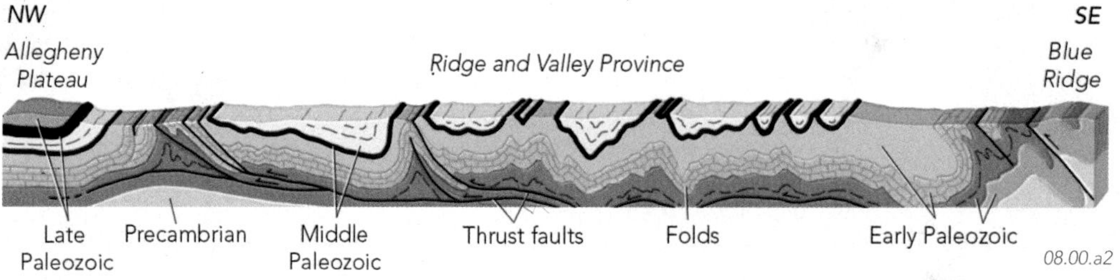

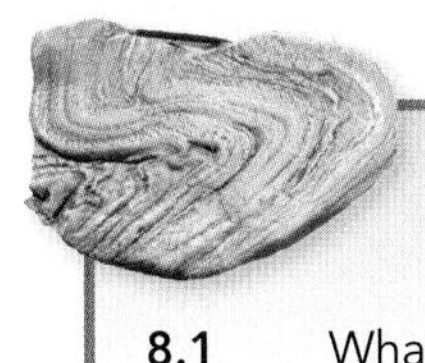

TOPICS IN THIS CHAPTER

08.00.a3

Large folds warp the rock layers, in this case folding the layers downward. Folds exist at almost all scales, from the large folds shown in the cross section to folds small enough to be picked up in a hand specimen. [Maryland]

How do folds form, and how do we describe their shapes?

08.00.a4

The cross section on the facing page has thicker lines that represent faults along which the folded layers have moved. These faults commonly stack one layer of rock on top of another, as shown in this photograph. [Tennessee]

Which type of faults are these, and what caused the faults to move? Are the faults and folds somehow related?

08.00.a5

During folding and faulting, imposed forces and slightly increased temperatures cause mineral grains to rotate, change shape, and partially dissolve. These changes formed a new structural feature called *rock cleavage*, which cuts steeply across these nearly horizontal layers.

How do new structural and metamorphic features form, and what do they tell us about the conditions that affected the rock?

Origin of the Ridge and Valley Province

The Appalachian Mountains have a complex geologic history that includes four main time periods when the region was subjected to tectonic forces that caused the rocks to deform. One of these mountain-building episodes formed the folds, faults, and cleavage of the *Ridge and Valley Province*.

A broad region of the eastern United States, including the Ridge and Valley Province, is covered by Paleozoic sedimentary rocks, which formed between 542 Ma and 251 Ma. These rocks were deposited in various environments, including shallow seas, shorelines, and rivers. Rocks in the Appalachian Mountains were folded, faulted, and heated several times during the Paleozoic. The deformation culminated with a *continental collision* between Africa and eastern North America approximately 300 million years ago. The collision uplifted the central part of the Appalachian range and shoved huge slices of rock up and over sedimentary rocks west of the mountains. The rock layers within and below the slices, such as those in the Ridge and Valley Province, responded to the forces by folding, faulting, and squeezing out of the way. Squeezing of the rocks was accompanied by increased temperatures because the rocks were buried. Minerals in the rocks changed their arrangement or changed to new minerals. This process of change is called *metamorphism*. As the region was uplifted, erosion was guided by the folds and faults. It carved away some rock layers faster than others to produce the region's distinctive ridges and valleys that reflect the shape of the folds.

8.0

What Processes Can Deform and Change a Rock After It Forms?

ROCK CAN BE SUBJECTED TO FORCES, high temperatures, or both at the same time. As a result, the rock moves, rotates, changes shape, and has minerals grow, form, or be destroyed. How does rock respond to forces and temperature changes, and what types of features form in these new conditions?

A What Are Force and Stress?

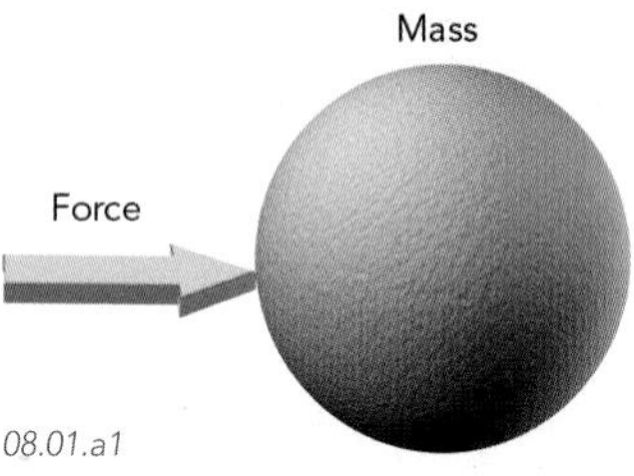

08.01.a1

◁ *Force* is a push or a pull that causes, or tends to cause, a change in the motion of a body. It is commonly expressed in terms of the amount of acceleration experienced by a mass.

08.01.a2

◁ The amount of force divided by the unit area to which the force is applied is called the *stress*. The force from a stone weight is distributed evenly across the top of a broad, wooden pillar.

08.01.a3

◁ If the same amount of weight is on a thinner pillar, the stress (force per unit area) on the pillar is greater. It might cause the pillar to splinter or break.

B How Do Rocks Respond to Force and Stress?

▷ Rocks within Earth are subjected to forces from the weight of overlying rocks, from tectonic forces pushing or pulling on the rocks, from cooling and heating, and from pressurized fluids, such as water and magma. In dealing with geologic structures, we talk about stress more than force because rocks, like the wooden pillar above, respond to whether some amount of applied force is spread out or is concentrated. These figures show stress with a blue arrow.

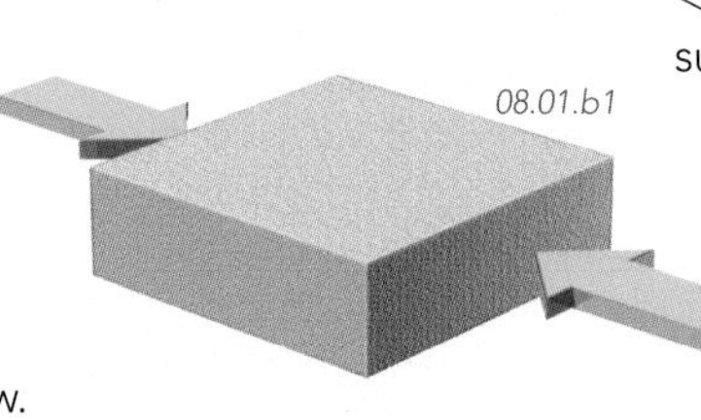

08.01.b1

◁ A volume of rock may remain unchanged if subjected to only a small amount of stress. If the imposed stresses are higher, three things can happen to a volume of rock. It can be *displaced* from one location to another, be *rotated*, or have its shape *modified*. All three responses may occur at the same time.

Displacement

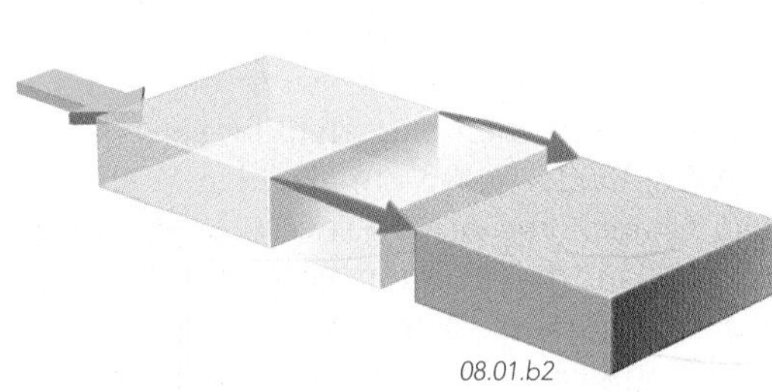

08.01.b2

△ In response to stress, a volume of rock may be moved, or displaced, from one place to another. During *displacement*, a rock can behave as a rigid object or can change shape as it moves. In the photograph below, the white layer has been displaced by movement along fractures.

08.01.b5

Rotation

08.01.b3

△ A volume of rock may be *rotated* in response to stresses. Rotation can tilt the volume of rock or spin it horizontally. The rock layers in the photograph below were deposited as horizontal layers, but the layers have since been rotated (in this case tilted). [Morrison, Colorado]

08.01.b6

Strain

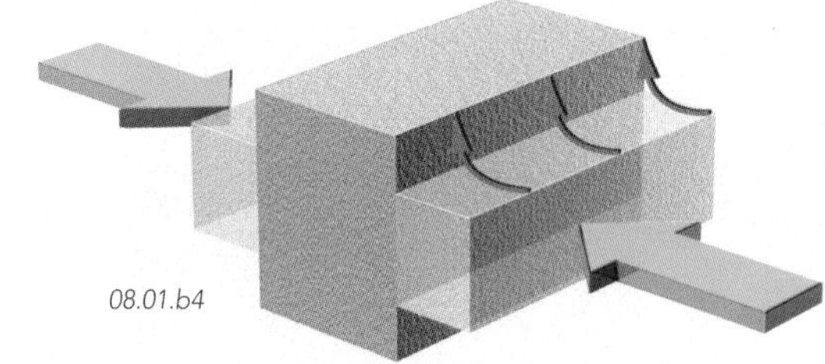

08.01.b4

△ A rock can respond to stress by deforming internally—changing size or shape. A change of size or shape is called *strain*. Stress is the cause, and strain is the effect. Below, pebbles in a conglomerate were *strained* in response to stress pushing on the rock.

08.01.b7

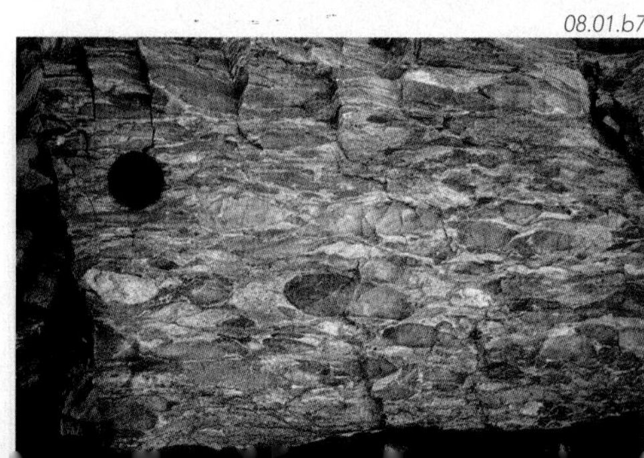

C What Processes Affect Rocks at Different Depths?

The structural behavior of rock changes as temperature and stress increase with depth. The changing conditions also influence the stability of minerals and the way rocks react to water and other fluids.

Conditions

08.01.c1

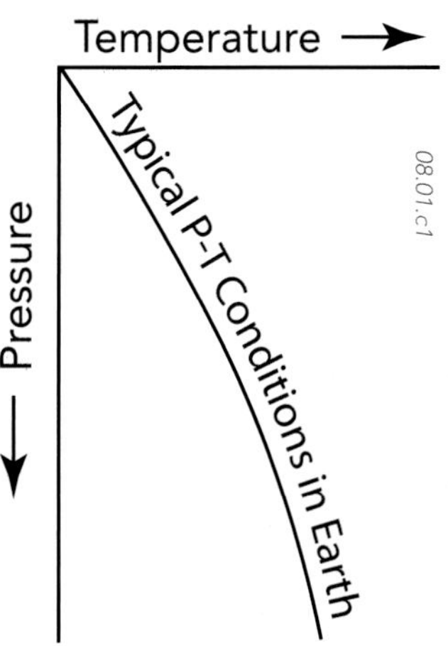

△ This graph shows that temperature (T) and pressure (P) increase with depth within Earth. The term *pressure* is similar to the term *stress* in that it refers to the amount of force per area. We use pressure to signify that the amount of force imposed on the rock is the same in all directions. When we dive into a swimming pool, the water exerts an equal pressure on us from every direction.

Structural Behavior

Rocks can respond to stress by brittle fracture or by flowing like soft plastic.

Shallow

08.01.c2

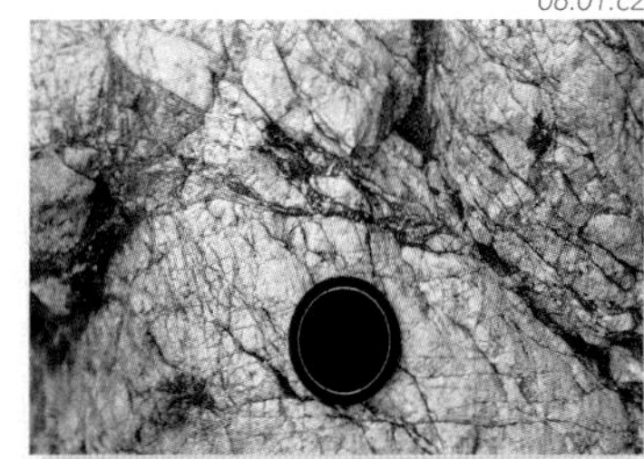

△ At cool, shallow levels of the crust, rocks usually exhibit *brittle* behavior by fracturing, in some cases into pieces. [Nevada]

Deep

08.01.c5

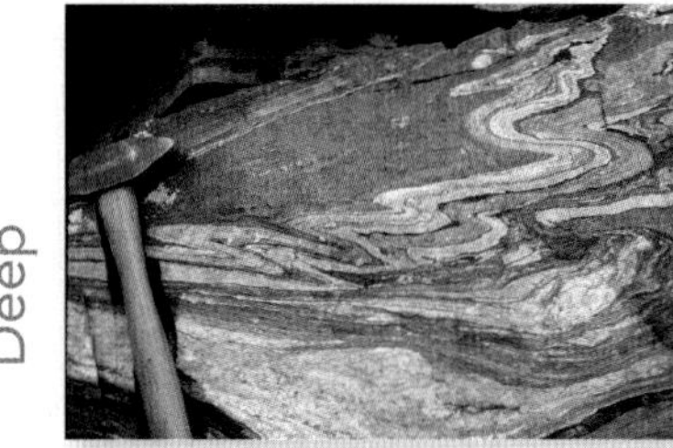

△ At deeper levels, where temperature and pressure are higher, rocks usually respond to stress by flowing in the solid state, which is called *ductile* behavior. [eastern Washington]

Mineral Response

A mineral may become unstable or may not be affected as temperature and pressure change.

08.01.c3

△ At low temperature, many minerals, like quartz in these cobbles, are stable, exhibiting little response to the conditions. [Arizona]

08.01.c6

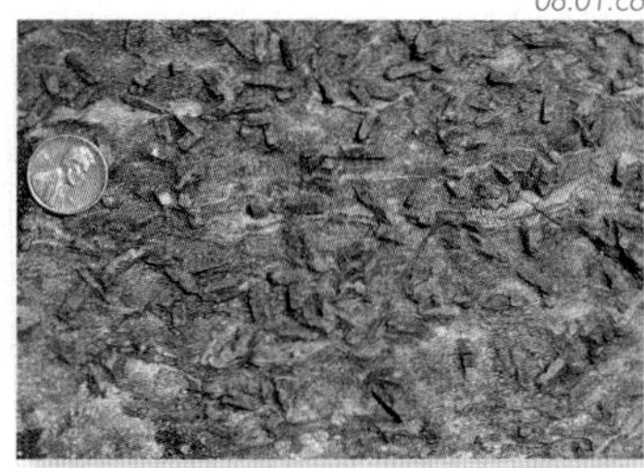

△ At high temperature and pressure, minerals commonly *recrystallize* into larger or smaller crystals, and new minerals may grow at the expense of existing minerals. [northern Idaho]

Effect of Fluids

Fluids, such as water, can help minerals grow or dissolve and can affect the strength of a rock.

08.01.c4

△ Low-temperature fluids have little effect on many rocks but may form mineral-filled fractures, called *veins*. [Nevada]

08.01.c7

△ At depth, high-temperature water and other fluids can mobilize chemical constituents, form high-temperature veins, and promote recrystallization of minerals. [North Cascades, Washington]

Determining the Conditions at Which a Mineral Is Stable

To better understand the interior of Earth, geologists use various approaches to investigate the conditions where different minerals are stable. One approach is to observe samples of rock retrieved from deep drill holes, but such observations are usually limited to about the upper five kilometers of Earth's crust. To investigate deeper environments, we observe exposures of rocks that once were at depth but then were uplifted by tectonics and exposed at the surface by erosion.

Geologists also investigate deep environments by doing laboratory experiments in which rocks are subjected to various temperatures, pressures, and fluids. Special laboratory devices, such as the one shown here, permit geologists to subject rocks and minerals to high temperatures and pressures, simulating conditions within the deep Earth. Geologists place a small sample inside the device and then raise the temperature and pressure. After a specific length of time, the sample is cooled, depressurized, and removed from the device. The geologist then examines the sample with various types of microscopes and analytical instruments to determine which minerals were stable under those conditions and which ones were not.

08.01.mtb1

Before You Leave This Page Be Able To

- ✓ Describe or illustrate the concept of stress.
- ✓ Summarize or sketch the three ways that a mass of rock can respond to stress.
- ✓ Sketch or describe how temperature and pressure vary with depth.
- ✓ Summarize the differences in structural behavior, mineral response, and effect of fluids between shallow and deep environments.
- ✓ Briefly describe how we study the conditions at which a specific mineral is stable.

How Do Rocks Respond to Stress, Temperature, and Fluids?

HOW ROCKS RESPOND TO STRESS depends on three main factors: the type and magnitude of stress, the pressure and temperature conditions, and the amount of fluid. The interaction of these factors creates a diversity of geologic structures, including fractures, folds, and veins.

A What Kinds of Stress Affect Rocks?

Rocks within Earth are subject to stress applied by the surrounding rocks. Any point within Earth is affected by stresses from all directions, and the entire array of stresses is called the *stress field*. We simplify the stress field by showing only the stresses applied from three mutually perpendicular directions. The size of the blue arrows corresponds to the amount of stress—larger arrows signify more stress.

Confining Pressure

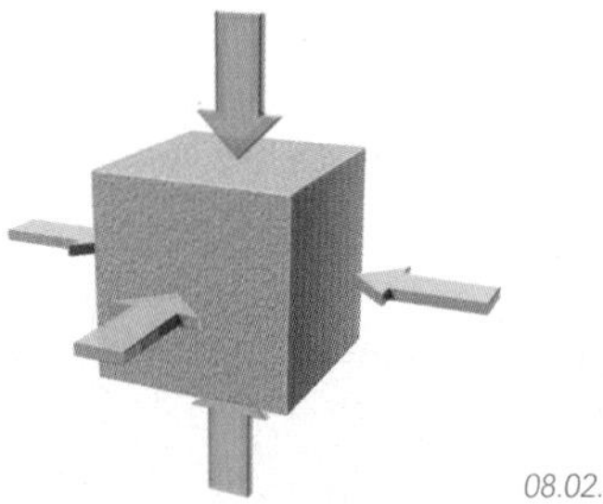

08.02.a1

Any point within Earth is pushed downward by the weight of overlying rocks. Adjacent rocks also experience this weight and so push outward in all directions against other rocks. The rock experiences the same amount of force from each direction, in other words *pressure* from being confined at depth.

Differential Stress

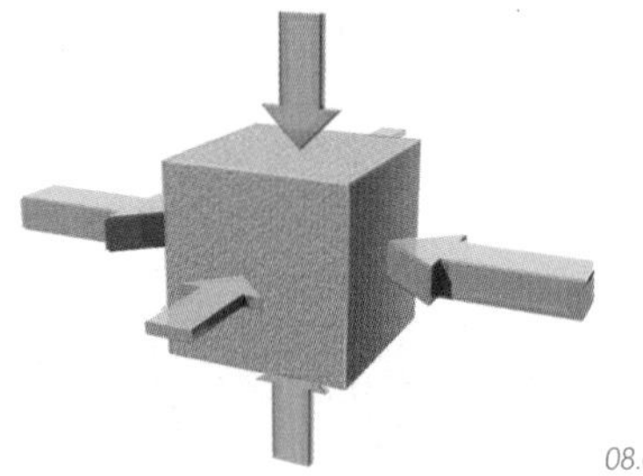

08.02.a2

If stress from another source, such as from tectonics, affects the rock, the imposed stress may add to or subtract from the confining pressure. As a result, the amount of combined stress will be greater in some directions than in others—the rock is subjected to *differential stress*.

Fluid Pressure

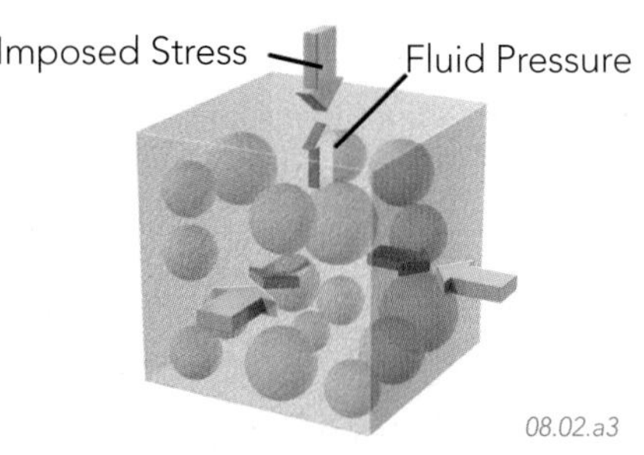

08.02.a3

Fluids, such as water, in the pore spaces of a rock exert a pressure that pushes outward in all directions. This *fluid pressure* opposes the imposed rock pressure and reduces the amount of overall pressure on the rock. This can cause rocks at depth to behave as if they were at low pressures very near the surface.

B What Is the Strength of a Rock and How Does It Vary with Depth?

Strength of Rock

08.02.b1

When a small amount of stress is applied to a rock, the rock may contract slightly like an elastic material but otherwise is strong enough to be undeformed.

08.02.b2

As stress increases, the rock remains essentially undeformed as long as its *strength* is greater than the amount of differential stress.

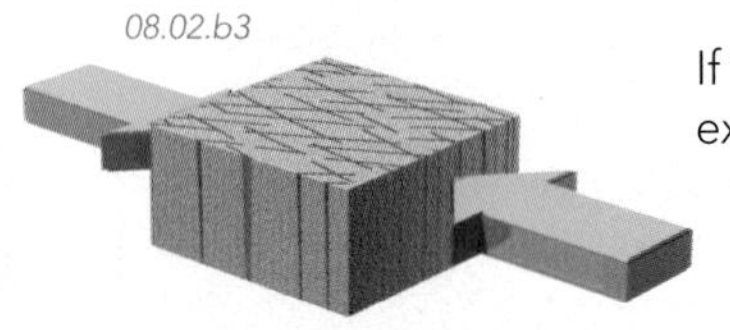

08.02.b3

If the imposed stress exceeds the strength of the rock, the rock fails structurally, such as by fracturing or by flowing as a ductile solid.

Strength of Continental Crust at Depth

The strength of continental crust varies as a function of depth because temperature and pressure both increase downward. ▽

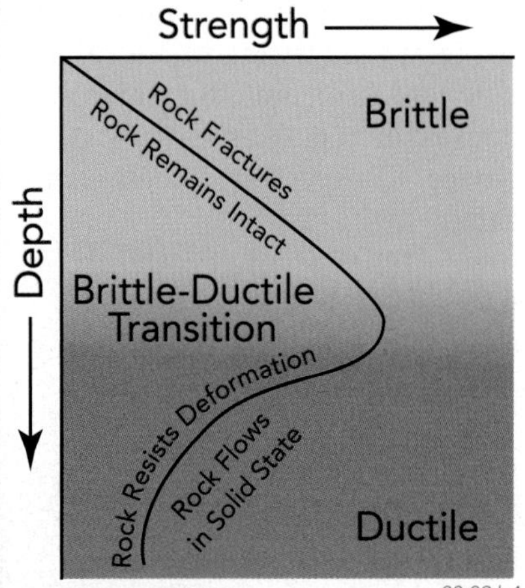

08.02.b4

At shallow levels of the crust, the rocks deform by fracturing and other types of *brittle deformation*. Rocks in Earth's crust initially become stronger with depth because increasing confining pressure acts to force rocks together and makes slip along any fractures more difficult.

With increasing depth, rocks gradually begin to deform (flow) by *ductile deformation*. There is a gradational boundary, or transition, between the *upper brittle* and *lower ductile* parts of the crust.

At greater depths, the effects of temperature dominate over the effects of pressure. Rocks become progressively weaker as they become hotter and can flow more easily in the solid state. The strength of the crust decreases rapidly downward.

C How Do Rocks Respond to Differential Stress?

Rocks may be subject to three types of differential stress: *compression*, *tension*, or *shear*. The way in which rocks respond to these stresses varies as a function of depth because temperature and pressure increase with depth.

Compression | Tension | Shear

Type of Stress

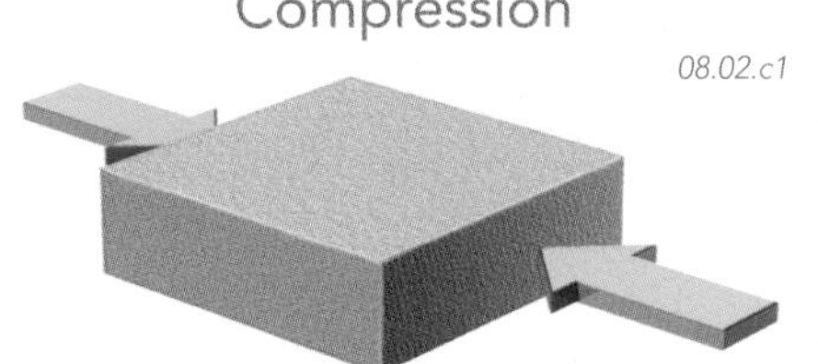

1. When a stress pushes in on rock, the stress is called *compression*. In this book, compression will be illustrated by inward-directed arrows.

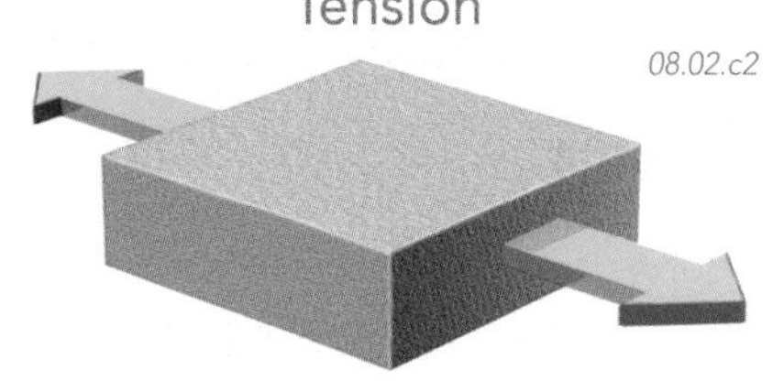

4. When stress is directed outward, pulling the rock, the stress is called *tension*. Tension is shown with stress arrows pointing away from the rock.

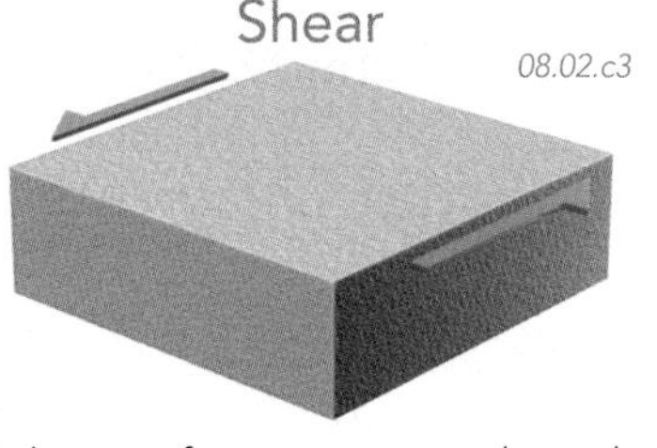

7. A third type of stress acts to *shear* the rock as if stresses on the edges of a block were applied in opposite directions.

Shallow Levels of Crust

2. Compression in shallow levels of the crust can cause rocks to deform by brittle processes, such as causing the rock to fracture and slip.

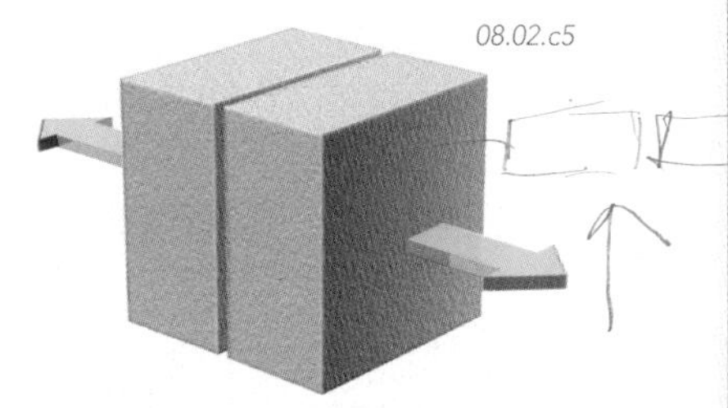

5. Tension can form fractures that help the rock stretch as it is pulled apart. Fluids, if present, can deposit minerals in the fracture, forming a vein.

8. Shearing in shallow parts of the crust usually forms a *fault*, which is a fracture along which two rock masses have slipped past one another.

Deeper Levels of Crust

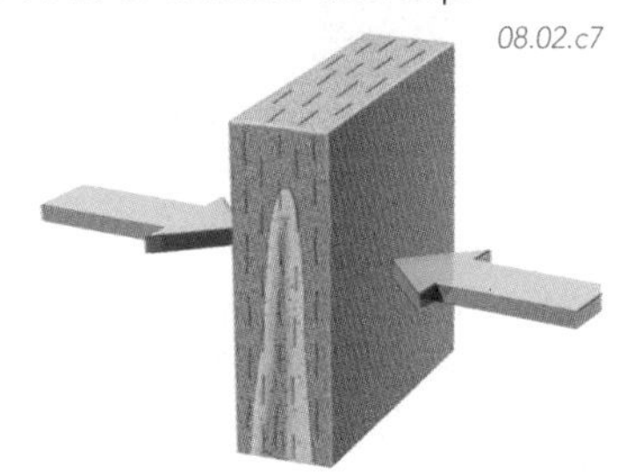

3. In deep parts of the crust, where rocks are hot enough to flow, compression can squeeze the rocks and form tightly squashed folds and new metamorphic structures.

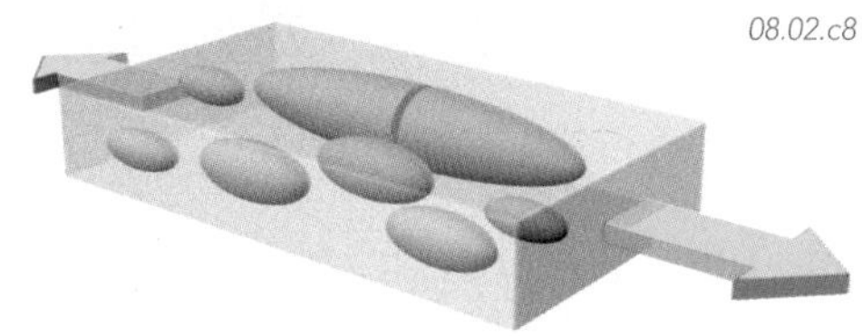

6. Tension is difficult to achieve deep in the crust because the high confining pressure pushes inward, but differential stress can stretch the rock. If accompanied by high fluid pressure, stress can fracture the rock and form veins.

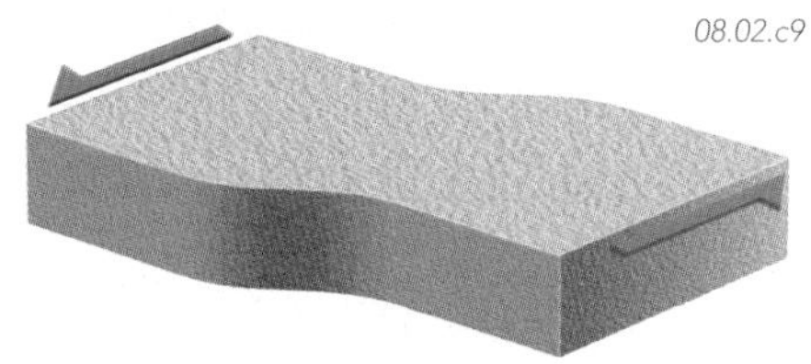

9. In deep, ductile environments, shearing commonly is distributed across a wide zone. Rocks within the zone of shearing deform and flow as weak solids. A zone of shearing is called a *shear zone*.

How We Determine the Strength of Rocks

The strength of any material, including a rock, is normally determined in the laboratory by gradually increasing the amount of stress on a sample of the material until it deforms. The samples in the photograph were compressed end-on until they bulged or fractured and slipped. Some rocks are stronger than others, so deformation experiments are performed on various rocks, including granite, marble, sandstone, and salt. These experiments are conducted under temperatures and pressures appropriate for different depths in the crust and mantle. The strength of rocks can also be investigated by examining how rocks, such as those in deep mines and drill holes, respond to natural stresses.

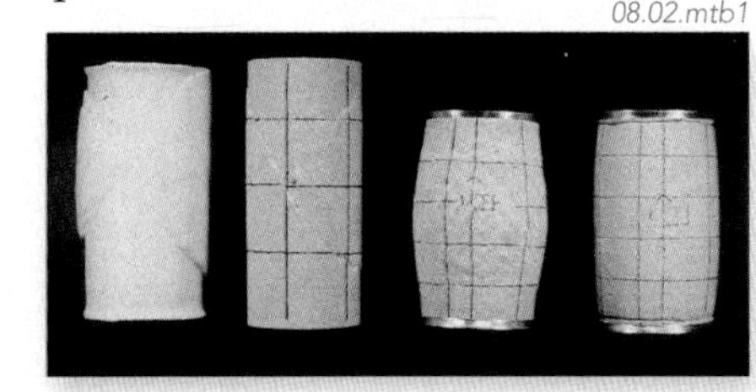

Before You Leave This Page Be Able To

- ✓ Sketch the difference between confining pressure and differential stress.
- ✓ Sketch and summarize how the strength of rocks varies with depth.
- ✓ Sketch and describe the three types of stress, and provide examples of the structures that each type forms at shallow and deep levels of the crust.

How Do Rocks Fracture?

FRACTURES ARE THE MOST COMMON geologic structure. They range from countless small cracks visible in any rock exposure to huge faults hundreds or thousands of kilometers long. What are the different types of fractures, and how do the different types of fractures form?

A In What Different Ways Do Rocks Fracture?

There are two main types of fractures: *joints* and *faults*. Joints and faults both result from stress but have different kinds and amounts of movement across the fracture.

Joints

08.03.a1

◁ **1.** Most fractures form as simple cracks representing places where the rock has pulled apart by a small amount. These cracks are called *joints* and are the most common type of fracture.

08.03.a3

2. These sandstone ledges are cut by a series of near-vertical joints. The layers are not offset by the joints but are simply pulled apart a small amount. [San Juan River, Utah]

Faults

▷ **3.** A *fault* is a fracture where rocks have slipped past one another. Rocks across a fault can slip up and down, as shown here, or can slip sideways or at some other angle. A fault displaces the rocks on one side relative to the other side.

08.03.a2

08.03.a4

4. The long fracture in the center of this photograph cuts across and offsets the rock layers. That is, the layers across the fault have been displaced relative to each other. [Shoshone, southern California]

B How Do Joints Form?

Joints form as stress pulls a rock apart. The orientation in which a joint forms is controlled by the amounts of stress imposed from different directions. In the diagrams below, the size of each arrow reflects the magnitude of stress in that direction. Larger arrows indicate that greater stress is being applied in the direction shown.

Stress Environments in Which Joints Form

08.03.b1

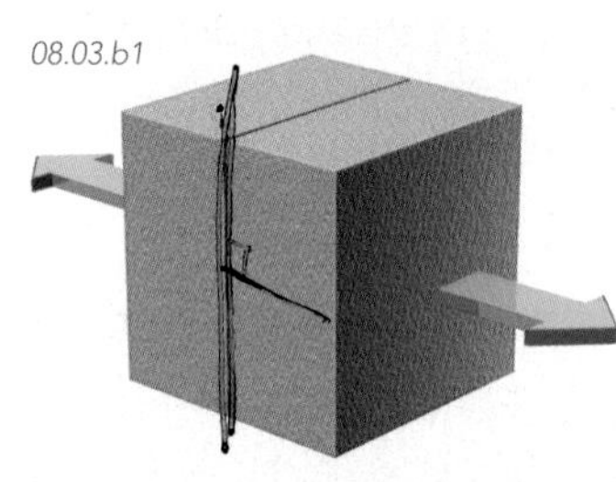

◁ The simplest way that a joint can form is by *tension*, where stresses pull on the rock. The joint forms as a plane that is perpendicular to the direction of tension. True tension is possible only in very shallow levels of the crust because deeper levels have too much confining pressure from the weight of overlying rocks.

08.03.b2

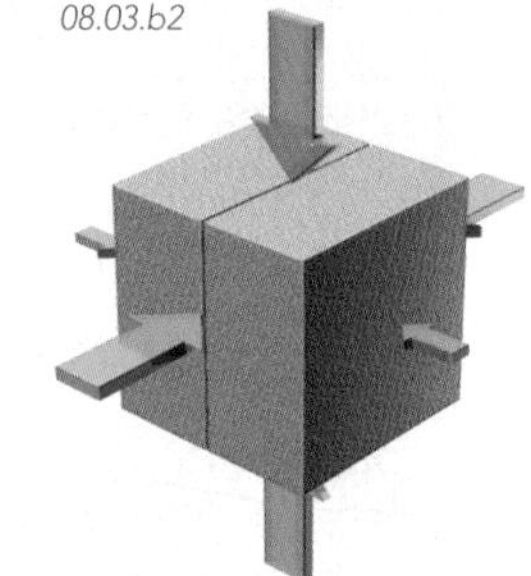

◁ At most crustal depths, joints form because fluid pressure opposes the inward push of confining pressure. The block shown here is subjected to deferential stress, with the least amount of compression being in the direction shown by the smallest arrows. Joints form perpendicular to this direction of least stress.

Stress Orientations Control the Orientations of Joints

08.03.b3

◁ *Vertical joints* form when the stress field allows the rock to be pulled apart in a horizontal direction. The vertical joints can form in any compass direction, depending on the orientation of the stresses. A rock pulled in a north-south direction, for example, will have joints oriented east-west.

08.03.b4

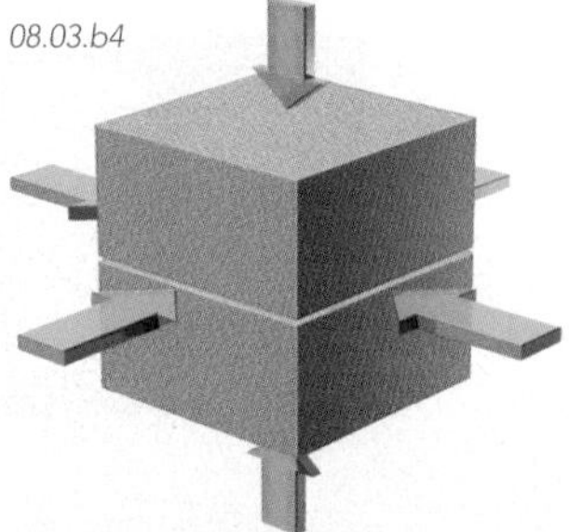

◁ *Horizontal joints* form if a rock is pulled apart in a vertical direction. This can occur when additional tectonic stresses push on the sides of the rock, which causes the vertical stress to be the smallest stress.

B How Is Cleavage Related to Folds?

Cleavage has a close and consistent geometric relationship to folds formed during the same event.

Cleavage typically is *parallel to the axial surface* of folds that formed during the same episode of deformation. Such cleavage generally cuts across bedding and other layering in a systematic way. Cleavage can form vertically, as shown here, horizontally, or somewhere in between.

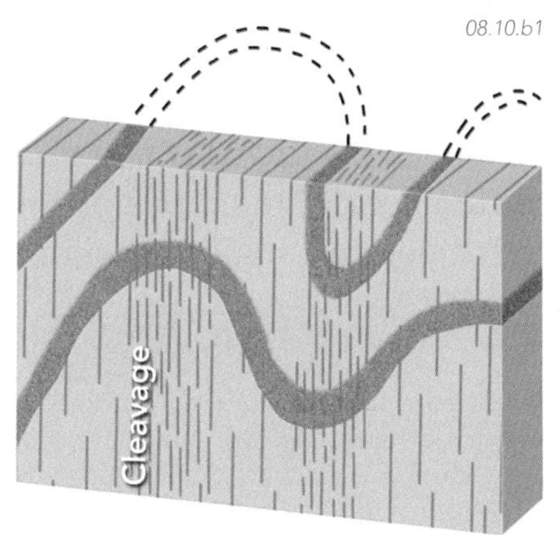

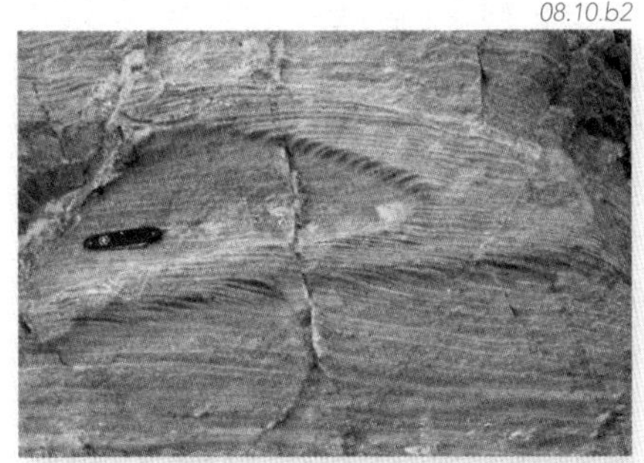

These two photographs show cleavage that is parallel to the axial surface of folds. The one on the left is looking down on a single fold with associated cleavage. The one on the right shows dark cleavage planes that formed parallel to the axial surfaces of a series of small folds.

Using the Relationship Between Bedding and Cleavage to Locate Large Folds

1. On this limb of the fold, cleavage cuts across beds that dip to the right.

2. In the hinge of a fold, cleavage and bedding are perpendicular to one another.

3. On this limb of the fold, bedding dips to the left, opposite to what is seen on the first hill (the other limb of a syncline).

4. Based on the relationship between cleavage and bedding in the hills to the left and the right, how would cleavage and bedding be oriented on this hill?

5. Cleavage cuts beds that dip to the right on this last hill.

C How Can Movement on Faults and Shear Zones Influence Metamorphism?

Reverse Fault

Movement on a fault may bury or uncover underlying rocks and directly influence the type of metamorphism. A thrust fault tectonically buries rocks beneath the fault where they can be metamorphosed at higher pressures and temperatures.

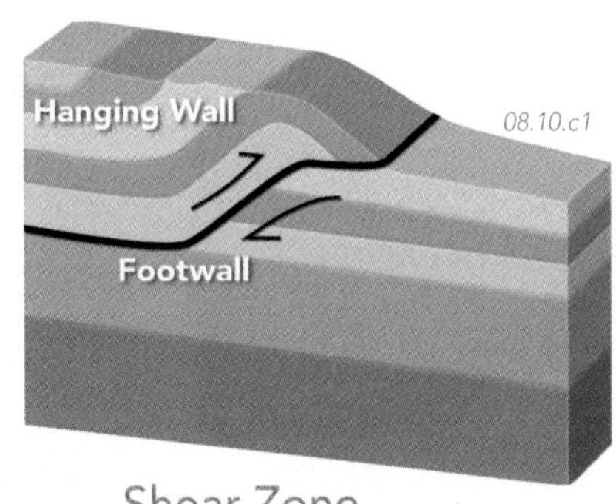

Normal Fault

Hanging Wall

Footwall

08.10.c2

In contrast, normal faulting uplifts and uncovers rocks in the footwall by faulting away the overlying rocks. The footwall will therefore cool as it nears Earth's surface, causing metamorphism to cease or to have lower pressures and temperatures over time.

Shear Zone

A shear zone is a zone of ductile shearing that displaces one rock mass past another. Rocks caught up in a shear zone can be very strongly deformed. A shear zone can overturn and greatly thin rock units that are smeared out within the zone. In some places, like the Alps of Europe and the Big Maria Mountains of southern California, a kilometer-thick section of rocks has been thinned to less than one percent of its original thickness. These are among the most spectacular geologic structures in the world.

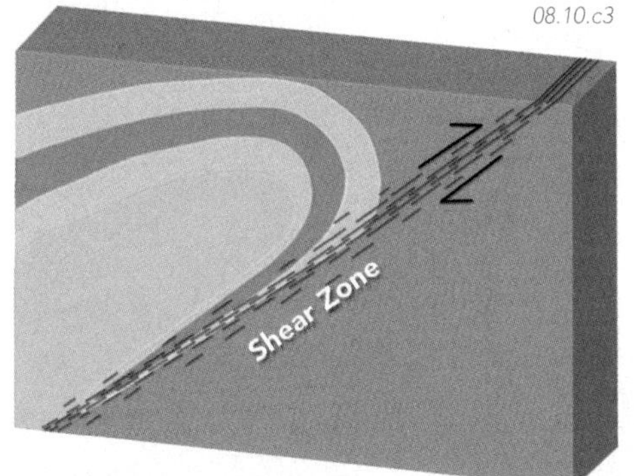

Before You Leave This Page Be Able To

- ✓ Sketch or describe how joints, faults, folds, cleavage, and shear zones can be associated with one another.
- ✓ Describe and sketch how cleavage relates to bedding in folds.
- ✓ Briefly describe how thrust faulting and normal faulting can influence the conditions of metamorphism.

What Types of Deformation and Metamorphism Occur in Compressional Settings?

INTENSE DEFORMATION AND METAMORPHISM occur in regions that undergo *compression*, such as in mountain belts along *convergent* plate boundaries. Different types of structures and metamorphic features form in different parts of a convergent system and reflect differences in the types of rocks involved, the way the rocks deform, the role of magma, and the metamorphic temperatures and pressures.

A What Structural and Metamorphic Features Form in a Fold and Thrust Belt?

Thrusts and folds commonly occur together in regional belts, called *fold and thrust belts*. These form where thrust faults cut through a thick sequence of layered rocks.

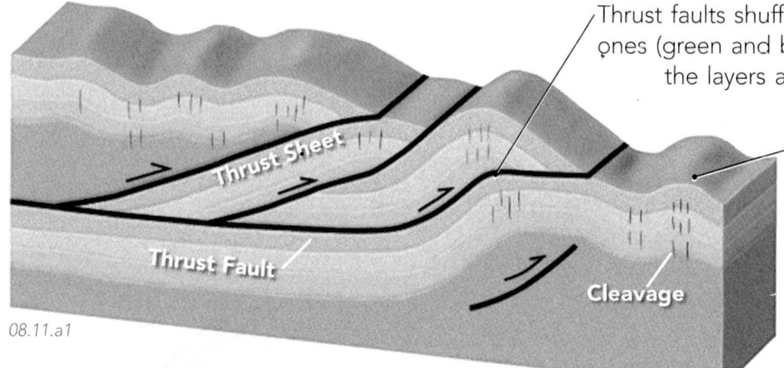

Thrust faults shuffle rock layers by displacing older rocks over younger ones (green and blue over tan in this image). Large folds form where the layers are forced up and over bends in the thrusts.

Other folds develop from overall shortening of rocks in the thrust belt or over thrust faults propagating to the surface.

Most thrust belts contain variably developed cleavage (shown with thin dashed lines) related to the folds and to shear along the thrusts. The deformed rocks also contain numerous joints.

▷ These anticlines and synclines, each several meters high, formed by shortening of layers of slightly metamorphosed black shale and tan sandstone in a Precambrian fold and thrust belt in Arizona.

◁ Larger folds, formed by thrusting and overall shortening, deform shale and sandstone layers during regional thrusting in the foothills of the Patagonian Andes in Argentina, South America.

B What Deformation and Metamorphism Occur Along Continental Collisions?

Thrust faults form during continental collisions when one continent is underthrust beneath another. Thrust sheets typically form in a broad zone between the two plates with sheets of rock sliced off from both the underthrusting and the overriding plates. The most intense metamorphism is shown in purple.

Rocks within the thrust sheets are strongly folded and sheared especially near major thrust faults and shear zones. This tectonic setting is not accompanied by magma that rises into the upper crust, so most metamorphism is *regional metamorphism*, occurring under typical crustal conditions of temperature and pressure.

Collisions are structurally complex with faults and folds forming over a broad region. [Tibet] ▷

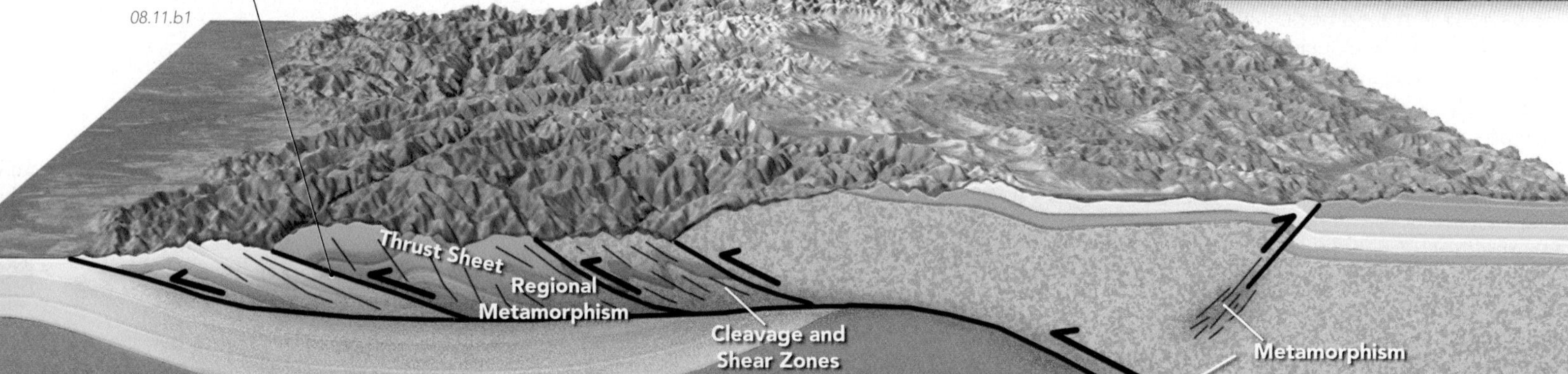

C Where Do Strike-Slip Faults and Shear Zones Form?

During strike-slip movement, one block of rock is sheared sideways past another block of rock. This can occur in various settings, including transform plate boundaries and within the interior of plates.

Rocks can be subjected to horizontal shear stresses, which act to shear the two sides of a block in opposite directions.

08.13.c1

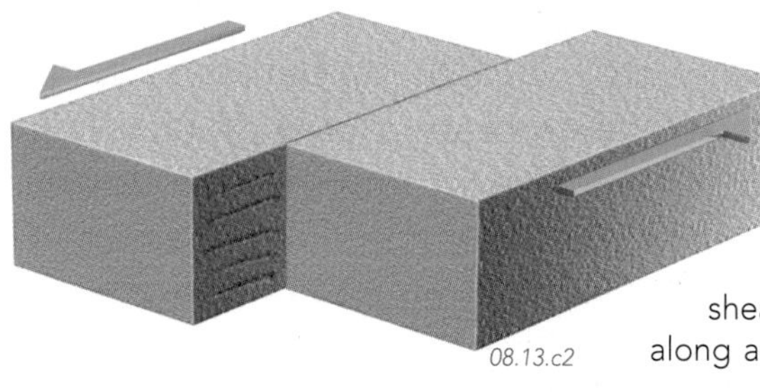

08.13.c2

As a result of the stresses, shearing moves rocks horizontally past one another. Shearing in the upper parts of the crust occurs along a fault, as shown here, but shearing at depth will occur along a metamorphic shear zone.

Shear Within Plates

▷ Stresses can form a strike-slip zone that functions as a plate boundary or that is totally within a tectonic plate, as shown here. A strike-slip zone *within a plate* may offset the rocks hundreds of kilometers or less than a meter. In many cases, a rock layer shows drag folds as it is tracked toward or across the shear zone.

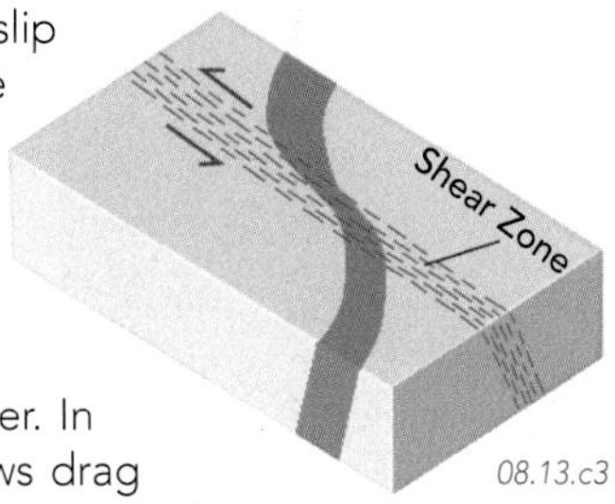

Shear Zone

08.13.c3

Transform Faults

▷ All transform boundaries are strike-slip faults that accommodate the lateral displacement of one plate past another. Most are a boundary between two oceanic plates, as shown here, but a transform fault can also separate two continental plates or can separate an oceanic plate from a continental one.

08.13.c4

D How Do Folds Form Along Strike-Slip Faults?

Strike-slip faults and metamorphic shear zones are accompanied by folds, which form where one block of rock is sheared past another or where the rocks are forced around a bend in the fault.

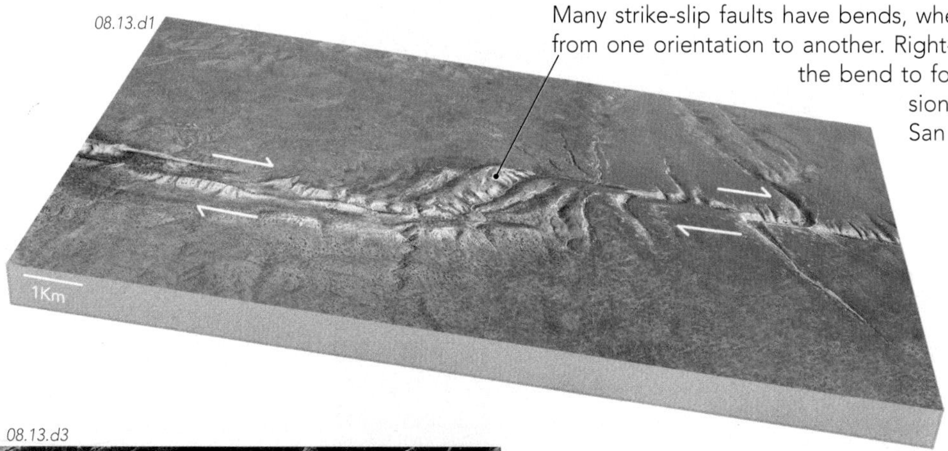

08.13.d1

1Km

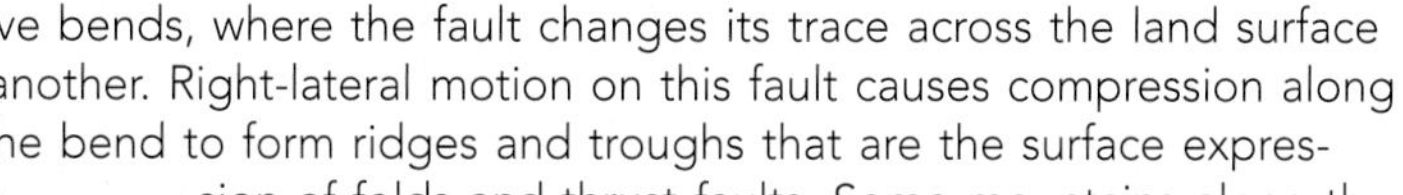

Many strike-slip faults have bends, where the fault changes its trace across the land surface from one orientation to another. Right-lateral motion on this fault causes compression along the bend to form ridges and troughs that are the surface expression of folds and thrust faults. Some mountains along the San Andreas fault are anticlines that are still forming. ▽

08.13.d2

08.13.d3

◁ Shearing within a strike-slip zone can form folds of various orientations. The folds here are formed in young (late Cenozoic) sedimentary rocks along the San Andreas fault. Note the geologist just left of center for scale. Folds can be used to determine the direction of shearing along some fault zones. [Mecca Hills, southern California]

Before You Leave This Page Be Able To

- ✓ Sketch or explain a strike-slip fault and a metamorphic shear zone, and describe some common features along each type of structure.
- ✓ Summarize or sketch how folds can form along a strike-slip fault zone.

How Are Geologic Structures and Metamorphic Rocks Expressed in the Landscape?

ROCKS AND GEOLOGIC STRUCTURES on Earth's surface are exposed to weathering and erosion, which remove weaker rocks faster than stronger ones. Joints, faults, and cleavage provide easy access into the rocks for water and other agents of weathering and erosion and so are preferentially worn away. Folding, faulting, and metamorphism can tilt and otherwise deform rocks, which are then eroded into distinctive landforms that provide clues about what type of structure is present and what events might have happened.

A How Are Joints Expressed in Landscapes?

08.14.a1

1. Joints generally occur in *joint sets* in which many joints have a similar orientation. Most rocks contain several joint sets, which cut rock layers into a series of rectangular blocks. In this photograph, the rock is cut by two sets of joints. [Tasmania]

08.14.a2

3. Joints strongly control weathering because jointed rocks weather faster than unjointed rocks. Here, weathering along joints carves grooves and notches, leaving rocks between the joints as small pillars. [Baby Rocks, northern Arizona]

08.14.a3

2. Joints are largely responsible for the appearance of many cliffs, ledges, and other outcrops of rock. In this cliff, a near-vertical joint set cuts across horizontal layers of sandstone. [Zion National Park, Utah]

08.14.a4

4. *Columnar joints* form by the cooling and contraction of solidified igneous rocks and are distinctive in outcrop. The size and orientation of the columns reflect how the rock cooled but most columns, like these, are steep. [Grand Canyon, Arizona]

B How Do Tilted and Folded Layers Erode?

08.14.b1

1. Erosion can rapidly strip off weak layers, but erosion slows upon encountering an underlying hard layer. Such erosion can carve a slope, called a *dip slope*, parallel to the planar or gently curving layers. [Lime Ridge, Utah]

08.14.b2

3. Where the continuity of a tilted layer is interrupted, such as where stream valleys cross a ridge, remaining parts of the layer can be eroded into large, triangular-shaped rock faces called *flatirons*. [Kayenta, Arizona]

08.14.b3

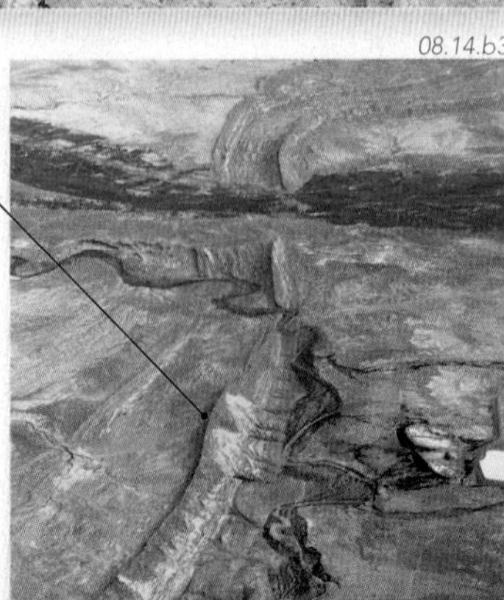

2. Erosion of dipping layers in a tilted fault block or on the limb of a fold can create a landscape with linear or curved ridges formed from more resistant rock layers. Such a ridge has a dip slope on one side and is called a *hogback*. [The Hogback, northern New Mexico]

08.14.b4

4. As layers change dip, the landscape expression changes too. These layers form a dip slope near the top of the mountain, but form steep fins of rock where the layers are nearly vertical near the base of the mountain. [Split Mountain, Utah]

C How Are Faults Expressed in Landscapes?

08.14.c1

1. We commonly recognize faults because of *offsets* or abrupt *terminations* of layers. Here, a curved fault truncates bedding in red sedimentary rocks, juxtaposing them against dark basalt. [Grand Canyon, Arizona]

08.14.c2

3. When fault movement offsets Earth's surface, it can cause a step in the landscape, called a *fault scarp*. This dirt-colored fault scarp formed during the 1983 Borah Peak earthquake in southeastern Idaho.

08.14.c3

2. Rocks along faults are highly fractured and are easily eroded. As a result, many fault zones erode into linear topographic notches or linear valleys. This fault truncates layers and forms a linear notch. [Echo Cliffs, Arizona]

08.14.c4

4. Faults that are currently active can offset streams, ridges, and other topographic features. The San Andreas fault in California is the linear feature cutting across drainages in the center of the photograph. [Carrizo Plain, California]

D How Are Metamorphic Rocks Expressed in Landscapes?

08.14.d1

1. Metamorphic rocks can be shiny even from a distance, because their mica minerals share a similar orientation and reflect light. This shiny phyllite preserves wavy ripple marks and a set of veins that weather as raised lines. [Pamour, Ontario, Canada]

08.14.d2

2. Metamorphic rocks have many different expressions in landscapes because different rock types, metamorphic histories, and structural orientations are involved. This folded rock is schist. [Patagonia, Argentina]

08.14.d3

3. Most metamorphic rocks have cleavage, foliation, or other layers that form platy outcrops and that weather into tabular slabs of rock. [Aurland Trail, Norway]

08.14.d4

4. Metamorphic rocks can include numerous dikes, sills, and pods of igneous rocks, such as these pink granites. [Black Canyon of the Gunnison, Colorado]

08.14.d5

5. Many metamorphic rocks display distinctive features, such as folded dikes, vertical foliation, and a jagged appearance. [Grand Canyon, Arizona]

Before You Leave This Page Be Able To

- ✓ Identify joints in a photograph of a landscape and describe or sketch how joint sets weather and erode.
- ✓ Summarize or sketch the features that form when tilted or folded layers are eroded.
- ✓ Summarize or sketch how you might identify a fault in the landscape.
- ✓ Describe some characteristics displayed by metamorphic rocks in the landscape.

How Do We Study Geologic Structures?

UNDERSTANDING STRUCTURAL AND METAMORPHIC FEATURES is a key step in reconstructing the history of Earth. Geologists usually begin the process by doing field studies, in which they collect field observations and other data. Field studies can be followed by laboratory studies to better understand the timing and conditions of the different structural and metamorphic events represented in the field.

 What Can We Learn from Field Studies?

The primary way in which geologists collect structural data is by doing field studies in the locality of interest. Such field studies involve hiking around the area while observing, describing, and measuring the various geologic and metamorphic features encountered.

08.15.a1

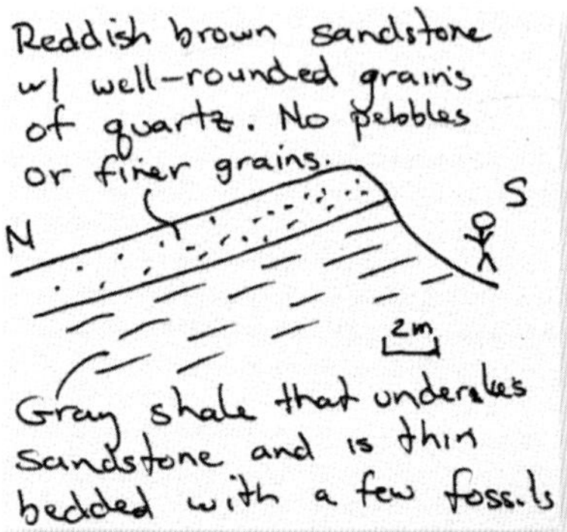

Reddish brown sandstone w/ well-rounded grains of quartz. No pebbles or finer grains.
N S
2m
Gray shale that underlies sandstone and is thin bedded with a few fossils

One of the first steps in the field is to carefully observe the exposed geology. Geologists hike across the area and describe the different rock units, geologic structures, and other features. They then record the observations and descriptions in a notebook, which typically includes sketches (◁) to document in pictures what is difficult to describe in words. Sketching is an important way to explore ideas and possible alternative explanations in the field.

08.15.a2

Geologists pay special attention to aspects that are diagnostic of a certain type of geologic structure, such as highly fractured and shattered rocks near or along a fault zone. ▷

08.15.a3

08.15.a4

Soil cover
sandstone dips to northeast 25 degrees
red sandstone
25
gray shale
Soil cover
Soil cover
fault cuts Soil
tan sandstone
red ss
28
gray shale
Soil cover
fault
brown cong.
Soil cover
Soil
tan Sandstone
Soil cover
brown conglomerate
Soil

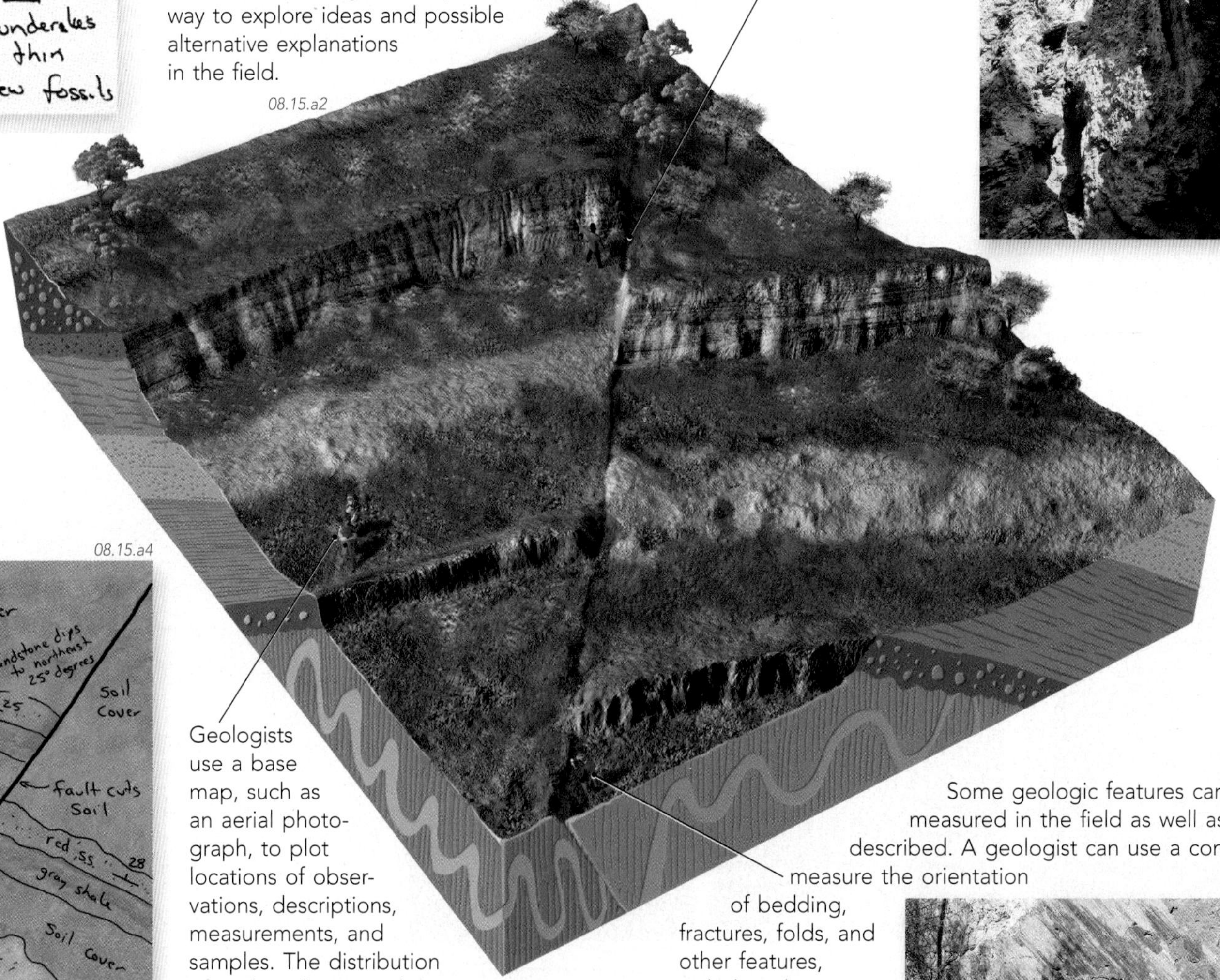

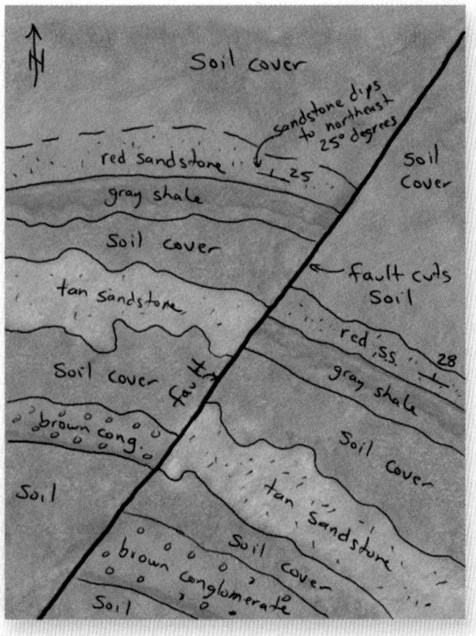

Geologists use a base map, such as an aerial photograph, to plot locations of observations, descriptions, measurements, and samples. The distribution of each rock unit and the orientations of beds and other structures are plotted on the base map to create a geologic map (◁) of the area.

Some geologic features can be measured in the field as well as described. A geologist can use a compass to measure the orientation of bedding, fractures, folds, and other features, including these scratches on a polished fault surface. ▷

08.15.a5

B How Can We Determine When a Fault Was Active?

To illustrate how a structural problem is approached, the figure below shows how we might determine the timing of movement on two faults. Ideally, we would like to know when a fault formed, through what time period it was active, and the age of the most recent movements. It is rarely possible to know all these ages.

Overlap—A lava flow can be erupted or a sedimentary unit can be deposited *across* a fault, overlapping it without showing any offset. Such a relationship demonstrates that the fault has not moved since the overlapping unit was emplaced.

Faulted Units—A fault must be younger than any units it cuts across, such as these sedimentary and volcanic layers. If the ages of such layers can be determined, then these ages help us infer the age of the fault.

Units Deposited During Faulting—If faulting displaces the land surface and forms a fault scarp, sedimentary and volcanic units may accumulate in down-dropped fault blocks. These might include coarse sediment derived from the fault scarp.

Intrusion—Dikes and plutons that cut across a fault and are themselves unfaulted indicate that the fault is older than the intrusion. From the relations shown here, the left fault is older than the gray intrusion and the dark lava flow.

Tilting—Some faulting is accompanied by tilting of the rock units, and so the age of a fault can be inferred from the age of tilting.

08.15.b1

C How Do We Investigate Metamorphic Rocks?

The approach we use to study metamorphic rocks is similar to the one outlined above for studying the age of faults. Many areas have experienced more than one episode of deformation and metamorphism, so field geologists look for key localities that demonstrate the relationships between deformation and metamorphism.

08.15.c1

Metamorphic minerals can grow before, during, or after a deformation event. These white crystals are oriented in all directions, show no foliation or lineation, and therefore grew after deformation had ceased.

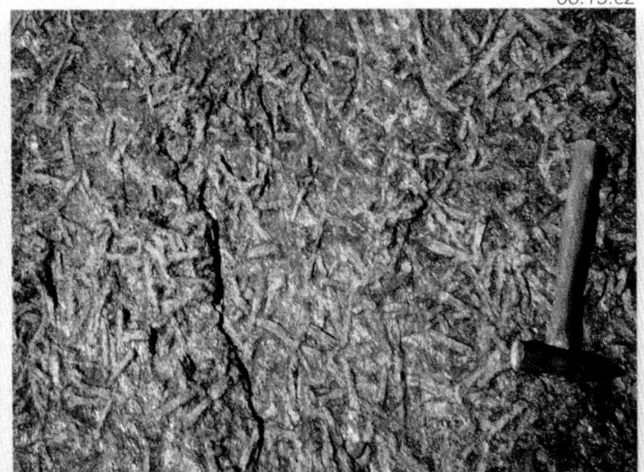

08.15.c2

Certain metamorphic minerals provide constraints on temperatures and pressures reached during metamorphism. These unusually large crystals of the mineral andalusite indicate a specific range of P-T conditions.

08.15.c3

Minerals and other features are sometimes large enough to be easily observed, but some require observing the rock with a small magnifying glass or hand lens. This metamorphic rock has crystals of garnet.

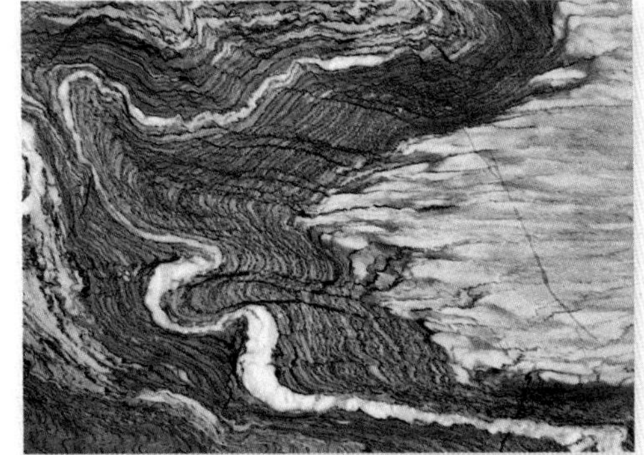

08.15.c4

Many studies focus on trying to reconstruct the sequence of events. This meter-wide marble slab has dark bands of horizontal cleavage cutting older metamorphic layering and folded white veins.

Before You Leave This Page Be Able To

- ✓ Summarize how we observe and measure geologic features and the ways we record this information.
- ✓ Sketch or describe aspects we would observe to infer the age of a fault.
- ✓ Summarize some aspects we might observe in a metamorphic rock to learn something about its history of metamorphism and deformation.

What Is the Structural and Metamorphic History of New England?

NEW ENGLAND CONTAINS A WEALTH of geologic structures and metamorphic rocks, ranging from domes of high-grade gneiss to slightly cleaved sedimentary and volcanic rocks. This area provides an opportunity to examine the regional context of structures and metamorphic features.

A What Rocks and Structures Are Exposed on the Surface?

08.16.a1

This figure is a moderately detailed geologic map of New England superimposed over topography. The different colors on the map show the distribution of different kinds and ages of rocks. The color patterns illustrate that the region contains belts of different rocks that trend north-south across the region.

Thin lines are contacts (boundaries) between rock units, whereas thick lines are faults.

Brown and olive-green colors show Precambrian rocks, such as those in the Adirondack and Green Mountains.

Purple, pink, and blue colors represent Paleozoic rocks, which are the most widespread rocks in the area. In the western part of New England, such as near the Catskill Mountains, rocks are relatively unmetamorphosed sedimentary rocks that are locally folded and cleaved. ▽

08.16.a2

Rock units form complex patterns in the eastern part of the geologic map, as is typical of a terrain that has been metamorphosed and deformed multiple times.

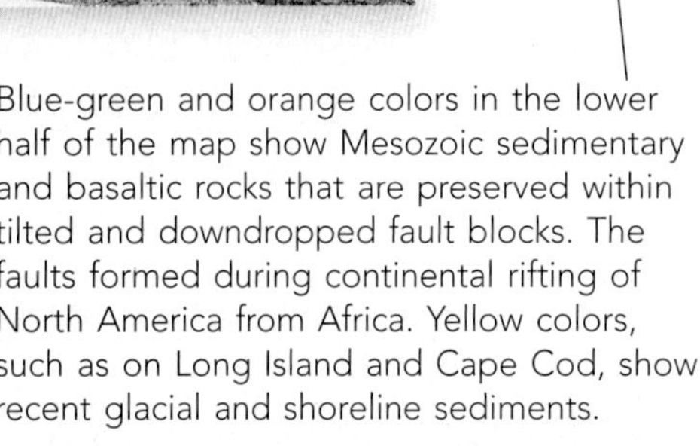

Blue-green and orange colors in the lower half of the map show Mesozoic sedimentary and basaltic rocks that are preserved within tilted and downdropped fault blocks. The faults formed during continental rifting of North America from Africa. Yellow colors, such as on Long Island and Cape Cod, show recent glacial and shoreline sediments.

Metamorphism is shown on this map with a stipple (dotted) pattern over the different colors. Paleozoic and older rocks in the center of the region were metamorphosed at high temperatures and pressures. They include garnet-bearing schist and banded gneiss. ▷

08.16.a3

TOPICS IN THIS CHAPTER

09.00.a3

△ **The geologic feature** for which Siccar Point is famous is a *contact* that separates two chapters in Earth's history. Below the contact are gray sandstone and shale, whose beds are nearly vertical. Above the contact are gently dipping beds of reddish sandstone and conglomerate, which contain angular pieces of the underlying gray sandstone. The contact between these two rock sequences with very different geologic histories is what inspired Hutton's profound insight.

How does a contact like the one exposed at Siccar Point form and what does it imply about the length of Earth history?

09.00.a4

Ruins of an Earlier World

As James Hutton explored the rocky coasts of Scotland in the late 1700s, he encountered the remarkable geologic exposures at Siccar Point. The insight he gained on that day in 1788 changed the world! James Hutton's profound realizations provided a new way to think about Earth, which is why he is known as the *father of geology*.

At Siccar Point, James Hutton's attention was drawn to the enigmatic contact, which even from a distance is striking, with vertical gray beds below and gently inclined red beds above. Hutton pondered what had happened to produce such an arrangement of rock types. He wondered if the ancient contact represented the same processes currently occurring on the beach next to the outcrop—modern beach sand was being deposited in horizontal layers over the vertical beds of gray sandstone. In other words, Hutton's insight was that you might be able to use modern processes to interpret events that had occurred in Earth's past. This principle, today called *uniformitarianism*, was the key step in the development of geology as a science. It is often stated as "*the present is the key to the past.*"

Following this new logic, Hutton realized that to explain the relationships at Siccar Point, the gray sandstones below the contact must have been tilted and eroded before the red sandstone was deposited over the upturned layers. In essence, Hutton realized that this contact represented an *ancient erosion surface*, which we now call an *unconformity*. Hutton concluded that the gray rocks below the unconformity represented a mountainous landscape that had been eroded away and called these rocks "the ruins of an earlier world."

Hutton noted that erosion and many other geologic processes could be observed to occur relatively slowly compared to the life span of a human, so he realized that the contact at Siccar Point required Earth to have a very long history, much longer than was perceived at the time. Hutton concluded that the history of Earth was very long and partially shrouded, with "no vestige of a beginning, no prospect of an end."

How Do We Infer the Relative Ages of Events?

DETERMINING THE AGE OF ONE ROCK relative to another is a key starting point for deciphering the geologic history of an area. Many of the techniques used to determine *relative* ages of rock units and geologic structures are based on logical common sense and are extensions of observations of how the world works today. The collection of strategies used is called *relative dating.*

Most Sediments Are Deposited in Horizontal Layers

Most sediments and many volcanic units are deposited in layers that originally are more or less horizontal, a principle called *original horizontality.* If layers are no longer horizontal, some event affected the layers after they formed. Some beds can be deposited with an initial dip, as on the front of sand dunes and deltas.

These canyon walls expose horizontal gray and reddish layers. These layers were deposited horizontally and have remained nearly so for 300 million years. [Goosenecks, southern Utah]

09.01.a1

09.01.a2

Just to the east, the same gray and reddish layers are folded. They are no longer horizontal, so something (deformation) must have happened. [San Juan River, Utah]

A Younger Sedimentary or Volcanic Unit Is Deposited on Top of Older Units

When a layer is deposited, any rock unit on which it rests must be older than the layer, a concept called the *principle of superposition.* This principle is illustrated below.

1. A layer of tan sediment is deposited over older rocks.

09.01.a3

2. A series of horizontal red layers are then deposited over the first layer.

09.01.a4

3. A third series of layers is deposited last and is on top. In this sequence, the oldest layer is on the bottom and the youngest layer is on the top.

09.01.a5

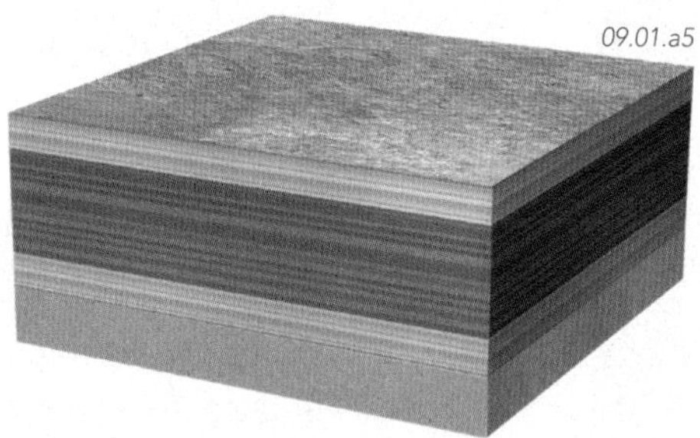

09.01.a6

◁ **4.** Observe all the different layers in this rock sequence. Which exposed layer is oldest, and where would you look to find the youngest rock layer? All of the sedimentary layers now exposed in the canyon walls had formed and been lithified into rock before the canyon was cut by erosion. [Dead Horse Point, Utah]

▽ **5.** Where is the oldest layer in this tilted sequence? It is on the left, in the lowest part of the section, as long as deformation was not so severe that it overturned the beds.

09.01.a7

A Younger Sediment or Rock Can Contain Pieces of an Older Rock

As any rock or deposit forms, pieces of older rock can be incorporated into it. A cobble eroded from bedrock and carried by a river cannot exist unless the bedrock already was there. The presence of clasts of an *older rock* in a *younger rock* clearly indicates the relative ages, even if you cannot see the two rock units in contact with one another.

09.01.a8

Tilted quartzite beds at the bottom were exposed on the surface and weathered into clasts, which became incorporated into the overlying conglomerate.

09.01.a9

The dark, lower basalt contributed clasts into an overlying layer of tan conglomerate. The conglomerate contains *clasts* of—and is therefore younger than—the basalt.

09.01.a10

A light-colored granite contains dark-colored pieces of older metamorphic rocks that fell into the granitic magma. The metamorphic rocks are older than the granite.

A Younger Rock or Feature Can Crosscut Any Older Rock or Feature

Many rocks are cross cut by fractures (joints and faults), so the rocks were there before the fractures formed. Dikes, sills, and veins can also intrude into or across preexisting rock units, showing *cross-cutting relations*.

09.01.a11

Several fractures cut across, or *crosscut*, the limestone layers, so they are younger. The fractures are said to be *crosscutting*.

09.01.a12

Magma was injected as a dark dike that crosscuts a red sedimentary rock. The dike is younger than the red rock.

09.01.a13

Light-colored dikes of granite crosscut darker igneous rocks. The dikes are younger than the dark igneous rocks.

Younger Rocks and Features Can Cause Changes Along Their Contacts with Older Rocks

Magma comes into contact with preexisting rocks when it erupts onto the surface and when it solidifies at depth. In either setting, the magma may locally bake or chemically alter any other rocks that it encounters, and these changes, called *contact effects,* indicate that the magma is younger than the altered rocks.

09.01.a14

A dike of basalt intrudes across a grayish conglomerate. Heat and fluids from the magma affected the older conglomerate, causing a reddish baked zone next to the dike.

09.01.a15

A volcanic dome collapsed, sending hot ash and rock fragments downhill. The heat from the hot volcanic materials baked and reddened the older underlying rocks.

Before You Leave This Page Be Able To

- ✓ Sketch and explain each of the five principles of relative dating.
- ✓ Apply the principles of relative dating to a photograph or sketch showing geologic relations among several rock units, or among rock units and structures.

How Are Ages Assigned to Rocks and Events?

DETERMINING THE RELATIVE SEQUENCE OF EVENTS is only one part of deciphering the geologic history of a location. We also strive to know *when* these events occurred and to assign them *ages* in hundreds, thousands, millions, or billions of years. This is done by using *analytical dating methods*, most of which involve chemically analyzing a rock for the products of natural radioactive decay.

A How Does Radioactive Decay Occur?

Some atoms of an element differ in the number of neutrons they contain in their nucleus and therefore have different *atomic weights;* these varieties of the same element are called *isotopes*. Some isotopes are unstable through time, changing into a new element or isotope by the process of *radioactive decay*.

1. This lattice shows atoms before any radioactive decay. These starting atoms are called the *parent atom* or *parent isotope*. Over time, some of the parent isotope will decay into a different isotope that is called the *daughter product*.

2. At a later time, *half* of the parent atoms (green) will have decayed into the daughter product (purple). The amount of time it takes for this to occur is called the *half-life*. After one half-life, the ratio of parent atoms to daughter atoms is 1.

3. After a time equal to another half-life has passed, half of the *remaining* parent atoms have decayed into the daughter product. After two half-lives, 3/4 of the parent atoms have decayed and 1/4 remain.

4. This table summarizes the radioactive decay shown in the figures above. If the number of parent atoms was initially 1,000, half of the parent atoms (500) will have decayed to atoms of the daughter product after one half-life. After two half-lives, only 250 parent atoms remain, alongside 750 atoms of daughter products.

	Before Any Decay	After One Half-Life	After Two Half-Lives
Atoms of Parent	1,000	500	250
Atoms of Daughter	0	500	750

5. The time it takes for an isotope to decay is constant, predictable, and measurable in the laboratory. Geologists, therefore, can calculate the age of a rock by measuring the ratio of parent to daughter atoms. Dating rocks using radioactive decay is called *isotopic dating*.

B How Do We Measure and Calculate Isotopic Ages?

1. Geologists, working alongside chemists and physicists, use an instrument called a *mass spectrometer* to measure the ratio of parent isotopes to daughter product in the rock or the mineral to be dated.

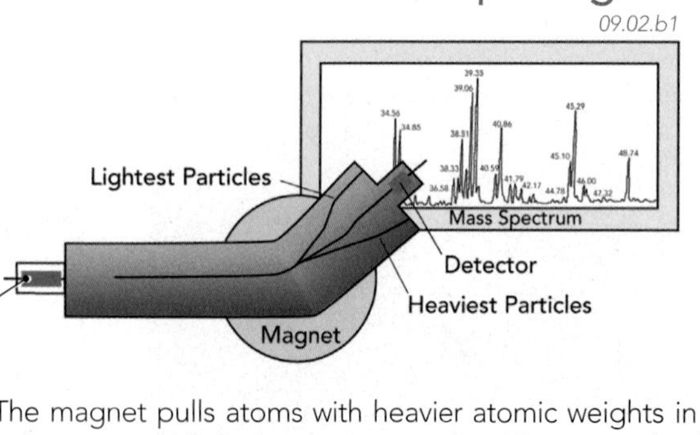

09.02.b1

Lightest Particles

Mass Spectrum

Detector

Heaviest Particles

Magnet

2. The rock or mineral is prepared and placed in a mass spectrometer where the sample is ionized and propelled down a tube toward a very strong electromagnet.

3. The magnet pulls atoms with heavier atomic weights in one direction and lighter atoms in another. The strength of the magnet can be altered by adjusting the amount of electric current passing through it. With the proper settings, only atoms of the desired atomic weight reach a collector at the end of the tube, which counts the number of atoms that arrive.

4. Mass spectrometers measure ratios of isotopes more easily than absolute amounts of isotopes, so most results and calculations use ratios between isotopes. The results are calculated using equations and commonly plotted on a graph.

5. There are limitations to calculating isotopic ages. Each method of dating has certain assumptions, such as the rock has not lost or gained parent isotopes or daughter product. Also, some methods require that the rock not have been heated above a certain temperature after it was formed. Geologists consider and evaluate each assumption before applying the determined age to a rock.

C What Can Isotopic Ages Tell Us?

Different isotopic systems can be used for isotopic dating, and all isotopic ages do not provide the same kind of information. Some record when the rock incorporated the parent isotope; others record later cooling.

09.02.c1

1. We date volcanic eruptions using a variety of isotopic systems. Volcanic rocks form on the surface and cool rapidly, so an age of the rock is typically the *age of eruption.*

09.02.c2

2. Hot plutons lose certain isotopes, so we determine the age of such bodies using only those minerals that retain isotopes and provide the *age of solidification.*

09.02.c3

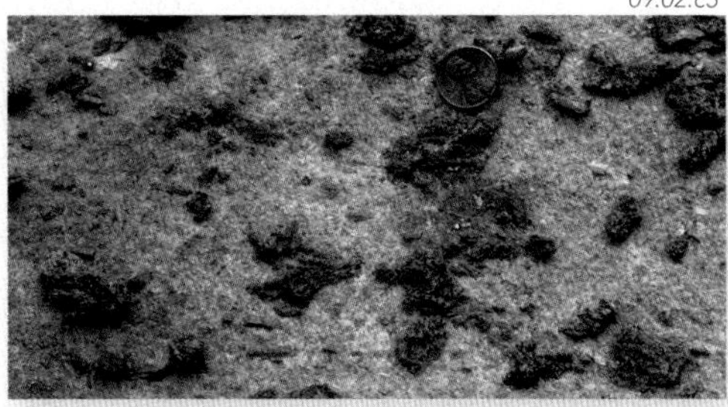

3. The *age of a metamorphic event* is investigated using minerals that formed during metamorphism or minerals that record certain metamorphic temperatures.

09.02.c4

4. Less-retentive minerals, such as biotite mica, are used to determine *when a rock cooled* through a specific temperature, such as while it was being uplifted to the surface.

09.02.c5

5. Dates from individual cobbles or even sand-sized grains in a sedimentary rock help infer the age of the *source rocks* from which the sediment was eroded.

09.02.c6

6. Black pieces of charcoal incorporated into recent sediment can be dated with carbon-14, which provides an age for when the recent sediment was deposited.

09.02.c7

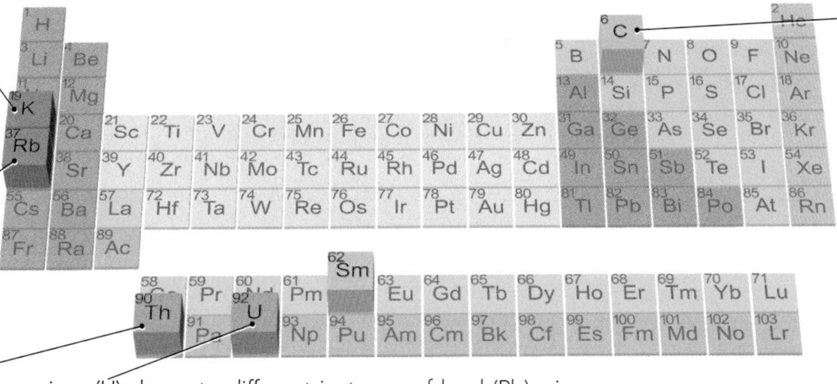

7. Potassium (K) decays to the noble gas argon (Ar) on the far right side of the periodic table), and to an isotope of calcium. K-Ar dating is used to date volcanic rocks and the cooling of deep rocks.

8. Rubidium (Rb) decays to strontium (Sr), and Rb-Sr dating provided some of the first ages for old granites and metamorphic rocks.

10. One isotope of carbon (C), called carbon-14, is used to date wood, charcoal, bones, shells, and carbon-rich rocks and water, but it has a relatively short half-life that makes it best suited for dating materials that are only hundreds to thousands of years old.

9. Thorium (Th) and uranium (U) decay to different isotopes of lead (Pb), via a series of steps. They are suited for dating many kinds and ages of rocks, including grains in sediments. Samarium (Sm) decays to Neodymium (Nd) and is used mostly to date old rocks and to investigate sources of magma.

▷ 11. In many cases, geologists use different methods on a single rock to obtain information about different parts of the rock's geologic history. For granite, a U-Pb age on zircon can provide the age when the magma solidified, and a K-Ar age on the mica mineral biotite provides the time when the rock cooled below 300°C. By using different dating methods on the same rock, we can show when the rock formed and how fast it cooled through time.

09.02.c8

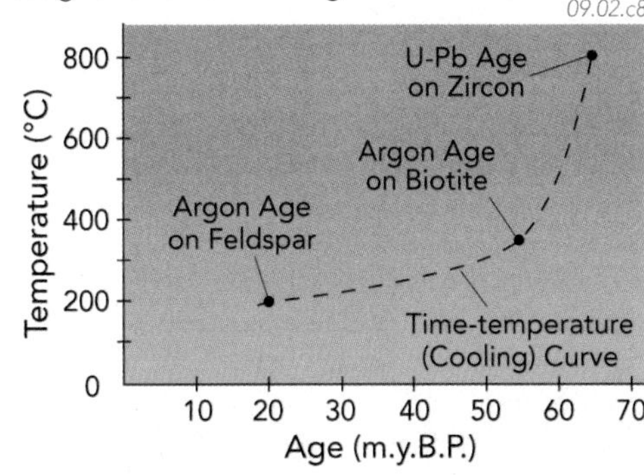

Before You Leave This Page Be Able To

✓ Explain how to determine how many half-lives have passed based on the ratio of parent to daughter atoms.

✓ Summarize how a mass spectrometer is used to determine isotopic ages.

✓ Describe the different ways that isotopic dating is used for dating geologic events.

How Do We Study Ages of Landscapes?

KNOWING THE AGES OF ROCKS AND STRUCTURES provides only one piece of the geologic story. We also need to understand when and how the landscape features, such as mountains and valleys, formed.

 How Does a Typical Landscape Form?

Most landscapes have a similar history—rocks form and then are eroded. The history of many regions typically includes the deposition of a sequence of sedimentary layers and then erosion down through the layers.

1. The sequence begins with water covering a rock unit that was already there, such as the preexisting layer shown above or older metamorphic and igneous rocks.

2. The first new sedimentary unit is deposited on top of the preexisting rocks. Most sediments, such as a layer of sand, are deposited as nearly horizontal layers.

3. Through time, the depositional environment changes and so a series of different sedimentary layers successively accumulate, with each younger layer being deposited on top.

4. At some point, all the layers that will be deposited are there. Deposition stops and erosion can begin.

5. If the region is uplifted or the seas withdraw, the area can begin to be eroded by rivers, streams, glaciers, and the wind. Erosion cuts downward with time.

6. Erosion by a river cuts downward, forming a canyon. The canyon widens as small drainages carve outward and as the steep canyon walls move downhill in landslides and slower movements.

7. Observe the photograph below and think about how the landscape formed before reading on. The canyon walls expose three or four different sedimentary units, including a thick, reddish-tan rock (sandstone) that forms the cliff. Older layers are below the cliff, and younger ones are above it. All the layers were deposited and became hard rock before the river cut the canyon. [Labyrinth Canyon, Utah]

8. What would you conclude about the sequence of events in the landscape below. For a typical section of rocks, the lowest layers are the oldest and the highest ones are the youngest. That is true for the section preserved here, but with a caveat. The top cliff has columnar joints and is an igneous rock. It could have been, and was, inserted (intruded) into the layers long after the layers had formed. [Scotland]

09.03.a7

09.03.a8

B How Do We Infer the Maximum and Minimum Age of a Landscape Surface?

To investigate when a landscape surface formed, we commonly try to find a rock unit or other geologic feature that was there before the surface formed or one that came after the surface already existed.

The age of a landscape surface must be *younger* than any rocks on which it is carved. In this example, erosion beveled across an older series of tilted layers, which were then covered by a thin veneer of sediment and soil.

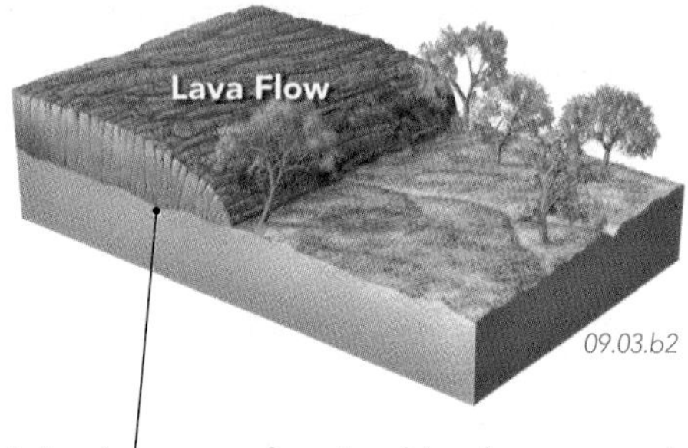

A landscape surface is *older* than any rock that is deposited on top of the surface. A lava flow is ideal for dating a surface because it formed during a short time and its age can usually be determined by isotopic dating.

09.03.b3

This landscape displays a cliff of tan sandstone overlooking a broad valley covered by a dark lava flow. The valley was already eroded down to its present level, and the sandstone cliff already existed, before the lava was erupted across the landscape.

C What Are Some Other Indications of the Age of a Landscape Surface?

1. A landscape surface progressively develops more soil if it remains undisturbed by erosion and deposition. A surface with well-developed soil, such as this one with thick red clay and white carbonate accumulations, is thousands of years old. Recent sediment along the stream has no soil.

2. Sometimes the age of a landscape surface cannot be dated directly, but we can infer its age relative to other features. Many rivers are flanked by raised, gentle surfaces called *terraces*. The terraces were formed sometime in the past, before the river eroded down to its present level.

09.03.c3

3. Observe this photograph and think about the relative ages of the different levels of the landscape. The oldest parts are the top of the high terraces, and these surfaces would likely have the best developed soil. Deposits along the active channel are the youngest and probably have no soil.

09.03.c4

4. In some settings, stones become concentrated on the surface through time, forming a feature called *desert pavement*. Exposed stones get coated with dark material called *desert varnish*. It takes more than ten thousand years for a well-developed pavement with darkly varnished stones to form.

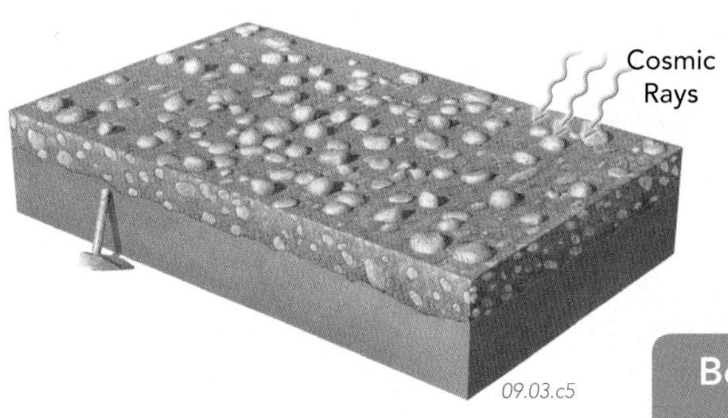

5. Stones on the surface progressively accumulate tell-tale amounts of certain isotopes produced when cosmic rays strike the stones. This form of isotopic dating is used to determine how long the stones have been on the surface. Geologists collect samples of the stones and analyze them in the laboratory by using a mass spectrometer.

Before You Leave This Page Be Able To

- ✓ Describe the sequence of events represented in a typical landscape of flat-lying sedimentary rocks.
- ✓ Summarize or sketch how you could assess the age of a landscape surface.

What Is the Significance of an Unconformity?

EROSION SURFACES CAN BE BURIED AND PRESERVED beneath later deposits. These buried erosion surfaces, called *unconformities*, can be billions of years old and can represent large intervals of time. They provide a glimpse of the shape and longevity of ancient landscapes. What do these features look like, how do they form, and what do they tell us about geologic events that occurred in the distant past?

A What Does an Unconformity Represent?

Erosion surfaces, formed in the past, can be buried and preserved within a section of rocks. Such buried erosion surfaces are called *unconformities*, of which there are several varieties.

1. A gray limestone is deposited under the sea in nearly horizontal layers. The blue in this figure represents water.

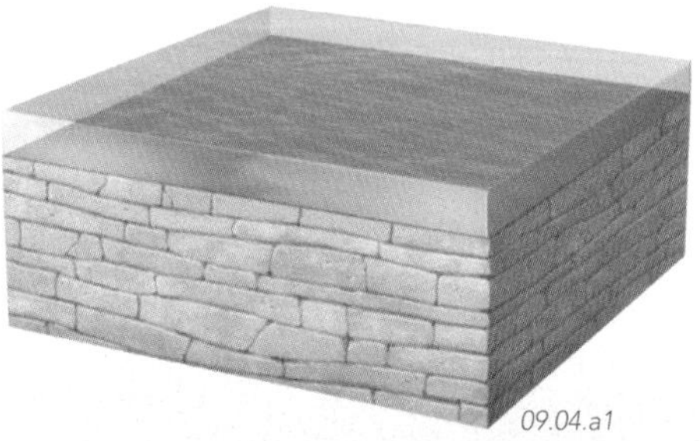

2. Later, the sea withdraws and the limestone beds are folded. As the folded beds are uplifted, they are beveled by erosion. This results in tilted beds being exposed on the surface.

3. Conglomerate is deposited over the eroded beds, forming an unconformity, a buried erosion surface. Where the underlying layers have been tilted, as in this example, it is an *angular unconformity.*

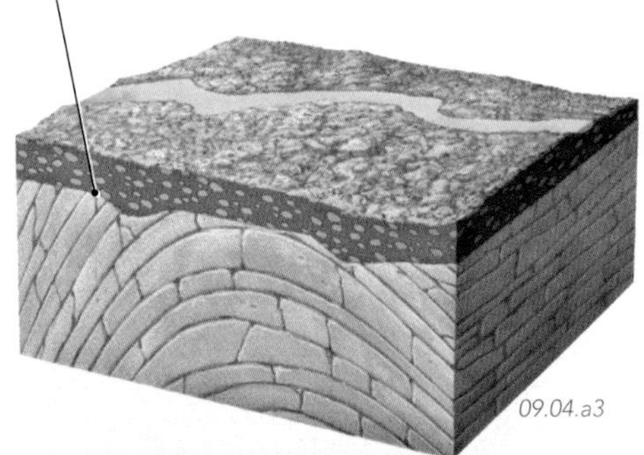

◁ 4. Examine this photograph of the eastern Grand Canyon. There is an obvious angular unconformity between tilted layers below and nearly flat-lying layers (beds) above. The rocks below the unconformity are approximately 1,100 million years (1.1 billion years) old, whereas those above are 540 million years old. There is a long time span represented by the unconformity, for which there is no record, except that there was tilting followed by erosion and then deposition of the upper layers. Other events could have—and did—occur, but we have no record of them at this site.

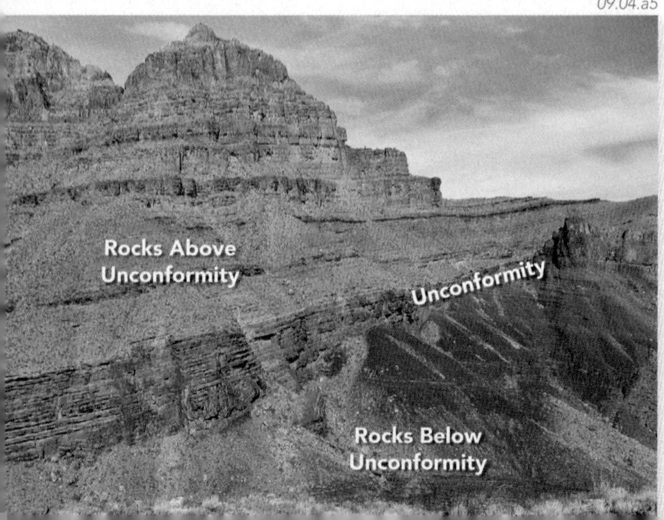

◁ 5. In this view of the same unconformity, the unconformity slopes from right to left as it bevels across layers in the underlying reddish rocks. A ledge of sandstone directly above the unconformity is thicker to the left and thins toward the right. This thinning indicates that the rocks below the unconformity formed a small hill, against which the sandstone was deposited. Such preserved ancient topography is common along unconformities.

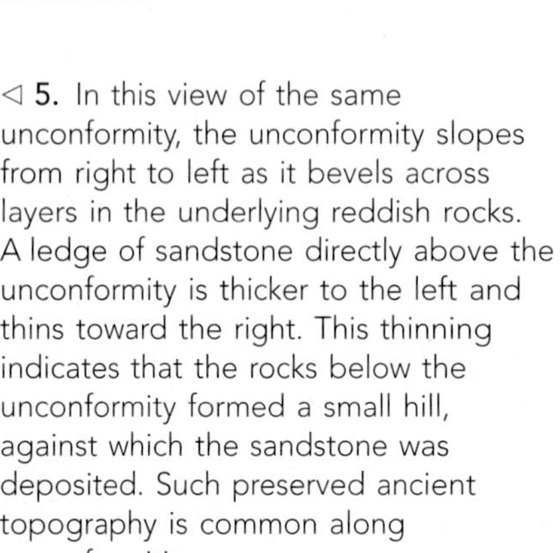

△ 6. The same unconformity is exposed across many parts of the United States and is called the *Great Unconformity.* In this view, gently tilted sedimentary layers unconformably overlie vertical layers within a metamorphic rock. The time represented by the unconformity is 1.4 billion years (from 1.7 billion years below to 400 million years above). [Ouray, Colorado]

B How Does a Nonconformity Form?

Some erosion surfaces form on top of rocks that are not layered, such as granite or massive metamorphic rocks. This type of unconformity is called a *nonconformity*.

The formation of a nonconformity begins as a granite, or other nonlayered rock, is uplifted to the surface, where an erosion surface is carved across it. The top of the granite is weathered into a reddish zone of sand, clay, and iron oxides.

As conditions change, the erosion surface is buried by sand and cobbles, perhaps derived in part from weathering of the granite. The contact at the top of the granite is a *nonconformity*.

△ This *nonconformity* has dark sandstone over tan granite. It is the same Great Unconformity as shown on the previous page, but here it overlies granite instead of layered rocks. [Grand Canyon]

C What Is a Disconformity and What Does It Indicate About an Area's History?

If rock layers are not tilted before they are overlapped by younger layers, but the boundary still represents millions of years of time, the contact is called a *disconformity*.

1. The first step in the development of a disconformity is deposition of essentially horizontal layers. These layers are limestone, but could be any type of sediment.

2. Next, the area is exposed at the surface because the rocks are uplifted or because sea level drops. Weathering takes place along the exposed land surface.

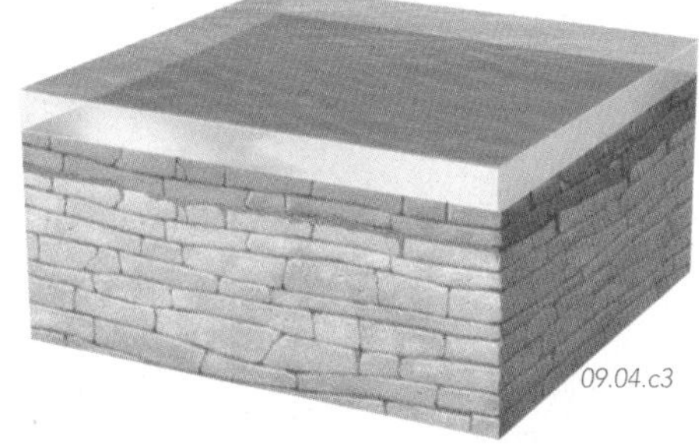

3. The surface is buried by an overlying layer, forming the *disconformity*. This new layer can be deposited by water or can be deposited on land, such as by sand dunes. A disconformity can form because of erosion, as illustrated here, or because there was little or no deposition for some time period.

4. This image, colorized for emphasis, shows two disconformities, one above the reddish lens and an older one below the lens.

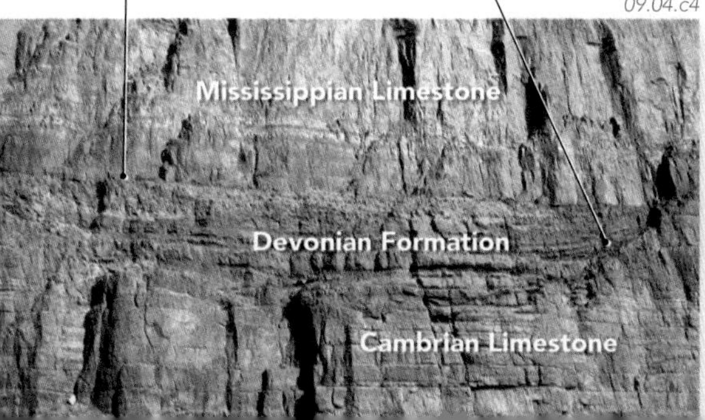

5. This close-up photograph shows a disconformity between two limestones that are millions of years different in age.

Before You Leave This Page Be Able To

- ✓ Summarize or sketch the three main types of unconformities and what sequence of events is implied by each type.

9.4

What Are Fossils?

ROCKS CONTAIN FOSSILS—EVIDENCE OF ANCIENT LIFE. Rocks of the appropriate age and type can preserve shells, coral, bone, petrified wood, leaf impressions, dinosaur tracks, and features created by burrowing worms. What kinds of creatures left these remains, and how are these traces of past life preserved in the rock record?

A What Are Fossils and How Are They Preserved?

Fossils are any remains, traces, or imprints of a plant or animal that are preserved in a rock or deposit. Fossils can be expressed in different ways depending on what type of life is represented, in what environment the plant or animal lived, and how the remains were buried and preserved.

09.05.a1

△ Most fossils encountered in the field represent the preserved *hard parts* or the replaced remains of marine organisms, such as shells and coral. The photograph above shows heads and stems of animals called *crinoids*.

09.05.a2

△ Vertebrate animals have hard parts, such as *bones*, that can be preserved. Most bones are found as fragments instead of complete skeletons because of the destruction and dispersal caused by scavengers and erosion.

09.05.a3

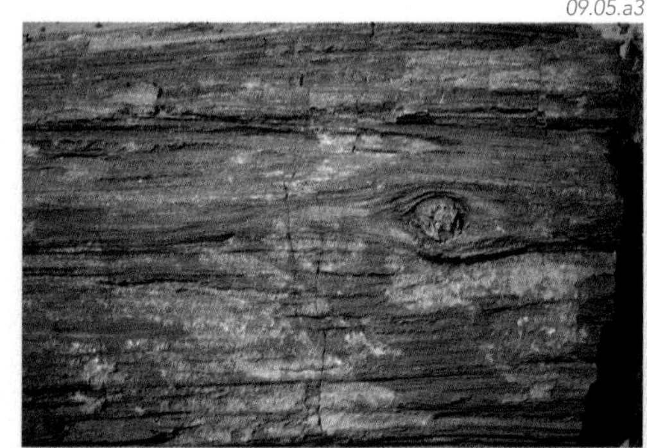

△ Some fossils are preserved because they have been *replaced* by silica, pyrite, or some other material. One example is wood from trees that is replaced by fine-grained silica, forming petrified wood.

09.05.a4

△ Another type of fossil forms when an animal is buried and decays. This leaves a cavity in the rock that mimics the animal's shape. The cavity is a *mold* if unfilled and is a *cast* if it is later filled by minerals.

09.05.a5

△ After burial, some carbon-rich plants and animals become *thin films of carbon* or other materials that preserve the original shape of the plant or animal. This fossil fern is almost 300 million years old.

09.05.a6

△ Fish and other soft creatures can be preserved as *impressions*, especially when the remains come to rest in quiet waters of a lake or deep sea. Such fossils can preserve amazing details, including fins.

▷ Animals can become fossils in other ways. Insects become trapped in tree sap, which through time loses its volatile substances and hardens into golden-brown *amber*.

09.05.a7

09.05.a8

◁ Some fossils do not preserve the actual organism, but instead represent something that the organism constructed. This mound-like feature, called a *stromatolite*, was built by ancient microscopic algae.

B What Traces Do Creatures Leave in the Rock Record?

In addition to preserved remains of the organism, rocks contain other features made by animals that moved across the surface or burrowed into soft sediment. These features are called *trace* fossils.

09.05.b1

Creatures that walk on land, such as reptiles, or on the sea bottom, like crabs, can leave *footprints* behind, such as this one from a dinosaur. Most footprints are indentions in sediment that get covered and filled by later sediment. A series of related footprints is called a *trackway*, from which geologists can infer how the creature moved, how much it weighed, and whether it traveled alone or in a group.

09.05.b2

Worms and other creatures wriggle, dig, or tunnel into the mud, forming cavities that can be filled by a different kind of sediment, such as sand. This type of trace fossil is a *burrow* or *worm burrow*. Burrows represent places where worms and other creatures went through the mud. The creatures were too soft to be preserved but still left behind a record.

C What Determines Whether a Fossil Is Preserved?

Most creatures are never preserved as fossils because fossil preservation requires certain favorable circumstances. The most important factors include the existence of hard body parts and rapid burial after death.

Hard Parts—Preservation as a fossil is much more likely if a creature, such as this crinoid, has a shell, bones, teeth, or some other hard part. Such animals are overrepresented in the fossil record compared to insects or jellyfish, which lack such hard parts. Also, the soft parts of creatures can be eaten by scavengers, crushed during sediment compaction, or otherwise destroyed.

09.05.c1

Rapid Burial—A fossil cannot be preserved unless it is buried. If a creature's remains are left on the surface, whether on land or in the sea, they can physically or chemically decompose or be scavenged by other creatures. Rapid burial means less opportunity for destruction. Preservation is easier beneath the sea than on land because of rapid burial and the lower content of oxygen, which accelerates decay on land.

What Features Look Like Fossils but Are Not

Some natural geologic features look like fossils but are not fossils. These features form through *inorganic* processes and do not represent the remains of any organism.

The most common features mistaken for fossils are the dark, branching mineral growths shown here. These growths, called *dendrites*, typically consist of manganese-oxide minerals that grow in this branching pattern.

Spherical features, called *concretions*, which grow in sediment during cementation, are also commonly mistaken for fossils. These weather out of sediment as small spheres or oddly shaped objects that can look organic, but are not.

09.05.mtb1

Before You Leave This Page Be Able To

- ✓ Describe the different ways in which a plant or animal can be preserved as a fossil.
- ✓ Describe two types of commonly encountered trace fossils.
- ✓ Describe the two main factors that influence whether a creature is preserved as a fossil.
- ✓ Describe a feature that can be mistaken for a fossil.

How and Why Did Living Things Change Through Geologic Time?

DIFFERENT FOSSILS OCCUR IN DIFFERENT ROCK UNITS. Some of these differences reflect variations in sedimentary facies, such as between reefs and rivers, but most reflect the documented, systematic way that fossils vary with age up through a section of rocks. Why did these changes in fossils occur, and what do they tell us about how life on our planet has changed through time?

A How Do Fossils Vary with Age?

The earliest geologists recognized that fossils change as one moves upward from one sedimentary layer to another. These systematic changes of fossils with age, called *faunal succession,* helped geologists identify time periods marked by major changes in life on Earth. Using the principles of relative dating and faunal succession, geologists subdivided geologic time into four major chapters, each with subdivisions. Later, results from isotopic dating provided numeric ages, in millions of years before present, for when each chapter started.

Subdivisions of geologic time.

1. The *Cenozoic Era,* meaning *recent life,* spans the last 65 million years. It is called the *age of mammals* because mammals, such as this mammoth (▽), became a dominant type of life.

09.06.a1

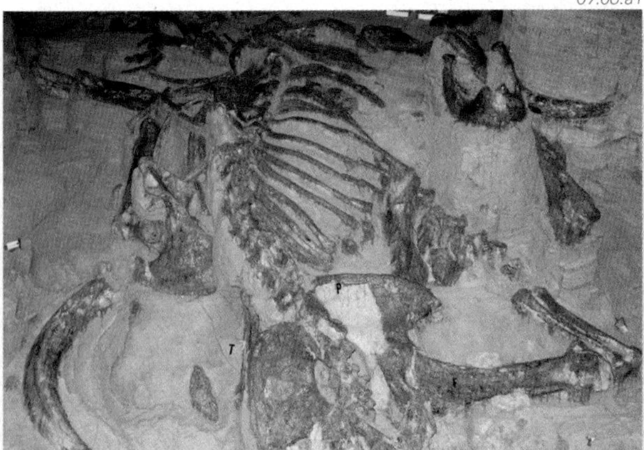

09.06.a2

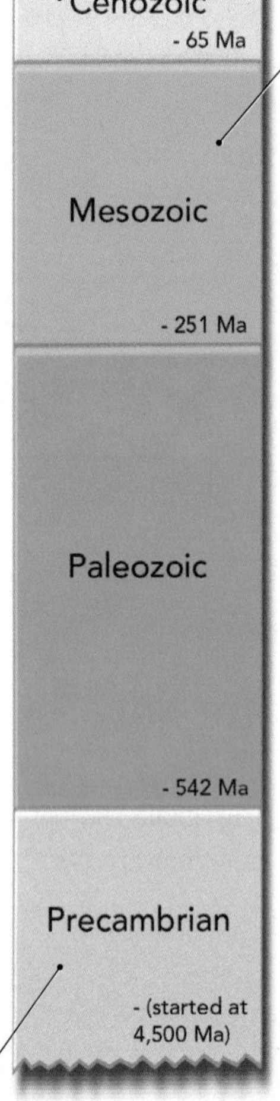

2. The *Mesozoic Era* (*middle life*) is known as the *age of dinosaurs* because dinosaurs (▽) rose to dominance during this era. The end of the Mesozoic Era is marked by the extinction of dinosaurs.

09.06.a3

3. The *Paleozoic Era* (*ancient life*) marked the rise of the major groups of marine animals, including coral, creatures like clams that had hard shells, and various types of fish (▽). Plants, insects, and amphibians also colonized the land during this era. The end of the Paleozoic Era is marked by a major extinction called the *Great Dying.* This extinction killed off many species of animals in the seas and on land.

09.06.a4

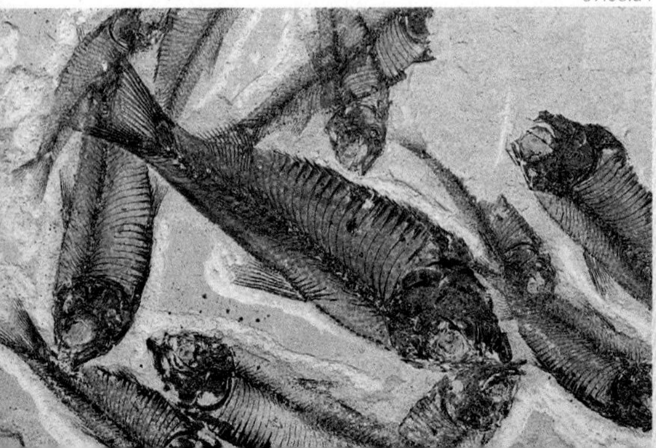

4. The *Precambrian* (before the Cambrian Period) comprises nearly 90 percent of geologic time. For most of this time, only simple life forms existed, such as bacteria and algae that formed stromatolites. ▷

09.06.a5

C What Life Existed During the Paleozoic Era?

During the Paleozoic Era, an extraordinary diversity of life evolved—both in the seas and on land. Artistic reconstructions of three times of the Paleozoic were produced by Karen Carr, whose work is featured in museums.

Early Paleozoic

09.10.c1

In the early Paleozoic, corals, crinoids (which look like platy underwater lilies), and mollusks were anchored to the seafloor, trilobites and snails moved across the seafloor, and shelled creatures with tentacles propelled through the water.

Middle Paleozoic

09.10.c2

In the middle of the Paleozoic, corals built large reefs, and pieces of crinoid stems littered the seafloor. Fish became diverse and abundant. On land, many forms of insects appeared, and plants included ferns and seedless trees.

Late Paleozoic

09.10.c3

Amphibians and early reptiles evolved during this time, with a dramatic rise of reptile groups and a continued diversity of marine life. Land plants, insects, and marine life continued to diversify until a major extinction at the end of the Paleozoic.

Possible Causes of the Great Dying

The end of the Paleozoic Era marks Earth's greatest extinction, called the Great Dying. On land, about 70% of all species, including many invertebrates, amphibians, and reptiles, went extinct. The event took a huge toll in the oceans, extinguishing almost 90% of marine species, including trilobites. Geologists are still actively investigating a number of possible causes.

A great outpouring of lava occurred at the end of the Paleozoic. Large volumes of basalt erupted in northern Asia, forming a unit called the *Siberian Traps* (*trap* is an old word used to describe basalt). Such eruptions expel volcanic ash and gases, such as water vapor, carbon dioxide, and sulfur dioxide. These products have the potential to warm or cool the planet and possibly lead to catastrophic effects, such as changing circulation patterns in the oceans.

There is some evidence, currently being debated, for a large meteorite impact at the end of the Paleozoic. Geologists have proposed that unusual carbon molecules with uncommon gases resulted from such an impact. Several suitably large impact craters have been proposed but not tied to the extinction. Such an impact could trigger the massive eruptions of the Siberian Traps, but this connection remains conjectural.

Throughout most of the Paleozoic, continents were separated by warm, shallow seas. By Permian time, the supercontinent *Pangaea* had formed. Its formation closed seas that had once fostered Paleozoic life. The supercontinent became more arid, and vast evaporite deposits formed and could have changed the salt concentrations in seawater. These many effects of the formation of Pangaea may have helped kill off specialized organisms and set the stage for a more dramatic event.

Another alternative is that conditions in the atmosphere and oceans led to a massive overturn of ocean water, causing deep, oxygen-poor water to be brought to the surface. This could have caused a dramatic change in shallow ocean temperatures and in the amount of CO_2 in the atmosphere, leading to sudden and catastrophic climate changes. Such changes could affect the entire planet, resulting in a mass extinction on the land and in the oceans. This theory, like the others, is not conclusive and one of geology's greatest mysteries remains unresolved.

Before You Leave This Page Be Able To

- ✓ Summarize the environments of early life and some important evolutionary events that took place during Earth's early history.
- ✓ Briefly describe what happened during the *Cambrian explosion*, including fossils of the Burgess Shale.
- ✓ Explain four possible causes for the *Great Dying*, the largest extinction event in Earth history.

What Were Some Milestones in the Later History of Life on Earth?

MASS EXTINCTION AT THE END OF THE PALEOZOIC ERA provided evolutionary opportunities for new life forms. The organisms that repopulated the early Mesozoic seas and lands were very different from many Paleozoic organisms. The end of the Mesozoic Era is defined by another major extinction event, which gave rise to yet another evolutionary chapter, the Cenozoic Era. The artwork of Karen Carr provides us with one interpretation of the scenes represented by the bones, shells, and other fossils.

A What Life Was Abundant During the Mesozoic Era?

Diverse life existed during the Mesozoic Era, but it is known as the *age of dinosaurs* because dinosaurs are the best known creatures of this time. The Mesozoic has three periods: Triassic, Jurassic, and Cretaceous.

Early Mesozoic: Triassic

09.11.a1

During the Triassic Period, small and nimble dinosaur-like creatures and mammals appear beneath the seed-bearing conifer forests. In the seas, shallow-sea niches left open by the Permian extinction were occupied by coiled ammonites and other marine animals.

Middle Mesozoic: Jurassic

09.11.a2

Dinosaurs diversified during the Jurassic Period, including *Stegosaurus* with plates on its back and the huge plant-eating *Apatosaurus*. Carnivorous predators, like *Allosaurus*, stalked the landscape. The Jurassic Period also featured *Archaeopteryx*, an early bird. The seas flourished with creatures, including ammonites, star fish, and large marine reptiles.

Late Mesozoic: Cretaceous

09.11.a3

During the Cretaceous Period, dinosaurs remained diverse, and included various plant-eating dinosaurs that walked on four or two legs, as well as predators like the raptors lurking in the bushes. Flying reptiles and birds graced the skies. Not shown is the fearsome *Tyrannosaurus rex*. For the first time, flowering plants, called *angiosperms*, became abundant on land. Insects remained a vibrant and diverse group, and most mammals continued a rather low-key existence.

09.11.a4

During the Cretaceous Period, animals similar to those of the Jurassic thrived in the seas, including fish of many kinds, straight and coiled nautiloids, large marine reptiles, and turtles. Not shown because of their tiny size are countless floating and free-swimming organisms called *plankton*.

B What Were Dinosaurs and What Wiped Them Out?

Dinosaurs evolved from Permian ancestors and existed on Earth for 165 million years. By the middle of the Mesozoic, they dominated the land, but they and many other animals went extinct at the end of the Mesozoic Era. The end of the Mesozoic also is called the *KT boundary* because it separates the Cretaceous (K) from a time that was called the Tertiary (T). Numerous hypotheses have been proposed to explain the KT extinction.

1. There were two types of dinosaurs that differed in their hip structure. One group of dinosaurs had a hip structure similar to lizards and included a diverse group of carnivores, such as *Tyrannosaurus rex*, and herbivores, such as *Apatosaurus*. Some walked slowly on four legs; others walked and ran on two legs.

09.11.b1

2. Another group of dinosaurs had a birdlike hip structure (but were not related to birds) and were herbivores. Some like *Stegosaurus* and *Triceratops* walked and grazed on four legs. Others, like duck-billed dinosaurs, could move on two legs.

09.11.b2

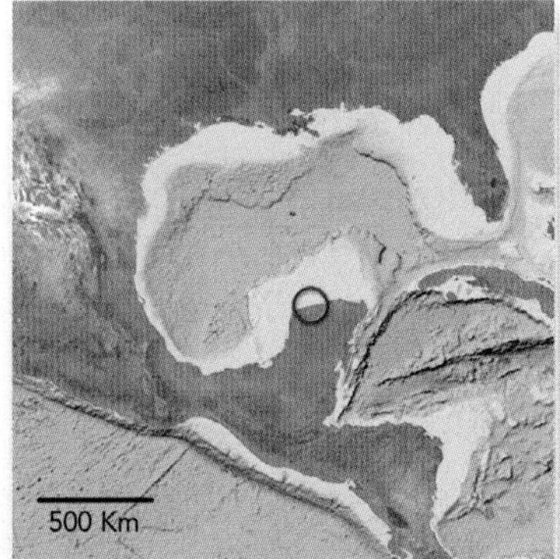

3. A widely known hypothesis for the KT extinction involves a huge comet or asteroid striking Earth, sending massive amounts of dust and gas into the atmosphere and blocking sunlight. The surface of the planet would have been cold for decades. The presumed impact at 65 million years ago is the Chicxulub crater on the Yucatán Peninsula in Mexico (shown by the red circle).

4. Another possible cause of the extinction was massive outpourings of basalt, represented by the *Deccan Traps* in India (not shown). Huge eruptions could have put enough sulfur-dioxide gas into the atmosphere to cause a volcanic winter that would have lasted decades.

C What Life Appeared During the Cenozoic Era?

The Cenozoic Era is called the *age of mammals*. After the dinosaurs went extinct, mammals were able to diversify rapidly and fill the niches left behind by the KT extinction.

1. By early Cenozoic time, the ancestors of modern mammals, including bats, rodents, primates, sloths, whales, hoofed animals, and carnivores were abundant and lived in a variety of niches. Marsupial mammals, represented by modern kangaroos, thrived on the isolated southern continents of South America and Australia.

09.11.c1

◁ 2. Although they lived 20 million years ago, many of the mammals shown here may be familiar to you because they are fairly similar to their modern descendants. Each type of mammal, however, underwent many changes between then and now. Horses, for example, changed dramatically in size. These changes are well recorded by bones and teeth of different species of horses found at thousands of sites around the world.

09.11.c2

◁ 3. Late in the Cenozoic, during the Ice Ages, a number of large mammals roamed the continents. Some of these animals, like the mammoth, went extinct as the Ice Ages ended and humans spread across the globe. The first humans (Homo sapiens) appeared before 300,000 years ago, based on fossil evidence. Human-migration data are still controversial, but by at least 50,000 years ago Homo sapiens populated several parts of the planet, having left their sites of origin in Africa. The details of human history are refined by discoveries of new archeological sites and even older ancestors.

Before You Leave This Page Be Able To

- ✓ Contrast the kinds of organisms that lived during the Mesozoic Era with those that lived during the Cenozoic Era.
- ✓ Describe some of the variety observed in dinosaurs, and summarize two theories for why dinosaurs became extinct.

How Do We Reconstruct Geologic Histories?

WE RECONSTRUCT THE SEQUENCE OF GEOLOGIC EVENTS by using the various strategies of relative dating, correlation of rock units, isotopic dating, and other geologic principles. Geologists commonly start by studying a single section of rocks and determining the sequence of events it represents. Understanding the causes and broader contexts of the events requires correlating several rock sections.

A How Do We Correlate Units and Events in Two Sections of Rocks?

There are various strategies for matching—or correlating—two sections of rocks. The general approach is to find units that match and develop a logical explanation for why other parts of the sections do not match. Read each principle and its example, and then compare that part of the two sections.

Principle	Example
Lateral Continuity—The surest form of correlation is to be able to physically trace a unit through the landscape from one place to another.	Both sections are capped by a reddish sandstone that can be traced through the landscape between the two sections.
Distinctive Rock Type—A unit may have distinctive characteristics that enable it to be matched between two sections.	Both sections have breccia that contains large blocks of banded gneiss.
Fossils—If two units contain the same assemblage of *index fossils*, they have the same age and are correlative.	The gray limestone in both sections contains late Triassic fossils of the same age.
Similar Sequence of Rocks—Two sections of rock may contain a similar sequence of layers.	Both sections contain a yellowish mudstone, overlain by gray limestone and underlain by gray shale.
Record the Same Event—Two units may record the same event, such as a change in sea level, even if they express the event in different ways.	Gray shales in both sections record rising sea level.
Isotopic Age—If datable units, such as volcanic layers, yield the same numeric age, they may be time correlative.	Basalt flows in both sections give similar isotopic ages (~230 Ma).
Magnetic Signature—Two units may record the same sequence of reversals of Earth's magnetic field, from reversed (R) to normal (N).	These units both yielded a similar paleomagnetic sequence, which allows but does not demonstrate a correlation of the two units.
Position in Sequence—Two different rock types may correlate if they are in the same position in the sections.	The beach sand in section 1 is in the same position as a tidal-flat-related mudstone in section 2.
No Correlative—Sometimes a unit in one section has no correlative unit in the other section.	A local landslide deposit in section 2 did not extend far enough to be present in section 1.
Relation to Unconformities—Two units may correlate if they have a similar relationship to the same unconformity.	Conglomerate at the base of both sections overlies metamorphic rocks along the same unconformity.

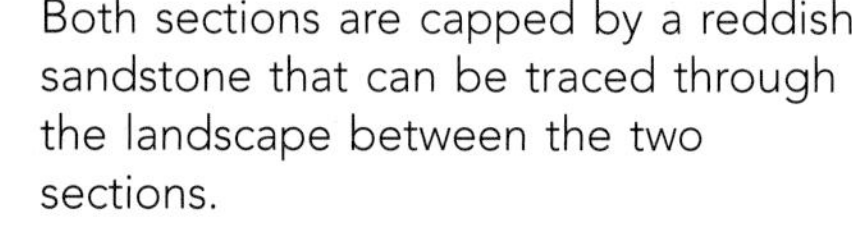

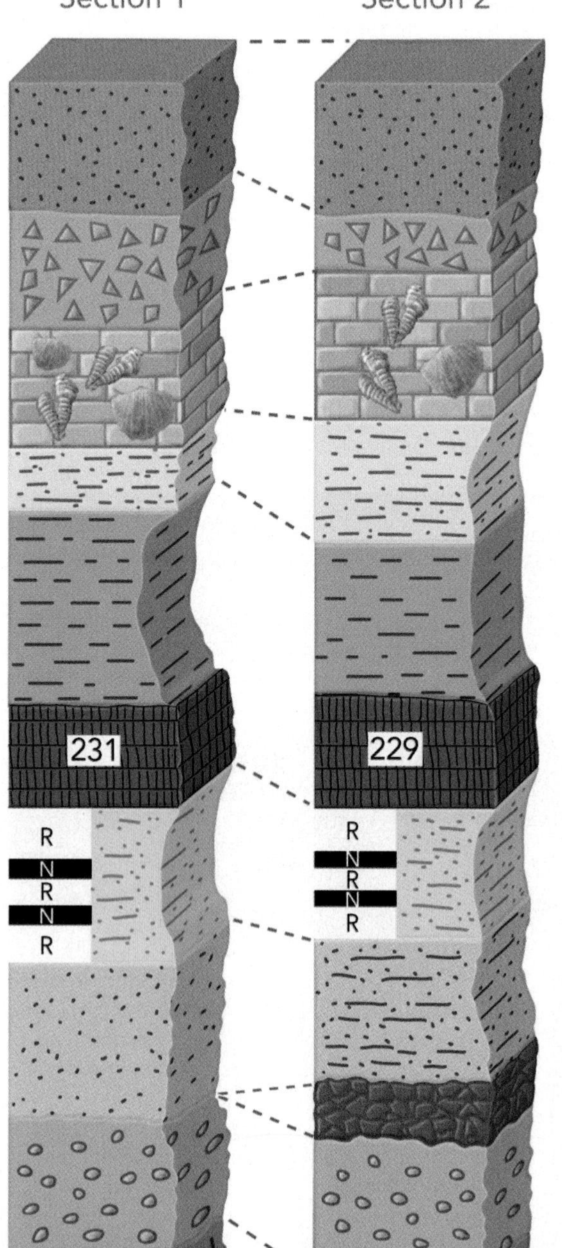

B Why Do Some Rock Units Change from One Section to Another?

When investigating causes of events, geologists seek to understand how and why the rock units change laterally. Even though two units are deposited at the same time, they may be different rock types. There are several explanations why the sequence of layers changes from one place to another.

Facies Change—The type of sediment deposited at the same time can change from one place to another because the *sedimentary facies* changed laterally, such as from a delta environment to a shallow marine one.

Restricted Event—A unit may not be present in nearby sequences because it simply was not deposited there. Many units are formed by a relatively small event and have a restricted aerial distribution, such as these sand dunes.

Change in Thickness—Accompanying some facies changes are variations in the thickness of sediment deposited. Thickness changes can also reflect variations in topography over which a unit is deposited, such as river deposits that are thickest in the center of the valley.

Eroded Away—Another explanation for why a unit is not present or is thinner is that it has been partly or completely eroded away in one place but is preserved in another.

C What Are Some Approaches to Investigate Geologic History?

Geologists rarely use a single approach to investigate geologic history. Instead, most geologists study sections of rocks in the context of the geologic relationships as expressed on geologic maps and cross sections. This is because nearly all geologic problems involve geometric relationships in three dimensions.

1. The view below, looking north, shows a geologic terrain with various rock units and other geologic features, numbered in the order in which they formed, from oldest (1) to youngest (11). The area has two distinct parts. The northern half largely consists of volcanic features and broad plains, whereas the southern half features folded rock units that form curved ridges and valleys. Examine this figure and identify the reasons why the units and features are interpreted to have formed in the relative order reflected by the numbering.

2. A recent-looking scoria cone (11) has erupted lava (also numbered 11) that flowed to the south. The lava flow poured over a fault (10), and is not faulted, so the fault is older.

3. South of the fault is a series of layers (2 to 5) that wrap around and everywhere dip outward away from a granite (unit 1). The lowest layer in the series is the oldest (2), and the highest layer is the youngest (5).

4. The contact between layer 2 and the granite (1) is depositional with pieces of granite in layer 2. So the granite is older than layer 1. Also there is no baking or other contact effects.

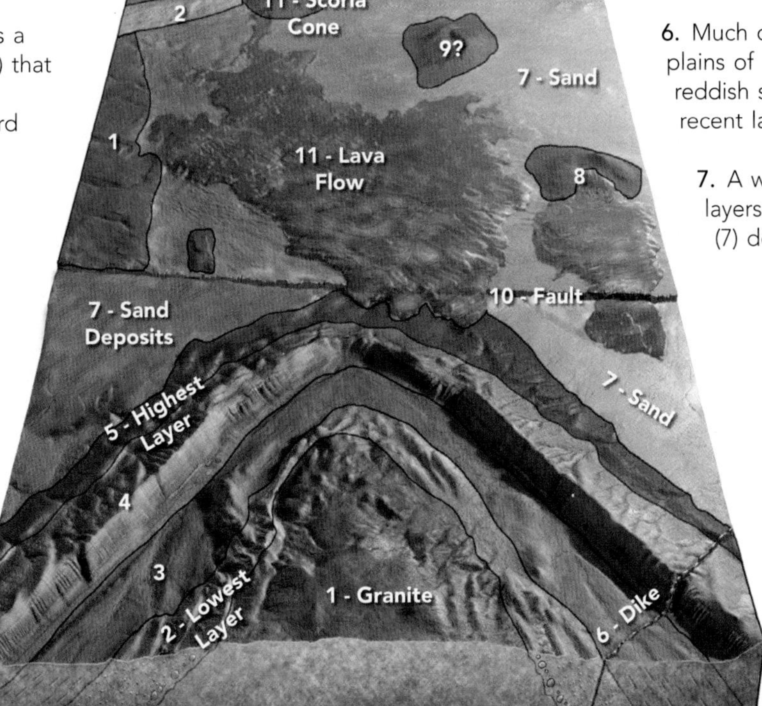

5. There are three scoria cones in all, and of the three, one cone is not eroded (11) and one is very eroded (8). A third cone (9) is intermediate in appearance and is probably between the two in age, but is not in contact with other geologic features so its age is uncertain.

6. Much of the northern part of the area consists of broad plains of unconsolidated sand (unit 7). The sand has some reddish soil developed on top of it and is overlain by the recent lava flows and scoria cones.

7. A weathered dike (6) crosscuts the granite and layers 2 to 5, so is younger. It is older than the sand (7) deposited across the area.

Before You Leave This Page Be Able To

- ✓ Summarize or sketch the principles by which two sequences of rocks can be correlated.
- ✓ Describe or sketch why layers can change from one sequence to another.
- ✓ Reconstruct the sequence of events from a cross section or block diagram.

Why Do We Investigate Geologic History?

INVESTIGATING WHEN AND IN WHAT ORDER geological events occurred is important to society. Geologic history helps us evaluate the potential for geologic hazards, explore for resources, understand the physical world around us, and understand changes in life and the environment over time.

A How Do Geologic Ages Help Us Evaluate Geologic Hazards?

When assessing the potential for geologic hazards, such as earthquakes and volcanic eruptions, we are most interested in knowing *when* these events last occurred and when such activity might happen again.

In this image, a dark, recent lava flow overlies a lighter colored, older one. When did each eruption occur, and what is the likelihood of another eruption? Evaluating such volcanic hazards involves dating the ages of the rocks using isotopic methods and estimating the ages of lavas and volcanic cones based on their amount of erosion.

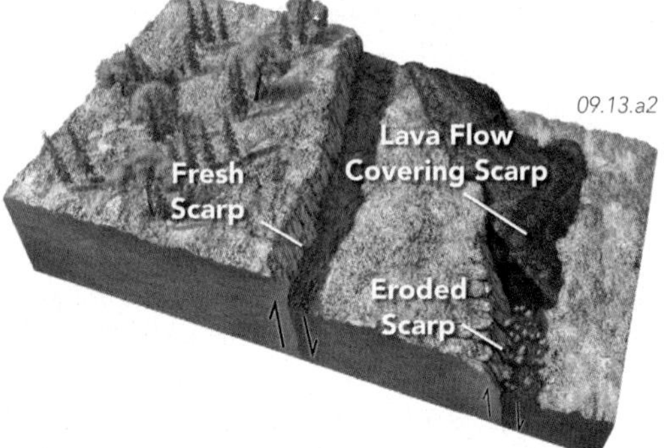

In this scene, the landscape is cut by two earthquake-related fault scarps. The upper scarp is recent and not eroded, and the lower one is partly eroded and covered by a lava flow. The age of the earthquakes can be investigated by examining soils and rocks that predate and postdate faults, and by dating sediments associated with faulting.

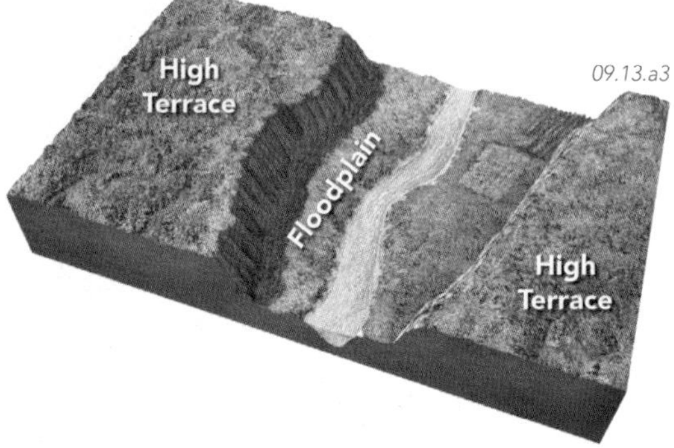

Flood potential is evaluated from records of stream flow, but these may only cover the last 100 or so years. Inferring the recurrence of larger, less frequent floods relies on geologic evidence, such as determining ages of terraces from soil development, carbon-14 ages on charcoal, and surface-dating methods.

B How Do Geologic Ages Help Us Explore for Natural Resources?

Understanding geologic history is an essential part of evaluating an area's potential for important natural resources, such as mineral, energy, and water resources.

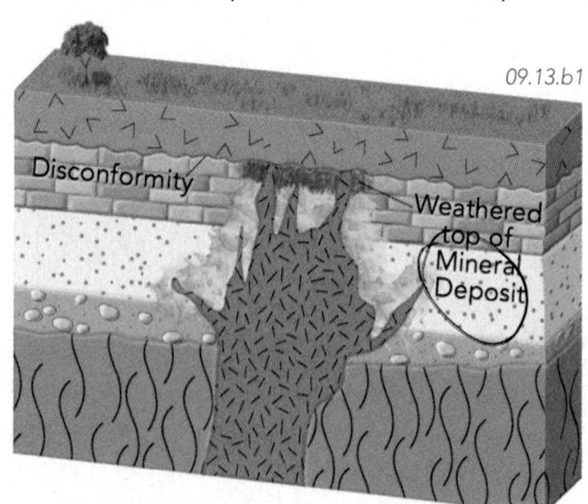

Here, a granite released metal-rich fluids that formed adjacent copper and gold deposits. Exploring for such mineral deposits involves knowing the age of events that contributed to mineralization. Dating the granite or minerals deposited by the fluids helps determine these ages. Also important are crosscutting relations between the granite, mineralized fractures, and earlier or later rocks and structures.

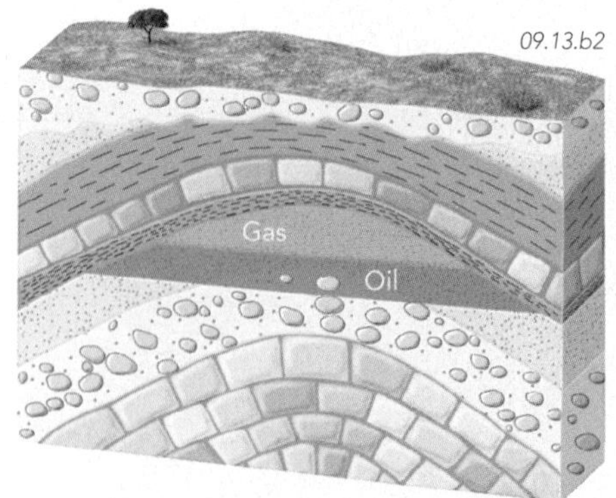

Oil and gas accumulate in the subsurface, such as near the tops of anticlines. Exploration for oil and gas involves a thorough investigation of the sequence of rock units, determined by relative dating and the fossils within the sedimentary rocks. Later events, such as folding and erosion, play a key role in determining whether oil is trapped at depth or could escape to the surface and flow away.

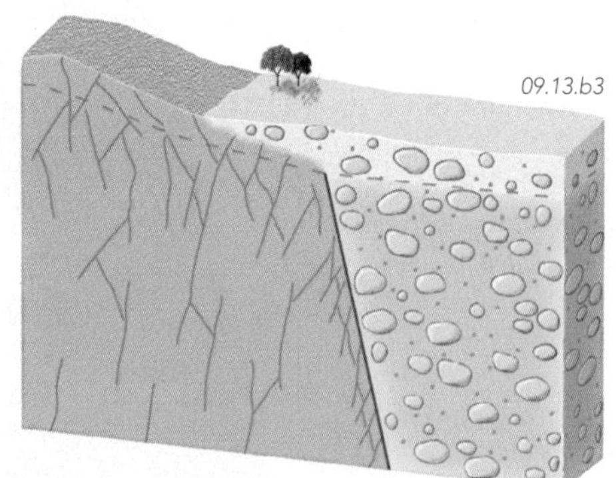

Sedimentary basins contain abundant groundwater. However granite may only contain water in fractures. We assess the formation age of sedimentary basins and their water-rich sediments using fossils, relative-dating techniques, and numeric ages on interbedded volcanic rocks. The age of some groundwater can also be dated using isotopic ages to better understand groundwater supplies.

C How Does the Sequence of Events Help Us Understand Our Physical World?

Geology is all around us. It is expressed in the shape of the landscape, the loose sediment on the surface, and the hard rocks beneath. The landscape we see today is the result of a long series of events.

1. This view to the north shows a geologic map superimposed on topography of the area around the state of North Carolina.

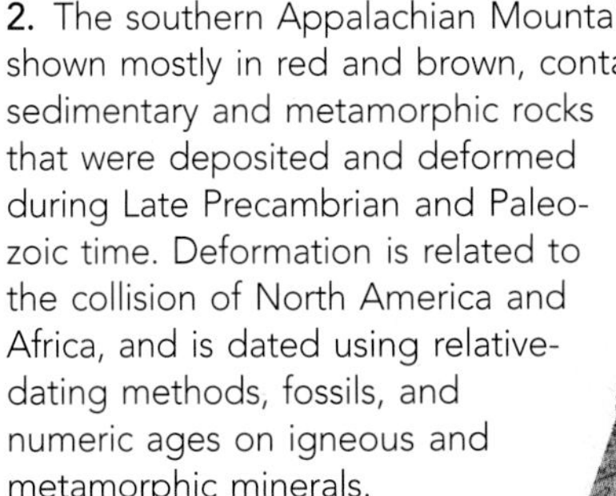

2. The southern Appalachian Mountains, shown mostly in red and brown, contain sedimentary and metamorphic rocks that were deposited and deformed during Late Precambrian and Paleozoic time. Deformation is related to the collision of North America and Africa, and is dated using relative-dating methods, fossils, and numeric ages on igneous and metamorphic minerals.

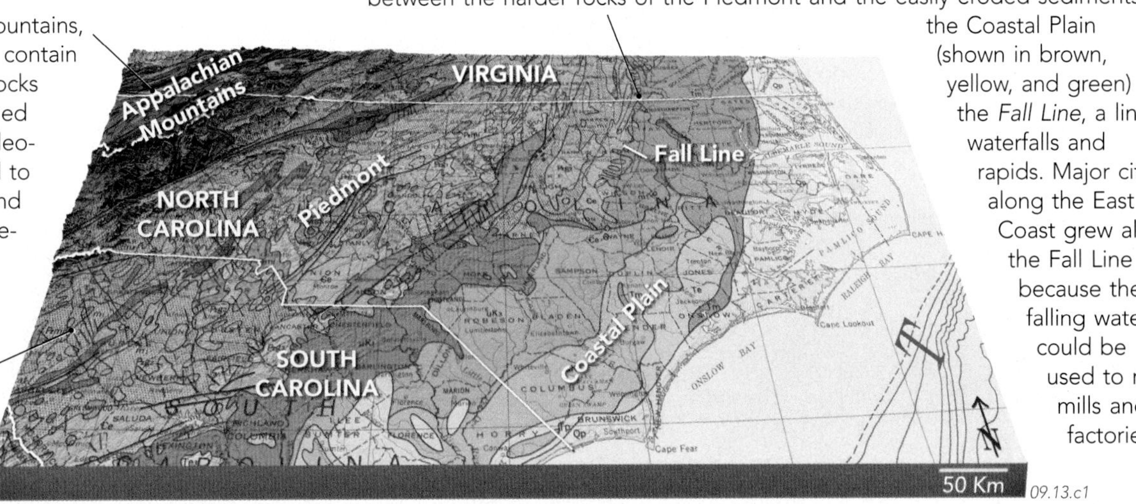

09.13.c1

3. East of the Appalachians is the Piedmont, which contains strongly deformed metamorphosed rocks and granite. The Piedmont has similar ages of rocks as the Appalachian Mountains, but the two series of rocks originated in very different places and settings. Cutting through the Piedmont are large faults (heavy lines on this map). Dating has determined that most of these faults are no longer active, so are unlikely to cause earthquakes.

4. In from the shoreline is the Coastal Plain, a low region underlain by Cenozoic and Late Mesozoic sedimentary rocks and sediment. The boundary between the harder rocks of the Piedmont and the easily eroded sediments of the Coastal Plain (shown in brown, yellow, and green) is the *Fall Line*, a line of waterfalls and rapids. Major cities along the East Coast grew along the Fall Line because the falling water could be used to run mills and factories.

D How Does Geologic History Help Us Investigate Our Origins?

Geologists investigate recent historical events and the origins of humans using many of the same strategies and techniques used to reconstruct the history of ancient rocks and structures. The record of historical events of the last several thousand years is commonly investigated by working with archeologists who excavate the ruins of ancient cities and other archeological sites.

1. Archeologists and geologists use the same relative-dating principles applied to older rocks. The oldest ruins generally are on the bottom, having been covered over by successive generations of younger habitations. Pieces of an older wall can also be contained within a younger wall if the prehistoric builders reused preexisting materials. This site represents the ancestral Woodland Cultures of Iowa.

09.13.d1

2. Investigating early history of humans relies heavily on the input of geologists. Bones from early human ancestors have been found in sedimentary units ranging from tens of thousands of years to five or six million years. These sequences are dated by interbedded volcanic rocks or by fossils of small mammals or other creatures. [Ethiopia, Africa]

09.13.d2

3. Events that are hundreds to tens of thousands of years old can be dated using carbon-14 ages on bones, wood, and charcoal preserved at the site. The relative positions of dated samples becomes an important check for consistency.

Before You Leave This Page Be Able To

- ✓ Describe or sketch how geologic ages help evaluate geologic hazards and mineral, energy, and water resources.
- ✓ Provide an example of how knowing the ages of rocks and structures helps us understand landscapes at a larger scale.
- ✓ Discuss dating techniques used to investigate early human sites.

What Is the History of the Grand Canyon?

GEOLOGICALLY, THE GRAND CANYON HAS IT ALL. It contains some of the best exposed and studied, as well as most beautiful, rock sequences in the world. It is discussed in almost every geology class because it so clearly expresses a history of geologic events over the last 1.7 to 1.8 billion years.

This north view of the Grand Canyon region shows the location of a geologic cross section from A to B. The Colorado River, which formed the canyon, flows from right to left, exits the canyon through high cliffs and enters Lake Mead.

The Grand Canyon is eroded into the Colorado Plateau, a region of broad plateaus, mesas, and deep canyons, which expose a mostly flat-lying sequence of Mesozoic and Paleozoic sedimentary rocks.

The river flows southwest across the area, cutting across horizontal to locally tilted layers. The deepest part of the canyon is where the Colorado River erodes through the uplifted Kaibab Plateau.

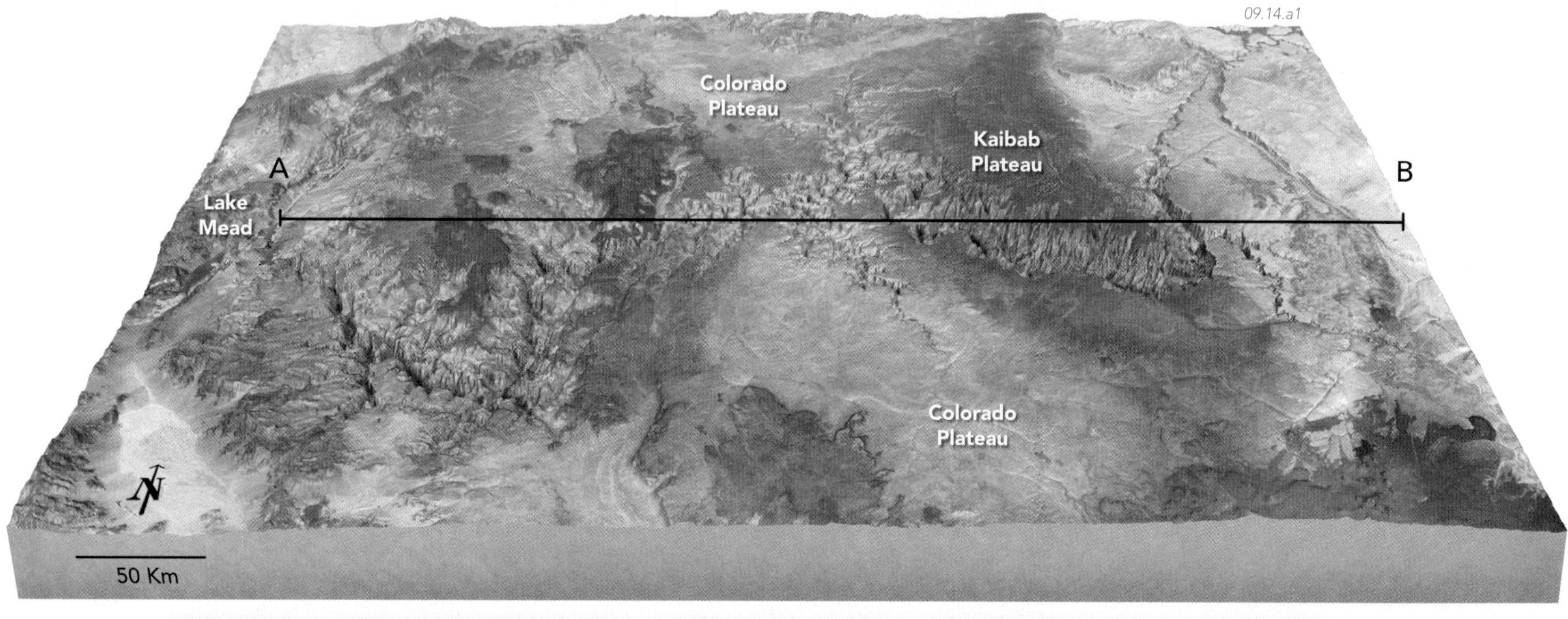

Basalt flows cap some plateaus and predate formation of the main canyon. They are dated by argon methods at 8 million years.

Large faults, like the Hurricane fault, cut across the region, downdropping rocks to the west. These faults cut basalt flows that are less than 1 to 2 million years old.

Some basalt flows flowed down into the already-carved canyon, demonstrating that much of the canyon is older than 4 to 5 million years.

Paleozoic sedimentary layers cap most plateaus and are warped over a few broad folds (monoclines).

Mesozoic sedimentary rocks are preserved on the down-folded sides of monoclines and contain famous dinosaur tracks and petrified wood in the Painted Desert.

A
Cenozoic Sedimentary Rocks
W
Grand Wash Cliffs
Late Cenozoic Basalts
Hurricane Fault
Late, Middle, and Early PZ Units
Kaibab Plateau
Monoclines
Mesozoic Sedimentary Rocks
E
B
09.14.a2
Precambrian Metamorphic Rocks
Precambrian Granite
Faults with Complex Histories
Precambrian Sedimentary and Igneous Rocks

Most of the Plateau exposes a flat-lying sequence of, from top to bottom, late, middle, and early Paleozoic rocks. These rocks compose the colorful upper walls of the canyon.

The oldest rocks are metamorphic rocks and granites that are nearly the same age (1.7 billion years). These are exposed in the bottom of the canyon.

The near-vertical basement rocks are overlain by tilted late Precambrian sedimentary and volcanic rocks, shown in purple. The contact is an angular unconformity, and is called the *lower unconformity.*

A separate, *upper unconformity* marks where gently dipping Paleozoic layers overlie the moderately tilted late Precambrian ones.

Sequence of Rocks

09.14.a3

09.14.a4

09.14.a5

Geologic History and Key Age Constraints

The sequence of events in the Grand Canyon has been reconstructed using relative dating, fossils, and many different isotopic dating methods. The geologic history resulting from these studies is summarized below, which should be read from bottom to top (oldest to youngest).

7. *Deformation, Uplift, and Erosion*—The Paleozoic strata largely have escaped deformation and remain nearly flat, except near a few faults and folds. The monoclines are bracketed, using relative-dating methods, to between 80 and 40 million years ago. The region was uplifted some at this same time, but the modern canyon was not carved until much later, mostly within the last 5 million years.

6. *Deposition of Late Paleozoic Layers*—Overlying sedimentary layers (shown in red, pink, tan, and blue-green) record a wide range of environments, including shallow marine, shorelines, rivers, and a dune-covered desert. These rocks are dated with marine and nonmarine fossils as late Paleozoic (Pennsylvanian and Permian). Disconformities separate some of the formations and represent time when the region was above sea level.

5. *Deposition of Early and Middle Paleozoic Units*—After erosion carved the upper unconformity, seas covered the land and deposited sandstone, shale, and limestone (shown in brown and blue). These deposits are dated by trilobites and other fossils as early and middle Paleozoic (Cambrian, Devonian, and Mississippian). Later, the seas left and in several instances formed disconformities within the limestones.

4. *Tilting and Upper Unconformity*—Layers in the Late Precambrian rocks were gently to moderately tilted and then beveled by erosion. This produced the *upper unconformity*. As this unconformity is followed west, it truncates the *lower unconformity* beneath the Kaibab Plateau (see the cross section A–B). This combined unconformity represents even more missing time (from 1.7 billion years to 540 million years, or more than 1.1 billion years); it is appropriately called the *Great Unconformity* and can be followed eastward to the Great Lakes region.

3. *Late Precambrian Rocks and Lower Unconformity*—In the Late Precambrian, sedimentary and volcanic rocks were deposited in horizontal layers across the upturned basement layers. This formed the *lower unconformity*. The lower parts of these late Precambrian rocks are dated by several isotopic methods at 1.1 billion years. Since the underlying basement rocks are 1.7 billion years, the lower unconformity represents 600 million years of time not recorded by any rocks!

2. *Uplift and Erosion of the Basement*—After the metamorphism, the basement rocks cooled as they were uplifted and eroded over a period that lasted for hundreds of millions of years. Erosion beveled across the steep metamorphic layers.

1. *Basement Rocks*—Metamorphic and plutonic rocks in the bottom of the canyon represent the oldest events. They were formed, metamorphosed, and deformed to near-vertical orientations, all between 1.76 and 1.70 billion years ago.

The Percentage of Geologic Time That the Canyon Records

Although the Canyon is a classic geologic locality with a thick sequence of formations, it represents a relatively small amount of geologic time. The oldest rocks are "only" about 1.7 billion years old, so the area contains no record for 2.8 billion years of Earth history (4.5–1.7 billion years). Next, the two unconformities together cut out another 700 to 800 million years of history.

Even the Paleozoic sequence is missing more time than it records! The formations only represent five out of the seven geologic periods (rocks of the Ordovician and Silurian Periods are not present), and none of the formations span an entire period. Finally, Mesozoic and Cenozoic rocks are mostly present outside the canyon, so yet more time is not represented by rocks in the canyon walls.

Before You Leave This Page Be Able To

- ✓ Summarize examples of how different methods of dating events and rocks were used to reconstruct the geologic history of the Grand Canyon.
- ✓ Summarize why the Canyon does not represent all of geologic time.

What Is the Geologic History of This Place?

This terrain exposes various geologic relationships that have been documented in the field and recorded as descriptions. Samples collected from the area were analyzed either for their isotopes or their characteristic fossils. You will use this information to reconstruct the sequence and ages of events that produced features exposed in the landscape today.

Goals of This Exercise:

- Observe the distribution of different rock types exposed in the terrain to characterize the sequence of rocks and the geologic features that are present.
- Use descriptions of units and of key contact relationships to infer the relative sequence of events.
- Assign the sedimentary units to ages, based on the kinds of fossils they contain.
- Calculate isotopic ages for key igneous rocks to help constrain when important events occurred.

Procedures

Use your observations to complete the following steps. Your instructor may provide you with rock or fossil specimens.

1. Observe the terrain to understand the overall pattern of rocks. Based on this pattern, use the associated descriptions to determine in what order the units formed and where in that sequence different geologic features, such as a fault, developed.
2. Examine the six fossils in the table below, and the geologic period to which each is assigned; complete the stratigraphic section on the worksheet, listing the units in the order in which they formed, from bottom to top in the section.
3. Use the table of isotopic measurements below to calculate the age of a sample of granite and a sample of the dike.
4. Summarize the geologic history by arranging the different events in their proper order on the worksheet.

Field Notes

The units and features are described below. Each unit or feature has a letter assigned to it, but these do not reflect the order in which the features formed. Some letters were skipped so that some features would have letters that were easy to remember, such as V for the volcano.

Unit A—Tan sandstone with land fossils, including plants of Permian age.

Unit B—Greenish shale with marine fossils, including Ordovician trilobites. The top of the unit was weathered and eroded prior to deposition of unit A, but the layers in the two units are parallel to each other and to their mutual contact.

Unit C—Coarse sandstone and beach conglomerate that contains Cambrian trilobites. The base contains clasts derived from the underlying granite (G).

Unit D—Finely crystalline dike that has baked units A, B, C, and G.

Feature F—Fault that cuts units B, C, and G. It does not cut some units.

Unit G—Coarse granite that is weathered near the contact with unit C.

Unit K—Gray limestone with marine fossils of Cretaceous age.

Units L and V—Unweathered lava flow (L) associated with a volcano (V).

Feature N—Narrow canyon.

Unit R—Partly consolidated river gravels with a thick, well-developed soil. Contains land mammals of middle Cenozoic age.

Unit S—Reddish and pinkish sandstone that was deposited by rivers and in lakes. It contains Jurassic dinosaur bones.

Identification of Fossils

Rock Unit	Fossil	Period
R	Mammals	Cenozoic
K	Fish	Cretaceous
S	Dinosaurs	Jurassic
A	Plants	Permian
B	Trilobites	Ordovician
C	Trilobites	Cambrian

Table of Isotopic Measurements

Rock Unit	Half-Life of Isotope	# Parent Atoms	# Daughter Atoms
G	1 Billion Years	250	750
D	50 Million Years	500	500

1. This view shows a landscape with various rocks and features. There is a central plateau (high flat area) flanked by several mountains, an obvious volcano, a canyon, and a number of lines and curved features that cross the landscape.

2. The geology in the subsurface is shown on the sides of the block. Any type of unconformity is shown with a squiggly line, reflecting some topographic relief along the erosion surface represented by the unconformity. Normal depositional contacts are shown by thin lines, and a fault is marked by a thicker line.

3. A section of layers forms a series of cliffs and slopes on three corners of the block. These were encountered first and so are lettered A, B, and C. Unit A is a brown sandstone that was deposited on land and contains Permian plant fossils. Unit B is greenish marine shale and contains Ordovician trilobites. Unit C is a coarse sandstone and beach conglomerate that contains Cambrian trilobites.

4. A dark dike (D) forms a linear wall across the landscape. It mostly is uninterrupted by other geologic features, except for one obvious gap near a belt of some tan-colored soils. The dike consists of dark basalt and was dated by isotopic methods.

5. An older series of river channels (R) cross the plateau and form low troughs in the topography. One channel goes all the way to the edge of the canyon. Along their lengths, the channels are partially filled by river gravels and are characterized by well-developed, tan soils. They contain bones of small horses and other fossils from the middle Cenozoic.

6. The top of one mountain in the area exposes higher layers than are preserved elsewhere. There is a red sandstone (S) that contains bones of Jurassic dinosaurs. The sandstone is overlain by a gray limestone (K) that has fish and other marine fossils from the last part of the Mesozoic (Cretaceous).

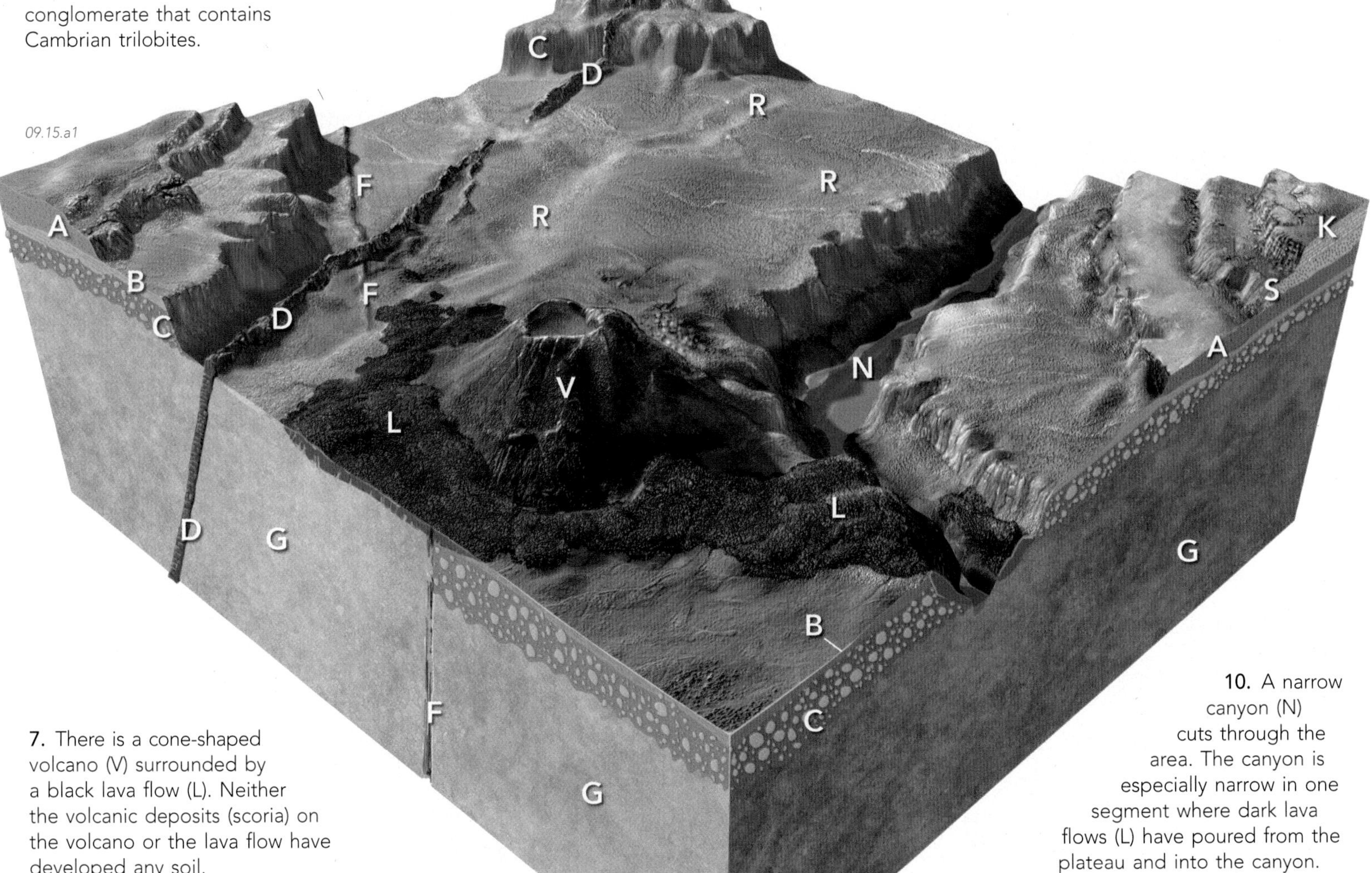

7. There is a cone-shaped volcano (V) surrounded by a black lava flow (L). Neither the volcanic deposits (scoria) on the volcano or the lava flow have developed any soil.

8. A fault (F) forms an obvious line across parts of the area, but is not continuous. It is also shown in cross section on the side of the block. It has not formed a fault scarp, but is expressed in the topography because it is the boundary between rock types that erode in slightly different ways.

9. The lowest unit in the area is a gray granite (G). Geologists determined an isotopic age on a sample of the granite, and these results are in the table on the previous page.

10. A narrow canyon (N) cuts through the area. The canyon is especially narrow in one segment where dark lava flows (L) have poured from the plateau and into the canyon.

11. Reconstruct the history using superposition, crosscutting relationships, and the relationship of different features to the landscape. Be systematic, focusing your attention on any pair of objects that are in contact. For example, does the dike crosscut the fault or vice versa? Is unit A above or below unit B? Some objects may not be in direct contact with each other, but their relative age can be determined by comparing their ages relative to some other feature.

CHAPTER

10 The Seafloor and Continental Margins

SEVENTY PERCENT OF EARTH'S SURFACE IS OCEAN. Beneath the oceans is an underwater landscape that includes broad plains, submarine mountains, and deep trenches and canyons. What clues do these features provide about how our planet operates? This chapter is about the surface and subsurface of the seafloor, how we study the seafloor, and what the various features tell us about Earth processes.

Beneath Monterey Bay, along the coast of central California, the seafloor displays a puzzling feature—a great submarine canyon. In this image, satellite data are shown for land, and computer-shaded and colored data show seafloor depths.

What features are present on the seafloor, and how do we explore the depths, rock types, and structures of the seafloor?

A broad continental shelf flanks the coast, with relatively shallow water (less than about 100 meters) extending out kilometers to tens of kilometers from the coast. The area is a prized marine ecosystem and is the site of the Monterey Bay National Marine Sanctuary.

What is a continental shelf and how does it form?

10.00.a1

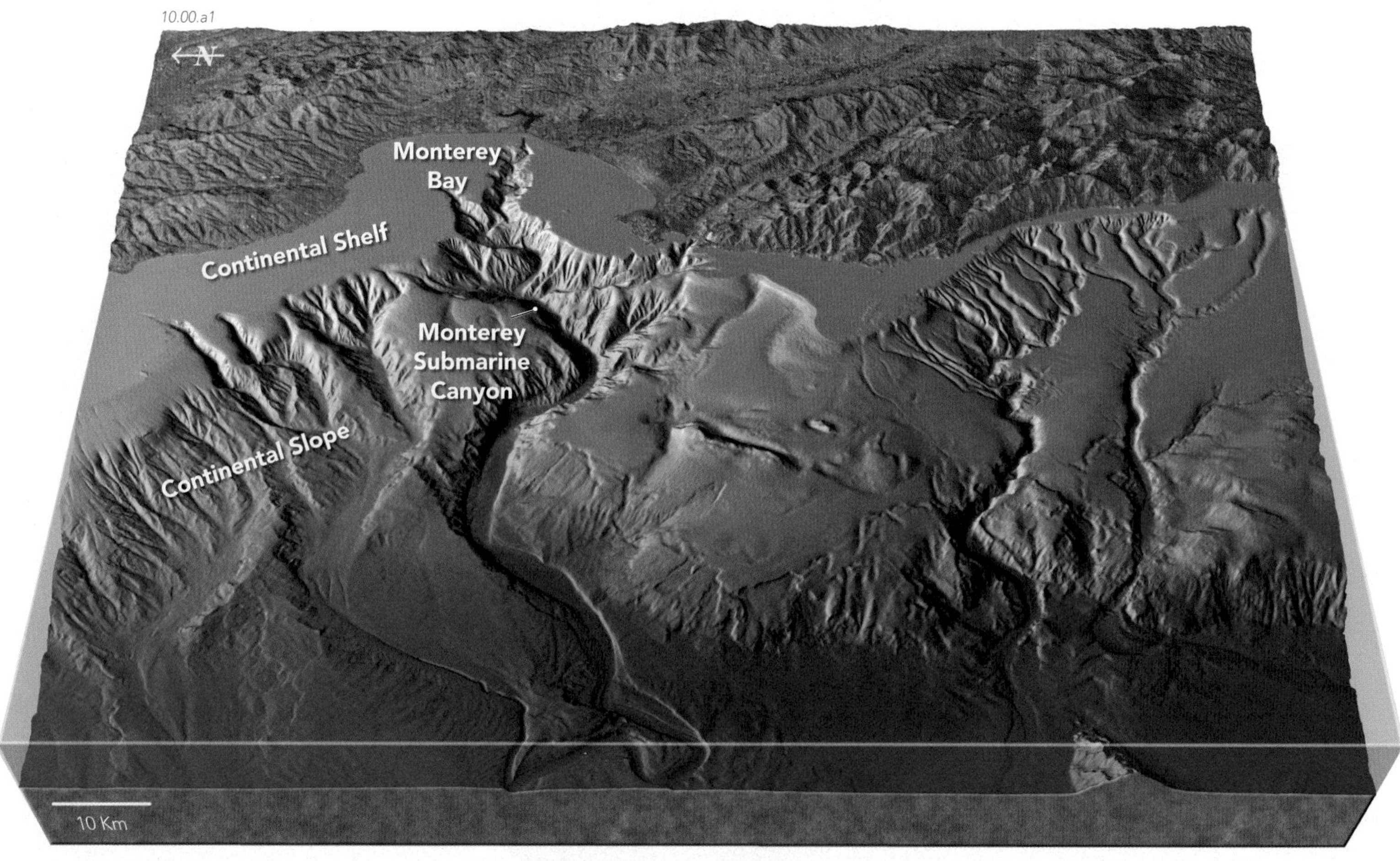

Monterey Submarine Canyon is enormous. It is similar in scale to the Grand Canyon. The canyon bottom is as much as 1,800 meters (nearly 6,000 feet) below the rim and, in this deep segment, the canyon is 20 kilometers (12 miles) wide. It resembles many valleys on land; it curves, goes from higher to lower areas, and has smaller side valleys that merge with the main channel.

What processes carve submarine canyons?

The continental margin near Monterey Canyon is heavily studied. Surveys done using ship-borne instruments provide detailed information about the canyons and other geologic features, such as landslides, bedrock ridges, and linear fault scarps.

What processes occur on the seafloor, and what types of features do they produce?

TOPICS IN THIS CHAPTER

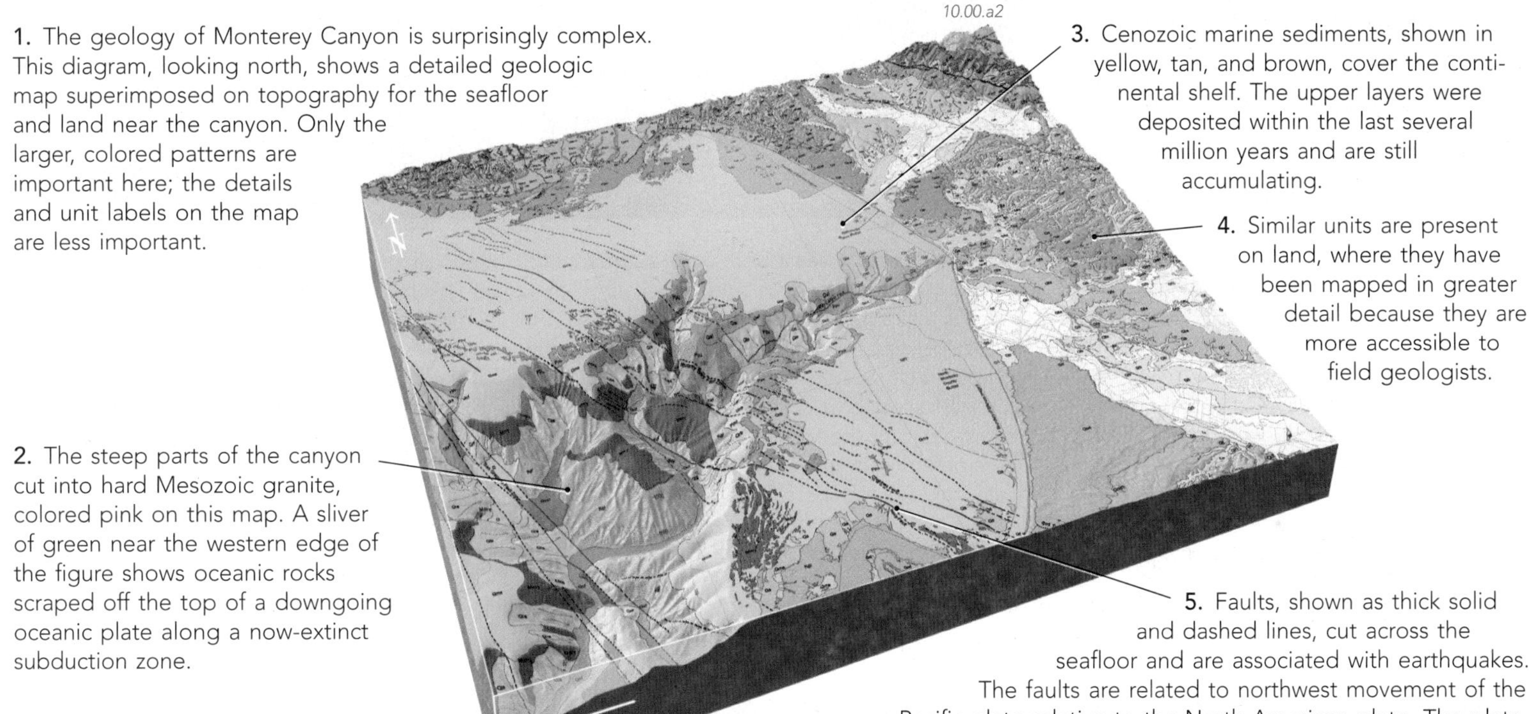

1. The geology of Monterey Canyon is surprisingly complex. This diagram, looking north, shows a detailed geologic map superimposed on topography for the seafloor and land near the canyon. Only the larger, colored patterns are important here; the details and unit labels on the map are less important.

2. The steep parts of the canyon cut into hard Mesozoic granite, colored pink on this map. A sliver of green near the western edge of the figure shows oceanic rocks scraped off the top of a downgoing oceanic plate along a now-extinct subduction zone.

3. Cenozoic marine sediments, shown in yellow, tan, and brown, cover the continental shelf. The upper layers were deposited within the last several million years and are still accumulating.

4. Similar units are present on land, where they have been mapped in greater detail because they are more accessible to field geologists.

5. Faults, shown as thick solid and dashed lines, cut across the seafloor and are associated with earthquakes. The faults are related to northwest movement of the Pacific plate relative to the North American plate. The plate boundary is on land in this area, near the far corner of the diagram.

Origin of Monterey Canyon

We do not expect to find huge canyons beneath the sea. When and how did Monterey Canyon form, and what processes are going on today in and around the canyon? Scientists explore the submarine canyon by bouncing sound waves off the seafloor, dredging and drilling rock samples from the bottom, and diving to the bottom in small submarines.

The formation and evolution of the canyon reflect the complicated plate tectonic events that have affected California in the last 20 million years. Geologists have concluded that the upper part of the canyon was originally carved by rivers when its granitic base was above sea level, before 10 million years ago. Strike-slip motion between the North American and Pacific plates shaved off this granitic slice and transported it northward up the coast of North America. During this movement, the canyon was submerged below sea level and filled by sediments, which were later eroded by underwater currents.

For the past several million years, dense slurries of sediment-rich water, called *turbidity currents*, have flowed down the canyon, scouring the channel and undercutting the canyon walls. The canyon widens as the steep, unstable walls collapse downward in underwater landslides and debris flows. The turbidity currents carry sediment more than 200 kilometers (120 miles) down the canyon and into deeper water, where the sediment is deposited in a broad feature called a *submarine fan*. The lower part of the canyon, like many submarine canyons, was never above sea level and has been carved entirely by turbidity currents. The position of the lower channel has shifted over time, as segments of the canyon have been offset by strike-slip faulting or buried by landslides.

How Do We Explore the Seafloor?

EXPLORING THE SEAFLOOR presents different challenges than from mapping and studying geology on land. Mapping the oceans requires methods for *remotely* observing the bottom of the sea, such as by bouncing sound waves off the seafloor. Geologists and oceanographers collect samples of rocks on and below the seafloor by going down in small submarines or by using ships to drill holes through the sediment and rock.

A How Do We Map and Investigate the Seafloor?

10.01.a1

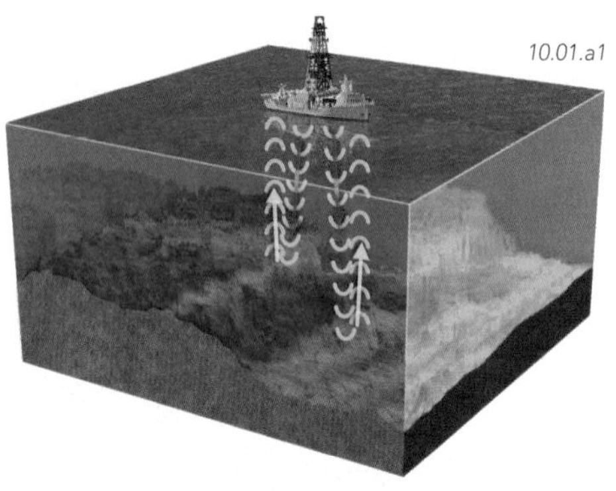

Scientists map parts of the seafloor by transmitting *sound waves* from a ship and then timing how long the waves take to bounce off the seafloor and return to sensors on the ship. The longer this takes, the deeper the seafloor. Using this technique, called *sonar*, one can direct sound waves straight down, as shown here, or at an angle to the seafloor.

10.01.a2

Scientists visit the seafloor in small submarines called *submersibles*, capable of carrying two or three people. Submersibles allow direct observation of geologic features and phenomena. Scientists can take photographs for later study and collect samples of rocks, seawater, and life forms. Smaller robotic versions of submersibles are operated remotely from ships.

10.01.a3

Specially equipped research vessels allow us to drill holes into the seafloor. We can retrieve samples of the sedimentary and volcanic rocks that make up the upper part of the oceanic crust. The layers preserved in drill cores allow geologists to reconstruct the sequence of events, the ages of the rocks, and the variations in seawater chemistry over time.

B What Can We Learn from Ocean Drilling?

Geologists and oceanographers drill holes into the seafloor to retrieve samples for later study. The drilling process yields cylinder-shaped samples, called *drill core*, which provide numerous types of data.

10.01.b1

1. *Type of Sediment or Rock*—Geologists cut open the drill core to identify the type of sediment or rock and to observe layers and other features. These observations yield interpretations of ocean-floor processes, environments, and past events.

10.01.b2

2. *Fossils*—Microscopic and larger fossils within the drill core help geologists assign sediment and rock layers to different parts of the geologic timescale. They also provide constraints on the environments in which the sediment formed. These microscopic fossils are called Foraminifera.

3. *Isotopic Ages*—Small samples of the core can be crushed in order to separate minerals for isotopic dating. This is mostly done on volcanic units, such as basalt flows, when studying the seafloor. The ages of the layers, combined with measurements of layer thickness, yield the rates at which the layers accumulated:

rate of deposition = thickness/time span

4. *Other Measurements*—Geologists analyze core samples in other ways to answer specific questions about past climates and seawater chemistry. Analyzing for different oxygen and carbon isotopes in a series of layers yields a detailed record of how ocean conditions varied over the past thousands to millions of years. Many such changes were global, so these measurements can be used to correlate layers between different parts of the world.

C How Do We Map the Seafloor from Space?

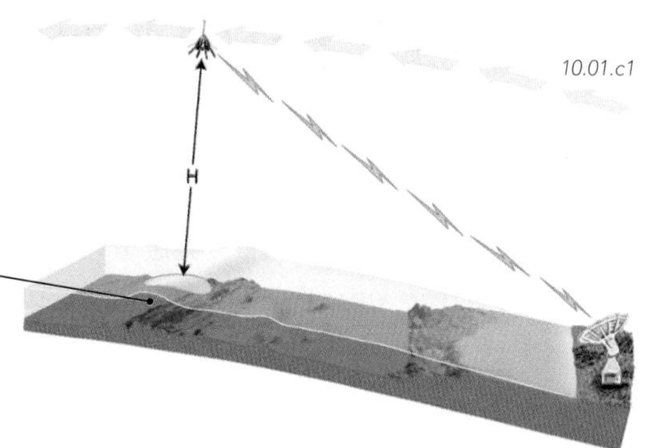

Most satellites are able only to observe the seafloor in very shallow water, and are primarily used for surface measurements, such as temperature of the upper water surface. Some satellites, however, use radar to measure the height of the sea surface.

In some areas, the sea surface can be tens of meters higher than average sea level. These high areas are caused by the gravitational forces associated with submarine mountains and mid-ocean ridges. These forces attract seawater and cause it to mound up. These variations in height of the sea surface can, therefore, be used to estimate the topography of the underlying seafloor.

D How Do We Image What Is Below the Ocean Floor?

To investigate the geometry of rock units *beneath* the sea, a government-research or industry ship tows a device that bounces sound waves off the seafloor and off rock layers in the subsurface. The sound waves are recorded by devices, called *geophones*, that are towed behind the ship. Waves reflected from shallower layers arrive back sooner than those reflected by layers deeper in the subsurface.

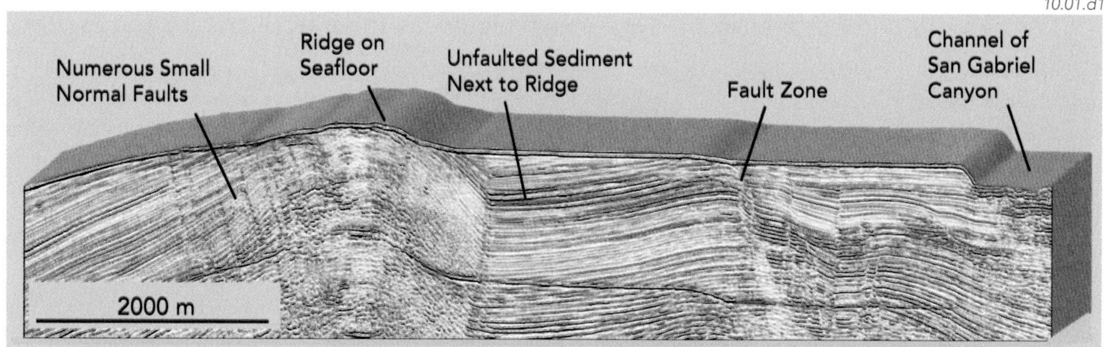

Geologists process the data using sophisticated computer programs that model the passage of the sound waves through the layers and back to the geophones. The data are plotted in a type of cross-section view called a *seismic-reflection profile* (△). This profile from offshore of Southern California shows tilted, folded, and faulted layers. The *seismic-reflection* technique is widely used in exploring for oil and natural gas beneath the seas, as well as on land.

Diving to the Deepest Parts of the Ocean

Ocean exploration is similar in many ways to exploring space, but not quite as expensive. Getting to the deepest parts of the ocean requires specialized submarines that can only accommodate a few passengers. Such travel is quite dangerous because of the high pressure in the deep oceans. Nevertheless, humans have explored very deep regions using remotely guided probes and by diving in submersibles. Currently, the world's deepest diving submersible is the Japanese vehicle *Shinkai*, shown below to the left. It can take humans to depths of 6,500 meters (more than 21,000 feet).

Among the features observed on the seafloor are *manganese nodules*, shown below. They form when manganese precipitates out of seawater, forming baseball-sized spheres. These are an important potential source of manganese and other metals, but the logistics and environmental issues associated with remote mining on the deep seafloor are still being investigated.

Before You Leave This Page Be Able To

- ✓ Summarize the four methods we use to explore the topography and rocks of the seafloor (sonar, submersibles, drilling, and satellites).
- ✓ Summarize the kinds of information we can obtain from cores drilled in the seafloor.
- ✓ Describe the seismic-reflection method for mapping the geometry of geologic units beneath the seafloor.

How Is Paleomagnetism Used to Study the Ocean Floor?

PALEOMAGNETISM IS THE ROCK RECORD OF PAST CHANGES in Earth's magnetic field. The magnetic field is strong enough to orient magnetism in certain minerals, such as magnetite and hematite, in the direction of the prevailing magnetic field. Magnetic directions preserved in rocks, such as volcanic rocks, intrusions, and some sedimentary rocks, provide an important way to investigate the origin of the seafloor.

A What Causes Earth's Magnetic Field?

Earth has a metallic iron core, which is composed of a *solid inner core* surrounded by a *liquid outer core*. The liquid core flows and behaves like a dynamo (an electrical generator), creating a magnetic field around Earth.

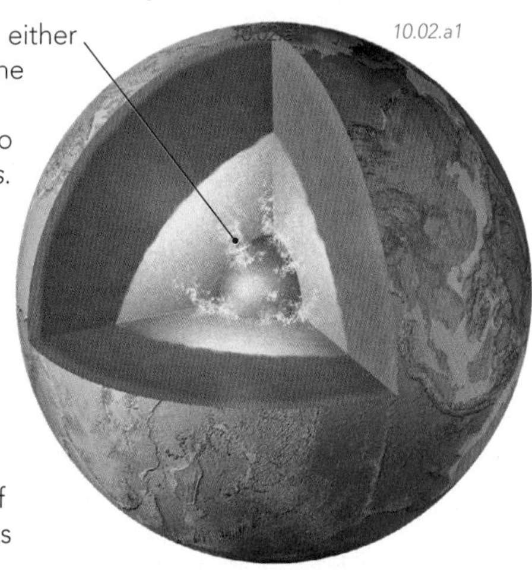

1. The inner core is transferring either heat or less dense material to the liquid outer core. This transfer causes liquid in the outer core to rise, forming *convection currents*. These convection currents are limited to the outer core and are not the same as those in the upper mantle.

2. Movement of the molten iron is affected by forces associated with Earth's rotation. The resulting movement of liquid iron and electrical currents generates the magnetic field.

▷ 3. Earth's magnetic field currently flows from south to north, causing the magnetic ends of a compass needle to point toward the north. This orientation is called a *normal* magnetic field.

10.02.a3

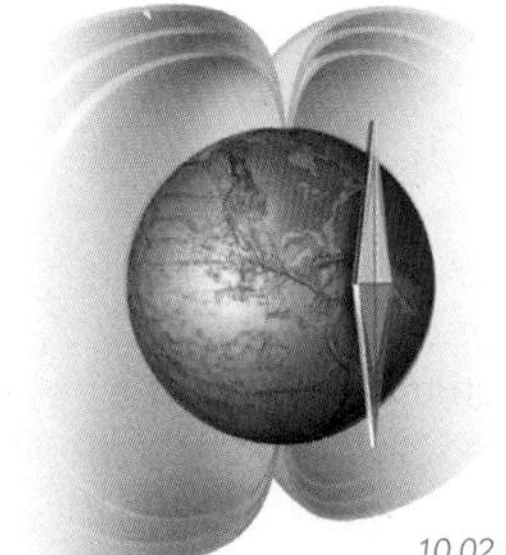

◁ 4. Many times in the past, the magnetic field has had a *reversed polarity*, so that a compass needle would point *south*. The switch between normal polarity and reversed polarity is called a *magnetic reversal*.

B How Do Magnetic Reversals Help Us Infer the Age of Rocks?

Earth's magnetic field typically remains either normal or reversed anywhere from 100,000 years to a few million years. Geologists have constructed a magnetic timescale by isotopically dating sequences of rocks that contain reversals. This timescale then serves as a reference to compare against other sequences of rocks.

1. Geologists measure the direction and strength of the magnetism preserved in rocks with an instrument called a *magnetometer*. With this device, geologists can tell whether the magnetic field had a *normal* polarity (compass would have pointed north) or a *reversed* polarity (compass would have pointed south) when the rock formed.

▷ 2. This figure shows the series of magnetic reversals during the last 10 million years, the most recent part of the Cenozoic Era. This time period is within the *Neogene Period*. In older geologic literature, this time period included the *Quaternary* and part of the *Tertiary*.

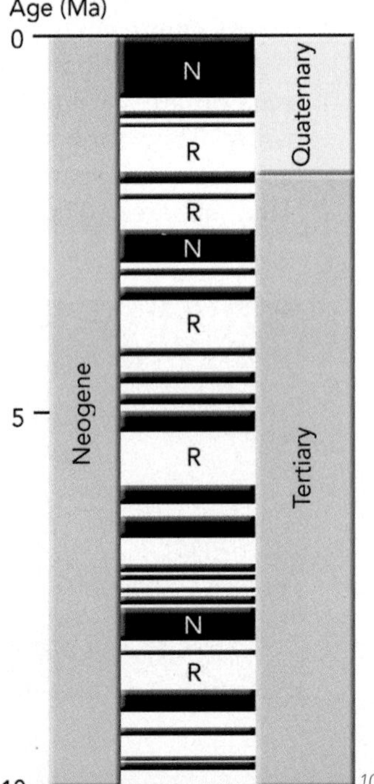

3. Periods of normal magnetization (N) are shown in black, and those of reversed magnetization (R) are shown in white.

4. Variability in the spacing and duration of magnetic reversals produced a unique pattern through time. The pattern of reversals can be measured in a rock sequence and compared to this magnetic timescale to see where the patterns match. This allows an estimate of the age of the rock or sediment. The magnetic timescale is best documented for the last 200 million years. Other age constraints, such as isotopic ages or fossils, are used to further refine the age of the magnetized rocks.

C How Are Magnetic Reversals Expressed at Mid-Ocean Ridges?

In the 1950s, scientists discovered that the ocean floor displayed magnetic variations in the form of matching stripes on either side of the mid-ocean ridge. Geologists interpreted the patterns to represent a magnetic field that had reversed its polarity. In the process, they developed the theory of plate tectonics.

10.02.c1

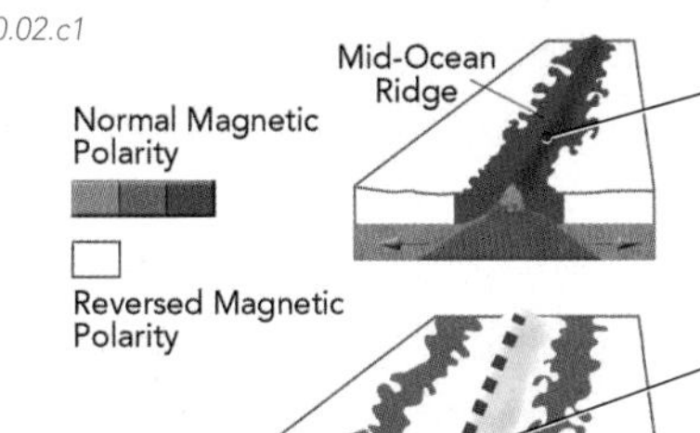

As an ocean basin pulls apart at a mid-ocean ridge, basaltic lava erupts onto the surface or solidifies at depth. As the rocks cool, the orientation of the Earth's magnetic field is recorded by an iron-rich mineral called *magnetite*. In this example, the magnetite records normal polarity (shown with a reddish color) at the time the rock forms.

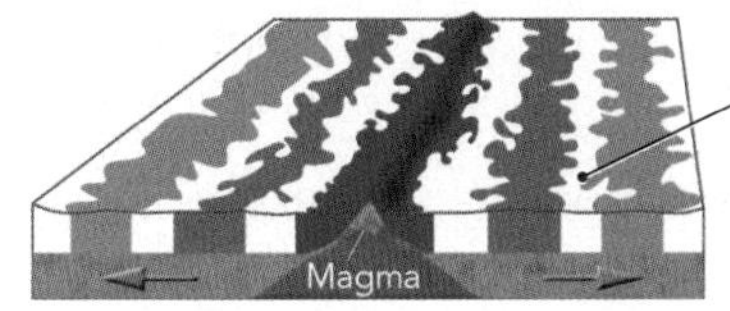

If the magnetic field reverses, a rock formed during that time will acquire a *reversed polarity* (shown in white). Rocks forming all along the axis of the mid-ocean ridge will have the same magnetic direction, forming a stripe of similarly magnetized rocks parallel to the ridge. Once the rocks have cooled, they retain their original magnetic direction, unless they are heated significantly or altered by certain types of fluids.

Magma

Eventually, with continued plate spreading, a pattern of alternating magnetic stripes is produced on the ocean floor. This pattern is strong enough to be detected by magnetometers towed behind a ship.

D How Do Magnetic Patterns on the Seafloor Help Us Study Plate Tectonics?

Magnetic patterns allow us to estimate the ages of large areas of seafloor and to calculate the rates at which two diverging oceanic plates were formed.

Positive Magnetic Anomaly
Earth's Magnetic field
Negative Magnetic Anomaly
4 3 2 1 Present 1 2 3 4
Age Before Present (Millions of Years)
Crest of Mid-Ocean Ridge
Normal
Reversed
Magnetic Stripes

10.02.d1

1. As magnetic instruments are towed behind a ship, the strength of the magnetic field is measured and plotted. Stronger measurements plot high on the graph and are called *positive magnetic anomalies*.

2. The magnetic signal is weaker over crust that was formed under a reversed magnetic field. The magnetic direction in such rocks is opposite to and works to counteract the modern magnetic field. The reverse magnetization of the rocks slightly weakens the measured magnetic signal and will plot low on the graph, forming a *negative magnetic anomaly*.

3. The seafloor patterns are compared with the patterns on the magnetic timescale to assign ages to each reversal. Geologists simplify and visualize these data as reversely and normally magnetized stripes on the seafloor and in cross section.

4. We calculate rates of seafloor spreading by measuring the width of a specific magnetic stripe in map view and then dividing that distance by the length of time the stripe represents:

rate of spreading for stripe = width of stripe / time duration

If a magnetic stripe is 60 km wide and formed over 2 million years, then the rate at which spreading formed the stripe was 30 km/m.y. This rate is equivalent to 3 cm/year. An equal width of oceanic crust was added to a plate on the other side of the mid-ocean ridge, so the total rate of spreading across the ridge was 60 km/m.y. (6 cm/year), a typical rate of seafloor spreading. At these rates, how wide is the area of seafloor shown in the bottom figure?

Before You Leave This Page Be Able To

- ✓ Describe how Earth's magnetic field is generated.
- ✓ Summarize how magnetic reversals help with dating rocks.
- ✓ Describe or sketch how magnetic patterns develop on the seafloor.
- ✓ Calculate the rate of seafloor spreading if given the width and duration of a magnetic stripe.

What Processes Occur at Mid-Ocean Ridges?

MID-OCEAN RIDGES FORM where two oceanic plates diverge. Magma ascending from the mantle erupts onto the seafloor or solidifies at depth, making new oceanic crust. Heat associated with the hot rocks and magma causes undersea vents of hot water that nourish unique life forms on the seafloor.

A What Happens When Plates Spread Apart?

As two oceanic plates move apart, solid rock and magma rise from the mantle to occupy the space between the plates. The cooling and solidifying magma forms new oceanic crust, which then gets transported away from the mid-ocean ridge as more magma rises to the surface. Slices of this rock sequence can be scraped off and preserved on land, allowing scientists to study it in detail without diving to the bottom of the ocean.

As the oceanic crust is stretched apart, basaltic lava erupts within the rift, forming *pillow basalts* on the seafloor. Some magma solidifies within large chambers and in magma-filled fissures (dikes) parallel to the mid-ocean ridge (perpendicular to plate movement).

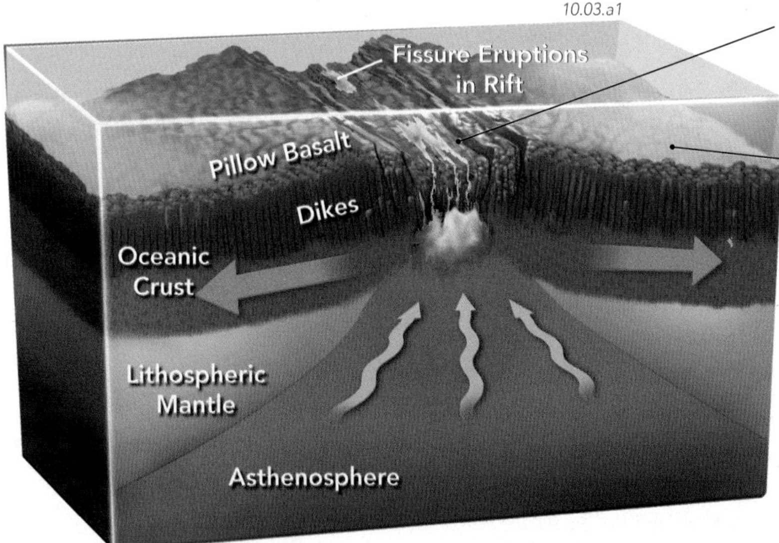

In many mid-ocean ridges, normal faults allow blocks of crust to be displaced down and inward toward the center, forming a fault-bounded rift.

As the cooled crust moves away from the ridge and becomes older, it is progressively covered with deep-sea sediment. Over time, the sediment tends to smooth over the rough topography formed in the rift. As a result, older oceanic crust tends to have relatively smooth topography. Below a depth of about 4,500 to 5,000 meters, sediment is dominated by clay and silica-rich materials because at these depths carbonate minerals dissolve into seawater as fast as they accumulate.

B What Accounts for Variations in the Shape of Mid-Ocean Ridges?

Many mid-ocean ridges possess the typical features shown above, but ridges vary in their width, ruggedness, and overall shape. These variations reflect differences in the rates of spreading, magmatism, and faulting. Compare the three topographic profiles below, each of which shows a detailed view of the center of a ridge.

1. Some ridges, such as parts of the East Pacific Rise, are broad and do not have a large, well-developed rift in the center. Such ridges are spreading apart at relatively fast rates (10 centimeters per year). These ridges are broad because rapid spreading allows the new oceanic crust to move far from the spreading center before it cools and subsides. Furthermore, such ridges are interpreted to have more underlying magma, which rapidly pours onto the surface out of fissures, rather than forming a rift.

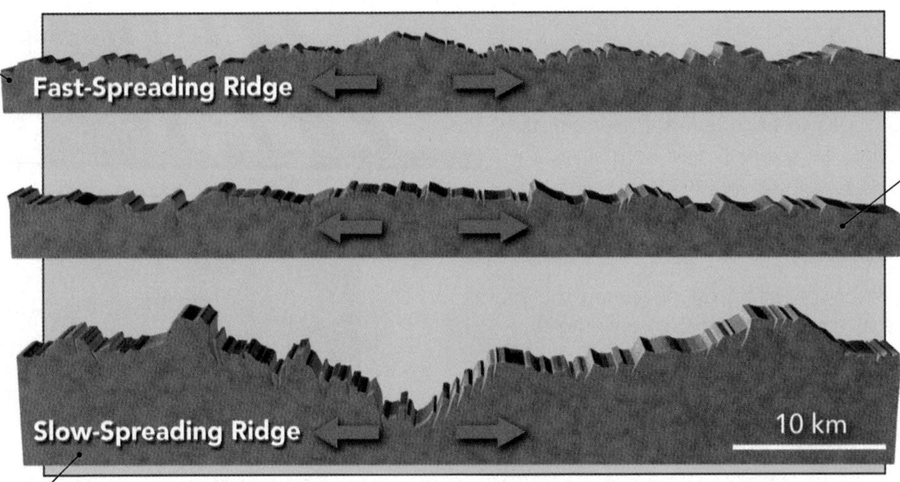

2. Other mid-ocean ridges, such as the Mid-Atlantic Ridge, have well-defined rifts that are 1 to 3 kilometers deep and are bounded by normal faults that dip inward toward the rift. These ridges have slower spreading rates (1 to 2 centimeters per year). This allows rocks near the ridge to cool and strengthen enough to form large faults.

3. Other ridges are intermediate in character between these two end members. Such ridges lack a high central area or a deep rift. They are intermediate in spreading rate, breadth, roughness, and degree of faulting. Parts of the East Pacific Rise have this shape.

C What Are Black Smokers and How Do They Form?

10.03.c1

◁ Mid-ocean ridges contain features called *black smokers*, shown here in a photograph taken from a submersible. Black smokers are *hydrothermal vents*, where hot water from within the rock jets out into the cold seawater. As the hot water cools, metals, sulfur, and other elements dissolved in the hot water form small crystals that make the water black and cloudy.

Sulfide and sulfate minerals precipitate around the vent, forming a hollow, circular column called a *chimney*. Some chimneys are more than 5 meters (16 feet) high and a meter across, and can grow tens of centimeters per day. Black smokers and sulfide-rich chimneys are interpreted to have formed on mid-ocean ridges and other submarine volcanoes in Earth's geologic past, forming mineral deposits rich in copper, zinc, and other valuable elements.

10.03.c2

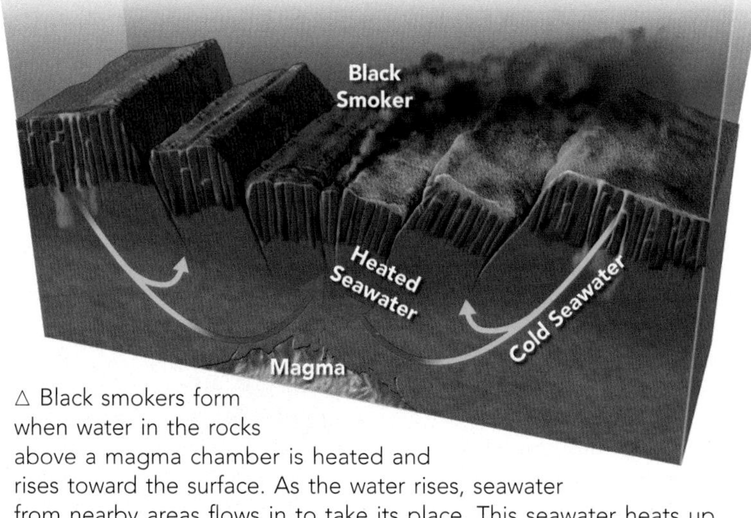

△ Black smokers form when water in the rocks above a magma chamber is heated and rises toward the surface. As the water rises, seawater from nearby areas flows in to take its place. This seawater heats up and becomes rich in dissolved chemicals. The heated seawater rises toward the surface along faults and other pathways, eventually venting in a black smoker. The water is very hot, commonly over 350°C, but it does not boil because of the pressure exerted by the deep water.

Life at Hydrothermal Vents

Deep-sea hydrothermal vents associated with black smokers support a unique and only recently discovered community of unusual creatures. Scientists are actively exploring the ecosystems of these vents, in part because such sites may have been where life originated on Earth. The photographs below were taken by scientists using a submersible to investigate these vents and their unusual inhabitants.

Sunlight is the energy source for green plants, which provide the bulk of food for animals living on Earth's surface. No sunlight reaches the deep seafloor. Instead, life around the hydrothermal vents utilizes a completely different energy source. Here, life is dependent on somewhat unusual bacteria that are able to break down hydrogen sulfide (H_2S), one of the chemical compounds common within black smokers. These bacteria produce sugars, which feed giant (meter-long) red tube worms. The worms can tolerate the hot water and live close to the vents, where the bacteria are abundant. In fact, many bacteria live within the worms' tissues. The worms in turn form the main food for an assembly of scavenging animals, including fish and white crabs, such as those shown below in the photo on the right. Large clams also live around hydrothermal vents and draw nutrients by extracting small bits of material from the water and from the bacteria. Fossils of tube worms in ancient hydrothermal vent deposits show that such communities have existed for millions of years.

10.03.mtb1

10.03.mtb2

Before You Leave This Page Be Able To

- ✓ Summarize or sketch the processes that accompany the formation of new oceanic crust at mid-ocean ridges.
- ✓ Summarize or sketch the differences between fast-spreading and slow-spreading mid-ocean ridges.
- ✓ Describe black smokers, how they form, and where the hot water originates and how it gets heated.
- ✓ Describe the type of life that exists around hydrothermal vents and where the different creatures derive their food.

What Are Major Features of the Deep Ocean?

BENEATH THE WORLD'S OCEANS lie rugged mountains, active rifts, gentle plains, broad plateaus, and deep trenches. The seafloor varies in depth and in thickness of sediment cover, largely because of differences in age. What is the shape of the deep seafloor and how do the various types of features form?

Topography of the Deep Seafloor

10.04.a1

1. Much of the ocean floor, called the *abyssal plain*, has a gentle slope and lies at depths below about 4.5 kilometers (2.8 miles). The abyssal plain and other old parts of an oceanic plate generally have a smooth topography because sediment evens out most original irregularities.

2. The elevation of the seafloor decreases from mid-ocean ridge to abyssal plain because oceanic lithosphere cools, becomes denser, and subsides as it moves away from the ridge. The start of the abyssal plain marks where oceanic crust reaches an equilibrium temperature and subsides very slowly or not at all.

Seamount
Abyssal Plain
Mid-Ocean Ridge
Oceanic Trench
Island Arc
Accretionary Prism

3. The abyssal plain locally contains isolated mountains, called *seamounts*, which vary from gentle submarine hills to mountains and steep, recently active volcanoes.

4. Trenches are the deepest parts of the ocean, but comprise a very small area. They are the surface expression of a subduction zone, where one oceanic plate is flexed or bent as it plunges beneath another plate. Within the trench, sediment and basaltic rocks are scraped off the top of the subducted plate and incorporated as slices into the *accretionary prism*.

Sediment Thicknesses

10.04.a2

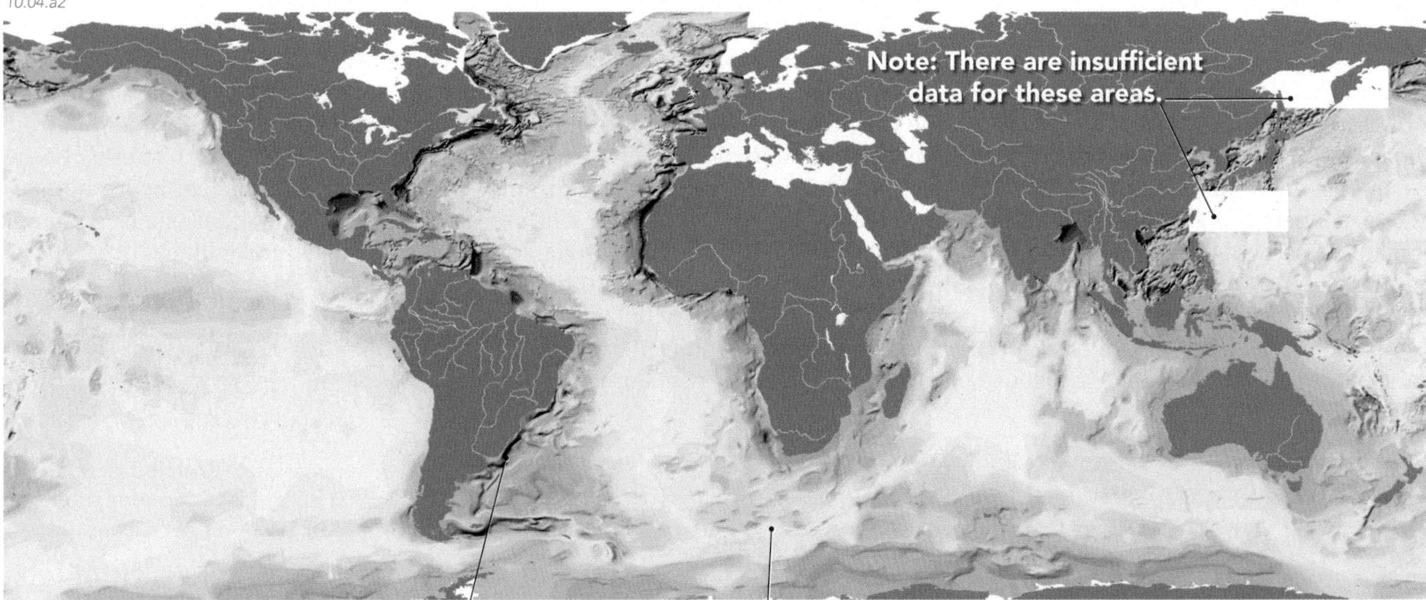

This map shows sediment thickness on the seafloor. It ranges from light blue where thickness is less than 200 meters, to orange and red, where sediment is over 5 to 10 kilometers (3–6 miles) thick. The white lines on the continents are rivers. What patterns do you observe on this map?

The thickest sediment is along continental margins, especially those that were formed by rifting. Seafloor sediment is also thickest near the mouths of rivers or where the oceanic crust is relatively old (see maps on the next page).

There is virtually no sediment cover over the youngest crust at the mid-ocean ridges, which here are along the belts of light blue.

Before You Leave These Pages Be Able To

- ✓ Sketch or describe some features of the deep seafloor.
- ✓ Describe how age of the seafloor relates to mid-ocean ridges, depths of seafloor, and sediment thicknesses.

Depth and Age of the Seafloor

This top map shows the depth of the seafloor. Dark gray areas are continents, and the blue and purple regions are oceans and seas. Dark blue shows the deepest areas and light blue-gray represents the shallowest depths. Letters indicate the positions of selected ridges (R), trenches (T) along convergent boundaries, and passive margins (P) that are not plate boundaries.

The deepest locations occur in trenches (T) along active margins, where one plate subducts beneath another plate. The shallowest parts of the seafloor are on continental shelves, many of the widest of which are along passive margins (P). Mid-ocean ridges (R) are intermediate in depth.

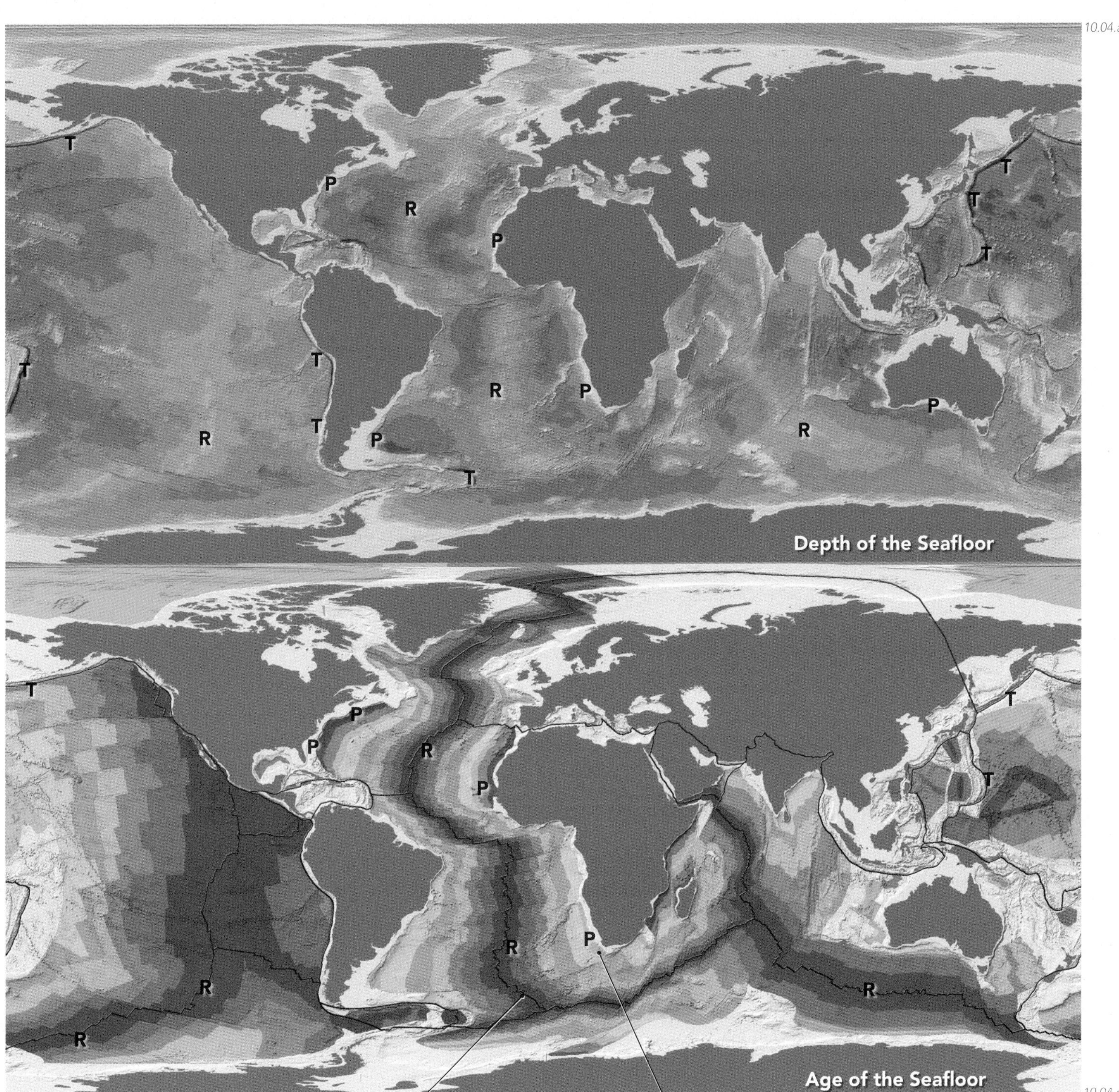

This bottom map shows the age of the seafloor. Purple represents the oldest areas (about 180 million years), and the darkest orange represents very young oceanic crust. Compare these two maps.

The youngest oceanic crust is near mid-ocean-ridge spreading centers (R). These areas are also higher than most of the ocean floor.

The oldest oceanic crust in any ocean is the most distant from mid-ocean ridges. None is older than about 180 million years, because all older oceanic crust has been subducted back into the mantle. The oldest seafloor is much younger than the oldest continental rocks.

How Do Oceanic Islands, Seamounts, and Oceanic Plateaus Form?

ISLANDS AND SUBMARINE MOUNTAINS, called *seamounts*, rise above the seafloor. These islands include exotic places, such as Tahiti and the Galapagos. Also on the seafloor are broad oceanic plateaus that cover huge areas. How are such islands, seamounts, and oceanic plateaus formed?

A How Do Some Oceanic Islands and Seamounts Form?

Most oceanic islands consist of mafic (basalt) to intermediate (andesite) volcanic rocks and are formed by a series of volcanic eruptions onto the seafloor. Many of these islands are not associated with an island arc, but form linear chains or irregular clumps of islands and seamounts, some of which are related to a *hot spot*.

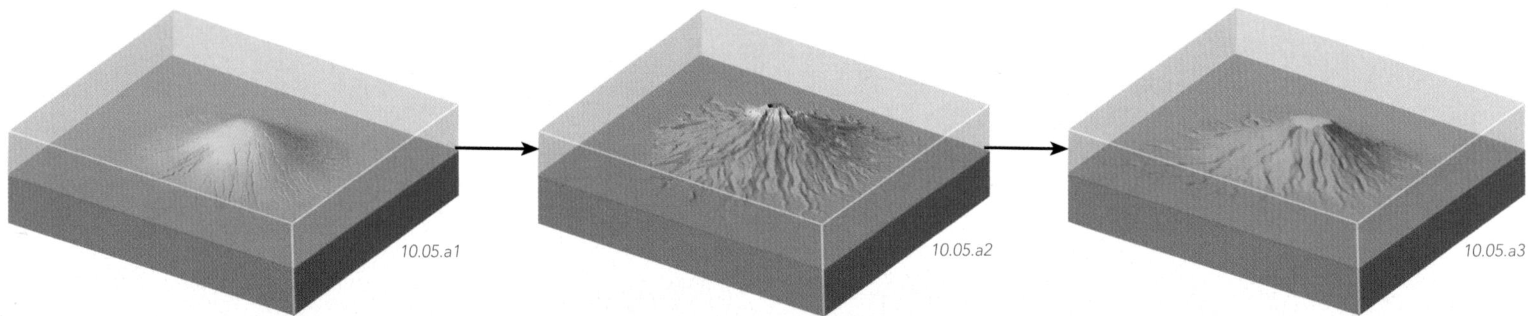

Magmatism, such as that caused by an underlying hot spot, begins building a submarine volcano by eruption of lava onto the seafloor.

Continued eruptions build up the volcano until it rises above the sea as an island. Once magmatism ceases, such as when the island moves off a hot spot, the oceanic plate cools and subsides.

The top of the mountain is leveled off by wave erosion and continues subsiding, becoming a submarine, flat-topped mountain. Over time, it is covered by marine sediments.

B How Do Oceanic Plateaus Form?

Some large regions of the seafloor rise a kilometer or more above their surroundings and are called *oceanic plateaus*. Such plateaus are largely composed of flood basalts and, like the seafloor in general, are mostly late Mesozoic and Cenozoic in age (mostly 130 million years ago to the present).

This perspective shows the Kerguelen oceanic plateau, which rises up above the surrounding seafloor in the southern Indian Ocean. The plateau is several thousand kilometers long, but only reaches sea level in a few small islands. The land in the lower right corner is part of Antarctica.

Oceanic plateaus are interpreted by most geologists to form at hot spots, above rising mantle plumes. The plumes travel through the mantle as solid masses, not liquids.

When the top of a plume encounters the base of the lithosphere, it causes widespread melting. Submarine flood basalts pour out onto the seafloor through fissures and central vents.

Immense volumes of basalt (as much as 50 million cubic kilometers) erupt onto the seafloor over millions of years. This volcanism creates a broad, high *oceanic plateau*.

C Where Are Hot Spots, Linear Island Chains, and Oceanic Plateaus?

Hot spots have created many Pacific islands that we associate with tropical paradises and exotic destinations. Hawaii is the most famous island chain formed by movement of a plate over a hot spot, but several other linear island chains, in both the Atlantic and Pacific, formed in the same manner.

On this map, red dots show the locations of likely hot spots, many of which are located at the volcanically active ends of linear island chains. There is however great debate about which areas really are hot spots and how hot spots form.

The dark gray areas in the oceans represent linear island chains, clumps of islands, and oceanic plateaus, such as this area around Iceland.

10.05.c1

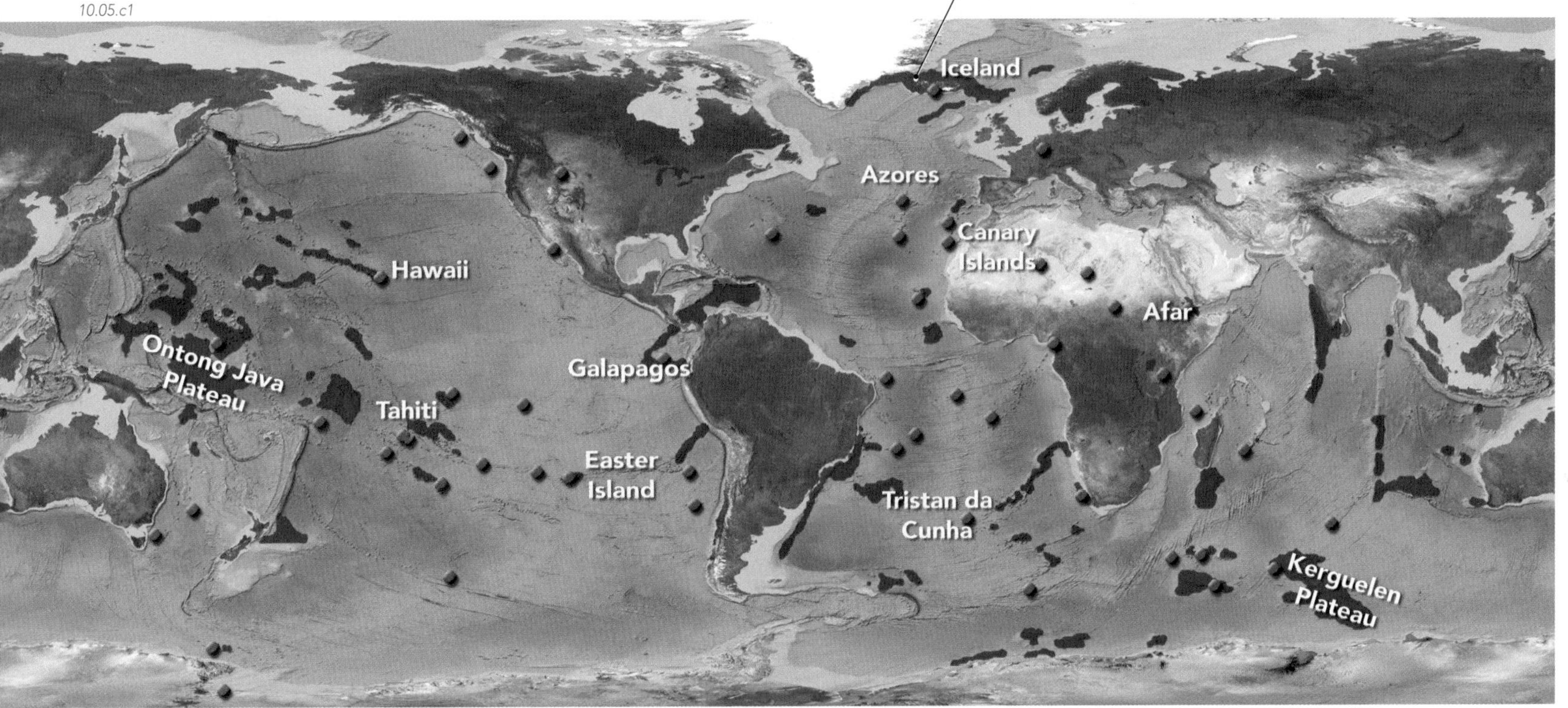

The *Ontong Java Plateau* is the largest oceanic plateau on Earth, covering millions of square kilometers or nearly 1 percent of Earth's surface area. It formed in the middle of the Pacific ocean 120 million years ago and is no longer near the hot spot that formed it.

Volcanic islands near *Tahiti* define northwest-trending chains that are forming over several hot spots. In each chain, the islands to the northwest are older than those in the southeast, indicating that the Pacific plate is moving to the northwest relative to the underlying source of magma.

The *Galapagos* is a clump of volcanic islands west of South America. The western islands, shown in the satellite image to the lower right, are volcanically active and have erupted within the past several years. Eruptions have created shield volcanoes and smaller cinder cones, both of which are shown in the photograph to the lower left.

Tristan da Cunha, a volcanic island in the South Atlantic Ocean, marks a hot spot just east of the Mid-Atlantic Ridge. Volcanism associated with the hot spot created a large submarine ridge (shown in gray) that tracks the motion of the African plate over the hot spot.

The *Kerguelen Plateau*, in the southern Indian Ocean, is the second largest oceanic plateau in the world. It mostly consists of basalt and was formed in several stages during the late Mesozoic (between 115 and 85 million years ago).

10.05.c2

10.05.c3

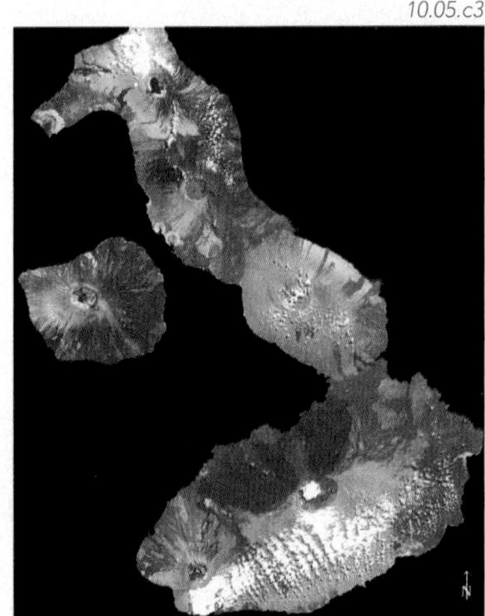

Before You Leave This Page Be Able To

- ✓ Summarize or sketch how a mantle plume can form oceanic islands, providing several examples.
- ✓ Summarize how oceanic plateaus are interpreted to have formed.

What Processes Form Island Arcs?

MANY ISLANDS OCCUR IN LONG ARCS that cross the seafloor and are associated with deep trenches. How do arcs form, why are they curved, and what processes occur in front of, within, and behind them?

A How Do Island Arcs Form?

Island arcs form where one oceanic plate is *subducted* beneath another. The subduction creates a trench and generates the magma that forms an arcuate belt of volcanic islands, such as the Aleutian Islands and Java.

1. An *oceanic trench* is formed where the subducted oceanic plate is *flexed* downward beneath the overriding plate. Many island arcs reside in the open ocean, away from large land masses. Therefore, the trench receives most of its sediment from eruption and erosion of the adjacent volcanoes.

2. As the oceanic plate is subducted, it heats up, causing metamorphic reactions that release water from the minerals. This water promotes melting in the asthenosphere above the subducted plate. The asthenosphere-derived magma rises into the overriding lithosphere, erupting onto the surface or solidifying in the crust. These magmatic additions thicken the crust over time.

3. As a new volcano begins to grow, volcanic eruptions first occur in deep water on the seafloor. Over time the eruptions may construct a mountain that rises above the sea. As the crust thickens, the magmas become intermediate in composition (andesite) and form dangerous, *composite volcanoes*. Submarine mountains and ridges lie between the islands.

Oceanic Trench
Island Arc
Oceanic Crust
Lithospheric Mantle
Asthenosphere
Lithospheric Mantle
Asthenosphere

10.06.a1

B What Happens in Front of and Behind an Island Arc?

Island arcs are not fixed in position. They can migrate across the surface of Earth over millions of years, depending on what happens in front of the arc (at the trench) or behind the arc.

As the dense oceanic plate is subducted into the asthenosphere, it sinks downward and tends to bend or roll back away from the island arc. The trench, which is simply the surface expression of the bend in the downgoing plate, follows the rollback of the slab. This process is called *trench rollback*. Subduction continues during trench rollback.

As the subducting slab and trench migrate, the island arc follows them because the position of the volcanoes is determined by the location of the subducting plate.

As the arc and trench both migrate, stretching of the crust can cause *rifting* within or behind the arc. The arc can be rifted (split) into two parts, one of which gets carried away from the trench and becomes inactive. Rifting can lead to seafloor spreading and formation of a new *back-arc basin* that can be hundreds of kilometers wide. One possible signature of back-arc spreading is oceanic crust in a back-arc basin that overlaps in age with volcanic rocks within the island arc.

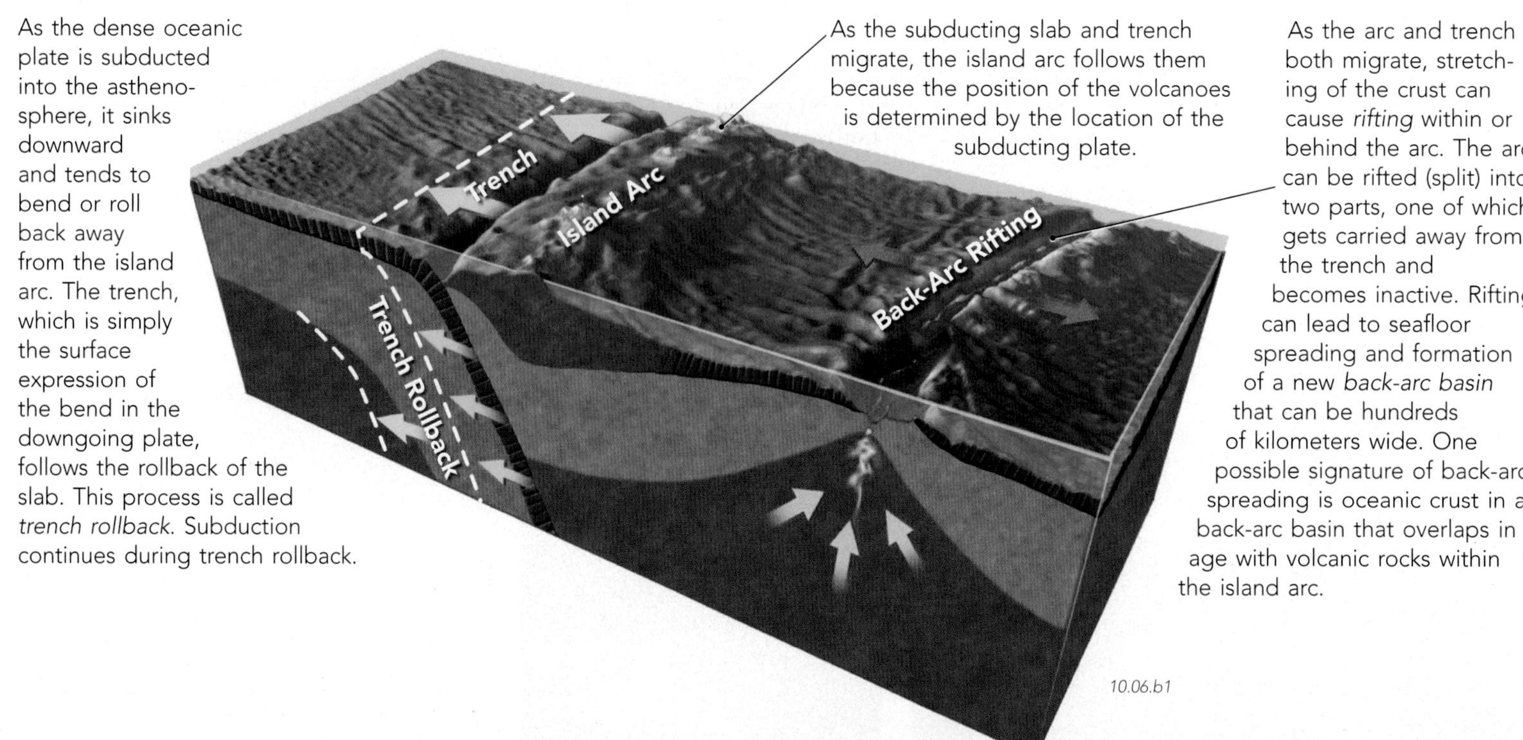

10.06.b1

C Why Are Island Arcs Curved?

1. In map view, island arcs have a distinctly curved or arcuate shape. This view of the Aleutian arc of Alaska shows the curved shape of the island arc and of the associated trench that lies in front of it.

10.06.c1

10.06.c2

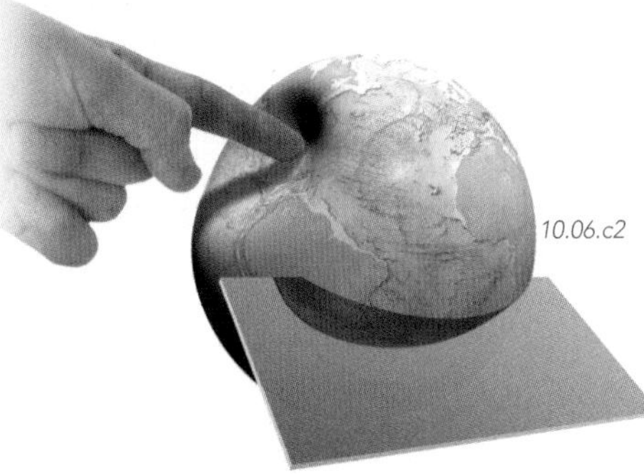

10.06.c3

△ 2. In thinking about why island arcs are curved, we need to consider that plates are interacting on a *spherical* Earth, not a flat plane. Cutting into a globe (like the red plane) or depressing the surface of a globe creates a curved feature (arc).

△ 3. Also, on a sphere, such as Earth, there is more surface area on the outside than at depth. A plate that is subducted into the interior becomes buckled as it is forced to fit into a smaller width. The arc and trench have an arcuate shape because the downgoing slab does too.

D Where Are the Main Island Arcs of the World?

Japan and the *Philippines* are both part of island arcs offshore of mainland Asia.

In the *Mariana arc*, the Pacific plate subducts westward, forming the world's deepest trench.

The *Aleutian arc* extends from mainland Alaska westward to the Kamchatka peninsula of Asia. It is formed where the Pacific plate is subducted northward beneath the North American plate.

The *Lesser Antilles arc* forms the eastern edge of the Caribbean Sea. It includes the islands of Montserrat, site of recent eruptions, and Martinique, site of the deadly 1902 eruption of Pelée.

10.06.d1

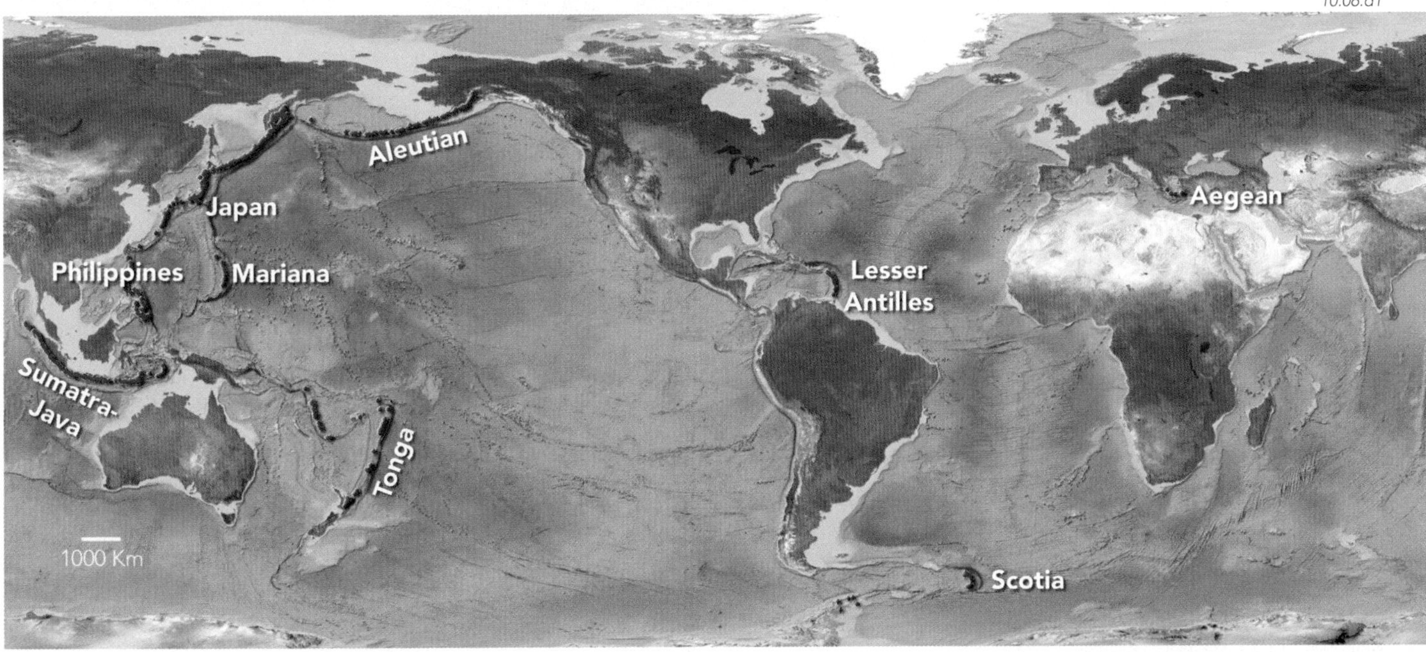

The *Sumatra-Java arc* is a typical island arc in the east, but in the west it lies upon a promontory of Asian continental crust. It is located where the Indian plate is subducted northward. This zone caused the deadly 2004 Indian Ocean tsunami.

Along the *Tonga* trench and island arc, the Pacific plate is subducted to the west. Spreading west of the arc has created several small back-arc basins. The Tonga subduction zone is the site of many large earthquakes each year.

The *Scotia arc* is a small island arc between South America and Antarctica. Beneath the arc, an oceanic section of the South American plate is subducted beneath another oceanic plate.

Before You Leave This Page Be Able To

- ✓ Summarize the processes that occur within, in front of, and behind island arcs.
- ✓ Summarize why island arcs and their associated trenches are curved.
- ✓ Describe some examples of island arcs.

How Did Smaller Seas of the Pacific Form?

A SERIES OF SMALL SEAS exist around the edges of the Pacific Ocean. They are separated from the main Pacific basin by chains of islands and slivers of continents. These include the Sea of Japan and the Gulf of California. Each sea has it own unique and interesting history, and together they illustrate the most important ways in which smaller seas in the Pacific formed.

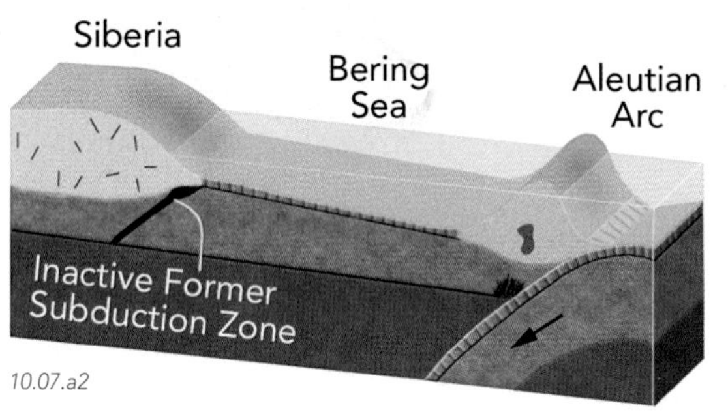

10.07.a2

1. The *Bering Sea* lies between mainland Alaska and the Aleutian island arc. The eastern part of the sea, near Alaska, is shallow and is underlain by continental crust. The western part of the Bering Sea is deeper because it is underlain by oceanic crust. The oceanic crust is part of the North American plate, so there is no plate boundary between Alaska and the oceanic crust beneath the Bering Sea. Instead, the western edge of the Alaskan mainland is a passive margin.

10.07.a1

△ 2. In the Mesozoic, oceanic plates were subducted directly beneath coastal Alaska and Siberia, but the site of subduction zone jumped offshore, trapping some old oceanic crust between the new Aleutian arc and the mainland.

3. The *Sea of Japan* is a moderately deep basin between Japan and mainland Asia. Before 20 million years ago, Japan was part of a volcanic arc along the coast of mainland Asia. Rifting within the arc split Japan away from Asia. This led to back-arc seafloor spreading, which formed the Sea of Japan. ▽

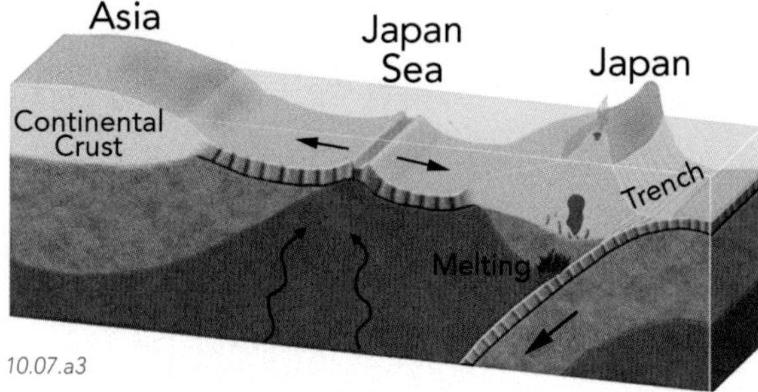

10.07.a3

4. The *China Sea*, between China and the Korea, is relatively shallow because it is mostly underlain by continental crust.

5. The *Philippine Sea* lies between the Philippines and the Mariana island arc. It contains several distinct basins separated by long, submarine ridges, the origin of which are discussed on the next page.

6. The *Java Sea* of Indonesia and Malaysia is shallow and is part of a continental platform between the larger islands of the region.

7. South of *Indonesia*, oceanic portions of the Indian plate are subducted northward beneath the Asian plate, forming a trench and the Sumatra-Java island arc. The continuation of this subduction zone to the northwest caused the huge earthquake and deadly tsunami that devastated coastlines around the Indian Ocean in 2004.

Origin of the Philippine Sea

11. The *Philippine Sea* is an example of how features on the seafloor reflect the geologic history of an area. The Mariana arc is active and is flanked by the Mariana trench, which contains the deepest seafloor in the ocean (nearly 11 kilometers below sea level).

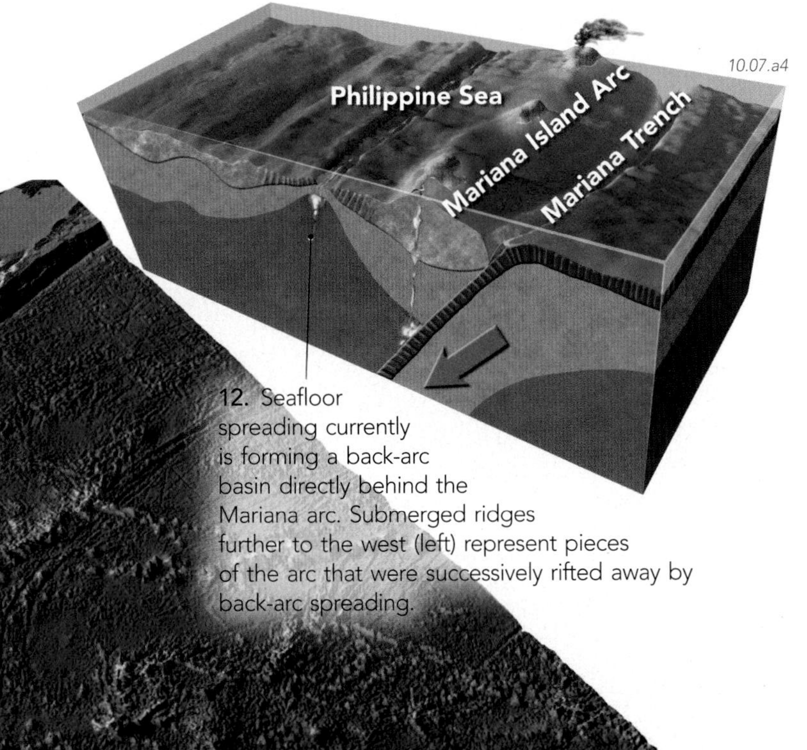

12. Seafloor spreading currently is forming a back-arc basin directly behind the Mariana arc. Submerged ridges further to the west (left) represent pieces of the arc that were successively rifted away by back-arc spreading.

Origin of the Gulf of California

13. Prior to 10 million years ago, Baja California was part of the mainland of western Mexico, and an oceanic plate subducted eastward beneath the land.

14. As North America, including Mexico, encountered the East Pacific Rise spreading center, the plate boundary became a transform boundary and later jumped inland, splitting off Baja California and shifting it northward along the coast.

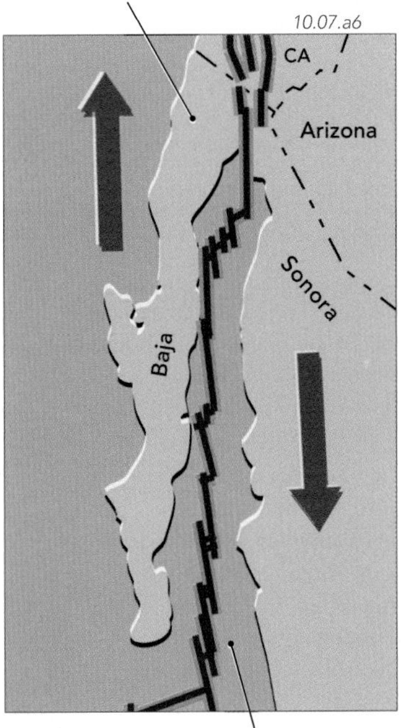

15. As Baja moved northward away from the mainland, the *Gulf of California* formed in the place Baja vacated. The gulf has transform faults linked by short spreading centers that are a continuation of the East Pacific Rise. Baja is now part of the Pacific plate and continues to move northward relative to the North American plate.

10. At the *Tonga trench*, the Pacific plate subducts westward beneath oceanic crust that is attached to Australia. The subduction zone forms the trench and associated island arc of the Tonga Islands. It is very active, being associated with numerous large earthquakes.

9. The seafloor east of Papua New Guinea and northeast of Australia is unusually complicated. It contains small basins, trenches, and island arcs, reflecting complex interactions between a number of small oceanic plates. From a plate-tectonic perspective, it is the most complex area of oceanic crust in the world.

8. The shallow seas between Australia and Papua New Guinea are underlain by a continuation of Australian continental crust.

Before You Leave This Page Be Able To

- ✓ Summarize or sketch the different ways in which smaller seas formed in the Pacific Ocean, providing an example of each.
- ✓ Summarize the history of the Gulf of California and how it is related to the boundary between the Pacific and North American plates.

How Did Smaller Seas Near Eurasia Form?

EUROPE AND ASIA HOST A NUMBER OF SEAS, including the Black Sea, North Sea, and Mediterranean Sea. The Arabian Peninsula, between Africa and mainland Asia, is flanked by the Red Sea to the west and the Persian Gulf to the east. Several seas were formed by present or past plate-tectonic activity. Others were valleys and low areas flooded by rising sea levels after the last Ice Age.

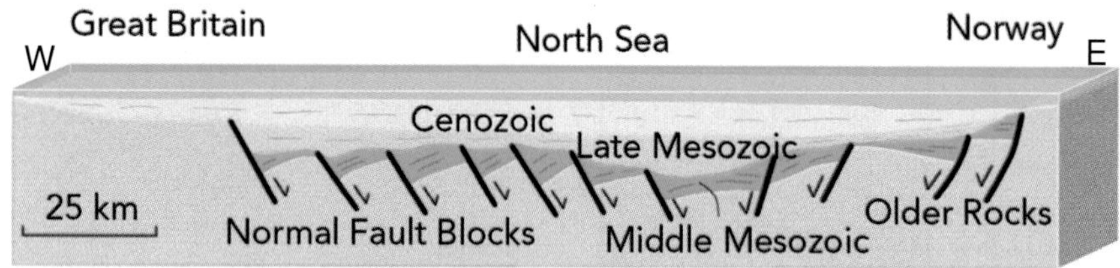

10.08.a2

△ The *North Sea*, between Great Britain and Norway, is underlain by continental crust that was thinned by extension and normal faulting when Europe rifted away from North America during the Mesozoic. The faulted sedimentary layers, shown in the cross section above, contain important oil fields that were discovered by drilling into the seafloor.

The *Baltic Sea* of Scandinavia is a shallow sea underlain by continental crust. It was originally a river valley and was further scoured by glaciers during the last Ice Age. As the glaciers retreated and sea level rose, the valley was flooded by seawater.

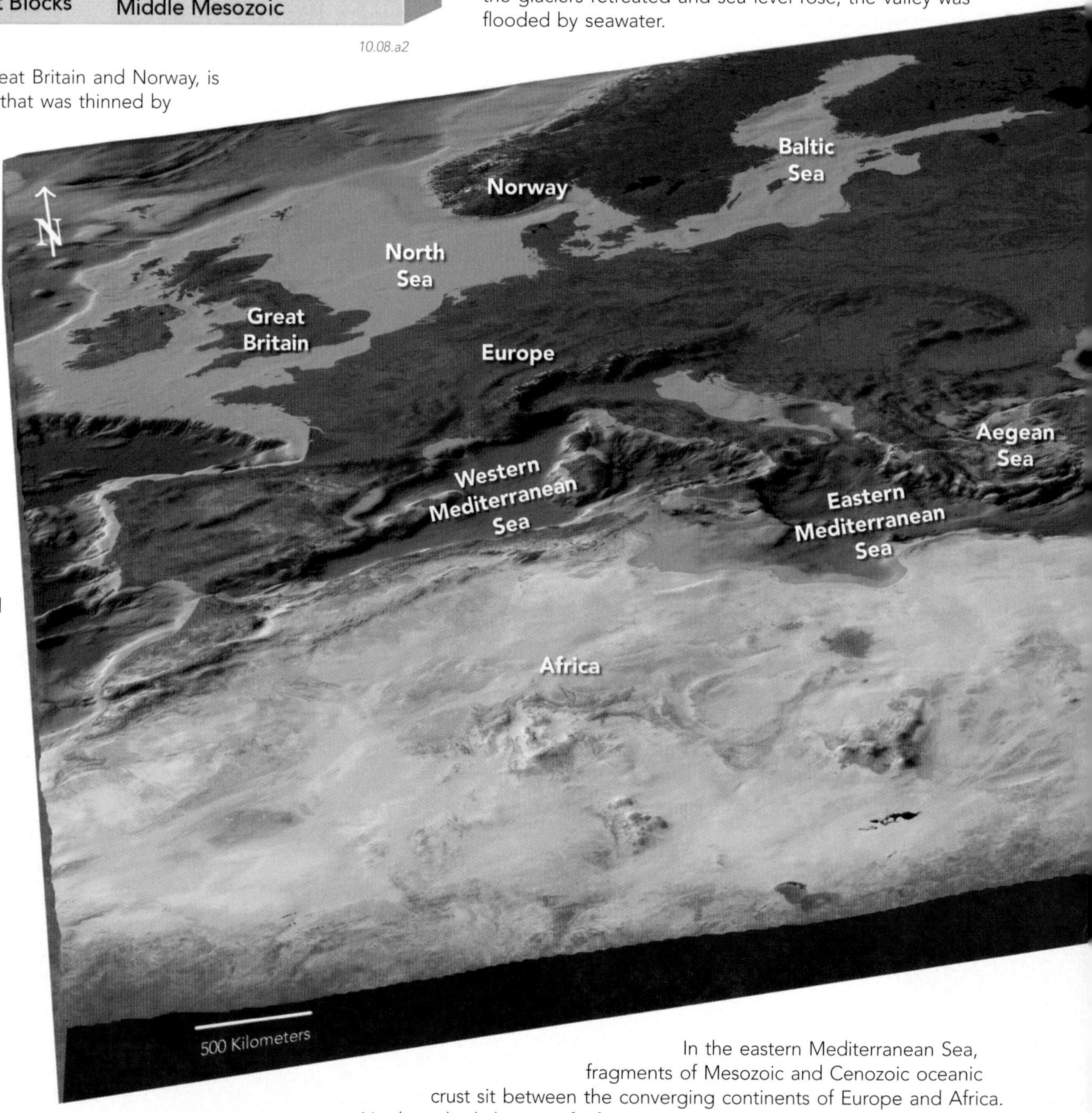

The *Mediterranean Sea* separates Africa from Europe. Most of it is more than several kilometers deep because it is underlain by oceanic crust or thinned continental crust. The western Mediterranean was formed primarily by Cenozoic rifting.

In the eastern Mediterranean Sea, fragments of Mesozoic and Cenozoic oceanic crust sit between the converging continents of Europe and Africa. Northward subduction of African oceanic crust forms volcanoes in the *Aegean Sea*, east of mainland Greece.

The crust beneath the *Black Sea* was part of a large open ocean, called *Tethys*, but was then trapped by the continental collision. It was reduced to a large lake, isolated from the sea. After the last Ice Age ended and sea level rose, seawater from the Mediterranean Sea overtopped a low divide near Istanbul, Turkey, flooding the Black Sea. Some geologists studying the origin of the Black Sea think that this flooding was catastrophic and may be the origin of ancient stories about a massive flood.

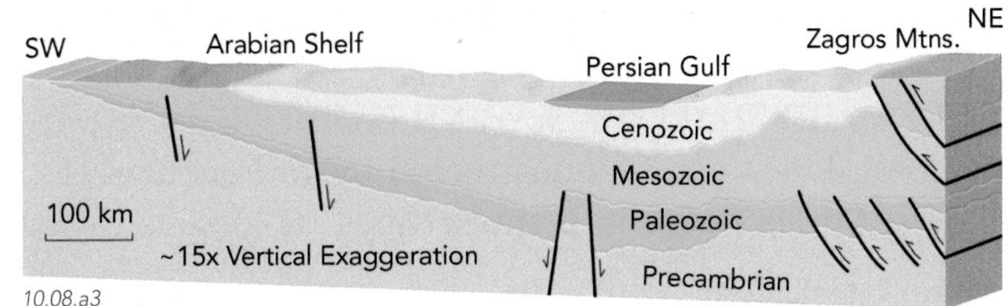

10.08.a3

△ The *Persian Gulf* lies between Saudi Arabia and Iran and is related to the collision of the Arabian plate and the Eurasian plate. As the Zagros Mountains of Iran are thrust over Arabia, the weight of the thrust sheets forces the Arabian plate downward. This has created the gulf, as shown in the cross section above. Oil is forced up along the layers toward the large oil fields of the Arabian shelf of Saudi Arabia, Kuwait, and adjacent countries.

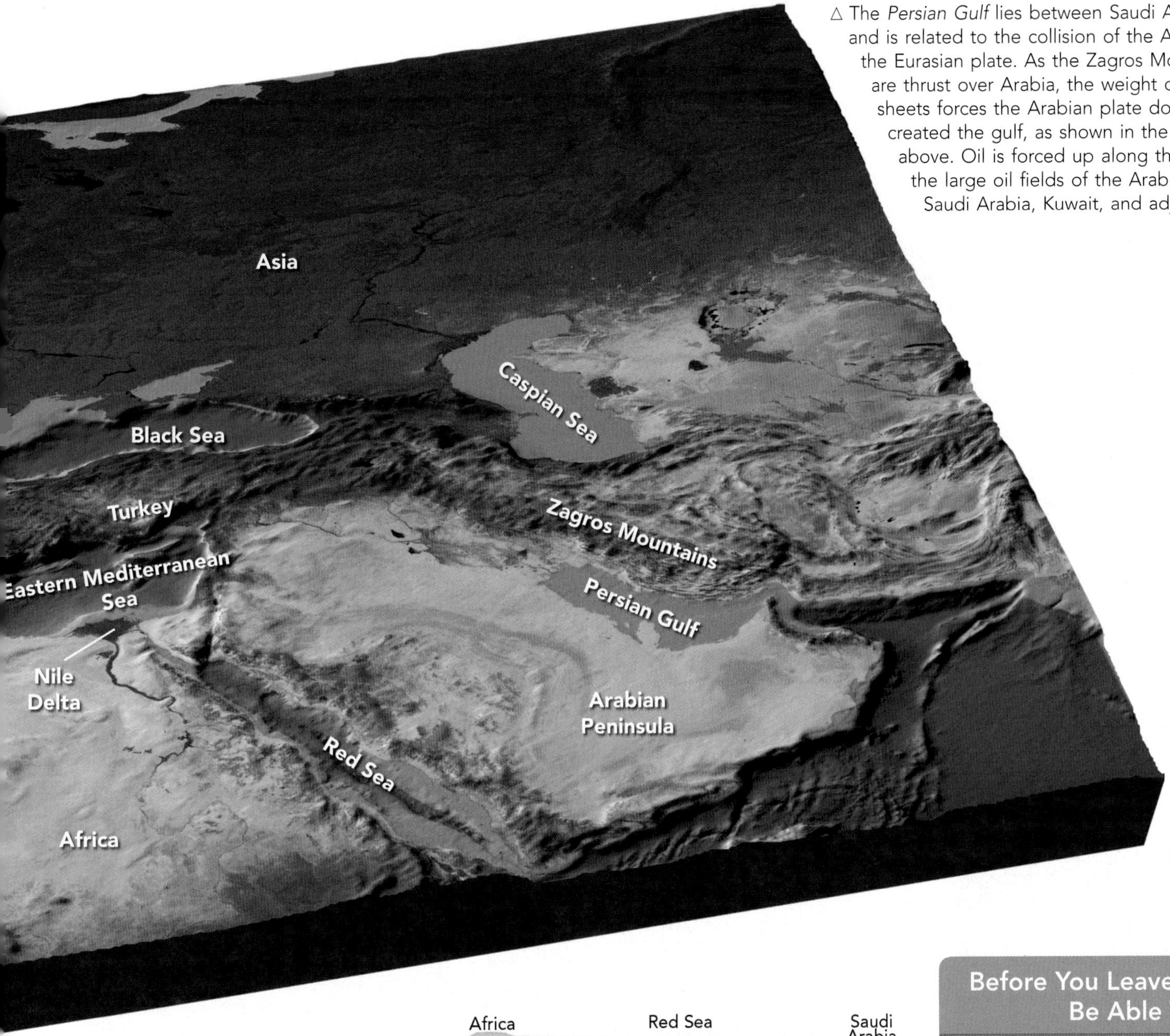

10.08.a1

▷ The *Red Sea* was formed at a divergent boundary, where the Arabian Peninsula split from Africa via continental rifting. New oceanic crust is being created by seafloor spreading within the southern Red Sea as the two plates move apart.

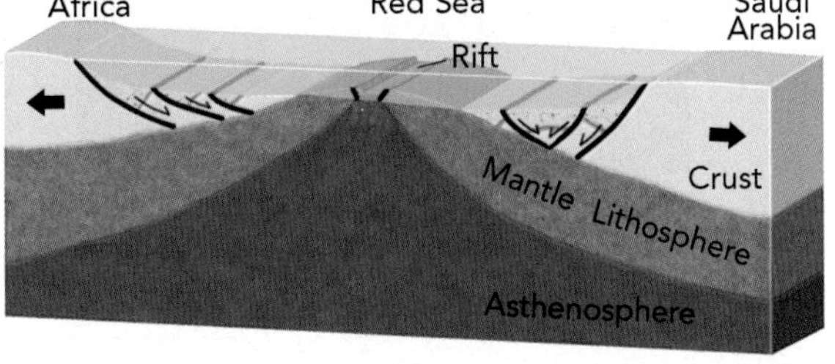

10.08.a4

Before You Leave This Page Be Able To

- ✓ Summarize or sketch the origin of the North, Baltic, Black, and Red Seas.
- ✓ Summarize or sketch how the Persian Gulf is related to the collision of Arabia and Asia.

How Do Reefs and Coral Atolls Form?

REEFS ARE SHALLOW MARINE LANDFORMS, mostly built by colonies of living marine organisms, such as coral. Reefs can be constructed by various marine organisms or from accumulations of shells and other debris. Corals thrive in many settings, as long as the seawater is warm, clear, and shallow.

A In What Settings Do Coral Reefs Form?

Corals are a group of invertebrate animals that form calcium-carbonate structures. Coral reefs most commonly form in shallow tropical seas, especially where the water is relatively free of suspended sediment. Many reefs are battered by large waves and provide carbonate sediment to adjacent parts of the seafloor.

1. Some reefs occur along the edges of continents, forming *barrier reefs* offshore from the main coastline. Erosion of the reefs forms low, sandy islands with white sand beaches. The reefs and islands protect the continent from large waves. They enclose a lagoon on the landward side, but have open ocean on the other side.

2. Reefs and other carbonate accumulations can form broad, shallow *platforms*, like the Bahama Islands east of Florida. These are places where older reef deposits and dunes rise slightly above sea level. Between most islands, the water is shallow and the seabed is composed of white, carbonate-rich sand derived from wave erosion of the reef and land.

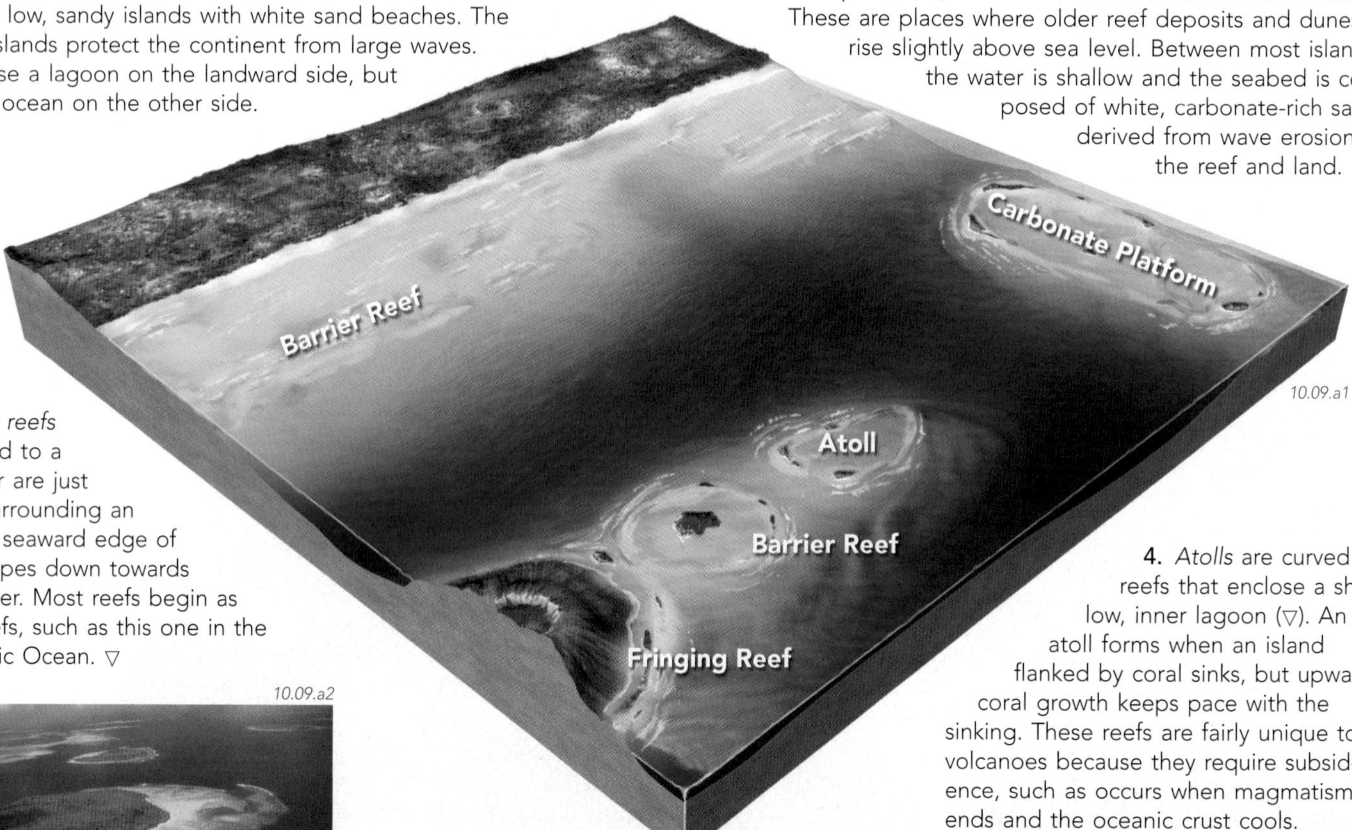

Barrier Reef
Carbonate Platform
Atoll
Barrier Reef
Fringing Reef

10.09.a1

3. *Fringing reefs* are attached to a shoreline or are just offshore, surrounding an island. The seaward edge of the reef slopes down towards deeper water. Most reefs begin as fringing reefs, such as this one in the South Pacific Ocean. ▽

10.09.a2

10.09.a3

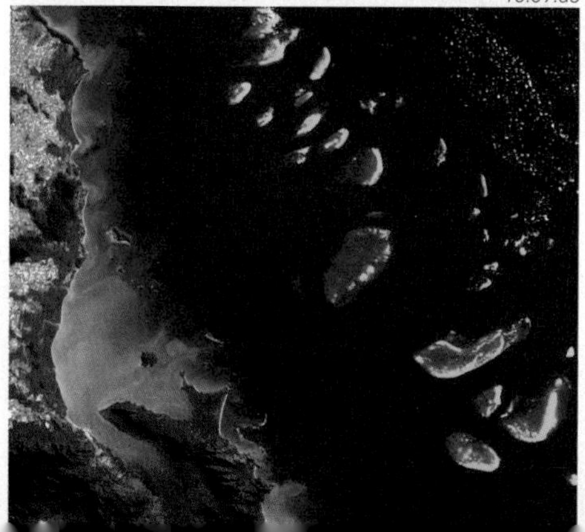

4. *Atolls* are curved reefs that enclose a shallow, inner lagoon (▽). An atoll forms when an island flanked by coral sinks, but upward coral growth keeps pace with the sinking. These reefs are fairly unique to volcanoes because they require subsidence, such as occurs when magmatism ends and the oceanic crust cools.

10.09.a4

10.09.a5

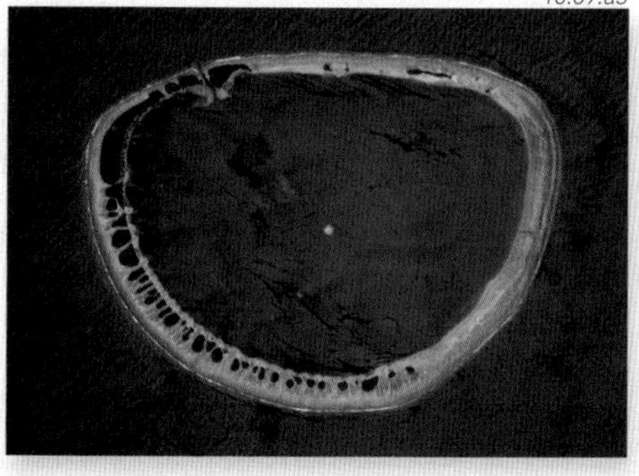

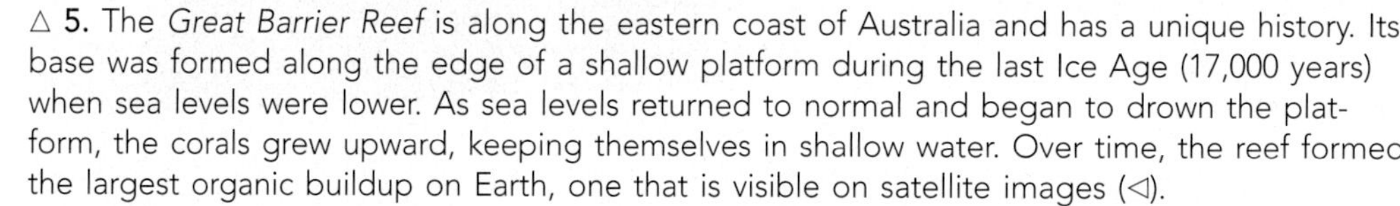

△ 5. The *Great Barrier Reef* is along the eastern coast of Australia and has a unique history. Its base was formed along the edge of a shallow platform during the last Ice Age (17,000 years) when sea levels were lower. As sea levels returned to normal and began to drown the platform, the corals grew upward, keeping themselves in shallow water. Over time, the reef formed the largest organic buildup on Earth, one that is visible on satellite images (◁).

B How Do Atolls Form?

Charles Darwin proposed a hypothesis for the origin of atolls after observing a link between certain islands and atolls during his research aboard the ship *Beagle* from 1831–1836. According to his model, shown below, atolls form around a sinking land mass, such as a cooling or extinct volcano. Another model (not shown) interprets some atolls as being the result of preferential erosion of the less dense center of a carbonate platform.

10.09.b1–3

Stage 1: A volcanic island forms through a series of eruptions in a tropical ocean, establishing a shoreline along which corals can later grow and construct a *fringing reef*.

Stage 2: After volcanic activity ceases, the new crust begins to cool and sink. Coral reefs continue building upward as the island subsides, forming a *barrier reef* some distance out from the shoreline.

Stage 3: The volcano eventually sinks below the ocean surface, but upward growth of the reef continues, forming a ring of coral and other carbonate material. This is an *atoll*.

C Where Do Reefs Occur in the World?

Most of the world's reefs are in tropical waters, located near the equator, between latitudes of 30° north and 30° south. Reef corals are more diverse in the Pacific, probably because many species went extinct in the Atlantic during the last Ice Age. Coral reefs are shown as red dots on the map below.

10.09.c1

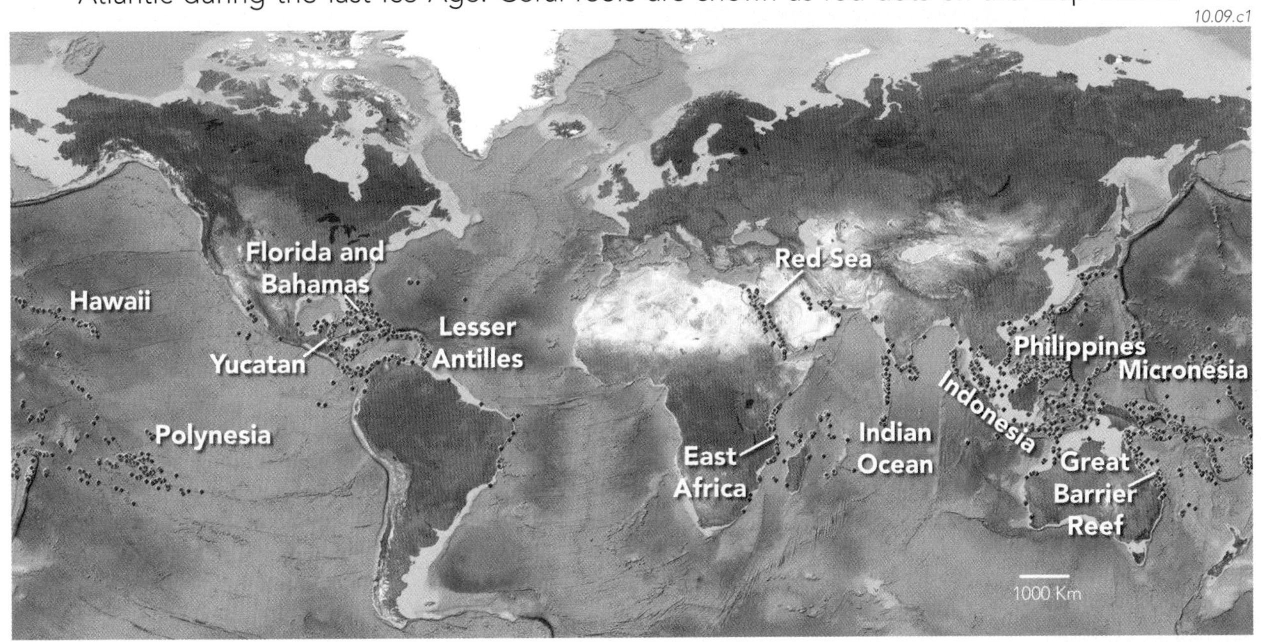

1. The *Great Barrier Reef*, along the northeastern flank of Australia, is the largest reef complex in the world. The world's second largest reef surrounds part of *New Caledonia*, a series of islands east of Australia and south of Micronesia.

2. The central and southwestern Pacific, including *Polynesia* and *Micronesia*, has numerous atolls and various reefs, including barrier and fringing reefs. Hawaii is also warm enough for reefs.

3. Well-known reefs are present throughout much of the *Caribbean* region, including *Florida*, the *Bahamas*, and the *Lesser Antilles*. The longest barrier reef in the *Caribbean* extends some 250 kilometers (150 miles) along the Yucatan Peninsula, from the northern border of Belize south to Honduras.

4. Reefs occur along the continental shelf of *East Africa*, such as in Kenya and Tanzania. Other reefs encircle islands in the *Indian Ocean* and the shoreline of the *Red Sea*.

5. Reefs in the *Philippines* cover an estimated 25,000 square kilometers and consist of fringing reefs with several large atolls. Reefs also flank *Indonesia* and nearby *Malaysia*.

Before You Leave This Page Be Able To

- ✓ Describe the different kinds of reefs and where they form.
- ✓ Summarize the stages of atoll formation.
- ✓ Name some locations with large reefs.

What Is the Geology of Continental Margins?

THE EDGES OF MOST CONTINENTS ARE HIDDEN beneath the seas, out some distance from the shoreline. The edge of a continent marks the transition between continental and oceanic crust, but this transition is typically concealed and smoothed out by thick layers of sediment. What features are present along the edges of continents, and how do these features form?

A What Features Are Typical of Continental Margins?

Some continental margins are *active plate boundaries*, such as the western coast of South America, where oceanic crust subducts beneath the edge of the continent. Many continental margins are not plate boundaries and are called *passive margins*. Both active and passive margins share some features.

10.10.a1

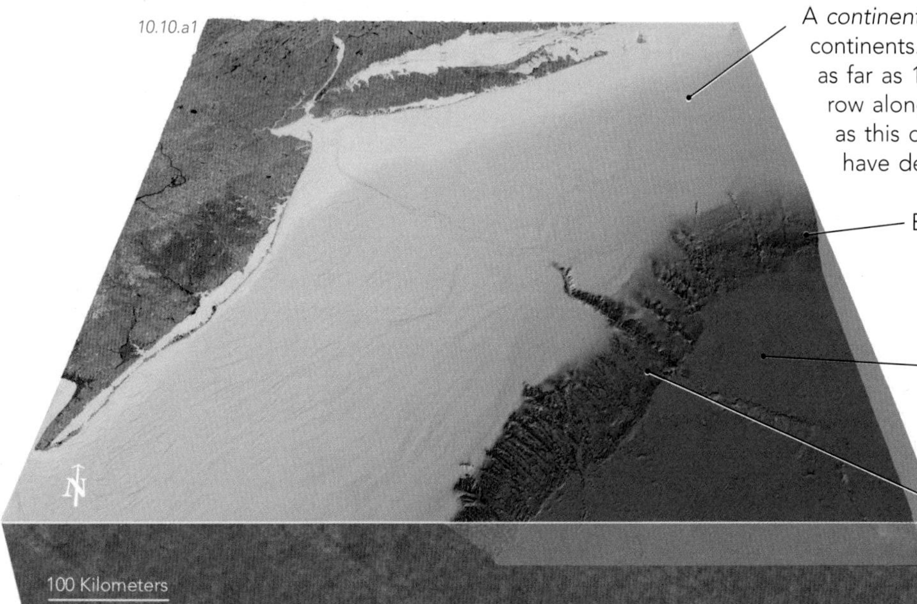

A *continental shelf* is a gently sloping surface that surrounds nearly all continents. On passive margins, it can extend out from the shoreline as far as 1,500 kilometers (930 miles) seaward, but it is typically narrow along active margins. The gentle slopes of most shelves, such as this one along the northeastern United States, are thought to have developed during the last Ice Age, when sea level was lower.

Extending out from the shelf is the *continental slope*, where deep ocean truly begins. Here the ocean floor slopes seaward at angles typically between 5° and 25°. The slopes are exaggerated in this figure.

Farther out from the continental slope is the *continental rise*. Sediment transported off the continental slope accumulates here, forming a broad, gentle slope.

The continental shelf and slope are locally cut by submarine canyons, such as the Monterrey Submarine Canyon in California. The submarine canyons shown here cut into the continental shelf and slope offshore of New York and New Jersey.

B What Rocks, Sediments, and Structures Characterize Continental Margins?

The transition from continent to deep ocean reflects progressively thinner continental crust and an abrupt change to oceanic crust. The thinned crust along most passive margins records rifting apart of the continent. Sediment on the continental margin varies greatly in thickness across the shelf, slope, rise, and abyssal plain.

1. Sediment is generally thinnest near the shoreline and on nearby parts of the continental shelf, which is underlain by continental crust with a close-to-normal thickness.

2. There are normal faults farther out, beneath the continental shelf and slope. These formed during the initial continental rifting that formed the margin. Movement along the faults has thinned the crust, causing the deeper seafloor.

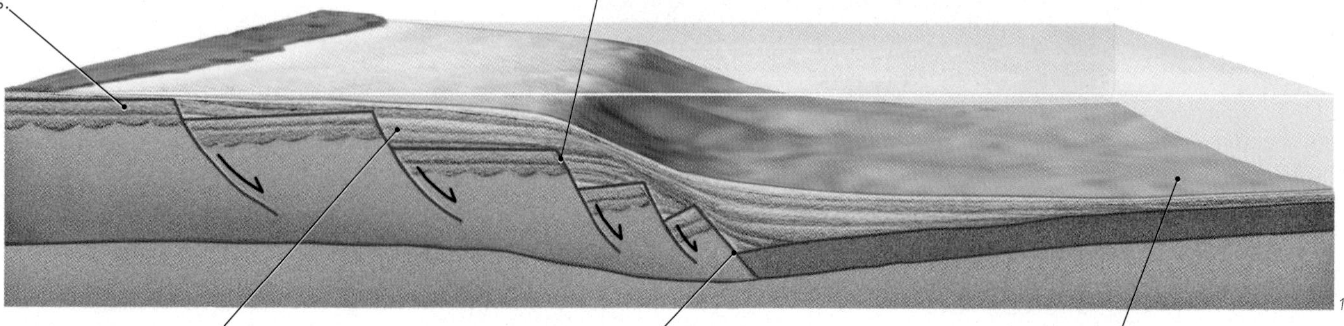

10.10.b1

3. Thick sediment has accumulated over the downdropped fault blocks beneath the shelf and slope. The sediment can host important oil and gas resources.

4. *The continental slope* marks the abrupt change from thinned continental crust of granitic composition to even thinner oceanic crust composed of basalt and gabbro.

5. The abyssal plain is farther from land and sources of land-derived sediment. It has a thin sediment cover composed of small particles, such as clays.

C What Settings Lead to Underwater Slope Failure?

Continental slopes are blanketed by sediments, most of which are unconsolidated and weak. The combination of weak materials and a steep angle causes some slopes to fail due to the force of gravity. Failure may be triggered by earthquakes, large storms, or overloading by newly deposited sediments.

Turbidity Currents

10.10.c1

◁ 1. As sediments collapse during a slope failure, they can break up and incorporate seawater between the grains. This forms a dense mixture of water and sediment (mud, silt, and sand), such as this mass produced in a laboratory. These mixtures are more dense than normal seawater and they flow downslope as fast-moving slurries, called *turbidity currents*. Turbidity currents have destructive potential and are capable of eroding rock, even under water.

2. The dense, cloud-like slurry of a turbidity current travels through the water until the current slows and the grains progressively settle. Larger grains settle out first, forming *graded beds* and alternating coarser and finer sediment. ▷

10.10.c2

Submarine Canyons

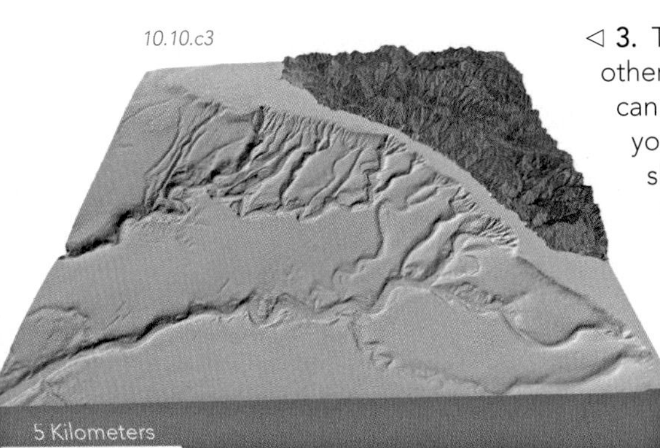

10.10.c3

◁ 3. Turbidity currents and other submarine movements can erode submarine canyons into the continental slope. This example is off the coast of central California. As the currents flow downhill, they erode the floor and walls of the canyon, making it larger over time.

▷ 4. The upper parts of some submarine canyons, such as this one on the continental shelf near the Hudson River in New York, were carved by rivers when sea levels were more than 100 meters (330 feet) lower than today. Once submerged by rising sea levels, such canyons can channel turbidity currents and be enlarged.

10.10.c4

Submarine Fans and Submarine Landslides

5. This diagram illustrates a turbidity current, shown in gray, beginning on the continental slope and flowing down a submarine canyon.

6. As a turbidity current exits the steep canyon, it spreads out and slows down. Sand grains can no longer be suspended by the turbulence and settle out. As the current slows further, silt is deposited, followed by clay particles. This process forms *graded beds* (coarser sediment at the base and finer sediment at the top).

7. As the turbidity current slows and spreads out across the continental rise, it deposits its load of sediment in a fan-shaped deposit, called a *submarine fan*. A submarine fan can be hundreds to more than a thousand kilometers wide and typically consists of shale and other deep-marine sediment that alternate with sandy turbidite deposits with graded bedding.

10.10.c5

8. Underwater slopes can also fail as *submarine landslides*. A landslide mass can contain large, fairly coherent blocks or can come apart as it detaches from the slope and moves downhill. A landslide commonly forms distinctive lumps on the seafloor and may leave behind a ragged scar on the slope. Repeated, large landslides were key to forming and widening some submarine canyons.

Before You Leave This Page Be Able To

- ✓ Summarize or sketch the features of a continental margin, such as the continental shelf, slope, and rise.
- ✓ Summarize or sketch the rocks, sediments, and structures that occur along a typical continental margin.
- ✓ Explain turbidity currents, submarine canyons, fans, and landslides.

How Do Marine Salt Deposits Form?

SALT DEPOSITS OF MARINE ORIGIN occur along many continental margins, forming layers and structural domes. Marine salt deposits are related to certain geologic settings and are important resources for salt, sulfur, and petroleum. How and where do marine salt deposits form?

A How Does Salt Occur Along Continental Margins?

10.11.a1

Natural salt is mostly composed of the sodium-chloride mineral *halite* (NaCl), the same mineral in common table salt. Halite can be associated with other salt minerals and with *gypsum*. Outcrops are very soluable.

10.11.a2

Most salt deposits can form layers, as is typical for any sedimentary rock. Salt layers can be thinner than a centimeter, or comprise a sequence several kilometers thick. In this outcrop, thin white salt layers are interlayered with gray shale. [Colorado]

10.11.a3

Salt is a very weak geologic material, flowing easily when subjected to the stresses associated with deep burial and tectonics. As a result, it commonly flows as solid but soft masses that dome up surrounding layers. [Colorado]

B How Does Salt Form Near Continental Margins?

Many salt deposits form when seawater evaporates, leaving behind a residue of salt that was dissolved in the water. Such evaporation is especially efficient in warm, dry climates and in water bodies with limited connection to the oceans.

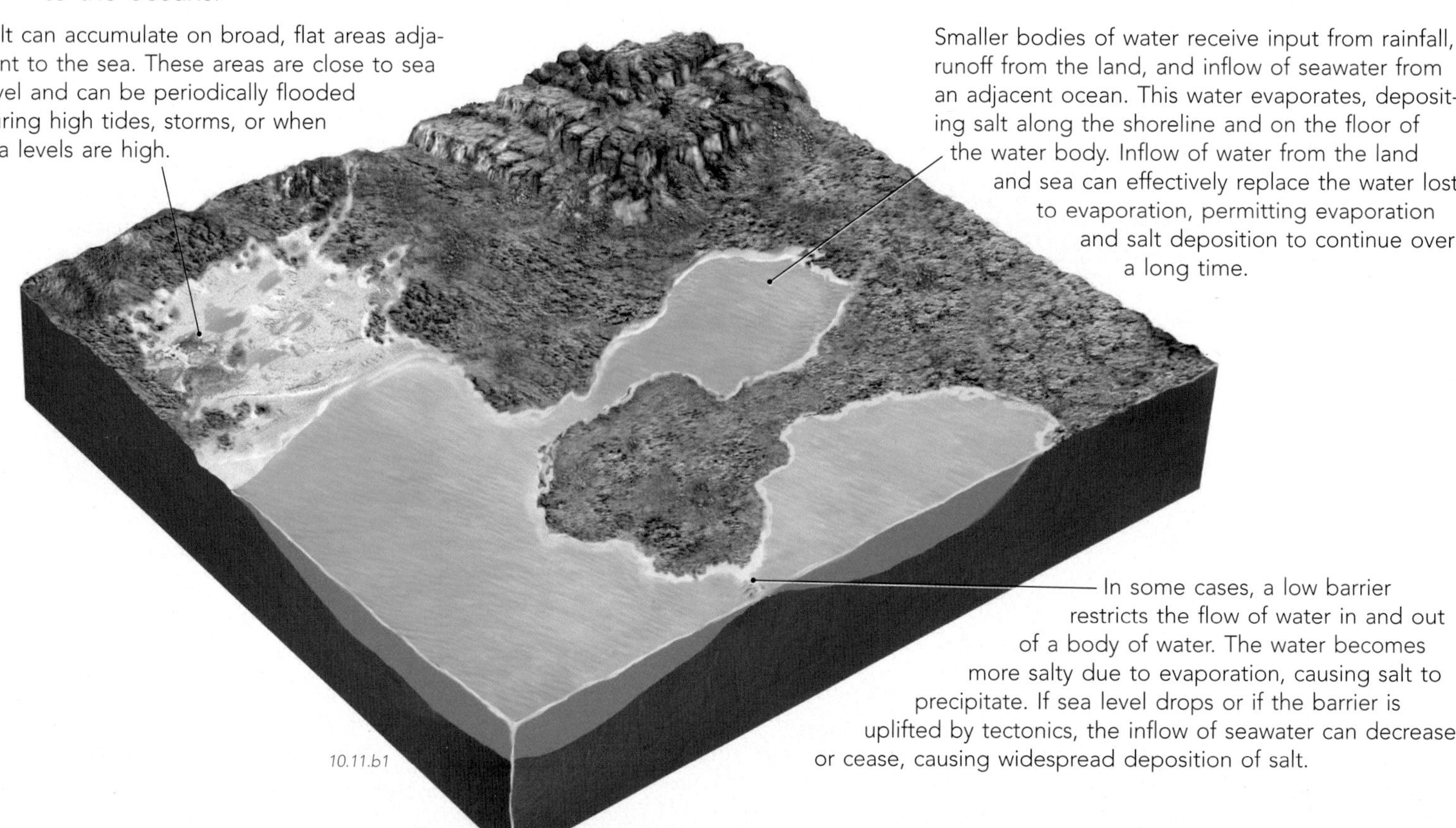

Salt can accumulate on broad, flat areas adjacent to the sea. These areas are close to sea level and can be periodically flooded during high tides, storms, or when sea levels are high.

Smaller bodies of water receive input from rainfall, runoff from the land, and inflow of seawater from an adjacent ocean. This water evaporates, depositing salt along the shoreline and on the floor of the water body. Inflow of water from the land and sea can effectively replace the water lost to evaporation, permitting evaporation and salt deposition to continue over a long time.

In some cases, a low barrier restricts the flow of water in and out of a body of water. The water becomes more salty due to evaporation, causing salt to precipitate. If sea level drops or if the barrier is uplifted by tectonics, the inflow of seawater can decrease or cease, causing widespread deposition of salt.

10.11.b1

C What Structures Do Salt Deposits Form?

Because it is a weak rock, salt can form its own unique kinds of geologic structures. It also can greatly influence how faults and folds develop in overlying rocks.

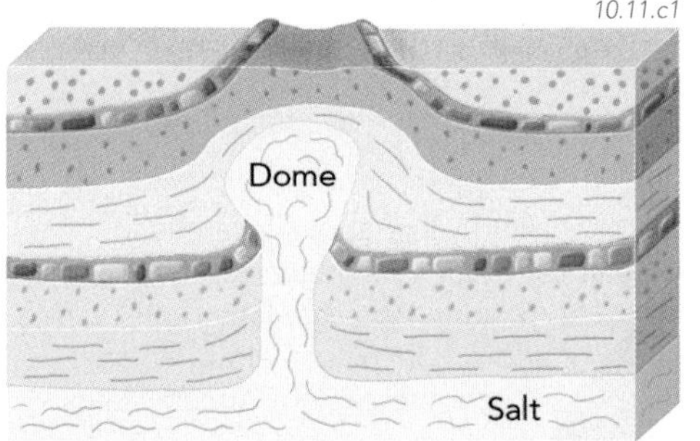

Salt is less dense than most rocks; when buried, it can buoyantly flow toward the surface in steep, pipe-like conduits. The resulting structure is a *salt dome*.

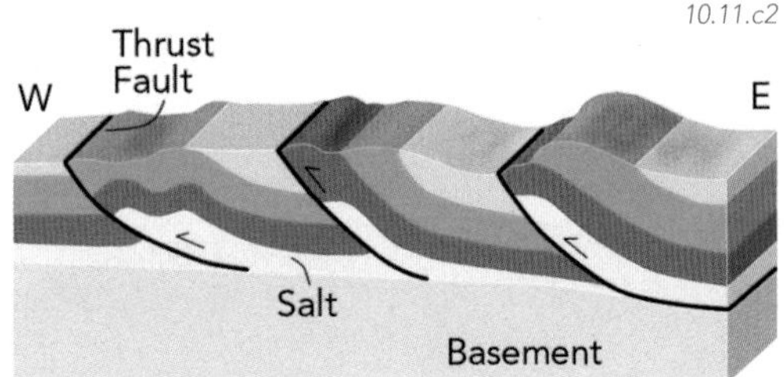

When a region containing a thick salt layer is deformed, the salt can slip and flow, allowing overlying rocks to fold and fault. This cross section shows part of the Jura Mountains near the French-Swiss border. The folds and faults are underlain by a layer of salt.

▷ Where salt reaches the surface, such as in a salt dome or an anticline, it can flow downslope under the influence of gravity and form a *salt glacier*, such as this one. [Iran]

D What Salt Structures Occur Along the Gulf Coast of the United States?

The Gulf Coast of the southern United States is world-famous to geologists for its numerous salt structures, both on land and offshore. The salt structures have played a key role in the formation of the region's large oil fields and provide important sources of salt and sulfur minerals. For these reasons, they have been extensively studied by seismic surveys and by expensive drilling, sometimes in thousands of meters of water.

1. This diagram shows the land and seafloor in the Gulf of Mexico offshore of the Texas-Louisianna coast. An interpretation of the subsurface geology is drawn on the sides of the block.

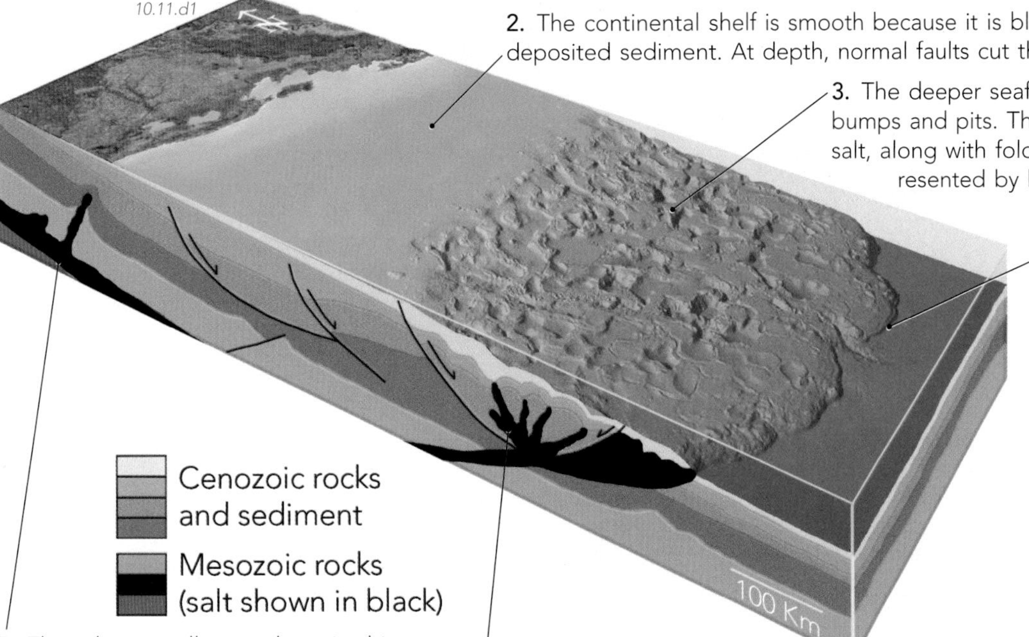

2. The continental shelf is smooth because it is blanketed by nearly flat-lying layers of recently deposited sediment. At depth, normal faults cut the layers and displace some blocks downward.

3. The deeper seafloor in this region has unusual and puzzling bumps and pits. These features are caused by subsurface flow of salt, along with folds and faults in the overlying layers. Salt is represented by black areas on the side of the diagram.

4. The unusual seafloor is bounded by a relatively steeper slope called the *Sigsbee Escarpment*. As shown on the side of the block, the escarpment marks the front of a large mass of salt in the subsurface. This salt flowed upward and sideways from depth, reaching all the way to the seafloor in places.

5. The salt originally was deposited in a thick layer, shown here in black, when continental rifting during the Mesozoic formed narrow basins. The basins at times had limited connection with the sea, causing extensive evaporation of seawater and deposition of the salt layer. The salt was later buried by sediments, shown in light green, orange, and yellow.

6. As the salt was buried and subjected to increased pressure, it flowed sideways and rose up through the overlying sedimentary layers. Movement of the salt folded and domed the layers. In places, it formed steep, pillar-shaped *salt domes*, shown here as finger-like black masses.

Before You Leave This Page Be Able To

- ✓ Summarize how salt forms near continental margins.
- ✓ Describe how salt can occur in salt domes, some folded mountain belts, and salt glaciers.
- ✓ Summarize how salt structures are expressed in the Gulf Coast region.

How Did Earth's Modern Oceans Evolve?

EARTH'S CONTINENTS AND OCEANS have changed over time, and their present configuration is simply a snapshot of a longer evolution. Two hundred million years ago, the continents were joined together in a supercontinent called *Pangaea*. This huge land mass has since separated, forming the modern oceans.

200 Million Years Ago – End of Pangaea

These artistic renditions by geologist Ronald Blakey depict the breakup of a supercontinent and the movement of the continental fragments during the last 200 million years. He used a type of oval map that can show the entire world. It is not just one side of a globe.

About 200 million years ago, all the continents were joined in the supercontinent of *Pangaea,* which was surrounded by an enormous ocean. Pangaea was assembled in the Late Paleozoic and, as shown here, had begun to break up via continental rifting by 180 to 200 million years ago (in the Early Mesozoic).

By 180 million years ago, Africa and South America had begun to rift apart from North America, forming rift valleys and a narrow sea along what is now the eastern United States.

As North America rifted away, the land masses of Africa, South America, India, Australia, and Antarctica remained linked in a southern supercontinent called *Gondwana.*

10.12.a1

150 Million Years Ago – New Oceans Open

By about 150 million years ago, in the middle of the Mesozoic Era, the breakup of Pangaea was well underway. The Central Atlantic Ocean had formed as North America separated from Africa. The Gulf of Mexico formed by rifting along the southern edge of North America.

Continental rifting began along the future borders of Africa and South America, but seafloor spreading had not yet started to form the South Atlantic Ocean.

The *Tethys Sea,* a tropical ocean, was a large wedge-shaped extension of the main global ocean. To the north, a series of collisions started to consolidate Asia into a larger continent.

The land masses south of Tethys started to rift apart and began to resemble the familiar shapes of South America and Africa. The rest of *Gondwana* remained intact, for now.

10.12.a2

120 Million Years Ago – Central and South Atlantic

During the Cretaceous Period (120 million years ago), North America, Europe, and Asia were still connected. The North Atlantic Ocean, between North America and Europe, had yet to open.

The southern parts of the Atlantic Ocean opened as South America began to separate from Africa. The rest of Gondwana began to rift apart, as India separated from Australia and Antarctica, forming the early stages of the Indian Ocean.

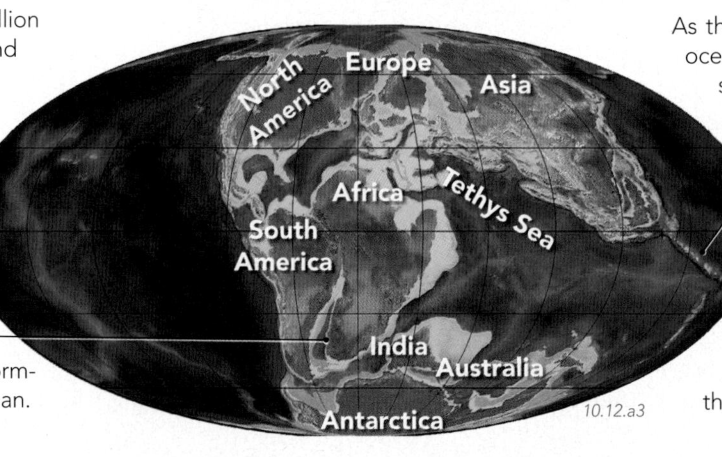

10.12.a3

As the new oceans grew, the large global ocean began to shrink because of subduction beneath North and South America, Asia, and island arcs within the ocean.

As the newly formed oceans continued opening, their waters moderated land temperatures and produced numerous shallow-marine environments. This led to an incredible diversity of life in the sea and on land.

90 Million Years Ago - Atlantic Ocean Fully Opens

The Central Atlantic Ocean was fully open between North America and Africa, with a spreading center down the middle of the ocean. The North Atlantic, between North America and Europe, was not yet rifted open.

The opening of the South Atlantic Ocean separated Africa and South America, isolating their land animals.

India was headed northward across the Tethys Sea toward an eventual collision with the southern flank of Asia.

Antarctica rifted apart from Africa and Australia, which allowed the South Atlantic to connect to the southern Indian Ocean.

10.12.a4

30 Million Years Ago - Closing the Tethys Sea

Greenland and the rest of North America began rifting apart from Europe at about 80 million years ago, opening up the North Atlantic Ocean.

The Pacific Ocean contained spreading centers (the belts of lighter blue on the seafloor), but has grown smaller over time as its oceanic plates are subducted beneath the Americas, Asia, and numerous island arcs.

The Tethys Sea was nearly closed when India collided with Asia to form the *Himalaya Mountains*, and when Africa and nearby continental fragments converged with southern Europe to form the *Alps* and other ranges.

Australia was completely isolated, allowing its collection of marsupials and other unusual animals to thrive and evolve. Antarctica remained over the South Pole.

10.12.a5

Present Day

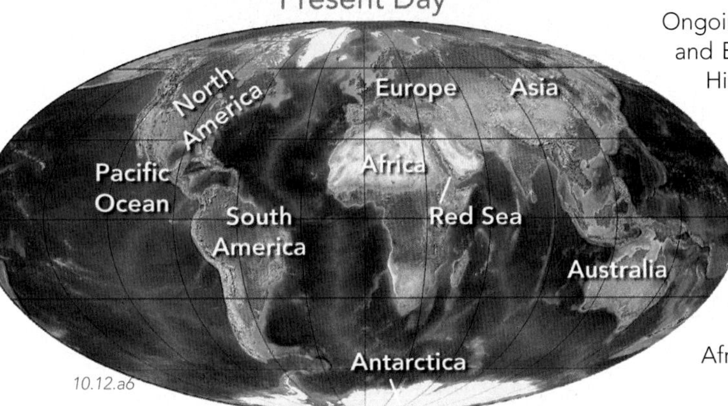

Today, the Atlantic Ocean continues to grow because it has a spreading center that adds to the oceanic plate but does not have subduction zones to consume any oceanic material. It has grown to its present size at the expense of the Pacific Ocean, which is the last remnant of Pangaea's global ocean.

Ongoing convergence between the Indian and Eurasian plates continues to form the Himalaya Mountains north of India. It may also be starting to form a new plate boundary further south, within the Indian Ocean.

The Red Sea is a developing rift and may continue to grow at the expense of another ocean. It is uncertain whether the nearby East African Rift will split off a piece of Africa.

10.12.a6

The Future of the Oceans

What will the oceans look like in 50 million years? Geologists calculate the likely future locations of the oceans and continents by using current plate velocities and by making assumptions about how plates act during collisions. One prediction is that Africa will collide with Europe and Asia, closing the Mediterranean Sea and forming a very large supercontinent, shown to the right. The Pacific will continue to shrink as spreading in the Atlantic Ocean pushes the Americas (not shown) farther to the west. Our present situation is not final. It is just one scene in a very long movie.

10.12.mtb1

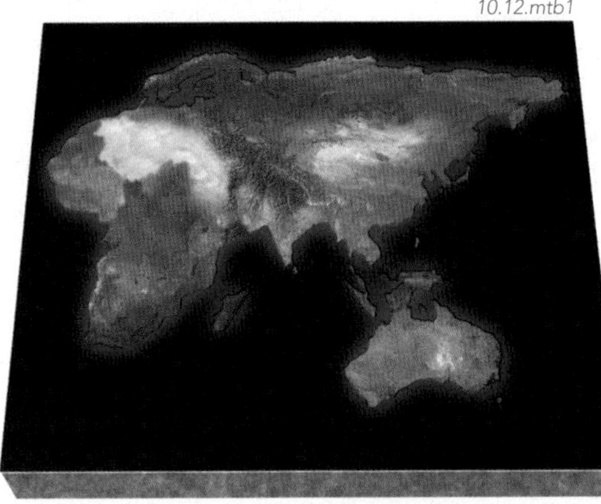

Before You Leave This Page Be Able To

- ✓ Summarize the major changes in the Earth's oceans since 180 million years ago, including approximately when the Central Atlantic, South Atlantic, North Atlantic, and Indian Ocean formed and which continents rifted apart to form each ocean.
- ✓ Describe or sketch why growth of the Atlantic Ocean must have caused the Pacific Ocean to shrink over time.

How Did the Gulf of Mexico and the Caribbean Region Form?

THE GULF OF MEXICO AND CARIBBEAN SEA display various features, including an island arc, several deep troughs, and numerous islands and smaller ocean basins. The present setting and recent geologic history of the region provide an opportunity to examine how continental margins are formed and how ocean basins evolve over time. Examine the map below, which shows seafloor depths and plate boundaries.

Present Setting

The *Gulf of Mexico* is nearly enclosed by Florida, the Gulf Coast of the United States, and Mexico. It is deepest in the center and is flanked by broad continental shelves offshore of the United States and Yucatan Peninsula.

Shallow seafloor, underlain by continental crust, flanks the Florida Peninsula and Bahama Islands. Deeper seafloor separates this region from the island of Cuba to the south and from the Yucatan Peninsula to the southwest.

A trench curves around the outside of the Lesser Antilles island arc. The trench and island arc are the result of westward subduction of Atlantic oceanic lithosphere beneath the Caribbean plate.

10.13.a1

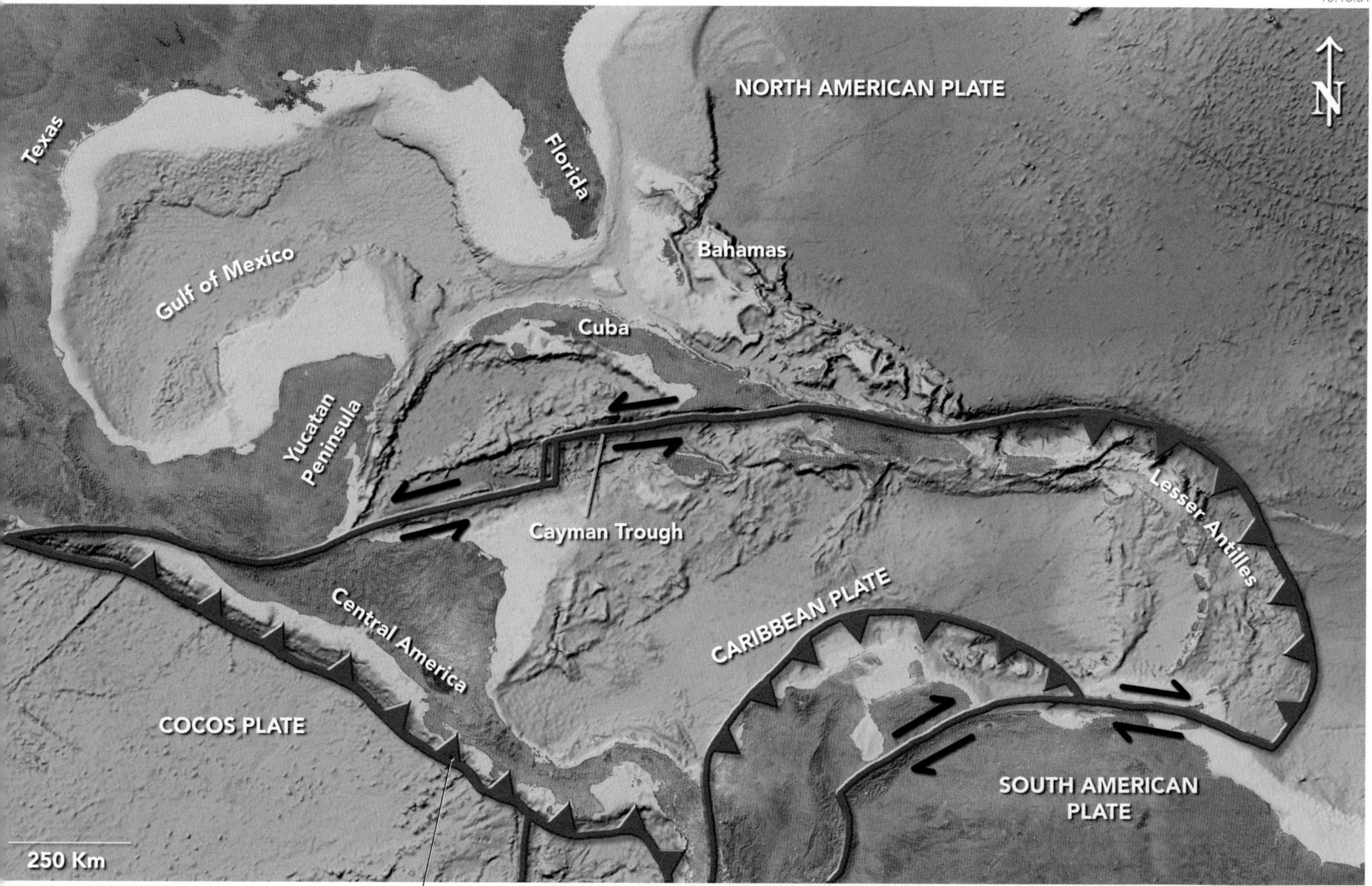

A deep trench marks where oceanic plates in the Pacific, including the Cocos plate, are subducted northeastward beneath Central America. Volcanoes and earthquakes are common in the overridding plate.

An east-west-trending escarpment, called the *Cayman Trough,* cuts across the seafloor and the southern end of Cuba. It is a transform boundary along the northern edge of the Caribbean plate.

The seafloor in some parts of the Caribbean plate is shallower than expected. Here, the oceanic crust is anomalously thick (up to 20 kilometers thick) and is composed of thick sequences of basalt.

Jurassic History (~200 to 145 Million Years Ago)

By the Jurassic Period, North America had begun to rift apart from Africa and South America. Continental rifts filled with sediment and salt, and the thinned continental crust became continental shelves.

10.13.a2

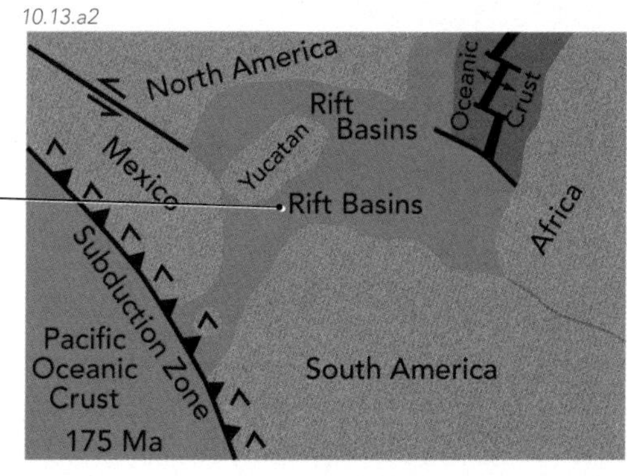

10.13.a3

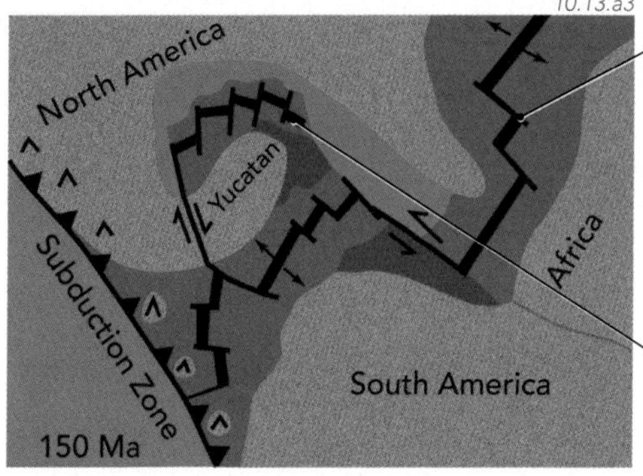

By the Late Jurassic, the continents had rifted apart and seafloor spreading produced new oceanic crust.

Spreading formed the Gulf of Mexico when the Yucatan pulled away from the Gulf Coast.

Cretaceous History (145 to 65 Million Years Ago)

In the Cretaceous Period, spreading in the Gulf of Mexico ceased, but sediment deposition continued.

Seafloor spreading moved North America farther from Africa and South America.

10.13.a4

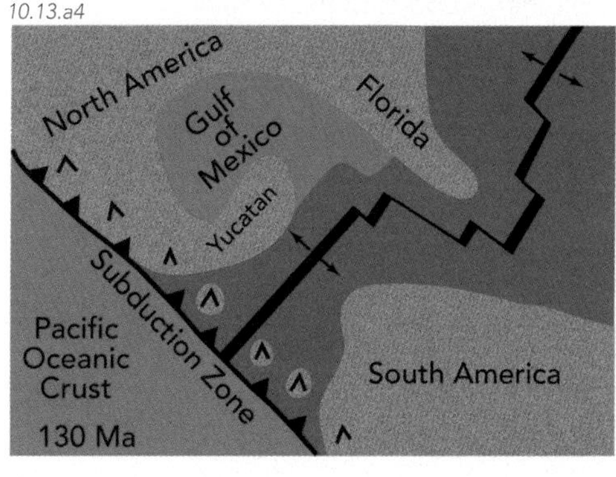

10.13.a5

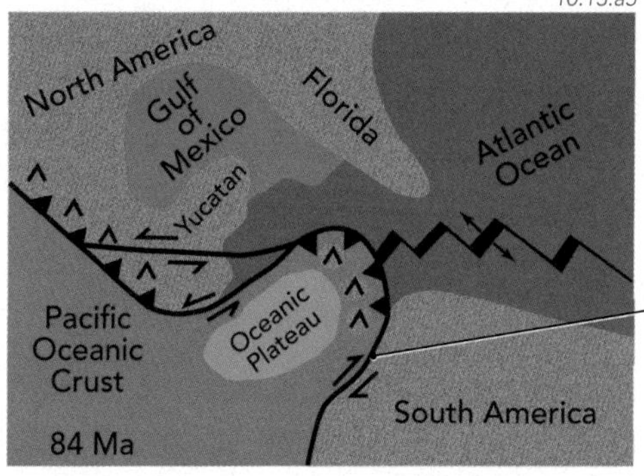

In the Late Cretaceous, an island arc and oceanic plateau, probably from somewhere in the Pacific, moved into the area.

Transform faults bounded the sides of the plate on which the oceanic plateau rode.

Tertiary History (65 to 5 Million Years Ago)

In the early Tertiary Period, the island arc and oceanic plateau collided with Florida near Cuba.

A new volcanic arc formed between South and Central America but did not connect these lands until about 5 million years ago.

10.13.a6

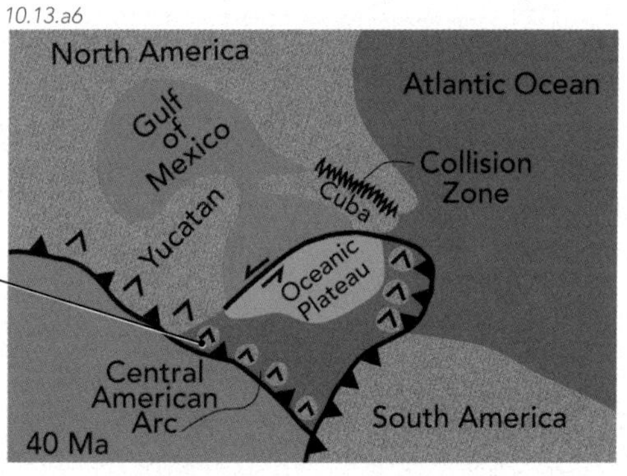

10.13.a7

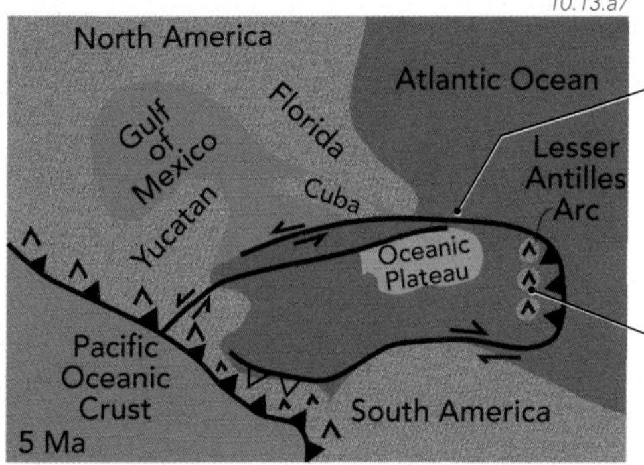

By 5 million years ago, long transform faults allowed the Caribbean plate to continue moving eastward.

The Lesser Antilles island arc formed above a west-dipping subduction zone.

How This Geologic History Was Reconstructed

The history summarized above was not easy to piece together, especially since much of the geologic record is undersea. The first studies were done on land, mapping the geology and determining the ages and sequences of rock units. The locations of volcanoes, faults, and other structural features helped identify tectonically active areas.

An understanding of the undersea geology was largely obtained by seafloor mapping and geophysical surveys conducted for petroleum exploration on the continental shelves. Such surveys were followed by very expensive drilling through the sedimentary layers in order to retrieve rock samples, calibrate the surveys, and determine whether petroleum is present. The local geology was then interpreted in the context of global reconstructions of plate motions, largely derived from paleomagnetism.

Many aspects of this geologic history are being actively investigated and debated by geologists. Perhaps the most controversial topic is whether the thick basaltic sequence in the Caribbean originated as an oceanic plateau in the Pacific, as shown above, or was formed locally, between North and South America.

Before You Leave This Page Be Able To

- ✓ Summarize the main physical features of the Caribbean and Gulf of Mexico, describing how they relate to modern plate-tectonic boundaries.
- ✓ Briefly summarize the main events that shaped the Gulf and Caribbean.
- ✓ Describe how the geologic history of the region was studied, both on land and beneath the sea.

How Did These Ocean Features and Continental Margins Form?

The terrain below contains various features on the seafloor, as well as parts of three continents. Some general observations of each feature provide clues about what that feature is. You will use this information to interpret how each feature formed, what the area was like in the past, and how it will look in the future.

Goals of This Exercise:

- Observe the terrain and make observations about the shape, size, and character of each feature.
- Use the general descriptions to determine which features are present in different parts of the terrain.
- Interpret how each feature formed and use this information to infer the present-day plate tectonics of the area.
- Use all the information to reconstruct what the area probably looked like 20 million years ago and what it will look like 20 million years into the future.

1. This figure shows a region approximately 1,000 kilometers (600 miles) wide. The seafloor is shaded according to depth, with lighter blue colors indicating shallower areas. Numbers indicate the isotopic ages of volcanic rocks in millions of years before present (labeled *Ma*). The view is toward the north.

6. A linear chain of islands and seamounts extends from the oceanic plateau toward the southeast. The islands and seamounts are shaped like volcanoes and are composed of volcanic rocks, mostly basalt. The ages of the volcanic rocks decrease to the southeast. The shield volcano at the southeast end of the chain is still active.

2. A broad oceanic plateau rises from deep water and locally forms small islands. Samples collected by drilling and dredging are mostly basalt and are dated at 40 million years.

3. A curved belt of volcanic islands is flanked on the east by a deep oceanic trench. Most of the volcanoes are composed of andesite and show evidence of recent explosive eruptions. Most islands have been volcanically active for more than 35 million years.

4. There is a narrow sea between the volcanic islands and a continent to the west. In the center of the sea is a low ridge, whose axis contains a rift valley and evidence of active submarine eruptions of basalt.

5. The western continent contains a narrow shelf offshore. There is no evidence of recent volcanoes, earthquakes, or mountain building along this edge of the continent. The oldest oceanic crust next to the continent is 20 million years old.

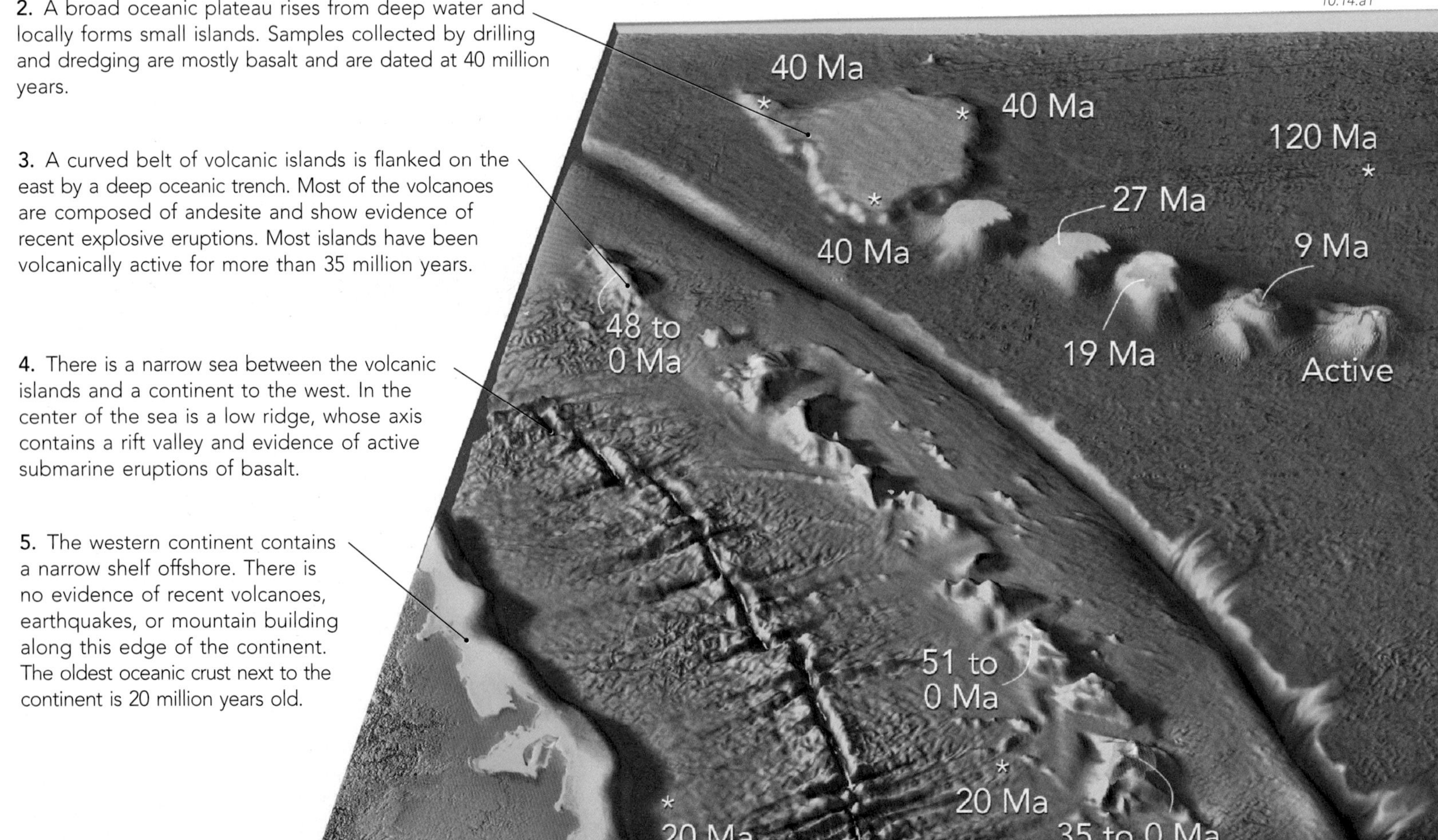

Procedures

Use your observations of this region to complete the following steps, entering your answers in the appropriate places on the worksheet.

1. Observe the terrain and determine which types of features are shown (e.g., mid-ocean ridge, continent, etc.).
2. Based on the descriptions, describe how each feature probably formed.
3. Interpret whether adjacent features are related to one another using their relative positions and ages.
4. Identify the main geologic features on the cross section along A–A′.
5. In the appropriate place on the worksheet, draw a cross section along the front of the terrain. Show your interpretations of the plate geometries and different types and thicknesses of crust.
6. Describe what the area might have looked like 20 million years ago based on the ages and relative motions of the plates. Draw a very simplified map of your interpretation on the worksheet.
7. Predict what the area will look like 20 million years into the future. Draw a simplified map of your interpretation on the worksheet.

7. The shelf surrounding the central continent is broad and shallow, extending several hundred kilometers out from the shoreline. The edge of the shelf shows no evidence of earthquakes or active faulting. Several large canyons are cut into the shelf and lead down to large piles of sediment on the abyssal plain. The continent has fairly subdued topography.

8. To explore for oil, the shelf of the central continent has been investigated using seismic surveys. A geologic cross section summarizing these results is presented below for the line A–A′ (shown on the map). All sedimentary layers are Cenozoic (younger than 65 million years).

10.14.a2

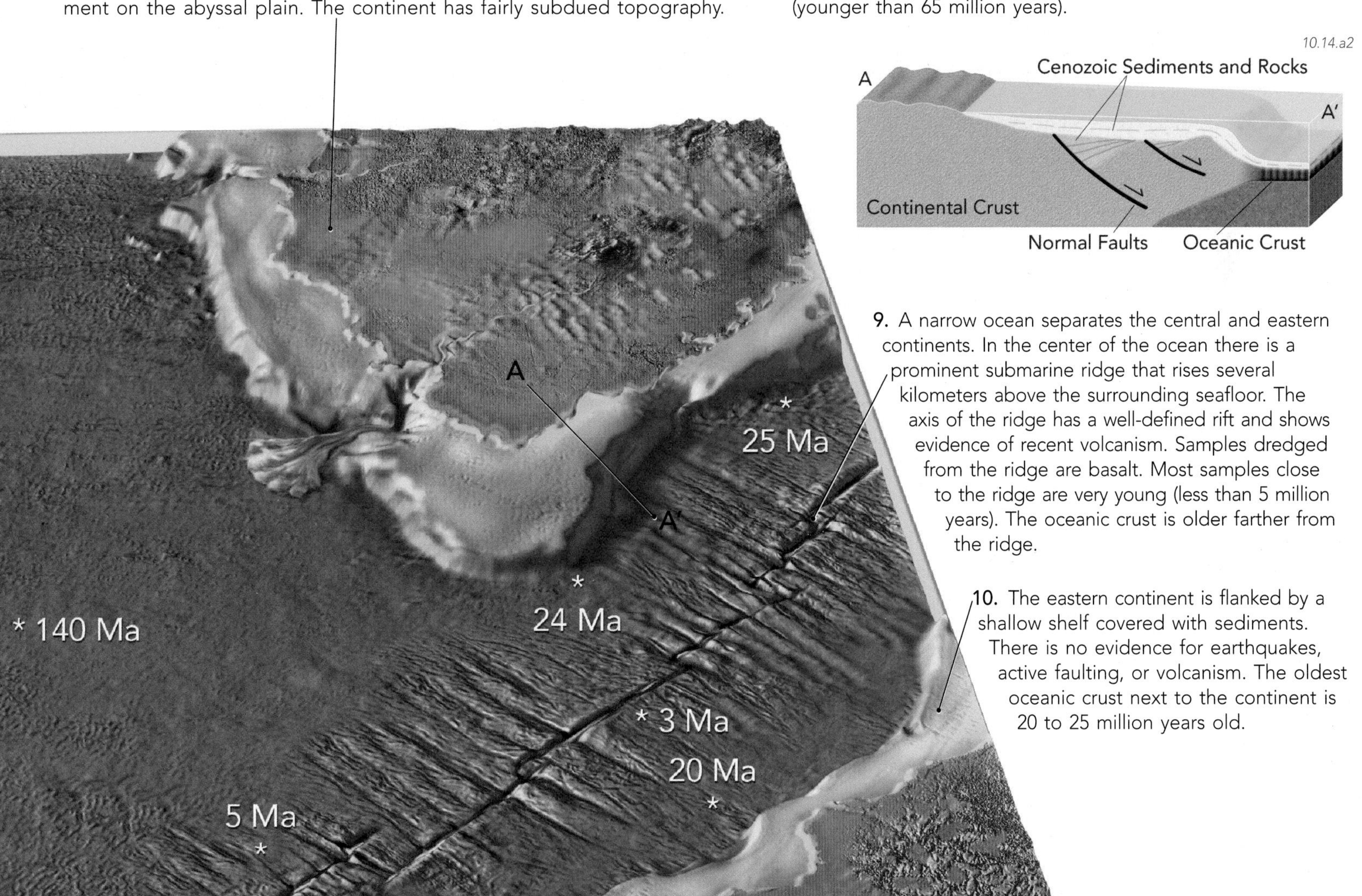

9. A narrow ocean separates the central and eastern continents. In the center of the ocean there is a prominent submarine ridge that rises several kilometers above the surrounding seafloor. The axis of the ridge has a well-defined rift and shows evidence of recent volcanism. Samples dredged from the ridge are basalt. Most samples close to the ridge are very young (less than 5 million years). The oceanic crust is older farther from the ridge.

10. The eastern continent is flanked by a shallow shelf covered with sediments. There is no evidence for earthquakes, active faulting, or volcanism. The oldest oceanic crust next to the continent is 20 to 25 million years old.

CHAPTER 11

Mountains, Basins, and Continents

THE SURFACE OF THE EARTH contains mountains, high plateaus, and low areas called *basins*, in which sediment accumulates. Earth's surface also contains continents that have grown and moved through time. How do mountains, basins, and continents form, and what factors control their elevations?

This view, looking north, shows satellite imagery superimposed on topography for the region around the Tibetan Plateau of southern Asia. A topographic profile across the region is shown on the next page.

What regional features can you observe in this perspective view and on the topographic profile?

The Tibetan Plateau is the largest, highest, and flattest plateau on Earth. Its average elevation is 5 kilometers (over 15,000 feet), which is higher than any peak in the United States, except for those in Alaska.

Why is the Tibetan Plateau so high, and what controls the elevation of a region?

The Tarim Basin is the large desert north of the Plateau. It is 3,000 meters lower in elevation than the plateau and is being filled by sediment derived from the adjacent highlands.

How do basins form, and why are they lower than their surroundings?

11.00.a1

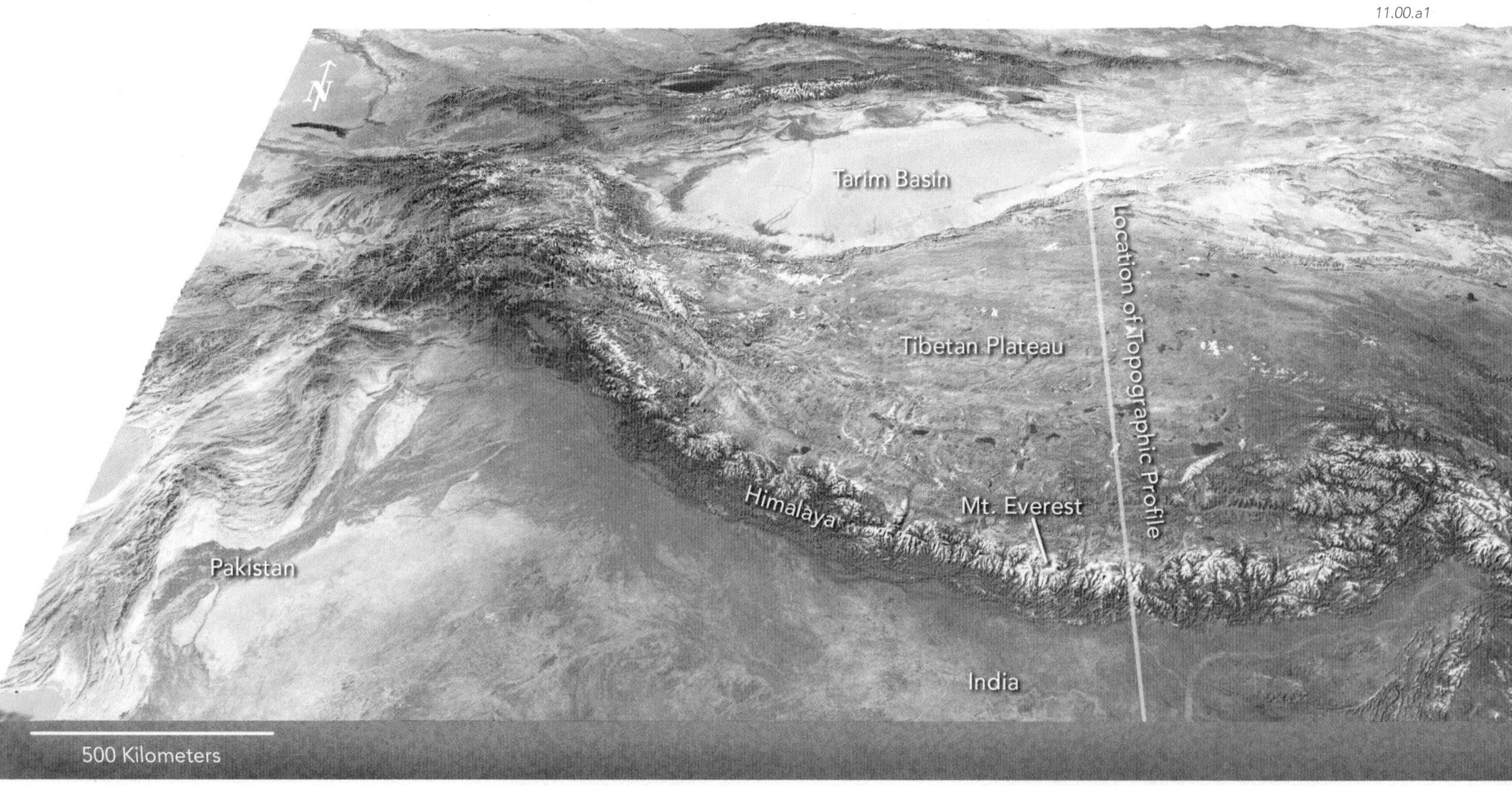

The Himalaya is a spectacular mountain range that rises along the southern edge of the Tibetan Plateau. It is the world's highest mountain range, with numerous peaks more than 8 kilometers (>26,000 feet) above sea level.

Why is this mountain range so high compared to all others on our planet?

Mount Everest is the world's highest mountain, rising 8,850 meters (29,035 feet) above sea level. It straddles the border between Nepal (to the south) and Tibet (to the north), and climbers can approach the mountain from either side.

What geologic processes form mountains, and what controls which areas have mountains versus which ones do not?

Most of India to the south has much lower elevation and relief, and mostly is tectonically stable away from the mountain front. Its oldest rocks are approximately 2.7 billion years old, representing the earliest period of Earth's history.

When did the first continents form, and how do continents change over time?

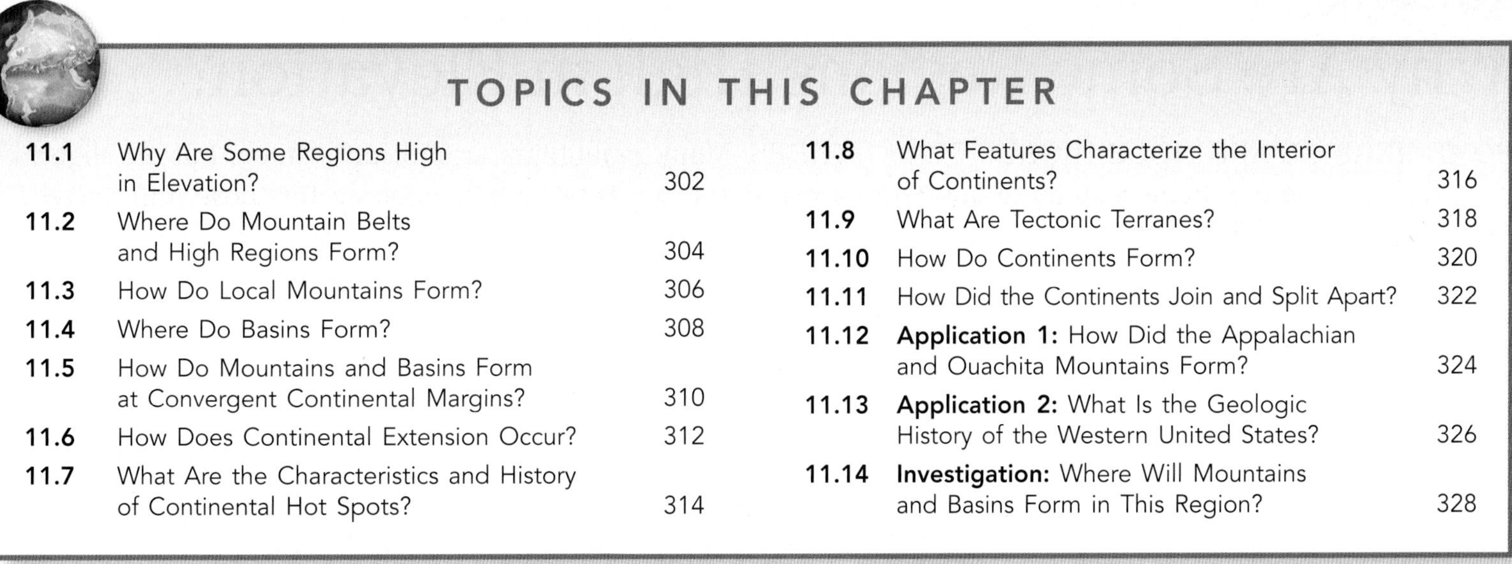

TOPICS IN THIS CHAPTER

11.00.a2

◁ This photograph, looking south, shows peaks of the high Himalaya. Part of the less rugged, but still high-elevation Tibetan Plateau is in the foreground.

China

11.00.a3

△ From the Tibet (north) side, Mount Everest stands as a rugged, imposing mountain. This view is taken from one of the base camps where climbers begin their arduous and dangerous climb to the top.

▽ The topographic profile below shows the high Tibetan Plateau viewed to the west. The high mountains on the left edge of the Plateau are the Himalaya, with the lowlands of India to the left (south). The sandy Tarim Basin is to the right (north). To depict the topographic features at this regional scale, the topography is vertically exaggerated by 10 times.

11.00.a4

100 Kilometers

Investigating the Timing of Uplift of a Region

Tibet and the Himalaya mountain range are high in elevation now, but when did they become so? Several approaches are used to determine when a region was uplifted (became higher in elevation). These include GPS, dating the uplift history of the rocks, and examining sediment in adjacent basins.

Uplift can be measured directly with a Global Positioning System (GPS), and studies indicate that parts of the Himalaya are rising a few centimeters per year. Another approach is to find rocks at high elevations that were deposited at low elevation. The top of Mount Everest contains a faulted slice of Paleozoic limestone with marine fossils; the limestone was deposited at sea level and the mountain was uplifted later.

Isotopic dating methods are an important way to determine the age of uplift. As deep rocks are uplifted toward the surface and uncovered by erosion, they cool, locking in daughter products from radioactive decay. Certain dating techniques tell us when deep rocks arrived to within 2 to 4 kilometers of the surface and so indicate the age of uplift. In the Himalaya, such methods yield ages as young as several million years, indicating recent and ongoing uplift.

As mountains and other high regions are uplifted and eroded, they contribute clasts to adjacent sedimentary basins. The age of uplift, therefore, can be inferred by determining when clasts derived from a mountain were added to the sedimentary sequence. Sediments along the foothills of the Himalaya indicate that debris originating from the mountain range first appeared around 45 million years ago.

Why Are Some Regions High in Elevation?

SOME REGIONS ARE MUCH HIGHER THAN OTHERS. Many mountains are not only steep, but are high in elevation. Elsewhere, huge regions of land are barely above sea level. What accounts for these differences? Regional variations in elevation primarily reflect the kinds of materials that lie at depth. A change in the subsurface can cause a region to be uplifted (rise in elevation) or to subside (drop in elevation).

A What Controls Regional Elevation?

Regional elevations are controlled primarily by the thickness of the crust, but can also be influenced by the temperature and density of materials in the crust and upper mantle.

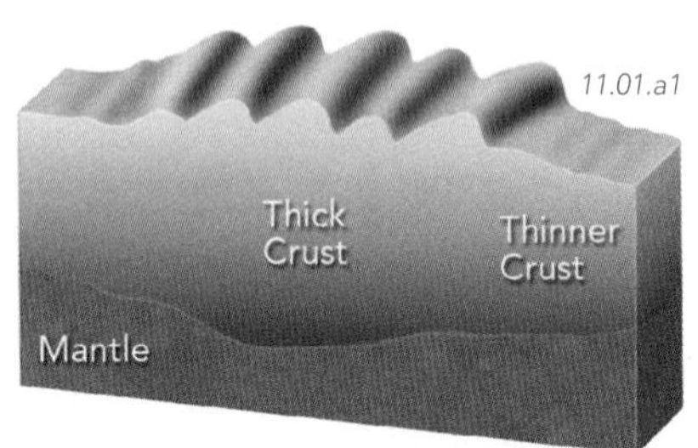

Regions with thick crust are *higher* than those with thinner crust. In other words, mountain ranges have deep crustal roots.

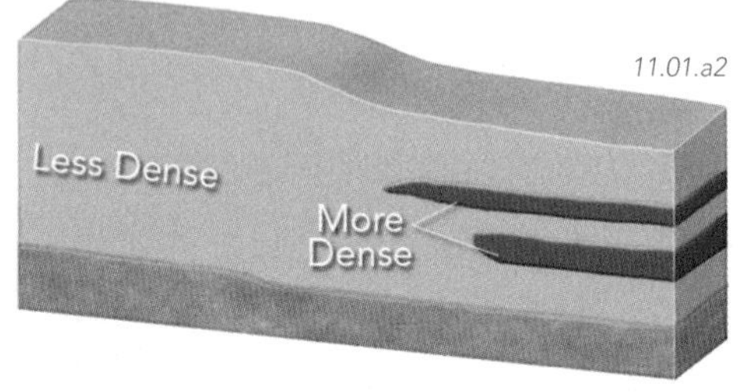

Regions underlain by less dense crust will be *higher* in elevation than areas with a similar thickness of denser crust.

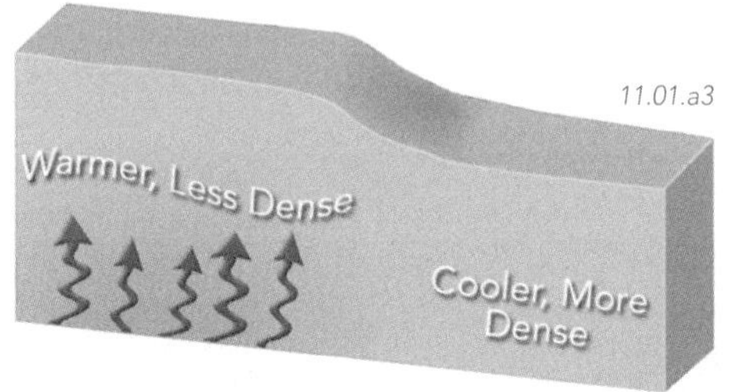

As rocks are heated they expand, taking up more volume and becoming less dense, both of which *increase elevation*.

B What Causes Variations in Crustal Thickness?

Variations in crustal thickness between regions reflect differences in their geologic histories. Such differences include whether the crust is continental or oceanic, and whether it has been deformed, eroded, or buried.

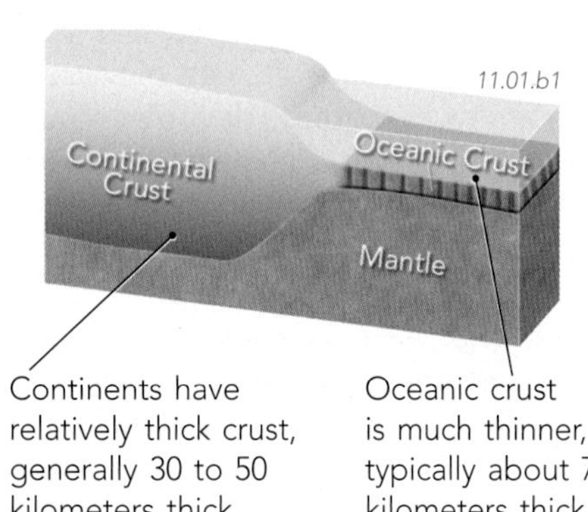

Continents have relatively thick crust, generally 30 to 50 kilometers thick.

Oceanic crust is much thinner, typically about 7 kilometers thick.

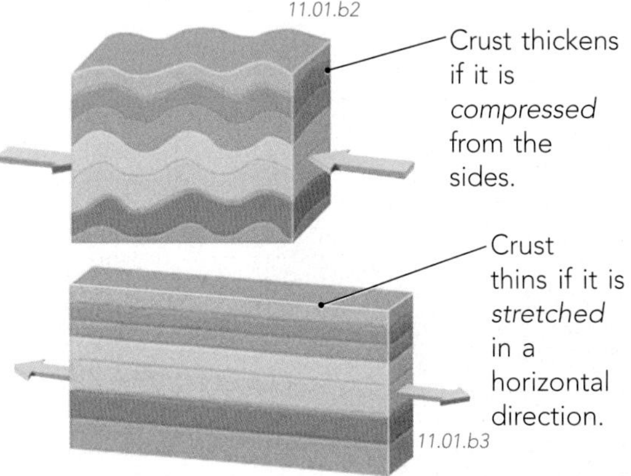

Crust thickens if it is *compressed* from the sides.

Crust thins if it is *stretched* in a horizontal direction.

Crust that *loses material*, such as by erosion, will become thinner.

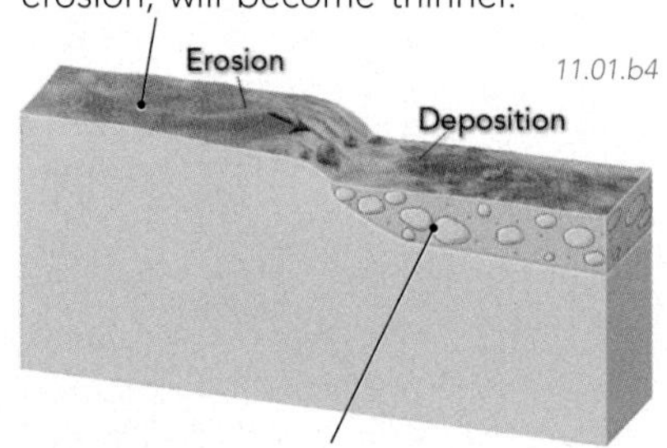

Crust that *gains material*, such as by deposition, will become thicker.

C What Is the Influence of the Thickness of the Lithosphere?

The lithosphere is, on average, about 100 kilometers thick, but varies in thickness from nearly zero at mid-ocean ridges to more than 150 kilometers beneath some ancient continental interiors. These variations greatly influence elevation because the mantle part of the lithosphere is denser than the asthenosphere.

Mid-Ocean Ridge

11.01.c1

Lithosphere

Continental Crust

Lithosphere

Asthenosphere

1. A region with *thin* lithosphere, such as a mid-ocean ridge, will be higher than an adjacent region with *thicker* lithosphere, even if they have the same crustal type and thickness.

2. As the new oceanic plate moves away from the ridge, the asthenosphere cools enough to become lithosphere. The plate, therefore, thickens, becomes more dense, and subsides as it cools.

3. The lithosphere is generally thicker in the central, ancient parts of continents, far away from modern plate boundaries.

4. The continental lithosphere can be thinned near plate boundaries by heating and other plate activity. The affected region can rise in elevation.

D How Is Regional Elevation Increased?

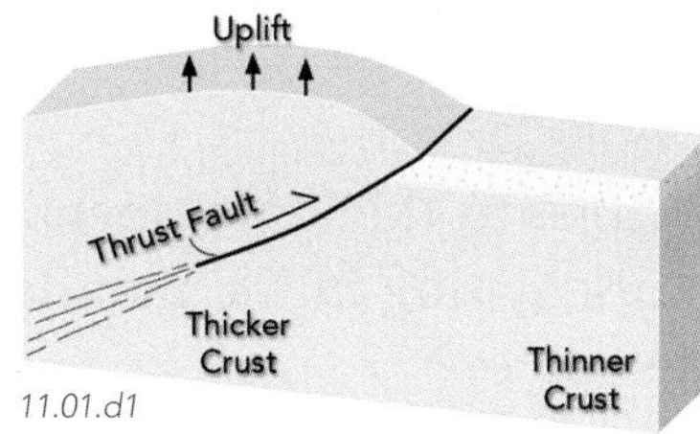

Crust that is *shortened*, such as by thrust faults, also thickens. This thickening causes the region to be uplifted.

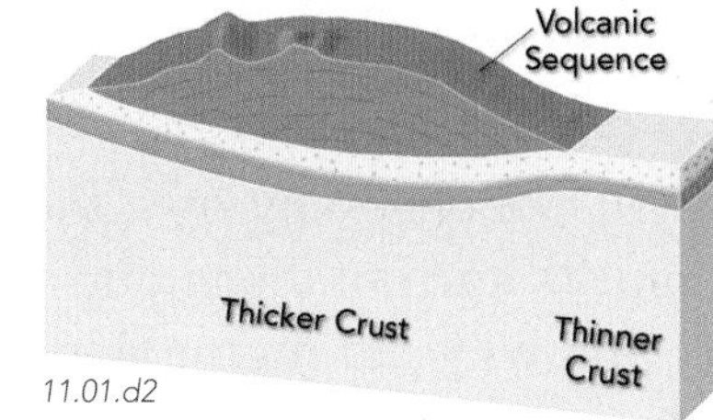

Crust can be thickened by adding material to the *surface*, such as by constructing huge volcanic fields.

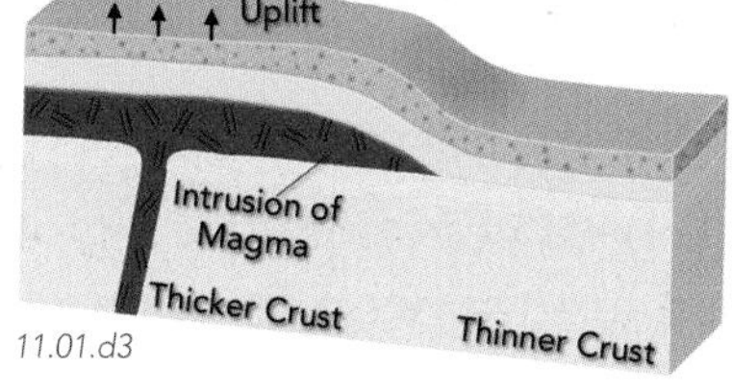

Magma can be added to the crust *at depth*, and this addition of material thickens the crust.

If the crust or mantle beneath a region is *heated*, the rocks expand and the region can increase in elevation.

Several processes may operate together. Magma can add material and heat up the crust. The region can be uplifted even more if the material added has a lower than average density.

E How Is Regional Elevation Decreased?

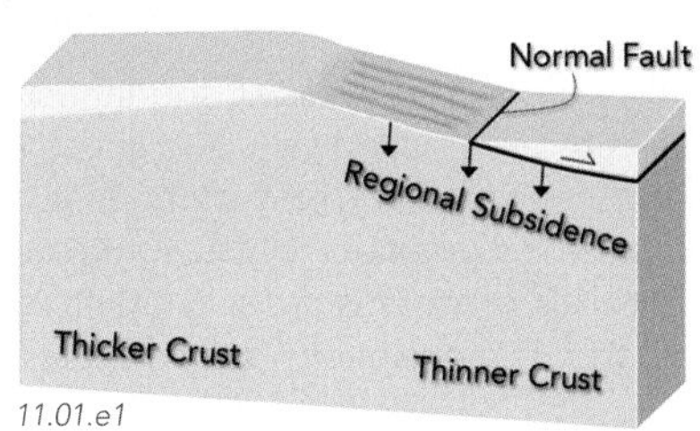

Thinning the crust, such as during normal faulting, decreases its thickness and causes the region to subside.

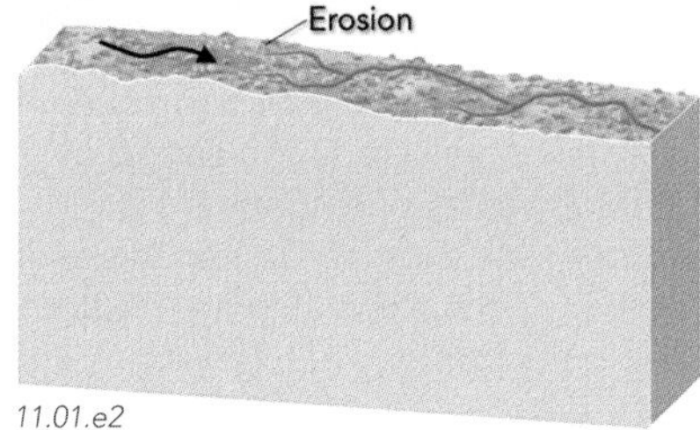

Crustal thickness can be reduced if material is eroded from the top, as is common in many mountain belts.

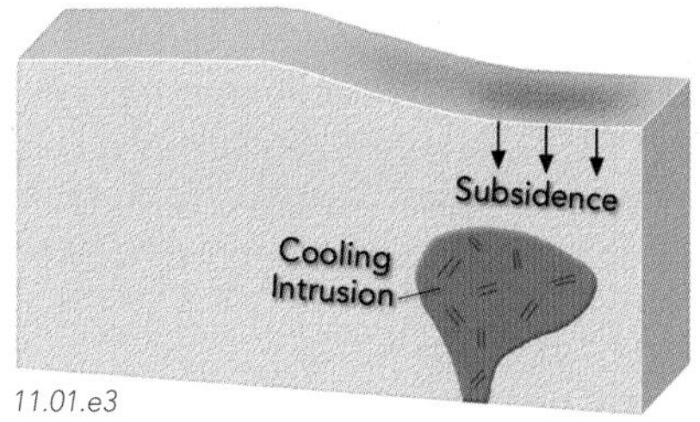

Rocks contract when they cool, so cooling of the crust or mantle causes subsidence.

The Discovery of Isostasy

Isostasy is the principle that regional elevations adjust to the types and thicknesses of rocks at depth. It was discovered because of observations made by George Everest while surveying India around 1850. Surveyors at the time understood that a weight suspended on a line (to level the surveying equipment) was deflected from vertical a very small amount by the gravitational attraction of nearby mountains. When taking this into account, Everest noted an unexplained discrepancy in positions on his survey.

To explain the discrepancy, a mathematician calculated the expected gravitational attraction of the Himalaya. He found that the deflection of the weight from vertical was less than predicted. Astronomer George Airy then used an analogy with common objects, such as floating icebergs, to suggest that higher mountains had thicker roots. According to this model, the lower density crustal material in the roots attracts the suspended weight less than would the denser mantle material that the crustal root has displaced.

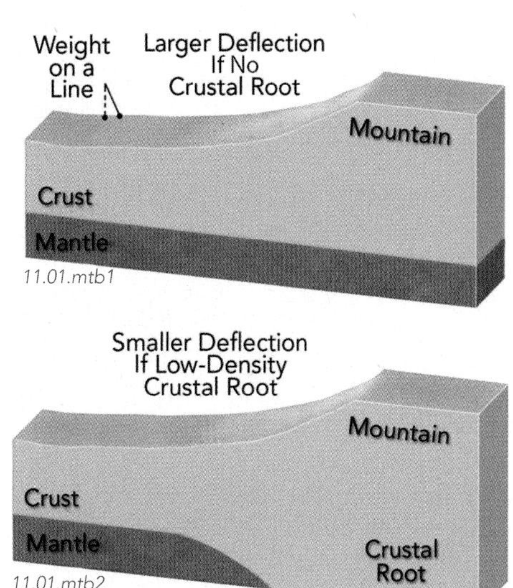

Before You Leave This Page Be Able To

- ✓ Summarize or sketch the factors that control regional elevation.
- ✓ Summarize or sketch what causes variations in crustal thickness.
- ✓ Summarize several ways to increase elevation and to decrease elevation.
- ✓ Explain the observation that led to the discovery of isostasy.

Where Do Mountain Belts and High Regions Form?

MOUNTAIN BELTS AND OTHER HIGH REGIONS generally owe their high elevations to thick continental crust. Less commonly, a region is higher than its surroundings due to processes originating in the mantle. Where are the world's main mountain belts and why did mountains form in these places?

A In Which Tectonic Settings Do Regional Mountain Belts Form?

Regional mountain ranges are hundreds or thousands of kilometers long. They are large enough that they can only be supported by major variations in the thickness of the crust and lithosphere. Most ranges occur near convergent plate boundaries or where there has been large-scale movement of material in the mantle.

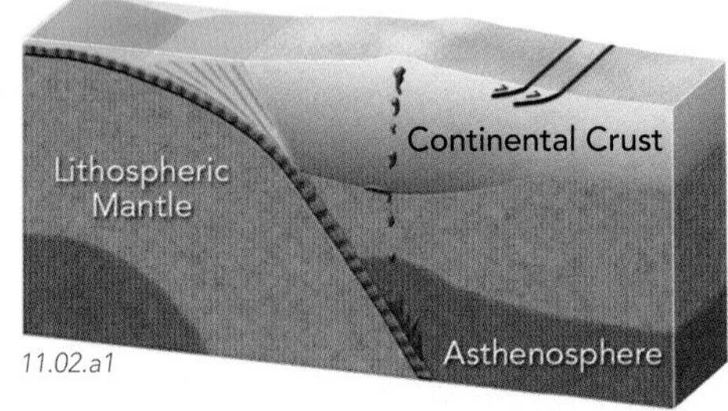

11.02.a1

Subduction Zones—Convergent margins are high in elevation largely because the crust is thickened by magmatic additions from the subduction zone and by crustal shortening. Furthermore, the lithosphere is thinned because it is heated.

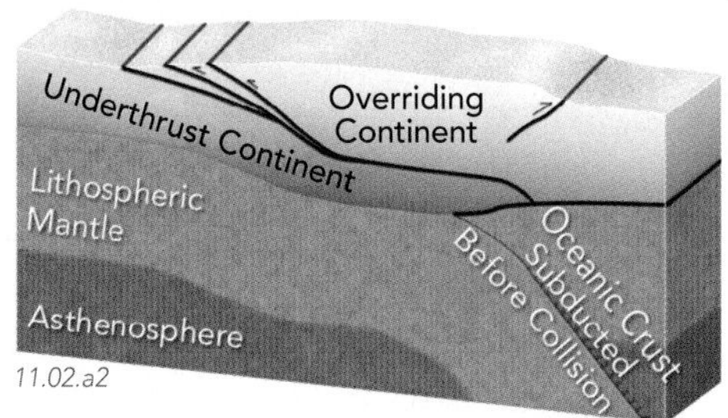

11.02.a2

Continental Collisions—Collision zones can be extremely high due to increasing of the crustal thickness as one continent is shoved beneath another. Crustal thickening also occurs by thrusting, folding, and other forms of deformation.

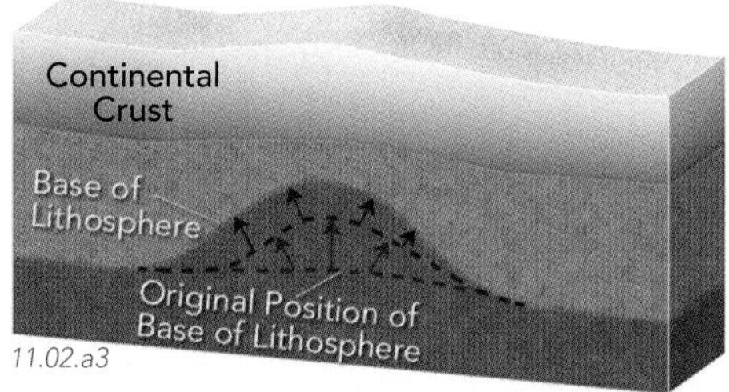

11.02.a3

Mantle Upwellings—Areas of the lithospheric mantle can be replaced by upwellings of less dense asthenosphere. This occurs near hot spots, plate boundaries, and in some other settings. This process is causing uplift in some parts of the western United States.

B What Causes These Regions to Have High Elevation?

Western Canada has been a convergent margin for most of the last 100 million years. Its mountain ranges overlie crust thickened from magmatic additions and from collisions with island arcs and pieces of continents.

The *Alps* mountain range of southern Europe is high because it has thick crust due to collisions between Europe and smaller continental blocks that came from the south.

The *Tibetan Plateau* and the *Himalaya* are extremely high because of thickened crust caused when the Indian continent collided with, and was partly shoved beneath, Asia.

11.02.b1

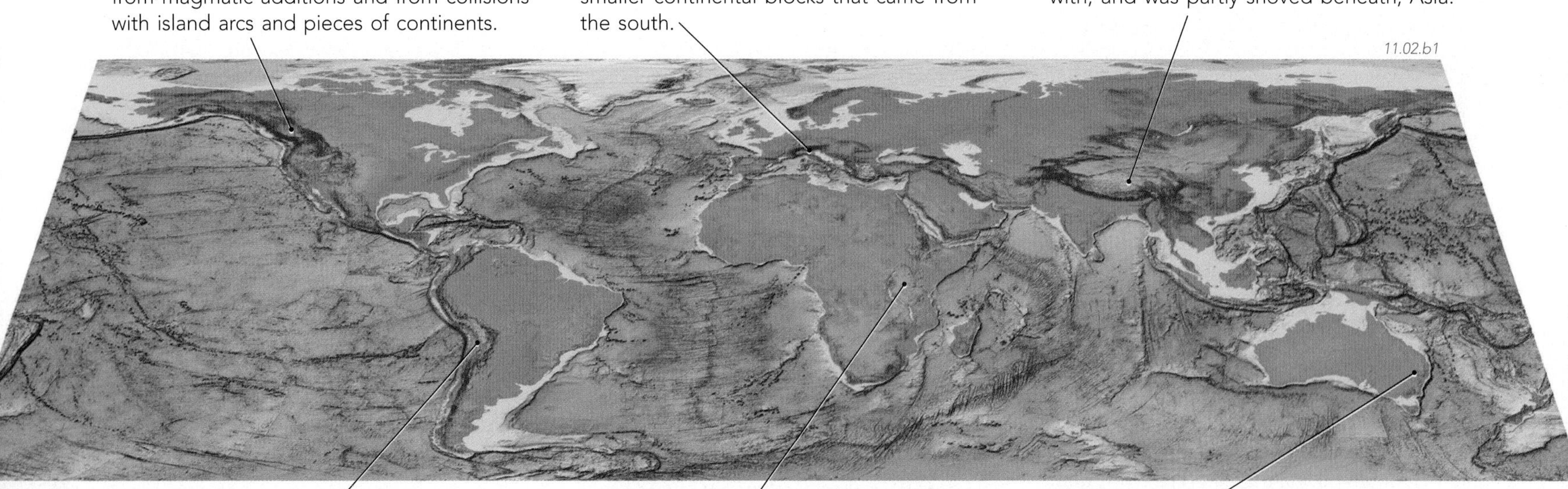

The *Andes* of South America are above a subduction zone. The underlying crust is hot and is thick because of magmatic additions and crustal shortening.

The *East African Rift* is higher than most of Africa because of magmatic heating of the crust, thinning of the lithosphere, and the presence of a hot spot (mantle upwelling).

The *Great Divide range* forms the eastern flank of Australia. There is currently no plate boundary here, and the age of uplift and origin of the range are not resolved.

C What Happens When Mountain Belts Are Eroded?

Mountains, once created, are subjected to weathering and erosion. These processes wear mountains down, but are countered by uplift related to *isostasy*. The uplift is driven by the root of thick crust beneath the belt.

Early Mountain Building

As a mountain belt forms, uplift is commonly faster than erosion, and the mountain becomes higher and more rugged over time. A high mountain belt results from rapid uplift and slow erosion.

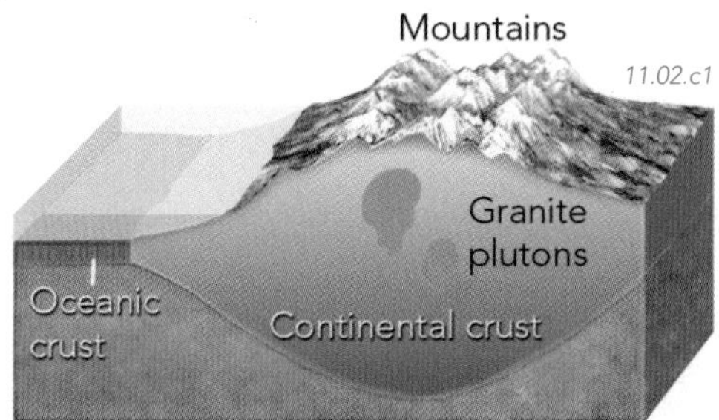

As soon as it starts forming, a mountain will be weathered and eroded, contributing sediment to streams and rivers. The sediment will be transported to adjacent low areas, such as a continental margin or other type of basin.

Erosion and Isostatic Rebound

As material is eroded from the top of a mountain, there is less weight holding down the thick crustal root. The buoyant crust can uplift, a process called *isostatic rebound*.

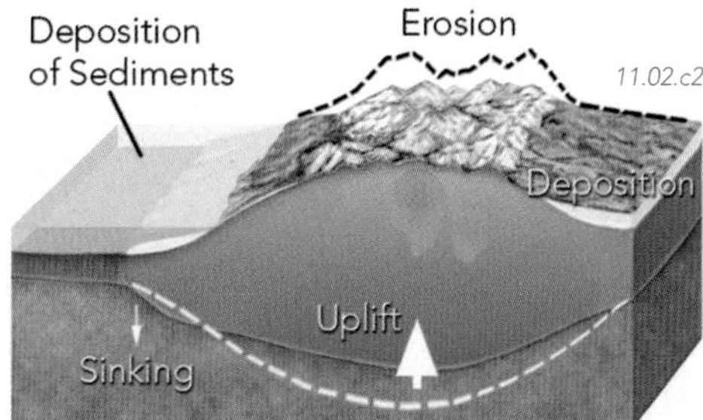

Sediment derived from the mountain is deposited in nearby basins, on both the sea and continental sides. The added weight of the sediment depresses the crust in these regional basins, making room for more sediment.

Late Stages of Evolution

Erosion and isostasy cause rocks deep within the crust to be uplifted and exposed at the surface. As a result, metamorphic and plutonic rocks are exposed in many mountain belts.

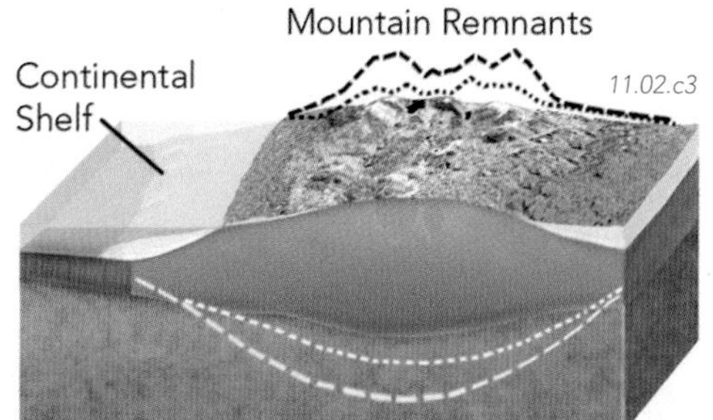

Through erosion and isostasy, the mountain is eroded down and the thick crustal root is gradually reduced in size. Material eroded from the mountain ends up in adjacent basins, increasing the crustal thickness beneath the basins.

D What Controls Regional Elevations in North America?

The topographic profile below illustrates how elevations vary from east to west across the United States. It does not show the full thickness of the crust, only the elevation of the land and the depth of the seafloor.

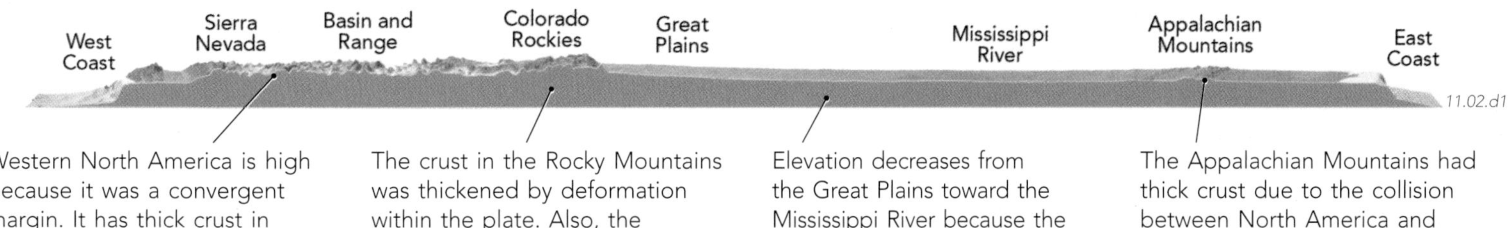

Western North America is high because it was a convergent margin. It has thick crust in some areas and a thin lithosphere in some areas.

The crust in the Rocky Mountains was thickened by deformation within the plate. Also, the lithosphere is being thinned beneath the Rio Grande Rift.

Elevation decreases from the Great Plains toward the Mississippi River because the lithosphere grows cooler and thicker in this direction.

The Appalachian Mountains had thick crust due to the collision between North America and Africa. Much of this thickness has been lost due to erosion.

Rule of Thumb for Elevations

Regional elevations rise from regions with thinner crust to those with thicker crust, but by how much? A rule of thumb is that increasing the thickness of the crust by 6 km will result in an increase in elevation of 1 km (~3,300 ft). Here's an example from Arizona.

Phoenix has an elevation of 300 m (1,000 ft), whereas Flagstaff is more than 2,100 m (7,000 ft). This difference is ~2 km, so the crust beneath Flagstaff should be 12 km thicker than the crust beneath Phoenix ($2 \times 6 = 12$). Geophysical measurements show that the crust beneath Phoenix is ~28 km thick, whereas crust beneath Flagstaff is ~40 km thick. The difference is 12 km.

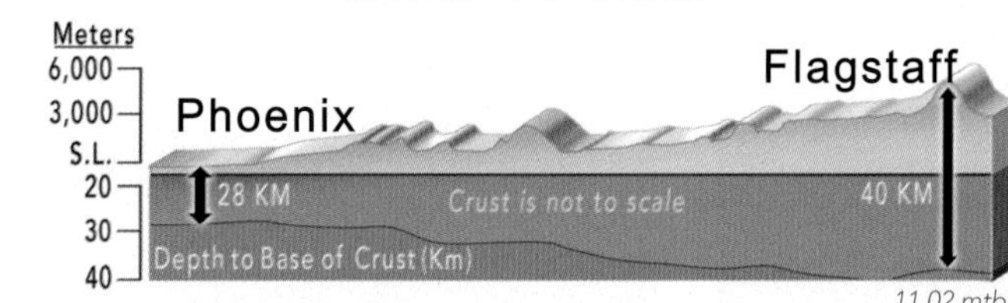

Before You Leave This Page Be Able To

- ✓ Summarize the main tectonic settings of high regions, explaining reasons for the high elevations in each setting.
- ✓ Summarize the explanation for high elevations in the world's main mountain ranges and plateaus.
- ✓ Summarize the differences in elevation across North America.

How Do Local Mountains Form?

THE DISTINCTION BETWEEN LOCAL MOUNTAINS and regional mountain ranges is important. Regional mountain ranges are hundreds to thousands of kilometers long, contain many peaks, and typically involve uplifted, thickened crust. Other mountains are *local* features, too small to be accompanied by regional increases in crustal thickness. Instead, such mountains simply rest upon—and are supported by—the crust.

A How Does Volcanism Form Local Mountains?

A local mountain may be formed by a volcanic eruption that piles lava, ash, and scoria onto the crust. Such mountains vary in size from small scoria (cinder) cones to large shield and composite volcanoes.

11.03.a1

Volcanism creates mountains by piling volcanic materials on a preexisting surface. Some of the smallest volcanic mountains and hills are scoria cones. They are clearly local features. [Flagstaff, Arizona]

11.03.a2

Composite volcanoes, composed of ash, lava, and mudflows, can make lofty and steep mountains. *Mt Fuji* in Japan is a symmetrical composite volcano constructed in the last 11,000 years via eruptions of lava flows, ash, and cinders.

11.03.a3

Prolonged volcanism can build even larger mountains. *Mt Kilimanjaro,* an active volcano in Tanzania, Africa, is over 5,800 meters (19,000 feet) in elevation. It was built in the last 2 million years from eruptions totalling 4,200 cubic kilometers in volume!

B How Do Faults Build Mountains?

Local mountains can also arise through faulting. Thrust faults create mountains by thrusting one fault block up and over another. Normal faults also form local mountains, even though they regionally stretch and thin the crust.

Mountains Formed by Thrust Faulting

Thrust faulting will make a mountain if the overthrust block is uplifted faster than it is eroded, or if it is composed of erosion-resistant rocks such as granite and other crystalline rocks.

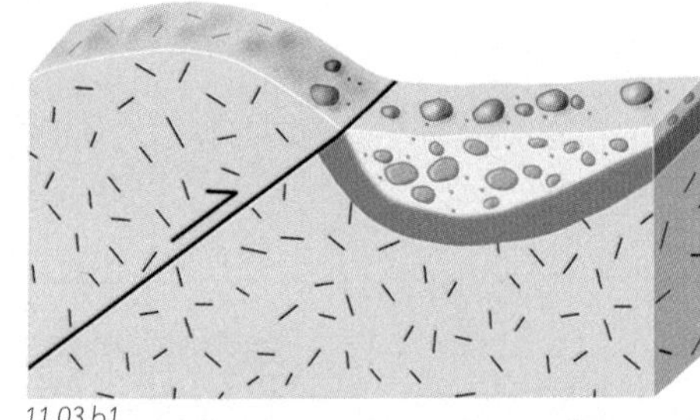

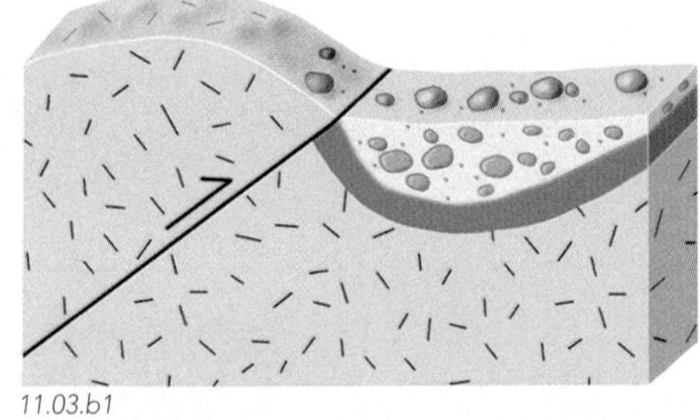

11.03.b1

Denali, the tallest peak in North America, consists of hard granite. It was uplifted along a thrust fault that formed due to localized compression along a nearby major strike-slip fault.

11.03.b3

Mountains Formed by Normal Faulting

During normal faulting, one block slips down, forming a basin. The other block remains high or is moved upward, and can form a local mountain if it is not eroded away.

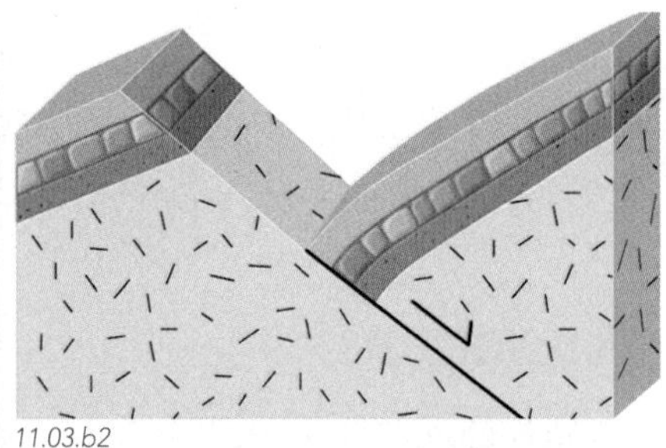

11.03.b2

11.03.b4

Normal faulting along the eastern side of Death Valley, California, forms rugged local mountains. The valley floor is displaced down relative to the mountains, forming a basin (Death Valley) that traps sediment eroded off the ranges. The floor of Death Valley is locally below sea level, but is not connected to the sea.

C How Does Folding Build Mountains?

Another way to make local mountains is by folding. Folding can warp the Earth's surface as well as the underlying rock layers. Also mountains may form from a folded, erosion-resistant layer.

1. *Folding* can form mountains and hills by deforming near-surface rocks, as is happening near Los Angeles, California.

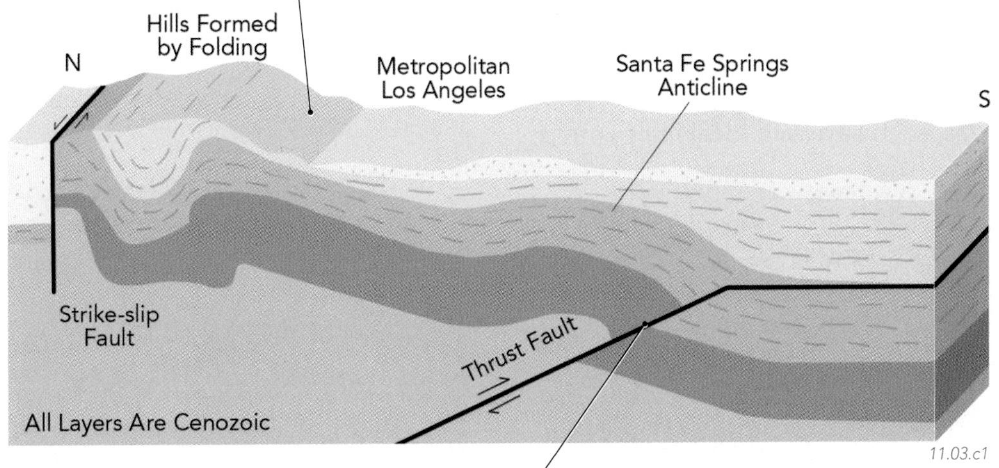

2. Some oil wells near Los Angeles were being crushed and bent beneath an anticline. No one knew why until the large 1994 Northridge earthquake revealed thrust faults in the area. One newly discovered fault, shown above, was breaking and bending oil wells and causing folding of sedimentary rocks beneath the L.A. area.

11.03.c2

△ 3. Some mountains, such as this one in Dinosaur National Park in Utah, owe their origin to folding followed by erosion. In this area, the folding ended more than 50 million years ago. Erosion downcut through the rocks until it encountered these folded layers of hard, light-colored sandstone. Soft rocks underlie the valley.

D How Can Differential Erosion Form a Local Mountain?

11.03.d1

1 Km

1. A *granite pluton* intruded into softer rocks commonly resists erosion and is left higher than its surroundings. A mountain or hill that remains when other rocks have been eroded down is called an *erosional remnant.*

2. A resistant rock layer can protect softer rocks beneath from erosion, forming a local hill or mountain. Such a feature can have a nearly flat top and is called a mesa. The photograph to the right is of a sandstone-capped mesa. [Moab, Utah] ▷

11.03.d2

3. A tilted resistant layer can outlast neighboring softer layers when the landscape is eroded. Along the Waterpocket Fold of Utah (▷), erosion of tilted layers of hard sandstone formed local mountains and linear ridges. The valleys are underlain by softer rock layers, such as shale.

11.03.d3

11.03.d4

△ 4. *Stone Mountain* in Georgia consists of granite that solidified at a depth of 10 kilometers. It was then uncovered by erosion, which removed the overlying rocks.

11.03.d5

△ 5. *Uluru*, also called *Ayers Rock*, is an erosional remnant of tilted sedimentary rock. It rises above an ancient and nearly flat landscape in central Australia.

Before You Leave This Page Be Able To

- ✓ Summarize how mountains are formed by volcanism.
- ✓ Describe or sketch how thrust faulting, normal faulting, and folding can each build mountains.
- ✓ Describe some ways that erosion can result in a mountain, ridge, or mesa.

Where Do Basins Form?

BASINS ARE LOW RELATIVE TO THEIR SURROUNDINGS and commonly trap sediment and water. They form in many tectonic settings, both on land and beneath the oceans, and can accumulate different kinds of sediment, depending on their geologic environments.

A In What Tectonic Settings Do Basins Form?

Basins form on both oceanic and continental plates and along plate margins. Some basins are as large as an ocean, while others are smaller, depressed areas, such as near a local fault or fold.

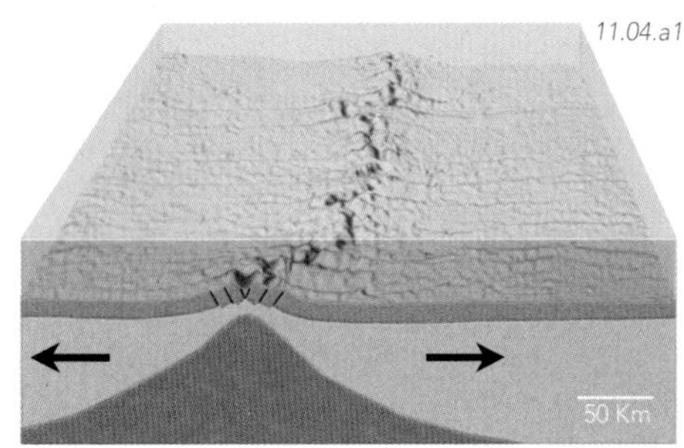

Oceans are the largest basins. They are initiated by rifting and may widen with time. Deep ocean sediments contain microscopic shells as well as fine sediment derived from continents and islands. This view shows a typical mid-ocean ridge.

A *passive margin* is the edge of a continent but is not a plate boundary. It generally is underlain by thin, previously rifted crust. It receives sediment from the continent and provides shallow-water environments, such as offshore of North Carolina.

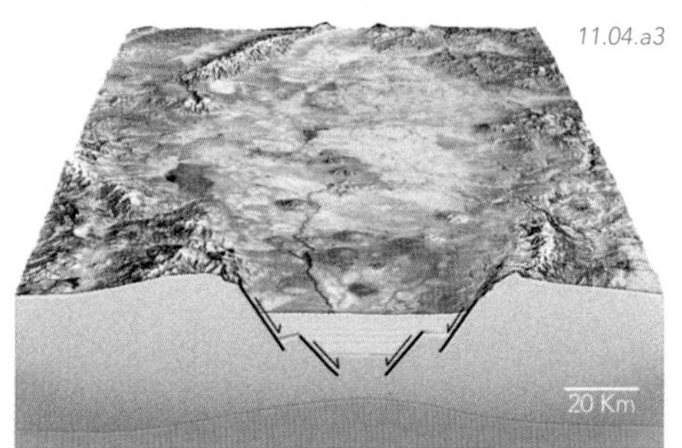

Continental rifts can accumulate coarse continental sediment and evaporite deposits. If the rifting progresses to seafloor spreading, a continental rift evolves into a passive margin. The rift shown here is similar to the Rio Grande Rift of the Southwest.

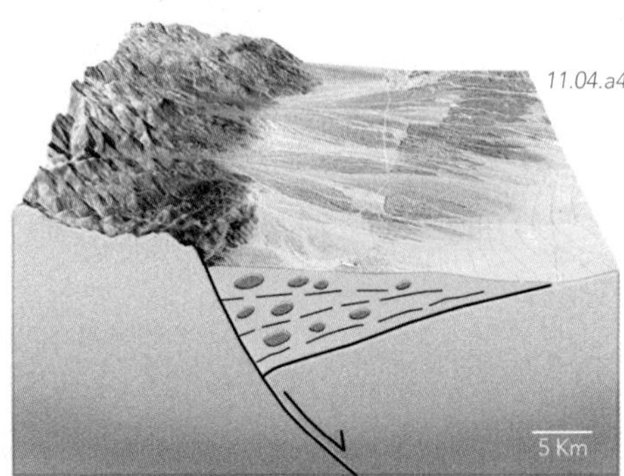

Normal faulting can downdrop a block, forming a basin that can be filled with sediment. Such basins can occur on land, such as in Death Valley, or along rifted margins, such as those flanking the Atlantic Ocean.

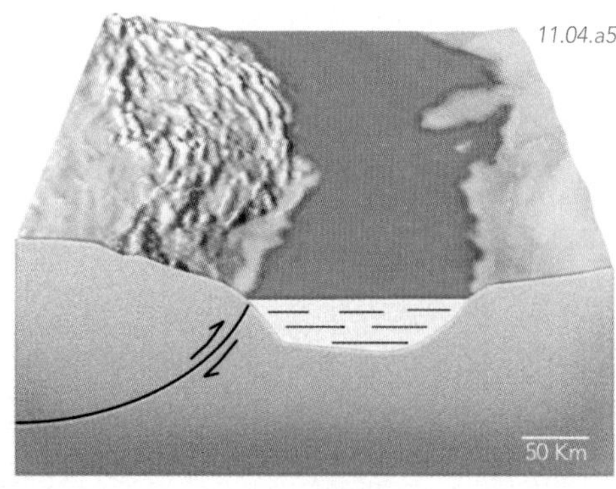

A *foreland basin* occurs when crust (either continental or oceanic) is warped by the weight of thrust sheets. The basin develops as a depression in front of the thrusts. The Persian Gulf is one such basin.

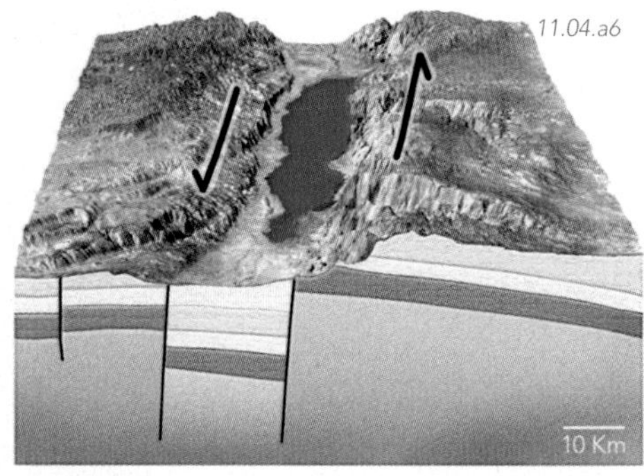

Basins can form along a strike-slip fault if motion along the fault downdrops one block relative to another, such as where the fault has bends or curves. The basin shown is like the Dead Sea of the Middle East.

Regional subsidence, where a broad region drops in elevation, may be due to several causes, including regional cooling or movement of the underlying mantle. Such basins are hundreds of kilometers wide.

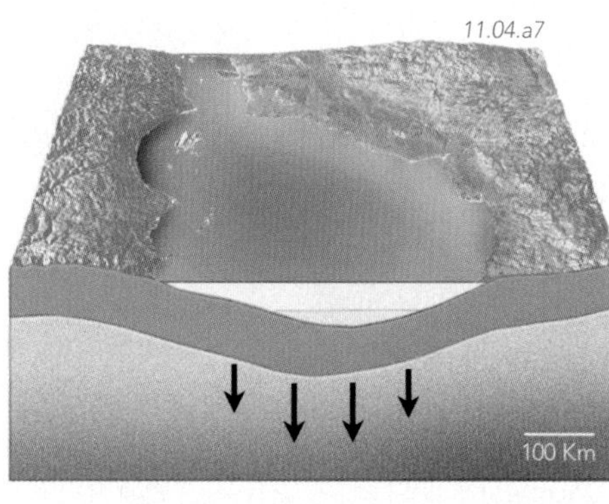

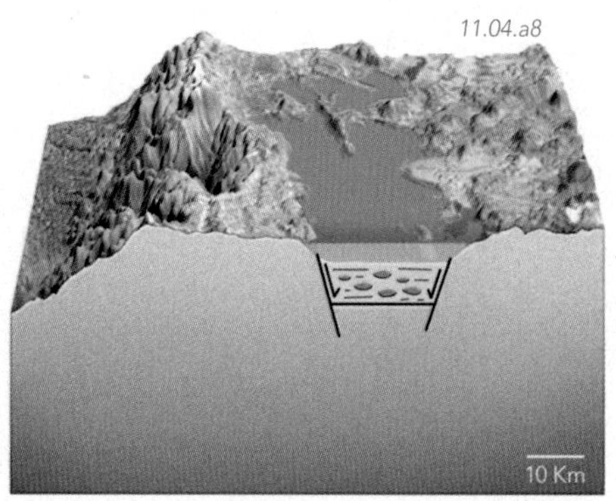

Basins are not restricted to low elevations, but can occur on plateaus or within mountain ranges. Closed basins, some containing lakes, are present in the Andes (shown here), Tibet, and some other high regions.

B What Formed These Basins?

This map shows basins of different ages that locally contain more than 5 kilometers of sedimentary and volcanic units.

A passive margin formed during the Paleozoic by rifting of the western edge of North America. It locally accumulated more than 10 kilometers (6 miles) of sediment.

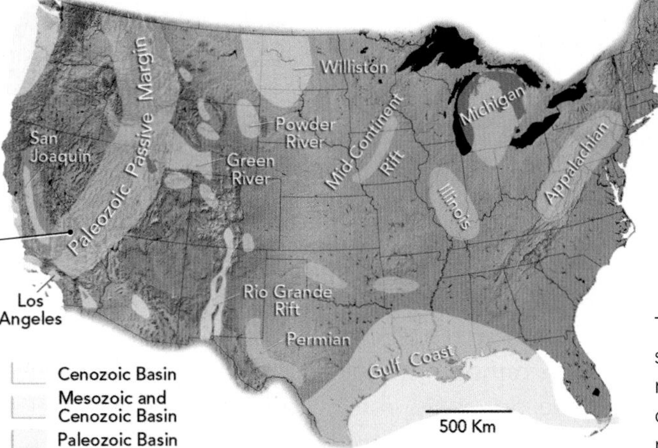

The Michigan and Illinois Basins formed within the continent, probably due to Paleozoic collisional tectonics in the Appalachians and from other deep processes.

The Appalachian Mountains and nearby areas contain thick sedimentary sequences deposited along the Paleozoic continental margin and in other basins before the Appalachian Mountains were formed.

The Gulf Coast contains thick sequences of sedimentary rocks, mostly related to Mesozoic rifting followed by subsidence of the edge of the continent. It is currently a passive margin.

C How Do We Determine the Age of a Basin?

Geologists use a variety of techniques to determine when a basin formed. They try to identify layers in the basin whose ages are known, perform isotopic dating of volcanic rocks, or find key fossils. Determining the age, thickness, and character of sediments tells us when and how fast the basin formed.

3. A unit *younger* than a basin may lie flat and may overlap the edge of the basin and its faults. It shows that the basin had stopped forming by the time the unit was deposited.

2. Units deposited *during* formation of a basin may be very thick and may contain coarse sediments that record steep slopes along the flanks of the basin.

1. Units *older* than a basin may have the same thickness across the area because the basin did not yet exist. These older units were then tilted and faulted when the basin formed.

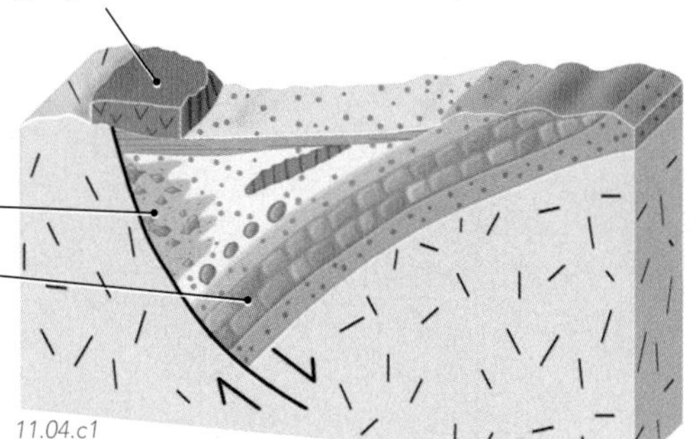

▷ 4. We can calculate the rate of deposition for each unit by dividing the thickness of the unit by the time during which the unit was deposited. This plot, for units in the left side of the basin, shows that the basin formed after 15 million years.

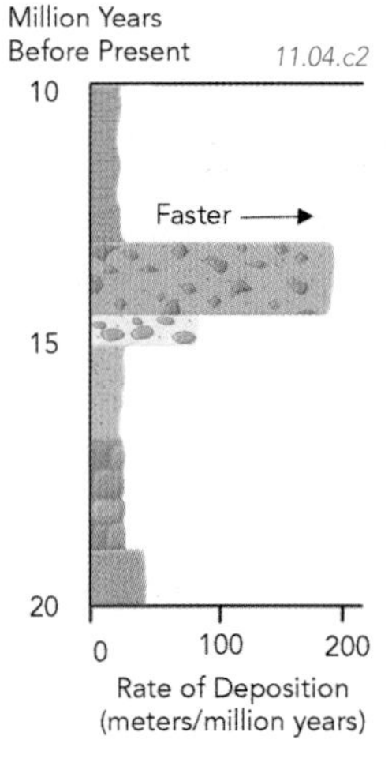

The Michigan Basin

Beneath the hills of Michigan is a deep basin, which contains a fairly complete column of sedimentary rocks deposited during the early and middle parts of the Paleozoic Era. On the geologic map shown here, rock layers form a bull's-eye pattern around the roughly circular basin, with the youngest layers (blues and greens) occurring in the center of the basin. A geologic cross section across the basin (below) shows how the layers are thicker in the center of the basin. This indicates that the basin was subsiding during deposition of the sediments. The origin of the basin is somewhat enigmatic and probably involves several causes. The basin probably formed during an episode of continental rifting, but may also have subsided partly because of flow, thinning, and cooling of the hot lower crust.

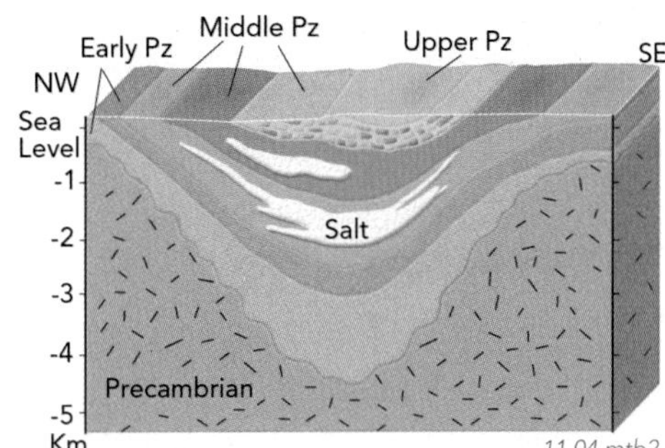

Before You Leave This Page Be Able To

- ✓ Describe the different ways that a basin can form.
- ✓ List some basins in the United States and describe what caused each to form.
- ✓ Describe how you might determine whether a unit was older, younger, or the same age as a basin.

How Do Mountains and Basins Form at Convergent Continental Margins?

SUBDUCTION ZONES BENEATH CONTINENTS are accompanied by various processes that form mountains and basins. Magmatic additions and crustal compression cause thickening of the crust and the formation of a central mountain belt. Basins can form in front of, within, and behind the mountain belt.

A What Processes Accompany Ocean-Continent Convergence?

Along *ocean-continent* convergent boundaries, an oceanic plate subducts beneath a continental plate. Subduction causes melting in the mantle beneath the continent and also leads to compression and thickening of the continental crust. Such margins are generally dominated by a regional mountain belt.

1. As the oceanic plate approaches the convergent margin, it flexes and bends downward into the inclined subduction zone. An *oceanic trench* forms as a result, and acts as a deep oceanic basin that traps sediment eroded from the adjacent mountain belt. The area between the trench and the mountain front is close to or below sea level because it is underlain by a combination of oceanic crust and thin continental crust.

2. Convergence of the two plates causes *horizontal compression* within the continent. This results in thrust faults and other structures that thicken the crust and cause further uplift of the mountain belt. A *fold and thrust belt* can form behind the continental arc as the mountain belt is thrust over the interior of the continent.

3. The weight of the thrust sheets causes the continent to flex downward, forming a basin in front of the thrust belt. This basin is called a *foreland basin* because it occurs in front of the mountain belt. It receives sediment from the mountain belt and from other parts of the continent.

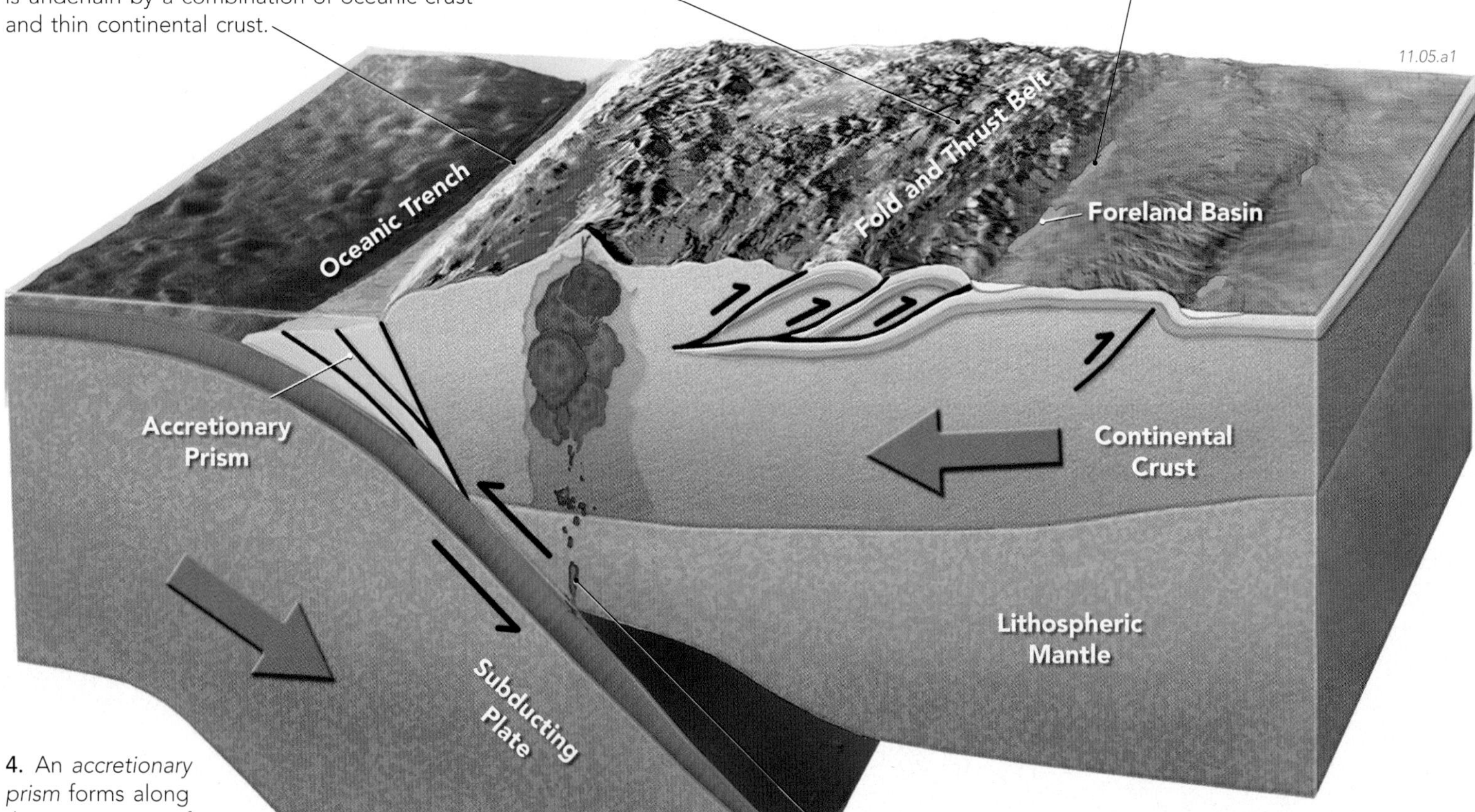

4. An *accretionary prism* forms along the upper parts of the subduction zone as sediment is scraped off the downgoing slab. It is a structurally complex zone of faults, folds, and rocks under various metamorphic conditions. As more material is stuffed under the prism, the prism thickens and is uplifted, but generally remains below sea level.

5. Magma generated along the subduction zone rises into the crust. It thickens the crust by erupting as volcanic rock on the surface and by solidifying at depth. The highest parts of most subduction-related mountain belts are near the areas with the greatest volcanic activity.

B What Features Accompany Continental Collisions?

Continental collisions involve the convergence of two tectonic plates that each carry continental crust. A continent is too buoyant to be subducted deeply, so one continent ends being shoved beneath the edge of the other continent. The collision transmits large stresses to the plates on either side, forming thrust faults and thickened crust.

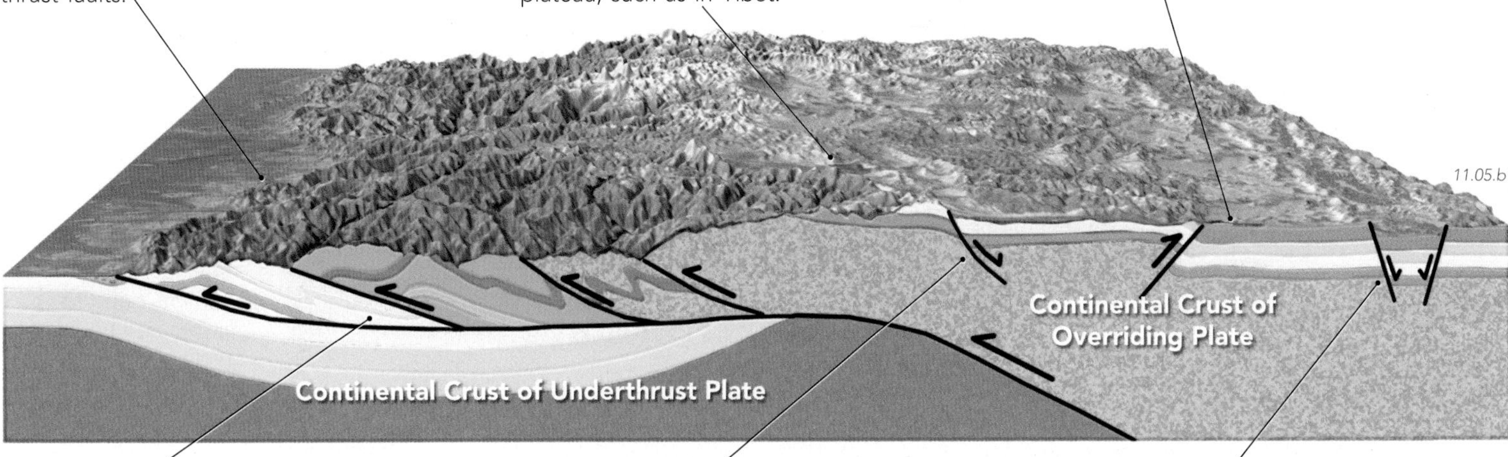

An Ancient Basin in New York

New York State contains a well-known, thrust-related basin of middle Paleozoic (Devonian) age. The basin, called the Catskill Delta, is related to a collision between eastern North America and a continental fragment that moved westward. During the collision, the stresses caused thrust faulting within the Appalachian Mountains. As the thrust sheets were pushed westward toward the interior of North America, their weight warped the crust, forming a foreland basin.

The cross section below plots the reconstructed thicknesses and rock types of the sedimentary layers in this basin. The sedimentary layers are thicker and coarser to the east because this area subsided more than areas to the west. In the oldest layers, coarse sediments are restricted to areas near the fault, indicating that faulting was active at this time. Later, coarse sediments from the mountains were deposited farther to the west, as faulting slowed and the basin was filled. Much later, the sedimentary basin was uplifted, tilted, and eroded. Its rocks form much of the scenery of central and western New York.

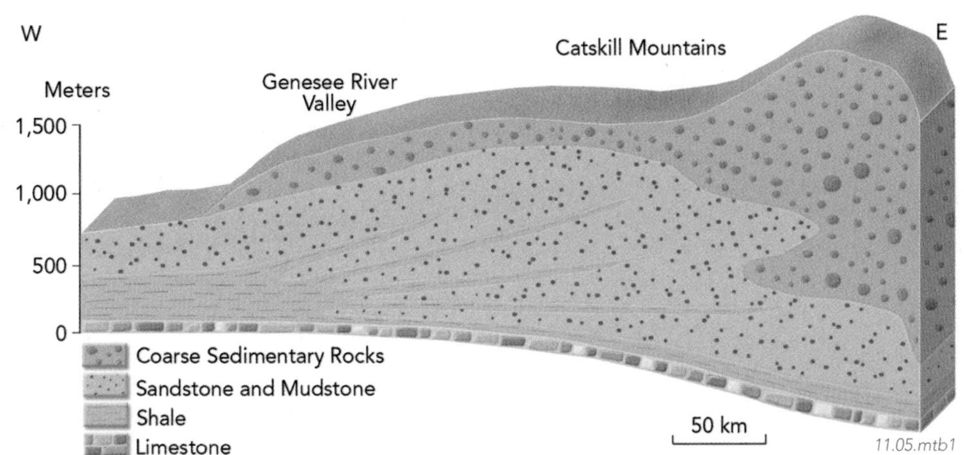

Before You Leave This Page Be Able To

- ✓ Summarize how mountains and basins form in an ocean-continent convergent margin.
- ✓ Summarize how mountains and basins form in a continental collision.
- ✓ Summarize how a basin formed in New York during the Paleozoic.

How Does Continental Extension Occur?

DURING CONTINENTAL EXTENSION, continental crust is stretched horizontally and thinned, typically causing the *region* to subside. Continental extension also breaks up the crust into faulted blocks, forming *local* mountain ranges and sedimentary basins.

A What Happens When Extension Accompanies Subduction?

Some regions experience crustal extension in spite of being near a convergent boundary. In these cases, the region may be fairly low in elevation, except for the large volcanoes.

1. Extension can occur in front of the arc, causing the crust to thin and the region to be below sea level.

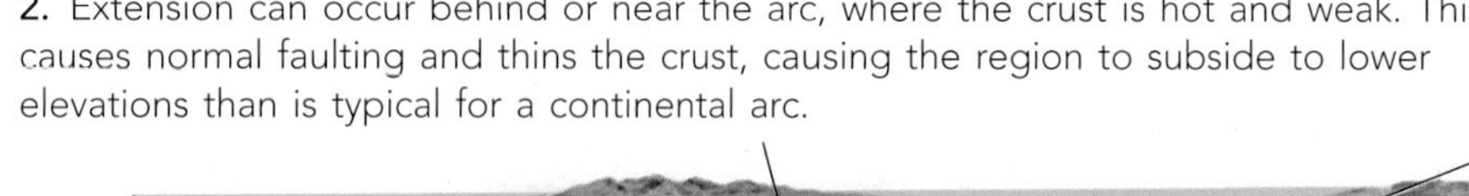

2. Extension can occur behind or near the arc, where the crust is hot and weak. This causes normal faulting and thins the crust, causing the region to subside to lower elevations than is typical for a continental arc.

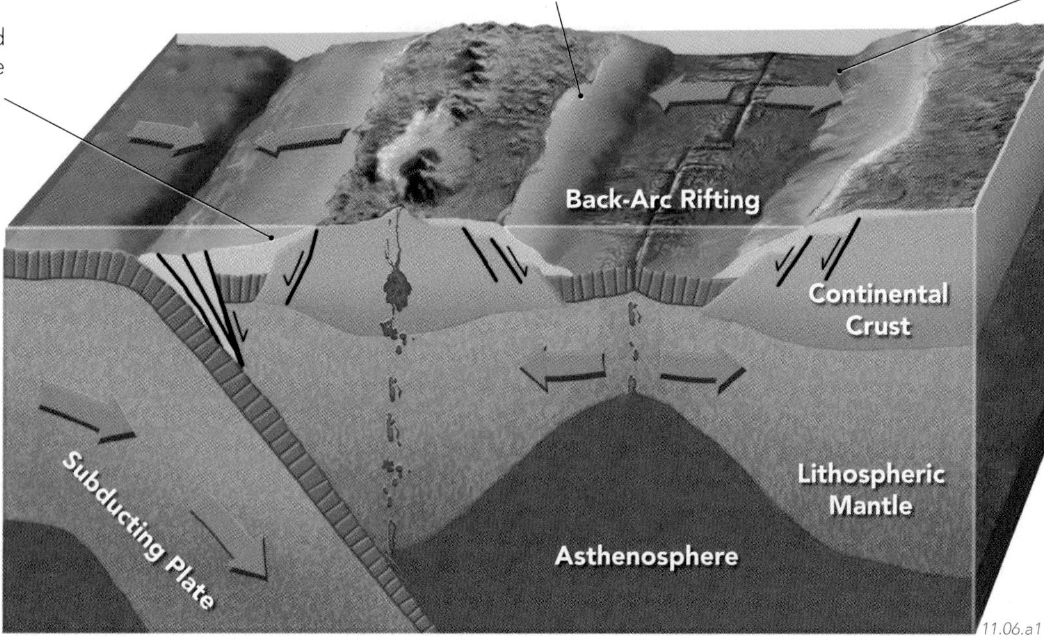

3. With enough extension, the arc can rift apart, forming a new ocean basin behind the arc. This *back-arc basin* will contain land-derived sediment along its margins and normal deep-ocean sediment in its center. Upward flow of underlying mantle continues to bring heat and material to the region, allowing the extension to continue.

B What Determines If the Overridding Plate Is Shortened or Extended?

A number of factors influence whether the plate above a subduction zone experiences compression and shortening, or extension. We present one of the most important factors here.

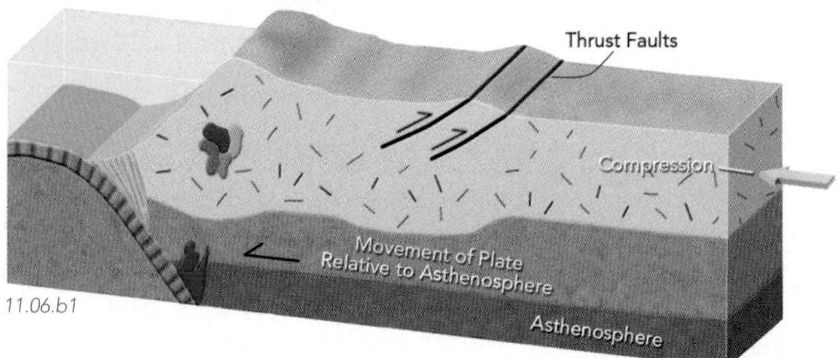

Compression and *horizontal shortening* are common in subduction zones where the continental plate moves toward the subduction zone. This movement pushes against the subducted slab, which is difficult to move sideways through the solid mantle. As a result, the continent experiences compression, as is occurring in parts of the Andes of South America.

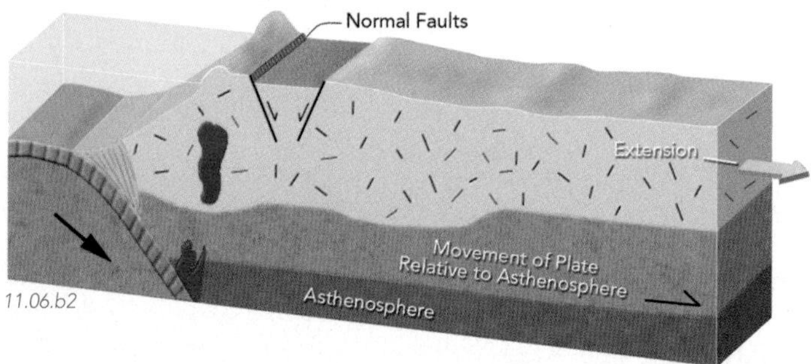

Extension is common when the overridding plate is not moving toward the slab *relative to the mantle*, or is even moving away. The slab tends to pull back by itself (trench rollback), and the continent extends as its edge is pulled by the trench toward the ocean. This is occurring along subduction zones in the western Pacific, such as near Japan and the Philippines.

C How Do Continents Accommodate Crustal Extension?

When continental crust is extended, the upper part responds by breaking into discrete blocks bounded by normal faults. If the fault blocks do not rotate during extension, only a small amount of extension can occur. If the blocks and faults rotate, greater amounts of extension can take place.

Non-Rotating Fault Blocks

In some extended areas, adjacent normal faults dip in opposite directions and cut the crust into wedge-shaped fault blocks.

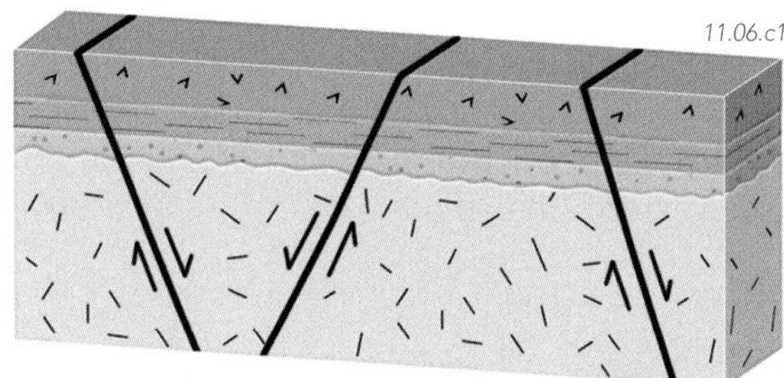

Movement along the faults downdrops some blocks, forming sedimentary basins.

The upthrown blocks become mountains. Their erosion contributes sediment to the basins.

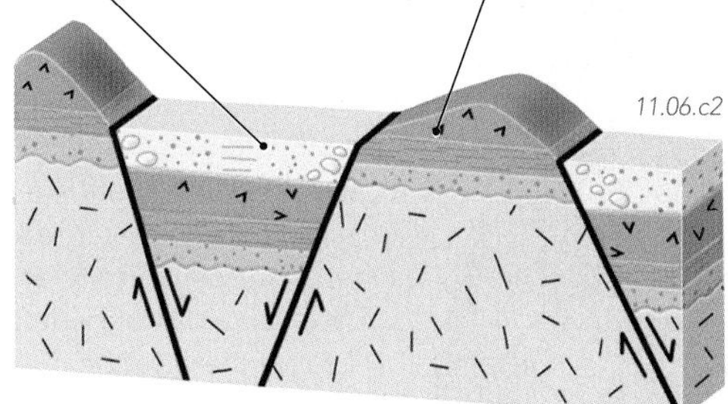

Over time, the basins fill with sediment unless rivers carry most of it away.

The mountains are gradually eroded down, and basin sediments may overlap the edges of the range.

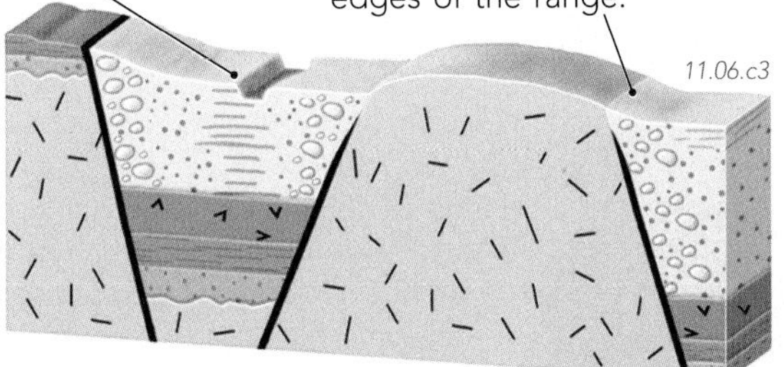

Rotating Fault Blocks

In some extended areas, adjacent normal faults all dip in the same direction and cut the crust into book-shaped fault blocks.

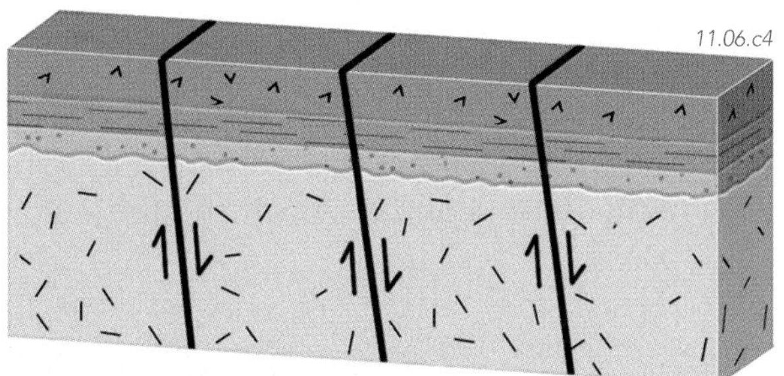

During fault movement, the blocks and faults both rotate, like books sliding on a shelf.

The corner of a block that is rotated down becomes a basin.

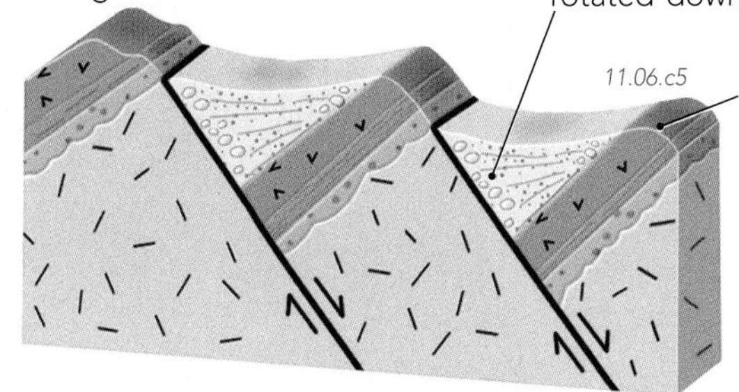

The corner that is rotated up becomes a mountain or ridge.

As faulting and extension continue, units are tilted to steep dips.

Faults are rotated to gentle dips and can have large displacements, allowing great amounts of crustal extension.

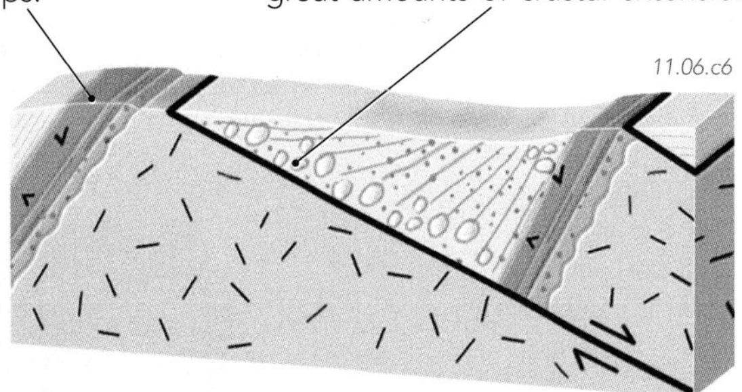

Crustal extension formed Death Valley, California, by a combination of normal faulting and tilting of a downdropped fault block relative to the mountain ranges.

Small normal faults offset layers in sandstone and siltstone. The normal faults record crustal extension in the western United States during the Late Cenozoic.

Before You Leave This Page Be Able To

- ✓ Summarize where extension can occur in a plate above a subduction zone.
- ✓ Summarize one factor that favors shortening versus extension in a plate above a subduction zone.
- ✓ Describe or sketch the formation of non-rotating and rotating fault blocks.

What Are the Characteristics and History of Continental Hot Spots?

A HOT SPOT WITHIN A CONTINENTAL PLATE is marked by high elevations, abundant volcanism, and continental rifting. Hot spots can facilitate complete rifting and separation of a continent into two pieces and can help determine where the split occurs. Several continental hot spots are active today.

A What Features Are Typical of Continental Hot Spots?

Hot spots are volcanic areas that are interpreted to be above rising mantle plumes. Continental hot spots are associated with several characteristics, including high elevations, volcanism, and the presence of rifts. Two examples are the Afar region of East Africa and the Yellowstone region of the western United States.

Afar Region, East Africa

Continental hot spots have high elevations largely because of heating and thinning of the lithosphere by a rising plume of hot mantle. A hot spot is interpreted to be beneath the Afar region of eastern Africa.

The East African Rift is *within* the African plate. It may never evolve into a full rift that fragments the continent into two parts and that leads to seafloor spreading.

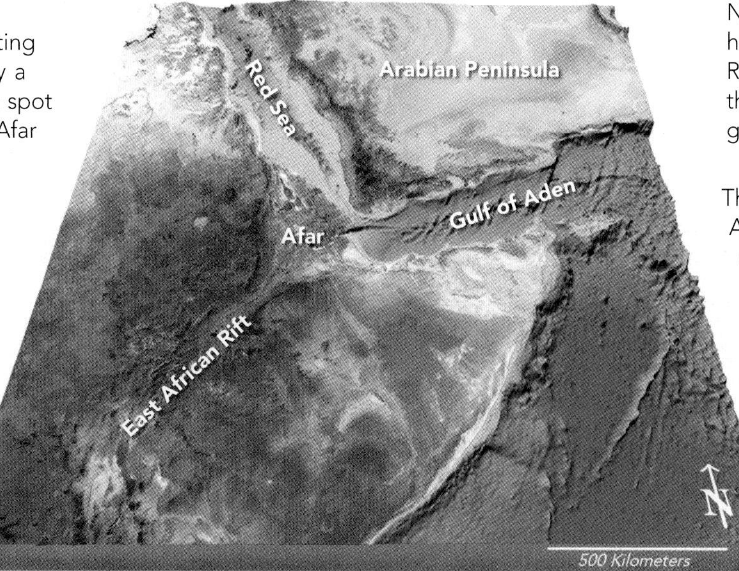

11.07.a1

Near the hot spot, the Arabian Peninsula has pulled away from Africa along the Red Sea and the Gulf of Aden. Beneath these seas, new oceanic crust is being generated by seafloor spreading.

The Red Sea, Gulf of Aden, and East African Rift come together in the Afar region, branching off like three spokes on a wheel. The Afar region is among the most volcanically active area on Earth and has experienced recent volcanic eruptions. Volcanism has been so prolific here that it has created a triangular area of new land in the corner of Africa where the Arabian peninsula pulled away.

Region Around Yellowstone National Park

Yellowstone is located in Wyoming and Idaho and sits in a region that is higher in elevation than surrounding areas.

The Snake River Plain of southern Idaho is underlain by thick sequences of basalt and other volcanic rocks. It is the site of recent eruptions at Craters of the Moon National Monument.

Montana
Yellowstone
Idaho
Snake River Plain
Wyoming
Basin and Range Province
Utah
Nevada
100 Km

11.07.a2

Large calderas, each issuing huge pyroclastic flows, have erupted in Yellowstone in the last 2.1 million years. Eruptions are not occurring today, but heat from hot volcanic rock and underlying magma drives the hot springs, geysers, and thermal pools for which Yellowstone is famous.

Yellowstone is believed to mark the present location of the hot spot, whereas the Snake River Plain records the track of North America as it moved southwest over the hot spot.

The Basin and Range Province of Utah and Nevada is a broad, continental rift adjacent to Yellowstone. It contains normal faults and fault blocks.

B How Do Continental Hot Spots Evolve?

Many continental hot spots have undergone a similar sequence of events. They start with doming and end with the formation of a new continental margin and a new ocean formed by seafloor spreading.

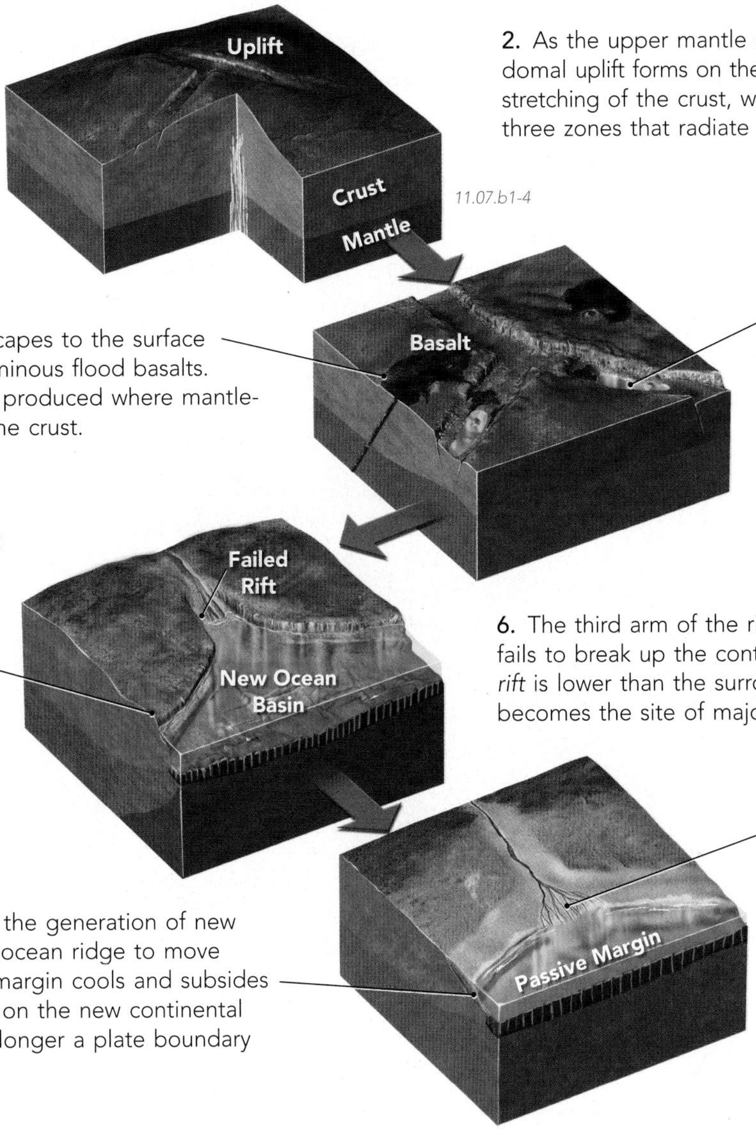

1. Hot spots are generated where a mostly solid mass rises from the lower mantle and encounters the base of the lithosphere. The rising material melts due to decompression and causes melting of nearby lithosphere.

2. As the upper mantle and crust are heated, a broad, domal uplift forms on the surface. Doming is accompanied by stretching of the crust, which commonly begins to rift along three zones that radiate out from the hot spot.

3. Some mantle-derived magma escapes to the surface and erupt as basalts, including voluminous flood basalts. More felsic magma (e.g., granitic) is produced where mantle-derived magma causes melting of the crust.

4. All three arms of the rift are outlined by normal faults, which downdrop long fault blocks. The downdropped blocks form basins that contain lakes and are partially filled by sediment and rift-related volcanic rocks.

5. Complete rifting of the continent occurs along two arms of the rift. This results in a new continental margin and seafloor spreading in the new ocean basin. The edge of the continent is uplifted because the lithosphere is heated and thinned due to the rifting.

6. The third arm of the rift begins to become less active and fails to break up the continent into more pieces. This *failed rift* is lower than the surrounding continent and commonly becomes the site of major rivers.

7. As seafloor spreading continues, the generation of new oceanic lithosphere causes the mid-ocean ridge to move farther out to sea. The continental margin cools and subsides and is covered by marine sediment on the new continental shelf. This continental margin is no longer a plate boundary and is now a *passive margin*.

8. Sediment transported by rivers down the failed rift will form a delta at the bend in the continent. This is currently occurring along the west coast of Equatorial Africa at the large inward bend in the coast (see the figure and text below).

Hot Spots and Continental Outlines

Hot spots have helped define the outlines of the continents by shaping the boundary along which continents separate from one another. The best example of this is the inward bend of the west coast of Africa. This bend occurs at the intersection of three arms of a rift, two of which led to the opening of the South Atlantic Ocean. The third *failed arm* cuts northeast into Africa and is the site of several large rivers. Large eruptions of basalt (flood basalt) occurred along the rifts, and active volcanism near the failed rift may mark the location of a hot spot. This figure shows what the area may have looked like 110 million years ago, after the continents started to rift apart.

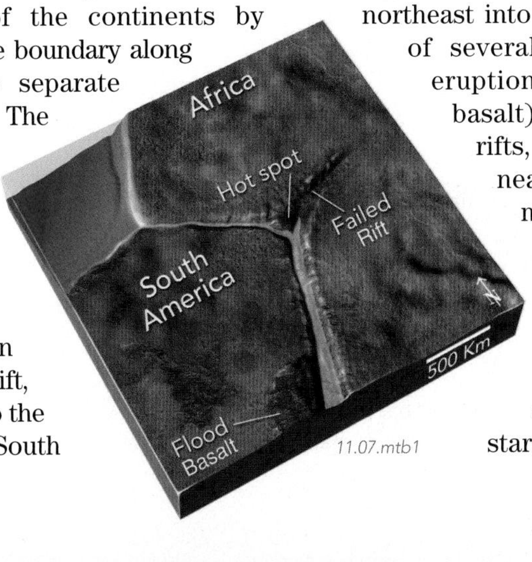

Before You Leave This Page Be Able To

- ✓ Summarize the features that are typical of continental hot spots, providing an example of each type of feature.
- ✓ Summarize or sketch how continental hot spots evolve over time.
- ✓ Describe or sketch how hot spots influence continental outlines, providing an example.

What Features Characterize the Interior of Continents?

THE INTERIORS OF CONTINENTS tend to be tectonically stable, largely because they are far from plate boundaries. Sedimentary rocks formed within continental interiors contain an important record of ancient rivers, lakes, wetlands, deserts, sea-level variations, and changes in global and regional climate.

A What Features Are Common in Continental Interiors?

Many continents display a similar pattern, with a central region of complexly deformed, older rocks surrounded by a relatively thin veneer of younger, nearly flat-lying sedimentary layers.

1. Many continents, including North America, have a central region called a *shield*. A shield is composed of old metamorphic and igneous rocks, commonly of Precambrian age. The crystalline (metamorphic and igneous) rocks exposed in the shield represent the kinds of rocks that underlie much of the continent, and are called the *crystalline basement*.

2. Surrounding the shield is a broad region, called the *continental platform*. It is characterized by nearly horizontal sedimentary rocks that were deposited on top of the basement. The sedimentary layers commonly contain broad basins and uplifts. Erosion across the gently dipping layers on the flanks of these structures exposes higher and lower rocks at the surface from place to place.

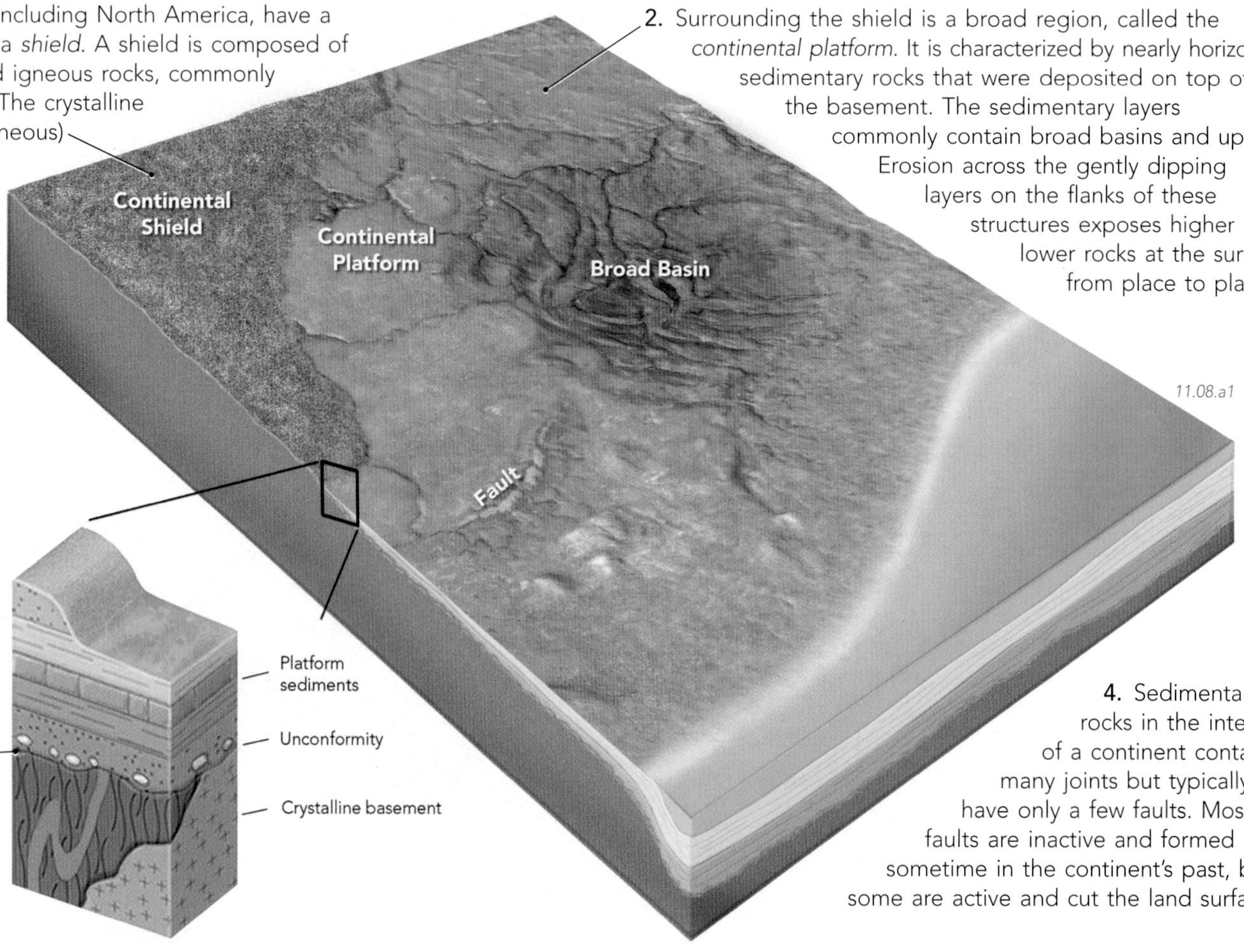

3. The boundary between the flat-lying platform sediments and underlying crystalline basement is a major *unconformity*. It separates rocks with very different ages, structural geometries, and geologic histories.

4. Sedimentary rocks in the interior of a continent contain many joints but typically have only a few faults. Most faults are inactive and formed sometime in the continent's past, but some are active and cut the land surface.

5. This geologic cross section across the state of Ohio is typical of the geology of central North America. In this area, Paleozoic sedimentary layers dip gently off the flanks of a dome, called the Findlay Arch. The section is vertically exaggerated, so true dips are less than shown here, and the thicknesses of the layers are greatly exaggerated.

Cross Section Across Ohio

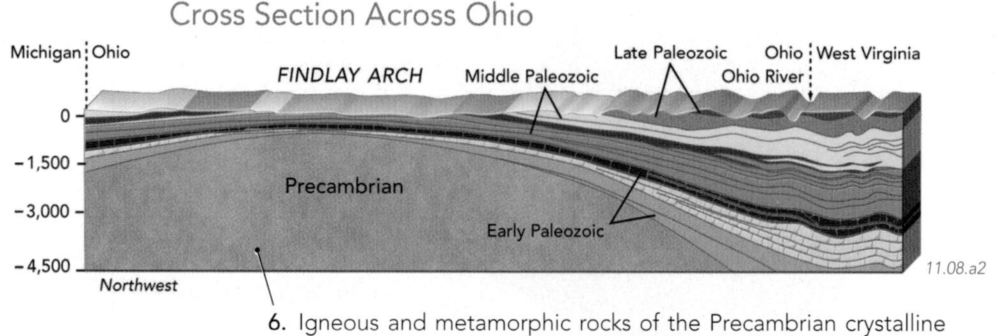

6. Igneous and metamorphic rocks of the Precambrian crystalline basement rest beneath the sedimentary layers, but are not exposed.

B What Regional Effects Influence the Geologic History of Continental Interiors?

The interiors of continents are relatively stable, but can be affected by tectonic events along distant plate boundaries. Continental interiors are more strongly influenced by global environmental fluctuations, such as climate change and the rise and fall of sea level.

1. Tectonic activity along the edges of a continent can cause broad uplifts and basins within the continent, such as in response to the loading of sediments along the margin. Thrust sheets emplaced onto the continent can create basins close to the thrusts and form uplifts farther inland as the continental plate flexes under the load.

2. Stresses can be transmitted from plate boundaries and from distant mountain belts to the interiors of continents. If the stresses are large enough, they may cause ancient faults in the crystalline basement to move, forming faults, folds, domes, and regional basins in the overlying sedimentary rocks.

3. Changes in global climate can cause sea level to rise and fall. Many continents have very low topographic relief and their edges are barely above sea level, so a sea level rise of tens of meters can cause significant flooding. Such changes in sea level dominate the history of many continents.

4. The climate of a continent can change in response to global effects, such as global cooling, or from regional effects, such as the rise of mountains along the coast. A continent also is subjected to different climates as the plate upon which it rides changes latitude (moves north or south).

Fold

Uplift

Fault

Dome

11.08.b1

Many continental interiors contain flat-lying layers of sedimentary rock. These are deposited in various environments, including shallow seas, beaches, rivers, lakes, and sand dunes. [Delaware River, Pennsylvania]

11.08.b2

11.08.b3

In many continental interiors, flat-lying sedimentary layers overlie Precambrian basement rocks along an unconformity. The older rocks were formed at depth, uplifted toward the surface, uncovered by erosion, and overlain by sediment. [Pikes Peak, Colorado]

Mountain Ranges in the Middle of Continents

Some mountain ranges occur in the middle of a continent, not along the edges. The Ural Mountains of central Russia, shown in this satellite image, are one such range, occurring far from any continental edge or plate boundary. Why is there a mountain range here?

In most such cases, such mountain ranges originally formed near a plate boundary along the edge of a continent. Subsequent collision between two continents causes the edge to become the center, in essence trapping the mountains within the center of a new, larger continent. In the case of the Ural Mountains, part of Europe collided with Siberia 200 to 300 million years ago.

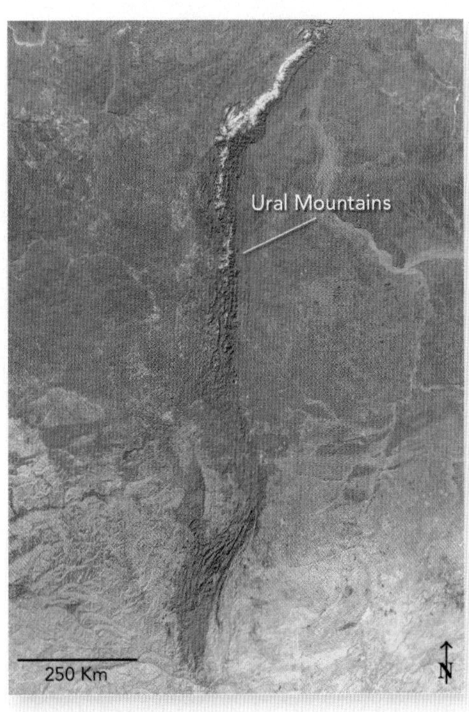

11.08.b4

Before You Leave This Page Be Able To

- ✓ Summarize or sketch what features are common in continental interiors.
- ✓ Describe what types of regional or global effects can influence the geology of a continental interior.
- ✓ Describe one way that a mountain can exist in the middle of a continent.

What Are Tectonic Terranes?

EMBEDDED WITHIN CONTINENTS are pieces of crust that have a different geologic history than adjacent regions. These exotic pieces, called *tectonic terranes*, originate in a variety of tectonic settings. Many are structurally added to the edges of continents during tectonic collisions.

A How Do We Recognize a Terrane?

A terrane is bounded by faults and has rocks, structures, fossils, and other geologic aspects that are unlike those in adjacent regions.

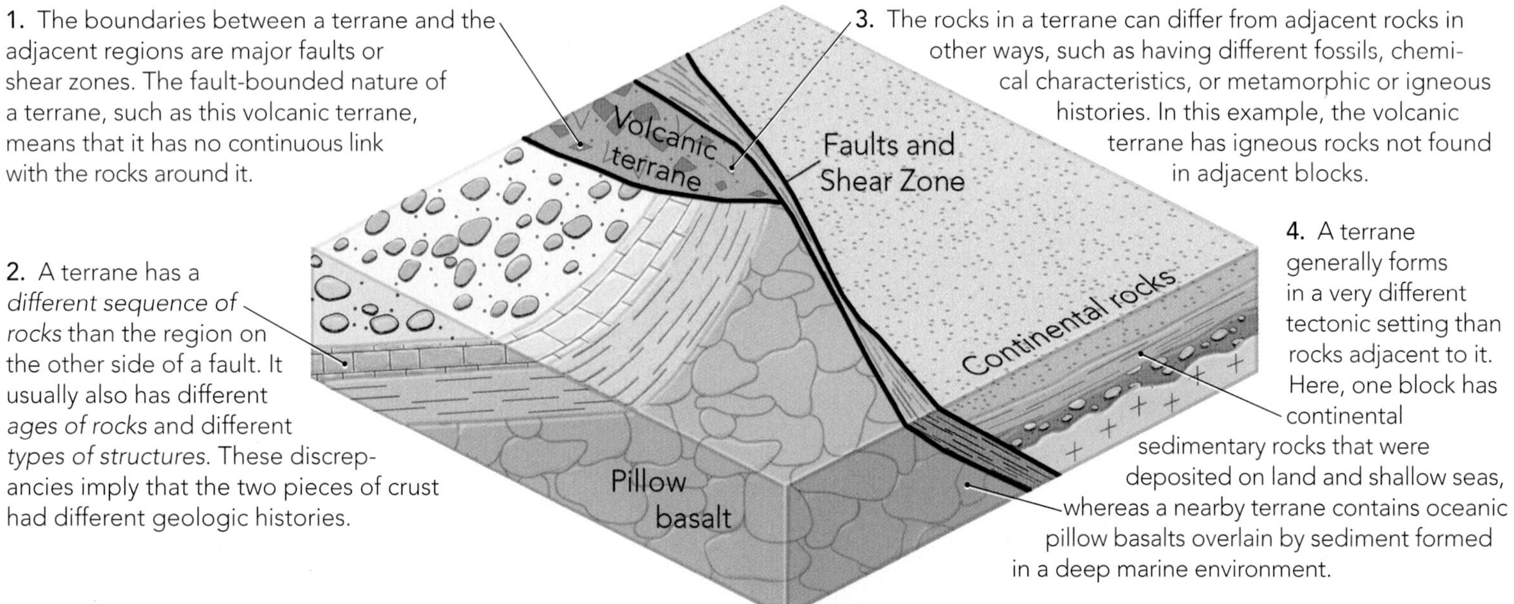

1. The boundaries between a terrane and the adjacent regions are major faults or shear zones. The fault-bounded nature of a terrane, such as this volcanic terrane, means that it has no continuous link with the rocks around it.

2. A terrane has a *different sequence of rocks* than the region on the other side of a fault. It usually also has different *ages of rocks* and different *types of structures*. These discrepancies imply that the two pieces of crust had different geologic histories.

3. The rocks in a terrane can differ from adjacent rocks in other ways, such as having different fossils, chemical characteristics, or metamorphic or igneous histories. In this example, the volcanic terrane has igneous rocks not found in adjacent blocks.

4. A terrane generally forms in a very different tectonic setting than rocks adjacent to it. Here, one block has continental sedimentary rocks that were deposited on land and shallow seas, whereas a nearby terrane contains oceanic pillow basalts overlain by sediment formed in a deep marine environment.

B Where Do Terranes Come From?

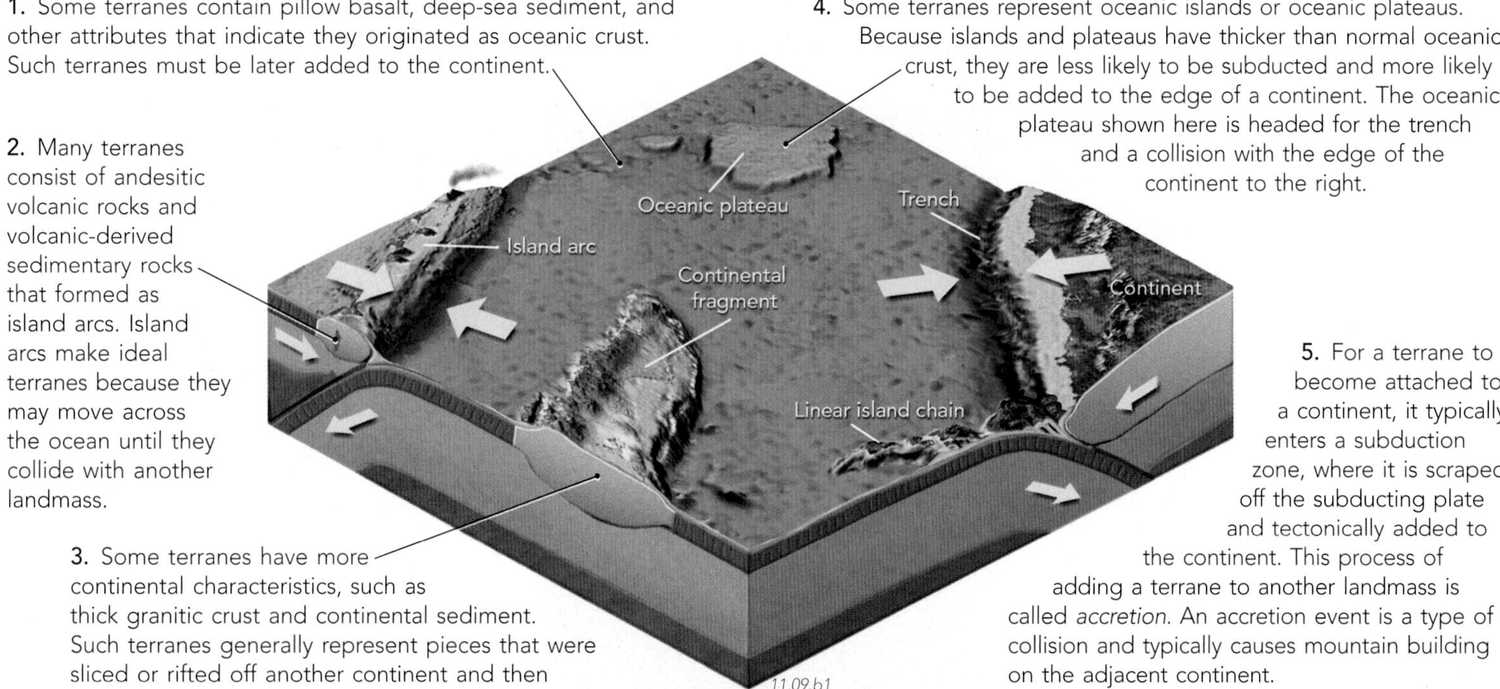

1. Some terranes contain pillow basalt, deep-sea sediment, and other attributes that indicate they originated as oceanic crust. Such terranes must be later added to the continent.

2. Many terranes consist of andesitic volcanic rocks and volcanic-derived sedimentary rocks that formed as island arcs. Island arcs make ideal terranes because they may move across the ocean until they collide with another landmass.

3. Some terranes have more continental characteristics, such as thick granitic crust and continental sediment. Such terranes generally represent pieces that were sliced or rifted off another continent and then tectonically transported to their present positions.

4. Some terranes represent oceanic islands or oceanic plateaus. Because islands and plateaus have thicker than normal oceanic crust, they are less likely to be subducted and more likely to be added to the edge of a continent. The oceanic plateau shown here is headed for the trench and a collision with the edge of the continent to the right.

5. For a terrane to become attached to a continent, it typically enters a subduction zone, where it is scraped off the subducting plate and tectonically added to the continent. This process of adding a terrane to another landmass is called *accretion*. An accretion event is a type of collision and typically causes mountain building on the adjacent continent.

C How Do We Infer the Origin of a Terrane and the Timing of Terrane Accretion?

Geologists investigate the origin of terranes, especially the tectonic setting in which a terrane formed and when the terrane was accreted to another landmass.

1. The origin of a terrane is revealed by the kinds of rocks it contains. For example, pillow basalts generally imply an oceanic origin. Geologists analyze the chemistry of volcanic rocks in order to compare them to modern-day volcanic examples, such as rocks from mid-ocean ridges, oceanic islands, and island arcs.

2. If rocks on opposite sides of a major fault are the same age but are otherwise dissimilar, it indicates that the rocks were not close to each other when they formed. In this example, the pillow basalts on the left are the same age as the metamorphic rocks on the right, so the terranes were probably not in close proximity then.

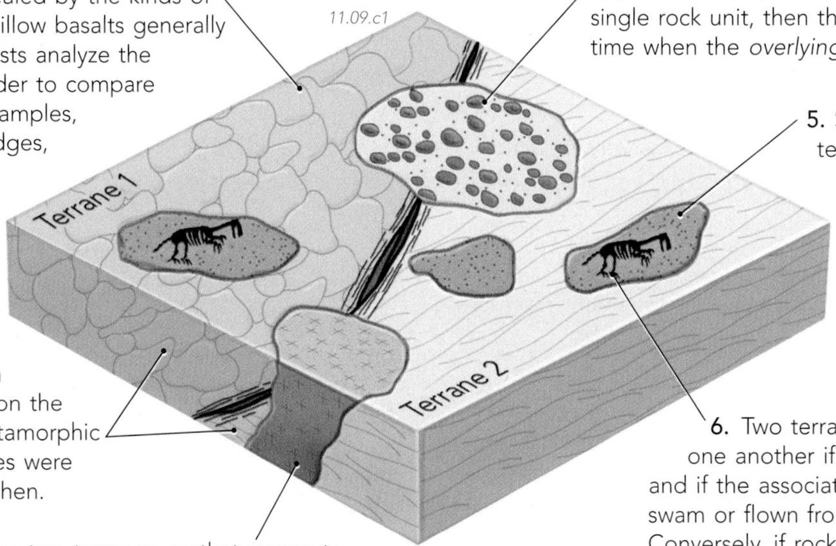

3. An intrusion that invades two terranes, or that crosscuts their boundary, indicates that the terranes were already together at the time when the intrusion invaded the crust.

4. If two terranes and their boundary are overlain by a single rock unit, then they were already together at the time when the *overlying* unit was deposited.

5. Sediment derived from one terrane can be deposited on top of adjacent rocks. The patches of brown sediment shown here contain cobbles of pillow basalt. This indicates that the two terranes were close to each other when the sediment was deposited.

6. Two terranes may have been adjacent to one another if they contain the *same fossils* and if the associated animals could not have swam or flown from one terrane to another. Conversely, if rocks from two terranes are the same age but have *different fossils*, then they probably originated in different settings and locations.

Terrane 1

Terrane 2

11.09.c1

Terranes of Alaska

Alaska, like most of western North America, is a mosaic of terranes. Some cover huge regions, but others are only kilometers long.

Terranes in Alaska, as simplified on the map here, are believed to have formed in various tectonic settings and places. On this map, the light gray area to the east is stable North America. Blue-colored terranes represent parts of North America and its continental margin that were sliced off and transported some distance. Purple and green terranes are exotic, representing slices of oceanic crust and accretionary prisms that were accreted to the continent during the Mesozoic and Cenozoic. The pink and red terranes were island arcs or continental magmatic belts. Yellowish areas depict rocks that *overlapped* the terranes after they were attached to the continent.

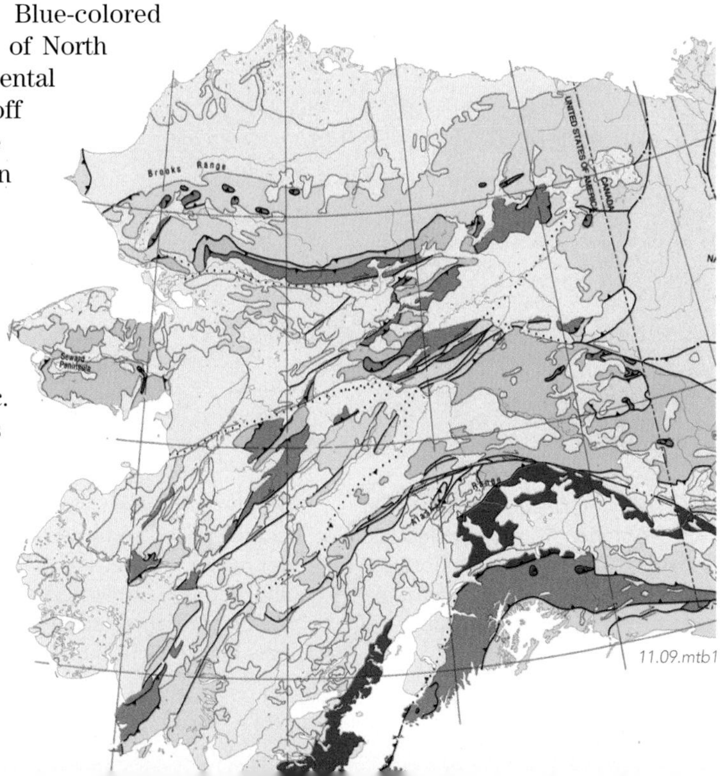

11.09.mtb1

A famous terrane, named *Wrangellia* for the Wrangell Mountains of southern Alaska, is colored red on this map. It is interpreted to have originated during the Late Paleozoic and Early Mesozoic (Triassic) as one or more island arcs that probably started south of the equator. It was then transported northward until it collided with the West Coast, after which pieces were sliced off by strike-slip faults and moved farther north along the coast. Pieces of Wrangellia are scattered from western Idaho northward to Alaska, but are considered to have been part of the same terrane because they have similar rock sequences (see below).

Alaska

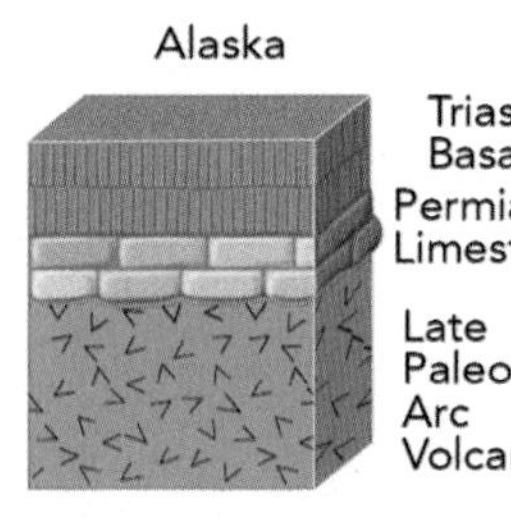

Triassic Basalt

Permian Limestone

Late Paleozoic Arc Volcanics

Western Idaho

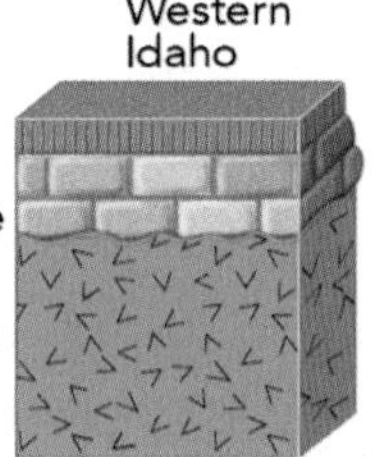

11.09.mtb2

Before You Leave This Page Be Able To

- ✓ Summarize the characteristics used to recognize a terrane.
- ✓ Describe a few of the main tectonic settings in which terranes originate.
- ✓ Summarize or sketch how we determine when two terranes were brought together.

How Do Continents Form?

CONTINENTS ARE CONSTANTLY BEING RESHAPED. Pieces can be removed by rifting or added by accretion of tectonic terranes. Continents are also internally rearranged, as areas are shortened during compression, stretched during extension, and shifted horizontally by strike-slip faults.

A How Old Is North America?

The crust of North America varies widely in age. The oldest parts of the Precambrian shield were formed as early as 3.8 to 4.0 billion years ago, but most of the continent was added later as a series of terranes.

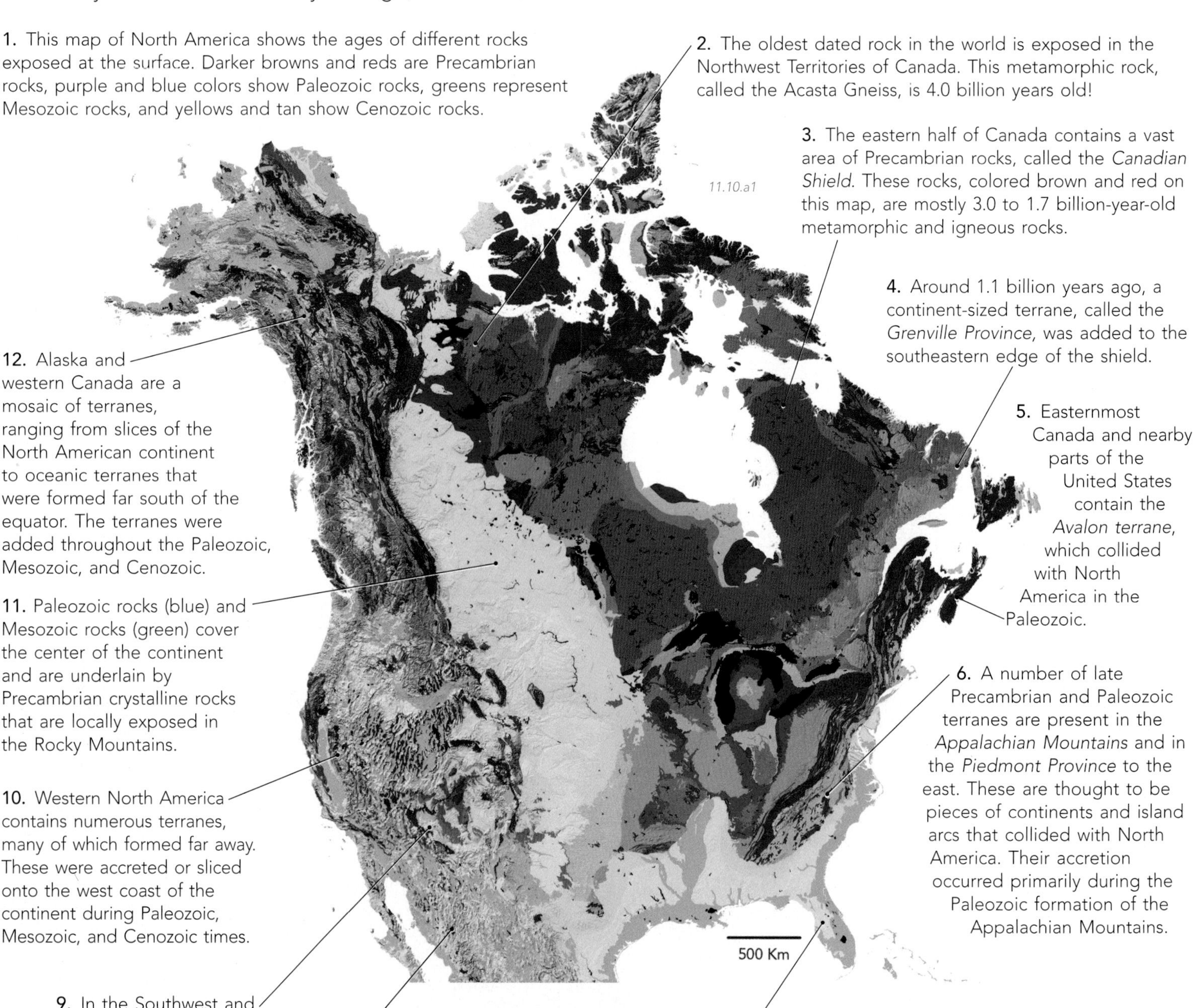

1. This map of North America shows the ages of different rocks exposed at the surface. Darker browns and reds are Precambrian rocks, purple and blue colors show Paleozoic rocks, greens represent Mesozoic rocks, and yellows and tan show Cenozoic rocks.

2. The oldest dated rock in the world is exposed in the Northwest Territories of Canada. This metamorphic rock, called the Acasta Gneiss, is 4.0 billion years old!

3. The eastern half of Canada contains a vast area of Precambrian rocks, called the *Canadian Shield*. These rocks, colored brown and red on this map, are mostly 3.0 to 1.7 billion-year-old metamorphic and igneous rocks.

4. Around 1.1 billion years ago, a continent-sized terrane, called the *Grenville Province*, was added to the southeastern edge of the shield.

5. Easternmost Canada and nearby parts of the United States contain the *Avalon terrane*, which collided with North America in the Paleozoic.

6. A number of late Precambrian and Paleozoic terranes are present in the *Appalachian Mountains* and in the *Piedmont Province* to the east. These are thought to be pieces of continents and island arcs that collided with North America. Their accretion occurred primarily during the Paleozoic formation of the Appalachian Mountains.

7. The tan, yellow, and green areas along the southern and southeastern edge of North America represent the *Coastal Plain*. This low-lying region was covered by Late Mesozoic and Cenozoic sediments that were deposited after this edge of North America was blocked out by early Mesozoic rifting.

8. Mexico is largely composed of terranes added to North America from the Paleozoic onward. The largest terranes are Paleozoic and Mesozoic island arcs that collided with the west coast of the Mexican mainland during the Mesozoic.

9. In the Southwest and the Southern Rockies, several large Precambrian provinces were added onto the southern edge of North America between 1.9 and 1.6 billion years ago.

10. Western North America contains numerous terranes, many of which formed far away. These were accreted or sliced onto the west coast of the continent during Paleozoic, Mesozoic, and Cenozoic times.

11. Paleozoic rocks (blue) and Mesozoic rocks (green) cover the center of the continent and are underlain by Precambrian crystalline rocks that are locally exposed in the Rocky Mountains.

12. Alaska and western Canada are a mosaic of terranes, ranging from slices of the North American continent to oceanic terranes that were formed far south of the equator. The terranes were added throughout the Paleozoic, Mesozoic, and Cenozoic.

B What Are the Ages of the Other Continents?

The other continents are similar to North America in that they contain Precambrian areas, called *shields*, that are flanked by belts of successively younger rocks. The younger rocks consist of either sedimentary and volcanic rocks deposited over the Precambrian basement, or terranes added to the edges of the shields.

1. On this map, Precambrian rocks are brown, with darker browns representing the oldest rocks. The other colors are blue for Paleozoic, green for Mesozoic, and yellow for Cenozoic.

2. Northern Europe contains a Precambrian shield, but the eastern and southern parts of the continent were added in the Paleozoic or in the Mesozoic and Cenozoic. The youngest additions are along the Mediterranean Sea, accreted during the collisions that formed the Alps.

11.10.b1

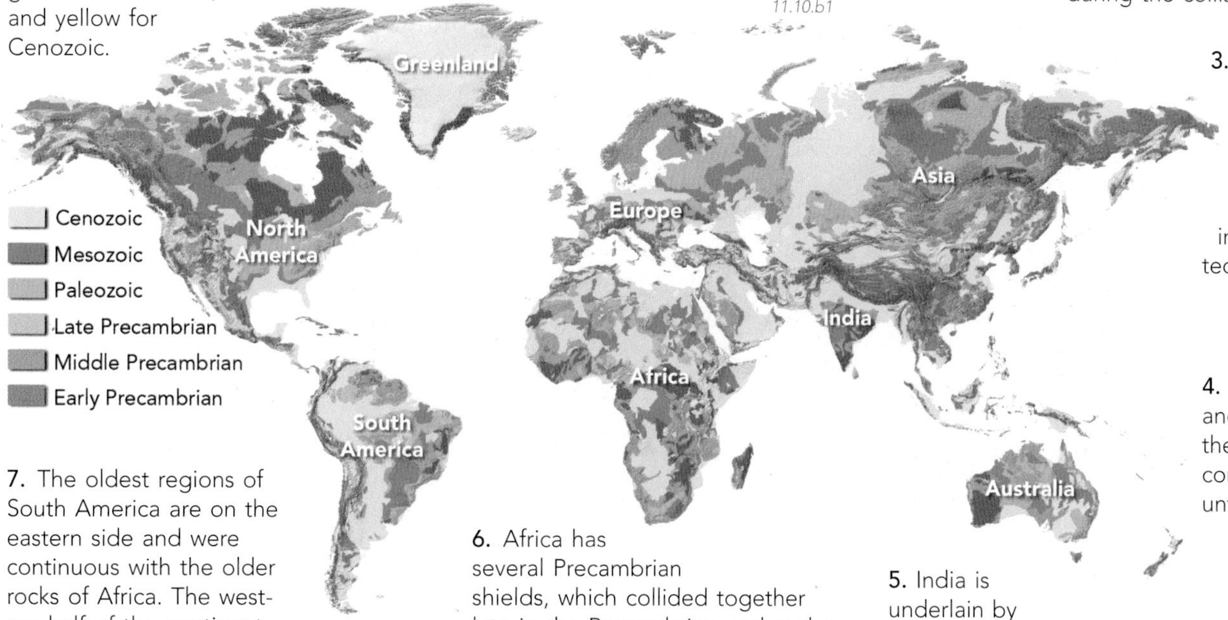

Greenland
North America
Europe
Asia
India
Africa
South America
Australia

Cenozoic
Mesozoic
Paleozoic
Late Precambrian
Middle Precambrian
Early Precambrian

3. Most of Asia was assembled from terranes during the Paleozoic and Mesozoic. The areas of Precambrian rocks (brown) are continental fragments incorporated into this tectonic jumble.

4. Western Australia contains ancient Precambrian rocks; the eastern part of the continent was not added until the Paleozoic.

5. India is underlain by Precambrian rocks that were once part of the ancient southern supercontinent of Gondwana. Rifting broke up the shield, dispersing the fragments of Gondwana around the Indian Ocean.

6. Africa has several Precambrian shields, which collided together late in the Precambrian and at the start of the Paleozoic. Younger rocks, shown in green and yellow, overlie these older rocks.

7. The oldest regions of South America are on the eastern side and were continuous with the older rocks of Africa. The western half of the continent was constructed more recently, mostly during the Paleozoic and Mesozoic.

In Suspect Terrain

California is the area many geologists think about when studying how continents grow from the accretion of tectonic terranes. John McPhee's popular books, *In Suspect Terrain* and *Assembling California*, provide an accessible account of how terranes were recognized and how they added new real estate to North America.

This map shows the various types of terranes added to the crust. The terranes include slices of Paleozoic and Mesozoic oceanic crust and sediment, Mesozoic island arcs, and an accretionary prism. The prism consists of oceanic material scraped off oceanic plates that were subducted eastward beneath the continental margin. The map does not show units that formed in place, such as granites in the Sierra Nevada.

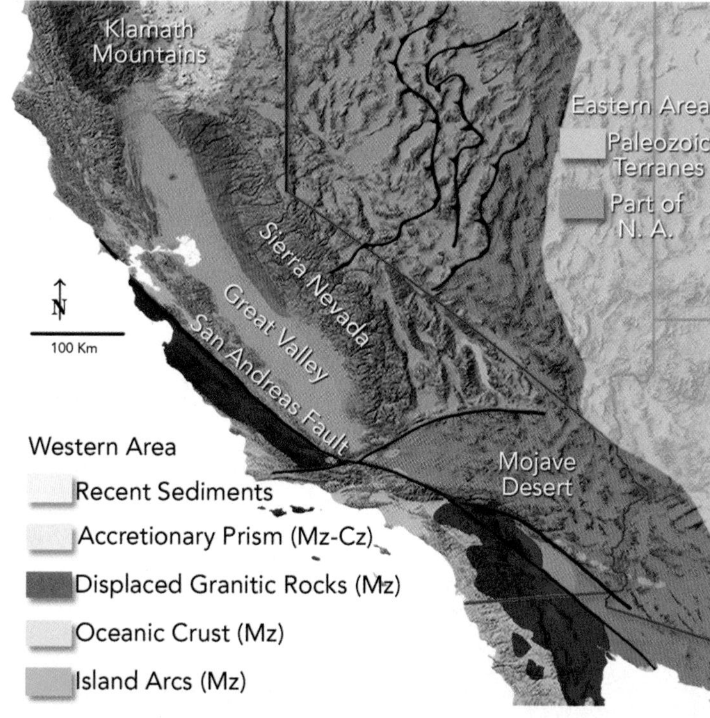

Before You Leave This Page Be Able To

- ✓ Identify the oldest (Precambrian) parts of North America and some areas that were added as terranes in Paleozoic or Mesozoic-Cenozoic times.
- ✓ Briefly describe why different parts of a continent can be different ages.
- ✓ List the types of terranes added to or displaced in California.

How Did the Continents Join and Split Apart?

CONTINENTS HAVE SHIFTED THEIR POSITIONS over time in response to plate tectonics. They have rifted apart and collided, only to rift apart again. Where were the continents located in the past, and which mountains resulted from their motions? The story of the movement of the continents is the same story as the origin of the modern oceans presented in the last chapter. But here we emphasize which continents were joined and how they separated. We start with the present and work backwards, but we recommend that, after reading both pages, you examine the maps again from oldest to youngest.

Present

The images on these pages show one interpretation of where the continents were located in the past. The artistic renderings of the continents, mountains, and oceans were created by geologist Ron Blakey. For most time periods, he created two views, one focused on the western hemisphere (image on the left) and one on the eastern hemisphere (image on the right), generally with some overlap.

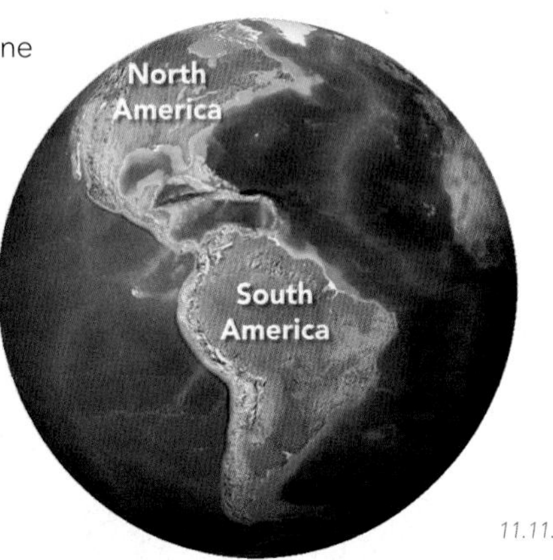

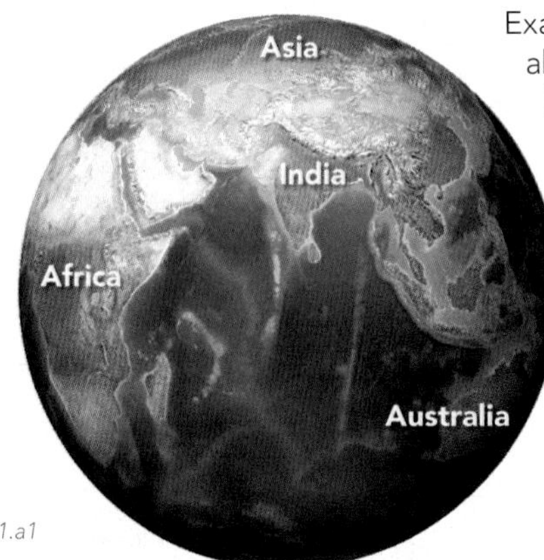

11.11.a1

Examine these globes and think about the *present-day* plate boundaries, envisioning which way the continents are moving relative to one another. Next, use the relative motions to predict where the continents were likely to have been in the past. Then check your predictions by examining the next two globes as you step backward in time.

150 Million Years Ago - Gondwana and Laurasia

The continents on either side of the Atlantic are currently moving away from each other due to seafloor spreading; in the past they would have been closer than they are today. The left globe is rotated so that the central Atlantic Ocean is in the center of the image.

At 150 million years ago, in the late Jurassic, North America had separated from Africa. It still was joined with Europe and Asia, forming the supercontinent of *Laurasia*. South America had not yet rifted away from Africa.

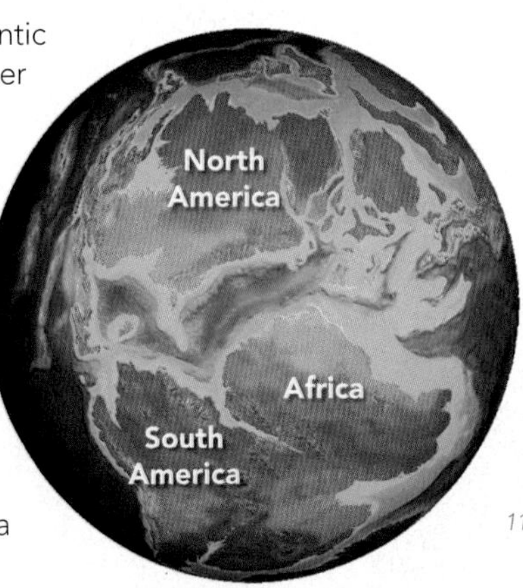

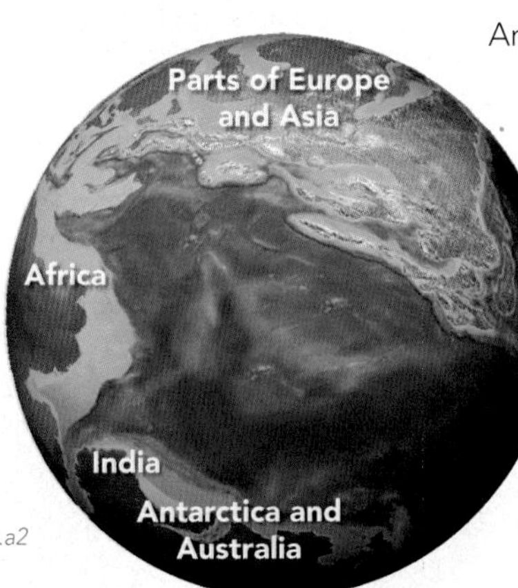

11.11.a2

Antarctica was still attached to the southern tips of Africa and South America. India was attached to the northern edge of Antarctica. The combined southern continents formed the supercontinent of *Gondwana*.

280 Million Years Ago - the Supercontinent of Pangaea

In the late Paleozoic, around 280 million years ago, all the continents were joined in a supercontinent called *Pangaea*.

The Appalachian and Ouachita Mountains had been formed by a continental collision between North America and the northern edge of Gondwana (South America and Africa).

11.11.a3

A wedge-shaped ocean, called the *Tethys Sea*, separated Asia from land masses farther to the south. Southern Africa, Australia, and Antarctica were close to the South Pole, and so at this time were partly covered with ice.

Early to Late Mesozoic Convergent Margin

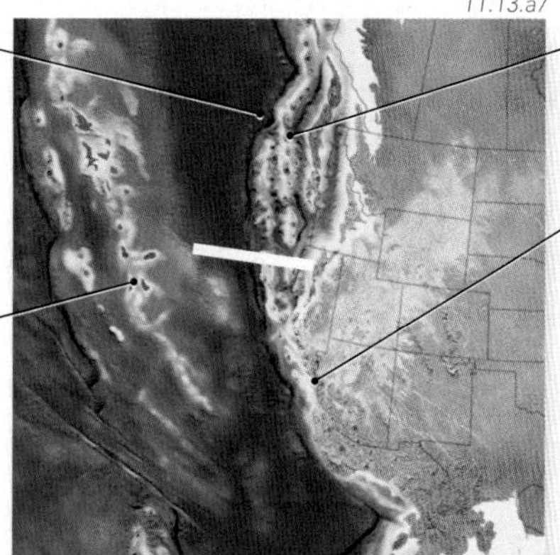
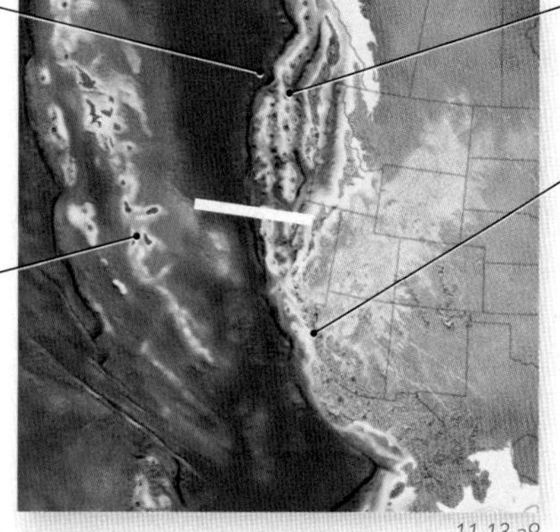
11.13.a7

1. The region's setting changed in the early Mesozoic (about 210 to 150 million years ago), when oceanic plates were subducted beneath western North America.

2. An exotic land mass, called *Wrangellia*, was located offshore and approached the trench, destined for a collision with the mainland.

3. Subduction caused magmatism within the overriding plate, forming an offshore arc from California northward.

4. Farther south in Arizona, southern California, and Mexico, volcanoes were built on the continent, partly within a large area of desert sand dunes (shown in orange).

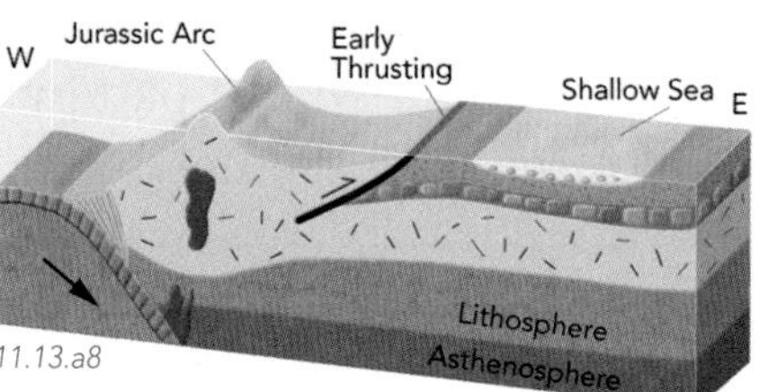

11.13.a8

5. This cross section, along the white line on the map, shows that at 160 million years ago there was an offshore arc and back-arc thrusting.

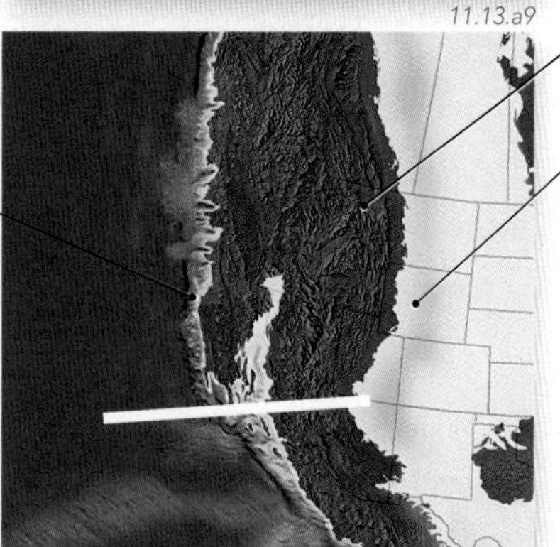
11.13.a9

6. This map shows the Late Mesozoic (about 85 million years ago). During this time, subduction continued beneath the western edge of North America. Inland from the coast, subduction-related magma erupted onto the surface and also solidified at depth forming large granitic batholiths in Washington, Idaho, and the Sierra Nevada.

7. Compression associated with the convergent boundary formed thrust faults and mountains from Canada to Mexico.

8. Inland, the continent was flexed down by the weight of the thrust sheets, allowing high sea levels at the time to flood the center of North America, from Mexico to the Arctic.

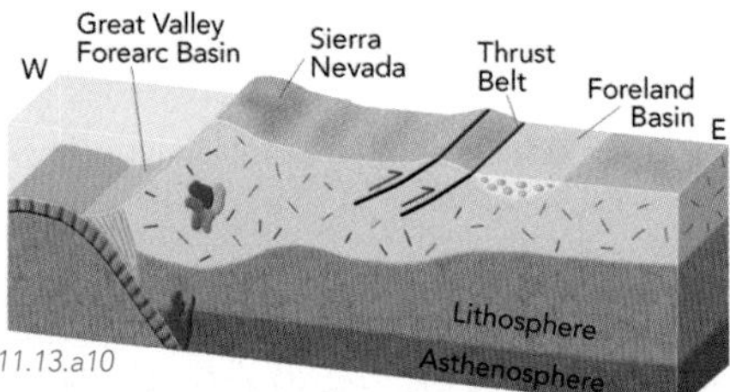

11.13.a10

9. This cross section, located along the white line, depicts the *Sierra Nevada batholith* and thrusting that formed a foreland basin.

Late Mesozoic and Early Cenozoic Laramide Orogeny

1. Late in the Mesozoic (about 80 million years ago), the oceanic plate started to subduct at a much lower angle and to scrape along the base of the overridding lithosphere. This change in the angle of subduction moved magmatism as far east as Colorado and shut off the supply of magma to the Sierra Nevada. Large slices of crust were transported up the coast in the Pacific Northwest (not shown).

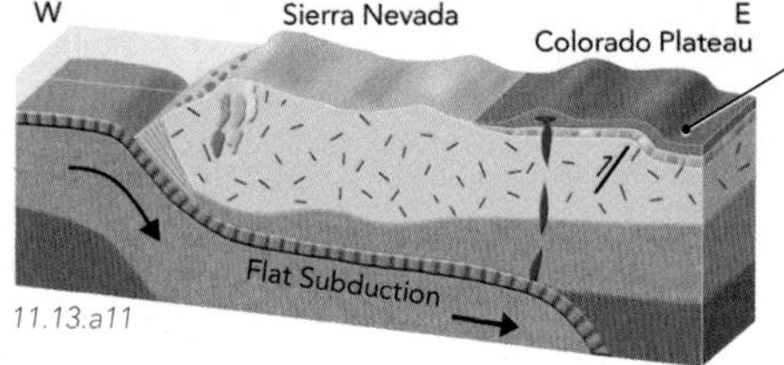

11.13.a11

2. The new subduction geometry caused compression farther into the continent, forming folds, uplifts, and basins in the Rocky Mountains and Four Corners region. This mountain-building event, called the *Laramide Orogeny*, lasted from 80 to about 40 million years ago.

Middle and Late Cenozoic Crustal Extension

1. Beginning at about 40 million years ago, the convergence between North America and the oceanic plates slowed. This helped end the compression of the *Laramide Orogeny*.

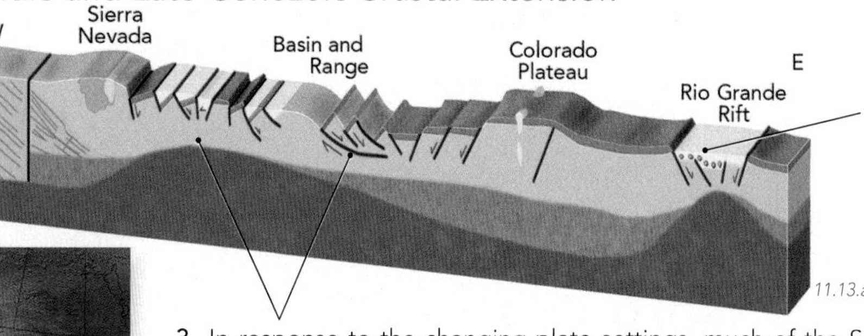

11.13.a12

4. Farther to the east, extension formed the *Rio Grande Rift* of New Mexico, West Texas, and south-central Colorado.

11.13.a13

2. By 15 million years ago, the Southwest had overrun a spreading center in the Pacific. This caused part of the convergent margin to be progressively converted into a transform boundary that eventually became the *San Andreas fault*.

3. In response to the changing plate settings, much of the Southwest was affected by crustal extension from about 30 million years ago to the present. Extension thinned the crust and lithosphere, stretching the crust to twice its original width. It shifted the Sierra Nevada hundreds of kilometers to the west and formed the *Basin and Range Province* of the Southwest.

Before You Leave This Page Be Able To

✓ Briefly summarize or sketch the main tectonic events that affected the western United States during Paleozoic, Mesozoic, and Cenozoic times.

11.13

Where Will Mountains and Basins Form in This Region?

The figure below shows part of a continent and adjacent ocean. There are no plate boundaries now, but a subduction zone will form along the western coast of the continent, and the eastern part of the continent will be rifted away. You will use the typical patterns that form along such boundaries to predict where mountains and basins will form once the new plate boundaries are fully developed.

Goals of This Exercise:

- Observe the continent and ocean below and read the descriptions of the types of features that will form in the future.
- Use your understanding of plate boundaries and of the settings in which mountains and basins develop to predict where mountains and basins will form. Sketch your predictions on a diagram of the region.
- Predict what the regional topography will be like and sketch it on a topographic profile.
- Predict and sketch how the thicknesses of the crust will vary across the region.

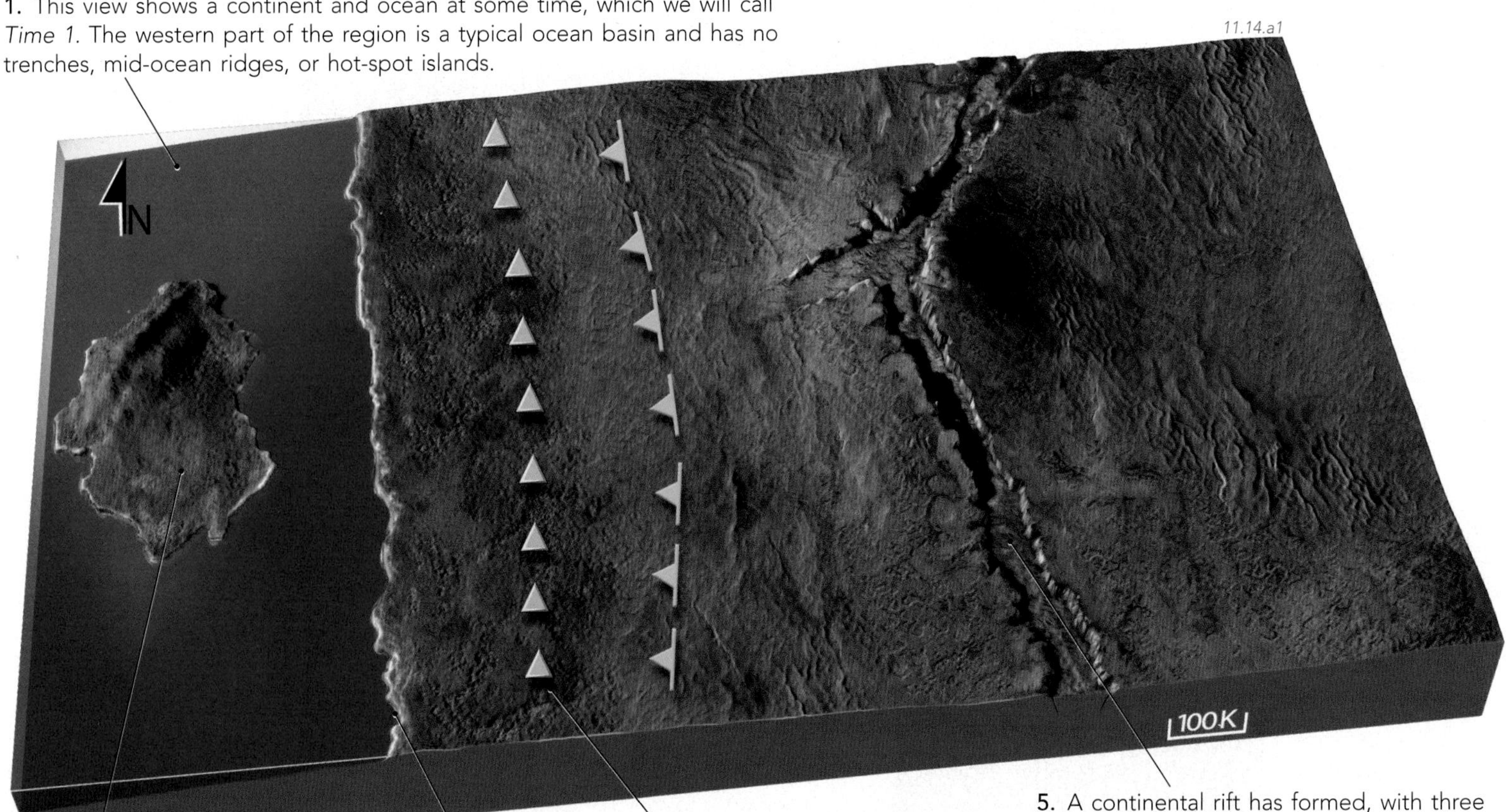

1. This view shows a continent and ocean at some time, which we will call *Time 1*. The western part of the region is a typical ocean basin and has no trenches, mid-ocean ridges, or hot-spot islands.

2. A small piece of continent lies offshore in the middle of the ocean. When the oceanic plate begins to move, this piece of continent will be carried toward and will collide with the main continent.

3. The ocean-continent edge is currently a passive margin, not a plate boundary. It will become an ocean-continent convergent boundary, and the oceanic material will be subducted eastward below the continent.

4. Once the convergence begins, a magmatic belt will form inland from the coast, near the position of the yellow triangles. Farther inland a thrust belt will form as shown by the blue dashed line with teeth. In the thrust belt, the western part of the continent will be thrust eastward over the central part of the continent.

5. A continental rift has formed, with three arms radiating out from a hot spot in the center of the continent. This rift will split the continent into two pieces. At some later time, the piece of continent to the right will break away completely and a new ocean basin will be formed by seafloor spreading. Even later, at a time we will call *Time 2*, the edge of the continent will have become a passive margin and the spreading center will be out of the region.

Procedures

Use the data to complete the following steps, entering your answers in the appropriate places on the worksheet.

1. Observe the regional features shown on the figure on the left page, which represents the situation at *Time 1*. Read the descriptions associated with that figure and decide what each statement implies about the *future topography* (elevations) of the area.
2. For each feature (subduction zone, thrust belt, etc.) that will form by *Time 2*, think about how that feature is typically expressed in the topography. Does it form a mountain range, a basin, or a mountain with a nearby basin?
3. On the worksheet, sketch your predictions about the area's topography for *Time 2* on the simplified figure below, which shows the same area as the figure on the previous page. The figure shows the overall shape of the continent but not the topography. Use the following letters: *O* for an oceanic trench, *A* for an accretionary prism, *M* for mountains, *V* for volcanoes in the continental magmatic belt, *B* for a basin, and *P* for a passive margin. Feel free to sketch some simple lines to portray the locations of the features. Your instructor may have you predict other features that might develop, such as a tectonic terrane **(T)** or collision event **(C).**
4. At the bottom of this page is a cross section, showing a topographic profile and the thickness of crust at *Time 1*. On the version of this section in the worksheet, draw a new profile of what the topography will look at *Time 2*. At the bottom on the cross section, draw the crustal thicknesses as you think they will be at *Time 2*, after the new features have formed. Use figures from this chapter and elsewhere in the book as examples of crustal thicknesses for different features.

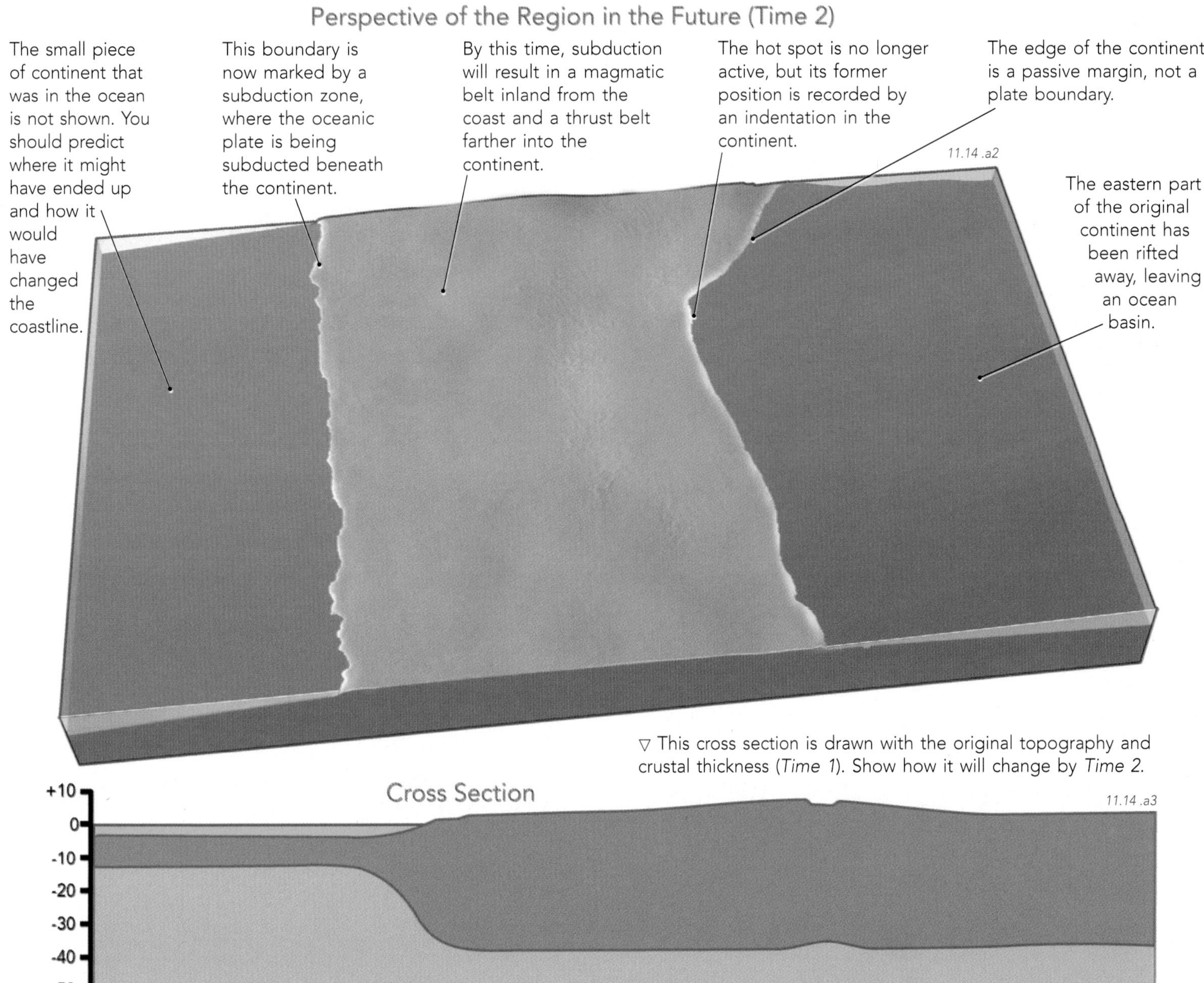

▽ This cross section is drawn with the original topography and crustal thickness (*Time 1*). Show how it will change by *Time 2*.

CHAPTER 12

Earthquakes and Earth's Interior

EARTHQUAKES ARE AMONG EARTH'S deadliest natural phenomena. Ground shaking during an earthquake can topple buildings, liquefy normally solid ground, and unleash massive ocean waves that wipe out coastal cities. A single earthquake can kill more than 100,000 people. What causes earthquakes, and how do we study them? In this chapter, we explore important aspects about earthquakes and Earth's interior.

The world's strongest earthquake in 40 years struck Indonesia on December 26, 2004. The magnitude 9 earthquake occurred west of Sumatra and was caused by movement on a fault, shown by the red line on this map. The fault is part of a plate boundary where the Indian-Australian plate is being subducted to the northeastward beneath the Eurasian plate. The red line shows the length of the fault that ruptured during the earthquake. Yellow dots nearby show the locations of smaller, related earthquakes.

What causes earthquakes and where are they most likely to strike?

The earthquake occurred beneath the ocean, where the Eurasian plate was thrust over the Indian-Australian plate. Movement on the fault abruptly uplifted the overriding plate, pushing up a large region of seafloor and displacing overlying seawater. This caused a massive wave, called a *tsunami*, that spread out across the Indian Ocean as a low wave, traveling at speeds approaching 800 kilometers/hour (500 miles/hour)! The curved gray lines show a model of the wave's position by hour.

What happens when an earthquake occurs under the sea, rather than on land?

The tsunami increased in height as it crashed into the coasts of Indonesia, Thailand, Sri Lanka, India, east Africa, and various islands. Low coastal areas were inundated by as much as 20 to 30 meters of water (65 to 100 feet) in Indonesia and 12 meters (40 feet) in Sri Lanka. Cities and villages were completely demolished along hundreds of kilometers of coastline, leaving more than 250,000 people dead or missing. The numbers below show casualties by location.

How does a tsunami form, how does it move through the sea, and what determines how destructive it is?

12.00.a1

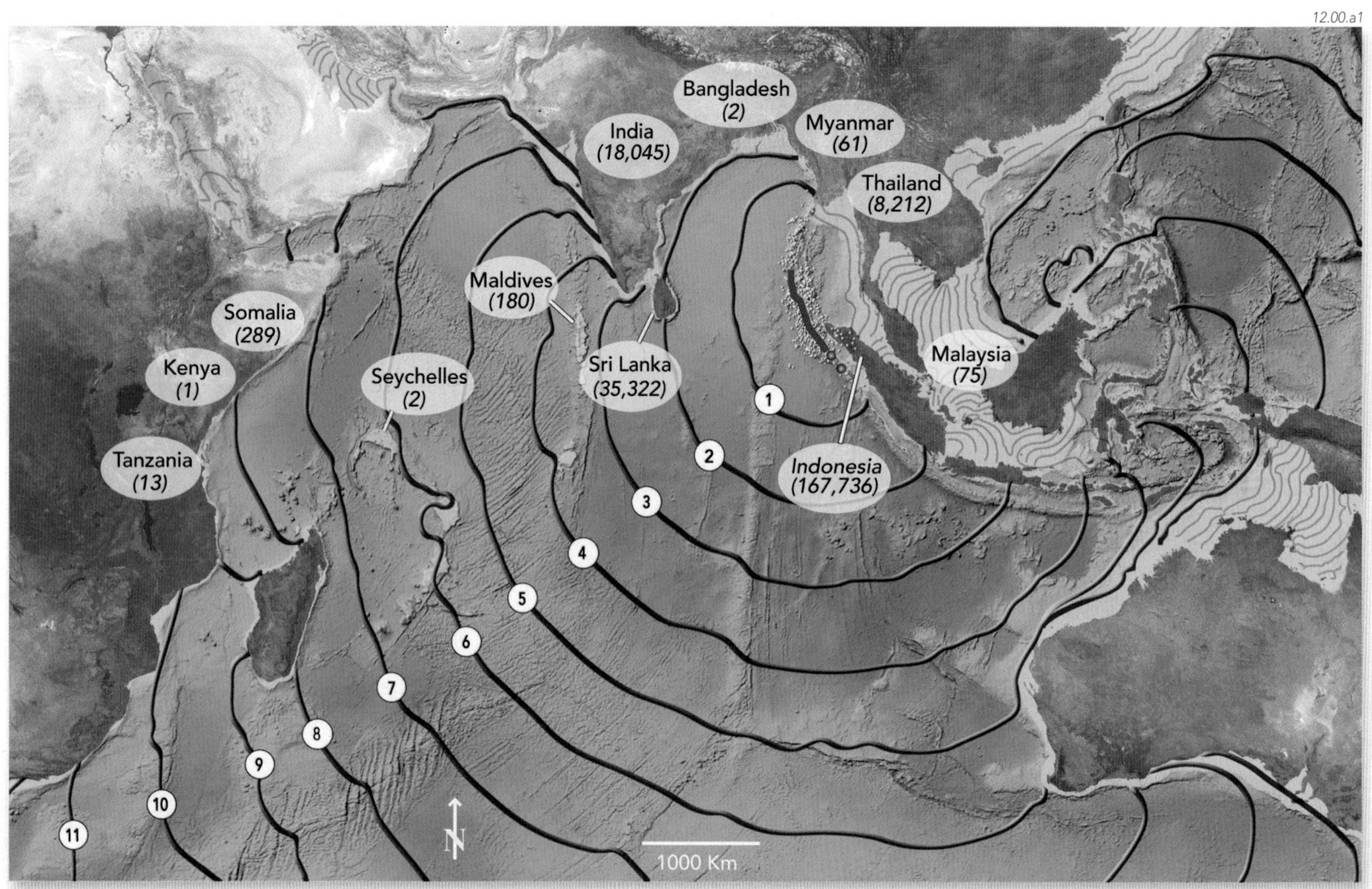

TOPICS IN THIS CHAPTER

12.00.a2

The destructive power of the tsunami is clear from this photograph of Banda Aceh, the regional capitol of Sumatra's northernmost province. This city of 320,000 people was reduced to rubble, and nearly a third of its inhabitants were killed or are missing. The tsunami inflicted damage to low-lying coastlines around the Indian Ocean, reaching as far away as the eastern coast of Africa.

The satellite images below show *Banda Aceh*, before and after the tsunami. The buildings and vegetation on the "before" image (left) were stripped bare by the water's rush onto the land and the subsequent retreat back to the sea. A slightly higher area to the north (top left) was largely untouched, retaining its forest.

Which areas along a coast are most at risk for a tsunami?

12.00.a3

12.00.a4

2004 Sumatran Earthquake and Indian Ocean Tsunami

The 2004 Sumatran earthquake struck on the morning of December 26, violently shaking the region and triggering the massive Indian Ocean tsunami. It ranks as one of the two or three largest earthquakes ever recorded. The magnitude of the earthquake is variably estimated at 9.0 to 9.3, depending on how the calculations are done. Large aftershocks followed the main quake, including one with a surprisingly large magnitude of 8.7. From the seismic records of the main quake and aftershocks, it is estimated that a fault surface 1,220 (760 miles) kilometers in length slipped by as much as 10 meters during the earthquake. The earthquake lasted over 8 minutes, an unusually long duration.

The earthquake started at a depth of 30 kilometers (19 miles) and ruptured all the way up to the seafloor. It lifted a large section of seafloor several meters in height, displacing tens of cubic kilometers of seawater, which spread out in all directions. The tsunami rose to heights of more than 30 meters (100 feet) when it came ashore, and in many places it washed inland for more than a kilometer.

As a result of the earthquake, parts of the Andaman Islands, northwest of Sumatra, were changed forever. Coral reefs, which had been undersea, were lifted above sea level. A lighthouse that was originally on land is now surrounded by seawater one meter deep. The changes to the land seem insignificant compared to the massive loss of life in this event, one of the deadliest disasters in world history.

What Is an Earthquake?

AN EARTHQUAKE OCCURS WHEN ENERGY stored in rocks is suddenly released. Most earthquakes are produced when stress builds up along a fault and causes the fault to slip. Similar kinds of energy are released by volcanic eruptions, explosions, and even meteorite impacts.

A How Do We Describe an Earthquake?

When an earthquake occurs, mechanical energy is released, some of which is transmitted through rocks as vibrations called *seismic waves*. These waves spread out from the site of the disturbance, travel through the interior or along the surface of Earth, and can be recorded by scientific instruments at *seismic stations*.

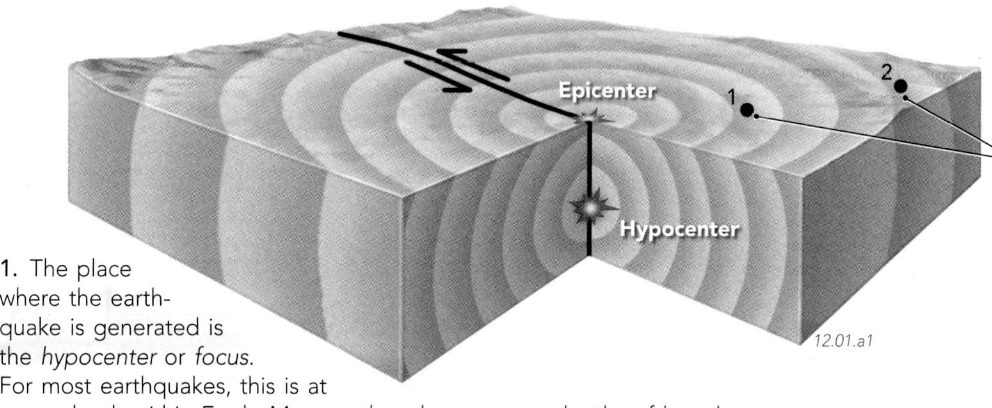

1. The place where the earthquake is generated is the *hypocenter* or *focus*. For most earthquakes, this is at some depth within Earth. Most earthquakes occur at depths of less than 100 kilometers (60 miles) and are as shallow as several kilometers. Those that occur in subduction zones may be as deep as 700 kilometers (430 miles).

3. Seismic waves, once generated, move out in all directions, as shown by the curved bands radiating out from the hypocenter. They can be measured by seismic stations (locations 1 and 2). Seismic stations closer to the hypocenter, such as station 1, will detect the waves sooner than those farther away, such as station 2.

2. The *epicenter* is the point on Earth's surface directly *above* where the earthquake occurs (directly above the hypocenter). If the seismic event happens on the surface, such as during a surface explosion, then the epicenter and hypocenter are the same.

B What Causes Most Earthquakes?

Most earthquakes are generated by movement along faults. When rocks on opposite sides of a fault slip past one another abruptly, the movement generates seismic waves as materials near the fault are pushed, pulled, and sheared. Slip along any type of fault can generate an earthquake.

Normal Faults

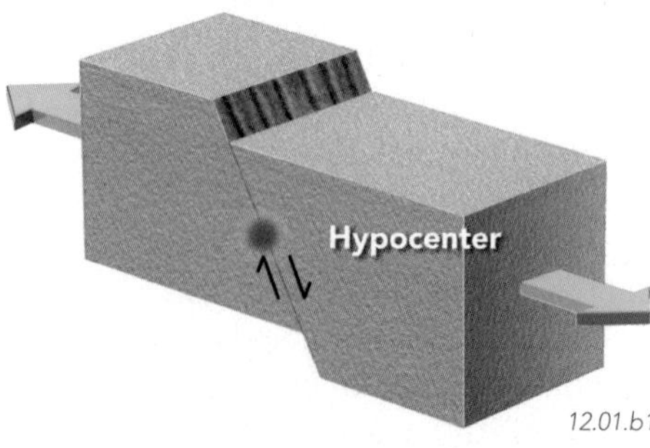

In a *normal fault*, the rocks above the fault (the *hanging wall*) move down with respect to rocks below the fault (the *footwall*). The crust is stretched horizontally, so earthquakes related to normal faults are most common along *divergent plate boundaries*, such as oceanic spreading centers, and in continental rifts.

Reverse and Thrust Faults

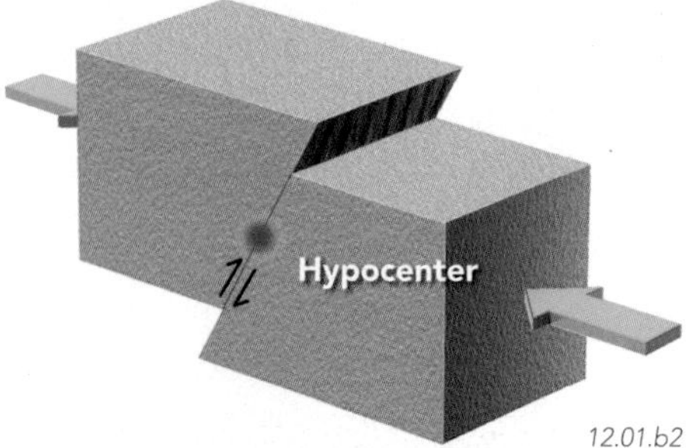

Many large earthquakes are generated along *reverse faults*, especially the gently dipping variety called *thrust faults*. In thrust and reverse faults, the hanging wall moves up with respect to the footwall. Such faults are formed by compressional forces, such as those associated with *subduction zones* and *continental collisions*.

Strike-Slip Faults

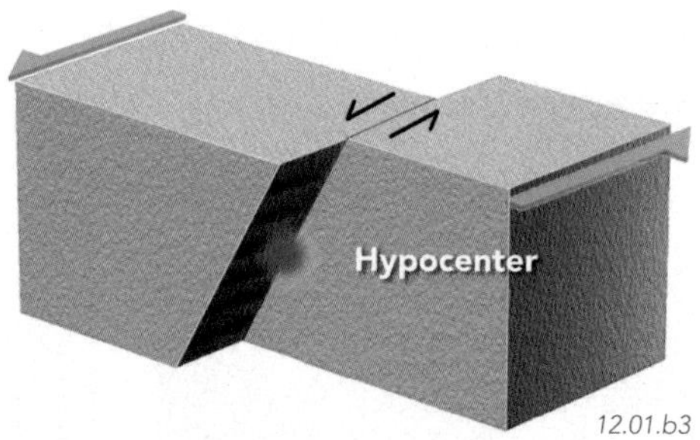

In *strike-slip faults*, the two sides of the fault slip horizontally past each other. This can generate large earthquakes. The largest strike-slip faults are *transform plate boundaries*, like the San Andreas fault in California and parts of the seismically dangerous Alpine fault, which cuts diagonally across the South Island of New Zealand.

C How Do Volcanoes and Magma Cause Earthquakes?

Volcanoes generate seismic waves and cause the ground to shake through several processes. An explosive volcanic eruption causes compression, transmitting energy through seismic waves (shown here with yellow lines).

Volcanoes add tremendous weight to the crust. This loading can lead to faulting and earthquakes. The fault shown here, which caused an earthquake at depth, has dropped the volcano relative to its surroundings.

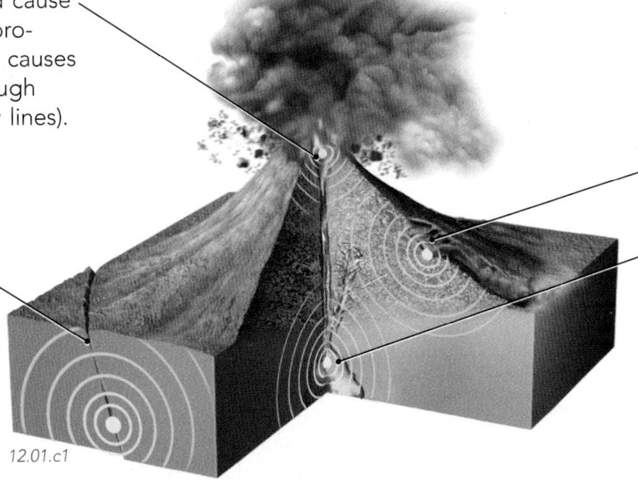

Many volcanoes have steep, unstable slopes underlain by rocks altered and weakened by hot water. The flanks of such volcanoes can fall apart catastrophically, causing landslides that shake the ground as they travel down the flank of the volcano.

As magma moves below the volcano, it can push rocks out of the way, causing a series of small, distinctive earthquakes. In some cases, the magma causes earthquakes as it opens space by inflating Earth's surface.

D What Are Some Other Causes of Seismic Waves?

Landslides

Catastrophic landslides, whether on land or beneath water, cause ground shaking. Lava flows forming new crust on the Big Island of Hawaii can become unstable and suddenly collapse into the ocean. Seismometers at the nearby Hawaii Volcanoes National Park often record seismic waves caused by such landslides.

Meteoroid Impacts

Ground shaking accompanies the impact of meteoroids on Earth's surface. The 100 kilometer-wide Manicouagan ring lake in Canada is one of Earth's largest meteoroid impact sites. The impact occurred about 200 million years ago, and would have resulted in an earthquake much larger than any recorded in history.

12.01.d2

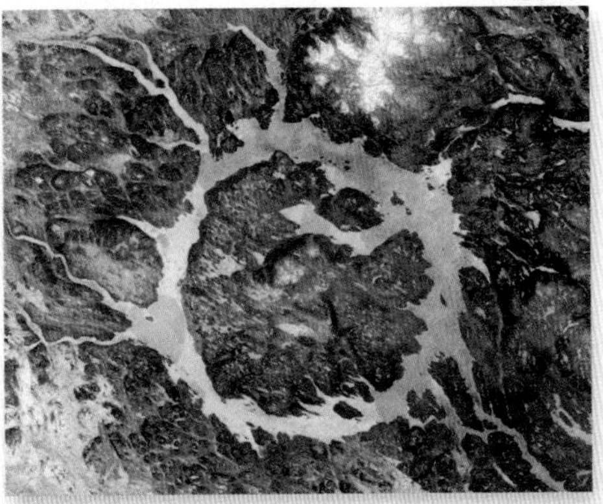

Explosions

Mine blasts and nuclear explosions compress Earth's surface, producing seismic waves measurable by distant seismic instruments. Monitoring compliance with nuclear test-ban treaties is done in part using a worldwide array of seismic instruments. These instruments recorded a nuclear bomb exploded by India in 1998. Seismic waves generated by a blast are more abrupt than those caused by a natural earthquake.

12.01.d3

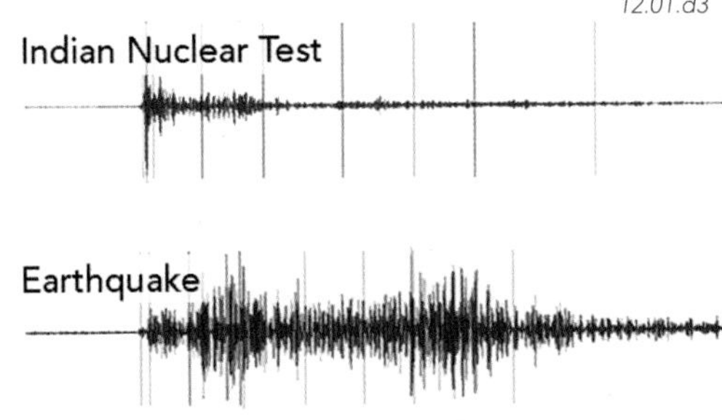

Earthquakes Caused by Humans

Humans can cause earthquakes in several ways. Reservoirs built to store water fill rapidly and load the crust, which responds by flexing and faulting. After Lake Mead behind Hoover Dam in Nevada and Arizona was filled, hundreds of moderate earthquakes occurred under the reservoir between 1934 and 1944. Similarly, very shallow (less than 3 kilometers deep) earthquakes occur near Monticello Reservoir in South Carolina. In China, there were fears that the filling of the Three Gorges Dam, the world's largest hydroelectric project, would trigger earthquakes in this seismically active area.

Humans have also caused earthquakes by injecting waste water underground into a deep well at the Rocky Mountain Arsenal northwest of Denver. This caused more than a thousand small earthquakes and two magnitude 5 earthquakes, which caused minor damage nearby. When the waste injection stopped and some waste was pumped back out of the ground, the number of earthquakes decreased.

Before You Leave This Page Be Able To

- ✓ Explain what a hypocenter and epicenter each represent.
- ✓ Sketch and describe the types of faults that cause earthquakes.
- ✓ Describe some other ways earthquakes or seismic waves are formed, including volcanoes and ways that humans cause earthquakes.

How Does Faulting Cause Earthquakes?

MOST EARTHQUAKES OCCUR because of movement along faults. Faults slip because the stresses applied to them exceed the ability of the rock to withstand the stress. Rocks respond to the stress in one of two ways—they either flex and bend, or they break. Breaking causes earthquakes.

 What Processes Precede and Follow Faulting?

Before faulting, rocks change shape (i.e., they *strain*) slightly as they are squeezed, pulled, and sheared. Once stress builds up to a certain level, slippage along a fault generally happens in a sudden, discrete jump. Faulting reduces the stress on the rocks, allowing some of the strained rocks to return back to their original shapes. This type of response, where rocks return to their original shape after being strained, is called *elastic behavior.*

Pre-Slip

▷ **1.** An active strike-slip fault has been offsetting a stream bed for hundreds of thousands of years, causing the stream to bend. The last earthquake occurred before people settled in the area. The straight section of the stream seemed a perfect place to put a bridge to provide a crossing for a road.

12.02.a1

2. The strike-slip fault is present at depth, but is not obvious at the surface because it is beneath the stream or covered with loose rocks, sand, and soil. The fault has some expression on the landscape, such as a very straight segment of the stream.

Stress Increase and Elastic Strain

▷ **3.** Tectonic stress continues to act upon the rocks along the fault. The rocks strain and flex, as shown by a slight warp in the right side of the block, but the stresses are not great enough to make the rocks break. The cement bridge, however, develops a few cracks.

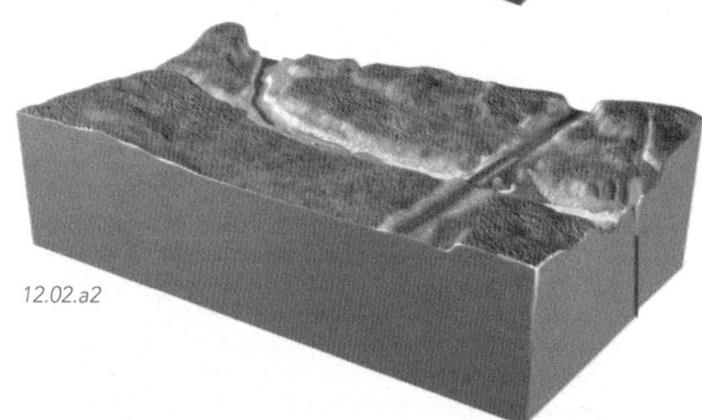

12.02.a2

4. As stress builds in the rocks along the fault, the rocks deform *elastically,* changing shape slightly without breaking. If the rocks are strong enough and there is sufficient friction along the fault surface, the rocks and fault hold the added stress.

Slip and Earthquake

▷ **5.** Finally the stress along the fault becomes so great that the fault slips and the rocks on opposite sides of the fault move past each other. A large earthquake occurs, generating seismic waves (not shown) radiating outward from the fault. Movement along the fault severs the bridge and offsets the road.

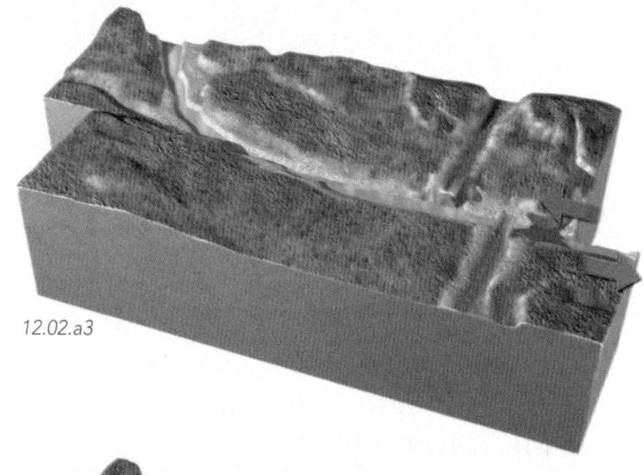

12.02.a3

6. With the stress partially relieved, the rocks next to the fault relax by elastic processes, many returning to their original, unstrained shape. The movement that has occurred along the fault, however, is permanent. It is not elastic.

Post-Slip

▷ **7.** After the earthquake, stress again begins to slowly build up along the fault. A new bridge is installed over the stream and the road is realigned. The abandoned part of the bridge, like the straight part of the stream, is a clue that something happened here.

12.02.a4

8. During this sequence, rock strains elastically before the earthquake, ruptures during the earthquake, and mostly returns to its original shape afterwards. This sequence is called *stick-slip behavior* because the fault sticks (does not move) and then slips.

B How Do Earthquake Ruptures Grow?

Most earthquakes occur by slip on a preexisting fault, but the entire fault does not begin to slip at once. Instead, the earthquake rupture starts in a small area (the hypocenter) and expands over time.

12.02.b1

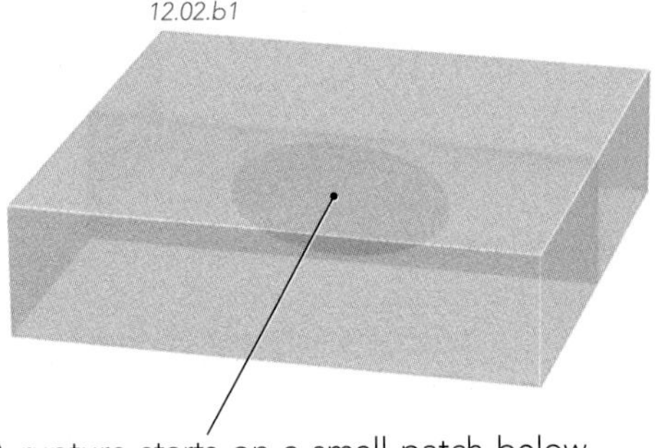

A rupture starts on a small patch below Earth's surface and begins to expand along the preexisting fault plane. Some rock breaks adjacent to the fault, but most slip occurs on the actual fault surface.

12.02.b2

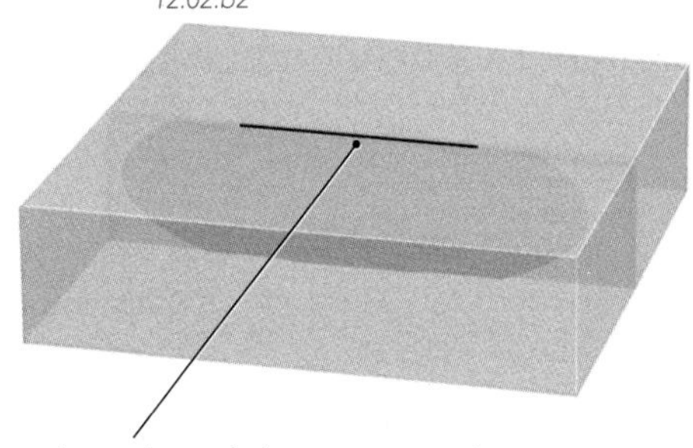

As the edge of the rupture migrates outward, it may eventually reach Earth's surface, causing a break called a *fault scarp*. Seen from above, the rupture migrates in both directions, but may expand more in one direction than in the other.

12.02.b3

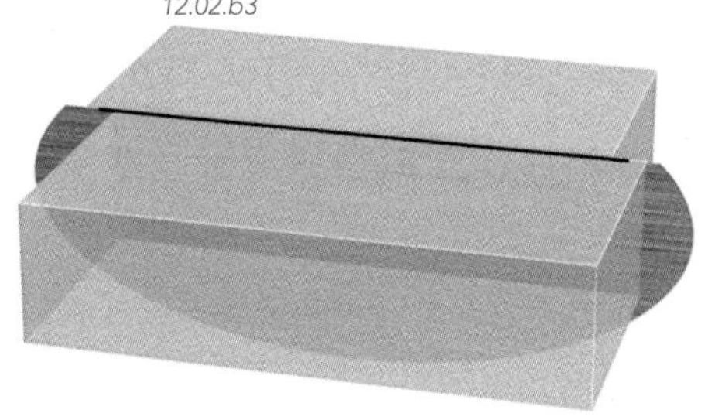

The rupture continues to grow along the fault plane and the fault scarp lengthens. The faulting relieves some of the stress, and rupturing will stop when the remaining stress can no longer overcome friction along the fault surface.

Earthquake Ruptures in the Field

12.02.b4

◁ The Landers earthquake of 1992 ruptured across the Mojave Desert of California, forming a *fault scarp*. In this photo, the scarp is cutting through granite. The fault had strike-slip movement, with some vertical movement.

▷ Movement along a normal fault offset the land surface during a 1983 earthquake, forming this fault scarp. [Borah Peak, Idaho]

12.02.b5

12.02.b6

△ The 1959 Hebgen Lake earthquake in southern Montana formed a several-meter-high *fault scarp*. The earthquake and fault scarp were generated by slip along a normal fault.

Build Up and Release of Stress

When a fault slips, it relieves some of the stress on the fault, causing the stress levels to suddenly drop. Gradually, the stress rebuilds until it exceeds the strength of the rock or the ability of friction to keep the fault from slipping. A conceptual model of how the amount of stress changes over time is shown below.

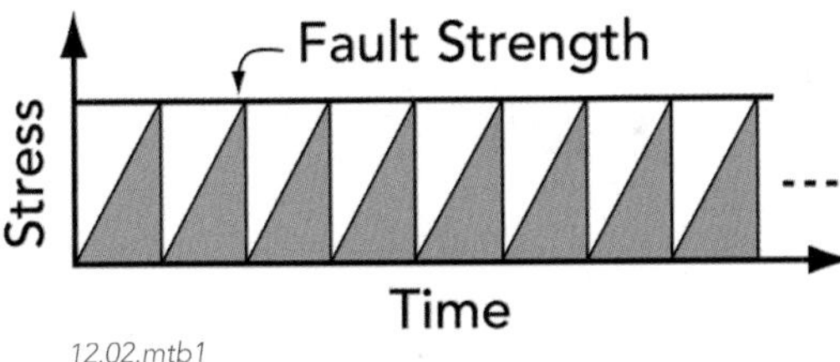

12.02.mtb1

On this plot, the magnitude of the stress imposed on the fault builds up gradually. When the *amount of stress* equals the *strength of the fault*, the fault slips, and the stress immediately decreases to the original level. In this manner, the amount of stress on a fault forms a zigzag pattern on the graph. It increases gradually (sloping line), and then decreases abruptly (vertical line) when an earthquake occurs. This process is called the *earthquake cycle*, and is one explanation for why some faults apparently produce earthquakes of a similar size. The time between repeating earthquakes is called the *recurrence interval*.

Before You Leave This Page Be Able To

- ✓ Describe or sketch how the buildup of stress can flex rocks, leading to an earthquake.
- ✓ Describe or sketch how a rupture begins in a small area and grows over time and ruptures Earth's surface.
- ✓ Describe some characteristics of fault scarps and ruptures.
- ✓ Describe how stress changes through time along a fault according to the earthquake-cycle model.

Where Do Most Earthquakes Occur?

MOST EARTHQUAKES OCCUR ALONG PLATE BOUNDARIES or in regions near plate boundaries, but some also strike in the middle of plates. Different tectonic settings generate different sizes and depths of earthquakes, with some types of plate boundaries being much more dangerous than others.

A Where Do Earthquakes Occur?

This map shows the world distribution of earthquake epicenters, colored according to depth. Yellow dots represent shallow earthquakes (0 to 70 km), green dots mark earthquakes with intermediate-depths (70 to 300 km), and red dots indicate earthquakes deeper than 300 km.

Examine this map and note how earthquakes are distributed. Note how this distribution compares to other features, such as edges of continents, mid-ocean ridges, subduction sites, and continental collisions.

12.03.a1

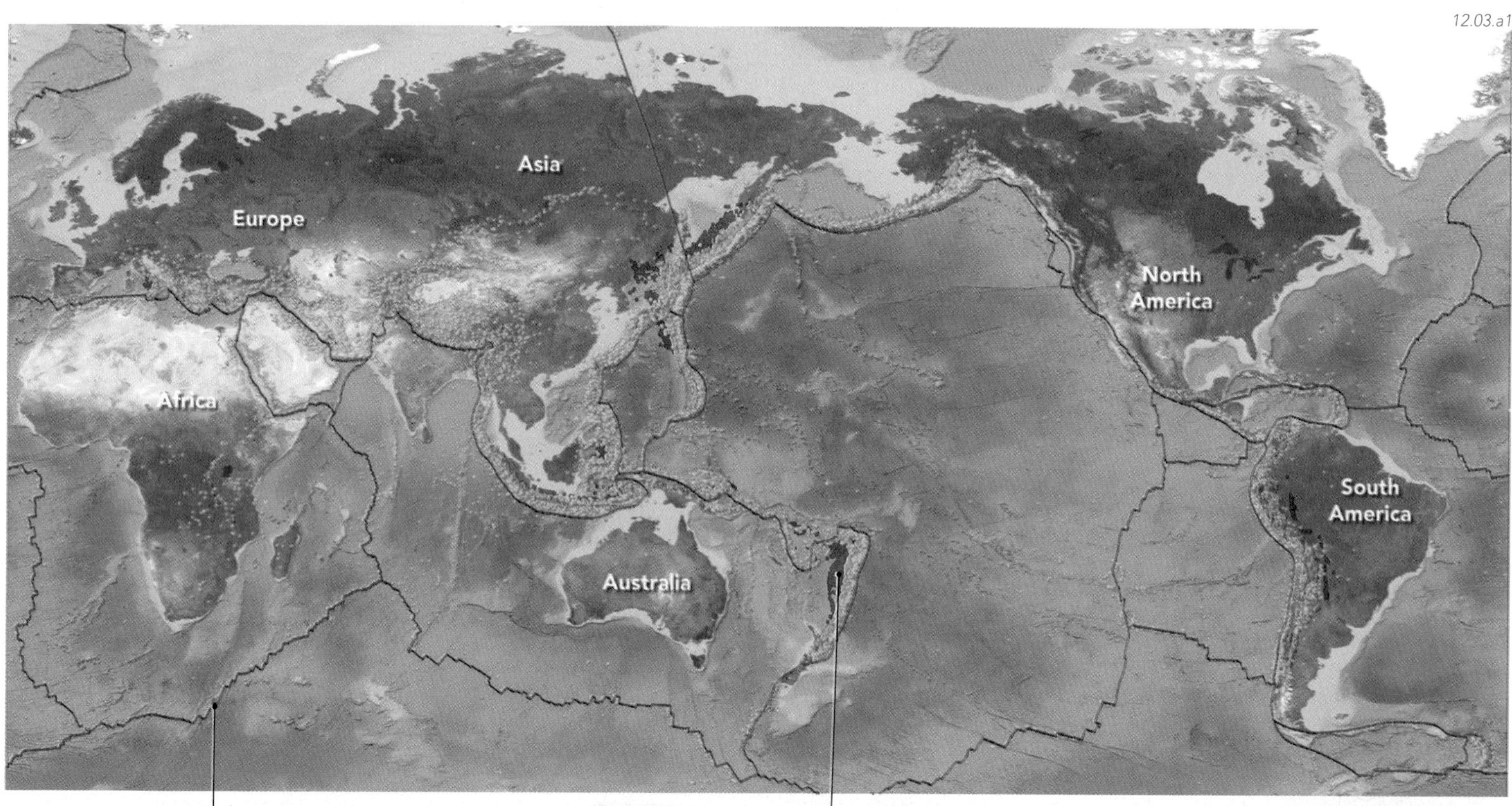

Most earthquakes occur in narrow belts that coincide with plate boundaries. Mid-ocean ridges, such as this one south of Africa, only have *shallow* earthquakes.

Deep- and *intermediate-depth* earthquakes occur only near subduction zones. There is a consistent pattern of shallow earthquakes close to the trench and progressively deeper earthquakes farther away. This pattern follows, and helps define, the position of the subducted slab, which is inclined from the shallow to the deep earthquakes.

B How Are Earthquakes Related to Mid-Ocean Ridges?

In mid-ocean ridges, seafloor spreading forms new oceanic lithosphere, which is very hot and thin. Stress levels increase downward in Earth, but the rocks in the lithosphere get too hot to fracture (they flow instead). As a result, earthquakes along mid-ocean ridges are relatively small and shallow, with hypocenters less than about 20 kilometers (12 miles) deep.

Many earthquakes occur along the axis of a mid-ocean ridge, where spreading and slip along normal faults downdrop blocks along the narrow rift. Numerous small earthquakes occur due to intrusion of magma into dikes.

Mid-Ocean Ridge
Transform Fault
Oceanic Crust
Lithospheric Mantle
Asthenosphere

12.03.b1

As the newly created plate moves away from the ridge, it bends and cools. The stress caused by the bending forms steep faults, which are associated with relatively small earthquakes.

Strike-slip earthquakes occur along transform faults that link adjacent segments of the spreading center. The typically thin lithosphere keeps earthquakes along these oceanic transform faults small.

C How Are Earthquakes Related to Subduction Zones?

A subduction zone, where an oceanic plate is underthrust beneath another plate, undergoes compression and shearing along the plate boundary. It can produce very large earthquakes.

1. As the oceanic plate moves toward the trench, it is bent and stressed, causing earthquakes in front of the trench.

2. Larger earthquakes occur in thrust faults formed in the *accretionary prism* as material is scraped off the downgoing plate.

3. Large earthquakes occur along the entire contact between the subducting plate and the overriding plate. The plate boundary is a huge thrust fault called a *megathrust*.

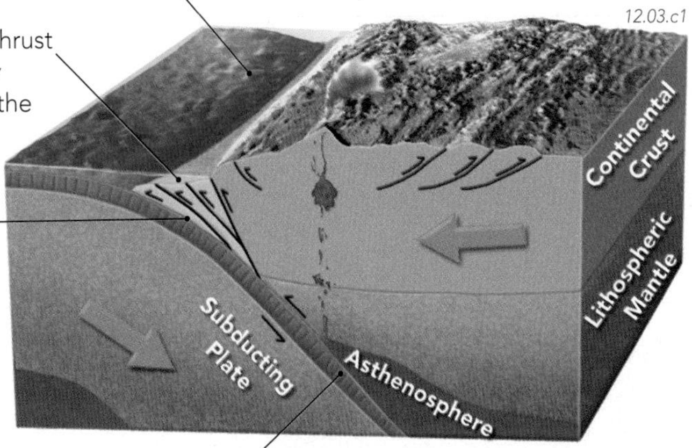

4. The downgoing oceanic plate continues to produce earthquakes from shearing along the boundary and from downward-pulling forces on the sinking slab. Subduction zones are typically the only place in the world producing deep earthquakes, as deep as 700 kilometers (430 miles). Below 700 kilometers, the plate is too hot to behave brittlely and fault.

5. Earthquakes can also be due to movement of magmas, volcanic eruptions, and thrust faulting behind the magmatic arc.

12.03.c1

Continental Crust

Lithospheric Mantle

Subducting Plate

Asthenosphere

6. A deep trench marks a subduction zone on the west side of South America.

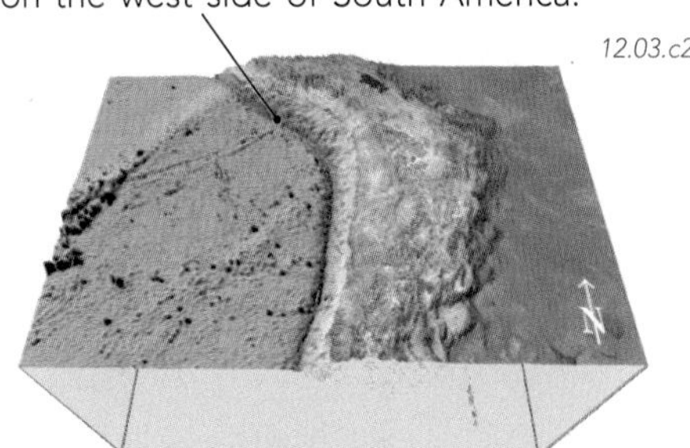

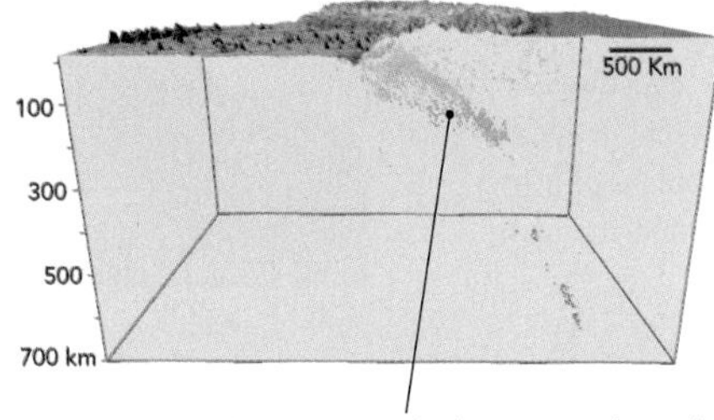

7. In a side view, subduction-related earthquakes, shown as dots, are shallower to the west (near the trench) and deeper to the east, recording the descent of the oceanic plate.

D How Are Earthquakes Related to Continental Collisions?

During continental collisions, one continental plate is underthrust beneath another. Large thrust faults form in both the overriding and underthrust plates near the plate boundary, causing large but shallow earthquakes.

Large, deadly earthquakes are produced along the plate boundary, or *megathrust*.

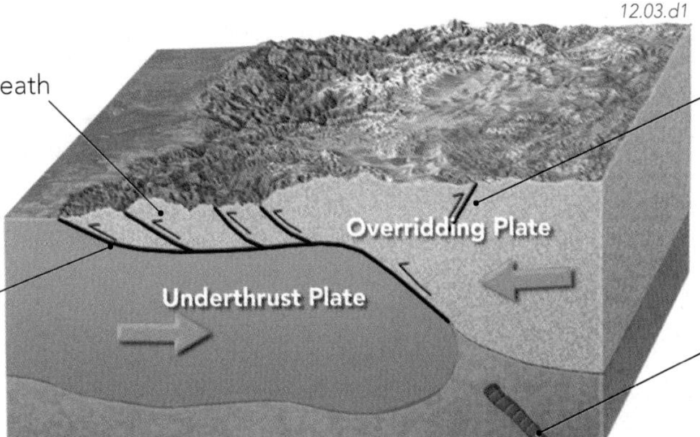

Thrust faults also form within both continental plates, causing moderately large earthquakes. The immense stresses associated with a collision can reactivate older faults within the interior of either continent. Strike-slip and normal faults may be generated as entire regions are stressed by the collision zone.

Any oceanic plate material that was subducted prior to the collision is detached, so subduction stops and deep earthquakes cease.

E How Are Earthquakes Generated Within Continents?

1. A transform fault, like the San Andreas fault, can cut *through* a continent, moving one piece of crust past another. The *strike-slip* motion causes earthquakes that are mostly shallower than 20–30 kilometers (10–20 miles), but that can be quite large.

2. Continental rifts generally cause *normal-fault* earthquakes, whether the rift is a plate boundary or is *within* a continental plate. Such earthquakes are typically moderate in size.

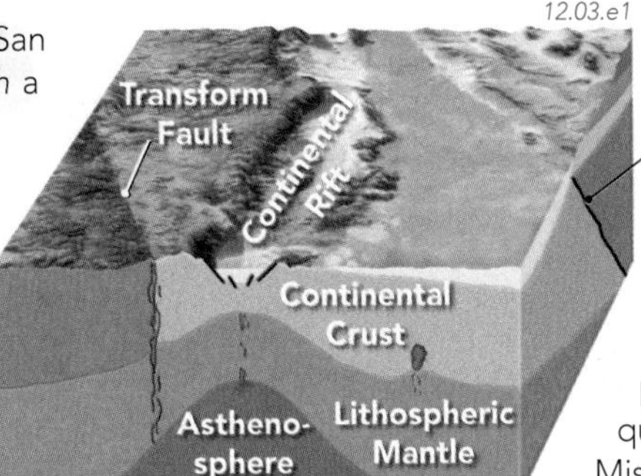

3. Intrusion of magma (shown here in red) within a plate can cause small earthquakes as the magma moves and creates openings in the rock.

4. Preexisting faults in the crust can readjust and move as the continental plate ages and is subjected to new stresses. These structures can produce large earthquakes, such as those in Missouri in 1811.

Before You Leave This Page Be Able To

- ✓ Explain why subduction zones have earthquakes at various depths, whereas mid-ocean ridges have only shallow earthquakes.
- ✓ Summarize how subduction and continental collisions cause earthquakes, identifying differences between these two settings.
- ✓ Describe how an earthquake can occur within a continental plate.

How Do Earthquake Waves Travel Through Earth?

EARTHQUAKES GENERATE VIBRATIONS that travel through rocks as physically distinguishable waves, called *seismic waves*. Geophysicists digitally record and process seismic waves in order to understand where and how the earthquake occurred. The word *seismic* comes from the Greek word for earthquake.

A What Kinds of Seismic Waves Do Earthquakes Generate?

Earthquakes generate different types of seismic waves. Those that travel *inside* Earth are called *body waves* and those that travel on the surface of Earth are *surface waves*. Scientists who study earthquakes are *seismologists*.

Shape of Waves

1. To describe seismic waves, we begin by defining waves in general. Most waves are a series of repeating crests and troughs.

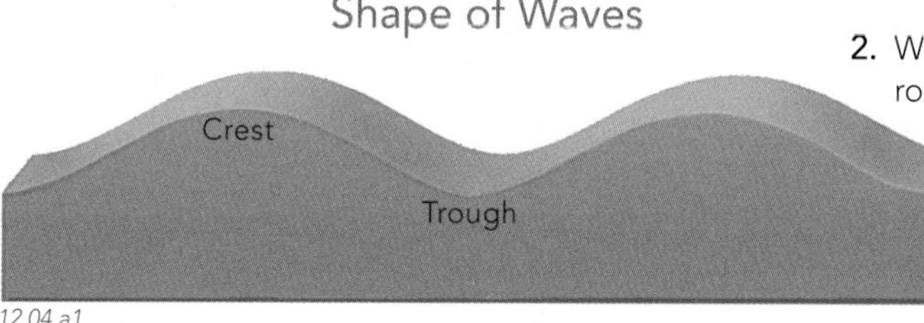

12.04.a1

2. Waves, whether moving through the ocean or through rocks, can travel, or *propagate*, for long distances. However, the material within the wave barely moves. Sound waves travel through the air and thin apartment walls, but the wall does not move much. Think of a wave as a pulse of energy moving *through* a nearly stationary material.

Types of Waves

3. An earthquake, as depicted by the red dot in the tan and gray block below, generates seismic waves. Most earthquakes occur at depth, so they first produce waves that travel through the Earth as *body waves*.

4. When body waves reach Earth's surface, some energy is transformed into new waves that only travel on the surface (*surface waves*). It is easier to visualize processes on the surface of Earth than within it, so we begin by discussing surface waves, of which there are two kinds.

Surface Waves

5. The first type of surface wave is a *vertical surface wave*. It is similar to an ocean wave, in that material moves up and down in an elliptical path. These earthquake waves propagate in the direction of the yellow arrow, or perpendicular to the crests of the waves.

6. The second type of surface wave is a *horizontal surface wave*, in which material vibrates horizontally and shuffles side to side. The motion of the material is *perpendicular* to the direction in which the wave travels.

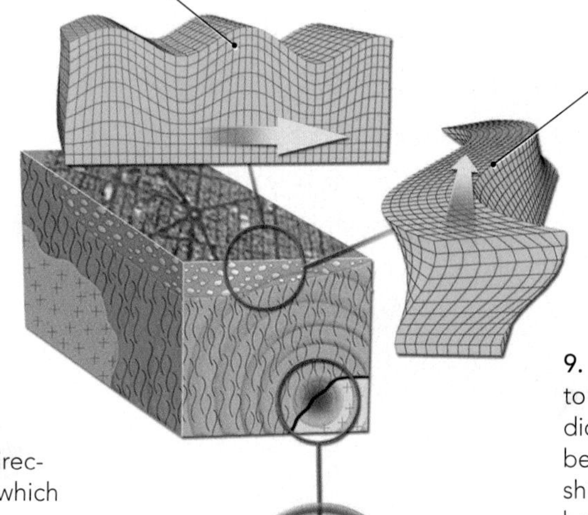
12.04.a2

Primary Body Wave

7. Body waves travel through Earth, and come in two main varieties. The *primary* or *P-wave* compresses the rock in the same direction it propagates. It is like a sound wave, which compresses the air through which it travels.

8. P-waves can travel through solids and liquids because these materials can be compressed and then released. The P-wave is the fastest seismic wave, and it travels through rocks at 6 to 14 kilometers/second depending on the properties of the rock.

Secondary Body Wave

9. The *secondary*, or *S-wave,* shears the rock side to side or up and down. This movement is perpendicular to the direction of travel. The wave shown below propagates to the right, but the material shifts up and down. It could also shift side to side, but the motion would still be perpendicular to the propagation direction of the wave.

10. S-waves cannot travel through liquids because liquids are not rigid (they cannot be sheared). If an area of the Earth's interior does not allow S-waves to pass, then it may be molten. S-waves are also slower than P-waves, travelling through rocks at about 3.6 kilometers/second.

B How Are Seismic Waves Recorded?

Sensitive digital instruments called *seismometers* are able to precisely detect a wide range of earthquakes. The recorded seismic data are uploaded to computers that process signals from hundreds of instruments registering the same earthquake. These computers calculate the hypocenter and magnitude, and produce digital maps showing the magnitude of ground shaking.

1. A *seismometer* detects and records the ground motion during earthquakes.

2. A large mass is suspended from a wire. It resists motion during earthquakes.

3. The mass hangs from a frame that in turn is attached to the ground. When the ground shakes, the frame shakes too, but the suspended mass resists moving because of *inertia*. As the ground and frame move under the mass, a pen attached to the mass marks a roll of recording paper that slowly rotates. As a result, the pen draws a line that records the ground movement over time.

4. This device only records ground movement parallel to the red arrows, so it only records a *single* direction *or component* of motion.

12.04.b1

5. Modern seismic detectors contain 3 seismometers oriented 90° from each other to record *three components* of motion (N-S, E-W, and up-down). From these three components, seismologists can better determine the direction and magnitude of the seismic signal.

6. Seismologists place seismometers away from human noise and bury them to reduce wind noise. Waves (in yellow) can come from any direction.

12.04.b2

C How Are Seismic Records Viewed?

1. Until the early 1990s, seismic waveforms were mostly represented as curves on a paper *seismogram*, which is a graphic plot of the waves. Seismologists developed this plot to better visualize the ground shaking caused by earthquakes. Today, most seismic data are displayed on computer screens, rather than on paper.

12.04.c1

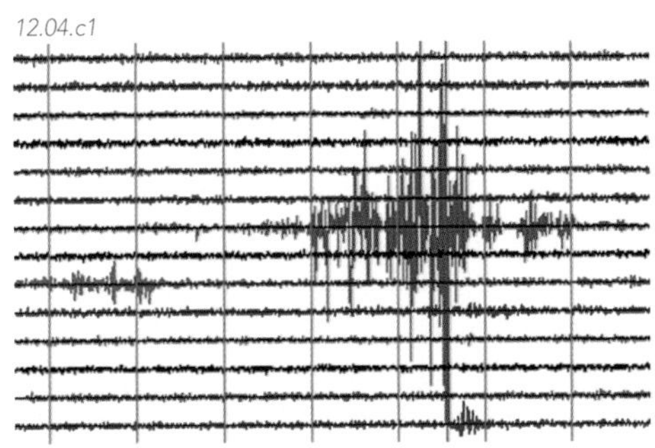

2. This diagram (seismogram) shows the record of an earthquake as recorded by a seismometer. It plots vibrations versus time. On seismographs, time is marked at regular intervals so that we can determine the time of the arrival of the first P- and S-waves.

3. Background noise commonly looks like small, somewhat random squiggles on seismograms.

4. After an earthquake, P-waves arrive first, marked by the larger squiggles. If the earthquake occurred at 8:00 am, the time of the P-wave's arrival was 2.5 minutes in this example.

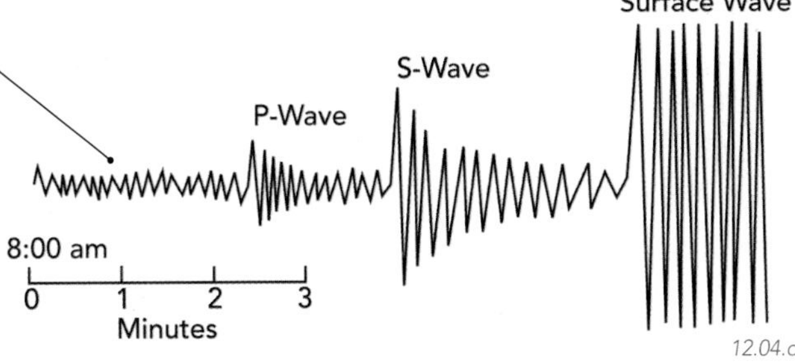

12.04.c2

5. The S-wave arrives later. The delay between the P-wave and the S-wave depends primarily on how far away the earthquake occurred. The longer the distance, the greater the delay.

6. Surface waves arrive last and cause intense ground shaking, as recorded by the longer squiggles on the seismograph.

Amplitude and Period

Seismic waves are characterized by how much the ground moves (*wave amplitude*) and the time it takes for a complete wave to pass by (*period*). Period is related to the wavelength and velocity of the wave. Both amplitude and wavelength can be measured from a seismogram. Amplitude is critical when estimating the magnitude and damage potential of an earthquake. The period can also be a critical component in assessing potential damage, because buildings vibrate when shaken by earthquakes. Every building has a natural period that can match, or *resonate* with, the earthquake wave. Resonance can cause intensified shaking and increased damage.

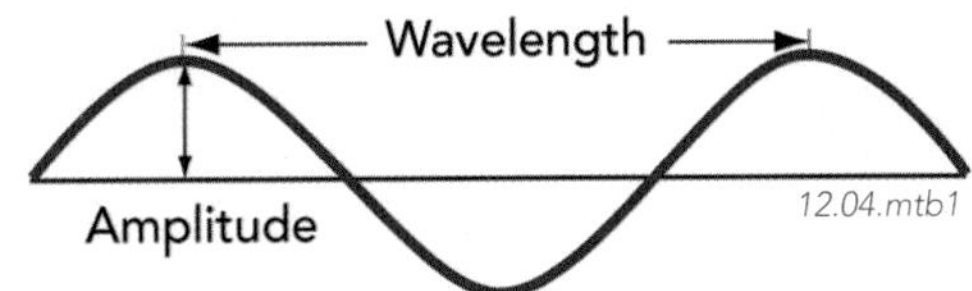

12.04.mtb1

Before You Leave This Page Be Able To

- ✓ Describe or sketch the characteristics of P-waves, S-waves, and surface waves, including the way motion occurs compared to the propagation of the wave.
- ✓ Sketch or describe how seismic waves are recorded, and the order in which they arrive at a seismometer.

How Do We Determine the Location and Size of an Earthquake?

EARTHQUAKES OCCUR DAILY AROUND THE WORLD, and a network of seismic instruments records these events. Using the combined seismic data from several instruments, seismologists calculate where an earthquake started and how large it was. The principal measurement of size is called *magnitude*.

A How Do We Locate Earthquakes?

Seismologists maintain thousands of seismic stations that actively sense and record ground motions. When an earthquake occurs, parts of this network can detect it. Large earthquakes generate seismic waves that can be detected around the world. Smaller earthquakes are detected only locally.

1. Seismometer Network Senses a Quake

Seismometers in the U.S. National Seismic Network (shown below) represent a fraction of all seismometers.

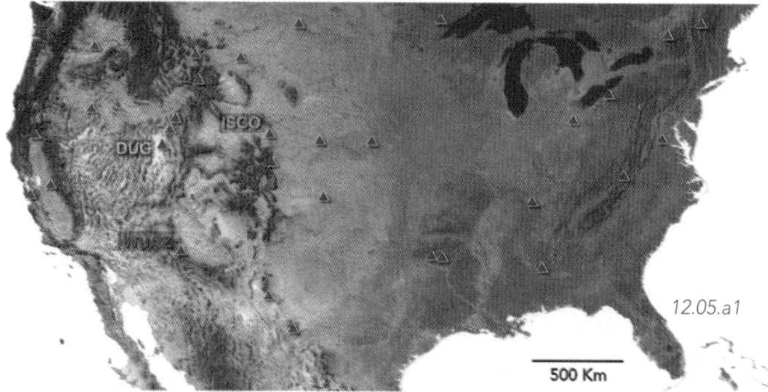

12.05.a1

On October 1, 2005, a moderate earthquake is felt in Colorado. Three stations (DUG, WUAZ, and ISCO) record wave arrivals and are chosen to locate the epicenter. Each station is given an abbreviation that reflects its location.

2. Select Earthquake Records

Records from at least three stations are normally compared when calculating an earthquake location.

P-waves travel faster than, and arrive before, S-waves, and the time interval between arrival of the P-wave and S-wave is called the *P-S interval*. The farther a station is from the earthquake, the longer the P-S interval will be. Picking the arrival of the P-wave and S-wave on these graphs is not always easy, but can be done by seismologists or by computer.

12.05.a2

The three seismograms show differences in the P-S interval. Based on the P-S intervals, ISCO is the closest station, followed by DUG and WUAZ.

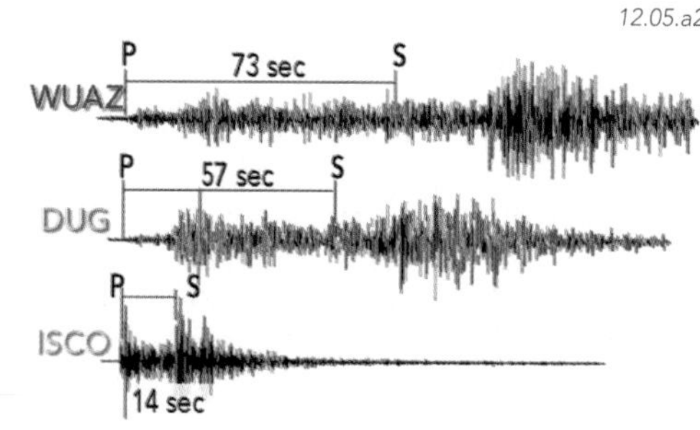

3. Estimate Station Distance from Epicenter

The P-S interval is mathematically related to the distance from the epicenter to the seismic station, factoring in the types of materials through which the waves pass. This relationship is shown on a graph as a *time-travel curve*.

12.05.a3

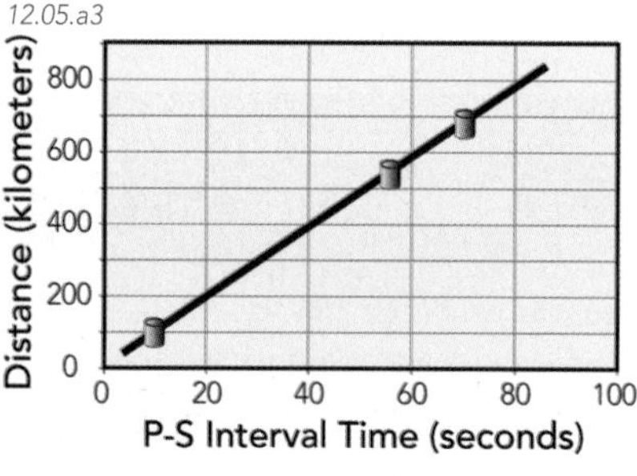

P-S intervals are measured from the seismograms shown in part 2 and then plotted on the graph. This gives the distance for each station.

Station	Distance (km)
WUAZ	670
DUG	540
ISCO	65

The distance from each station to the epicenter is now known, but not the direction.

4. Triangulate the Epicenter

The distance from each station to the earthquake can be compared graphically to find the epicenter.

12.05.a4

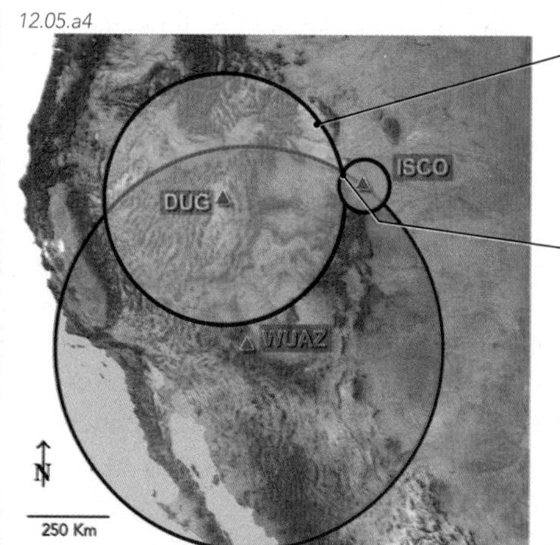

A circle is drawn around each station, with a radius equal to the distance calculated from the P-S interval.

The intersection of the three (or more) circles is the *epicenter* of the earthquake.

We calculate the depth of the earthquake's *hypocenter* in a similar way, using the interval between the P-wave and another compressional wave that forms when the P-wave reflects off Earth's surface near the epicenter. Again, we use multiple stations.

B How Do We Measure the Size of an Earthquake?

The *magnitude* of an earthquake is a measure of the released energy and is used to compare the sizes of earthquakes. There are several ways to calculate magnitude, depending on the earthquake's depth. The most commonly used scale, called the "Richter" or "Local" magnitude (Ml), is illustrated here.

Measuring Amplitude

The maximum height (amplitude) of the S-wave is measured on the seismogram. It is proportional to the earthquake energy. This measure is used for shallow earthquakes.

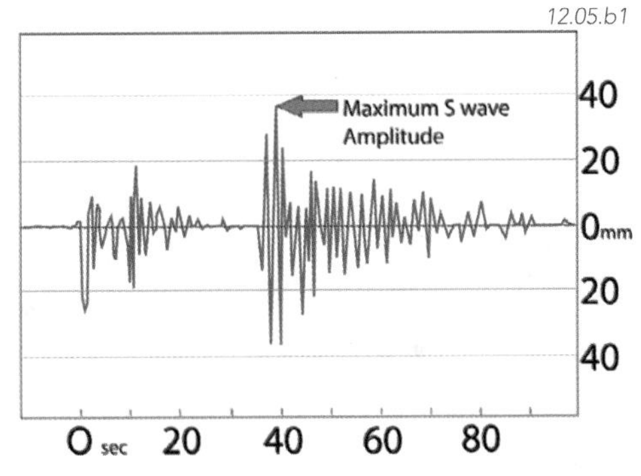

Seismographs are calibrated so that the measurements made by two different instruments are comparable.

Magnitude

This graph, called a nomograph, represents the mathematical relationship between distance, magnitude, and S-wave amplitude.

For each seismic station, a line is drawn connecting the distance and amplitude.

The earthquake's magnitude is read where each line crosses the center column. These three lines for the 2005 Colorado earthquake all agree, and yield a 4.1 Ml Local magnitude.

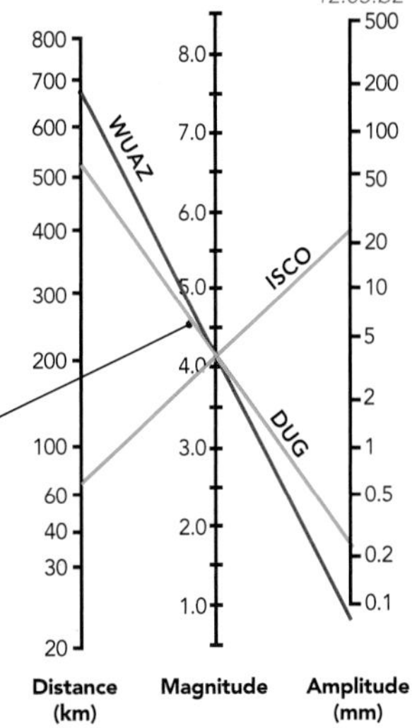

C What Can the Intensity of Ground Shaking Tell Us About an Earthquake?

Some of the most damaging earthquakes occurred before seismometers were in place. Reports of damage and shaking *intensity* are another way to classify earthquakes.

The Modified Mercalli Intensity Scale, abbreviated as *MMI*, describes the effects of shaking in everyday terms. A value of "I" reflects a barely felt earthquake. A value of "XII" indicates complete destruction of buildings, with visible surface waves throwing objects into the air!

A series of very large earthquakes in 1811 and 1812 shook Missouri, Arkansas, and the surrounding areas. Shaking was felt over a wide region. The magnitudes on this map, numbered from III to XI, indicate what you would feel and see if the earthquake happened today.

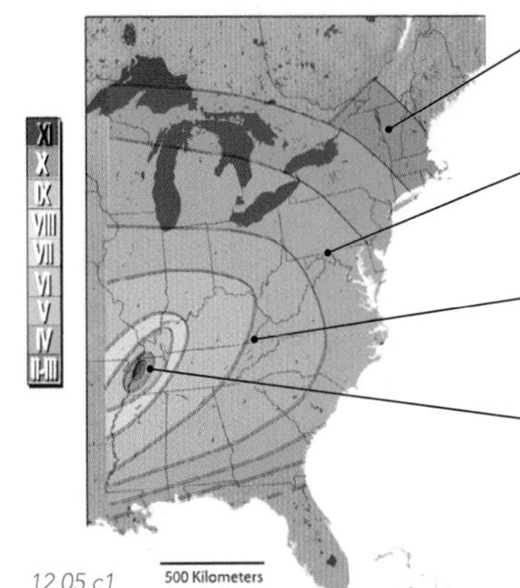

III. Felt strongly by persons indoors, especially on upper floors of buildings.

V. Felt by nearly everyone; many awakened. Some dishes and windows broken. Unstable objects overturned.

VI. Felt by all, many frightened. Some heavy furniture moved. Some plaster cracks and falls. Damage slight.

X and XI. Some well-built wooden structures destroyed. Most masonry and frame structures destroyed, along with foundations. Bridges destroyed. Rails bent. Damage extensive.

Energy of Earthquakes

Richter or local magnitude measures the amount of ground motion, but the scale is logarithmic so the ground motion increases by a factor of 10 from a magnitude 4 to a 5, from a 5 to a 6, and so on. The amount of energy released increases more than 30 times for each increase in magnitude, so a magnitude 8 releases more than 30 times more energy than a magnitude 7.

Another common measure of earthquake energy is *moment magnitude* or *Mw*, which is calculated from the amount of slip (displacement) on the fault and the size of fault area that slipped. Moment magnitude is useful for both large and small earthquakes. How do earthquakes compare to other energy releases with which we are familiar? An average lightning strike (*Mw* ~2) is miniscule compared to a small earthquake. However, an average hurricane is larger than the energy released by the largest historic earthquake (Chile, 1960).

Before You Leave This Page Be Able To

- ✓ Observe different seismic records of an earthquake and tell which one was closer to the epicenter.
- ✓ Describe how to use arrival times of P- and S-waves to locate an epicenter.
- ✓ Explain or sketch how we calculate local magnitude.
- ✓ Explain what a Modified Mercalli intensity rating indicates.

How Do Earthquakes Cause Damage?

MANY GEOLOGISTS SAY that "earthquakes don't kill people, buildings do." This is because most deaths from earthquakes are caused by the collapse of buildings or other human structures. Destruction can occur during the earthquake or later from fires, floods, and large ocean waves caused by the earthquake.

A What Destruction Can Arise from Shaking Due to Seismic Waves?

Direct damage from an earthquake results from ground shaking during the passage of seismic waves, especially surface waves near the epicenter of the earthquake. Damage can also be due to *secondary effects*, such as fires and flooding, that are triggered by the earthquake. The area below received mostly direct damage.

1. Mountainous regions that undergo ground shaking may experience *landslides*, *rock falls*, and other earth movements.

2. The ground can rupture along parts of the fault that slip during an earthquake or from shaking of unconsolidated materials. The cracks can destroy buildings and roads.

3. *Damage to structures* from shaking depends on the type of construction. Concrete and masonry structures are rigid and do not flex easily. Thus, they are more susceptible to damage than wood or steel structures, which are more flexible. In this area, a flexible, metal bridge in the center of the city survived the earthquake.

4. A concrete bridge farther downstream was too rigid and collapsed. Furthermore, it was built upon delta sediments that did not provide a firm foundation against shaking. In general, loose, unconsolidated sediment is subject to more intense shaking than solid bedrock.

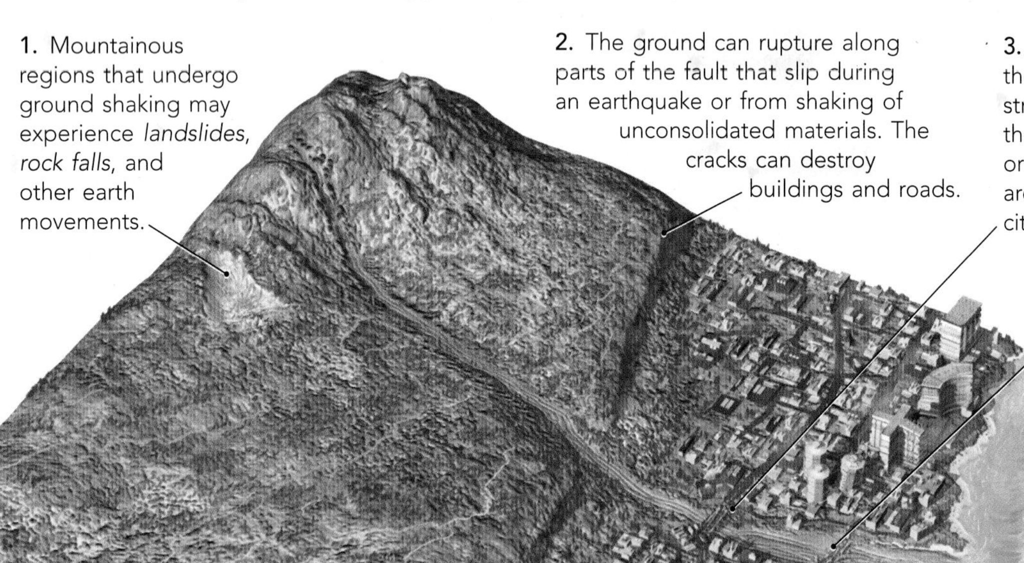

12.06.a1

5. *A tsunami* is a giant wave that can rapidly travel across the ocean. An earthquake that occurs undersea or along coastal areas can generate a tsunami, which can cause damage along shorelines thousands of kilometers away.

6. *Aftershocks* are smaller earthquakes that occur after the main earthquake, but in the same area. Aftershocks occur because the main earthquake changes the stress around the epicenter, and the crust adjusts to this change with more faulting. Aftershocks are very dangerous because they can collapse structures already damaged by the main shock. Aftershocks after a tsunami can cause widespread panic.

7. Ground shaking of unconsolidated, water-saturated sediment causes grains to lose grain-to-grain contact. When this happens, the material loses most of its strength and begins to flow, a process called *liquefaction*. This can destroy anything built on top (▽). [San Francisco, California, 1989]

8. Historically, most deaths from earthquakes are due to collapse of poorly constructed houses and buildings, such as those with mud, loosely connected blocks, and earthen walls. Even modern reinforced concrete can fail (▽). [Oakland, California, 1989]

12.06.a2

12.06.a3

B What Destruction Can Happen Following an Earthquake?

Some earthquake damage occurs from *secondary effects* that are triggered by the earthquake.

12.06.b1

Fire is one of the main causes of destruction after an earthquake. Natural gas lines may rupture, causing explosions and fires. The problem is compounded if water lines also break during the earthquake, limiting the amount of water available to extinguish fires. [Northridge, California]

12.06.b2

Earthquakes may cause both uplift and subsidence of the land surface by more than 10 meters (30 feet). Subsidence, such as occurred during the 2004 Sumatra earthquake, can cause areas that had been dry land before the earthquake to become inundated by seawater. [Sumatra]

12.06.b3

Flooding may occur due to failure of dams as a result of ground rupturing, subsidence, or liquefaction. Near Los Angeles in 1971, 80,000 people were evacuated because of damage to nearby dams during the 1971 San Fernando earthquake (Mw 6.7).

C How Can We Limit Risks from Earthquakes?

The probability that you will be affected by an earthquake depends on where you live and whether or not that area experiences tectonic activity. The risk of earthquake damage depends on the number of people living in the region, how well the buildings are constructed, and individual and civic preparedness.

▽ **1.** Earthquake hazard maps show zones of potential earthquake damage. Near Salt Lake City, Utah, the risk is greatest (reds) near active normal faults along the *Wasatch Front*, the mountain front east of the city.

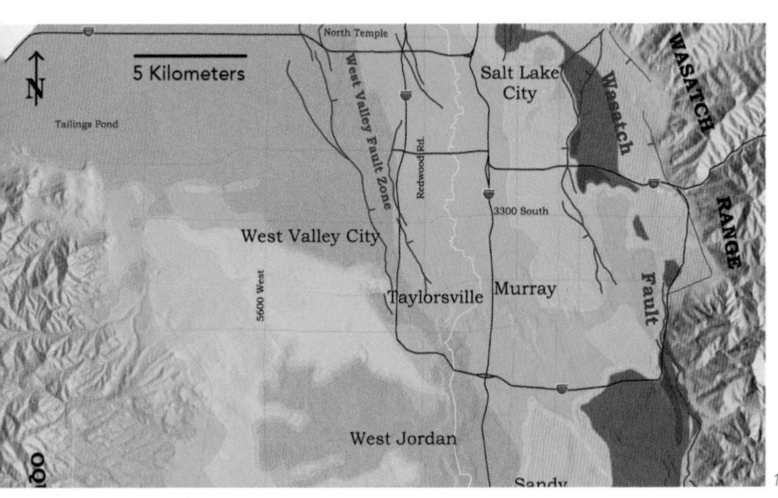

12.06.c1

2. Some utilities and hospitals have computerized warning systems that are notified of impending earthquakes by seismic equipment. The system will automatically shut down gas systems (to avoid fire) and turn on back-up generators to prevent loss of electrical power.

12.06.c2

3. Earthquakes have different periods, durations, and vertical and horizontal ground motion. This makes it difficult to design earthquake-proof buildings. Some rest on sturdy wheels or have shock absorbers (△) that allow the building to shake less than the underlying ground.

What to Do and Not Do During an Earthquake

There are actions you can take during an earthquake to reduce your chances of being hurt. If an earthquake strikes, you can seek cover under a heavy desk or table, and protect your head. You can also stand under door frames or next to inner walls, as these are the least likely to collapse. If possible, stand clear of buildings, especially those made of bricks and masonry.

During the shaking, stay away from glass and heavy objects that could fall, such as bricks or other loose debris. Always keep a battery-operated flashlight handy. Avoid using candles, matches, or lighters, since there may be gas leaks. Earthquakes may interrupt electrical and water service. Keeping 72 hours worth of food and water in an easily-carried backpack is a prudent plan for any natural disaster.

Before You Leave This Page Be Able To

- ✓ Describe how earthquakes can cause destruction, both during and after the main earthquake.
- ✓ Describe some ways to limit earthquake risk.
- ✓ Discuss ways to reduce personal injury during an earthquake.

What Were Some Major North American Earthquakes?

LARGE AND DAMAGING EARTHQUAKES have struck North America since written and oral records have been kept. We discuss seven important earthquakes here.

This map of the conterminous United States has yellow dots showing earthquakes that occurred in the last 15 years and that were larger than magnitude 4. The red lines on the map are faults that are interpreted to have slipped during the last 2 million years. Compare the distribution of earthquakes and recently active faults. Most active faults are in the western states, and most large earthquakes are in these same areas. Earthquakes have occurred elsewhere in the country, but most of these were too small to break the surface and form a fault scarp.

12.07.a1

Alaska, 1964

12.07.a2

◁ A magnitude (Mw) 9.2 earthquake, one of the two or three largest earthquakes ever recorded, struck southern Alaska in 1964. It killed 125 people and triggered landslides, and collapsed neighborhoods and the downtown of a nearby city. This event was caused by thrusting along the subduction zone. Most deaths and much damage were from a tsunami generated when a huge area of the sea floor was uplifted. This earthquake is not shown on the map.

San Francisco, 1906

12.07.a3

A huge earthquake occurred when 470 kilometers (290 miles) of the San Andreas fault ruptured near San Francisco. The earthquake was likely a magnitude (Ml) ~8. It ruptured the surface, leaving behind a series of cracks and open fissures. Within San Francisco, ground shaking destroyed most of the brick and mortar buildings. More than 300 people were killed and much of the city was devastated by fires that broke out after the earthquake.

Northridge, Los Angeles Area, 1994

This magnitude (Mw) 6.7 earthquake was generated by a thrust fault northwest of Los Angeles. The earthquake killed 57 people and caused $20 billion in damage. A section of freeway buckled, crushing the steel-reinforced concrete slabs. The thrust is not exposed on the surface, but when it ruptured it lifted up a large section of land. Geologists are concerned about a similar fault causing a similar earthquake right below downtown Los Angeles.

12.07.a4

Mexico City, 1985

A magnitude (Ml) 8.0 earthquake occurred on a subduction zone along the southwestern coast of Mexico, well west of Mexico City (not shown on this map). It damaged or destroyed many buildings in Mexico City and killed at least 9,500 people. Destruction was so extensive partly because Mexico City is built on lake sediments deposited in a bowl-shaped basin, which amplified the seismic waves. This geologic setting caused intensified and highly destructive ground shaking. Surface waves, which caused the most damage, traveled 200 kilometers (120 miles) from their source!

12.07.a5

Hebgen Lake, Yellowstone Area, 1959

◁ This magnitude (Ml) 7.5 event was generated by slip along a normal fault northwest of Yellowstone National Park. Ground shaking set loose the massive Madison Canyon slide, which buried 26 campers and formed a new lake, aptly named *Earthquake Lake.*

New Madrid, 1811–1812

New Madrid, Missouri experienced a series of large (Ml ~8) earthquakes generated over an ancient fault zone in the crust. The 1811–1812 earthquake death toll was probably low because of the sparse population at the time. The New Madrid zone has a high earthquake risk and, as shown on this earthquake-hazard map, is one of two areas in the eastern United States that are predicted to experience strong earthquakes in the future. Memphis lies in this zone, yet most of its buildings are not constructed to survive large earthquakes.

Charleston, 1886

This earthquake occurred at the other high-risk area along the East Coast, near Charleston, South Carolina. It had an estimated magnitude of Ml 7.3, the largest ever recorded in the Southeast. Buildings incurred some damage (▷), and 125 people died. The tectonic cause for this earthquake is still debated among geologists.

Earthquakes in the Interiors of Continents

Why do large earthquakes occur in the middle of continents, such as New Madrid, Missouri? Although the interior of North America is not near a plate boundary, the region is subjected to stress generated along far-off plate boundaries. Such stress includes compression from the Mid-Atlantic Ridge, called *ridge push.* These stresses can reactivate ancient faults that lie buried beneath the cover of sediment. In the case of New Madrid, there is seismic and other geophysical evidence to suggest that the area is underlain by an ancient rift basin that formed about 750 million years ago during the breakup of the supercontinent of Rodinia. Modern-day stress related to the current plate configuration is interacting with the ancient faults, occasionally causing them to slip and trigger earthquakes.

Before You Leave This Page Be Able To

- ✓ Describe some large North American earthquakes and how they were generated.
- ✓ Summarize the various ways these earthquakes caused damage.
- ✓ Describe why the eastern United States has earthquake risks.

What Were Some Major World Earthquakes?

THE WORLD HAS ENDURED a number of large and tragic earthquakes. These earthquakes have struck a collection of geographically and culturally diverse places, causing many deaths and extensive damage. Most large earthquakes occur along or near plate boundaries, especially along subduction zones.

12.08.a1

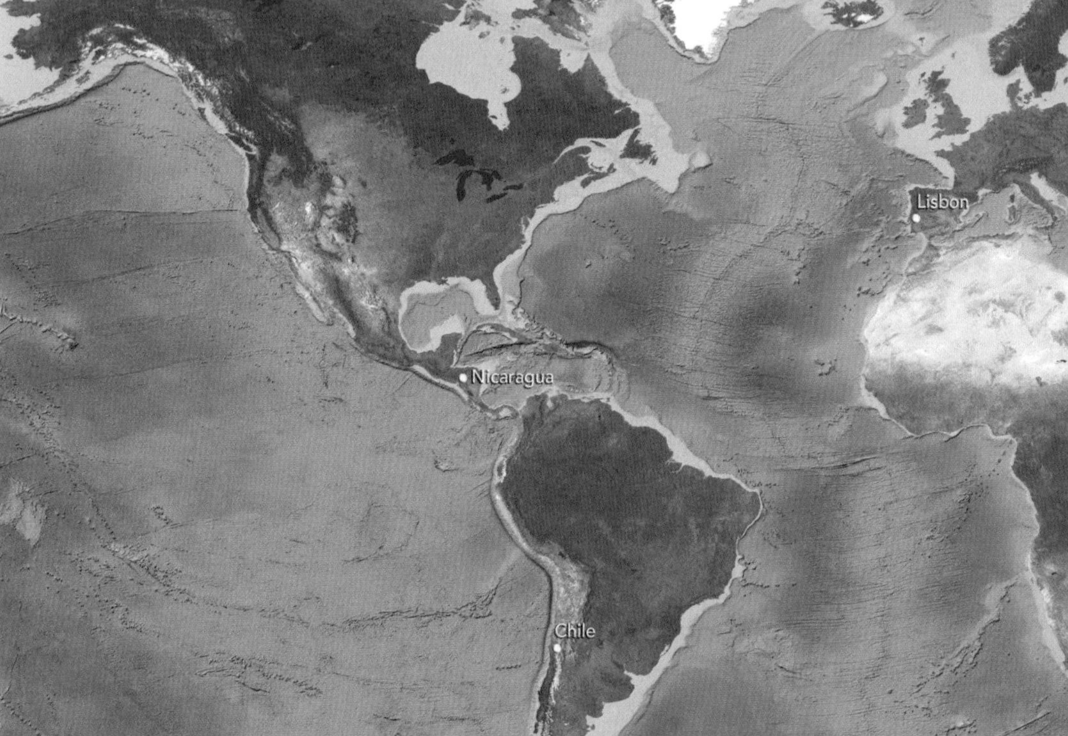

Nicaragua, 1972

On December 23, 1972, a magnitude (Mw) 6.2 earthquake killed about 6,000 people in central America. In the capital city of Managua, wood and adobe structures were leveled and fractures opened in the street (◁). The earthquake was caused by strike-slip along a boundary of the Caribbean plate.

12.08.a2

Chile, 1960

This huge, magnitude (Mw) 9.5 earthquake occurred offshore along a *megathrust* and triggered a destructive, Pacific-wide tsunami. At least 3,000 people died and $550 million of damage was done to infrastructure and buildings, such as in this city in Chile. ▽

12.08.a3

12.08.a4

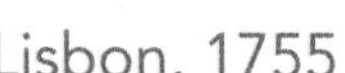

Lisbon, 1755

On November 1 (All Saints Day) in 1755, a large earthquake, estimated at magnitude (Mw) 8.5, shook Lisbon, Portugal. The earthquake demolished the city and triggered tsunamis, which sank ships in Lisbon's famous harbor. Photography was not yet invented, but the destruction was portrayed by artists (◁). The event caused an upheaval in religious and scientific thought, as people began to think that such catastrophes must be due to natural causes, since this one struck on such a holy day.

Turkey, 1999

In 1999, a large quake (Mw 7.4) generated along a transform fault zone killed more than 17,000 people and severely impacted the economy. The earthquake destroyed many buildings, including these multi-story apartment complexes. ▽

12.08.a5

12.08.a6

Armenia, 1988

◁ Old masonry structures were shaken apart in Leninakan, Armenia, in December 1988. This earthquake killed 25,000 people and was caused by an on-land transform fault related to lateral movement of pieces of southwestern Asia.

Kobe, Japan, 1995

▷ A magnitude (Mw) 7.2 thrust earthquake, also called the *Great Hanshin* earthquake, was the most damaging to strike Japan since the *Great Kanto* earthquake of 1923. The Kobe earthquake killed 6,500 and left 300,000 homeless. Damage was due largely to ground shaking and liquefaction.

12.08.a7

Chi Chi, Taiwan, 1999

This megathrust rupture generated a magnitude (Mw) 7.6 earthquake that was felt across Taiwan, killing 2,400 and displacing 600,000 people. The ShihKang Dam, 50 kilometers (30 miles) from the epicenter, was breached (▽), shutting off the local water supply.

12.08.a8

Sumatra, 2004

This magnitude (Mw) 9 earthquake was along a subduction zone (megathrust) west of the island of Sumatra. It offset the sea floor and unleashed a deadly tsunami that struck coasts around much of the Indian Ocean. The earthquake left more than 250,000 people dead or missing, mostly in Indonesia. It is discussed in more detail at the start of this chapter.

Deadly Earthquakes

Mortality due to earthquakes averages about 10,000 per year. Most earthquake-related deaths are due to collapse of poorly built structures in cities and villages. Earthquake-generated tsunamis also account for a large part of the yearly average. The table to the right shows some deadly earthquake events. The highest death tolls are due to a deadly combination of high population densities, substandard construction practices, and being situated along subduction zones or other high-risk areas.

Fatalities	Mw	Year	Location
8130,000	8	1556	Shaanxi, China
11,000	6.9	1857	Naples, Italy
70,000	7.2	1908	Messina, Italy
200,000	8.6	1920	Ninxia, China
143,000	7.9	1923	Kanto, Japan
200,000	7.9	1927	Tsinghai, China
32,700	7.9	1939	Erzincan, Turkey
66,000	7.9	1970	Colombia
23,000	7.5	1976	Guatemala
242,000	7.8	1976	Tangshan, China
31,000	6.6	2003	Bam, Iran

Before You Leave This Page Be Able To

- ✓ Briefly describe some of the world's most significant earthquakes and the tectonic settings in which these deadly earthquakes occurred.
- ✓ Summarize some ways that these earthquakes caused deaths.

How Does a Tsunami Form and Cause Destruction?

AN EARTHQUAKE BENEATH THE OCEAN can cause a large wave called a *tsunami*, which can wreak havoc on coastal communities. Most of Earth is covered by oceans, so many earthquakes, landslides, and volcanic eruptions occur beneath the sea. Each of these can generate a tsunami.

A How Are Tsunamis Generated?

Tsunamis are waves that affect an entire body of water from top to bottom. They are generated by abrupt changes in water level in one area relative to another. This occurs when a mass is dropped into the water, such as a landslide, or when the ocean floor is unevenly uplifted or downdropped by an earthquake.

1. A tsunami forms when a sudden change in sea level accompanies fault movement. The most common way this occurs is when a fault beneath the sea uplifts or downdrops an area of the seafloor.

2. A tsunami is a wave, or series of waves, that radiates away from the disturbance. It travels at speeds between 600–800 kilometers per hour (370–500 miles per hour) away from the source. In deep water, the wave energy is distributed over the entire water depth, forming a wave only a meter or so high but more than 700 kilometers across. It is so low relative to its width that it may be barely noticeable; it is much smaller in height out at sea than can be shown here.

3. As the wave approaches the shore, its energy is distributed over a smaller depth. The velocity of the front of the wave decreases to 30–40 km/hour, causing following water to pile up in a higher wave. Near shore, the tsunami becomes a massive, thick wave, like the front wall of a plateau of water. It may be a series of such waves.

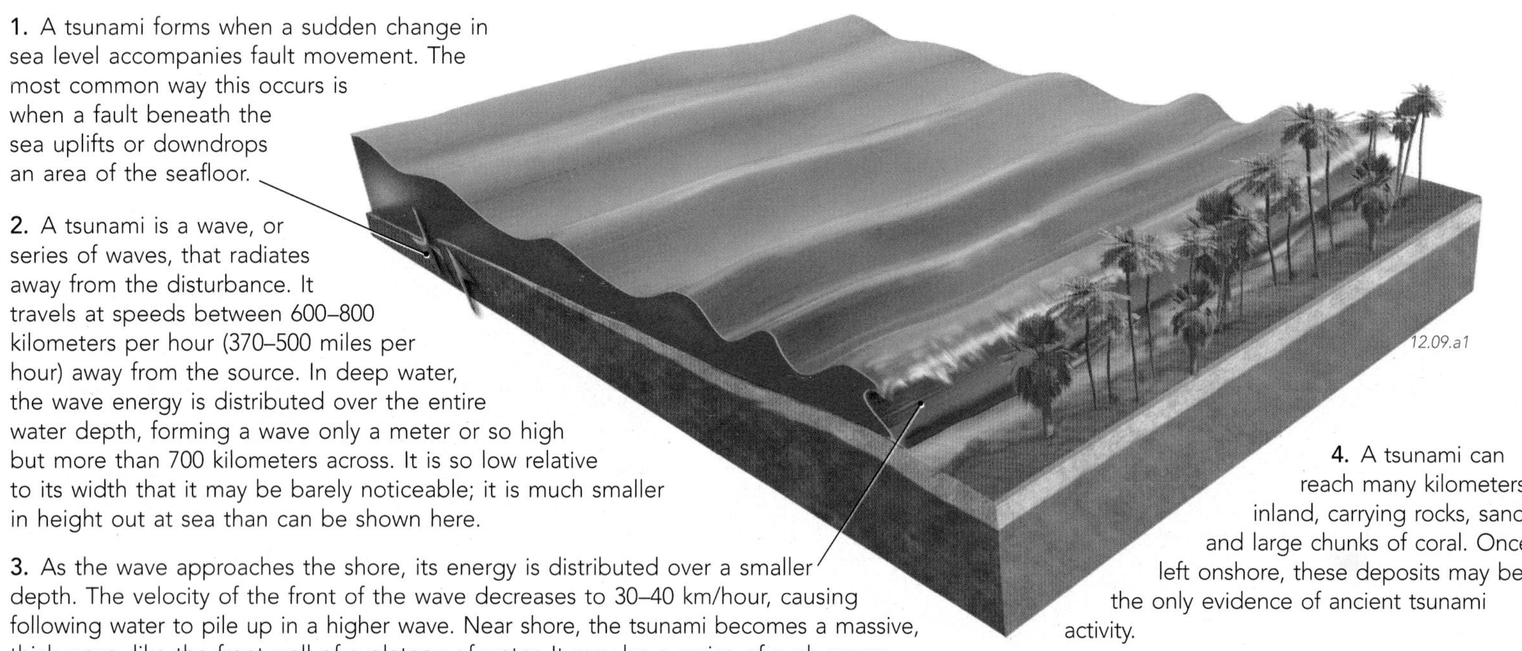

4. A tsunami can reach many kilometers inland, carrying rocks, sand, and large chunks of coral. Once left onshore, these deposits may be the only evidence of ancient tsunami activity.

Tsunamis Triggered by Landslides

A large mass of rock entering the water can catastrophically displace the water, generating a tsunami that radiates outward.

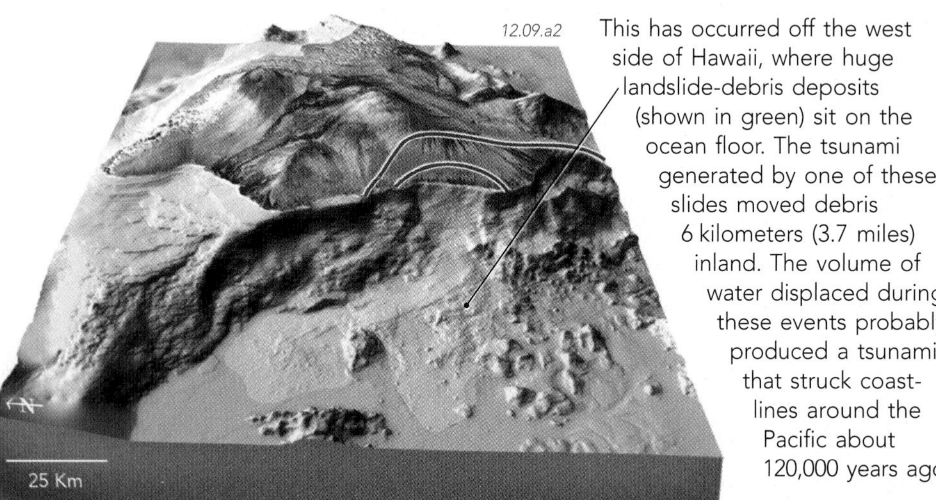

This has occurred off the west side of Hawaii, where huge landslide-debris deposits (shown in green) sit on the ocean floor. The tsunami generated by one of these slides moved debris 6 kilometers (3.7 miles) inland. The volume of water displaced during these events probably produced a tsunami that struck coastlines around the Pacific about 120,000 years ago.

Tsunamis Caused by Eruptions

The 1883 eruption of Krakatau in Indonesia, and the collapse of its immense caldera, generated a series of huge tsunamis that killed 36,000 people. A single catastrophic volcanic explosion produced the loudest sound ever heard, and most of Krakatau Island was demolished. The tsunami was as high as 40 meters (more than 130 feet), and some effects of the tsunami were recorded 7,000 kilometers away!

B What Kind of Destruction Can a Tsunami Cause?

Tsunamis cause death and destruction along coastlines where human populations are concentrated. On May 22, 1960, the largest earthquake ever recorded on a seismograph (Mw 9.5) occurred in the subduction zone (megathrust) offshore of southern Chile. The tsunamis that followed flattened coastal settlements in Chile, and traveled across the Pacific to devastate coastlines in Hawaii and Japan.

Chile, May 22, 1960

12.09.b1

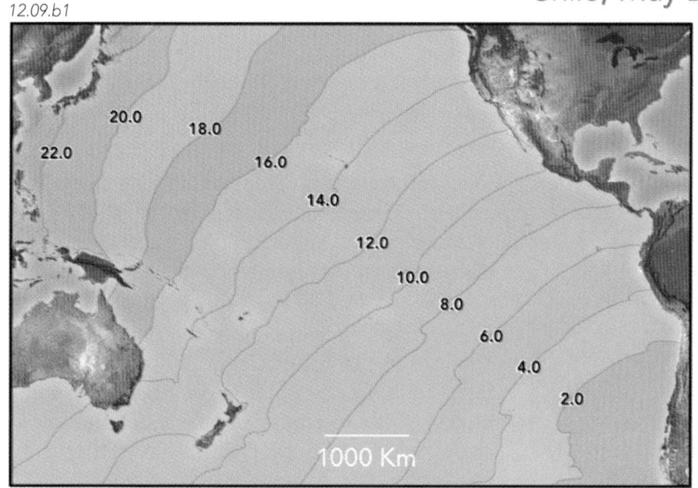

◁ During this earthquake, tsunamis were generated parallel to the coast. One headed in toward the shoreline, quickly striking Chile and Peru. Another set of tsunamis swept out across the Pacific Ocean at 670 kilometers (420 miles) per hour! Each stripe equals two hours of travel time.

12.09.b2

△ In Chile, the tsunamis struck 15 minutes after the earthquake. On Isla Chiloe, a 10-meter-tall wave swept over towns. The waves killed at least 2,000 people along the Peru-Chilean coast.

Hawaii, May 23, 1960

12.09.b3

About 15 hours after the earthquake in Chile, the tsunami related to the earthquake hit Hilo and other parts of Hawaii (△). A wave 11 meters (36 feet) high killed 61 people, damaged buildings, and caused $23 million in damage. Seven hours later, the tsunami killed 140 people in Japan.

Hokkaido, Japan 1993

12.09.b4

In 1993, a magnitude 7.8 earthquake occurred off the west coast of Hokkaido and within five minutes a tsunami struck the coastline. The tsunami killed at least 100 people and caused $600 million in property loss. It swept these boats inland across a concrete barrier built along the shoreline.

Papua New Guinea, 1998

12.09.b5

In 1998, a magnitude 7.1 earthquake and associated underwater landslides generated three tsunami waves that destroyed villages along the country's north coast, killing 2,200 people. A 10-meter-high wave destroyed a row of heavily populated houses along the coast shown here.

Tsunami Warning System

In an international effort to save lives, the United States National Oceanic and Atmospheric Administration (NOAA) maintains two *tsunami warning centers* for the Pacific Ocean. Twenty-six nations participate in this effort. Informed by worldwide seismic networks, these centers broadcast warnings based on an earthquake's potential for generating a tsunami. Since the huge loss of life after the Sumatran earthquake and accompanying tsunami in 2004, the United Nations has begun implementing a warning system in the Indian Ocean. Scientists are deploying warning buoys, like the one shown below, which can relay tsunami data by satellite. These buoys detect small changes in sea level as a tsunami passes underneath.

12.09.mtb1

Before You Leave This Page Be Able To

- ✓ Describe the different mechanisms by which tsunamis are generated.
- ✓ Summarize the kinds of damage tsunamis have caused.
- ✓ Briefly describe how tsunamis are monitored to provide an early-warning system.

How Do We Study Earthquakes in the Field?

GEOLOGISTS USE A VARIETY of tools and techniques to study evidence left behind by recent and ancient earthquakes. They examine and measure faults in natural exposures and in trenches dug across faults. Satellites and other tools allow faults to be studied in new and exciting ways.

A How Do We Study Recent Earthquakes in the Field?

Where a fault is visible at Earth's surface, it can be scrutinized in order to understand how it moves during an earthquake. Geologists investigate numerous features, some of which are shown below.

1. Faulting during an earthquake commonly is accompanied by smaller structures such as cracks and smaller faults. The geometry of these and other features can indicate the direction of fault movement.

2. When a fault moves, it can offset natural and human-made features. Streams and gullies, as well as roads, fences, and telephone lines, provide pre-earthquake reference points. Geologists can measure how much and in what direction the fault has offset these features.

12.10.a1

3. When a fault ruptures the surface, geologists carefully measure its location, dimensions, and orientation. Detailed drawings and photographs are essential for documenting features along the fault.

4. Faulting is commonly accompanied by changes in the topography of the land surface. Faulting can uplift linear ridges or form new hills. It can create ponds and other low areas by downdropping areas along the fault.

5. Rocks and soils, both in natural exposures and in trenches dug to study a fault, preserve a history of motion. They give clues to the magnitudes and recurrence of past earthquakes.

B How Do We Study Faults with Satellites and Geology-Based Models?

Ground Displacement

12.10.b1

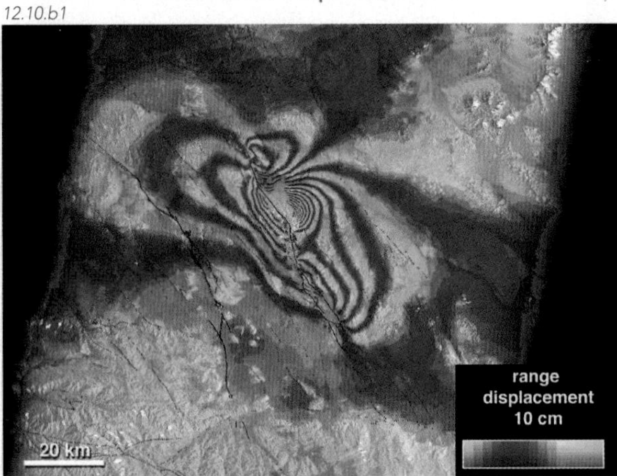

△ The topography around a fault changes when the fault moves. Very small changes in elevation can be detected through laser surveying or by comparing satellite radar data sets before and after faulting. To use the satellite method, an area is mapped before and after the earthquake. The two maps are combined into an *interferogram*, which shows how Earth has deformed near the fault rupture. In this image, color bands or fringes indicate strike-slip movement associated with the 1999 Hector Mine earthquake (Mw 7.1) in southern California. The fault is cutting diagonally northwest through the view.

Geologic Control of Damage

Damage from earthquakes can be compared to underlying geology. This helps geologists understand the patterns of damage and help plan for future earthquakes. This geologic map of San Francisco shows weak sediments as gray and light yellow.

The map below shows a geology-based model estimating the acceleration of the ground during the 1906 San Francisco earthquake. Dark reds indicate the most intense ground movement. Notice that many areas underlain by weak sediments experienced high accelerations.

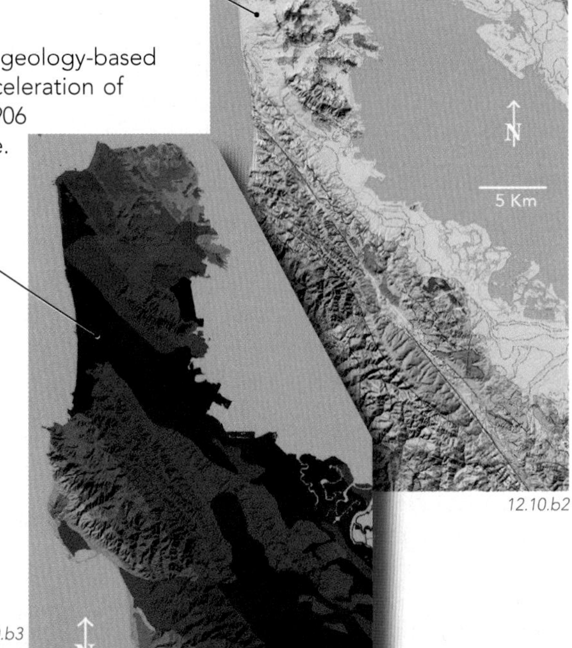

12.10.b2

12.10.b3

Features Along the San Andreas Fault

The San Andreas fault generally has a clear expression in the landscape. It is marked by a number of features that are common along active faults. Some of these features also can form in ways unrelated to active faulting.

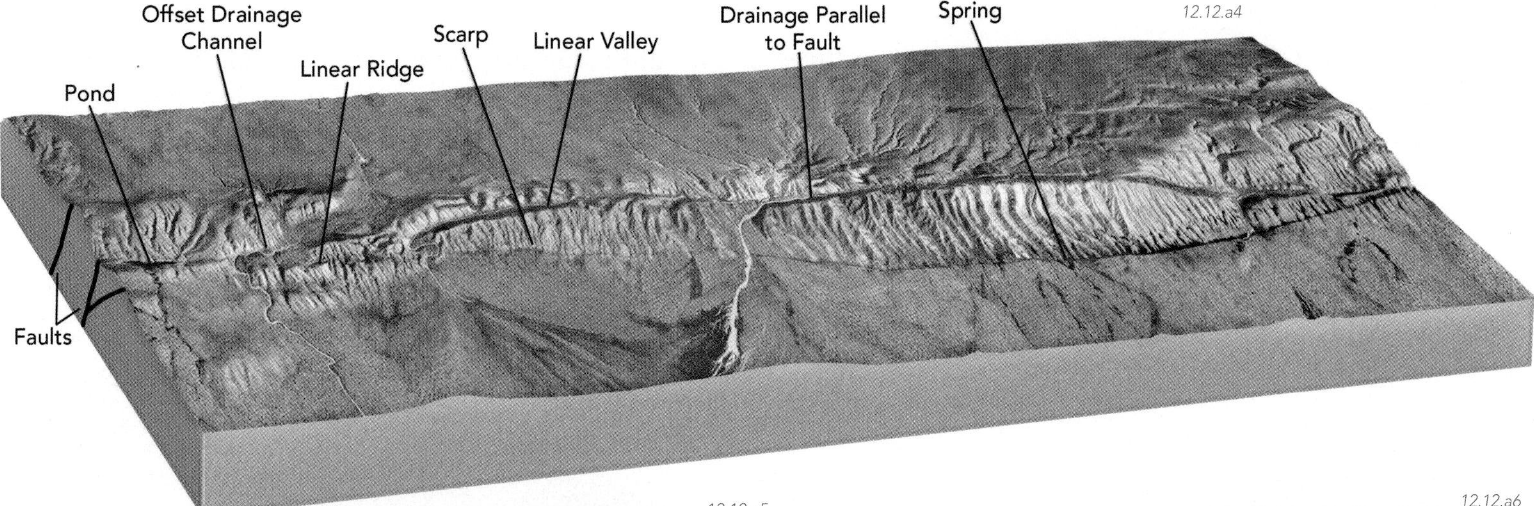

▷ Geologists explore the fault to find localities that have preserved a record of past faulting. Detailed studies of trenches dug across the fault help geologists unravel hundreds or thousands of years of the fault's history.

▷ The aerial photograph to the right shows the same part of the San Andreas fault as depicted in the figure above. Can you match some of these features between the two images?

7. North and east of the San Andreas is a series of faults, called the East California Shear Zone. This zone caused several >7 magnitude earthquakes in the 1900s and the large 1872 Owen Valley earthquake on the eastern side of the Sierra Nevada. The zone continues from the eastern side of the Sierra Nevada southward through the Mojave Desert, where it unleashed the 1992 Landers earthquake (Mw 7.3) and the 1999 Hector Mine earthquake (Mw 7.1).

8. On the map, note that the San Andreas fault has a distinct curve or *bend* in the middle of the southern locked (orange) segment. The bend causes regional compression and thrust faults, some of which are not exposed at the surface. These thrust faults caused the 1994 magnitude (Mw) 6.7 *Northridge earthquake* in metropolitan Los Angeles, and have uplifted the large mountains north and northeast of the city.

9. East of Los Angeles, the San Andreas branches southward into several faults. Some of these experienced several moderate-sized earthquakes in the 1900s, including some near important agricultural areas. The fault scarps for these events are colored pink and red on this map.

Before You Leave This Page Be Able To

- ✓ Briefly summarize the main segments of the San Andreas fault and whether they have had major earthquakes.
- ✓ Summarize features that might help you recognize the fault from the air.

How Do We Explore What Is Below Earth's Surface?

OUR VIEW OF GEOLOGY is typically limited to those rocks and structures that are exposed at the surface. In deep canyons we can glimpse subsurface rocks and structures. How else do we determine what lies beneath the surface?

1. The region shown here has a few hills of granite and a dark lava flow, but is otherwise covered by soil and vegetation. There are few clues as to what types of rocks and structures lie below the surficial cover. There are two general approaches for investigating subsurface geology: *obtaining samples* of rocks at depth, and performing *geophysical surveys* that measure the subsurface magnetic, seismic, gravity, and electrical properties.

12.13.a2

◁ 2. As magma rises to the surface, it can incorporate pieces of the rock through which it passes. Geologists study such pieces, called *inclusions*, to determine the types of rocks that lie beneath volcanoes.

3. We can gain a sense of what is below the surface by examining rocks and geologic structures that have been uplifted and are exposed at the surface. Geologists study rocks under the microscope to constrain the temperature and pressure conditions under which the rocks formed and to infer the geologic processes that created the rocks.

12.13.a3

◁ 4. Mines provide a more detailed subsurface view because the tunnels provide continuous exposures of rocks and structures. Some South African mines, such as the one in this photo, are deeper than 5 kilometers (3 miles).

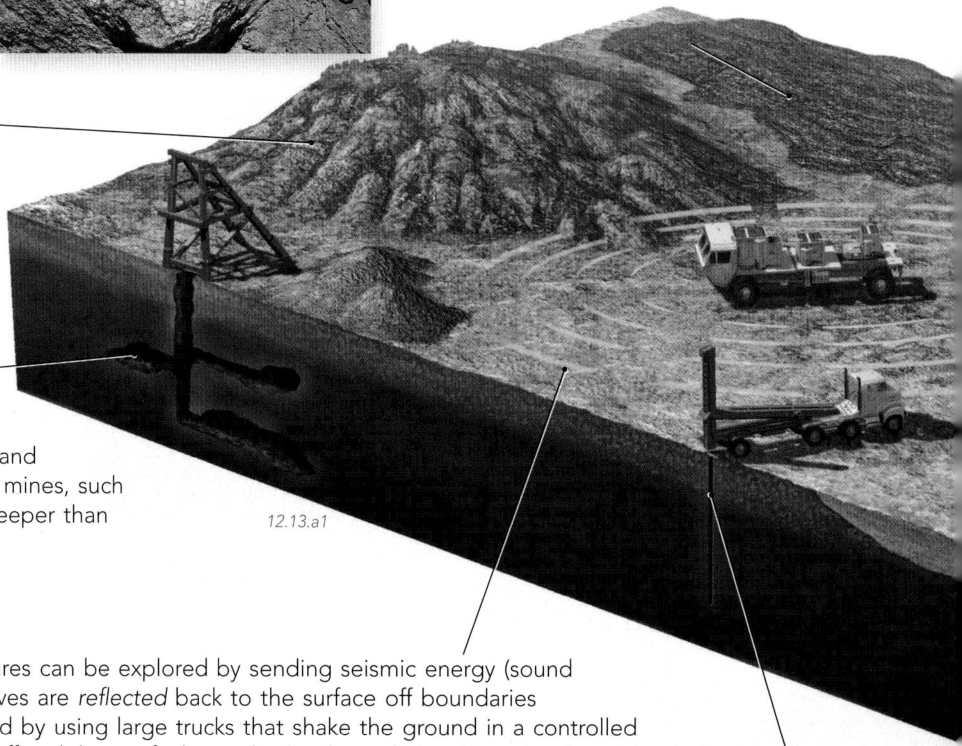

12.13.a1

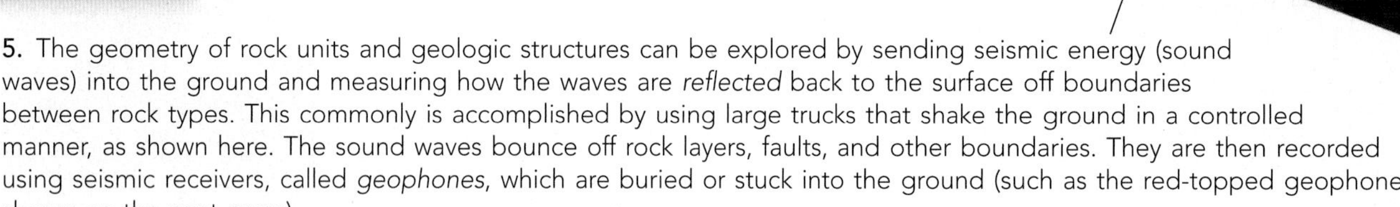

5. The geometry of rock units and geologic structures can be explored by sending seismic energy (sound waves) into the ground and measuring how the waves are *reflected* back to the surface off boundaries between rock types. This commonly is accomplished by using large trucks that shake the ground in a controlled manner, as shown here. The sound waves bounce off rock layers, faults, and other boundaries. They are then recorded using seismic receivers, called *geophones*, which are buried or stuck into the ground (such as the red-topped geophones shown on the next page).

6. Seismic-reflection data are processed using sophisticated computer programs and allow geologists to draw interpretive line drawings (▷) that show the geometry of the rock units, along with any faults and folds.

12.13.a4

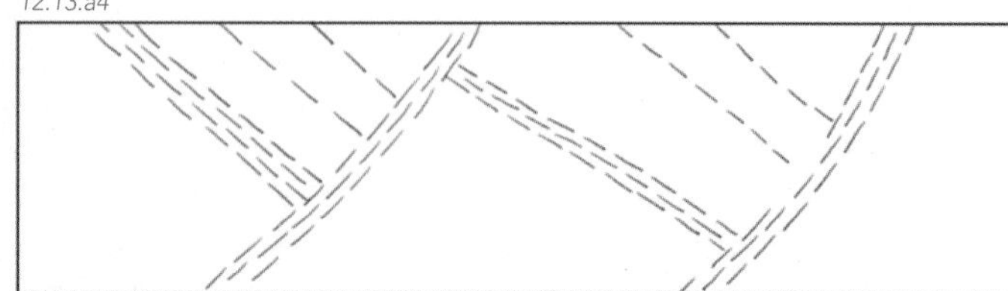

▷ 7. The geometry of the reflections, as expressed on the seismic profile, is integrated with information about the area's rock sequence and structures. We can then construct a geologic cross section representing an interpretation of the subsurface.

12.13.a5

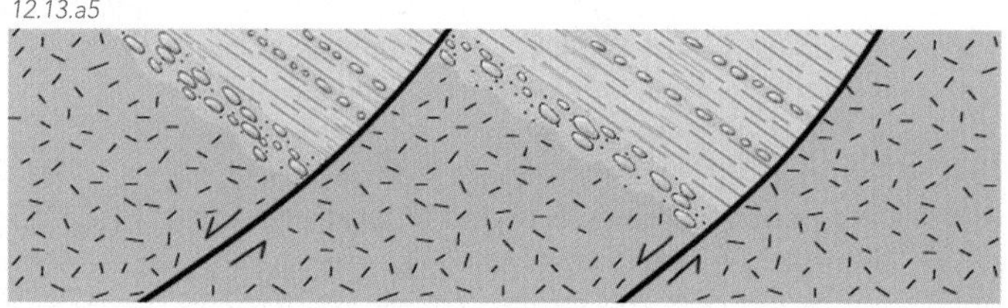

8. Geologists and engineers drill holes to search for petroleum, minerals, groundwater, and scientific knowledge. Most drill holes are less than several hundred meters deep, but some reach depths of 5 kilometers (3 miles) or more. Cylinder-shaped samples of rock, called drill cores, can be retrieved during the drilling process to provide samples of rocks from depth. ▽

12.13.a6

9. Instruments that measure the intensity of the Earth's magnetic field can be used to determine the subsurface distributions of magnetic rocks. The equipment can be carried on foot or towed behind a plane. Earth scientists who measure and interpret magnetic, seismic, gravity, and other types of physical data are *geophysicists*. Such data is called a *geophysical survey*. Many geology graduates are involved with geophysical surveys at some point in their careers.

▷ 10. Magnetic data are generally portrayed as a map, with warmer colors (reds) representing more magnetic rocks and cooler colors (blues) representing areas with lower magnetism.

▷ 11. The red and orange areas mark the dark lava flow and hills of gray granite, which are more magnetic than the sediments that cover the rest of the area.

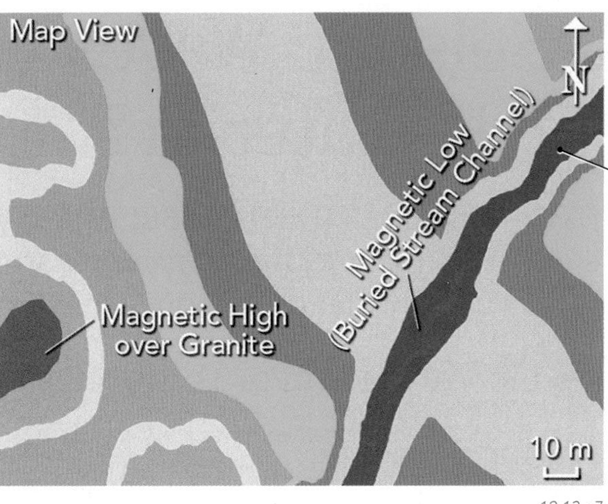

12.13.a7

12. A curving magnetic low, represented by the darker blue colors, coincides with a buried stream channel. In the figure below, the channel forms a band of gray soil where the two teams of geophysicists are standing.

13. The strength of gravity varies slightly from one place to another on Earth's surface. This is because some rocks, such as basalt, are more dense and cause a stronger pull than less dense materials, such as sediment. The variations in gravity can be measured using sensitive *gravity meters*.

▽ 14. In this area, the team of geophysicists measured gravity across the buried stream channel and plotted the data on a profile relative to the average value of gravity for the area. The plot shows a gravity minimum caused by low-density sediment within the buried channel.

12.13.a8

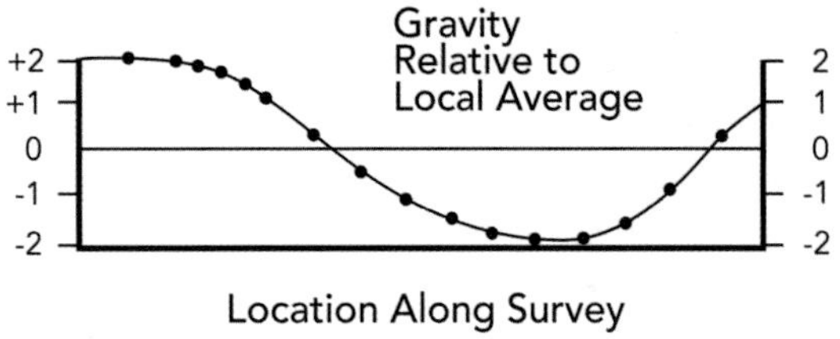

Gravity Relative to Local Average

+2, +1, 0, -1, -2 | 2, 1, 0, -1, -2

Location Along Survey

▽ 15. From the gravity profile, computer programs can model possible density configurations that are consistent with the data.

12.13.a9

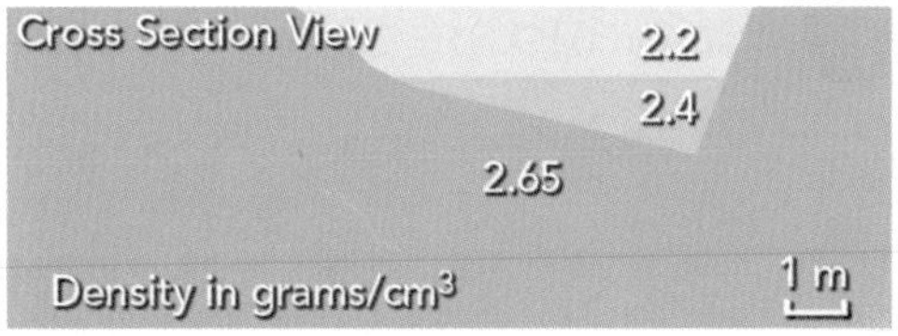

Cross Section View

2.2

2.4

2.65

Density in grams/cm^3

1 m

16. Some rocks, such as clays, conduct electrical currents better than other rocks. Rocks containing groundwater conduct an electrical current better than dry rocks. Geologists and geophysicists use these principles to explore for mineral deposits and groundwater. An electrical transmitter runs current into the ground, and one or more electrical receivers some distance away measure how much current reaches the surface. ▽

17. The results of an electrical survey across the buried stream channel are plotted in cross section and contoured, with warmer colors for rocks with higher conductivity, such as those with more water. Geologists compare all the various types of data to infer the subsurface geology.

12.13.a10

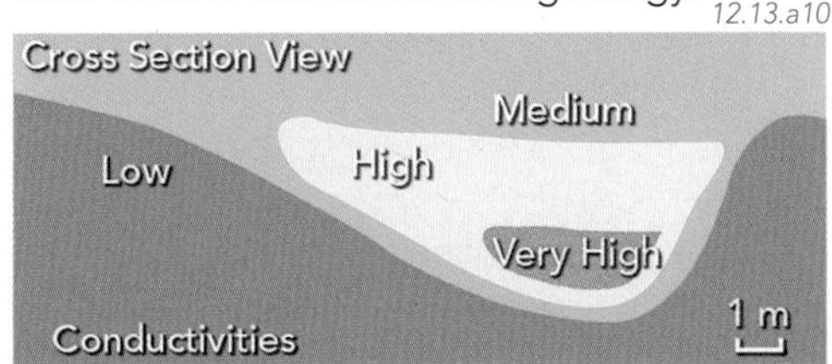

12.13.a11

Before You Leave This Page Be Able To

- ✓ Summarize how volcanic inclusions, exposed geology, drill holes, and mines provide observations of the subsurface.
- ✓ Briefly summarize what is measured by the various types of geophysical surveys (seismic, magnetic, gravity, and electrical).

What Do Seismic Waves Indicate About Earth's Interior?

EARTHQUAKES, EXPLOSIONS, AND OTHER SEISMIC EVENTS generate seismic waves that can be used to interpret Earth's internal structure. The way seismic waves travel through Earth enables us to identify distinct layers and boundaries within the interior, including the crust, mantle, and core.

A How Do Seismic Waves Travel Through Materials?

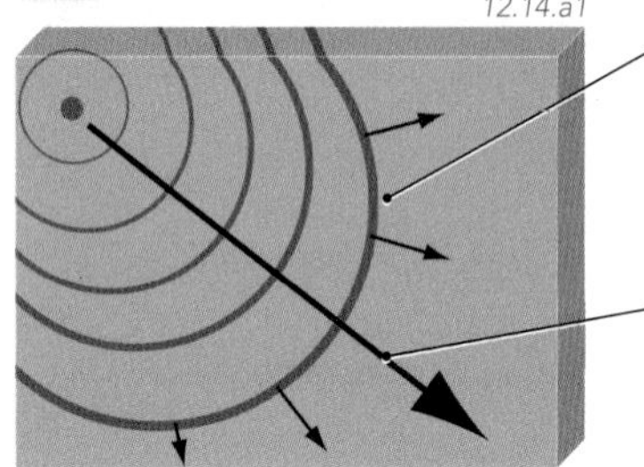

12.14.a1

An earthquake or other source of seismic energy generates seismic waves, which radiate out from the source in all directions.

The path that any part of the wave travels is a *seismic ray*. If the physical properties of the material do not change from place to place, then a seismic ray travels in a *straight line*. In this case, a family of straight rays diverges outward from the source.

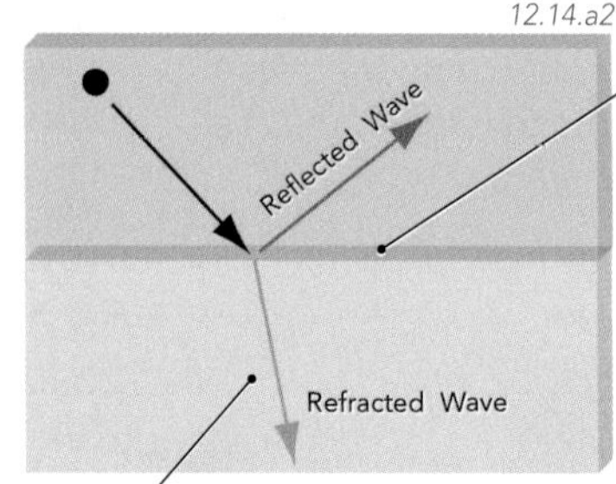

12.14.a2

Most seismic waves encounter boundaries between materials with different physical properties, causing the waves to reflect, speed up, or slow down. Some of the energy is *reflected* off the interface as a reflected wave.

Some of the energy is *bent* as it crosses the boundary. This process is called *refraction*.

How Seismic Waves Refract Through Different Materials

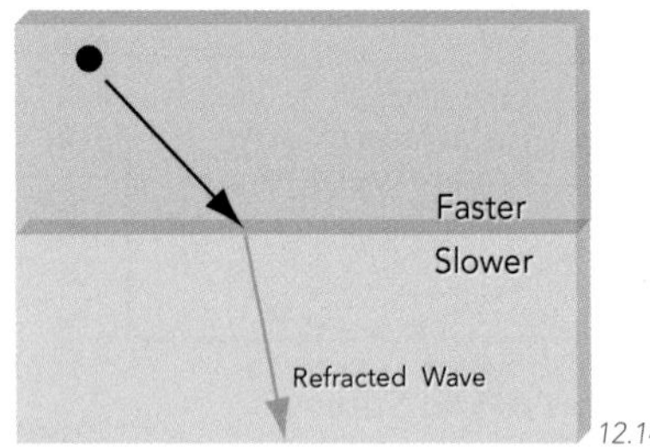

12.14.a3

If a seismic wave passes into a material that causes it to slow down, it will be refracted *away* from the interface at a steeper angle.

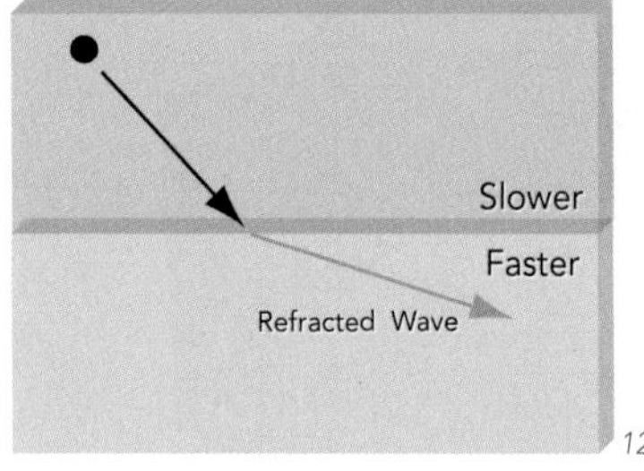

12.14.a4

If a descending seismic ray passes from a slow material to a faster one, it will be refracted to a shallower angle.

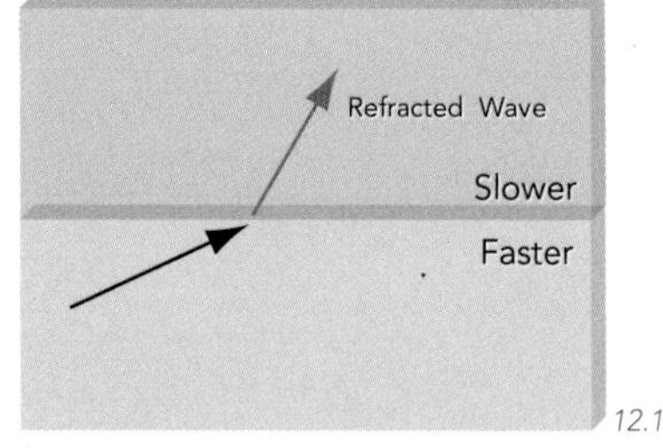

12.14.a5

If a rising seismic ray passes from a fast material to slower one, it will be refracted upward toward the surface.

B How Do Seismic Waves Travel Through Earth's Crust and Mantle?

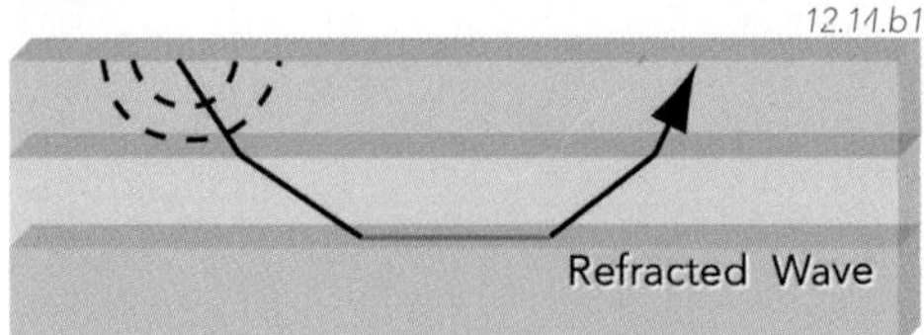

12.14.b1

△ 1. Refraction causes seismic waves to take curved paths through the Earth. Steeply descending rays will first be refracted to shallower angles as they encounter faster and faster material at depth. The waves will then be bent back toward the surface as they pass back through slower material.

2. In the figures below, an earthquake sends seismic waves into the crust and mantle. Both waves are refracted back toward the surface. Waves in the mantle travel faster than those in the crust, resulting in an interesting and useful phenomenon.

3. Close to the earthquake, waves that travel through the crust arrive sooner than those from the mantle because the crustal waves travel a shorter distance.

4. Farther from the earthquake, waves that travel through the mantle arrive at the surface first because the faster velocity lets them overtake the crustal waves.

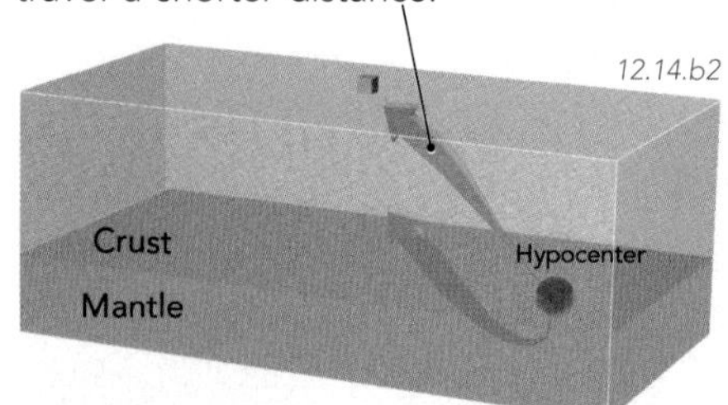

12.14.b2

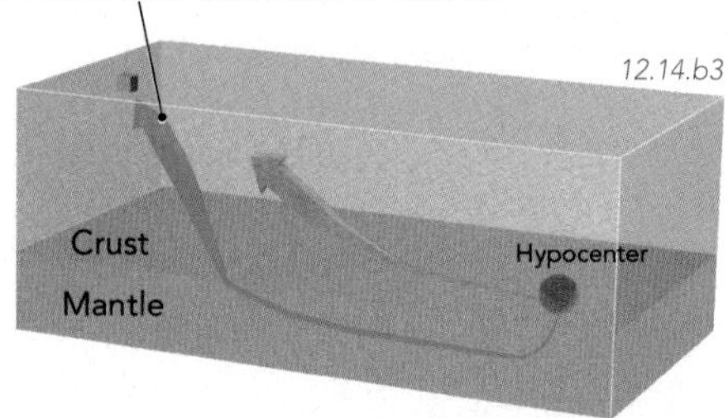

12.14.b3

5. Seismologists observe at what distance from the hypocenter the mantle waves begin to arrive first. They then use simple computer models of velocities, crustal thicknesses, and ray paths to calculate the depth to the crust-mantle boundary.

C How Are Seismic Waves Used to Examine Earth's Deep Interior?

Seismologists recognize distinct boundaries within Earth, largely based on changes in seismic velocities. Such changes reflect the physical and chemical properties of the rock layers through which the seismic waves pass. Not all seismic waves make it through every part of Earth. Observing where particular kinds of waves are blocked helps determine which parts of Earth are molten.

1. As P-waves travel through Earth, they speed up and slow down as they pass through different kinds of material. Their velocity depends upon three factors: (1) how easily the rocks are compressed; (2) how rigid the material is; and (3) the density of the material. Based on these factors, seismologists conclude that faster velocities indicate denser rocks.

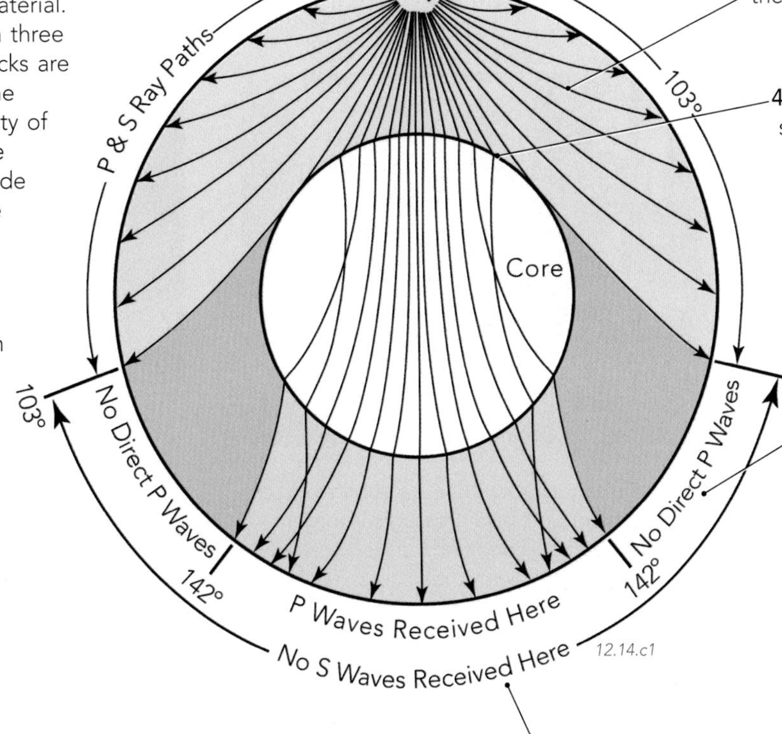

3. As P-waves and S-waves travel through Earth, many follow curved paths that return them to the surface.

4. Along the core-mantle boundary, some P-waves are *refracted inward* because the outer core has a *slower* velocity than the adjacent mantle. These P-waves pass through the core and out toward the other side of Earth.

5. There is a zone, called the *P-wave shadow zone*, that receives no direct P-waves. This is because the P-waves are either refracted upward before they reach this area, or are refracted inward through the core. Some weak P-waves reach the surface in this zone, but they took indirect routes by reflecting and refracting around Earth's core.

▽ 2. This graph plots P-wave velocity as a function of depth. Overall, P-wave velocity increases with depth in the mantle and in the core because the rocks in each part become more rigid and dense downward.

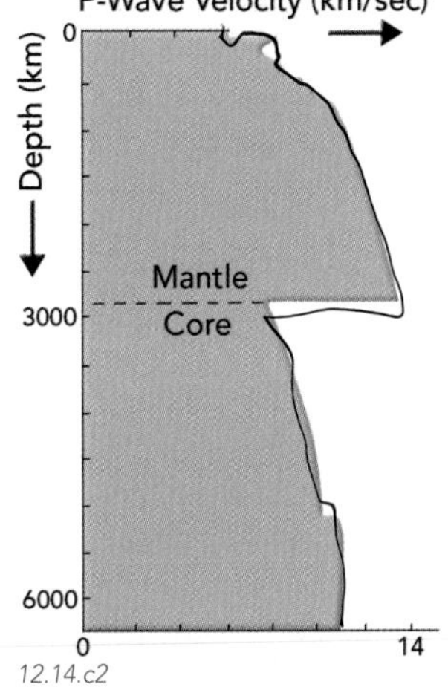

7. From the sizes and locations of the P-wave and S-wave shadows, we can determine the diameter and depth of Earth's core. Seismologists also learn about Earth's interior by studying *indirect waves*. These are waves that have reflected off boundaries or have changed wave type as they crossed a boundary (e.g., mantle to core).

6. On the opposite side of Earth from the seismic source, there is also an *S-wave shadow zone*, that receives no direct S-waves. This implies that S-waves cannot pass through the core. From this and other observations, seismologists conclude that the outer part of the core is molten and blocks S-waves.

The Moho

The boundary between the crust and mantle is named the *Mohorovicic Discontinuity* after the last name of the Croatian seismologist who discovered it. Most geologists simply call it the *Moho*.

Much effort is expended trying to determine the depth to the Moho because this tells us how thick the crust is. Geophysicists investigate this problem using various approaches. Some observe the arrivals of seismic waves from naturally occurring earthquakes, whereas others use mine blasts as the seismic source. The depth to the Moho can sometimes be identified as reflections on seismic-reflection profiles. Since seismic waves travel through the crust at ~6 km per second, it takes 10 seconds for a wave to travel 30 km down to the Moho, bounce off, and travel 30 km (19 miles) back up. It takes less time if the crust is thin and more time if it is thick.

Before You Leave This Page Be Able To

- ✓ Sketch or describe reflection and refraction of seismic waves.
- ✓ Sketch and explain how seismic waves pass through the crust and mantle.
- ✓ Describe how seismic waves are used to identify the diameter of the core and to show that the outer core is molten.

How Do We Investigate Deep Processes?

ROCK PROPERTIES, SUCH AS DENSITY, temperature, pressure, and composition, change through Earth. Seismologists use observations of seismic-wave velocities to determine how rock properties change with depth and how material moves in Earth's mantle and at the core-mantle boundary.

A How Do We Investigate Deep Conditions?

Much of what we know about Earth's interior comes from our knowledge of seismic-wave velocities and how they vary within Earth's interior.

12.15.a1

One way to constrain the conditions deep within Earth is to examine rocks that have resided at great depths. Some metamorphic rocks in Norway and China contain high-pressure minerals, which indicate that they were buried at ultra-high pressures and depths of 60 to 100 kilometers. Documenting the minerals and structures that formed under these conditions provides insight into what processes and conditions occur at depth.

12.15.a2

In the laboratory, rocks can be subjected to high temperatures and pressures in order to determine the conditions under which they melt, solidify, or flow in the solid state. Many minerals change into another mineral at high temperatures, high pressures, or both. The conditions under which these changes occur are then inferred for equivalent depths and temperatures within Earth's interior.

12.15.a3

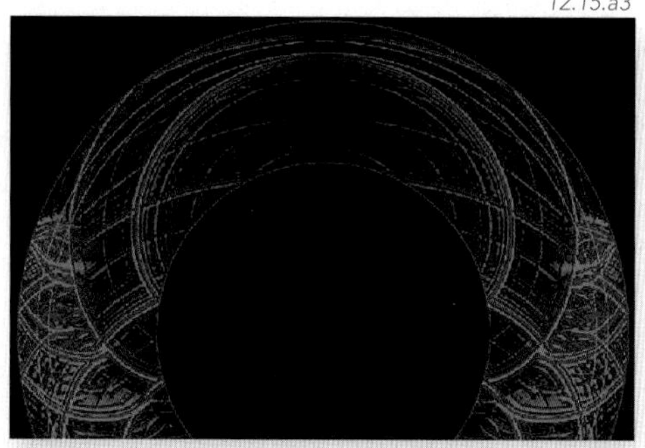

Computers and sophisticated numerical models are used to model processes that are too deep to observe directly. Such models can illustrate how seismic waves travel through the mantle, as shown here, or how the mantle might flow upward, downward, or laterally if there are lateral variations in density. Such density variations are caused by differences in temperature and in the types of minerals that are present.

B How Does Seismic Tomography Help Us Explore the Earth?

Seismologists examine Earth using earthquakes in much the same way that medical doctors examine the internal parts of the body with CT scans and other types of imaging technologies. The technique seismologists use is called *seismic tomography,* where "tomography" means an image of what is inside.

Seismic Observations

12.15.b1

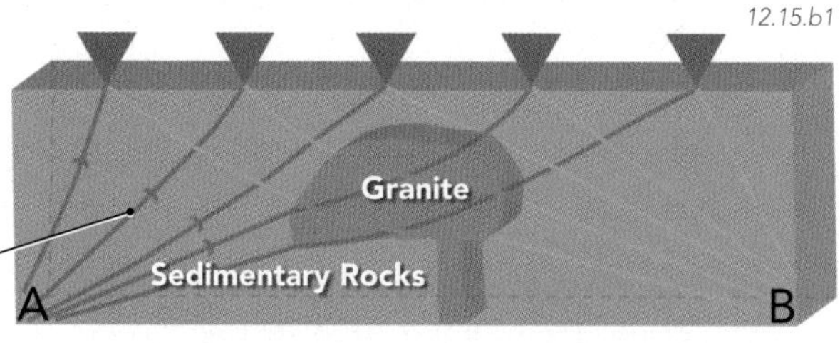

1. In seismic tomography, one examines a number of earthquake waves that have passed through the same subsurface region, but from different directions. In this diagram, the directions along which the seismic waves passed through the region are shown as a series of lines called *ray paths.*

2. Ray paths coming from points A and B are recorded on a number of seismometers, shown as triangles.

3. If part of the crust or mantle has a higher seismic velocity than other areas, then waves passing through that area will arrive *sooner* than expected. Those that travel through slow regions will arrive *later* than expected.

Seismic Interpretation

12.15.b2

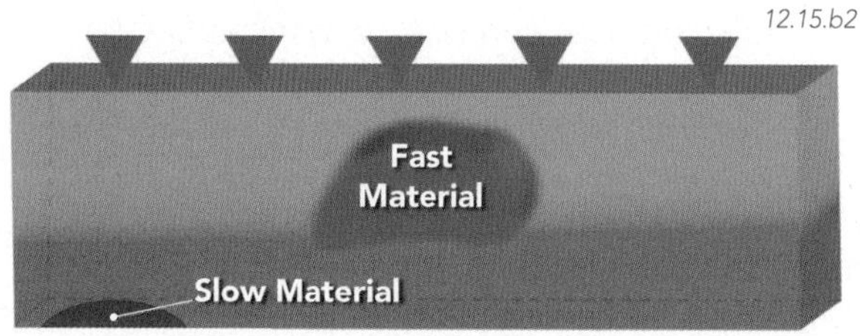

4. This figure models the velocities in the same region using seismic tomography. Areas that are slower than expected are shaded red and may represent areas that might be hotter than normal.

5. Some areas, such as the granite body, will be faster than expected and so are shaded blue. Fast areas might be abnormally cool or composed of stiff, dense rocks. Earthquakes do not come from every direction, so many details cannot be resolved and remain a little fuzzy.

C Why Do Global Wind Patterns Develop?

The Sun warms equatorial regions of Earth more than the poles, setting up a flow of warm air toward the poles, which is balanced by a flow of cold air from the poles toward the equator. Earth's rotation complicates the wind, producing curving patterns of circulating wind.

Rotation and Deflection

1. Earth is a spinning globe, with the equatorial region having a higher spin velocity than polar regions. As a result, air moving north or south is deflected sideways by the rotation, a response called the *Coriolis effect*.

2. Sunlight strikes equatorial regions more directly than it does areas closer to the poles and so preferentially heats the equatorial regions. As Earth rotates, the Sun's heat forms a band of warm air that encircles the globe and is re-energized by sunlight each day.

3. Warmed equatorial air rises and flows north and south, away from the equator. Air at the surface flows toward the equator to replace the air that rises. The Coriolis effect deflects this surface wind toward the west.

4. These flows of air combine into huge, tube-shaped cells of circulating winds, called *flow cells*. Some flow cells have surface winds flowing toward the poles. Others have winds toward the equator.

Flow Cells

5. Wind direction is referenced by the direction from which it is coming. A wind coming from the west is said to be a "west wind." A wind that generally blows from the west is a *westerly*.

6. Polar regions receive the least solar heating and are very cold. Surface winds move away from the poles, carrying cold air with them. *Polar Easterlies* blow away from the North Pole and are deflected toward the west by Earth's rotation.

7. *Westerlies* dominate a central belt across the United States and Europe, so weather in these areas generally moves from west to east.

8. *Northeast Trade Winds* were named by sailors, who took advantage of the winds to sail from the *Old World* to the *New World*.

9. *Southeast Trade Winds* blow from the southeast toward the equator.

10. *Westerlies* also occur in the Southern Hemisphere and are locally very strong because of the lack of continents, which disrupt winds.

11. *Polar Easterlies* flow away from the South Pole and deflect toward the west.

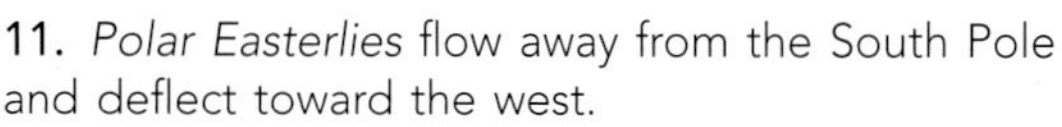

13.01.c1

Why the Coriolis Effect Occurs

How does the *Coriolis effect* deflect air (and water) movement on Earth's surface? Air, like the surface, is being carried around Earth by rotation. The surface has a faster velocity near the equator than at the poles because it has to travel a greater distance in 24 hours.

Therefore, as air moves toward the poles, it is rotating faster toward the east than the land over which it moves. It appears, from the surface, to be deflected to the east.

The opposite occurs as air moves toward the equator and encounters areas with a faster surface velocity. The air appears to lag behind, deflecting to the west as if it were being left behind by Earth's rotation.

13.01.c2

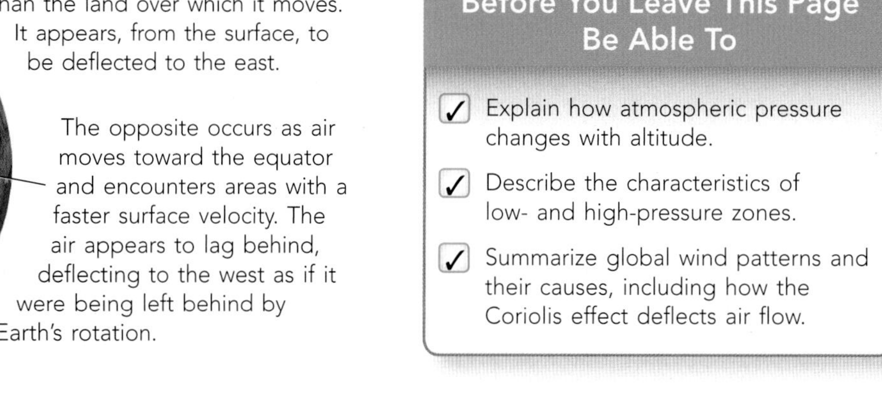

Before You Leave This Page Be Able To

- ✓ Explain how atmospheric pressure changes with altitude.
- ✓ Describe the characteristics of low- and high-pressure zones.
- ✓ Summarize global wind patterns and their causes, including how the Coriolis effect deflects air flow.

How Does Wind Transport Material?

WIND CAN PICK UP, TRANSPORT, AND DEPOSIT MATERIAL. It moves sand and finer particles by rolling them, bouncing them, or lifting them up and carrying them. The incorporation of sediment and other material into wind results in dust storms, soil erosion, and the formation of sand dunes and other wind-related features.

A How Does Wind Transport Sediment and Other Materials?

Wind is capable of moving sand and finer sediment, as well as pieces of plants and other materials lying on the surface. It generally moves material in one of three ways.

13.02.a1

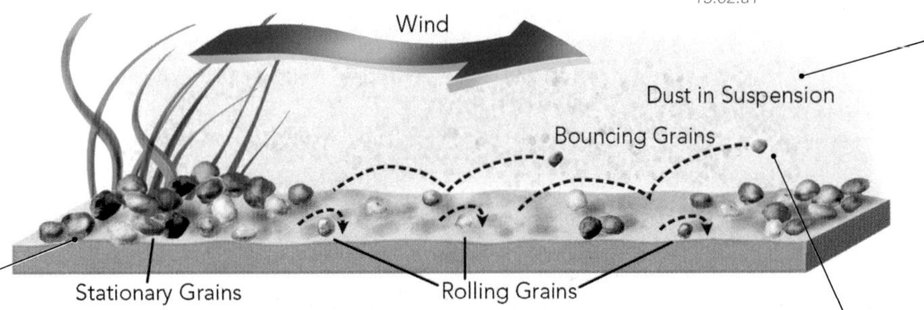

Most materials on Earth's surface are not moved by the wind because they are too firmly attached to the land (such as rock outcrops), are too large or heavy to be moved, or are both.

Wind can pick up and carry finer material, such as dust, silt, and salt, as particles drifting or sailing in air currents. This mode of transport is called *suspension*, and wind can keep some particles in the air for weeks or longer, transporting them long distances.

If wind velocity is great enough, it can roll grains of sand and silt and other loose materials across the ground.

Very strong winds can lift sand grains, carry them short distances, and drop them. This process is akin to bouncing a grain along the surface and is called *saltation*.

B What Are Dust Storms and Whirlwinds?

Strong winds can pick up large volumes of dust, forming a cloud that travels across the surface as a *dust storm*. At a smaller scale, *whirlwinds* are columns of rapidly rotating air common in open fields and deserts.

13.02.b1

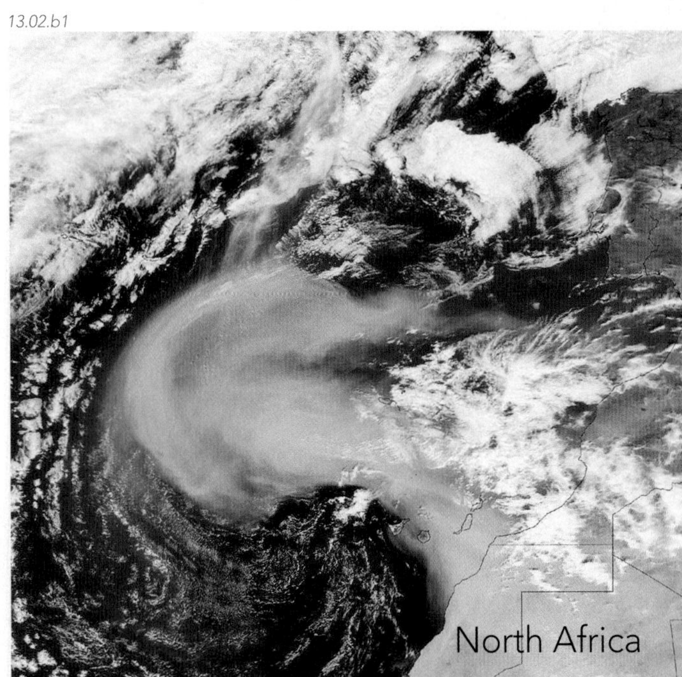

In this satellite image, a massive sand storm blows sand and dust west from the deserts of North Africa 1,600 kilometers (1,000 miles) into the Atlantic Ocean. Dust storms are most common in desert areas and near cultivated fields because these often lack sufficient vegetation to bind the soil to the land. Low-pressure cells can lift dust high into the atmosphere, where it can drift across an entire ocean.

13.02.b2

The front edge of a dust storm can be sharp and well defined, appearing as an ominous, turbulent cloud of fast-moving dust. Such storms are common around thunderstorms in dusty regions. [Arizona]

13.02.b3

Whirlwinds are rotating columns of wind that are typically tens to hundreds of meters across and less than 1 kilometer high. The fast-moving winds pick up loose dust and sand, and are commonly called *dust devils*. They last only minutes, do not move much material, and do not carry material very far. Spacecraft on Mars have observed dust devils and the twisting tracks they leave on the Martian surface. [Peru]

C Why Are Rain Forests Disappearing?

Deforestation of rain forests is occurring at about 2.5 acres per second, or 80 million acres per year. Some ecologists estimate that most rain forests will be destroyed by 2040. The main cause is economic pressure.

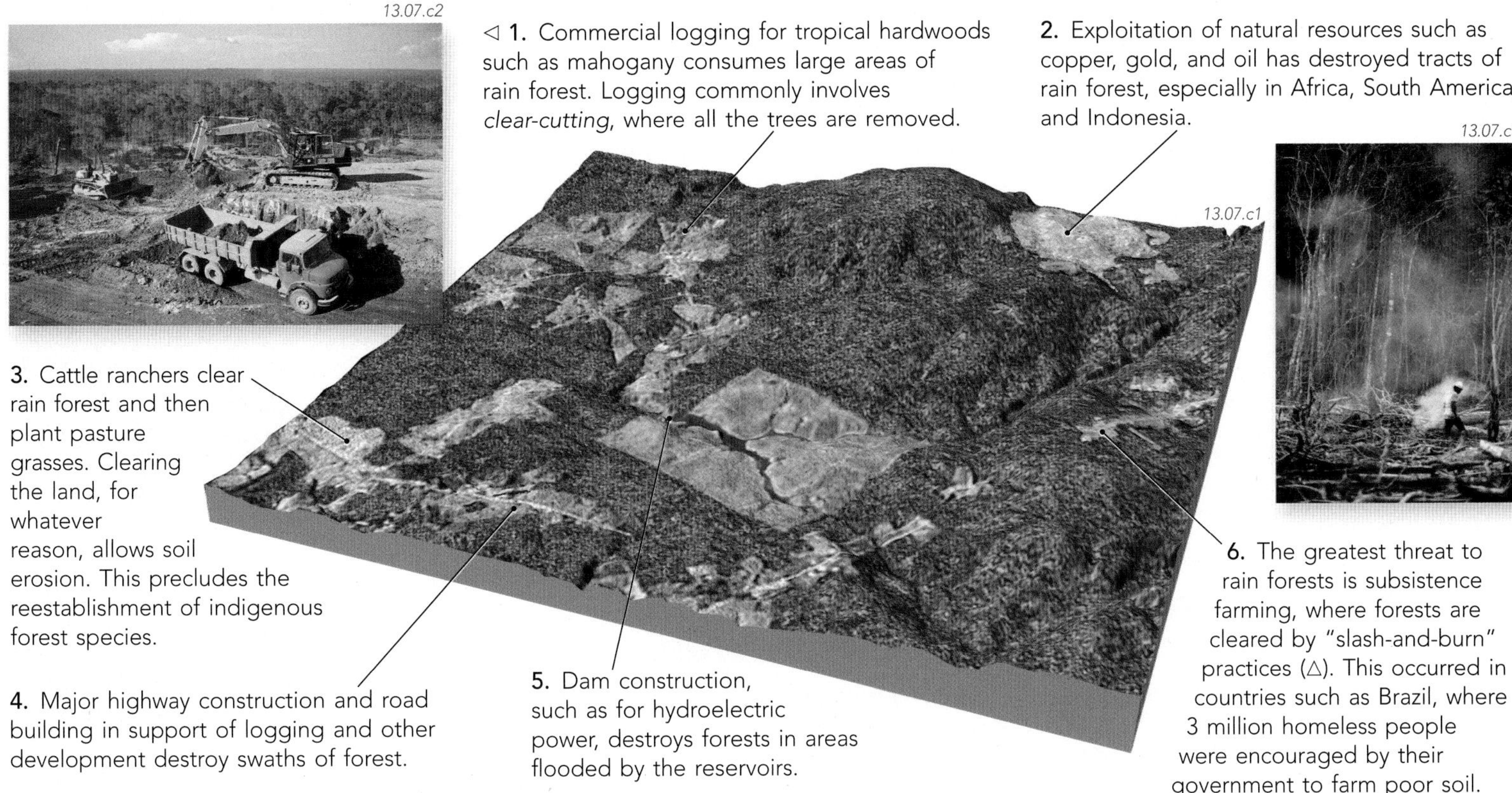

◁ **1.** Commercial logging for tropical hardwoods such as mahogany consumes large areas of rain forest. Logging commonly involves *clear-cutting*, where all the trees are removed.

2. Exploitation of natural resources such as copper, gold, and oil has destroyed tracts of rain forest, especially in Africa, South America, and Indonesia.

3. Cattle ranchers clear rain forest and then plant pasture grasses. Clearing the land, for whatever reason, allows soil erosion. This precludes the reestablishment of indigenous forest species.

4. Major highway construction and road building in support of logging and other development destroy swaths of forest.

5. Dam construction, such as for hydroelectric power, destroys forests in areas flooded by the reservoirs.

6. The greatest threat to rain forests is subsistence farming, where forests are cleared by "slash-and-burn" practices (△). This occurred in countries such as Brazil, where 3 million homeless people were encouraged by their government to farm poor soil.

D What Role Do Rain Forests Play in Ecology?

Rain forests are critical ecosystems with key ecologic niches. Extreme rainfall (150–400 centimeters per year) and infiltration into the ground leaches nutrients from the soil. The rain forest ecosystem of plants, animals, insects, and microbes is nutrient recycling—organisms die, rapidly decompose, and return nutrients to the system.

Carbon Dioxide Uptake

Rain forests are responsible for about 30 percent of the photosynthetic activity on Earth. If CO_2 levels in the atmosphere increase because of volcanic activity (▽) or other natural or human-related causes, rain forests can increase their CO_2 uptake. Thus, they act as a buffer against climate change.

Diversity Storehouse

Rain forests are estimated to contain at least 5 million species. They are the genetic storehouse for the world's ecology, including lush environments like the trees and plants below.

Local Climate

Rain forests intercept and use solar energy that would otherwise strike the ground. As a result, rain forest trees and plants provide shade and help to keep the land underneath cooler and sheltered during the day.

Before You Leave This Page Be Able To

- ✓ Describe the characteristics and vertical structure of a rain forest.
- ✓ Summarize where rain forests occur and what conditions produce enough precipitation to form a rain forest.
- ✓ Explain threats to rain forests and why rain forests are ecologically and genetically important.

What Are Deserts and How Do They Form?

DESERTS are dry lands, often with little vegetation, that cover many parts of Earth's landmasses. Most deserts are not barren sand dunes; they contain plants and animals adapted to life in a dry environment. What conditions create deserts, what controls their locations, and are deserts growing over time?

 What Is a Desert?

An *arid region* receives less precipitation than it loses to evaporation and other processes. Arid regions that have less than 25 centimeters (10 inches) of rainfall per year are known as *deserts*. Vegetation is sparse in desert ecosystems, commonly covering less than 15 percent of the ground. Many deserts lack permanent streams.

Sahara and Sahel

This satellite image shows the northern half of Africa. Most of the region is tan colored because it consists of sand and rock, with very sparse vegetation. This tan region is the *Saharan Desert*, stretching from Morocco to Egypt.

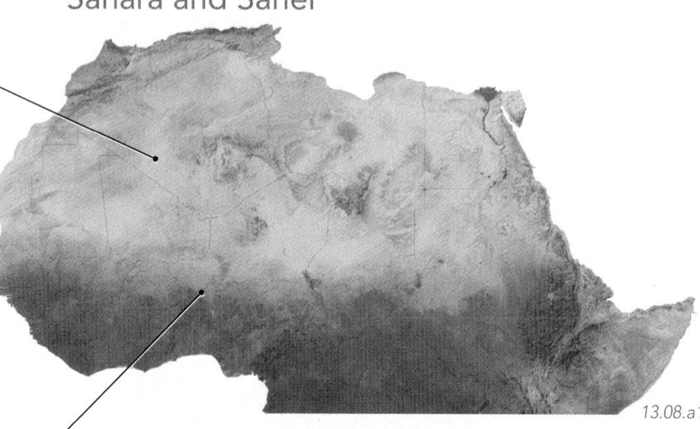

13.08.a1

South of the Sahara is the *Sahel*, a region that is relatively dry but not quite a desert. Regions intermediate between a true desert and a more humid climate are called *semiarid*. The Sahel is currently threatened by the encroachment of deserts from the north. The region has recently experienced a number of devastating droughts.

Mojave and Sonoran Deserts

13.08.a2

The Mojave Desert of Southern California has rocky mountain slopes with very few plants. The valleys can be sandy or rocky. [Big Maria Mtns., California]

13.08.a3

▷ The Sonoran Desert of Arizona receives more rain than the Mojave Desert, much of it during a summer monsoon. It has more cactus and other heat-adapted plants. [central Arizona]

B Where Do Deserts and Other Arid Lands Occur?

Examine the map below, which shows arid and semiarid regions in orange. Where are most of the world's arid lands? Look for patterns where dry lands occur. Compare the distribution of these lands with the (1) locations of atmospheric cells of rising and descending air; (2) locations and directions of ocean surface currents; (3) directions of prevailing wind; and (4) locations of mountains between the deserts and oceans.

13.08.b1

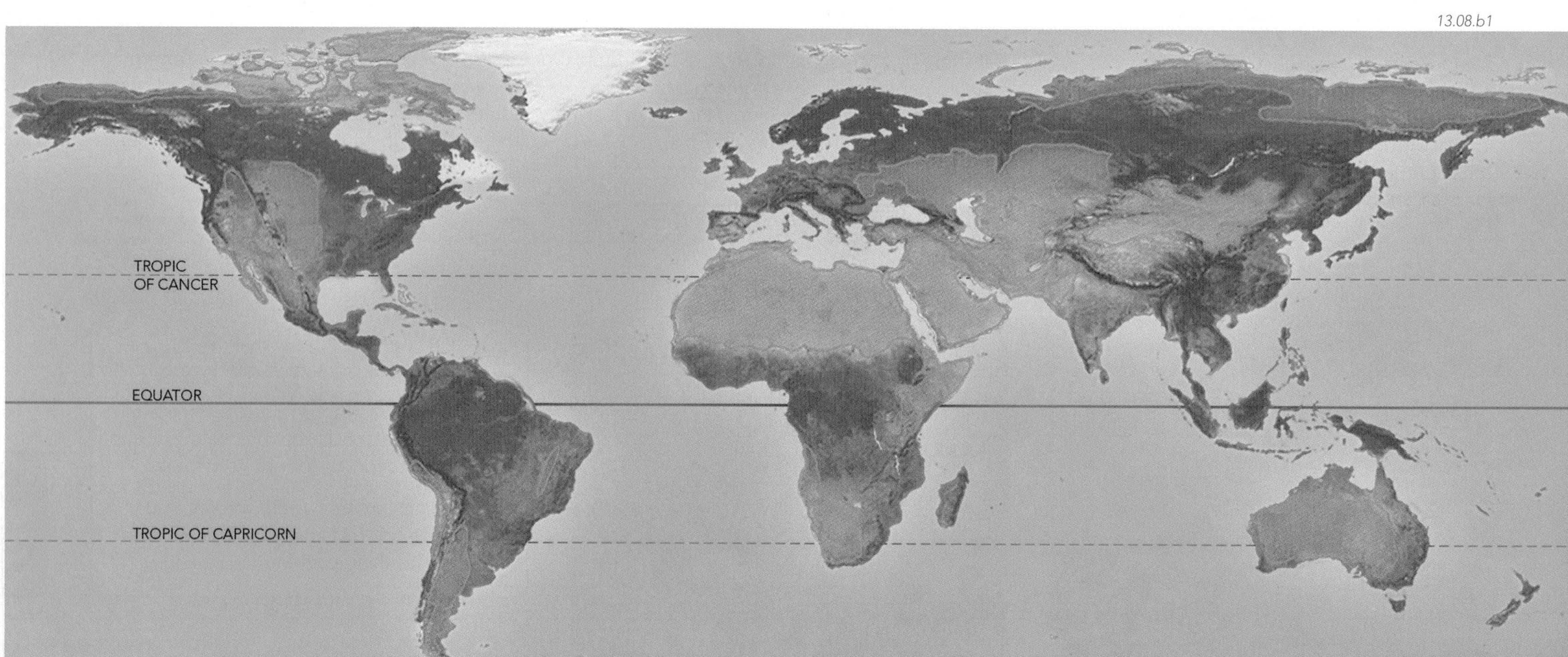

C In What Settings Do Deserts and Other Arid Lands Form?

Earth's large deserts result primarily from descending air in subtropical belts and from airflow associated with cold ocean currents. Deserts also form in rain shadows associated with mountain ranges and in cold, dry polar regions.

Coastal deserts form where cold, upwelling ocean currents cool the air and decrease its ability to hold moisture. This applies to the Atacama Desert, one of the driest places on Earth.

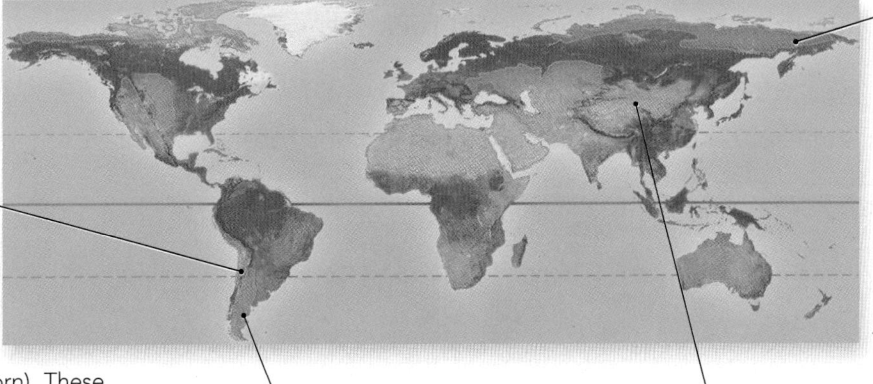

13.08.c1

Polar deserts form where cold, dry air prevails. Any available moisture is frozen for almost the entire year. Examples of a polar desert are the ice-free areas of the Arctic and the Dry Valleys of Antarctica.

Subtropical deserts form where the general atmospheric circulation brings dry air into the subtropics (near the Tropic of Cancer and the Tropic of Capricorn). These areas are situated beneath descending flow cells and are associated with high pressure. This is the primary cause of the world's large deserts, such as the Sahara and the Australian deserts.

13.08.c2

When moist air rises up over mountains, it rains. As the air descends on the downwind side of the mountain, it dries, forming a *rain-shadow desert*. The Andes mountains extract moisture from westerly winds, forming the Patagonia Desert to the east. A typical rain shadow is shown below.

13.08.c3

Continental deserts form in the interiors of some continents, far from sources of moisture. In many such settings, summers are hot and winters are very cold, such as in the Gobi Desert of Mongolia. The flow of relatively dry air from a continental interior can form deserts in adjacent lands. ▽

13.08.c4

D What Is Desertification?

Extended periods of drought, overgrazing by livestock, poor farming techniques, and diversion of surface water can cause soil loss and change grasslands to desert. Converting other lands to desert is called *desertification.*

This map shows the global risk of desertification. Red and orange indicate areas with high to very high risk. Areas most at risk include central Africa (the Sahel), the Middle East, and parts of the United States, including fertile farmland of the Great Plains.

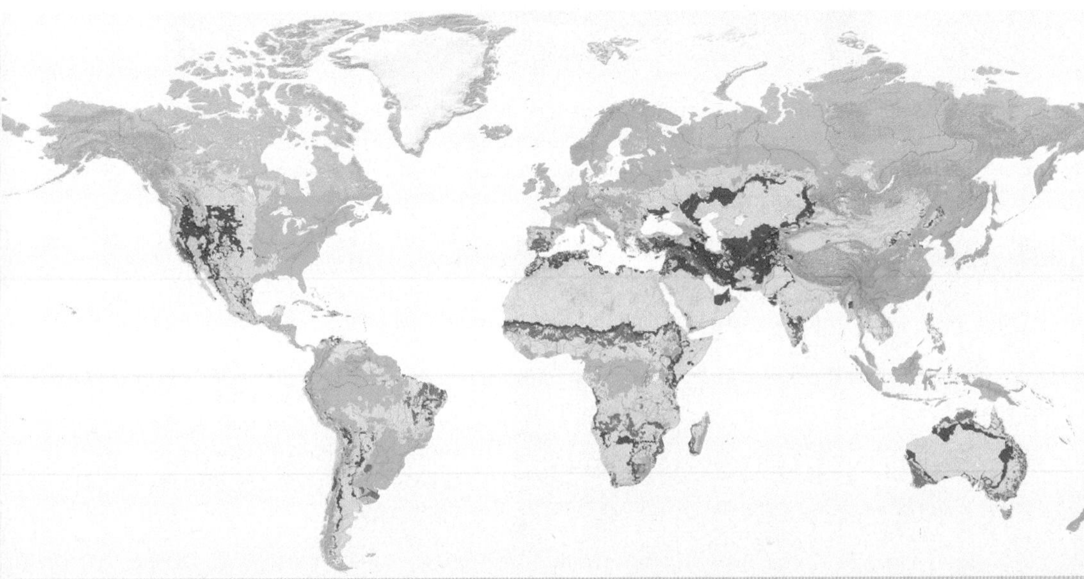

13.08.d1

13.08.d2

Plants shield the ground surface from erosion and hold the soil in place. Loss of plants and soil due to draught promotes erosion and desertification.

Before You Leave This Page Be Able To

- ✓ Describe what deserts and other arid lands are and where they occur.
- ✓ Summarize how and where different kinds of deserts form.
- ✓ Describe desertification.

What Features Are Common in Deserts?

THE DRYNESS OF DESERTS is expressed in the sparseness of vegetation and a lack of well-developed soil on many hillsides. Deserts have a number of other characteristic landscape features, such as sand dunes, channels that are normally dry, and dark coatings on rock faces.

A What Landscape Features Are Characteristic of Deserts?

The low rainfall, dry atmosphere, and sparse vegetation in most deserts result in several common features.

13.09.a1

1. *Alluvial Fans*—Loose rocks accumulate on rocky mountain slopes, are transported by streams and debris flows, and are deposited as fan-shaped aprons called *alluvial fans*. Alluvial fans form where steep, confined channels encounter more level terrain at the mountain front. [Death Valley, California]

13.09.a2

2. *Desert Washes*—Deserts contain sand and gravel-rich channels called *washes* or *arroyos*, which are normally dry. During intense rain, such as summer thunderstorms, a wash can rapidly fill with water draining off the land and rocky hillslopes. This causes a rapid rise in water (a *flash flood*). [Mojave Desert, California]

13.09.a3

3. *Playas*—Shallow, closed basins are *playas*. They receive runoff from washes and rivers, but the water has no outlet. Many playas partially or totally dry up, forming *salt flats*. [Atacama Desert]

13.09.a4

4. *Dunes*—Some sand in deserts is not held down by vegetation. It will form dunes if there is a sufficient supply of sand, such as from washes, and the wind is strong enough to pick up and concentrate material. [Death Valley, California]

13.09.a5

5. In this satellite image of Death Valley, there are *alluvial fans* that build out from the mountain front and down to the less steep basin floor.

6. The flat parts of the basin floor contain dunes and lake beds deposited when the playa lakes were more extensive.

7. The gray streaks running across the alluvial fans are *desert washes*, which branch and spread out over the fans in a network of channels.

8. The blue-green area on this image is a salt flat that formed by evaporation of water from a lake. The salt flat is actually light gray.

B What Features Develop over Time on Desert Landscapes?

Landscapes develop a number of features, some of which are more common in deserts than in other environments. Some of these features reflect the relative lack of rainfall in deserts.

13.09.b1

Pediment—Erosion and weathering of bedrock can carve a gently sloping erosion surface called a *pediment*. The broad dome in the distance is a gently sloping pediment. Pediments are visible in many deserts, especially where erosion has occurred over a long time, but pediments are not confined to deserts. [Mojave Desert, California]

13.09.b2

Desert Pavement—Over time, many desert surfaces become armored by rocks, forming a natural *pavement*. Rocks become concentrated on the surface because finer materials blow away, wash away, or move down into the soil. Desert pavement takes thousands of years to develop. [western Arizona]

13.09.b3

Caliche—Over time, soil in many environments accumulates soluble minerals, such as calcium carbonate (calcite). These minerals dissolve in rainwater, which percolates down into the soil. When the water evaporates, the dissolved components precipitate as coatings on clasts or as a distinct, hard layer called *caliche*. In wetter environments, dissolved material is flushed completely through the soil by descending waters.

13.09.b4

Desert Varnish—Exposed surfaces of resistant rocks can get coated with iron and manganese oxides and other materials, forming dark, natural *rock varnish*. The material in the varnish is largely wind-derived clays, oxides, and salts. In the example above, Native Americans carved into the varnish to create larger-than-life petroglyphs. Varnish takes thousands of years to form. [Butler Wash, Utah]

13.09.b5

Natural Stains—Many desert cliffs display vertical streaks of red, brown, and black. These colors are mostly coatings of iron oxides. They were deposited by water that flowed down the rock face and evaporated. Such stains are most obvious in deserts but can develop in other environments. [Yampa River, Utah]

Before You Leave This Page Be Able To

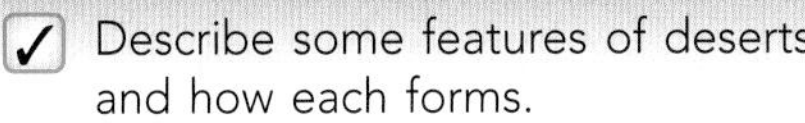

✓ Describe some features of deserts and how each forms.

What Is the Evidence for Global Warming?

OVER THE LAST 150 YEARS, people have measured atmospheric temperatures. This record, albeit short in geological terms, shows an increase in temperatures—*global warming*. There is currently much scientific and political discussion of this topic. What is the actual evidence?

A What Is Global Warming?

Global warming means *increasing* global atmospheric and oceanic temperatures from some point in the past to the present. Scientists examine various records of Earth's temperature history to investigate whether warming has occurred and whether it is related to natural events or human activity. In addition to direct temperature measurements, past temperatures are inferred from other types of observations, called *proxy evidence*. Most of the graphs below are from a recent report by the National Academy of Sciences and indicate how temperature is interpreted to have varied, relative to an arbitrary mean global temperature (averaged from 1961 to 1990).

Thermometer Record

Thermometers provide a direct measurement of air temperature. This record shows an average variation in temperature for the last 140 years. According to this record, it appears that average air temperatures have increased over the last century. Before the 1940s, the data show a relatively cool period.

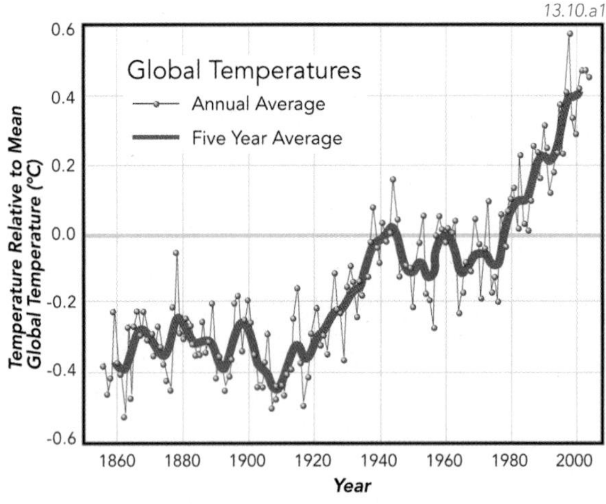

Borehole Temperature

Earth's shallow subsurface is heated by two sources: the atmosphere and heat from the deeper subsurface. If *surface* temperatures have changed over time, then the shallow *subsurface* temperature should reflect those changes. There are thousands of bore (drill) holes around the Earth, and measurements of temperature with depth are compared to predicted subsurface temperatures to infer changes in air temperature.

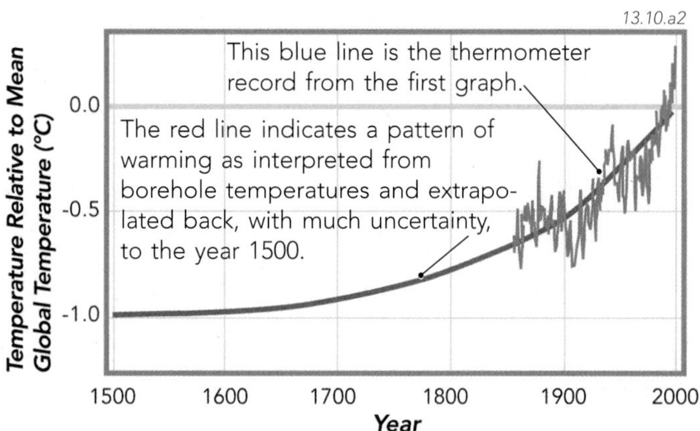

Glacier Length

Glaciers flow from areas of snow accumulation to lower elevations. The dynamics and energy flow of glacier movement and retreat are well understood. Most scientists interpret changes in the lengths of glaciers to be related to changes in atmospheric temperature and in the amount of precipitation. The combined data from glaciers around the world are interpreted as a warming trend beginning around the turn of the century.

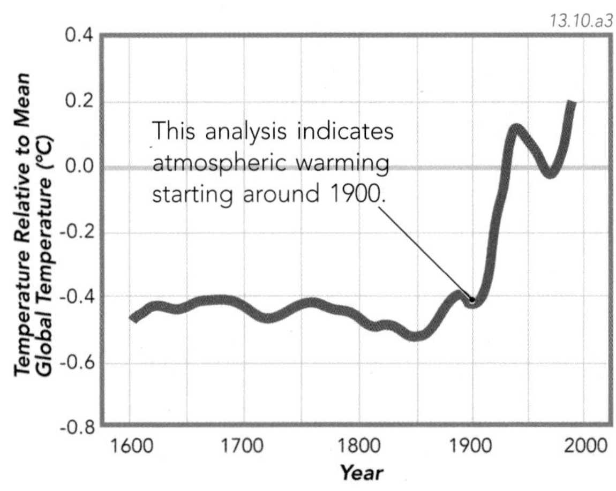

Tree-Ring Proxy

Tree-ring growth is partly dependent on climate. Some trees can grow for 300 years or more, and density of tree rings of successive populations from temperature-sensitive forests can be correlated around the world. This provides a climate record going back more than 1,200 years. This record shows an increase in air temperature between 1850 and 1998.

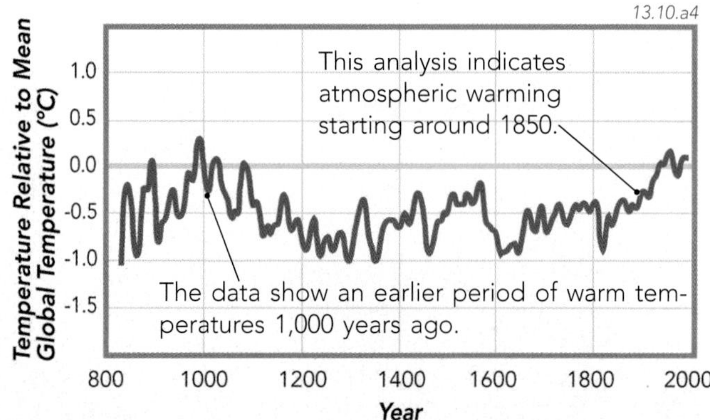

Ice Core Proxy

13.10.a5

Continental glaciers on Greenland and Antarctica have yearly layers that record winter precipitation and summer dust accumulations. Scientists extract ice cores by drilling, and then chemically analyze gases and ice in refrigerated laboratories (◁). Air trapped in tiny bubbles provides samples of the atmosphere back to at least 100,000 years ago. Oxygen and hydrogen isotopes in ice provide a proxy for temperature.

Coral Proxy

Corals grow in yearly cycles and secrete carbonate minerals to form their skeletons. Isotopic compositions of strontium and calcium incorporated in the carbonates depend on the surface temperature of seawater. Thus, layered corals provide a temperature record for a few hundred years. The coral data provide mixed results. Data from the Great Barrier Reef show a slight increase in sea surface temperature from about 1920–1980, but that even warmer periods occurred earlier. From these data, the period around 1900 was relatively cool, and we are coming out of that perhaps unusual period. Besides this, current temperatures are not unusual on this record.

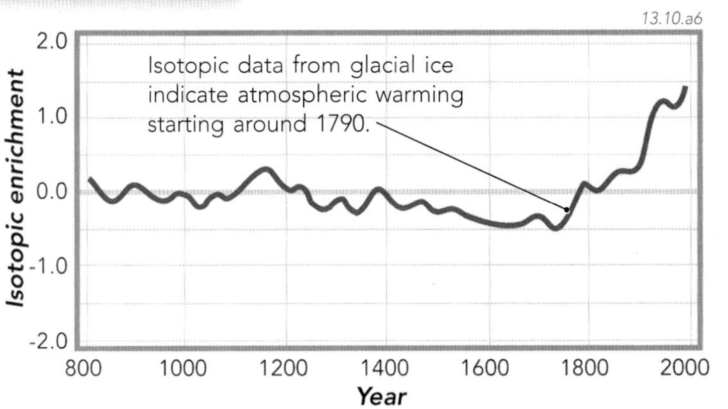

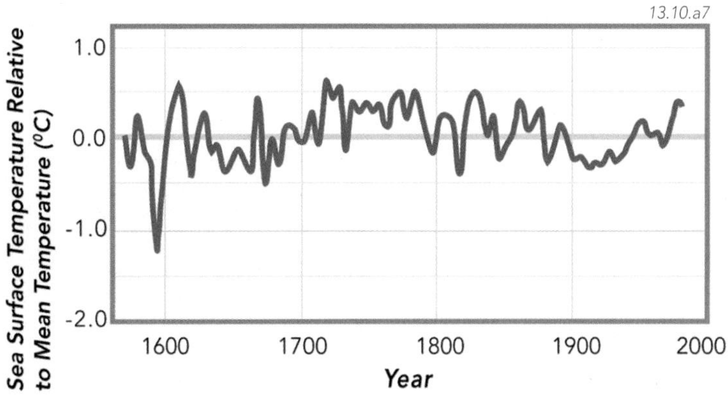

Comparing Temperature Reconstructions

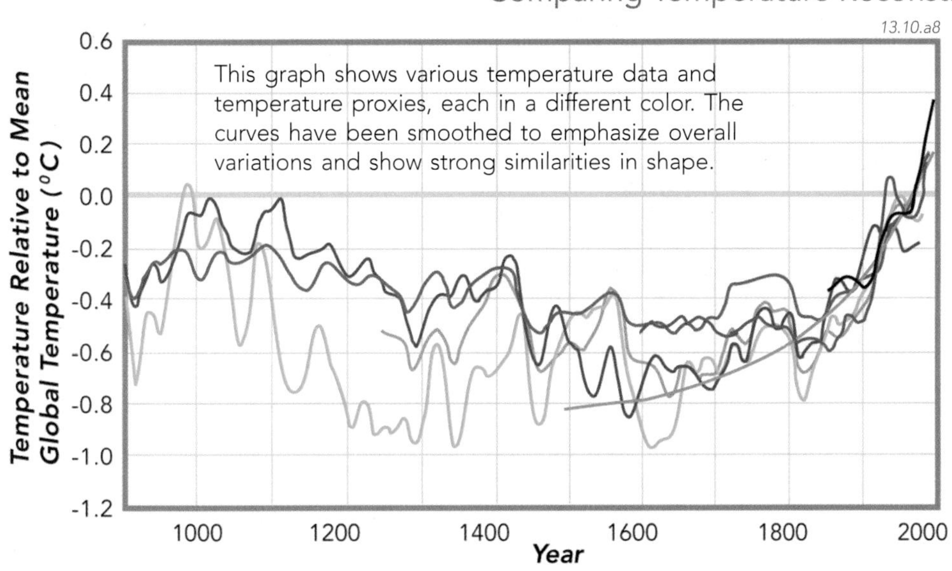

The National Academy of Sciences (NAS) published the chart to the left as part of a 2006 report to Congress. The report summarizes and compares the various temperature records shown on earlier graphs. The plot shows direct measurements, such as the temperature record and borehole data, as well as the various types of proxy data. Comparing the different types of data strengthens the case that some global warming has occurred. The NAS concluded that Earth's atmosphere (1) has warmed 0.6°C in the last 100 years, and (2) was hotter the last few decades than anytime in the last 400 years.

A Mystery Involving Atmospheric Satellite Data

One of the most convincing arguments *against* warming was that it was not detected in the extensive satellite data. Microwave instruments on NOAA satellites can look into different layers of the atmosphere and detect temperatures. Climate models, as well as actual air measurements, indicated temperature increases in the troposphere, but satellite data showed no such change. The inconsistency was resolved when researchers discovered that the instruments were also measuring the colder, higher stratosphere. Recalibration of the data now supports the temperature increase in the atmosphere.

13.10.mtb1

Before You Leave This Page Be Able To

- ✓ Describe what global warming means and explain how it might be measured.
- ✓ Summarize the major lines of direct measurement and proxy evidence indicating global warming in the last one hundred years.
- ✓ Explain the NAS conclusions using the combination of different temperature reconstructions.

What Factors Influence Global Warming?

MOST DATA INDICATE THAT SOME GLOBAL WARMING is occurring. Many scientists propose that human activities, including the burning of fossil fuels and the clearing of forests, contribute greenhouse gases to the atmosphere. Astronomical factors, such as Earth's orbit around the Sun and an increase in sunspot activity, can also contribute to warming. Other factors may lead to *global cooling*, such as ash from large volcanic eruptions and an increase in certain aerosols in the atmosphere.

A What Processes Influence Atmospheric Temperature Change?

Earth's surface temperatures are dominated by energy from the Sun. Sunlight heats the oceans, land, and atmosphere, but several factors influence how much of this energy reaches the surface and how much is retained.

Interaction of Sunlight with Earth's Atmosphere, Oceans, and Land

13.11.a1

1. Most of Earth's surface heating comes from sunlight, which heats the atmosphere, land, and oceans. Some of this energy escapes back into space. The rest is trapped by interactions with Earth, keeping the planet warm by a process called the *greenhouse effect*.

2. The amount of solar radiation hitting Earth varies regularly due to orbital fluctuations and changes in the Sun's energy output.

3. Some sunlight is *absorbed* by the atmosphere (shown as a yellow glow in the figure), and some is *reflected* off the atmosphere. The reflected sunlight returns to space without heating Earth, as depicted by the yellow arrows.

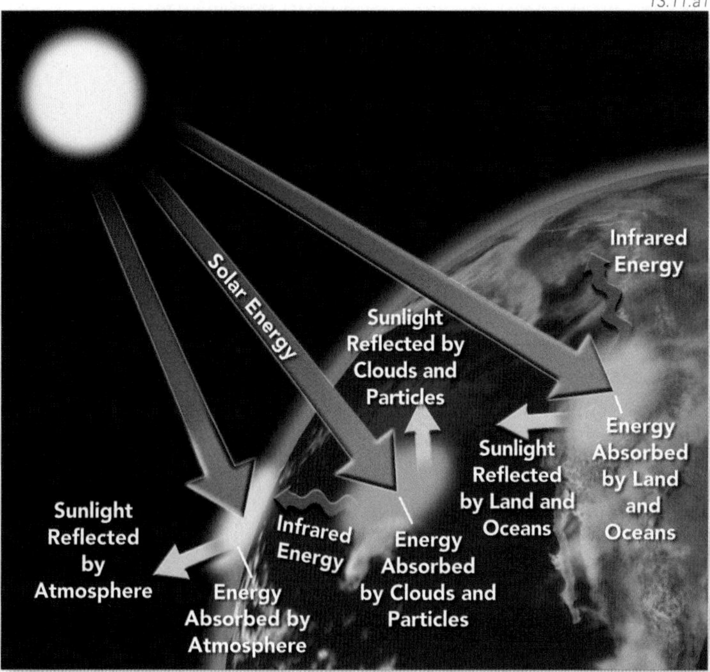

4. Light is *absorbed* by clouds, by soot from burning, and by fine particles called *aerosols*, which are produced by volcanoes, industry, and automobiles. Some of this absorbed energy radiates back into space as *infrared energy* (shown by the wavy red arrows). Clouds and particles also *reflect* some sunlight.

6. Some light is absorbed by the land and the oceans, both of which then radiate *infrared energy* back into the atmosphere. Some of this infrared energy is absorbed by atmospheric gases, such as water (H_2O). Infrared energy is also absorbed by gases such as carbon dioxide (CO_2), methane (CH_4), and nitrous oxides (NO_x), which are called greenhouse gases. Some portion of these gases is produced naturally and some is produced by human activities. More CO_2, for example, is produced naturally than is produced by humans.

5. Some light is *reflected* back into space from the land surface. Ice in continental glaciers is an effective reflector. As glaciers melt, darker land or ocean is uncovered. This increases the amount of energy absorbed by Earth or subsequently re-radiated back to the atmosphere.

13.11.a2

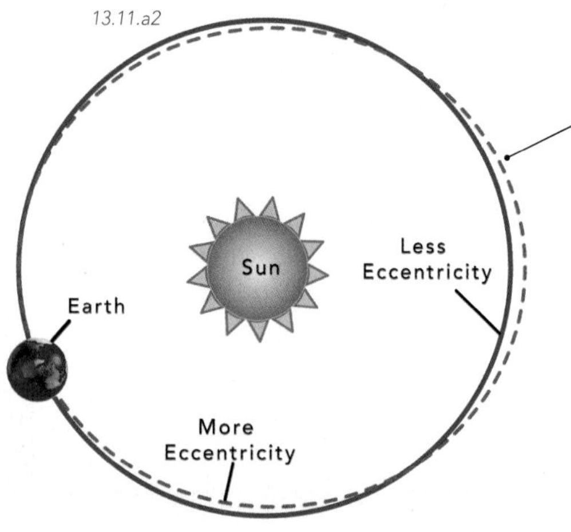

Orbital Variation

Earth's orbit is affected by the Moon, Sun, and the other planets, causing Earth's orbit to vary from more elliptical to more circular paths over time periods of ~100,000 years. This influences Earth's climate, but the current global warming has occurred in less than 200 years. This is too short a time period for orbital changes to have caused the observed warming. The tilt of Earth's spin axis also changes, but the effects occur over thousands of years. Variations in Earth's orbit and tilt are discussed in the next chapter.

Variations in Solar Radiance

Where can we look for a record of how the Sun shined in the past? Cosmogenic isotopes in ice cores and tree rings serve as proxies. These isotopic abundances are related to the strength of the solar wind, and in a complex way, to the Sun's radiance. Using these proxies it is estimated that the Sun's energy emission has varied by about 1 percent over the last 2,000 years. It is currently debated whether changes in solar output are enough to cause the temperature changes of the late twentieth century.

C What Are Some Geologic Controls on Ecology?

The materials and geologic structures determine what kind of life can survive in a given place—*the ecology.*

13.13.c1

Rocky mountains may catch more rainfall, but have little soil on their steep slopes. This limits the types of plants and animals that can thrive. Gentler mountains in wet climates develop thick soils that allow a greater abundance of plants and animals. [Banff, Alberta, Canada]

13.13.c2

Topography, which is controlled by geology and climate, determine where water flows and whether it runs off or soaks into the ground. Some rocks allow more water to run off into river systems, sustaining stream-side ecosystems and lakes. [Conway, New Hampshire]

13.13.c3

Flat areas, especially if underlain by soft rock or unconsolidated materials, allow water to soak in and develop better soils. The water and soils allow more vegetation to grow. Where there are plants, there are also animals. [Hudson River Valley, New York]

13.13.c4

Volcanic rocks can weather quickly, releasing nutrients such as magnesium and sulfur into soils. But some rocks, such as these mineralized volcanic rocks, release acids and other compounds detrimental to plants or animals. [Silverton, Colorado]

D How Does Geology Influence Agriculture?

Geologic structures and erosional processes provide environments, such as broad valleys, suitable for growing crops. Decomposed rocks provide all the inorganic nutrients needed by plants except for CO_2, which plants extract from the air to make their carbon-rich leaves, stems, and wood. Examine this conceptualized terrain.

Groundwater from this granite recharges local streams, whose water is used to irrigate crops.

Fault movement downdropped a fault block, forming a valley that filled with sediments eroded from the uplifted mountain range and adjacent areas.

This mountain range of fractured granite sheds debris rich in sodium, calcium, magnesium, and phosphorus. These are important nutrients for plants and animals.

Soils, which are composed of weathered rock and organic material, provide the foundation for crops and other plants.

5 Kilometers

13.13.d1

This river valley contains rich soils derived in part from fine sediment deposited on the floodplain during floods.

Wave action during the last glacial period planed off this uplifted coastal bench, which is now used for cropland. The bench only exists because of uplift.

Water-vapor-rich air from warm ocean currents provides abundant rainfall and keeps plants cool during foggy, coastal summer days.

Before You Leave This Page Be Able To

- ✓ Explain the factors that control where life can exist.
- ✓ Describe some ways that geology influences ecology, the distribution of ecosystems, and agriculture.

13.13

What Occurred During the Hurricane Season of 2004?

NOAA HURRICANE FORECASTS FOR 2004 predicted that the Atlantic would see slightly above-average activity, with 6 to 8 hurricanes. Of these 2 to 4 would become major hurricanes. The conditions, however, were more favorable for hurricanes than estimated. The Atlantic had 15 hurricanes, 6 of them major. Storms and hurricanes killed more than 3,000 people and caused over 40 billion dollars in damage.

A Where Did the 2004 Hurricanes Hit and How Strong and Damaging Were They?

Hurricanes are the most energetic of common short-term Earth events. An average hurricane expends more energy than was released in the largest recorded earthquakes. Meteorologists measure hurricane intensity based on wind speed, air pressure, and surge potential. On this scale, the 2004 hurricanes were slightly above average. However, since so many made landfall in vulnerable places, they were very destructive.

13.14.a1

During the 2004 season, four major hurricanes (*Jeanne*, *Frances*, *Ivan*, and *Charley*) made landfall along the southeastern United States. This map shows the paths they traveled.

One reason why so much damage occurred is that Florida became very popular with retirees in the period between 1966 and 2003, when only a few major hurricanes made landfall. Until 2004, the population had no experience with multiple giant storms in one season.

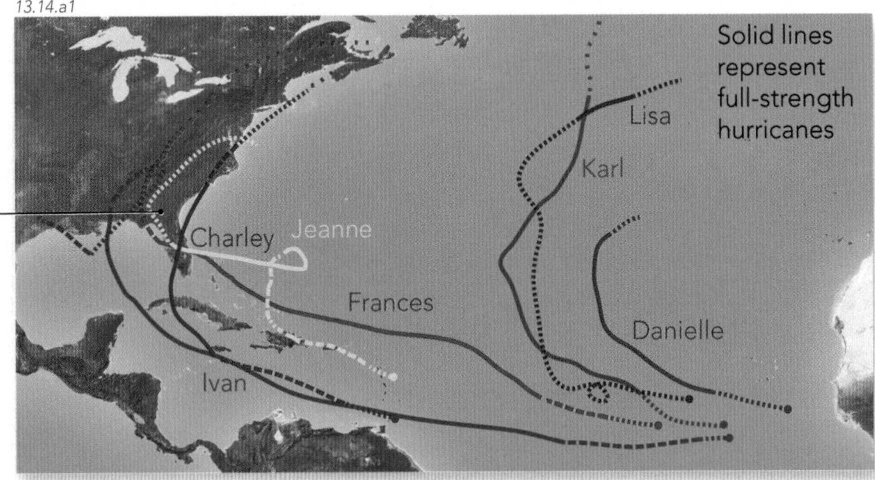

Major hurricanes were born off of west Africa. They traveled thousands of miles, gaining or maintaining strength before crashing onto a very small landing zone in Florida and adjacent areas. Luckily, other hurricanes missed the land.

Comparing Intensity and Damage of Hurricanes

13.14.a2

▷ The Saffir–Simpson Scale is a classification system based on wind velocity. It ranks hurricanes from 1 to 5, with 5 being the potentially most destructive. Ivan, shown to the left, was rated as a category 3 when it came ashore west of Florida (the land to the right of the storm).

Saffir–Simpson Category	Maximum Sustained Wind Speed (miles/hour)	Minimum Central Pressure (millibars)	Storm Surge	
			(feet)	(meters)
1	74–95	> 980	3–5	1.0–1.7
2	96–110	979–965	6–8	1.8–2.6
3	111–130	964–945	9–12	2.7–3.8
4	131–155	944–920	13–18	3.9–5.6
5	156+	less than 920	19+	5.7+

▷ This table compares actual damage for six recent hurricanes. The year 2004 was the worst hurricane season on record for damage until Hurricane *Katrina* destroyed parts of New Orleans in 2005. The amounts have not been adjusted for inflation.

Rank	Hurricane	Year	Category	Damage (billions of dollars)
1	Andrew	1992	5	26.5
2	Charley	2004	4	15
3	Ivan	2004	3	14,2
4	Frances	2004	2	8.9
5	Hugo	1989	4	7
6	Jeanne	2004	3	6.9

Before You Leave These Pages Be Able To

- ✓ Describe the Saffir-Simpson scale for hurricanes.
- ✓ Summarize or sketch conditions that caused 2004 to be a bad year for hurricanes.

B What Factors Affect the Appearance of a Shoreline?

Shorelines around the world have diverse appearances, from sandy white beaches to dark craggy cliffs that plunge vertically into the sea, with no beach at all. These differences are controlled by a number of factors.

Factors on the Water Side

1. From the water side, the appearance of a shoreline is greatly influenced by the strength of the waves and tides that impact the shore. Stronger waves will typically be associated with greater erosion and larger clasts in sediment along the shoreline.

2. The size and intensity of storms influence the appearance of a coast because storms bring with them large waves, strong winds, and intense rainfall. Some coasts are ravaged by hurricanes, whereas others rarely experience the erosive effects of powerful storms.

3. Slope of the seafloor is also a factor. Steep slopes can allow large waves to break directly against rocks along the shore, whereas more gentle slopes cause waves to break a short distance offshore.

4. The orientation of a coastline is also important, because waves typically approach from specific directions in response to prevailing winds. The dominant wave direction may change with the season (summer versus winter or dry versus rainy season). Also, some parts of the coast will receive less wave action because they are sheltered in a bay or are protected by an offshore feature, such as an island or barrier reef. The coastline below contains large boulders, most more than a meter across. It is directly exposed to waves along the stormy coast of Iceland, and some boulders probably were brought into the area by glaciers.

14.01.b4

14.01.b1

Factors on the Land Side

5. On the land side, the appearance of the shoreline reflects the hardness of the bedrock along the coast. Hard rock that resists erosion tends to form rocky cliffs, such as those at Big Sur, California (▷), whereas softer sediment and rock can be eroded into more gentle slopes and hills.

14.01.b2

6. Coastal landscapes also reflect the amount and size of sediment available. A coast cannot be rocky if the only materials present are soft and fine-grained. Rivers provide a fresh influx of sediment into the shoreline environment.

7. Coastlines undergoing uplift have a different appearance than those where the land has dropped relative to water level. A rise in sea level flooded river valleys along the North Carolina coast, producing a coastal outline marked by long, narrow bays. ▽

14.01.b3

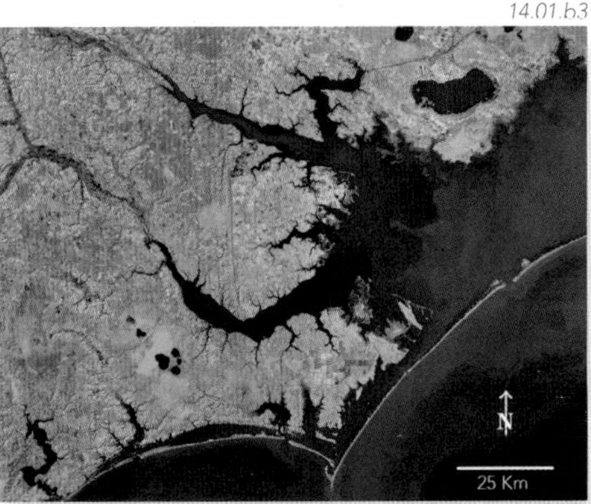

8. Climate is a major factor influencing coastal landscapes. Wet climates provide abundant precipitation for erosion and promote the growth of vegetation and formation of soil. Vegetation stabilizes soil and limits the amount of material that can be picked up by wind and water. Dry climates result in less vegetation and less soil.

Before You Leave This Page Be Able To

- ✓ Summarize or sketch the types of processes that affect shorelines.
- ✓ Summarize or sketch how different factors affect the appearance of a shoreline.

What Causes High Tides and Low Tides?

THE SEA SURFACE MOVES UP AND DOWN across the shoreline twice each day. These changes, called *tides*, are observed in the oceans and in bodies of water, such as bays and estuaries, that are connected to the ocean. What causes the tide to rise and fall, and why are some tides higher than others?

A What Are High and Low Tides?

Tides are cyclic changes in the height of the sea surface, generally measured at locations along the coast. The difference between high and low tide is typically 1 to 3 meters, but can be more than 12 meters or almost zero.

High Tide

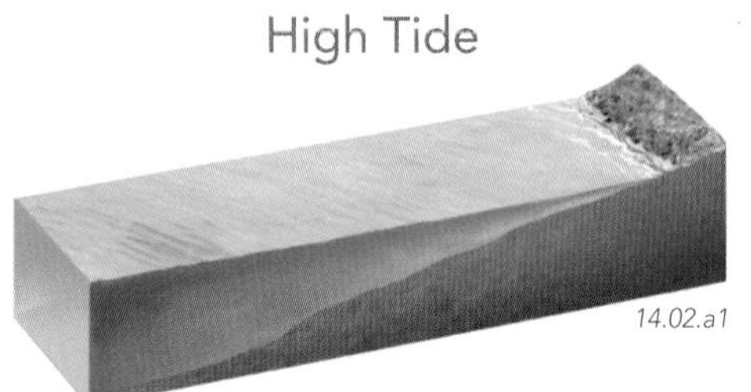

During *high tide*, the height of water in the ocean has risen to its highest level relative to the land. At this point, the water floods into low-lying areas. In most areas, high tide occurs every 12 hours and 24 minutes.

Average Sea Level

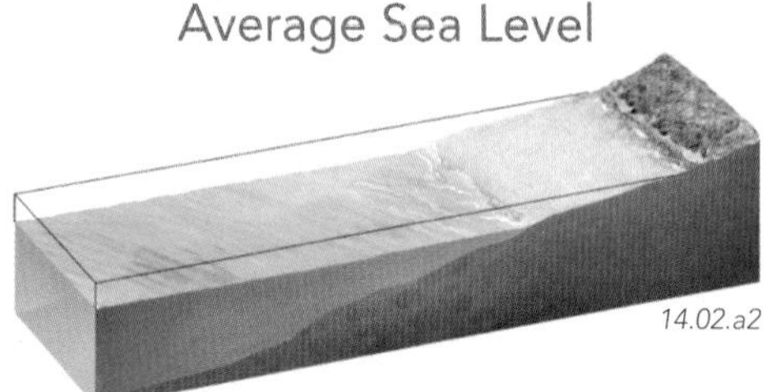

Following high tide, the water level begins to *fall* relative to the land—the tide is *going out*. At some time, water level will reach the *average sea level* for that location, but it keeps falling on its way to low tide.

Low Tide

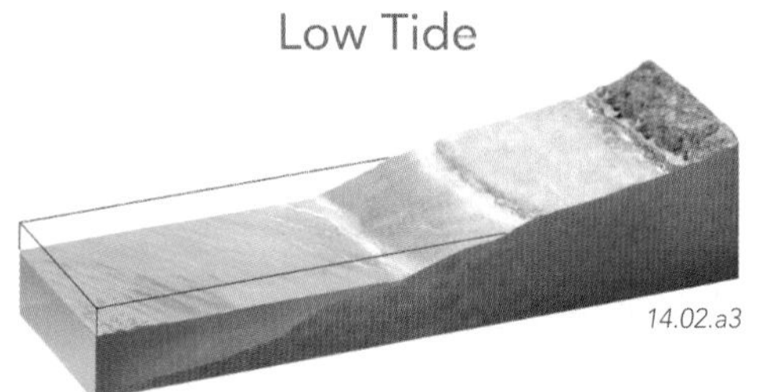

When the water level reaches its lowest level, it is *low tide*. Low tide in most areas also occurs every 12 hours and 24 minutes. Water level begins to rise again after low tide, and the tide is *coming in*. Rising tide spreads water across the land.

B What Causes High and Low Tides?

Tides rise and fall largely because water in the ocean is pulled by the gravity of the Moon and to a lesser extent the Sun. As Earth rotates on its axis, most shorelines experience two high and two low tides in each 25-hour period.

1. This figure depicts Earth and Moon as if looking directly down on Earth's North Pole. It shows the Moon much closer to Earth than it would be for the size of the two bodies. Earth rotates (spins) counterclockwise in this view and relative to the Sun completes a full rotation once every 24 hours.

2. The Moon orbits (travels around) Earth once every 29 days, also counterclockwise in this view. Due to this motion, it takes 24 hours and 50 minutes for a point on Earth that is facing the Moon to rotate all the way around to catch up with and again face the Moon.

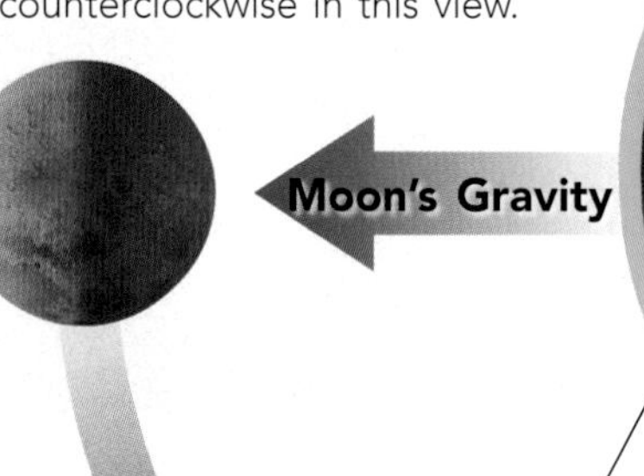

3. The Moon exerts a gravitational pull on Earth and its water. This pulls the water in the ocean toward the Moon, causing it to mound up on the side of Earth nearest to (i.e., facing) the Moon. Coastal areas beneath the mound of water experience *high tide*. On this figure, the thickness and mounding of the water are greatly exaggerated.

4. On parts of Earth that are facing neither toward or away from the Moon, sea level is lower as water is pulled away from these regions toward areas of high tide. Coastal areas here experience *low tide*.

5. On the side of Earth opposite the Moon, the Moon's gravity pulls the Earth more than the overlying water. The water is farther from the Moon and so feels less of the Moon's gravitational pull (recall that the force of gravity decreases with distance). The side of Earth *facing directly away* from the Moon, therefore, experiences *high tide*.

6. Earth rotates (spins on its axis) much faster than the Moon orbits Earth, so it is best to think of the mounds of water—but not the water itself—as remaining fixed in position relative to the Moon as Earth spins. During a complete rotation of Earth, a coastal area will pass through both mounds of water, causing most shorelines to have two high tides and two low tides in each 24-hour and 50-minute period.

C Why Are Some High Tides Higher Than Others?

From week to week, not all high tides at any location reach the same level—some are higher and others are lower than average. Similarly, some low tides are very low and others are less so. Such variations follow a predictable pattern that repeats about every 14 days and are related to the added influence of the Sun's gravity.

Spring Tides

1. Like the Moon, the Sun exerts a gravitational pull on Earth and its water. The Sun is larger and more massive than the Moon, but is farther away from Earth, so the Sun's gravitational effect on Earth is weaker than the Moon's.

2. The Sun's gravity attracts a mound of water on the side of Earth facing the Sun, and causes another mound on the side facing away from the Sun. These thin mounds, shown in dark blue, are fixed in position relative to the Sun, and locations on Earth rotate through each mound once every 24 hours.

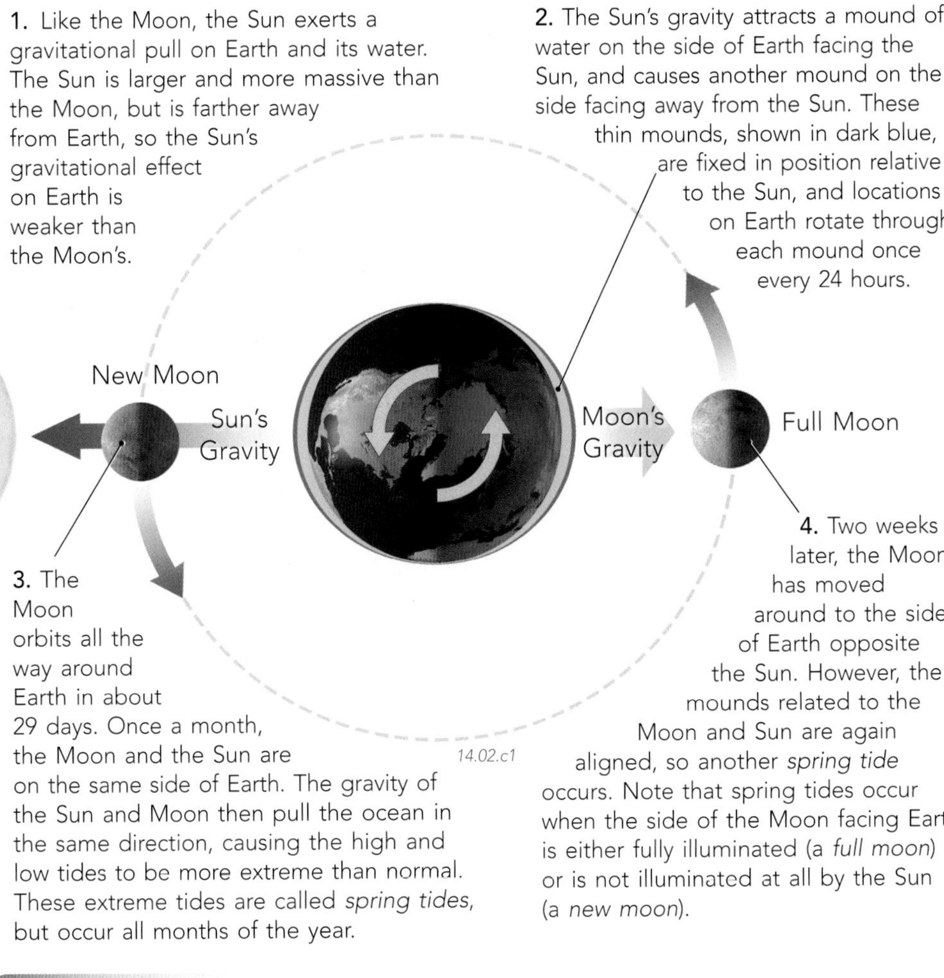

3. The Moon orbits all the way around Earth in about 29 days. Once a month, the Moon and the Sun are on the same side of Earth. The gravity of the Sun and Moon then pull the ocean in the same direction, causing the high and low tides to be more extreme than normal. These extreme tides are called *spring tides*, but occur all months of the year.

4. Two weeks later, the Moon has moved around to the side of Earth opposite the Sun. However, the mounds related to the Moon and Sun are again aligned, so another *spring tide* occurs. Note that spring tides occur when the side of the Moon facing Earth is either fully illuminated (a *full moon*) or is not illuminated at all by the Sun (a *new moon*).

Neap Tides

5. Seven days after each spring tide, the Moon has journeyed 1/4 of the way around Earth so that Moon and Sun are at right angles to one another, relative to Earth. In this position, the Moon's and Sun's gravity are pulling at right angles to one another.

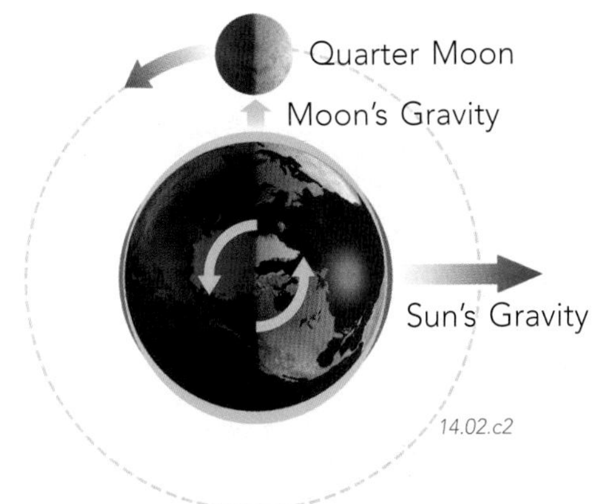

6. At these times of the month, the mounds of water related to the Sun and Moon do not coincide, so the differences between high tide and low tide are less than average. These lower-than-average high tides and higher-than-average low tides occur every two weeks and are called *neap tides*, from an Old English word for "lacking." They happen when the Moon, as viewed from Earth, is only half illuminated by sunlight (called a *quarter moon*).

The Most Extreme Tides in the World

Some places have higher tides than others. The difference between high and low tide can be so small as to be nearly undetectable, or can be so extreme as to be dangerous. The Mediterranean has almost no tide, and much of the Caribbean has only one tide each day. The world's highest tides are in the *Bay of Fundy* along the Atlantic Coast of Canada. In this place, the geometry of the shoreline and sea bottom funnel water in and out of the bay at just the right rate to cause a tidal range of as much as 16 meters (52 feet)! The large tides are used to generate electricity for Nova Scotia. These two photographs illustrate the extreme tidal range within the Bay of Fundy.

14.02.mtb1

14.02.mtb2

Before You Leave This Page Be Able To

- ✓ Describe or sketch what tides are.
- ✓ Sketch and describe how tides relate to the position of the Moon and why.
- ✓ Sketch or summarize how the gravity of the Moon and Sun cause spring tides and neap tides.

How Do Waves Form and Propagate?

WAVES ARE THE MAIN MECHANISM by which shorelines are eroded. Waves also transport and deposit material along the shoreline. Most ocean waves are formed by wind and are limited to the uppermost levels of the water. How do waves form, how do they propagate across the water, and what causes waves to break?

A What Is an Ocean Wave and How Is the Size of a Wave Described?

1. *Waves* are irregularities on the surface of a body of water. They vary in expression from a series of curved ridges and troughs to more irregular bumps and depressions to dramatic breaking curls perfect for surfers. ▽

14.03.a1

2. Out in the deep open water, many waves occur in *sets* of individual waves that are similar in size and shape to one another. The lowest part of a wave is called the *trough* and the highest part is the *crest*, and the vertical distance between the two is the wave *height*.

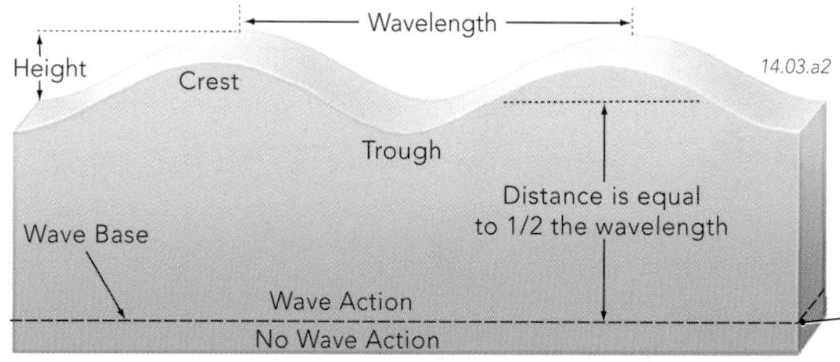

14.03.a2

3. The horizontal distance between two adjacent crests in a set of waves is called the *wavelength*, and indicates how far apart individual waves are. The wavelength and the vertical *height* of the wave are used to describe the size and spacing of a set of waves. The average wave is several meters high.

4. Most waves are near-surface features. They affect only the surface of the water and depths typically down to several tens of meters. Below some level, called the *wave base*, the wave ceases to have any effect.

5. The depth of the wave base is about half the wavelength:

depth of wave base = wavelength/2

If wave crests are 40 m apart, then the wave base will be about 20 m deep.

B How Do Waves Propagate Across the Water?

A set of waves can propagate a long distance across the ocean or other body from deep water, but the water through which the wave passes moves only a short distance.

▷ 1. Water waves propagate in a manner similar to seismic waves in rocks, where each wave is a zone of *compression* moving through the water. These compressive forces cause water within the wave to push against the water in front of it. These three figures are snapshots of a set of waves propagating to the right. Examine the motion of water particles A, B, and C within the wave.

▷ 2. Here, both waves have propagated through the water to the right, but the individual reference points have moved only a short distance. Point A, which is close to the surface, moves more than point B, which is deeper. Point C is below the wave base and does not move.

▷ 3. Later, points A and B are beneath the wave trough and have nearly returned to their positions on the reference line. As the next wave crest approaches, they will continue their motion to the left and upward, moving toward their starting points.

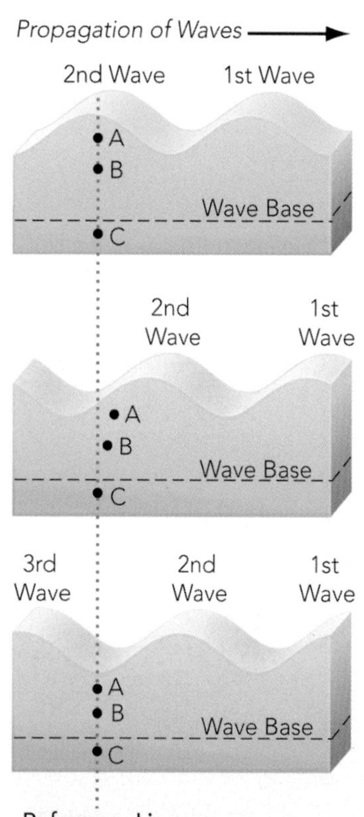

14.03.b1–3

▽ 4. During passage of an entire wave (crest to crest), points A and B will each have followed a small, circular path. The figure below shows the circular paths of 20 different points at different depths and positions along the wave.

5. Water particles on the surface of the water travel the most, going up and forward and then down and back, in a circular motion, as the wave passes.

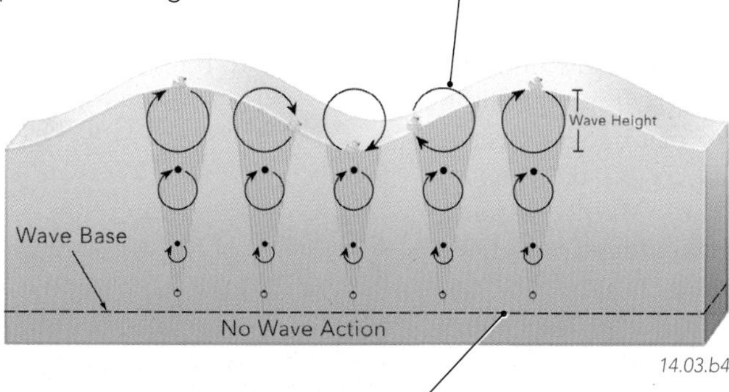

14.03.b4

6. Deeper within the wave, water particles travel smaller circular paths, and the paths become smaller with depth.

7. Water that is right above the wave base barely moves, and water below the wave base does not move at all as the wave passes.

B What Shoreline Features Are Associated with Deposition?

As sediment is transported along a coastline, it can preferentially accumulate (be deposited) in certain settings, forming a variety of low, mostly sandy features.

14.05.b1

Sandbar—Offshore of many shorelines is a low, sandy area, called a *sandbar*. Bars are typically submerged much of the time and can shift position as waves and longshore currents pickup, move, and deposit the sand. [Queensland, Australia]

Spit—Along some coasts, a low ridge of sand and other sediment extends like a prong off a corner of the coast. Such a feature, called a *sand spit* or simply a *spit*, can be easily eroded, especially by storm waves. [Puget Sound, Washington]

14.05.b2

14.05.b3

Barrier Island—Many shorelines have long, low islands offshore. These islands act as barriers, keeping most large waves and rough seas from reaching the coast. Most barrier islands are dominated by sand, including sand dunes. [North Carolina]

Baymouth Bar—This feature forms where a sandbar or spit blocks a bay from the sea. The bay may remain as a trapped body of water or may get filled in with sediment from land or with sediment washed off or over the bar. [Baja California, Mexico]

14.05.b4

Formation of a Spit, Baymouth Bar, and Barrier Island

A *spit* forms when waves and longshore currents transport sand and other beach sediment along the coast, building a long but low mound of sediment that lengthens in the direction of the prevailing longshore current.

If a spit grows long enough, it may cut off a bay, becoming a *baymouth bar*. This bar shelters the bay from waves and may allow it to fill in with sediment, forming a new area of low-lying land, like a marsh.

If sea level rises, former spits and bars may become long, sandy *barrier islands*. Alternatively, the pile of sediment in the barrier island was originally deposited by rivers when sea level was lower and later became islands as sea level rose.

The Shape of Cape Cod

Cape Cod sticks out into the Atlantic Ocean from the rest of Massachusetts like a huge, flexed arm. The "curled fist" is mostly a large spit, and other features are bars and barrier islands. Much of the sediment was originally deposited here by glaciers, which retreated from the area 18,000 years ago. As the glaciers melted, global sea level rose, flooding the piles of sediment and causing them to be reworked by waves and longshore currents.

14.05.mtb1

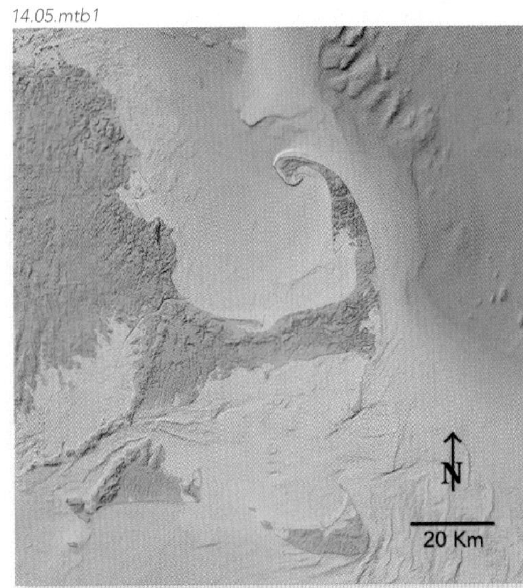

Before You Leave This Page Be Able To

- ✓ Describe the different types of shoreline features.
- ✓ Sketch and summarize one way that a sea stack, spit, baymouth bar, and barrier island can each form.
- ✓ List the types of features that are present on Cape Cod, and discuss how these types of features typically form.

What Are Some Challenges of Living Along Shorelines?

SHORELINES, BY THEIR DYNAMIC NATURE, can be risky places to live. Destruction of property and loss of life can result from waves, storm surges, and other events that are integral to the coastal environment. Beaches and other coastal lands can be totally eroded away, along with poorly situated buildings. How can home buyers identify and avoid such unsuitable and potentially risky sites?

A What Hazards Exist Along Shorelines?

Shoreline hazards mostly are related to the interaction between water and land, but also include strong winds.

14.06.a1

Waves are constantly present, but are a variable-intensity threat to land, buildings, and people along shorelines. The most damage from waves occurs during extreme events, such as hurricanes and other storms. Waves can erode land and undermine hillsides and structures, causing them to collapse into the water.

14.06.a2

A *storm surge* is a local rise in the level of a sea or large lake during a hurricane or other storm. A storm surge is caused by strong winds that pile up water in front of an approaching storm, inundating low-lying areas along the coast. Surges can be accompanied by severe erosion, transport, and deposition of material.

14.06.a3

Strong winds and rain commonly accompany storms that strike the coast. Coastal communities are especially susceptible to these hazards because they often lack a windbreak between them and the open water. Also many structures have been built along low-lying areas that are prone to rainfall-related flooding.

Before

14.06.a4

1. These images document damage caused by Hurricane Fran in 1996 along the beach on Topsail Island, North Carolina. The left photograph shows the area before the hurricane. White numbers mark two houses in both photographs.

After

14.06.a5

2. This photograph, taken after the hurricane, shows the loss of beach and destruction of houses caused by waves, storm surge, and erosion.

14.06.a6

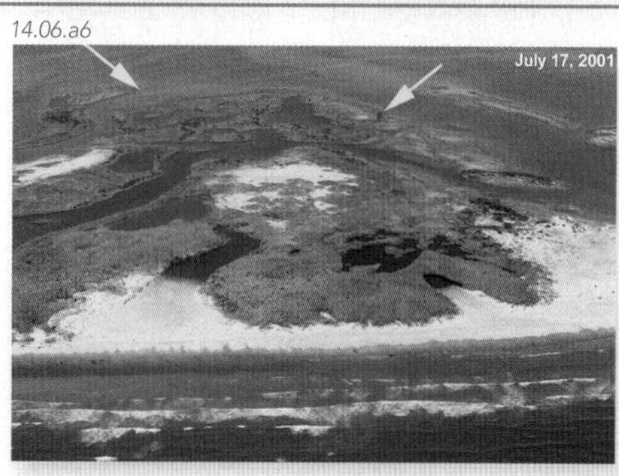

3. These images show an area of the Chandeleur Islands of Louisiana before and after Hurricane Katrina in 2005. Before the hurricane, the islands had sandy beaches, dunes, and marshes. Arrows point to the same location in each image.

14.06.a7

4. After the hurricane, the islands, marshes, and beaches were much smaller because of erosion by large waves and because the area was overtopped by the storm surge.

B What Approaches Have Been Tried to Address Shoreline Problems?

Various approaches have been tried to minimize the impacts of natural shoreline processes, including erecting barriers, trying to reconstitute the natural system, or simply not building in the most hazardous sites.

14.06.b1

▷ **1.** One way to limit the amount of erosion is to construct a *sea wall* along the shore. Such walls are made of concrete, steel, or some other strong material. Building a sea wall results in loss of beach in front of the wall. Large rocks and other debris are locally dumped along the shoreline to armor the coast in an attempt to protect it from erosion. Material used in this way is called *rip rap.* [Galveston, Texas]

14.06.b2

◁ **2.** Another type of wall, called a *jetty,* juts out into the water, generally to protect a bay, harbor, or nearby beach. Jetties are usually built in pairs to protect the sides of a shipping channel. Jetties and other walls can focus waves and currents on adjacent stretches of the coast. These directed waves and currents, deprived of their normal load of sediment by the wall, erode these adjacent areas as they try to regain an equilibrium amount of sediment. [Yaquina Bay, Oregon]

14.06.b3

3. A wall, called a *breakwater,* can be built out into the water to bear the brunt of the waves and currents. These breakwaters were built parallel to the coast to protect the beach from severe erosion and to cause sand to accumulate on the beach behind the structures. [Presque Isle State Park, Pennsylvania]

4. Low walls, called *groins,* are built out into the water to influence the lateral transport of sand by longshore currents and waves that strike at an angle to the coast. A groin is intended to trap sand on its up-current side, but has the sometimes unintended consequence of causing the beach immediately down-current of the groin to receive less sand and to become eroded.

14.06.b4

△ **5.** Some communities bring in sand to replenish what is lost to storms, currents, and nearby engineering projects. This procedure, called *beach nourishment,* can be expensive and may last only until the next storm. [Florida]

Avoiding Hazards and Restoring the Shoreline System to Its Natural State

One approach preferred by some people, including many geologists, is simply not to build in those places that have the highest likelihood of erosion, coastal flooding, coastal landslides, and other shoreline hazards. Geologists can map a shoreline and conduct studies to identify the most vulnerable stretches of coastline. With such information in hand, the most inexpensive approach—in the long run—is to forbid the building of houses or other structures in those areas identified as high risk. In the wake of the destruction of New Orleans and nearby communities by Hurricane Katrina, there is a debate about whether to rebuild those neighborhoods that are at highest risk, such as those below sea level.

Such geologic concerns often are either ignored or are overruled by financial and aesthetic interests of developers, communities, and people who own the land. Beach-front property is desirable from an aesthetic standpoint and so can be expensive real estate, which some people think is too precious to leave undeveloped.

Another approach is to try to return the system to its original situation, or at least a stable and natural one, rather than trying to "engineer" the coastline. Engineering solutions can be expensive, may not last long, or may have detrimental consequences to adjacent beaches. Returning the system to a natural state may involve restoring wetlands and barrier islands that buffer areas further inland from waves and wind. Examining the balance of sediment moving in and out of the system can help identify non-natural factors, such as dammed rivers, which if restored to original conditions would bring more sediment into the system and stabilize beaches, dunes, and marshes.

Before You Leave This Page Be Able To

- ✓ Summarize some of the hazards that affect beaches and other coastlines.
- ✓ Describe the approaches used to address coastal erosion and loss of sand, including not building in high-risk areas and trying to restore the system to a natural state.

How Do Geologists Assess the Relative Risks of Different Stretches of Coastline?

UNDERSTANDING THE GEOLOGY AND DYNAMICS of a shoreline is the first step in assessing the potential risks posed by waves, currents, coastal flooding, and other shoreline processes. Geologists study coastlines using traditional field methods and new methods that involve lasers and satellites.

A What Field Studies Do Geologists Conduct Along Shorelines?

To investigate potential shoreline hazards, geologists map and characterize the topographic and geologic features of the land, shoreline, and near shore sea bottom. They combine this information with an understanding of the important shoreline processes to identify those areas with the highest hazard.

1. To assess shoreline hazards, geologists document the *elevation of the land surface* above sea level. High areas clearly have less risk of being flooded by the sea. Precise elevations of the land are measured with various surveying tools, some using satellites (Global Positioning System, or GPS) or lasers that scan the ground surface from a plane. These surveys identify areas, such as this high marine terrace, that are too high to be flooded, even during a hurricane.

2. Areas that are close to sea level may be subjected to flooding by storm surges and storm-related intense rainfall that causes flooding along coastal rivers. Vulnerable low-elevation areas may extend far inland, such as along this low river valley.

3. Mapping the bedrock geology, as well as the loose sediments along the beach, guides geologists when assessing how different areas will erode. Coasts backed by resistant bedrock, such as along a cliff, will be less likely to be eroded by strong waves and currents.

4. Geologists also map the distribution and height of coastal dunes. Dunes, especially those that are large or are stabilized by vegetation, decrease the risk inland for surge and erosion. Marshlands, such as those on a delta, also help buffer areas farther inland from waves, storm surges, and strong coastal winds.

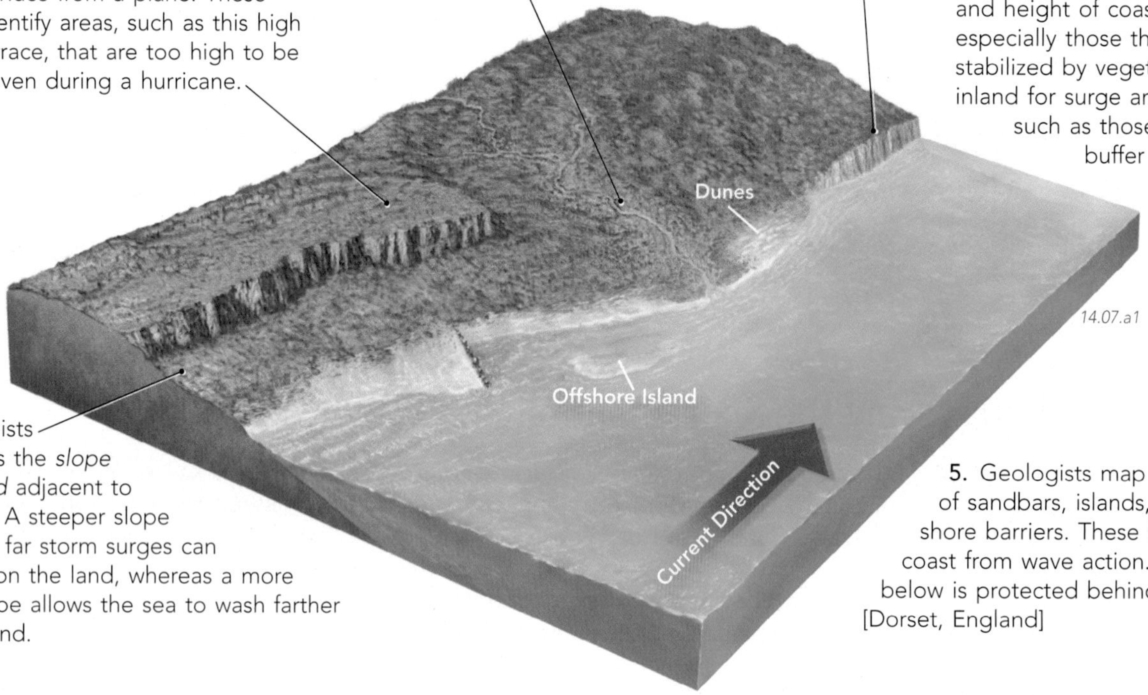

Dunes

Offshore Island

Current Direction

14.07.a1

5. Geologists map the location and height of sandbars, islands, reefs and other offshore barriers. These barriers can protect the coast from wave action. The coastline shown below is protected behind an offshore bar. [Dorset, England]

6. Geologists also assess the *slope of the land* adjacent to the shore. A steeper slope limits how far storm surges can encroach on the land, whereas a more gentle slope allows the sea to wash farther into the land.

7. Geologists document the width of beaches. An area with a wide beach between the shoreline and houses is generally less risky than an area where houses sit right behind a narrow beach. Sea walls, groins, and other constructed features can greatly affect beach width and therefore potential risk. A sea wall can limit the amount of landward erosion and may protect buildings from storm surges, but sand in front of the sea wall may be lost. A groin affects the width of the beach differently on either side. It may decrease the risk of storm erosion to the beach on the up-current side, which gains sand and becomes wider, but increase the risk of storm erosion to the beach on the down-current side, which loses sand and becomes narrower. Beach width is also affected by offshore sand bars. ▷

14.07.a2

B What Can New High-Resolution Elevation Data Tell Us About Shorelines?

Satellite data (GPS) and other new methods of mapping elevation now allow scientists to more accurately characterize shoreline regions and track in detail how a shoreline changes during storms. One new method is called *LIDAR*, which is an acronym for LIght Detecting And Ranging.

In the LIDAR method, a laser beam is bounced off the ground, detected back at the plane, and timed as to how long it takes to travel to the surface and back to the instrument. The shorter the time, the shorter the distance between the plane and the ground—and therefore the higher the elevation of the land. LIDAR elevations are accurate to within about 15 centimeters (6 inches) and can be quickly collected over larger areas than is practical to cover with conventional surveying.

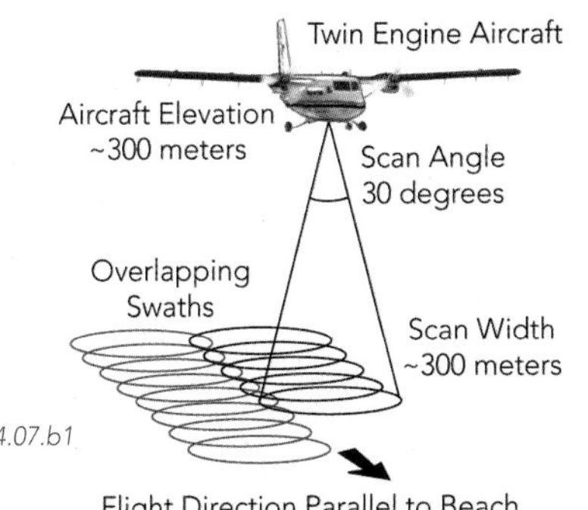

14.07.b1

As the plane flies forward, mirrors direct the laser toward different areas beneath the LIDAR sensor. Thousands of data measurements are recorded each second in a narrow belt, called a *swath*, across the land. The plane flies back and forth over the area, overlapping adjacent swaths to ensure that there are no gaps in the data. A GPS unit mounted on the plane allows technicians to accurately register the LIDAR data with geographic map coordinates and to match the data to features on the ground.

Mapping Hurricane-Related Changes in the Shoreline of Alabama

14.07.mtb1–3

May 2004 (pre-Ivan) A

Gulf of Mexico

Mississippi Sound

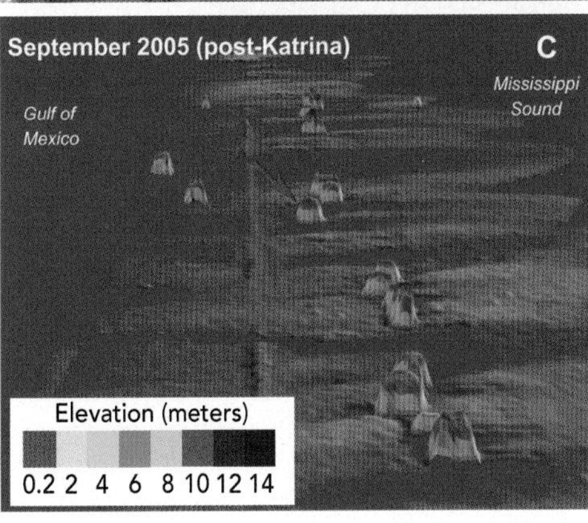

Coastal Alabama has been hit by a series of powerful hurricanes, most recently by Hurricane Ivan in 2004 and Hurricane Katrina in 2005. The U.S. Geological Survey (USGS) and other federal agencies have been using LIDAR to investigate the changes that such large hurricanes inflict on the coastline. One detailed study was of Dauphine Island, an inhabited barrier island along the Gulf Coast of Alabama.

The three images to the left show perspective views of detailed LIDAR elevations taken at three different times (before Ivan, after Ivan, and after Katrina). The first image shows a central road with houses (the colored "peaks") on both sides. In the second image, the storm surge from Hurricane Ivan has washed over the low island from left to right, eroding the left beach, covering the road with sand, and redepositing some of the sand on the right.

The aftermath of Hurricane Katrina, bottom left, is even more dramatic. All but a few houses were totally washed away. Both the width and height of the island decreased, leaving the remaining houses even more vulnerable to the next storm surge.

The two images to the right show the calculated changes caused by each storm. Each image was produced by comparing the *before* and *after* data sets, and computing the difference. Features shown in reds and pinks represent losses due to erosion, green areas show where deposition occurred, and whitish features were unchanged by that storm.

14.07.mtb4

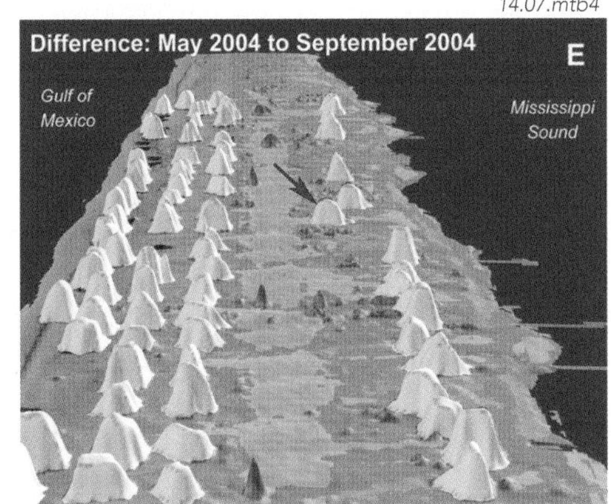

14.07.mtb5

Difference: September 2004 to September 2005 F

Gulf of Mexico

Mississippi Sound

Difference

Erosion Sedimentation

Before You Leave This Page Be Able To

- ✓ Describe how studying the geologic features along a coast can help identify areas of highest hazard.
- ✓ Summarize how LIDAR data are collected and provide an example of how they can be used to document changes in a shoreline.

What Happens When Sea Level Changes?

SEA LEVEL HAS RISEN AND FALLEN many times in Earth's history. A rise in sea level causes low-lying parts of continents to be inundated by shallow seas, whereas a fall in sea level can expose previously submerged parts of the continental shelf. Such changes in sea level are recorded in features formed along the shoreline and farther inland, and by marine sediments deposited on what is normally land.

A How Much Has Sea Level Risen or Fallen in the Past?

In the past, global sea level has been more than 200 meters higher and more than 120 meters lower than today. Such changes resulted in drastic changes in the location of shorelines and the outlines of continents.

This map shows one interpretation of shoreline positions in the eastern United States for various times in the past 120 million years. For each position, the line marks the boundary between sea to the southeast and land to the northwest. There is some disagreement about the shoreline positions at some of these times, such as 35,000 B.P.

Global sea level was high in the late Mesozoic, specifically in the mid-Cretaceous about 120 to 115 million years ago. Sea level was 200 meters higher than today, inundating the Atlantic coastal plain and a broad region inland from the present Gulf Coast. Farther west, these inland seas divided North America into two separate areas of land.

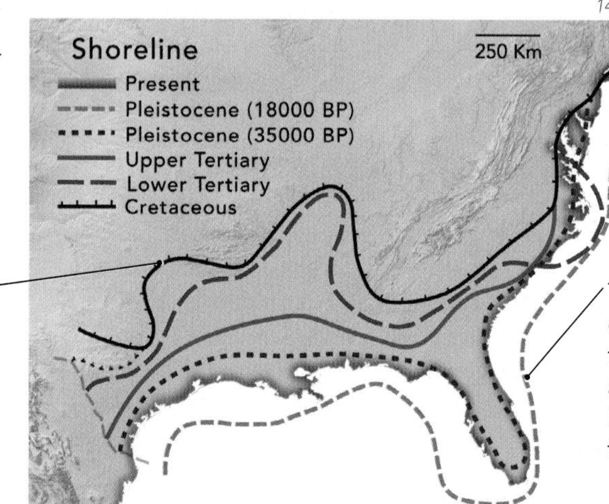

After the mid-Cretaceous, sea level fell overall, but with many smaller increases and decreases along the way. As sea level fell, progressively more of the continent emerged from the sea, and the land began to resemble the present-day outline of the continent.

The lowest sea level occurred more recently, during the most recent ice age about 18,000 to 20,000 years ago. Sea level then was about 125 meters lower than today, exposing large areas of the continental shelf, such as the area outlined offshore of Florida.

B What Features Form If Sea Level Has Risen or Fallen Relative to the Land?

Shorelines adjust their appearance, sometimes dramatically, if sea level rises or falls relative to land. A relative change in sea level can be caused by a global change in sea level or by tectonics, which causes the land to subside or be uplifted relative to the sea. Changing sea level produces two kinds of coasts.

Submergent coasts form where the land has been inundated by the sea because of a rise in sea level or subsidence of the land.

Preexisting deltas and barrier islands may become totally submerged by rising sea, forming offshore bars and other features.

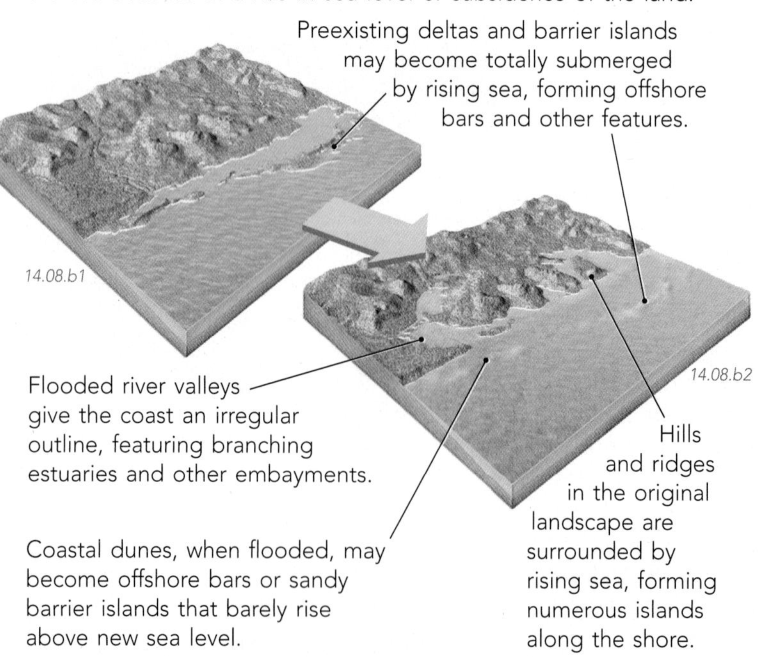

Flooded river valleys give the coast an irregular outline, featuring branching estuaries and other embayments.

Hills and ridges in the original landscape are surrounded by rising sea, forming numerous islands along the shore.

Coastal dunes, when flooded, may become offshore bars or sandy barrier islands that barely rise above new sea level.

Emergent coasts form where the sea has retreated from the land due to falling sea level or uplift of the land relative to the sea.

Sandbars that originally formed offshore can become coastal dunes if the sea retreats from the land.

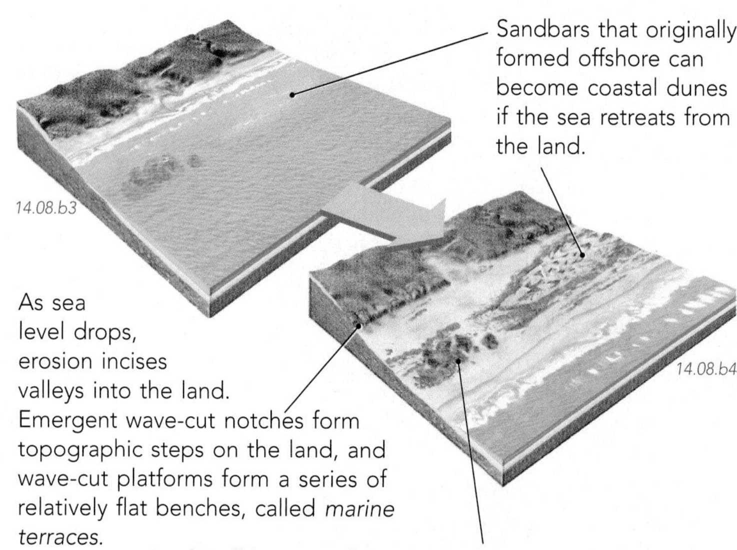

As sea level drops, erosion incises valleys into the land.

Emergent wave-cut notches form topographic steps on the land, and wave-cut platforms form a series of relatively flat benches, called *marine terraces*.

Reefs, offshore sandbars, and other once-submerged features may be exposed by falling sea level. Coral reefs exposed on land is a sign that a coast has emerged.

C What Are Characteristic Features of Submergent Coasts?

A rise in global sea level or subsidence of the land can result in several distinctive features.

14.08.c1

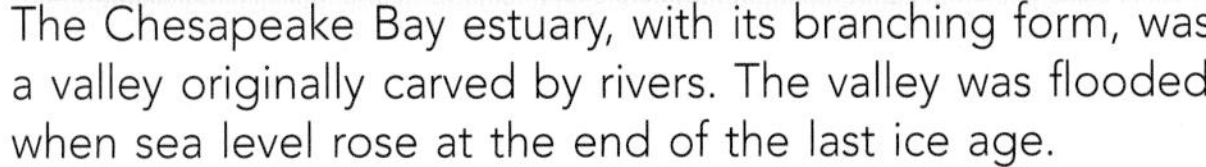

The Chesapeake Bay estuary, with its branching form, was a valley originally carved by rivers. The valley was flooded when sea level rose at the end of the last ice age.

14.08.c2

The coasts of Norway, Greenland, Alaska, and New Zealand all feature narrow, deep embayments called *fjords*. The steep-sided valleys were carved by glaciers and later invaded by the sea as the ice melted and sea level rose. [Norway]

14.08.c3

Many barrier islands are interpreted to have been formed by rising sea level. Some barrier islands began as coastal dunes or piles of sediment deposited by rivers. As sea level rose, the piles of sediment were surrounded by rising water. [North Carolina]

D What Are Characteristic Features of Emergent Coasts?

Some features reflect a fall in sea level, or uplift of the land by tectonics or by isostatic processes.

14.08.d1

A notch is cut by persistent wave erosion at sea level along a shoreline. Here, a wave-cut notch is now several meters above sea level because of tectonic uplift of the coast. [Crete]

14.08.d2

Wave-cut platforms form within the surf zone along many rocky shorelines and, when exposed above sea level, form relatively flat terraces on the land. The surface of such marine terraces may contain marine fossils and wave-rounded rocks. [California]

14.08.d3

Coral reefs and other features that originally formed at or below sea level can be exposed when seas drop relative to the land. These coral reefs, now well above sea level, provide evidence of uplift of the land. [Galapagos]

Before You Leave This Page Be Able To

- ✓ Summarize how much sea level has risen or fallen in the past 120 million years and how such changes affected the position of the coastline.
- ✓ Summarize what a *submergent* coast is and what types of features can indicate that sea level has risen relative to the land.
- ✓ Summarize what an *emergent* coast is and what types of features can indicate that sea level has fallen relative to the land.

What Causes Changes in Sea Level?

SEA LEVEL HAS VARIED GREATLY IN THE PAST. What processes caused sea level to be higher or lower than it is today? The large variations in past sea level resulted from a number of competing factors, including the extent of glaciation, rates of seafloor spreading, and global warming and cooling.

A How Does Continental Glaciation Affect Sea Level?

The height of sea level is affected by the existence and extent of continental ice sheets and glaciers. At times in Earth's past, ice sheets were more extensive than today, and at other times, they were absent.

The ice in glaciers and continental ice sheets accumulates from snowfall on land. When glaciers and ice sheets are extensive, they tie up large volumes of freshwater, causing sea level to drop.

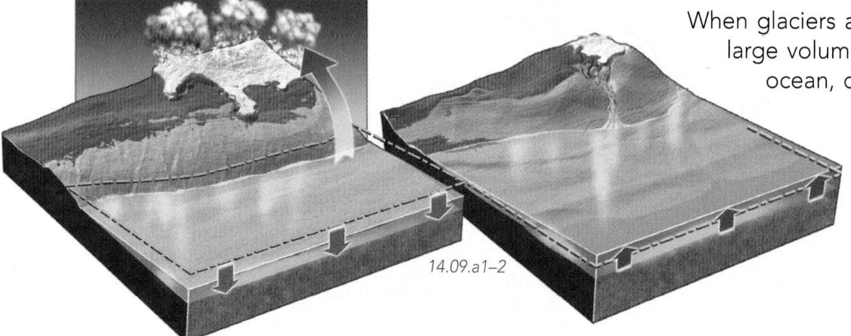

When glaciers and ice sheets melt, they release large volumes of water that flow back into the ocean, causing sea level to rise.

The growth and shrinkage of ice sheets and glaciers is the main cause of sea level change on relatively short timescales (thousands of years).

B How Do Changes in the Rate of Seafloor Spreading Affect Sea Level?

At times in Earth's history, the rate of seafloor spreading was higher than it is today, and at other times it was probably lower. Such changes cause the rise and fall of sea level.

The shape and elevation of a mid-ocean ridge and adjacent seafloor reflects the rate of spreading. As the plate moves away from the spreading center, it cools and contracts, causing the seafloor to subside, creating space for seawater.

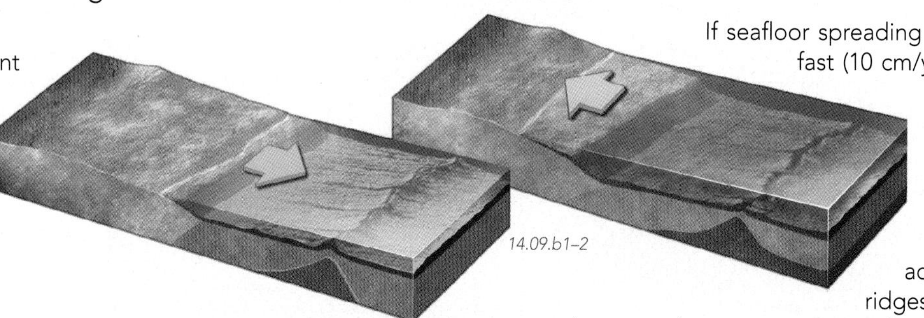

If seafloor spreading along a ridge is relatively fast (10 cm/year or faster), the ridge is broad because still-warm parts of the plate move farther outward before cooling and subsiding. So an increase in seafloor spreading rate is accompanied by broader ridges that displace water out of the ocean basins, causing sea level to rise. In other words, faster spreading yields more young seafloor, and young seafloor is less deep than older seafloor.

If seafloor spreading along a ridge is slow, the ridge is narrower because the slow-moving plate has time to cool before getting very far from the ridge. Slow seafloor spreading and narrow ridges leave more room in the ocean basin for seawater. So over time, a decrease in the spreading rate causes sea level to fall.

C How Do Changes in Ocean Temperatures Cause Sea Level to Rise and Fall?

Sea level is also affected by changes in ocean temperatures, which cause water in the oceans to slightly expand or contract. Such effects result in relatively moderate changes in sea level.

Water, like most materials, contracts slightly as it cools, taking up less volume. The amount of contraction is greatly exaggerated in this small image.

When ocean temperatures fall, water in the ocean contracts, causing sea level to fall.

14.09.c1–2

Water expands slightly when heated, taking up more volume. Again, the amount of expansion is exaggerated here.

When ocean temperatures increase, water in the ocean expands slightly, causing a small rise in sea level. The percentage of expansion is small, but can cause a moderate rise in sea level.

D How Does the Position of the Continents Influence Global Sea Level?

Continents move across the face of the planet, sometimes being near the North or South poles and at other times being closer to the equator. These positions influence sea level in several ways.

Glaciers and continental ice sheets form on land, and so require a landmass to be cold enough to allow ice to persist year round. This occurs most easily if a landmass is at high latitudes (near the poles) or is high in elevation.

At most times in Earth's past, widespread glaciation occurred when one or more continents were near the poles. By allowing glaciers to exist, a high-latitude position of a continent can cause a drop in global sea level.

A times in Earth's past, the larger continents were not so close to the poles. This lower latitude position minimizes or eliminates widespread glaciation, tending to keep sea levels high.

14.09.d1

E How Do Loading and Unloading Affect Land Elevations Relative to Sea Level?

Weight can be added to a landmass, a process called *loading*. A weight can also be removed, a process called *unloading*. Loading and unloading can change the elevation of a region relative to sea level.

Weight loaded on top of a region imposes a downward force, which if large enough, can downwarp the land surface beneath the load and in adjacent areas. Loading, such as by continental ice sheets, lowers the loaded region relative to sea level. This can allow seawater to inundate regions near the ice sheets. The ice in this figure and the amount of subsidence are very stylized and vertically exaggerated.

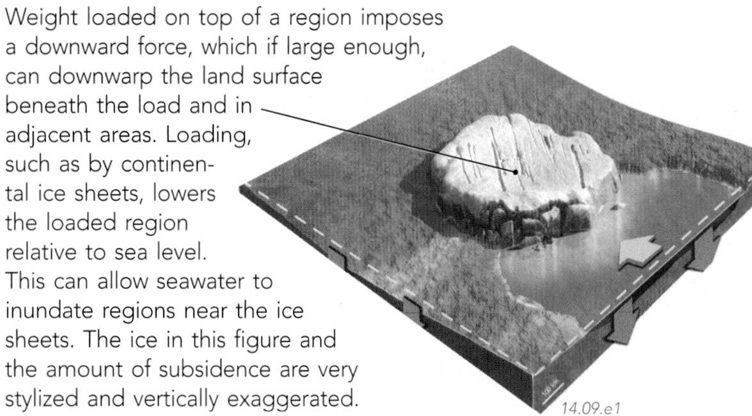

14.09.e1

If the weight is unloaded from the land, the region flexes back upward, a process called *isostatic rebound*. The uplifted, rebounding region rises relative to sea level.

Unloading and isostatic rebound can occur when continental ice sheets melt. Rebound begins as soon as significant amounts of ice are removed, but can still be occurring thousands of years after all the ice is gone.

14.09.e2

Ongoing Isostatic Rebound of Northeastern Canada

The northern part of North America has been covered by glaciers off and on for the last two million years. The weight of these continental ice sheets loaded and depressed this part of the North American plate. When the ice sheets melted from the area, beginning about 15,000 years ago, unloading caused the land, especially in Canada, to begin to isostatically rebound upward.

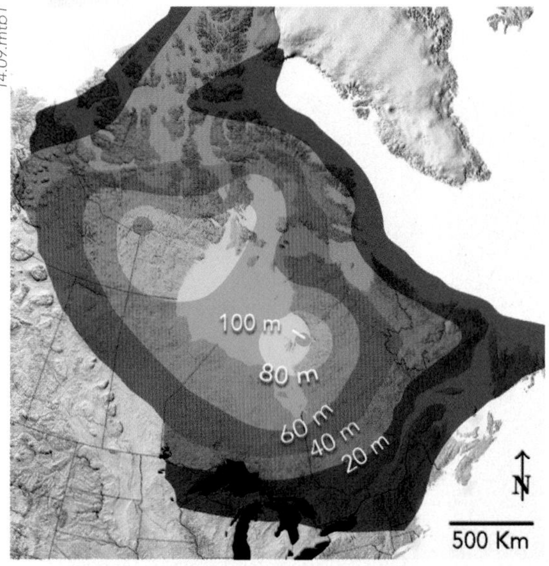

14.09.mtb1

The amount of rebound has been measured both directly and indirectly. Uplift can be measured directly by making repeated elevation surveys across the land and then calculating the amount of uplift (rebound) between surveys. Rates of rebound are typically millimeters per year, which is enough to detect with surveying methods. GPS is also sensitive enough to measure such changes. The amount and rate of rebound can also be inferred more indirectly by documenting how shorelines and other features have been warped and uplifted. Contours on this map indicate the amount of rebound interpreted to have occurred in northeastern North America over the last 6,000 years.

Before You Leave This Page Be Able To

- ✓ Summarize how continental glaciation, rates of seafloor spreading, ocean temperatures, and position of the continents affect sea level.
- ✓ Summarize how loading and unloading affect land elevations using the example of northeastern Canada.

What Is the Evidence for Past Glaciations?

WHAT EVIDENCE IS THERE THAT HUGE ICE SHEETS once covered the land? This evidence is directly expressed as features and deposits within the landscape, and is recorded indirectly in unusual marine deposits and in the isotopic signature of marine fossils in sediment deposited during glaciations.

A What Types of Evidence Do Glaciers Leave Behind in the Landscape?

To be able to recognize past glaciations, geologists visit active glaciers to observe what types of landforms and deposits are diagnostic of glaciation. Modern-day glaciers occur in *continental ice sheets* in Antarctica and Greenland and in smaller *mountain glaciers,* such as in the Alps, Andes, and Canada. Geologists observe both types of modern-day examples to explain prehistoric features—another example of "the present is the key to the past."

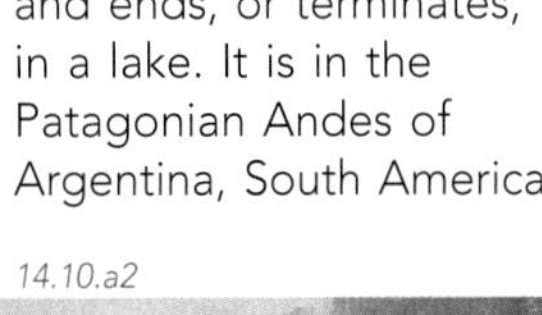

▽ This glacier of blue ice flows down a steep valley and ends, or terminates, in a lake. It is in the Patagonian Andes of Argentina, South America.

14.10.a2

Mountain glaciers produce distinctive landforms, including bowl-shaped basins flanked by steep ridges, and U-shaped valleys carved by the moving ice. The photograph to the right shows a glacially carved U-shaped valley in Norway.

14.10.a3

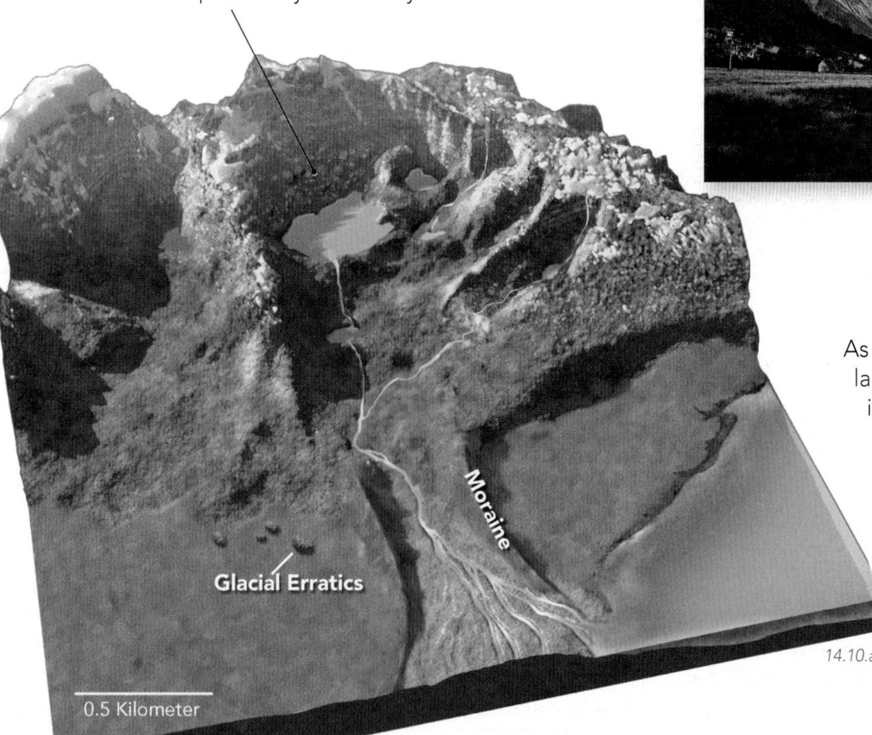

14.10.a1

As glaciers grind across the land surface, rocks contained in the moving ice tend to *polish* the underlying bedrock and to gouge scratch marks called *glacial striations.* Such polished and scratched bedrock is a clue to the former existence and extent of glaciers. [Duluth, Minnesota] ▽

Glaciers can carry huge rocks, some as big as a house, and leave them scattered about the landscape. Glaciers may transport large blocks hundreds of kilometers, taking them to places where such rock types are not present in the bedrock. Such out-of-place blocks (▽) are called *glacial erratics.* [Yellowstone National Park, Wyoming]

14.10.a4

Glaciers pick up and carry sediment as they move, literally dumping this sediment onto the landscape when the ice melts away. Such glacial sediment is called *till,* and till can accumulate in sheets, piles, and ridges, forming landforms called *moraines.* This sediment is a glacial till from Iceland. ▷

14.10.a5

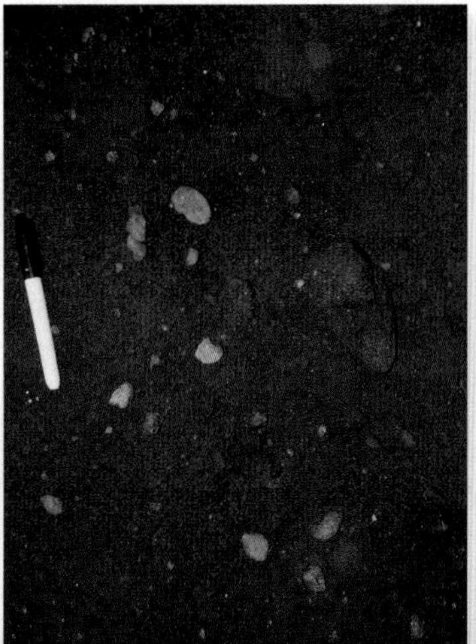

14.10.a6

B What Record Does Glaciation Leave in the Ocean?

A glaciation, because it greatly affects sea level and the influx of freshwater into the ocean, can also be inferred by examining the nature and chemistry of marine fossils and sediment.

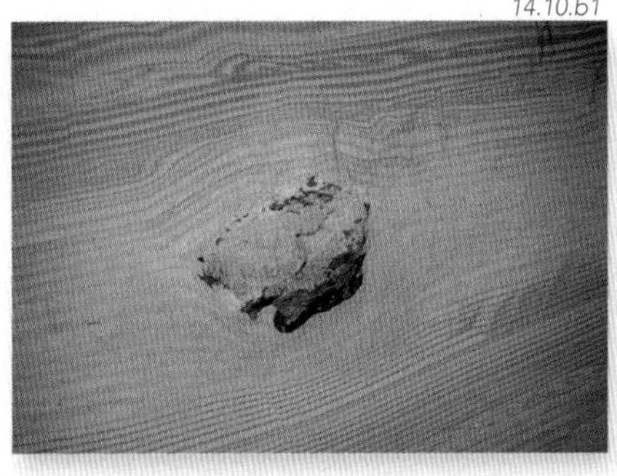

14.10.b1

An unusual feature of some marine and lake sediment is the presence of scattered stones in an otherwise fine-grained, clastic sediment. These stones are called *dropstones* because they have been carried within icebergs floating out into the ocean or lake and then dropped into the fine sediment on the seafloor or lake bottom.

14.10.b2

As they grow, some marine organisms, like these corals, build shells of calcium carbonate by extracting the necessary chemicals from seawater. As the chemistry and temperature of the water changes, so does the chemical composition of the shells formed in that water. Geologists analyze oxygen and carbon isotopes in fossils to infer changes in seawater temperature and chemistry over time. Such changes are used to infer the times of glaciation or times when melting of glaciers released freshwater into the ocean.

C How Do We Determine Where and When the Most Recent Ice Age Occurred?

From diverse lines of evidence, geologists reconstruct which areas were once covered by ice and which ones were not. They then use fossils and isotopic dating methods to determine when glaciers were most widespread; such a time is called a *glacial maximum* and causes sea level to fall. A time during the last ice age when glaciers became smaller is an *interglacial* and is marked by rising seas.

This stretched-out view of the Earth shows one interpretation of the maximum extent of ice sheets and glaciers (in white) during the last ice age, some 20,000 years ago.

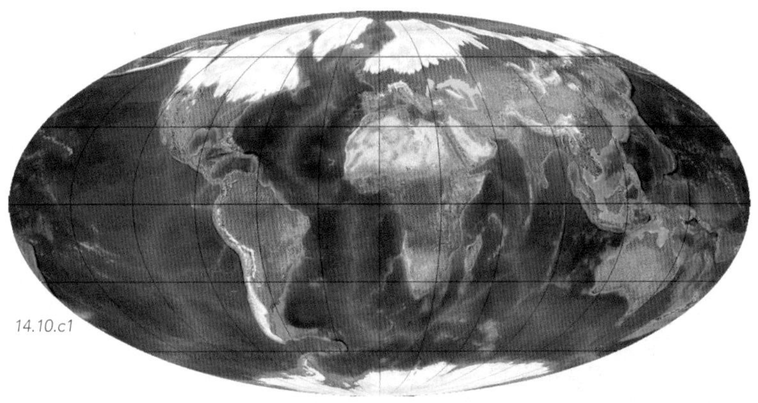

14.10.c1

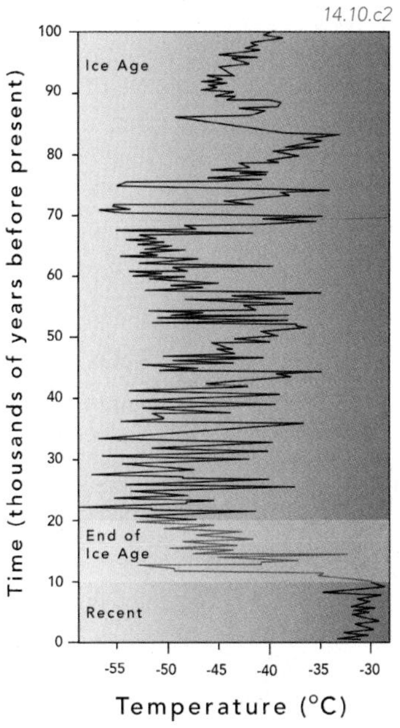

14.10.c2

This graph, based on the oxygen-isotope compositions of marine shells, shows how global temperatures are interpreted to have varied over the last 100,000 years. Geologists infer that data points to the right mean that temperatures were warmer, as they are today, and glaciers were less widespread (*interglacial* periods). Data points to the left indicate that glaciers were more widespread (*glacial* periods). From these and other data, the glaciers decreased and increased many times during the last 100,000 years.

The Ice-Age Hypothesis

The idea that huge ice sheets covered the land was not intuitive, but arose to explain an ever-growing number of hard-to-understand observations made in Europe and North America. As discussed in A. Hallam's book *Great Geological Controversies*, these observations included bones of reindeer and Arctic birds in southern France, large out-of-place boulders (erratics) scattered across much of Europe, and the presence of scratched and polished bedrock far beyond the existing glaciers in the Alps. European naturalists of the time debated how to explain these curious features. In the 1820s, they hypothesized that widespread, prehistoric glaciations had occurred in the Alps and northern Europe. Today, we call this interval of time the *ice ages*, and recognize that glaciers grew and shrank many times during the last two million years.

Before You Leave This Page Be Able To

- ✓ Describe evidence used to infer that glaciers once covered a landscape.
- ✓ Discuss how glaciations can be expressed in the ocean, and how we can use this record to interpret when glaciation occurred.

How Do Glaciers Form, Move, and Vanish?

GLACIERS ARE MOVING MASSES OF ICE, which can be small enough to be restricted to a single mountainside or large enough to cover large parts of a continent. How does a glacier form? Once formed, how does a glacier move across the landscape, and how does it eventually melt away?

A How Do Snow and Ice Accumulate in Glaciers?

Glaciers, such as the one below, form by the accumulation of snow and ice. The snow is derived from snowfall, but can be moved around by the wind or by *avalanches*, which are masses of snow and ice that fall, slide, or flow downhill. [Swiss Alps]

14.11.a1

Snow falls as individual flakes. Once on the ground, flakes are pressed together by the weight of other snowflakes on top. Loose snow can contain 90 percent air between the flakes.

As more snow accumulates on top, snowflakes farther down are compressed, forcing out more than 50 percent of the air. The snowflakes become compressed into small irregular spheres of more dense snow.

With increasing depth and pressure, the snow begins to recrystallize into small interlocking crystals, forming solid ice. Ice is a crystalline material and is considered to be a type of rock. Crystalline ice commonly appears bluish because it contains less air.

14.11.a2

B How Does a Glacier Form and Change as It Moves Downhill?

Glaciers form when the amount of snow and ice accumulating from snowfall exceeds the amount lost by various processes. In this situation, the snow and ice will begin to pile up and may start to move, forming a glacier.

1. In the place where a glacier forms, the amount of snow and ice that accumulates exceeds the amount lost due to melting and other processes—this upper part of the glacier or ice sheet is the *zone of accumulation*.

2. As the glacier moves downhill, it loses more and more ice and snow by melting, by wind erosion, and by loss of ice molecules directly to the air, a process called *sublimation*. At some point along the glacier, the losses of ice and snow exactly balance the amount of accumulation; this boundary is called the *equilibrium line*. The equilibrium line is typically, but not always, marked by a gradational boundary between snow-covered ice upslope on the glacier and bluish ice downslope. The bluish ice formed at depth and became exposed at the surface as upper levels of ice and snow are removed. In many cases, a glacier remains covered with snow until the very end, but blue ice can be observed at depth in fractures that cut the glacier's upper surface.

14.11.b1

3. The mountain glacier to the right has an upper, snow-covered part (zone of accumulation) and a lower area of blue ice below the equilibrium line. [Morteratsch, Swiss Alps]

4. At lower elevations, ice melts away faster than it can be replenished by downward movement of ice within the glacier and by snowfall. This causes the glacier to end or *terminate*, either on land or in the sea. The end is called the *terminus*.

14.11.b2

C How Do Glaciers Move?

Glaciers move downhill because the ice is not strong enough to support its own weight against the relentless downward pull of gravity. As glaciers spread downward, they move by internal shearing and flow of the solid ice, by simply sliding across the bedrock, or some combination of these.

14.11.c2

1. Most glaciers move in part because of internal shearing and flow of the ice crystals. As gravity pulls the ice downhill, friction along the base of the glacier causes this part of the glacier to lag behind the upper, less constrained parts (▽). The upper part of a glacier therefore flows faster than the lower part, causing internal shearing within the glacier.

4. As a glacier moves, internal stresses cause the upper surface of the ice to break, forming fractures, each of which is called a *crevasse*. This image (▷) is looking down on a glacier cut by numerous crevasses. [Mont Blanc, France]

14.11.c1

14.11.c3

◁ **2.** If the interface between the glacier and the underlying bedrock is very irregular and is relatively dry, the base may become locked to the bedrock and not move at all. Only the coldest glaciers are completely frozen at their bases.

14.11.c4

◁ **3.** If the bedrock-glacier interface is less irregular (i.e., smoother) or contains water from melting ice, the glacier may be able to slide over the bedrock. It can also internally shear and flow.

D What Happens When a Glacier Encounters the Sea or a Lake?

14.11.d2

When a glacier reaches the ocean or a lake, it may begin to float on the water if the floor of the sea or lake is deep enough. Ice, even the dense blue variety within glaciers, floats because it is less dense than either freshwater or saltwater.

14.11.d1

Glacier
Ice Block
Icebergs
Sea
Sea
Falling Stones

As the ice floats, it tends to spread or be pulled apart, forming large fractures (*crevasses*) within the ice. These allow large blocks of ice to collapse off the front of the glacier, a process called *calving*. ▷

As the blocks of ice fall into the water, they float, forming *icebergs*. As much as 90 percent of an iceberg is beneath the water. As icebergs melt, rocks and other sediment within them drop into the water. Larger rocks fall to the floor of the sea or lake, forming *dropstones*.

Glaciers, Snowfields, and Sea Ice

Not every large mass of ice on Earth's surface is a glacier. Some accumulations of snow and ice never move, and are simply called *snowfields*. Large masses of ice also form when the upper surface of a lake or the sea freezes. In the ocean, such ice is called *sea ice*. In all but the coldest places, like parts of the Arctic Ocean, sea ice freezes in the winter and thaws in the spring or summer. Freezing excludes salt from the crystalline structure of ice, so sea ice melts to form freshwater, not saltwater. In the photograph below, broken sheets of sea ice surround a rocky island in Antarctica.

14.11.mtb1

Before You Leave This Page Be Able To

- ✓ Sketch and describe how snow is transformed by pressure into ice.
- ✓ Summarize or sketch the differences in a glacier above and below the equilibrium line.
- ✓ Describe how glaciers move and what happens when they encounter a lake or the sea.

14.11

What Are Mountain Glaciers and What Landscape Features Do They Form?

MANY GLACIERS BEGIN IN MOUNTAINS and flow as ribbons of ice down valleys. Many geologists call this type of glacier a *mountain glacier*. As mountain glaciers move, they scour underlying rock and pick up and carry sediment toward lower elevations. Mountain glaciers produce distinctive landforms that we can use to recognize landscapes carved by mountain glaciers, even after the ice melts away.

14.12.a1

1. These mountain glaciers on Mont Blanc in the French Alps flow down the hillside, melting before they reach the resort town of Chamonix. Earlier this century, the glaciers extended farther and reached the valley floor, but their fronts have since melted back.

14.12.a2

2. Mountain glaciers in the Alaska Range accumulate in huge snowfields atop the range and flow downhill into valleys. Some people use the term *valley glacier*, instead of *mountain glacier*, to emphasize the flow of such glaciers within valleys.

14.12.a3

3. As a glacier moves past bedrock, it plucks away pieces of rock and is covered by rock pieces derived from nearby steep slopes. It grinds up some pieces into a fine rock powder. The glacier carries away this material, depositing the sediment where the glacier melts. Such sediment is called *moraine*. This glacier has dark fringes of rocky moraine on both sides and along its terminus. [Morteratsch, Swiss Alps]

14.12.a4

4. This view shows the terminus of a glacier. Piles of gray *moraine* flank the valley on both sides, left behind when the glacier melted. The bottom of the valley is also covered with moraine, which is being reworked by the meltwater stream emanating from the end of the glacier. Looking down past the end of the glacier shows a valley with a rounded U-shaped profile. Such *U-shaped valleys* are typical of glacial landscapes, as opposed to V-shaped valleys that are characteristic of erosion by streams and rivers. [Swiss Alps]

14.12.a5

◁ **5.** Near the source area of a mountain glacier, snow and ice accumulate in less steep areas, such as those below these higher peaks near Mont Blanc, France. The ice plucks pieces from the bedrock, excavating a bowl-shaped depression, called a *cirque*. When the ice melts, the cirque is exposed. The depression usually contains one or more small lakes. The figure below and the photograph to the right from the Yukon of Canada both feature a bowl-shaped cirque.

14.12.a6

14.12.a7

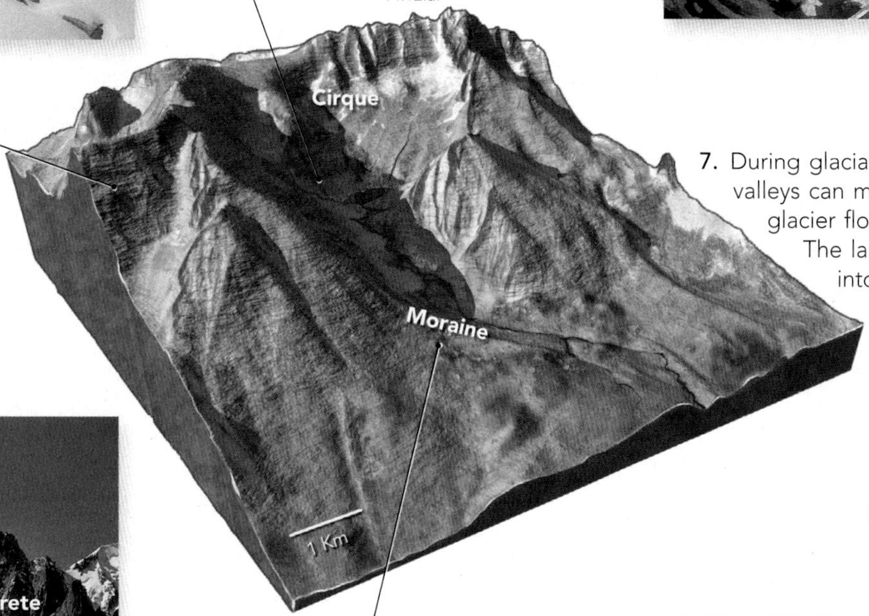

Cirque

Moraine

1 Km

6. Hard bedrock ridges that flank cirques are commonly narrow, sharp, and jagged, like the ridges shown below. Such a ridge is called an *arete* and is jagged because it has been glacially eroded from both sides and because physical weathering is very intense in such settings. [Chamonix, France] ▽

14.12.a8

Arete

Arete

Arete

7. During glaciation, glaciers from smaller valleys can merge with a larger, thicker glacier flowing down a main valley. The larger glacier scours deeper into the bedrock, so the main valley is deepened more than the side valleys. When the glaciers melt away, the side valleys are higher than the main valley; they therefore are called *hanging valleys*. [Alaska] ▽

8. A glacier carries large rocks and finer sediment, especially along its edges and base. It deposits the sediment as moraine, and we subdivide moraine into different types according to its position in the present-day or past glacier. ▽

14.12.a9

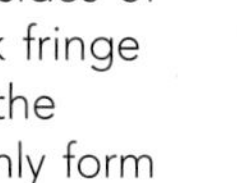

9. A *lateral moraine* forms along the sides of the glacier and is expressed as a dark fringe along the edge of the glacier. When the glacier melts, lateral moraines commonly form low ridges along the edges of the valley. ▷

10. A *medial moraine* is a sediment-rich belt in the center of the glacier. It represents where two glaciers joined, trapping their lateral moraines within the combined glacier. ▷

11. A *terminal moraine* forms at the termination of a glacier and generally marks the glacier's farthest downhill extent. Some terminal moraines are large and conspicuous because the end of the glacier deposited sediment there for an extended period of time. [Mount Spur, Alaska] ▷

14.12.a10

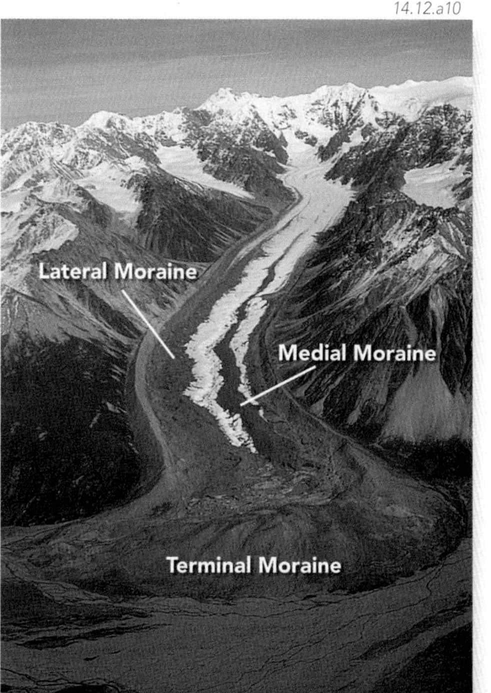

Before You Leave This Page Be Able To

- ✓ Summarize the landscape features associated with mountain glaciers and how each type of feature formed.
- ✓ Recognize a landscape formed by mountain glaciers.

What Are Continental Ice Sheets and What Record Do They Leave Behind?

CONTINENTAL ICE SHEETS ARE ENORMOUS FEATURES, covering huge areas of land. They currently occur in high-latitude places (close to the poles), such as Antarctica and Greenland. In the past, continental ice sheets and glaciers covered most of the northern parts of Europe, Asia, and North America, including most of Canada and the northern parts of the United States. What evidence did they leave?

◁ 1. Antarctica currently has the largest ice sheets. Most of the continent is covered with ice and snow, in many places thousands of meters thick. Rocks are exposed mostly along the coasts and in some of the mountain ranges, which become the focus of studies by field geologists, such as those in the camp above.

2. When ice sheets flow across the surface, they smooth and polish rocks over broad areas, and deposit glacial sediments (till) on the surface. [South Africa]

3. Some ice sheets and glaciers flow into the sea, where they can form a large ice shelf that floats on seawater. Most ice is derived from the land.

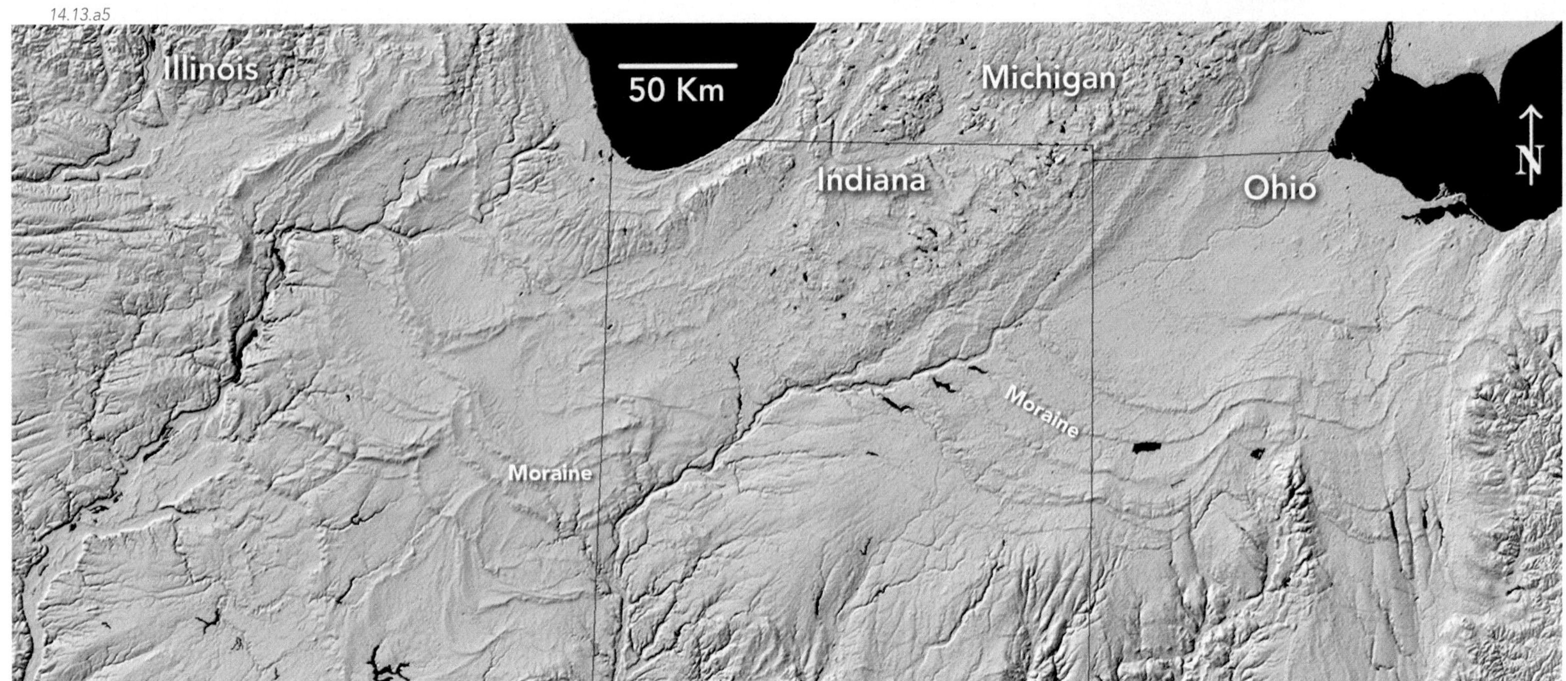

4. As continental ice sheets and glaciers melt away, they leave a veneer of glacial sediment on the landscape. South of the Great Lakes, ridges of glacial sediment mark the position of the front of the ice sheet as it melted back toward Canada. A ridge of moraine deposited as the front of a glacier melts back is a *recessional moraine*.

5. Most smooth areas were once covered by ice, but some areas with rougher topography, such as those in the lower right corner of this map, were never glaciated.

C How Do We Calculate the Rise in Sea Level If West Antarctica's Ice Melts?

To evaluate how melting of ice sheets would affect our shorelines, we can make some simple calculations to determine how much sea level would rise in an unlikely scenario—melting of *all* the ice from West Antarctica.

1. Examine the situation below. A rectangular tub of water has one block of ice floating in it and two blocks on land that will add water to the tub if the blocks melt. The ice blocks and the grids on the side of the tub are 10 cm on a side for easy measuring.

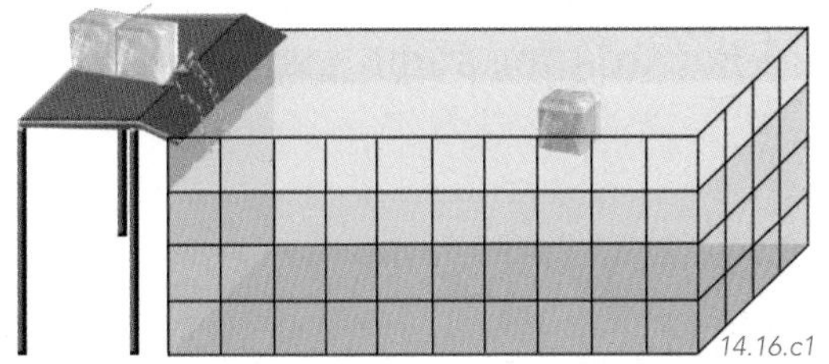

2. The block floating in the water is 10 cm on all sides, or 10 cm by 10 cm by 10 cm. We simply multiply these three dimensions to get the volume, which is 1,000 cm^3 (10 × 10 × 10 = 1,000). The two blocks on the table total 20 cm (two blocks wide) by 10 cm (two blocks deep) by 10 cm (one block) high. If we multiply 20 cm × 10 cm × 10 cm, we get 2,000 cm^3.

3. The Floating Block—Most of the floating block is below the surface. As ice melts, it yields a volume of water that is less than the volume of ice, because water is more dense than ice. As a result, melting ice that is floating in freshwater does not appreciably raise the level of the water. It does cause a slight rise in saltwater because melting ice yields freshwater, which is less dense, but we'll ignore this factor to simplify things.

4. Blocks on the Table—If the blocks on the table melt, all of the water helps raise the level in the tub. To see how much, we need only to worry about the surface area of the water, not how much water is already there at depth within the tub. Also, a volume of ice produces about nine-tenths that volume of water, or a ratio of 0.9 (volume water produced/volume ice melted).

5. To get the surface area of a rectangle of water, we just multiply the dimensions of its two sides. For the tub, it is 100 cm long by 40 cm wide, yielding a surface area of 4,000 cm^2. To figure out how much our melting blocks will raise the water level in the tub, we spread our volume of water over this surface area. The calculation is as follows:

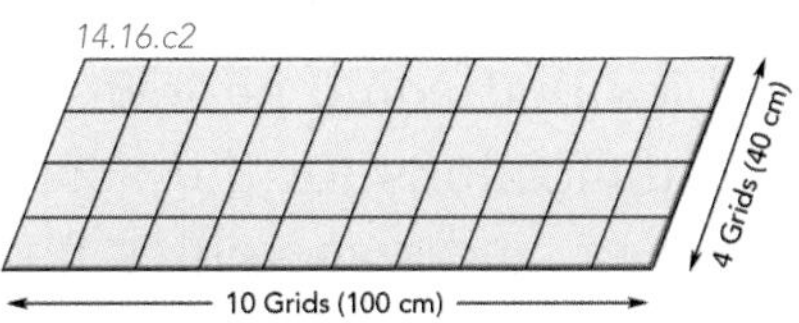

2,000 cm^3 (volume of the ice blocks on table)	×	0.90 (to convert the ice to water)	/	4,000 cm^2 (surface area of the water)	=	0.45 cm (rise in level of water)

6. So melting an ice block floating in the water does not appreciably change sea level, but melting ice on land does. The larger the amount of ice on land that is melted, the larger the rise. But the larger the *surface area* of the tub, the smaller the rise. For West Antarctica and our modern seas the calculation is:

3,000,000 km^3 (volume of all the ice)	×	0.90 (ice to water)	/	361,000,000 km^2 (surface area of the world's oceans)	×	0.0075 km (rise in sea level)

To get meters, we multiply 0.0075 km × 1,000 m/km = 7.5 m (25 feet)

7. This calculation does not take into account that as we add water and raise sea level, the ocean spreads out over the land and so the surface area increases. The number calculated when considering this factor is more like 6 meters (20 feet). Recall that this is a worst-case scenario that would occur only under a huge change in climate.

D What Impact Would Raised Sea Levels Have on the East Coast?

Think about all the shoreline photographs in this chapter and imagine those areas if sea level were 6 meters (~20 feet) higher. To plan for such contingencies, the USGS conducted a detailed assessment of the relative risk to sea-level rise for each part of the East Coast of the United States. On this map, each segment of coast was investigated for various geologic factors, such as elevation, slope of the land, hardness of the rocks, barrier islands, and various other aspects described in this chapter. From this analysis, each area was assigned a risk, from low to very high. The most vulnerable settings include the eastern coast of Florida, the barrier islands of Virginia and North Carolina, especially Cape Hatteras, and coastlines around Maryland, Delaware, and New Jersey. How risky is your favorite part of the East Coast?

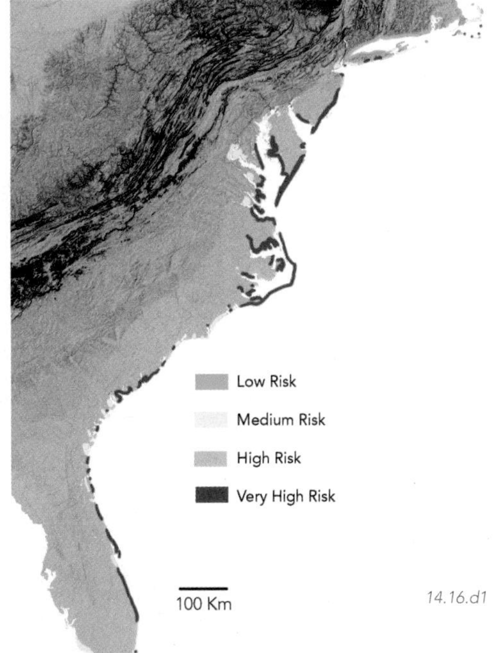

Before You Leave This Page Be Able To

- ✓ Briefly summarize the settings where ice occurs in West Antarctica.
- ✓ Calculate how much melting a block of ice will raise water levels in a tub, if you know the dimensions of the block and tub.
- ✓ Discuss why calculations about West Antarctica are important to people living along coasts everywhere in the world, including the East Coast of the United States.

How Could an Episode of Global Warming or a Glacial Period Affect North America?

On this shaded-relief map of North America, the land surface and seafloor are colored according to elevation above and below present sea level. These elevations represent possible levels to which the sea could rise or fall if the climate dramatically warms or cools and causes a change in the extent of glaciers. You will use estimates of the amount of ice that could be lost or gained to calculate how much sea level could rise or fall, and then evaluate the implications for the economy, transportation, and hazards for some major cities on the coasts of North America, and some cities farther inland.

Goals of This Exercise:

- Observe the shaded relief map of North America to identify areas that are close to sea level (above it and below it).
- Use estimates of the current amounts of ice on the planet to calculate how high sea level would rise if all the ice melted.
- Use estimates of the amount of ice that was present during the last glacial maximum (20,000 years ago) to calculate how much sea level would drop if these conditions returned today.
- Use your results to identify how such rises and falls in sea level would affect some major cites of North America.

14.17.a1

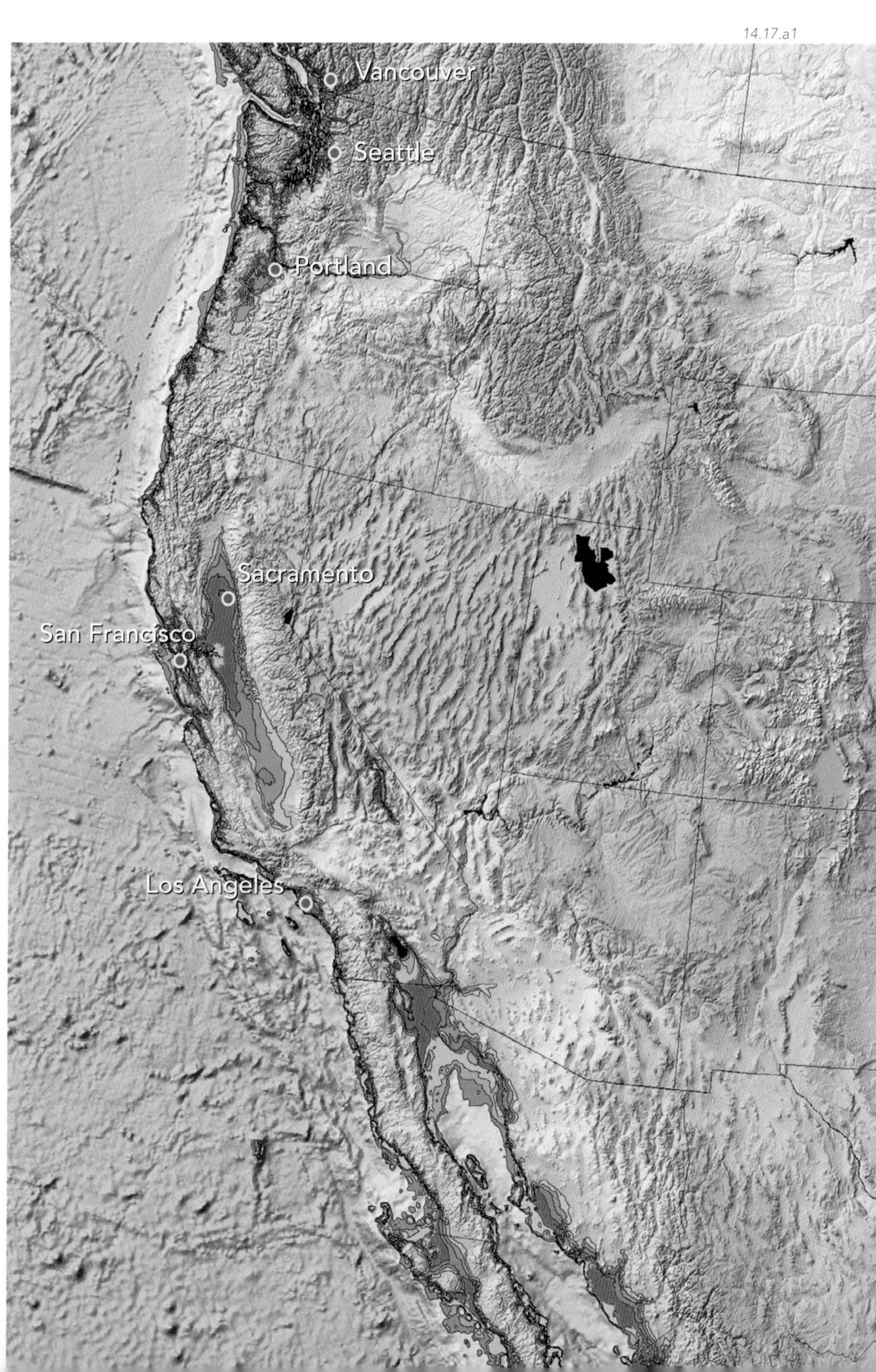

Data

Listed below are data about the present surface area of the oceans and estimates for the amount of ice that (1) is present today and (2) was present when glaciers were at a maximum 20,000 years ago. Use these data to complete the calculations on the next page.

1. The present surface area of the oceans is 361,000,000 km^2.
2. The total amount of ice (ice sheets, ice caps, glaciers) currently present on the planet is estimated to be 32,000,000 km^3.
3. During the last glacial maximum, 20,000 years ago, the amount of ice is estimated to have been 52,000,000 km^3 more than is present today.
4. When ice melts, the volume of water produced is ~0.9 times the volume of the ice. That is, the volume of water produced is only 90 percent of the volume of ice.
5. When water freezes into ice, the volume of the ice is ~1.1 times the volume of the water.

Procedures

Follow the steps below, entering your answers for each step in the appropriate place on the worksheet.

1. Calculate how much water would be released if all the ice on the planet melted (an unlikely scenario for any time soon). For this calculation the equation is: the volume of water gained = the volume of ice × 0.9 (converting ice to water).
2. Calculate the volume of water that would be tied up in ice if the glaciers returned to the same volume as 20,000 years ago. The equation is: the volume of water lost = the volume of additional ice 20,000 years ago × 1.1 (converting water to ice).
3. Calculate how much sea level would rise for the water volume gained in step 1 or sea level would fall for the water volume lost in step 2. Ignoring many important complications, the much simplified equation is: the change in sea level = change in water volume / surface area of the oceans.
4. Describe how each of the cites shown on this map would be affected by the two extremes. Would it be flooded, not flooded but much closer to the shoreline, or much farther from the shoreline? Discuss how this would affect that city's transportation, vulnerability to coastal flooding, economic livelihood, and any other factors you can think of.

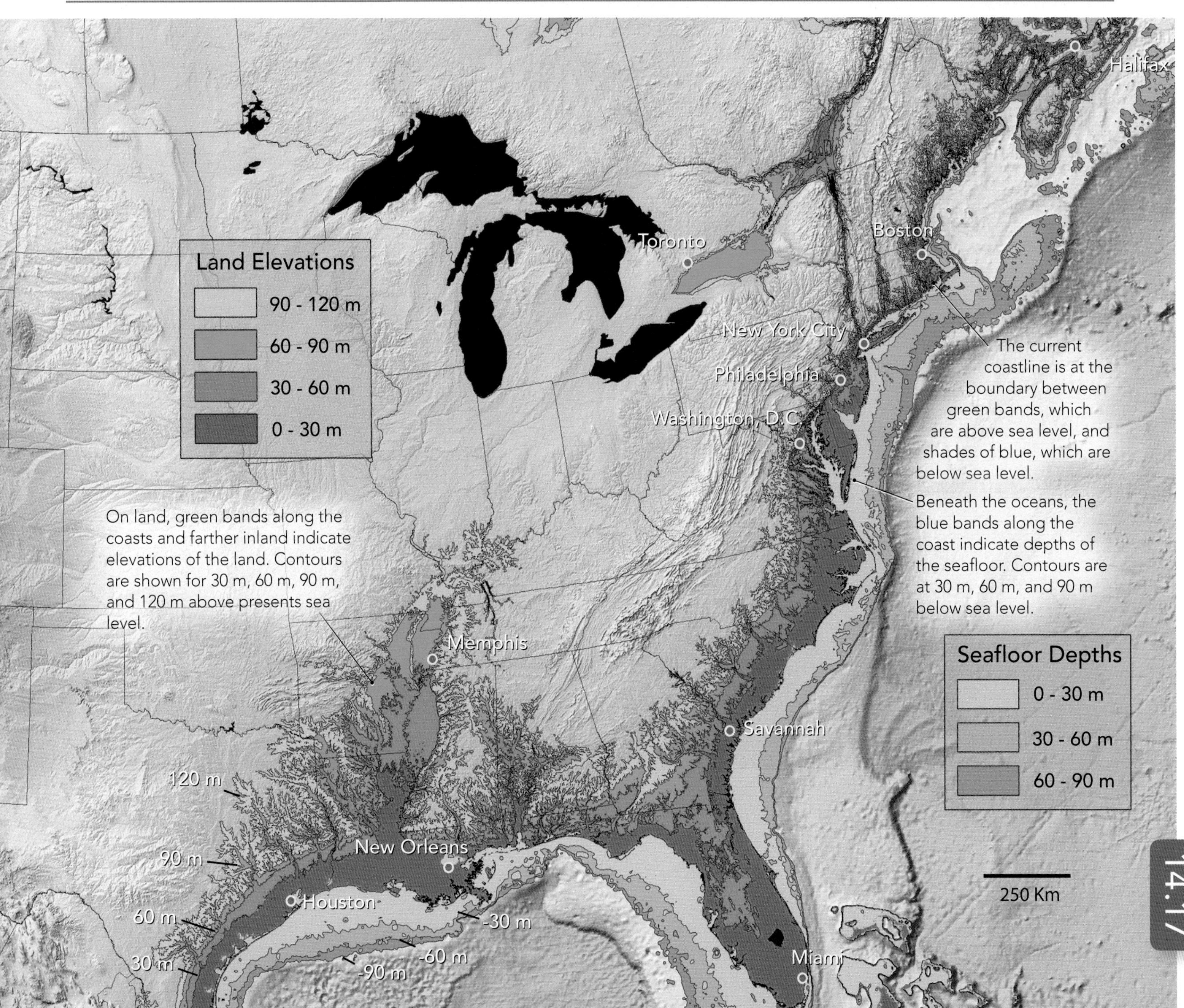

CHAPTER 15

Weathering, Soil, and Unstable Slopes

SLOPES CAN BE UNSTABLE, and slope failure can unleash catastrophic landslides and thick slurries of mud and debris. Such events have killed thousands of people, and destroyed houses, bridges, and entire cities. Where does this loose material come from, and what determines if a slope is stable? In this chapter, we explore slope stability and the processes that make soil, one of our more important resources.

The Cordillera de la Costa is a steep 2-kilometer-high mountain range that runs along the coast of Venezuela, separating the capital city of Caracas from the sea. This image, looking south, was constructed by overlaying topography with a satellite image taken in 2000. The white areas are clouds and the purple areas are cities. The Caribbean Sea is in the foreground.

In December 1999, torrential rains in the mountains caused landslides, and mobilized soil and other loose material as debris flows and flash floods that smothered parts of the coastal cities. Some landslides are visible in this image as light-colored scars on the hillsides.

How does soil and other loose material form on hillslopes? What factors determine whether a slope is stable or is prone to landslides and other types of downhill movement?

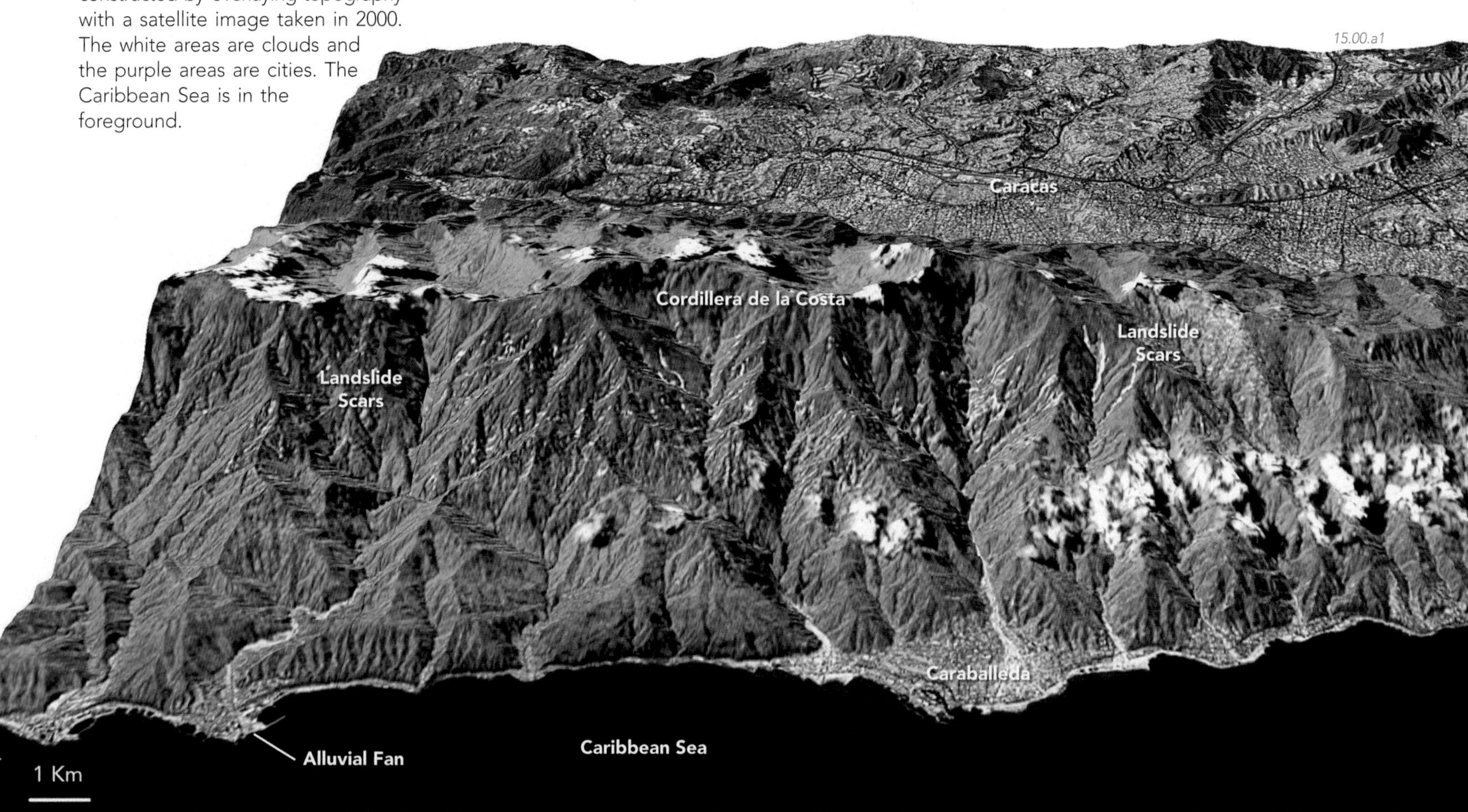

The mountain slopes are too steep for buildings so the coastal cities were built on the less steep fan-shaped areas at the foot of each valley. These flatter areas are alluvial fans composed of mountain-derived sediment that has been transported down the canyons and deposited along the mountain front.

What are potential hazards of living next to steep mountain slopes, especially in a city built on an active alluvial fan?

The city of Caraballeda, built on one such alluvial fan, was especially hard hit by the 1999 debris flows and flash floods that tore a swath of destruction through the town. The damage is visible as the light-colored strip through the center of town. Landslides, debris flows, and flooding killed more than 19,000 people and caused up to $30 billion in damage in the region.

How can loss of life and destruction of property by debris flows and landslides be avoided, or at least minimized?

TOPICS IN THIS CHAPTER

Caracas is situated among rolling hills on the south side of the mountains. As the city expanded, houses and businesses were built right up to the steep hills and mountains.

▽ In Caraballeda, huge boulders smashed through the lower two floors of this building and ripped away part of the right side. The mud and water that transported these boulders is no longer present, but the boulders remain as a testament to the fury of the event.

◁ This aerial photograph of Caraballeda, looking south up the canyon shows the damage in the center of the city caused by the debris flows and flash floods.

1999 Venezuelan Disaster

In December 1999, two storms dumped as much as 1.1 m (42 in.) of rain on the coastal mountains of Venezuela. The rain loosened soil on the steep hillsides, causing numerous landslides and debris flows that coalesced in the steep canyons and raced downhill toward the cities built on the alluvial fans. A *debris flow* is a slurry of water and debris, including mud, sand, gravel, boulders, vegetation, and even cars and small buildings. Debris flows can move at speeds up to 16 meters/second (36 mph).

In Caraballeda, the debris flows carried boulders up to 10 m (33 ft) in diameter and weighing 300 to 400 tons each. The debris flows and flash floods raced across the city, flattening cars and smashing houses, buildings, and bridges. They left behind a jumble of boulders and other debris among the ruins of the city.

After the event, USGS geologists went into the area to investigate what had happened and why. They documented the types of material that were carried by the debris flows, mapped the extent of the flows, and measured boulders (▽) to investigate the processes that occurred during the event. When the geologists examined what lay beneath the foundations of destroyed houses, they discovered that much of the city had been built on older debris flows. This should have provided a warning of what was to come.

What Physical Processes Affect Rocks Near the Surface?

ROCKS AT AND NEAR EARTH'S SURFACE are subjected to processes that break rocks apart and alter their components. These processes may change the color, texture, composition, or strength of the materials. Such processes constitute physical and chemical *weathering*. *Physical weathering* breaks rocks into smaller fragments that can then be moved from the original site by the process of *erosion*.

A What Is the Role of Joints in Weathering?

Joints are fractures, or cracks, in rocks that show no significant offset. Joints help break up rocks into smaller pieces because they permit water and roots to penetrate into the rock, thereby promoting weathering.

Most joints form at depth and then are uplifted, along with the rocks, to the surface. The orientation and spacing of *preexisting* joints and faults help determine the rates of physical and chemical weathering at the surface. More closely spaced joints promote more rapid weathering.

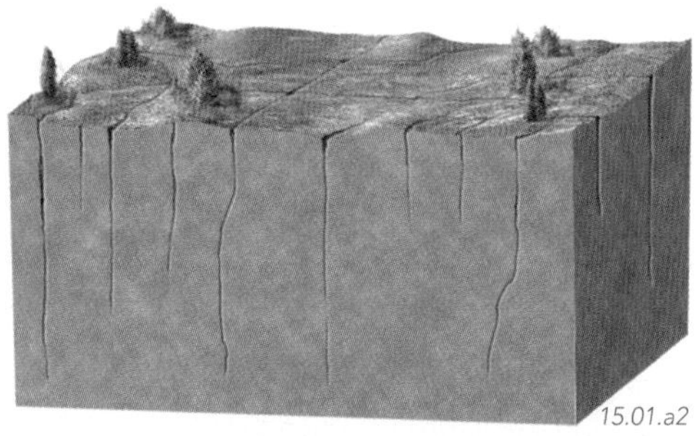

Some joints form as a result of expansion due to cooling or to a release of pressure as the rocks are uplifted toward the surface. These *expansion joints* can be difficult to distinguish from preexisting joints, that formed by other processes.

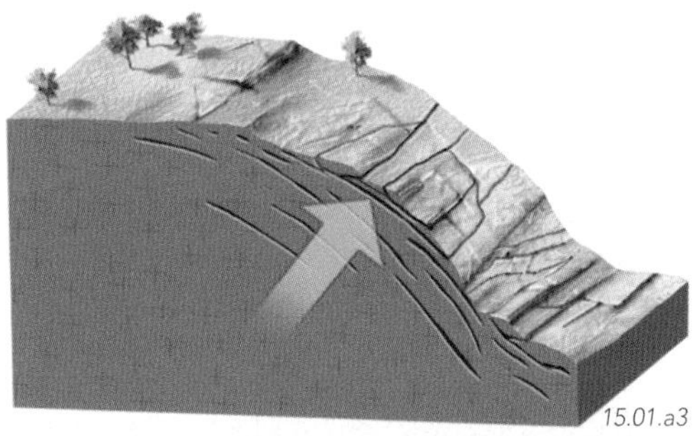

As Earth's topography is sculpted by erosion, the shape of the topography influences the stresses that build up as the weight of overlying rocks is *unloaded*. In some rock types, expansion joints form that mimic the topography in a process called *exfoliation*.

B How Are Joints Expressed in the Landscape?

Joints greatly influence how a landscape develops. Joints affect the strength of a rock, its resistance to weathering and erosion, and influence how pieces of rock are pried loose from the landscape.

15.01.b1

△ Joints are the dominant features of this roadcut, but the amount of jointing varies. The less-jointed areas would be more resistant to weathering than highly jointed ones. [Connecticut Valley, Massachusetts]

15.01.b2

△ The spacing and orientation of joints, along with rock type, determine how fast the rock will weather and which parts of the landscape will be most easily eroded. Joints play a prominent role in the weathering of rock layers in this photograph. [Capitol Reef National Park, Utah]

Exfoliation can shave off thin, curved slices of rock parallel to the surface, leaving curved, cone-shaped landforms. [Yosemite National Park, California] ▽

15.01.b3

How Do Mafic Rocks Weather?

Dark-colored rocks are usually composed of *mafic* (Mg- and Fe-bearing) minerals. Mafic minerals generally form at high temperature and at least moderate pressures. When these minerals are brought to the surface, they are unstable and convert to more stable minerals, such as clay minerals and iron oxides.

Gabbro is a coarse-grained mafic rock that is the intrusive equivalent of fine-grained basalt. Gabbro crystallizes at depth and contains calcium-rich plagioclase feldspar and dark, mafic minerals, such as pyroxene and amphibole. Feldspar and the mafic minerals in gabbro commonly weather to clay minerals by *hydrolysis*.

15.03.c1

The mafic silicate minerals in gabbro, along with magnetite, an iron oxide, can also *oxidize* when exposed to Earth's atmosphere and water. Magnetite (Fe_3O_4), a dark gray to black magnetic mineral, can be oxidized to hematite (Fe_2O_3), which commonly is reddish in color and has a reddish streak.

What Controls How Different Minerals Weather?

Many factors determine how minerals weather, including the climate, how much time the mineral has been exposed to weathering, and the chemical composition and atomic structure of the mineral.

Chemical Bonding

15.03.d1

15.03.d2

What happens when you put salt crystals into a glass of water? The salt begins to dissolve, but the glass—which is created from melted quartz—does not. Salt dissolves because its ionic bonds can be broken by water molecules, whereas quartz has strong, covalent bonds and does not dissolve in cold water.

Reactivity

15.03.d3

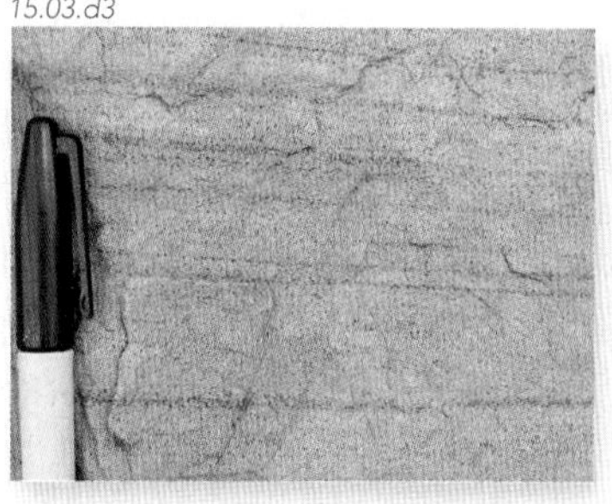

Sandstone is composed mainly of quartz. Most grains weather by *physical* processes, rather than chemical processes.

15.03.d4

Limestone is very soluble and prone to *chemical* weathering, especially dissolution. It also weathers physically.

Relative Resistance of Minerals to Weathering

15.03.d5

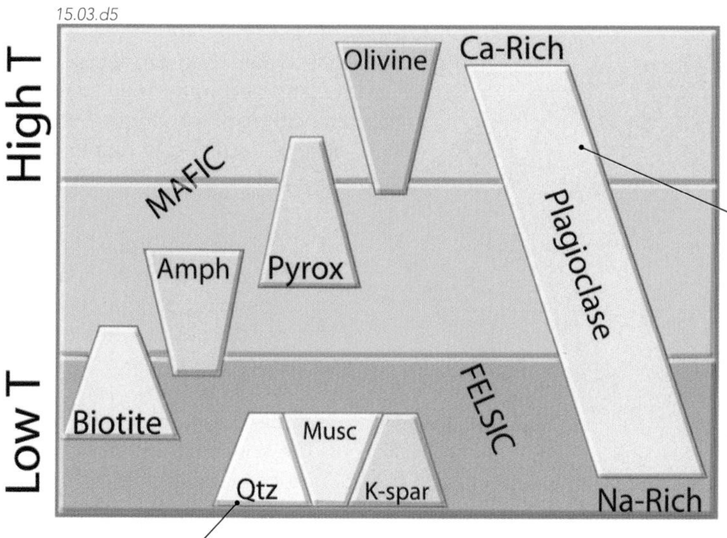

1. The stability of minerals is in a very general way related to the order in which the minerals commonly crystallize from a magma. According to *Bowen's Reaction Series*, mafic minerals and Ca-rich feldspar crystallize first, followed by K- and Na-rich feldspar, muscovite, and quartz. In this illustration, minerals are arranged in their general crystallization order, from top to bottom.

2. The first minerals to crystallize as the magma cools—at the highest temperatures—are typically those that are *least* stable when subjected to weathering at the low temperatures of Earth's surface.

3. Quartz crystallizes late according to Bowen's Reaction Series and is the silicate mineral most resistant to weathering. Although Bowen's Reaction Series is very idealized and not always applicable, it is one way to think about stabilities of minerals during weathering.

Before You Leave This Page Be Able To

- ✓ Describe differences in the weathering of quartz, feldspar, mafic minerals, and calcite.
- ✓ Explain the origin of the three main weathering products (sand, clay minerals, and dissolved ions).
- ✓ Summarize the factors that control how different minerals weather.

What Factors Influence Weathering?

DIFFERENT ROCKS WEATHER to different appearances. Some rock units are boldly exposed or weather to large, loose blocks, while others disintegrate to smooth, soil-covered slopes. What causes these variations? Weathering is largely controlled by the properties of the rock and the setting where weathering occurs.

 How Does the Character of a Rock Influence Weathering?

Differences in rock properties, such as mineral composition and particle size, play an important role in how a rock responds to weathering. Equally important are fractures, bedding planes, and other discontinuities.

Composition—Weathering of a rock is controlled by the types of minerals it contains. Most sandstone, such as in this cliff, consists largely of quartz, a mineral that is very stable on Earth's surface; it mostly weathers by physical processes. In contrast, the recesses below the cliff contain fine-grained sedimentary rocks that are more easily weathered and eroded.

Variation in Composition—Some rock units display large contrasts in their susceptibility to weathering. The more susceptible parts of the rock will weather faster than the more resistant parts. Such *differential weathering* can form alternating ledges and slopes, as shown here, or rocks with holes where less resistant material has been removed.

15.04.a1

Discontinuities—Fractures, bedding planes, and other discontinuities provide pathways for the entry of water into a rock body. A rock with lots of these features will weather more rapidly than a massive rock containing few such discontinuities. For example, highly fractured parts of a rock unit weather faster than less fractured parts. [Bluff, Utah]

Surface Area—Rock that is already broken into small pieces, such as these loose pieces, provides more surface area on which chemical weathering can act. Solid, unfractured bedrock provides less surface area and weathers more slowly.

B What Other Factors Influence Weathering?

How rocks weather also is controlled by factors not related to the rock itself but to its outside setting. Such external factors include location, topography, and aspects of the climate such as temperature and moisture.

Climate—Abundant precipitation and higher temperatures generally cause chemical reactions to proceed faster. Thus, warm, humid areas generally have more highly weathered rock. Chemical weathering is faster in warm, wet climates than in cold or dry climates. Elevation influences an area's climate, so is another important factor.

Hillslope Orientation—The orientation of the slope on which a rock occurs is an important factor in weathering. Many mountain ranges receive more rain on one side than the other. Slopes that are more sheltered from sunlight are cooler, can better retain their moisture, and may have more plants. Moisture, soil, and plants promote chemical weathering.

Slopes facing the Sun receive more light and heat than those facing away. Thus, sunny surfaces (south-facing slopes in the Northern Hemisphere) tend to be warmer and drier, to have more evaporation, and to have less chemical weathering, soil, and plants than slopes facing away from the Sun.

Steepness of Slopes—On steep slopes, rainfall runs off faster and weathering products may be quickly washed away by runoff.

On gentle slopes, weathering products can accumulate, and water may stay in contact with rock for longer periods of time, resulting in higher weathering rates.

Time—A crucial factor in weathering is time. Depending on the other factors discussed, rates of weathering can range from rapid to extremely slow. The more time available, the more weathering will occur. The speed of weathering and the volume of material affected in a given time will depend on slope, climate, hillslope orientation, and the composition and structural condition of the rock or sediment.

15.04.b1

C Why Does Weathering Produce Rounded Features?

Weathering processes usually work inward from an exposed surface. This commonly results in rounded shapes in weathered outcrops, and in some situations weathering generates loose, partially rounded blocks. The three figures below illustrate what can happen to a rock that has fractures, but lacks other type of discontinuities.

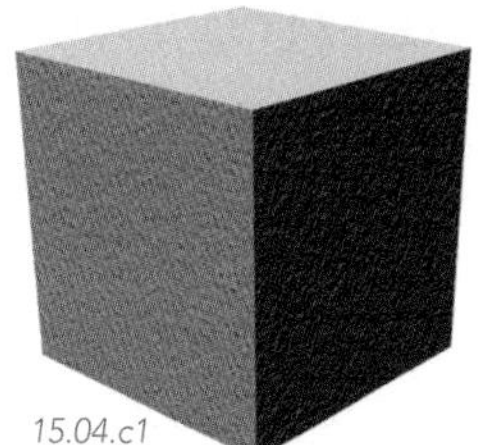

Rock that is newly fractured generally has sharp, angular edges. Weathering attacks edges from *two* sides and corners from *three* sides. These edges and corners wear down faster than a single smooth surface.

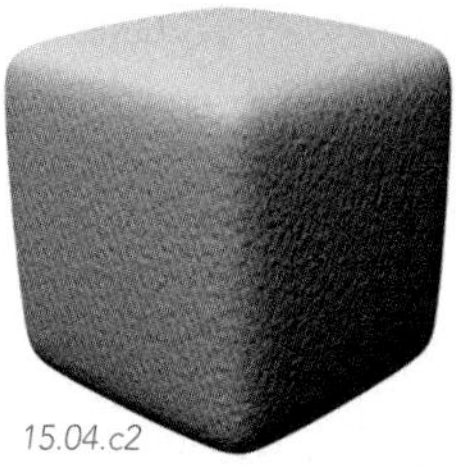

Over time, faster weathering of the edges and corners of the rock will begin to smooth away any sharp features, such as corners, edges, and any other parts of the rock that stick out.

Weathered rocks can become moderately rounded, losing their sharp edges and angular features. Rocks dislodged from the bedrock will get smaller with time as they are weathered from all sides. Most rounding occurs while the rock is being transported.

D How Is Weathering Expressed in the Landscapes?

Weathering exerts a huge control over the appearance of landscapes. Weathering helps define differences in appearance from one region to another, from one side of a hill to the other, or between different rock types.

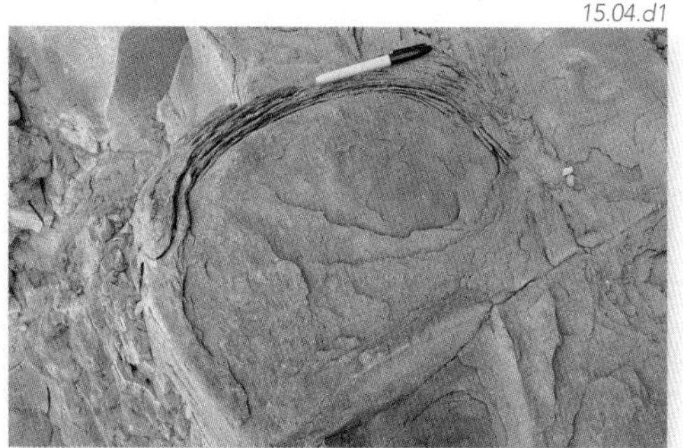

1. Weathering affects rocks from the outside in, so weathered rocks can have an outer weathered zone and an inner unweathered zone. The outer zone is called a *weathering rind*. As weathering continues, the weathering rind thickens and can sometimes be used to infer how long the rock has been on or near the surface and exposed to weathering. [San Juan River, Utah]

2. As weathering attacks a fractured rock, preferential weathering along the fractures can cause the fracture-bounded blocks to become rounded. The outer weathered rind of the blocks splits away from the stronger, less weathered rock, forming rounded shapes in a process called *spheroidal weathering*. [Boulder, Montana]

3. If conditions favor extensive weathering, such as exposure of a chemically reactive rock unit for a long time, thick soil can develop (△), obscuring the underlying bedrock. In a drier climate, on a steep slope, or for a more resistant rock, weathering is slower, so soil may be less developed and hillslopes may be rockier and more barren. [Australia]

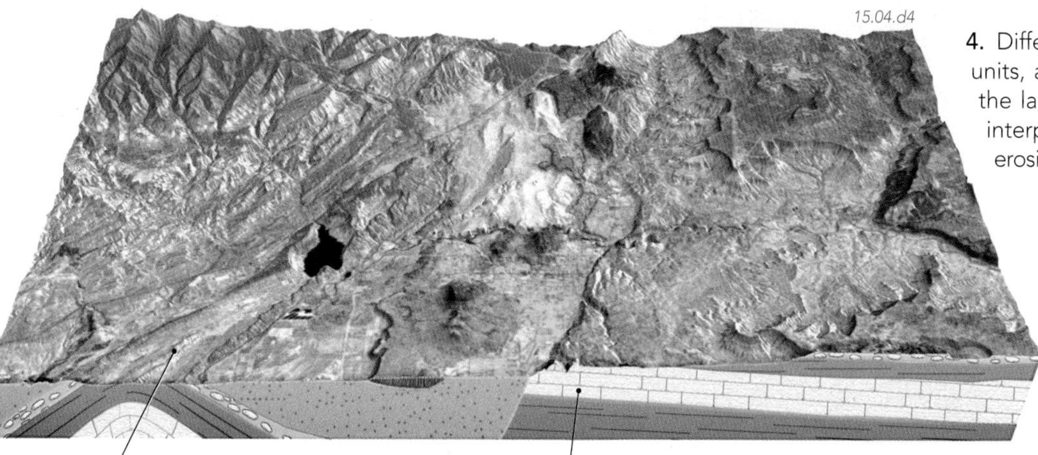

4. Differential weathering and erosion of different rock units, and of different parts of a single rock unit, produce the larger landscapes we see. This image illustrates the interplay between rock type, structure, weathering, and erosion. [loosely modelled after the St. George area, Utah]

5. Less resistant rock units, or parts of units that are highly fractured, occupy the valleys, slopes, and gently rounded hills.

6. Rock layers, such as this limestone, that are more resistant to weathering may form cliffs, ledges, hills, and mountains. What can you infer about the weathering of each rock unit in this scene?

Before You Leave This Page Be Able To

- ✓ Summarize or sketch how weathering is affected by the properties of a rock.
- ✓ Summarize or sketch how weathering is affected by the setting of a rock.
- ✓ Describe ways that weathering is expressed in the landscape, including its role in rounding off corners.

How Does Soil Form?

SOIL BLANKETS MUCH OF THE LAND SURFACE, providing a place for plants, animals, microbes, and humans to live. What is soil and how does it form? Soil is affected by geologic, biologic, and hydrologic processes, and thus represents the interplay between the lithosphere, biosphere, hydrosphere, and atmosphere. The processes and factors that influence weathering also control how soil forms in different climates.

What Is Soil?

Soil consists of weathered rock, plus material from the atmosphere, decaying plants, animals, and microbes.

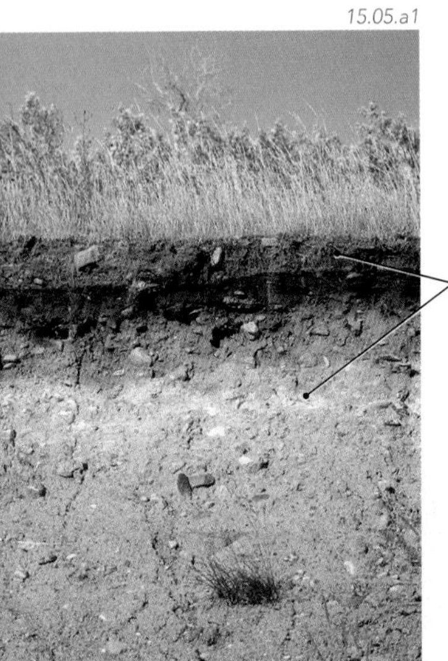

◁ What do you observe in this photograph of a cut through soil?

There are different zones or layers, called *horizons*, with rather gradational boundaries. These different layers are not the same as beds formed by sedimentation; instead each horizon forms and grows *in place* by weathering of rock and sediment, and by the addition of material from plants, animals, and the atmosphere.

▷ In this idealized soil profile, each soil horizon is assigned a letter to denote its position or its character.

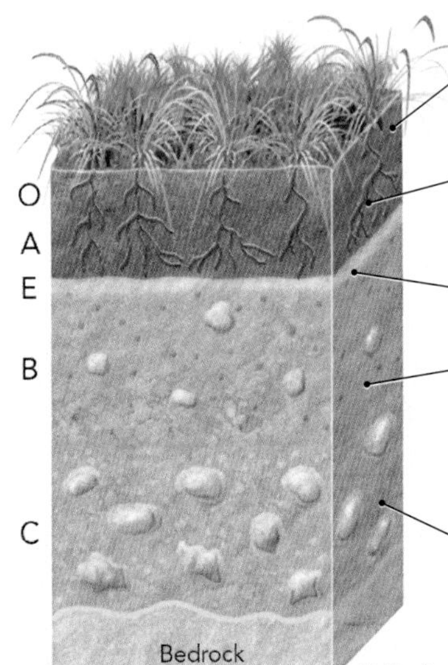

O horizon is a surface accumulation of organic debris, including dead leaves and other plant and animal remains.

A horizon is topsoil, composed of dark gray, brown, or black organic material mixed with mineral grains.

E horizon is a light-colored, leached zone.

B horizon contains little organic material, but can have a red color due to the accumulation of iron oxide. In dry climates, the B horizon can be whitish due to accumulated calcium carbonate.

C horizon is composed of either weathered bedrock or unconsolidated sediment, and grades downward into unweathered bedrock or sediment.

B What Processes Occur During Soil Formation?

What happens to form soil from rock and other materials? Soil forms gradually over thousands of years and involves some of the same processes as weathering, including dissolution, oxidation, hydrolysis, and root wedging. Soil formation also involves the vertical transport of dissolved material up and down through the profile.

Where Material Comes From

Soil material is mostly derived from weathering of underlying rock and sediment, but some is introduced by water, wind, and atmospheric gases.

Soil receives several types of material from the land surface. Leaves, pine needles, twigs, and other plant parts accumulate on the surface and are worked downward into the soil. Sediment washes onto the surface from adjacent hillslopes, during floods, or from windblown dust and salts. Roots give off CO_2 gas, other gases come from the atmosphere, and water mostly arrives in rainfall and snowmelt.

Weathering weakens and loosens underlying bedrock, providing starting material with which to make soil. This material can be worked up into the soil, or the soil can gradually affect deeper and deeper levels of the bedrock. Some residual material remains in place at depth.

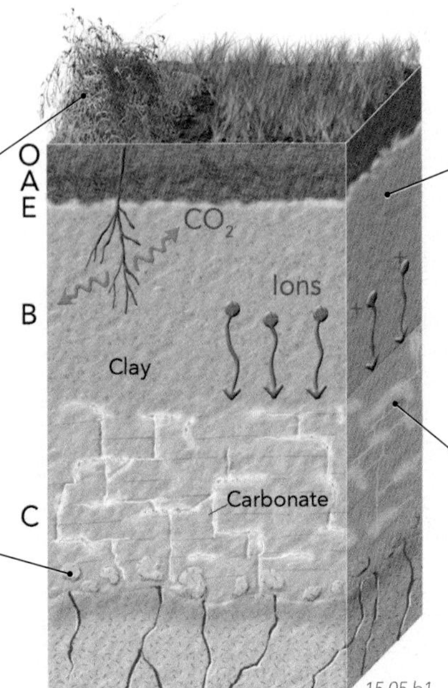

How Material Moves

Soil material moves both down and upward as it is carried by water, plants, animals, and gravity.

Zone of Leaching—The upper part of soil *loses* material downward. Water soaking into the soil from above leaches soluble ions liberated by chemical weathering, taking them deeper into the soil. Plant parts and other organic material are also worked downward into the soil. Clay minerals and other fine particles are carried downward by infiltrating water.

Zone of Accumulation—Chemical ions leached from above *accumulate* in the underlying zone, especially if the water evaporates rather than carries ions through the soil and into groundwater. Clay minerals, iron and aluminum oxides, salt minerals, and calcium carbonate can all accumulate in the soil, depending on how much oxygen and water pass through the soil.

C What Soil Profiles Are Typical for Different Climates?

Climate, especially temperature and moisture, strongly affects the type and rate of weathering, the abundance of plants, and the type of soil that results. There are many distinct types of soil, and one very useful soil classification uses names such as oxisols and ultisols. Here, we limit this discussion to three major soil types defined by climate. The top two diagrams represent thicker sequences of soil than does the one for arid climates. The accompanying photographs only show the upper parts of the soil profiles.

Tropical Climates

1. In humid, tropical climates, there is abundant rainfall and associated plant growth. Such areas include rain forests and swamps, both of which are characterized by dense plant growth.

15.05.c1

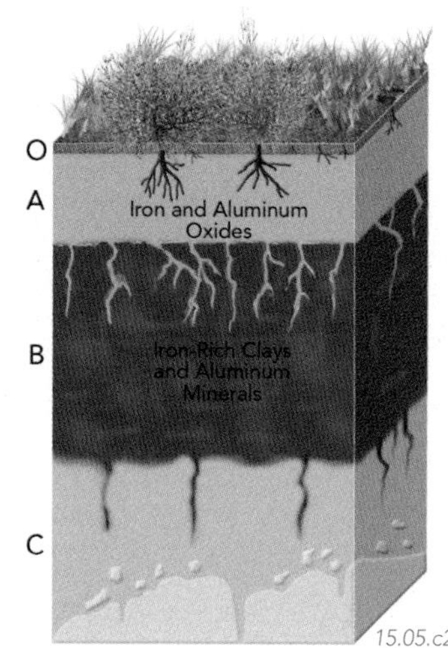

15.05.c2

2. In tropical climates, intense weathering and abundant soil moisture cause severe chemical leaching, leaving behind a soil rich in iron (Fe) and aluminum (Al) oxides, sometimes giving the soil a deep red color. This extremely leached type of soil is called a *laterite*.

Temperate Climates

3. Temperate climates are cooler, and generally have less rainfall than tropical climates. Such areas contain savannas, grasslands, farms, or lush forests of leafy, deciduous trees or pines.

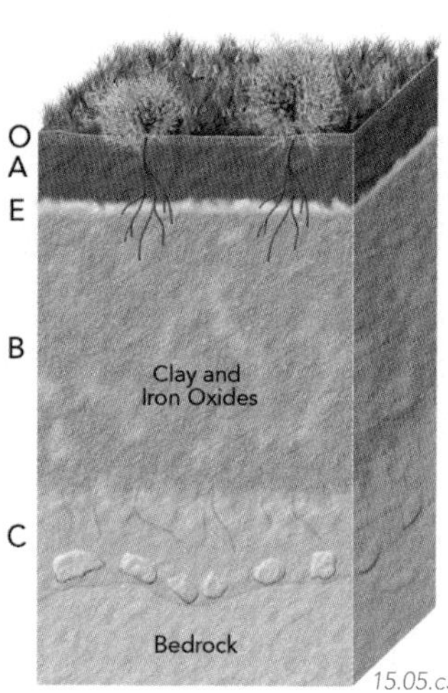

15.05.c3

15.05.c4

4. In cooler areas with moderate to high rainfall, the A and B horizons contain abundant insoluble minerals, such as quartz and clay, as well as iron-oxide minerals. More soluble minerals like calcium carbonate ($CaCO_3$) are absent. Informal names for such soils are *grassland soil* or *forest soil*, depending on the type of vegetation.

Arid Climates

5. Arid climates are dominated by overall dryness due to sparse precipitation. They can be very hot, as in subtropical deserts, very cold, as in the Dry Valleys of Antarctica, or moderate in temperature, but still dry. Plants and animals are sparse.

15.05.c5

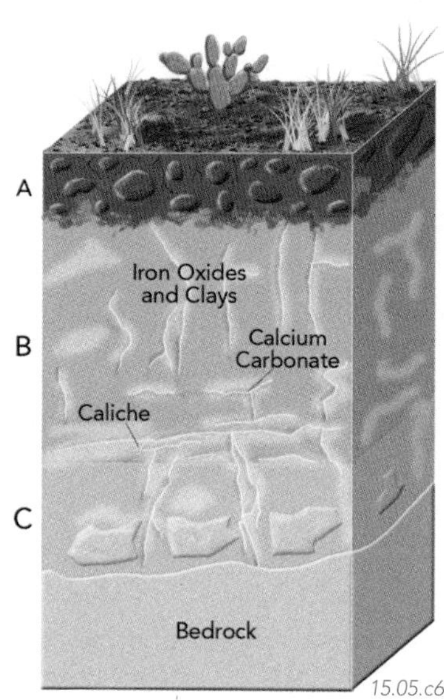

15.05.c6

6. In arid climates, there is limited vegetation so there is little or no O horizon, and usually only a thin A horizon. Clay, iron oxide, and salts, all partly derived from windblown material, accumulate at various levels in the soil. Ca^{2+} and CO_3^- ions are dissolved from upper soil horizons and chemically precipitated downward as calcite ($CaCO_3$). The amount of water passing through the soil is not enough to completely remove these ions, and so the amount of calcium carbonate increases with time, first coating clasts and eventually forming a discrete layer, called *caliche*. Such soil is a *desert soil*.

Before You Leave This Page Be Able To

- ✓ Describe what a soil is and the processes by which it forms.
- ✓ Sketch and describe the main soil horizons and the processes and materials that occur in each horizon.
- ✓ Discuss the different soils formed in different climates and the factors responsible for these differences.

Why Is Soil Important to Society?

SOIL IS ONE OF OUR MOST IMPORTANT NATURAL RESOURCES. It provides nutrients for the diversity of plants, animals, and microbes that inhabit our planet. Soil provides a necessary foundation for grasslands, forests, and the crops that feed and clothe our growing population. The loss of soil can be catastrophic for individual communities and regions, and for the plants and animals that depend on the ecosystem.

A How Do Soil and Vegetation Interact?

Soil and plants have a mutually beneficial relationship—soil permits most plants to grow, while plants contribute material to the soil and help bind the soil together, helping it develop and protecting it from erosion. Bacteria, fungi, and other microbes also play an important role in the development and health of soils and plants.

15.06.a1

The relationship between soil and plants is appreciated by farmers, who try to nurture and retain the soil, while harvesting a bounty of grains, fruits, vegetables, and cotton.

15.06.a2

The grass cover of this hillslope helps the underlying soil in many ways. It buffers the soil from rain and wind, captures water that helps the soil develop, and protects the soil from being eroded by runoff.

B What Activities Threaten Soil?

In most climates it takes 80 to 400 years to form about 1 centimeter of topsoil. Soil that is eroded due to poor farming practices or other detrimental activities is lost and cannot easily be replaced.

15.06.b1

Much land is used to raise livestock, such as cattle and sheep. Overgrazing by livestock or indigenous animals, such as deer, removes vegetation, which leaves soil vulnerable to wind and water erosion. It also removes the nutritionally rich upper layers of soil. Overgrazing can be especially devastating in times of drought and in some countries. [Namibia]

15.06.b2

Soil and other weathering products can be washed away on steep slopes, but can accumulate on more gentle slopes. Shaping steep terrain to provide flat areas suitable for farming, a practice called *terracing*, can better protect soil from erosion, promote soil formation, and provide a more level place on which to farm. [Peru]

15.06.b3

Cutting down forests and removing vegetation to provide lumber, grazing, or farmland can result in massive soil erosion. Severed roots decay and can no longer hold the soil in place. Eventually, the soil is unable to regenerate vegetation, which will ultimately lead to an increase in runoff, accelerated soil erosion, and possibly flood disasters. The loss of soil can defeat the activity (farming) for which the land was cleared. [Washington]

15.06.b4

Soil can become polluted near farms that use pesticides, herbicides, fungicides, and fuel oils. Soil can also become contaminated by salt from irrigation water that has acquired a high salinity from evaporating and passing over croplands. Some industries use pollutants that find their way into the soil, and some mining operations contaminate soil with chemicals and with elements, such as arsenic, that occur naturally in mineralized rocks. [Colorado]

C What Are Some Problems Related to Soil?

In addition to being a valuable resource, some soil can cause humans difficulty because of its low strength and how it behaves when shaken, wetted, dried, or compacted. Such soil can be recognized by geologists, builders, and home buyers so that building on, or buying, such risky sites can be avoided.

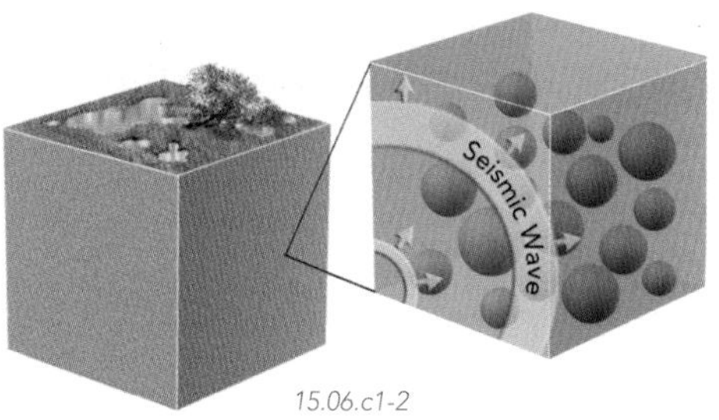

15.06.c1-2

Liquefaction occurs when loose sediment becomes oversaturated with water and individual grains lose grain-to-grain contact with one another as water squeezes between them. Quicksand is an example of liquefaction. Liquefaction is especially common when loose, water-saturated sediments are shaken, such as during an earthquake. The houses below, destroyed during the 1989 Loma Prieta earthquake, sank into artificial fill that liquefied.

15.06.c7

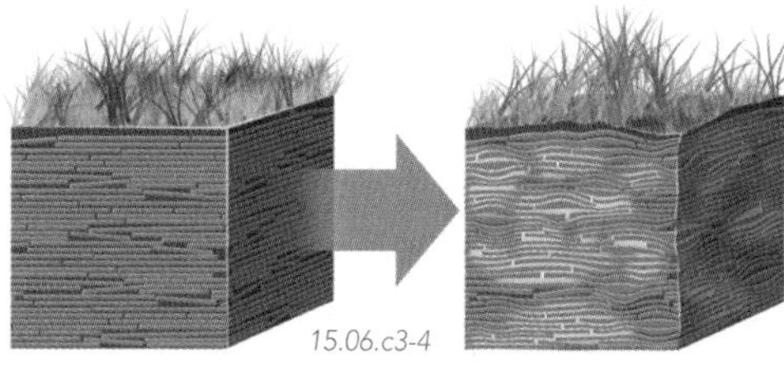

15.06.c3-4

Soil that contains a high proportion of certain clay minerals, called *swelling clays,* increases in volume when it becomes wet, expanding upward or sideways. When clays dry out, they decrease in volume, causing the soil to shrink or compact. Repeated expansion and compaction during wet-dry cycles can crack foundations, make buildings unsafe, and ruin roads, such as the one below. [Boulder, Colorado]

15.06.c8

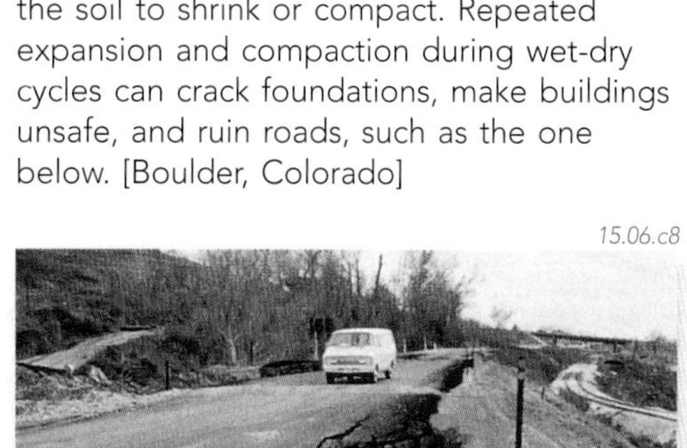

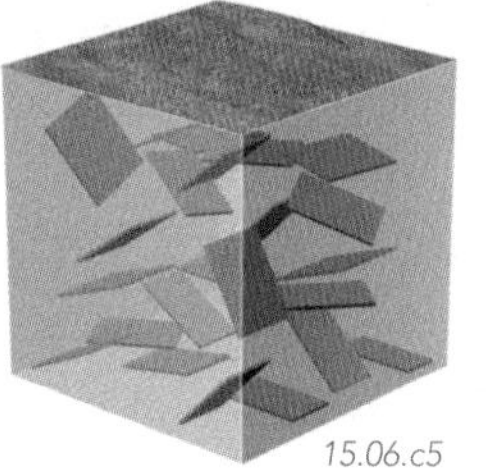

15.06.c5

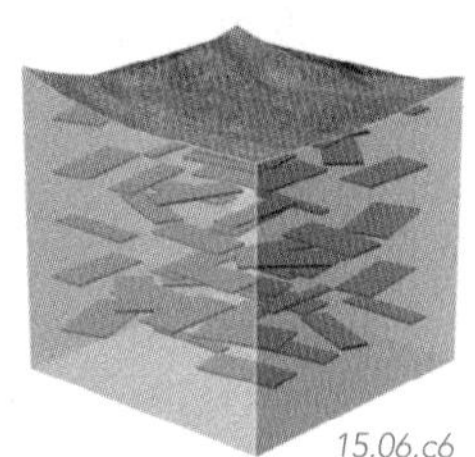

15.06.c6

In some soil, clay minerals start out arranged randomly, with much pore space between individual sheets. As water infiltrates the pore spaces, the clay minerals tend to lay flat, reducing open spaces and thereby compacting the soil. Such *soil compaction* typically does not occur uniformly, because some parts of the soil have more clay than others. Differential compaction can crack walls, foundations, and roads. ▽

15.06.c9

Mineral Deposits Formed by Weathering

Weathering processes move chemical elements, leaching them from some areas and concentrating them in others. During chemical weathering, for example, a body of soluble rock can be greatly reduced in volume by dissolution and leaching. Elements that are not leached from the rock can become concentrated enough to become valuable. The most important ore of aluminum, a rock called *bauxite* being mined below, forms in wet, tropical climates where high air temperatures and abundant water produce soil rich in aluminum. The bauxite results from the breakdown by weathering of clay minerals, which were also largely formed by weathering.

Weathering also plays a role in concentrating metals in near-surface mineral deposits. In many large copper mines, including the one shown below, weathering and down-flowing groundwater have leached copper and sulfur from the top hundred meters or so of the copper deposit, reprecipitating copper- and sulfur-rich minerals farther down, in many cases making the deposit rich enough in copper to mine.

15.06.mtb1

15.06.mtb2

Before You Leave This Page Be Able To

- ✓ Summarize activities that can threaten soil and its protective cover of vegetation.
- ✓ Describe some problems associated with certain soil types.
- ✓ Describe two ways that weathering can enrich a mineral deposit enough so the deposit can be mined.

15.6

What Controls the Stability of Slopes?

GRAVITY PULLS MATERIAL DOWNHILL, and some rocks, soil, and other loose material are not strong enough to resist this persistent force. Downward movement of material on slopes under the force of gravity is called *mass wasting* and occurs to some degree on all slopes. Mass wasting can proceed very slowly or can be rapid, sometimes with disastrous results. Mass wasting is an important part of the erosional process, moving material downslope from higher to lower elevations.

A What Role Does Gravity Play in Slope Stability?

The main force responsible for mass wasting is *gravity*. The force of gravity acts everywhere within Earth and on the surface, tending to pull everything downward toward Earth's center.

1. On a flat surface, the force of gravity acts on the weight of the block to push vertically down against the base of the block. The block will not move under this force.

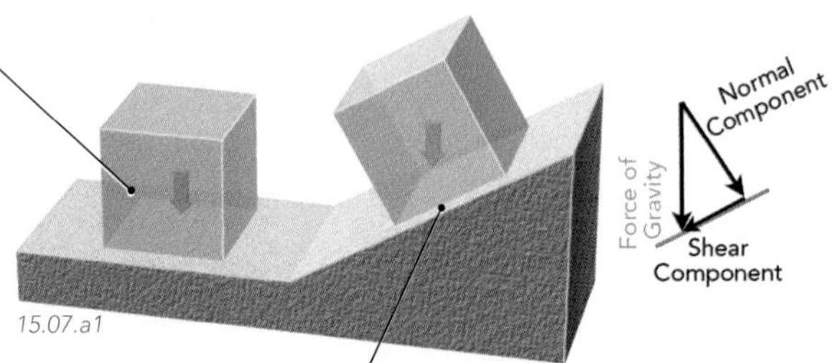

15.07.a1

2. On a slope, the force of gravity acts at an angle to the base of the block. Part of the force pushes the block *against* the slope and another part pushes the block *down* the slope. These two parts of the force are called *components*.

3. The part of the force pushing the block *against* the slope is called the *normal component*.

4. The other component acts parallel to the slope, trying to shear the block down the slope. It is called the *shear component*.

5. As the angle of slope becomes steeper, the *shear* component becomes larger while the *normal* component becomes smaller. As the slope angle steepens, the shear component becomes enough to overcome friction and causes the block to slide.

B How Steep Can a Slope Be and Still Remain Stable?

The steepest angle at which a pile of unconsolidated grains remains stable is called the *angle of repose*. This angle is controlled by frictional contact between grains. In general, loose, dry material has an angle of repose between 30° and 37°. This angle is somewhat higher for coarser material, for more angular grains, for material that is slightly wet, and if the material is partly consolidated. It is lower for material with flakes or rounded grains and for material that contains so much water that adjacent grains lose contact.

15.07.b1

1. Dry unconsolidated sand grains form a pile in which the angle of the resulting slope is equal to its angle of repose. If more sand is added, the pile becomes wider and higher, but the angle of repose remains the same. If part of the pile is undercut and removed, the grains slide downhill until the pile returns to a stable slope at the angle of repose. If sand is slightly wet, surface tension between the grains and a thin coating of water enables the sand to be stable on steeper slopes.

15.07.b2

3. Loose rocks and other material accumulate on some slopes and at the base of cliffs. Such material is called *talus* and commonly forms slopes that are at the angle of repose for the particular sediment. The smooth *talus slope* shown here became too steep in places and so locally slid downward. [Tibet]

15.07.b3

2. Most slopes of sand dunes reflect the angle of repose for dry sand. Slopes can be more gentle than the angle of repose, but if they begin to exceed the angle of repose then the slope fails, slumping downhill. Walking up a sand dune causes the barely stable sand to slide. [Morocco]

15.07.b4

4. The slopes of a scoria (cinder) cone reflect the angle of repose, because they are typically composed of loose, volcanic scoria. The angle of repose will be steeper for coarser scoria and for material that partially fused together during the eruption. [Arizona]

C What Factors Control Slope Stability?

The main control on slope stability is the angle of repose for that material. Intact rock can form steep slopes, but soil, sediment, and strongly fractured rock can form slopes reflecting their angle of repose.

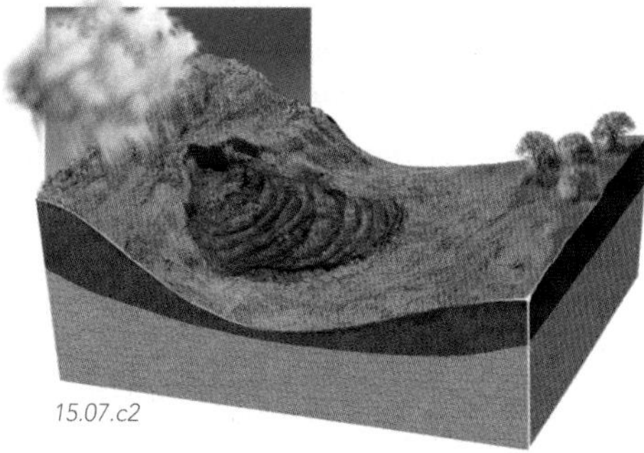

The addition of minor amounts of water increases the strength of soil, but oversaturation pushes grains apart and weakens the soil. Materials with high clay-mineral content can flow downhill when they become wet.

Fractures, cleavage, and bedding reduce the overall mechanical strength of a rock, and such discontinuities may allow rocks to slip if they are inclined downhill, as shown on the left side of this illustration.

D What Triggers Slope Failure?

Slope failure occurs when a slope is too steep for its material to resist the pull of gravity. Some slopes slide or creep downhill continuously, but others fail because some event caused the previously stable slope to fail.

1. Precipitation can saturate sediment, weakening an unconsolidated material by reducing grain-to-grain contact. A slope that was stable under dry conditions may fail when wet. Slopes can also fail after wildfires, which destroy plants that help bind and stabilize the soil.

2. Hillslopes can fail when the load on the surface exceeds its ability to resist movement. Humans sometimes build heavy structures on slopes, overloading the slope and causing it to fail. Areas with gentle slopes, such as near this town, are less risky for slope failures.

3. Modification of a slope by humans or natural causes can increase a slope's angle so that it becomes unstable. Erosion along river banks, such as shown here, or wave action along coasts can undercut a slope, making it unsafe.

4. Oversteepening of cliffs or hillslopes during road construction can cause them to fail, especially if fractures or layers are inclined toward the road.

5. A sudden shock, such as an earthquake along this fault scarp, may trigger slope instability. Minor shocks, such as from heavy trucks or human-caused explosions can also trigger failure.

6. Volcanic eruptions can shake, fracture, and tilt the ground, unleashing landslides from oversteepened slopes. Eruptions can cover an area with ash and other loose material, causing melting of ice and snow. This can rapidly release large amounts of water and mobilize volcanic material in destructive debris flows.

15.07.d1

Slope Stability in Cold Climates

In cold climates, water is frozen much of the year, and ice, although solid, does have the ability to flow. Freeze-thaw cycles, where ice freezes and then thaws repeatedly, also contribute to mass movement.

When water-saturated soil freezes, it expands, pushing rocks and boulders on the surface upward (▷). When the soil thaws, the boulders move down again. This process, called *frost heaving*, is a large contributor to downslope movement of material in cold climates. In addition, when the upper layers of soil thaw during the warmer months, the water-saturated soil may move downslope more easily.

Before You Leave This Page Be Able To

- ✓ Describe or sketch the role that gravity plays in slope stability.
- ✓ Describe the concept of the angle of repose and its landscape expressions.
- ✓ Describe some factors that control slope stability, and events that trigger slope failure.

How Do Slopes Fail?

THE RAPID DOWNSLOPE MOVEMENT of material, whether bedrock, soil, or a mixture of both, is commonly referred to with the general term *landslide*. Movement can occur by falling, sliding, rolling, slumping, or flow. Slope failures are classified by how the material moves and the type of material involved.

A What Are Some Ways That Slopes Fail?

Most people have seen images of slope failure when hiking in the hills, driving past a road cut, or watching television news, nature shows, or movies. A slope failure can be as subtle as a small pile of rocks at the base of a hill or as dramatic as a mudflow that has destroyed a neighborhood in China or California. The photographs below show images of various types and sizes of slope failures.

15.08.a1

△ **1.** This rocky cliff failed after being undercut by a river. Large sandstone blocks, one the size of a building, collapsed downward. The falling block detached along a prominent joint surface, which has since accumulated a brown rock varnish. [Yampa River, Utah]

15.08.a2

◁ **2.** During an earthquake, brown masses of rock and soil slid down these steep slopes in Alaska, smashed apart, and flowed as *avalanches* of rock, soil, and ice across a white glacier in the valley below. Parts of the avalanche flowed across the valley and partway up hillsides on the other side of the valley.

15.08.a3

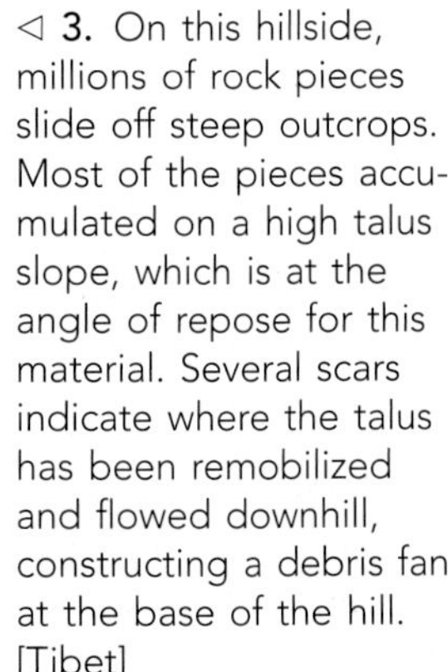

◁ **3.** On this hillside, millions of rock pieces slide off steep outcrops. Most of the pieces accumulated on a high talus slope, which is at the angle of repose for this material. Several scars indicate where the talus has been remobilized and flowed downhill, constructing a debris fan at the base of the hill. [Tibet]

15.08.a4

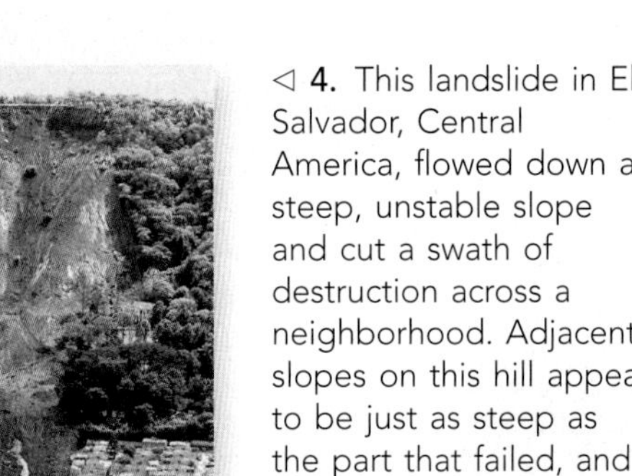

◁ **4.** This landslide in El Salvador, Central America, flowed down a steep, unstable slope and cut a swath of destruction across a neighborhood. Adjacent slopes on this hill appear to be just as steep as the part that failed, and may pose a hazard to the remaining homes.

▷ **5.** Undercutting of hillsides by coastal erosion and highway construction have made many slopes steep and unstable. Rocks and soil slumped downward, covering and blocking the highway along the Pacific Palisades, California. Such slope failures commonly occur during or after intense rainstorms.

15.08.a5

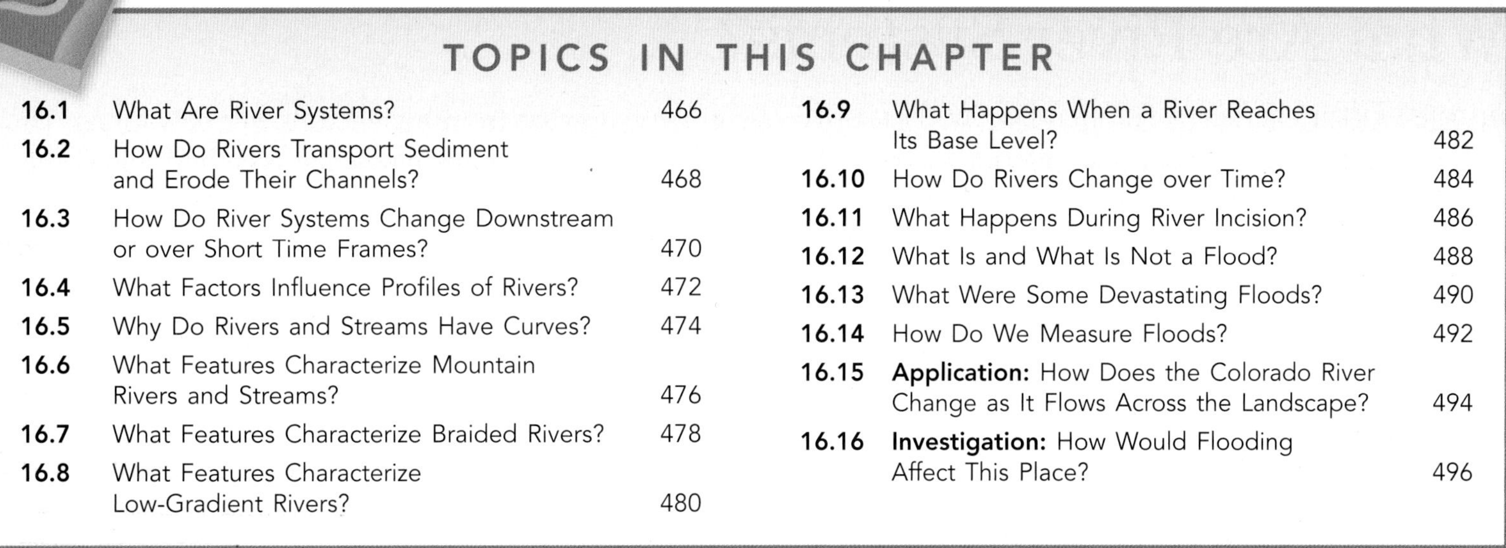

TOPICS IN THIS CHAPTER

A Variety of Rivers

Each river, like the *Yukon River*, has its own characteristics and history, which are specific to its geographic and geologic setting. Some rivers are steep and turbulent, moving large boulders, whereas others are slow and tranquil, transporting only silt and clay. Some rivers *meander* in huge looping turns, while others distribute their flow in a network of channels that split off and rejoin in a *braided* pattern. Certain principles govern the behavior of all rivers, such as whether a river erodes into its banks or deposits sediment, and what type of river it is. The processes involved with rivers cause them to change downstream and over time, producing a characteristic suite of landforms that dominate most landscapes. Rivers can flood huge tracts of land and transport enormous volumes of sediment. The Amazon River in South America (shown below) dumps millions of cubic meters of sediment-laden water into the ocean every minute.

16.00.a2

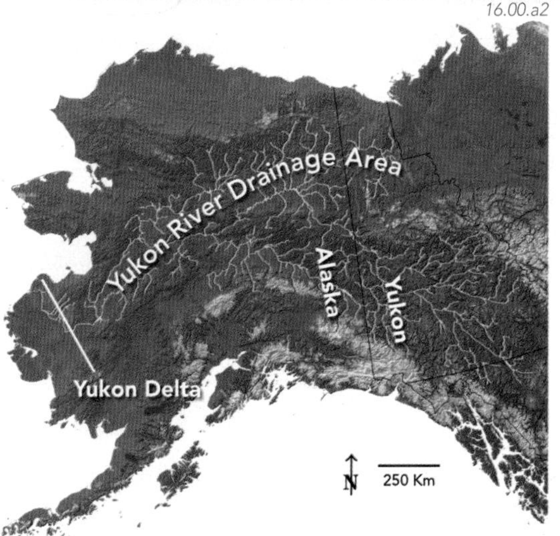

◁ **The Yukon River** collects water from a large region of Alaska and Canada's Yukon Territory. It drains an area of 840,000 km^2 (324,000 mile^2). Periodically, water volume in the river exceeds the confines of its channel, causing flooding.

How is the size of a river related to the size of the area it drains, what causes a flood, and what information do we need to predict flooding events?

16.00.a3

◁ **Many Alaskan Rivers** are full of sediment derived from weathering and erosion of the mountains and lowlands. This river in Denali National Park is choked with coarse gravels, sands, and fine sediment.

What types of sediment do different kinds of rivers carry?

16.00.a4

◁ **During the summer,** lush vegetation grows on the strips of land between the delta waterways.

What effect does vegetation have on rivers, and what effects do rivers have on vegetation?

16.00.a5

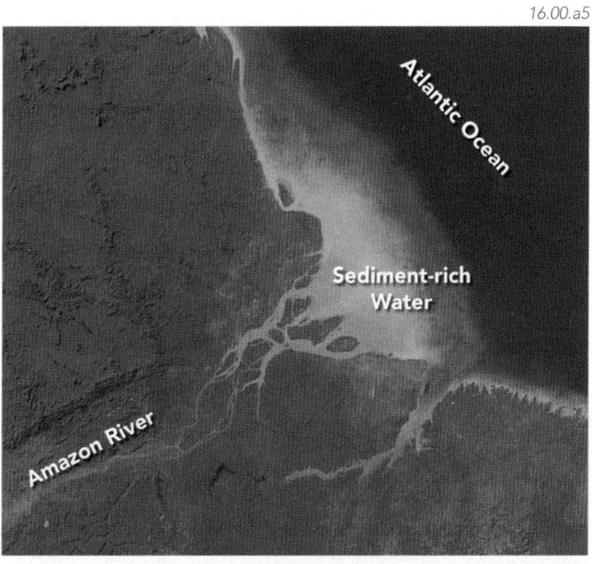

What Are River Systems?

RIVERS ARE CONDUITS OF MOVING WATER driven by gravity, flowing from higher to lower elevations. The water in rivers comes from precipitation, snowmelt, and springs. A river drains a specific area and joins other rivers draining other areas, forming a *network* of rivers that drains a large region.

A What Is a River?

Rivers and streams carry flowing water through a single channel or through a number of related, interconnected channels. Such channels vary in size from small streams to wide rivers.

1. The Brahmaputra River in India, shown in this satellite image (▽), is a main conduit for water falling on or melting off the Himalaya. The sediment load in this river is enormous, reflecting the ongoing uplift and erosion of the region.

2. Water flowing in rivers and streams is able to move rock fragments and dissolved minerals from high to low elevations. Note the varied sizes and styles of rivers and streams in this one image. ▷

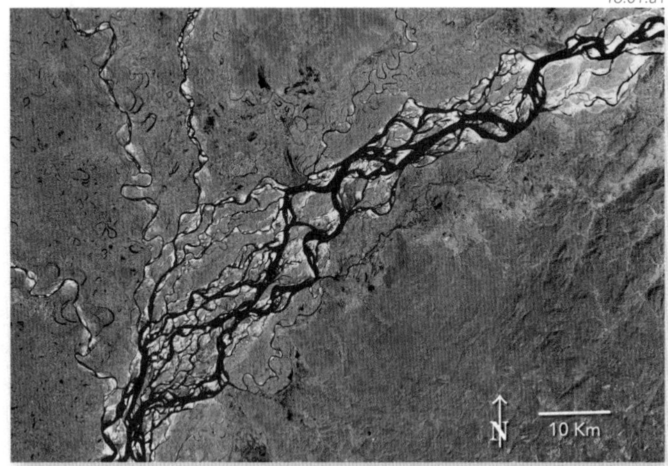

▽ 3. The amount of water that flows through a channel over a given amount of time is called the *discharge* (units of cubic meters per second or m^3/sec). A graph showing the change in the amount of flowing water (discharge) over time is called a *hydrograph*.

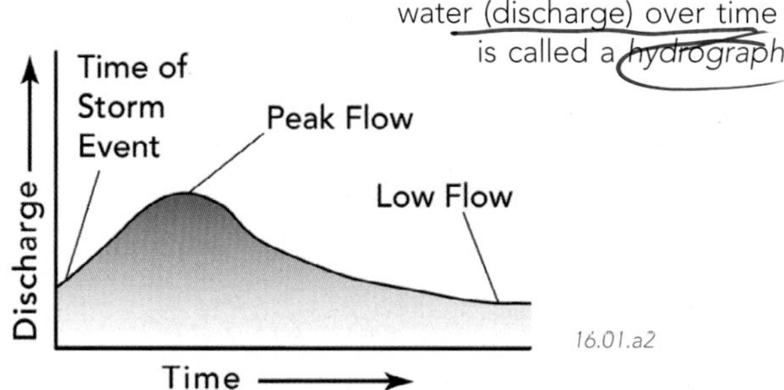

△ 4. This hydrograph shows that discharge increased and then decreased over time in response to a storm. The shape of the graph reflects how the river responds to precipitation and can tell us important information about the river and the area it drains.

B Where Does a Stream or River Get Its Water?

Each stream or river has a naturally defined area that it drains, called a *drainage basin*. A basin slopes from higher areas, where the stream or river begins, to lower areas, toward which the stream or river flows. Runoff from rainfall, snowmelt, and springs will flow out of the drainage basin at its low point.

▽ *Drainage Basin*—In this figure, each of two adjacent streams has a *drainage basin*, shaded in different colors. Runoff from the red area drains into the stream on the left; runoff from the blue area drains into the stream on the right. The ridge between the two drainage basins is the boundary between water flowing into different drainage basins, and is referred to as a *drainage divide*.

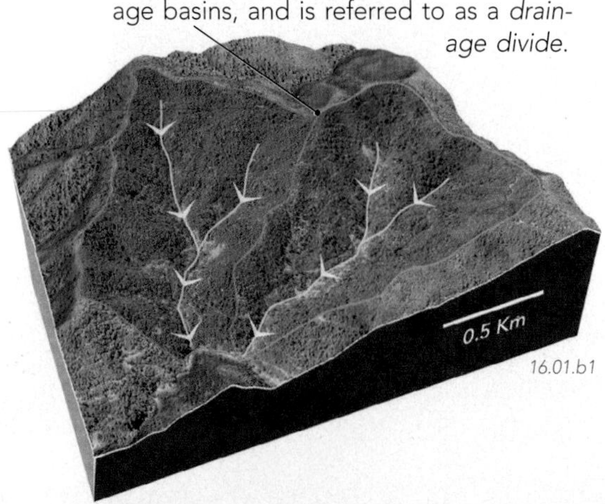

▽ *Basin Slope*—Overall slope of a drainage basin helps determine how fast water in the basin empties after a heavy rain or after snowmelt, as shown by the graph below.

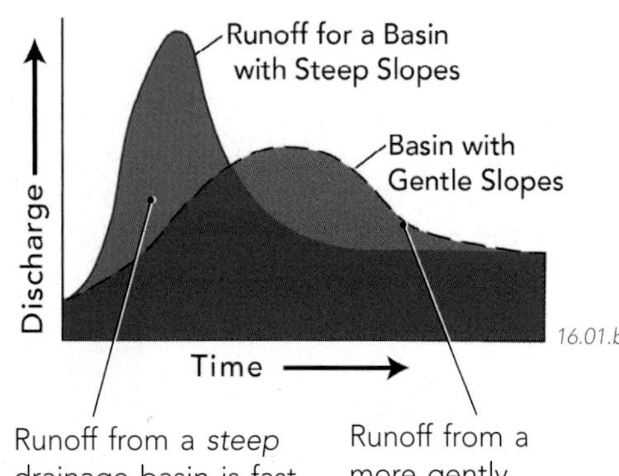

Runoff from a *steep* drainage basin is fast, and much water arrives downstream at about the same time, yielding higher discharge values.

Runoff from a more gently sloped basin is spread out over time, leading to lower peak discharge values.

▽ *Basin Shape*—A drainage basin's shape influences its flow response to rainfall. These plots show hydrographs for a single storm event, along with a simplified map of each basin's shape.

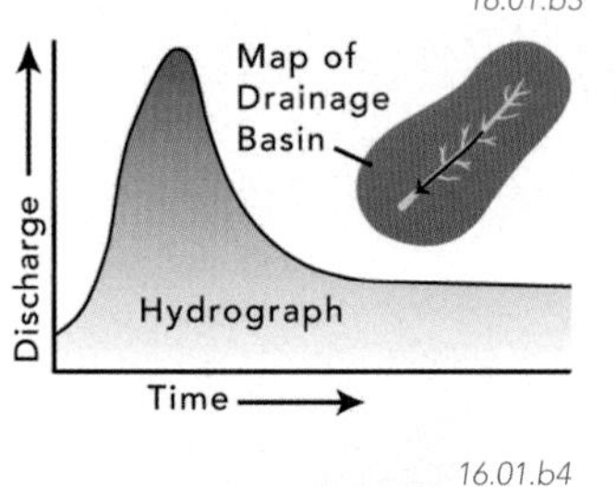

Following a storm event, a simple basin shows a single-peak increase in discharge with a gradual decrease.

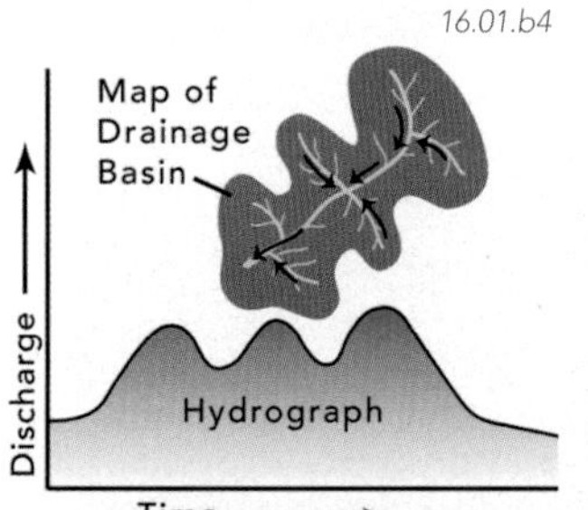

A complex, three-part drainage basin may show a three-peak response, even to a single event.

C What Are Tributaries and Drainage Networks?

Rivers and streams have a main channel fed by smaller subsidiary channels called *tributaries*. Each tributary drains part of the larger drainage basin, but a tributary can have higher flows than the main river. The combination of tributaries and the main river forms a *drainage network*. The response of a river to precipitation is influenced by the number and size of its tributaries.

▽ In this river system, smaller tributaries join to form larger drainages, which join to form even larger drainages. The drainage network has a branched appearance, like a tree.

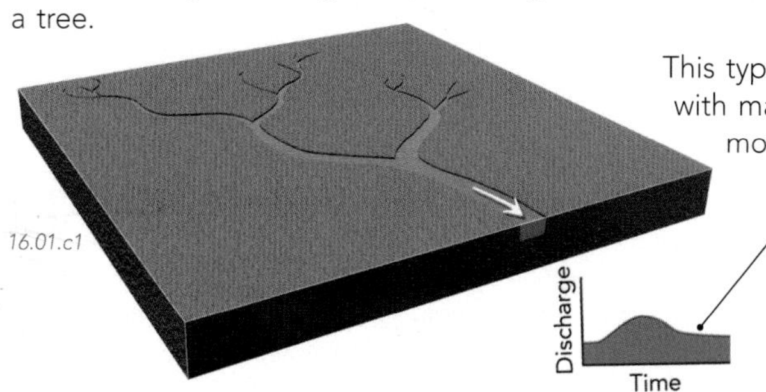

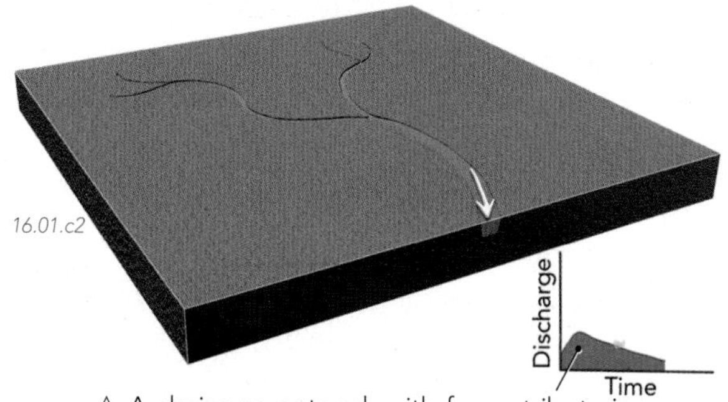

△ A drainage network with fewer tributaries responds faster to an event. The area tends to lose more sediment in response to increased flow.

D How Does Geology Influence Drainage Patterns?

The patterns that river systems carve across the land surface are strongly influenced by the geology. Channels form preferentially in weaker material and so reflect differences in rock type and the geometry of structural features such as faults, joints, and folds.

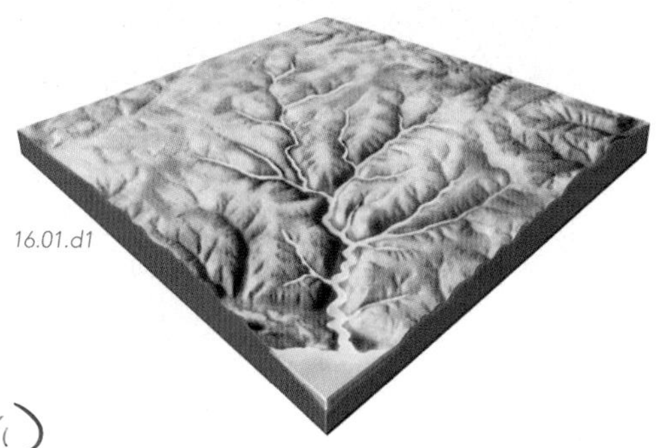

Dendritic Drainage Pattern—Where rocks have about the same resistance to erosion, or if a drainage network has operated for a long time, rivers can form treelike, or *dendritic*, drainage patterns.

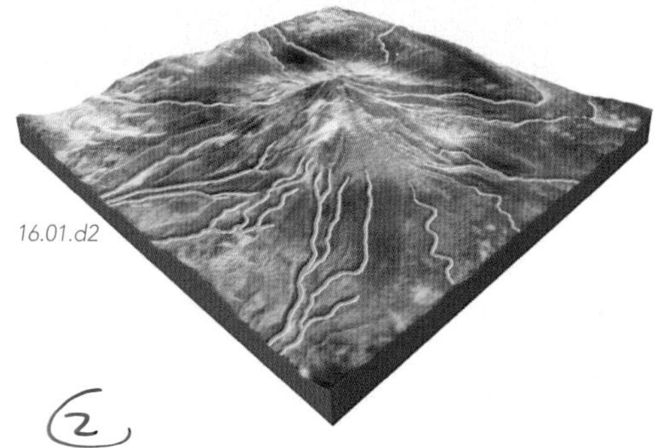

Radial Drainage Pattern—On a fairly symmetrical mountain, such as a volcano or resistant pluton, drainages flow downhill and outward in all directions (i.e., radially) away from the highest area.

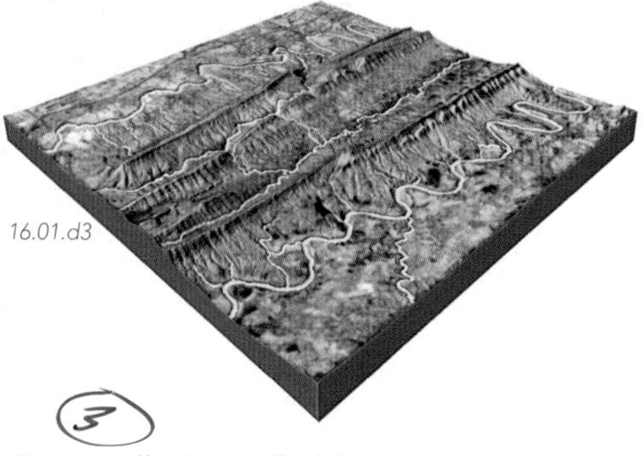

Structurally Controlled Pattern—Erosion along faults, other fractures, or tilted and folded layers can produce a drainage that follows a layer or structure, and then cuts across a ridge to follow a different feature.

North American Drainages

Colors on this map show areas of the land that drain into different parts of the sea. Boundaries between colors are drainage divides, the best known of which is the *continental divide*, separating drainages that flow westward into the Pacific Ocean from those that flow east and south into the Gulf of Mexico. Other drainages flow into the Arctic Ocean, and some drainages in the western United States have interior drainage (they do not reach the sea).

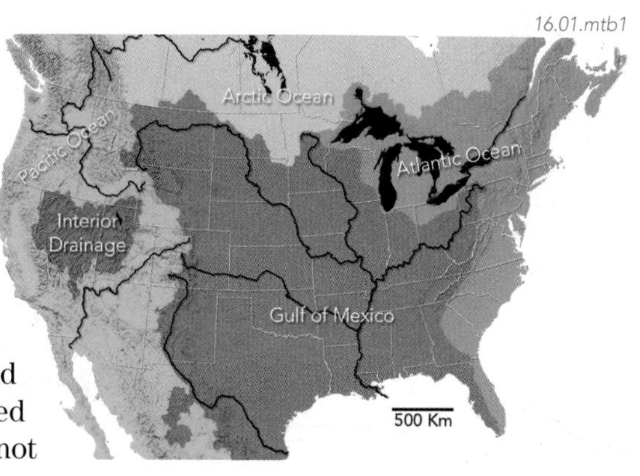

Before You Leave This Page Be Able To

- ✓ Sketch and describe the variables plotted on a hydrograph and what this type of graph indicates.
- ✓ Describe how the shape and slope of a drainage basin affects discharge.
- ✓ Sketch or describe how the distribution of tributaries influences a river's response to precipitation.
- ✓ Sketch three kinds of drainage patterns and discuss what controls each type.

How Do Rivers Transport Sediment and Erode Their Channels?

RIVERS AND STREAMS ERODE BEDROCK and loose material, transporting these as sediment and as chemical components dissolved in the water. The sediment is deposited when the river or stream can no longer carry the load, such as when the current slows or the sediment supply exceeds the river's capacity.

A How Is Material Transported and Deposited in Streambeds?

Moving water applies force to a channel's bottom and sides and is able to pick up and transport particles of various sizes: clay, silt, sand, cobbles, and boulders. The amount of sediment carried by the river is the *sediment load*. Water also transports material chemically dissolved in solution.

1. Fine particles, such as silt and clay (or collectively referred to as mud), can be carried *suspended* in the moving water, even in a relatively slow current.

5. Some chemically soluble ions, such as calcium and sodium, are *dissolved* in and transported by the moving water.

2. Sand grains can roll along the bottom or be picked up and carried down-current by bouncing along the streambed—the process of *saltation*.

3. Larger cobbles and boulders generally move by rolling and sliding, but only during times of high flow. Some of these large clasts can be briefly picked up, but only by extremely high flows.

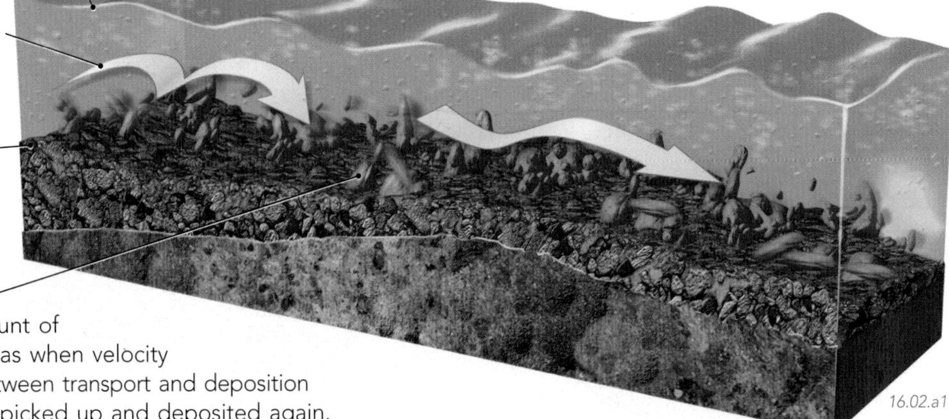

4. Material that is pushed, bounced, rolled, and slid along the bed of the river is the *bed load*. If the amount of sediment exceeds the river's capacity to carry it, such as when velocity drops, then the sediment is *deposited*. The balance between transport and deposition shifts as conditions change, and grains are constantly picked up and deposited again.

16.02.a1

B What Processes Erode Material in Rivers and Streams?

Moving water and the sediment it carries can erode bedrock or softer material that it flows past. Erosion occurs along the *base* and *sides* of the channel and can fragment and remove sediment *within* the channel. The silt, sand, and larger clasts carried by the water enhance its ability to erode.

1. Sand and larger clasts are lifted by low pressure created by water flowing over the grain tops. They can also be pushed up by turbulence. Once picked up, the grains move downstream and collide with obstacles, where they chip, scrape, and sandblast pieces off the streambed by the process of *abrasion*. Abrasion is concentrated on the upstream side of obstructions, such as larger clasts or protruding bedrock.

4. Soluble material in the streambed, such as salt, can be removed by *dissolution*. Most dissolved material in streams, however, comes from groundwater that has leaked into the stream.

16.02.b2

◁ 2. Concentrated erosion can also occur when water and sediment swirl in small depressions, carving bowl-shaped pits called *potholes*. [Susquehanna River, Pennsylvania]

3. Turbulent flow loosens and lifts material from the streambed, especially pieces bounded by fractures, bedding planes, and other discontinuities.

16.02.b1

C How Does Turbulence in Flowing Water Affect Erosion and Deposition?

Water, like all fluids, has *viscosity*—resistance to flow. Viscosity and surface tension are responsible for the smooth-looking surface of slow-moving streams and rivers. As the water's velocity increases, the flow becomes more *chaotic* or *turbulent*, and the water can pick up and move material within the channel.

1. All streams and rivers have turbulent flow to some degree. In very slow-moving streams, the viscosity limits more chaotic flow.

2. Moving water has *inertia* and so tends to keep moving in the same direction unless its motion is perturbed. In many cases, water is able to flow smoothly over somewhat uneven surfaces.

3. As water velocity increases, viscosity is no longer able to dampen chaotic flow, and the water flow becomes more complex, or *turbulent*. As turbulence increases, swirls in the current, called *eddies*, form in both horizontal and vertical directions.

4. Fast-moving water has numerous eddies where flow is not downstream.

5. Near the bottom of the river, upward-flowing eddies can overwhelm gravitational force and lift grains from the channel. Turbulence, in general, increases the chance for grains to be picked up and carried in the flow.

16.02.c1

D How Do Erosion and Deposition Occur in Rivers Confined Within Bedrock?

Many rivers and streams, especially those in mountainous areas, are carved into bedrock. If the bedrock is relatively hard, the shape of the river channel is controlled by the geology. If bedrock consists of softer material, such as easily eroded shale, then it will have less control on the shape and character of the river channel.

Erosion

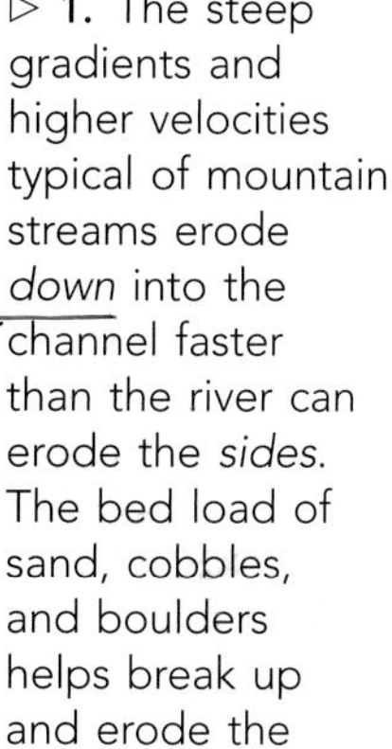

16.02.d1

▷ **1.** The steep gradients and higher velocities typical of mountain streams erode *down* into the channel faster than the river can erode the *sides*. The bed load of sand, cobbles, and boulders helps break up and erode the bedrock channel. Rapid changes in gradient, such as waterfalls, increase water velocity, turbulence, and erosion.

16.02.d3

▷ **2.** As a result, steep bedrock rivers commonly incise deep channels. They can have relatively straight sections, initially controlled by the location of softer rock types, faults, or other zones that are more easily eroded than surrounding rocks. Once formed, such hard-walled canyons are difficult for the river to escape.

Deposition

16.02.d2

◁ **3.** Deposition in bedrock channels occurs where the water velocity decreases, such as occurs along the river banks during flooding or in pools behind rocks or other obstacles. Rocks and sediment constrict this river, forming a pool of less turbulent water upstream. During floods, sediment is deposited in slow-moving eddies on the flanks of this pool, but such sediment is vulnerable to later erosion and is therefore very transient. [Grand Canyon, Arizona]

Before You Leave This Page Be Able To

- ✓ Sketch and describe how a river or stream transports solid and dissolved material.
- ✓ Sketch and explain the processes by which a river or stream erodes into its channel and which sites are most susceptible to erosion.
- ✓ Sketch and describe turbulent flow.
- ✓ Describe some aspects of erosion and deposition in bedrock channels.

How Do River Systems Change Downstream or over Short Time Frames?

RIVER SYSTEMS BECOME LARGER as more tributaries join the drainage network. As a river flows downstream, it generally increases in size, velocity, discharge, and the amount of sediment it carries. A river changes over short time spans, such as after a storm, and from winter to summer and from year to year.

A How Do River Systems Change Downstream?

A river changes in many ways as it flows downhill from its *headwaters*, where it starts, to its *mouth*, where it ends. The flow of a river from high elevations to lower ones is referred to as being *downstream*.

Gradient

1. The profile of most river systems is steep in the headwaters, gradually becoming less steep downstream toward the mouth. The steepness is also called the *gradient*, which is defined as the change in elevation for a given horizontal distance.

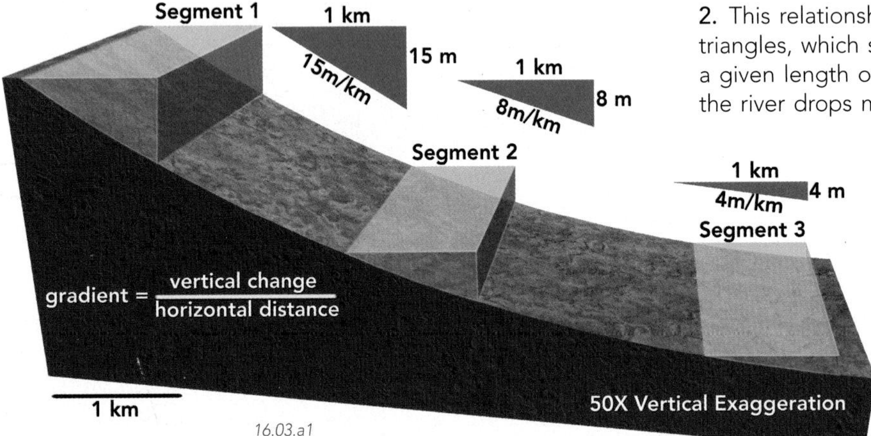

16.03.a1

2. This relationship is represented by the blue triangles, which show how much the river drops for a given length of river. A steeper gradient means the river drops more over the same horizontal distance. Gradient is expressed as meters per kilometer, feet per mile, degrees, or as a percentage (e.g., 4%). Here, gradient is calculated for three segments. It varies from 15 m/km to 4 m/km and decreases downstream. The vertical scale of the triangles is not the same as the horizontal scale.

Channel Size, Water Velocity, Discharge, and Sediment Load

3. Rivers erode bedrock and other materials and then transport the sediment down the river. Sediment can be deposited anywhere along the way or can be carried all the way to the mouth of the river. The river system shown here has a main river fed by three main tributaries, labeled T1, T2, and T3. Small graphs around the map plot how parameters change down the river, from the headwaters (H), past each tributary (T1, T2, and T3), to the start of a delta (D), and the river's mouth (M).

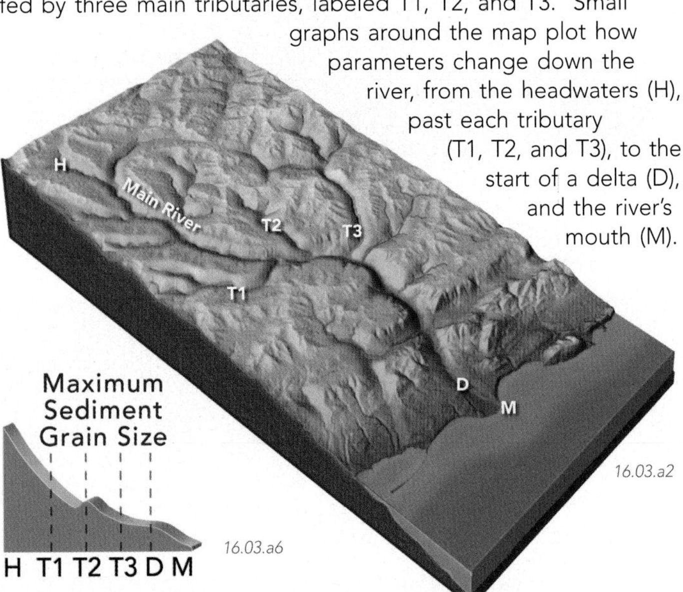

16.03.a2

16.03.a6

△ 4. As the gradient of the river decreases from the headwaters to the mouth, the *maximum size of sediment* that the river carries *decreases*. In other words, coarse material is more common in the headwaters than it is near the mouth.

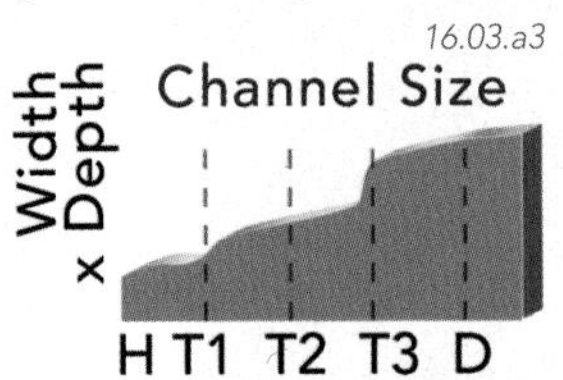

16.03.a3

◁ 5. There is an *increase in the overall size* of the channel, as represented by a cross section from side to side across the channel. Specifically, size means the *cross-sectional area* of the channel, obtained by multiplying the channel's width times its depth.

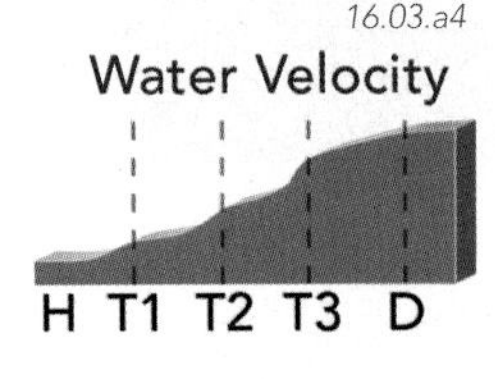

16.03.a4

◁ 6. The *velocity* of water flow *increases* downstream, as a higher volume of water allows the water to flow more easily and faster through the channel.

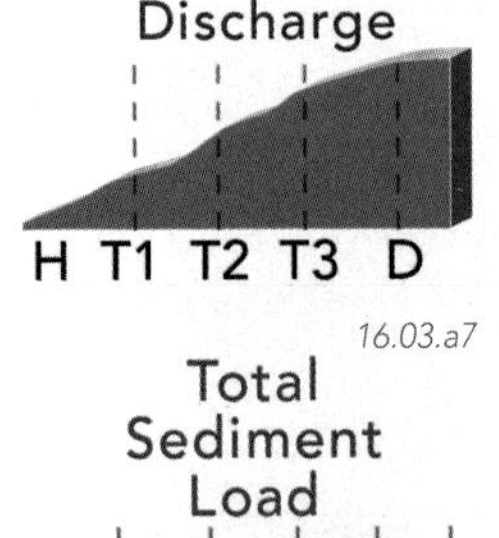

16.03.a5

◁ 7. Since the cross-sectional area and velocity of the channel both *increase*, so does the total *volume* of water flowing through the river. The volume of water flowing through any part of the river per unit of time is called the *discharge* and is calculated by multiplying the velocity times the cross-sectional area.

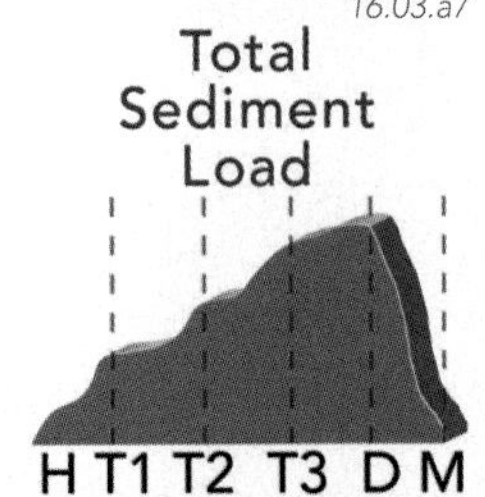

16.03.a7

◁ 8. The total amount of sediment that the river is carrying, the *sediment load*, *increases* downstream, until large amounts of sediment begin to be deposited within the delta and at the mouth of the river.

B What Is the Relationship Between Water Flow and Transported Sediment?

A river or stream can carry sediment only up to a certain size. Also, at a given flow rate, a river is capable of transporting only a certain *amount* of sediment, which is called its *capacity*. Normally, a river is carrying far less sediment than its capacity. As velocity decreases, so does capacity—a river can carry less sediment as it slows down.

▷ This graph shows the relationship between stream velocity and the size of the particles that can be carried by different modes of transport. The vertical bands of color indicate different grain sizes, and the inclined lines indicate whether sediment of that size is being carried in suspension, is being transported on the bottom of the river bed, or is being deposited.

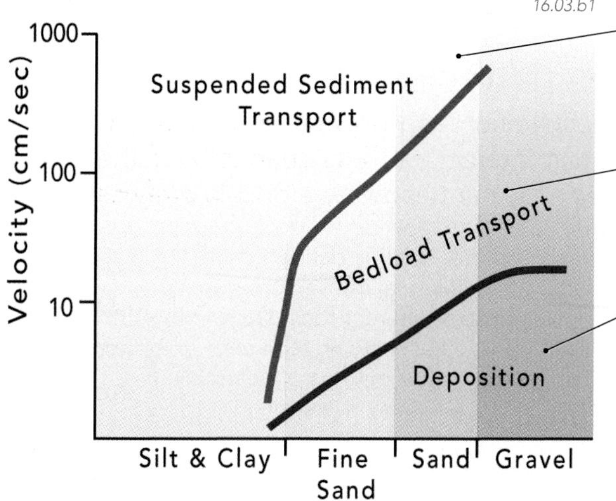

At high velocities (above 100 cm/sec), clay, silt, and sand can be carried *suspended* (floating and drifting) in the water. Those grain sizes extend above the red line.

At moderate velocities (100 cm/sec), silt and clay remain suspended, but sand and gravel slide, roll, or bounce along the riverbed, a mode of travel called *bedload transport*.

At low velocities (below 10 cm/sec), gravel and sand remain at rest on the riverbed or are deposited if a sediment-carrying river slows down to these velocities. These grain sizes plot below the blue line. Only silt, clay, and fine sand are transported, with silt and clay in suspension and fine sand in the bedload.

C How Does River Behavior Vary over Time?

The amount of precipitation, snowmelt, and influx from springs and groundwater varies, both during a single year and over longer timescales of decades to centuries. A river needs to be viewed at all these timescales.

▷ 1. The amount of water flowing in a river or stream can vary throughout the year. For this river and for the year shown, discharge is highest during spring snowmelt. The highest value on the plot is the *peak discharge*.

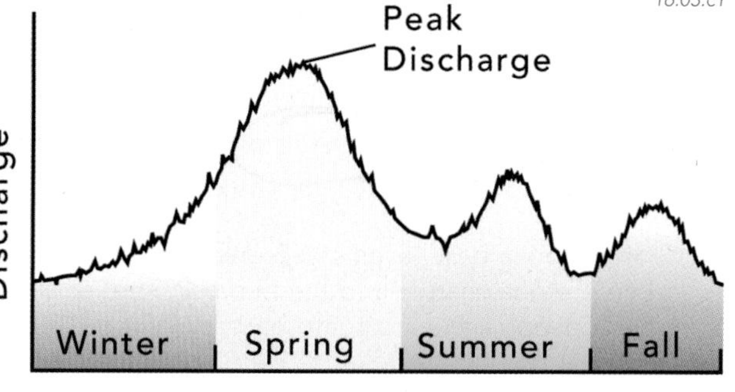

2. A stream or river that flows all year, like the one represented by the graph to the left, is a *perennial stream*. Because no place has rain all of the time to keep a stream flowing, some water in a perennial stream must be supplied by groundwater flow into springs, by a melting snowpack, by a lake, or by some combination. Some streams do not flow during the entire year, but only during rainstorms and spring snowmelt. Such a stream is an *ephemeral stream*, like the one shown above that flows a few weeks a year.

3. One way to examine stream behavior is by the amount of sediment transported past a certain stream segment over time.

▷ 4. In the short term, the amount of sediment a river transports, represented by the jagged line, varies rapidly and increases during storms and other short-duration events.

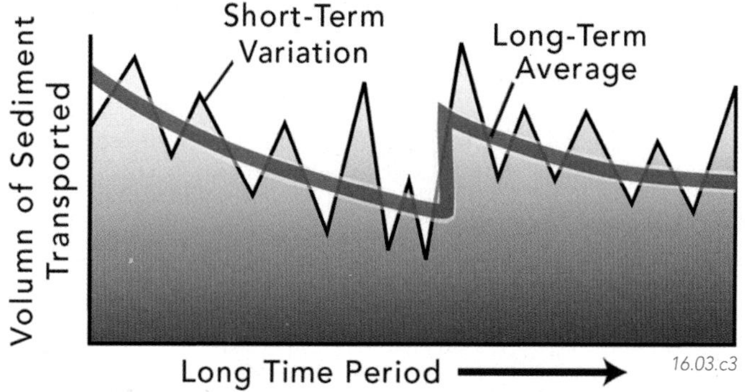

5. Over the long term (represented by the blue line), the amount of sediment transported on this river typically decreases with time, perhaps due to slow climate change or tectonic shifts. Interrupting these slow decreases is a sudden increase in the amount of sediment (the upward jump in the blue line) due to rapid climate change, tectonism, or human interference that modifies or resets long-term trends.

Before You Leave This Page Be Able To

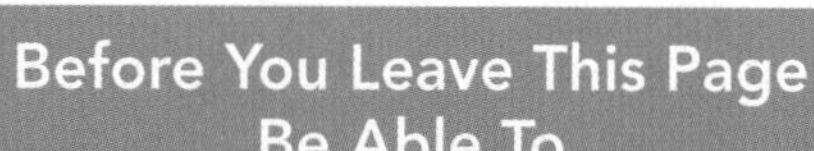

- ✓ Describe and sketch how to calculate a gradient for a river.
- ✓ Describe how gradient and other parameters change downstream.
- ✓ Describe how velocity relates to sediment size and capacity.
- ✓ Describe why discharge, or any other parameter, might exhibit changes at several different timescales.

What Factors Influence Profiles of Rivers?

RIVER SYSTEMS HAVE DIFFERENT GEOMETRIES, both in map view and when viewed from the side or in profile. Rivers have diverse settings, origins, and ages, and they respond to perturbations in their environment by eroding their channels and banks, by depositing sediment, and by changing their gradient.

A What Is the Shape of a River's Profile?

Rivers are *dynamic systems* driven by precipitation and gravitational forces. They respond to many factors that influence how the river operates and how it interacts with its channel and the adjacent landscape. Over time, most rivers attain a profile that is steeper near the headwaters and is progressively less steep downstream.

1. The idealized profile of a river is represented by the side of this block. The profile is steeper (has a higher gradient) near the headwaters of the river.

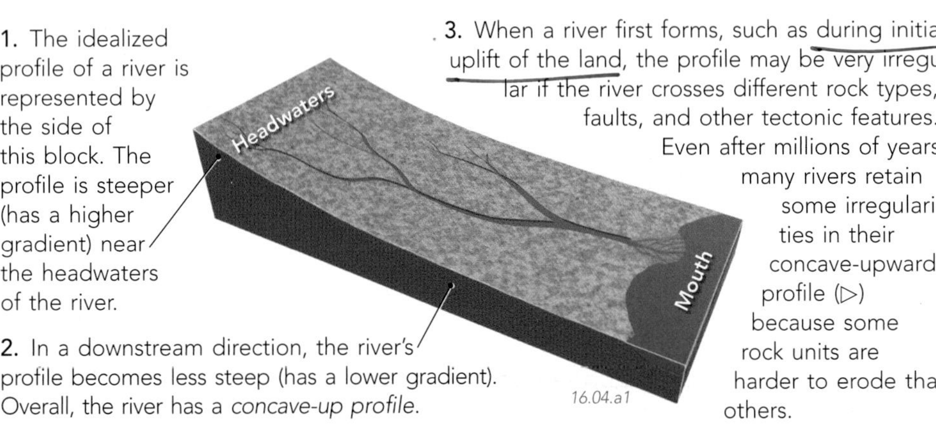

16.04.a1

2. In a downstream direction, the river's profile becomes less steep (has a lower gradient). Overall, the river has a *concave-up profile*.

3. When a river first forms, such as during initial uplift of the land, the profile may be very irregular if the river crosses different rock types, faults, and other tectonic features. Even after millions of years, many rivers retain some irregularities in their concave-upward profile (▷) because some rock units are harder to erode than others.

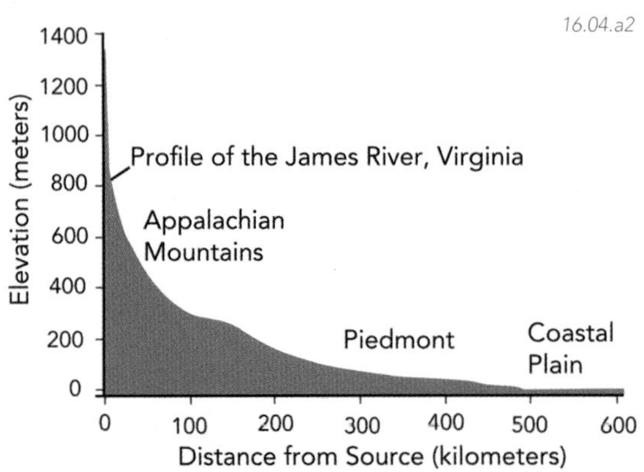

B What Controls the Profiles of Streams and Rivers?

Rivers and other processes erode mountains and carry the sediment downhill, eventually depositing it in a basin or along the sea. The lowest level to which a river can erode is its *base level*. The base level controls the topography along a river, how a river develops over time, and how it responds to change.

1. This terrain shows a typical drainage system consisting of mountainous headwaters, mid-elevation foothills, and a broad, low-elevation plain, ending at a shallow inland sea.

2. High above base level, steep gradients in the mountains cause streams and rivers to erode sharply into the bedrock. The terrain appears rough and may include deep canyons cut into bedrock.

3. Foothills in front of the mountains also experience erosion, but have intermediate gradients and generally appear less rough.

4. Closer to base level, rivers and streams on the broad plain have a much lower gradient, and the surrounding landscape has less relief and appears relatively smooth. This plain has low relief because either it has been eroded down or its low parts have filled with sediment. In this case, it is some of both.

Headwaters
Foothills
Broad Plain
River
Sea

5. A stream or river cannot erode below sea level. In this terrain, sea level represents the *base level*. In general, base level for a river is the ocean, a lake, or the bottom of a closed land basin (with internal drainage). For most river systems, the *ultimate base level* is the ocean.

6. As shown by the side of the block, variations in roughness of the landscape reflect the decrease in gradient from the mountains to the broad plains. A profile down the channel of any stream or river in this area is less irregular than the rough topography defined by the ridges and canyons. Most rivers and streams have a fairly smooth, *concave-up* profile.

16.04.b1

C What Factors Influence or Change Stream Profiles?

Streams and rivers generally do not achieve equilibrium because some rocks are more difficult to erode than others and because Earth is a dynamic planet, with frequent changes in tectonics, sea level, and climate.

Rock Type

As a river flows over different kinds of rocks, its ability to erode a channel is influenced by the type of rock over which it flows. Soft rocks erode more easily than hard rocks.

16.04.c1

◁ In unconsolidated sediment and easily eroded rocks, such as shale, the river can create a smooth, equilibrium-like profile because there are no major obstructions. The profile has a concave-upward shape.

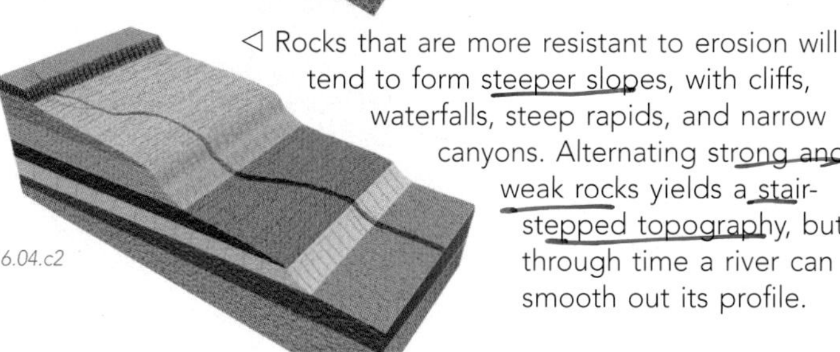

16.04.c2

◁ Rocks that are more resistant to erosion will tend to form steeper slopes, with cliffs, waterfalls, steep rapids, and narrow canyons. Alternating strong and weak rocks yields a stair-stepped topography, but through time a river can smooth out its profile.

Tectonics

Tectonic forces can cause *uplift* and *subsidence* of an entire region or can occur differentially, affecting one part of the region more than other parts.

16.04.c3

◁ Differential *subsidence* can flatten or steepen gradients, depending on where it occurs. In this example, subsidence occurred beneath the mountains, flattening the gradient and causing widespread deposition as stream velocity decreased and the river lost capacity.

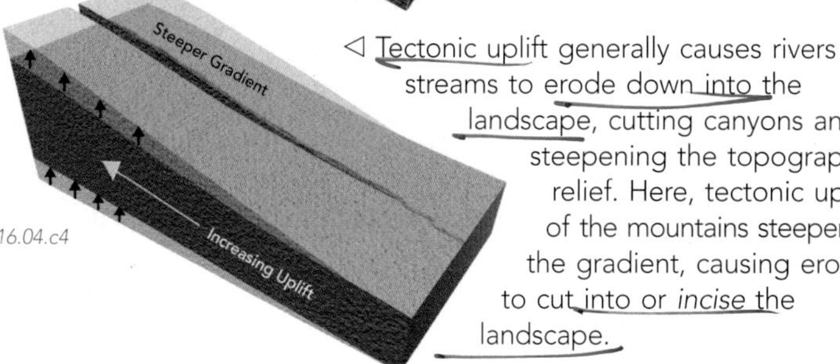

16.04.c4

◁ Tectonic uplift generally causes rivers and streams to erode down into the landscape, cutting canyons and steepening the topographic relief. Here, tectonic uplift of the mountains steepened the gradient, causing erosion to cut into or *incise* the landscape.

Sea Level

Sea level is the ultimate base level for rivers that empty into the ocean. Changes in sea level will change the location of the shoreline and the elevation of base level.

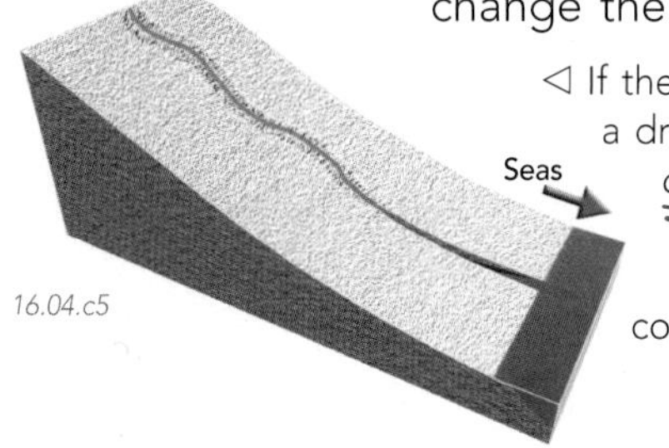

16.04.c5

◁ If the base level is *lowered*, such as by a drop in sea level, the river will *downcut* to try to match the new base level. In this example, erosional incision begins at the coast and works its way upstream.

16.04.c6

◁ If the base level *rises*, such as during a rise in sea level, the river will erode inland but deposit sediment along the coastline's new position, as the river tries to achieve a new equilibrium profile.

Climate

Rivers respond to changes in climate, such as an increase or decrease in rainfall or temperature. Under wet conditions, slopes will have more vegetation and so can hold soil, but increased discharge allows streams to carry sediment away, beveling the hills more than during dry periods.

16.04.c7

Stability of Conditions

If conditions, such as climate, remain stable, a river may approach an *equilibrium profile*. When a river is in steady state, there is a balance between the supply of sediment and the amount the river can carry. The channel becomes stable, neither eroding nor depositing material. Such a stream or river is called a *graded stream*.

16.04.c8

Before You Leave This Page Be Able To

- ✓ Sketch and describe the typical profile of a river.
- ✓ Describe the concept of base level and how it is expressed in a typical mountain-to-sea-landscape.
- ✓ Sketch or describe factors that influence a river's profile.

Why Do Rivers and Streams Have Curves?

ALL RIVERS HAVE CURVES OR BENDS, ranging from gentle deflections to tightly curved, but graceful, meanders. Why are rivers curved? What is inherent in the operation of a river that causes it to curve? Curves and bends are unavoidable because of processes that shape rivers.

A What Is the Shape of River and Stream Channels in Map View?

All rivers have curves or bends, but not all bends are the same. Some are gentle, open arcs, where the river veers slightly to one side and then the other, whereas others are tight loops. The shape of a river in map view can be thought of as having two main variables: whether there is single versus multiple channels and how curved the channel is. The amount that a channel curves for a given length is its *sinuosity*.

16.05.a1

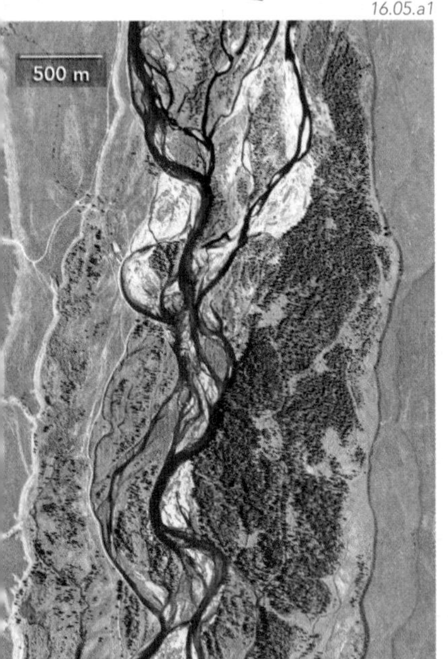

16.05.a2

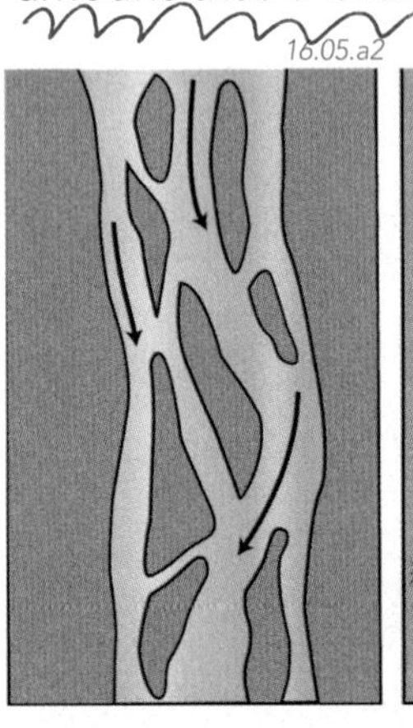

△ *Braided rivers* are characterized by a network of interweaving, sinuous channels, but the river can be fairly straight overall.

16.05.a3

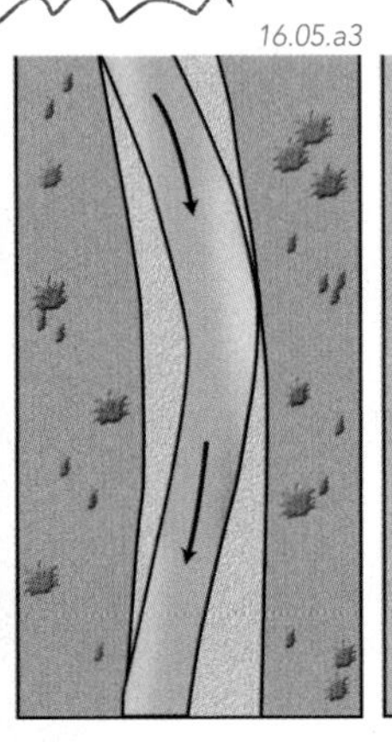

△ Many rivers consist of a single channel that is gently curved. This type of river is referred to as having *low sinuosity*.

16.05.a4

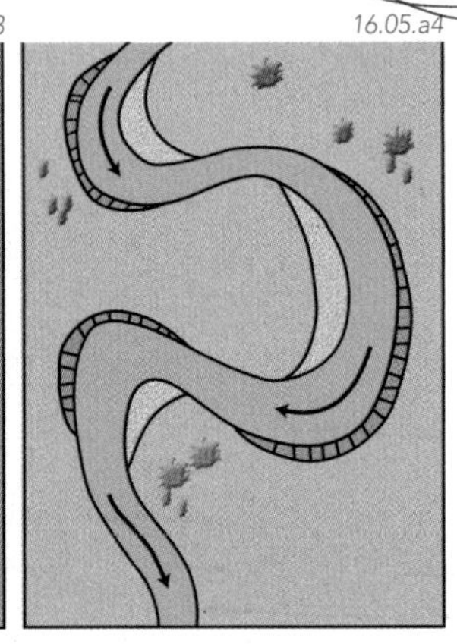

△ *Meandering rivers* have channels that are very curved, commonly forming tight loops. Such rivers have high sinuosity, and this type of bend is a *meander*.

16.05.a5

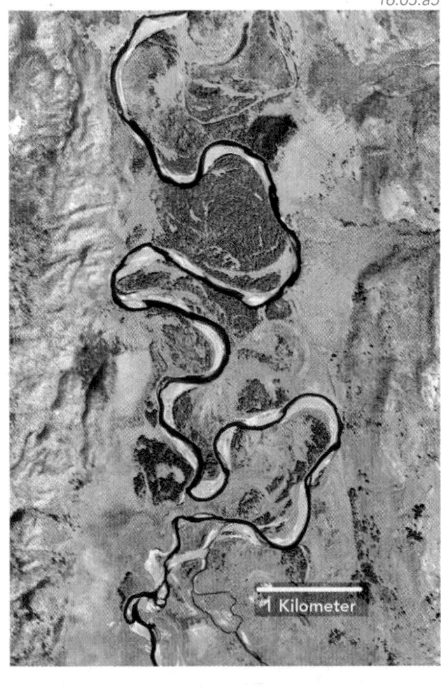

B What Processes Operate When a River Meanders?

River channels in alluvium and other soft materials generally do not have long straight segments but instead flow along sinuous paths. Curves or bends, called *meanders*, cause differences in water velocity in the channel and reflect a balance between deposition and erosion, as illustrated below for a meandering river.

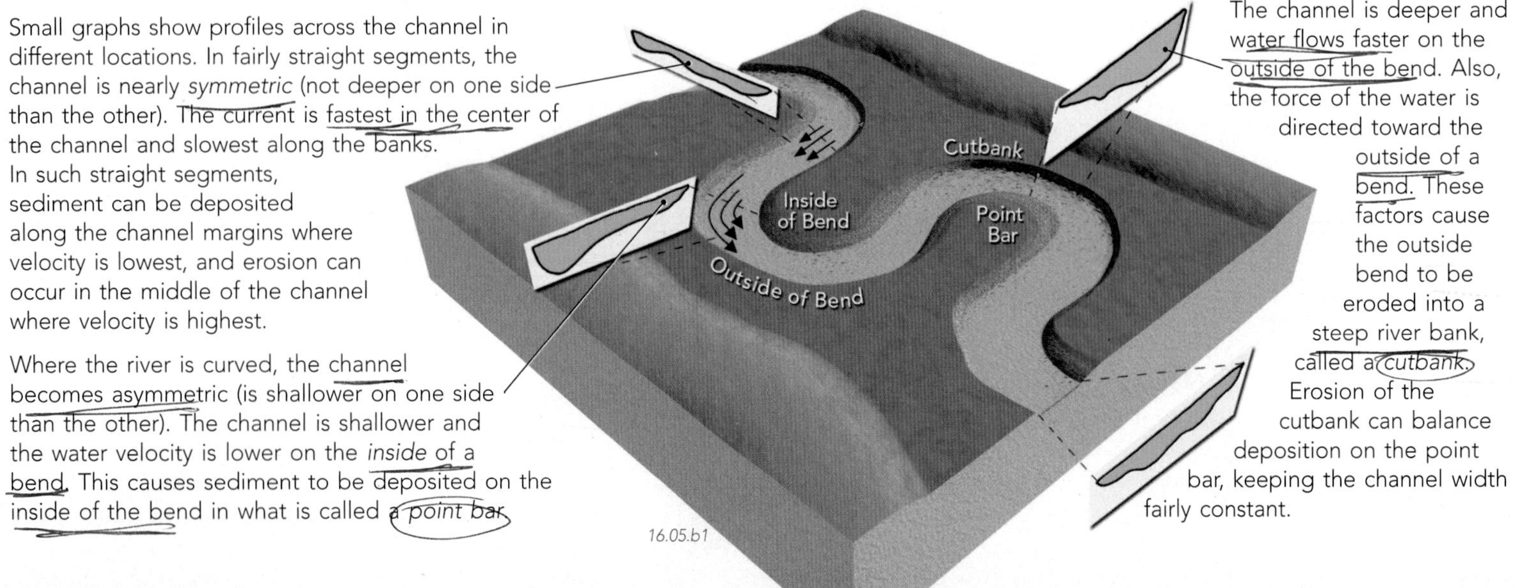

Small graphs show profiles across the channel in different locations. In fairly straight segments, the channel is nearly *symmetric* (not deeper on one side than the other). The current is fastest in the center of the channel and slowest along the banks. In such straight segments, sediment can be deposited along the channel margins where velocity is lowest, and erosion can occur in the middle of the channel where velocity is highest.

Where the river is curved, the channel becomes asymmetric (is shallower on one side than the other). The channel is shallower and the water velocity is lower on the *inside of a bend*. This causes sediment to be deposited on the inside of the bend in what is called a *point bar*.

The channel is deeper and water flows faster on the outside of the bend. Also, the force of the water is directed toward the outside of a bend. These factors cause the outside bend to be eroded into a steep river bank, called a *cutbank*. Erosion of the cutbank can balance deposition on the point bar, keeping the channel width fairly constant.

16.05.b1

C How Do Meanders Form and Move?

Meanders are landforms produced by migrating rivers and are extremely common in rivers that have low gradients. Meanders have been extensively studied in the field and simulated in large, sand-filled tanks. In the laboratory, water is initially directed down a straight channel in fine sand. Almost immediately, the water begins to transform the straight channel into a sinuous one, similar to the sequence shown below.

16.05.c1–6

1. A curve starts to form when a slight difference in roughness on the channel bottom causes water to flow faster on one side of the channel than on the other.

2. The side of the channel that receives faster flow erodes faster, creating a slight curve. The faster moving current slightly excavates the channel bottom, deepening the *outside* of the bend, forming deeper areas called *pools*.

3. The overall discharge in the river is constant, so the deeper channel on the *outside* of a bend takes more water, leaving less water for the other side. The water on the *inside* of the bend becomes shallower and slower.

4. The sediment carried by the slower water on the inside of the bend is dropped and deposited on a *point bar*.

5. Erosion scours the opposite (outside) band of the channel, forming a *cutbank*.

6. Through this process, each meander begins to preferentially erode its banks toward the outside. This causes the river to migrate toward the *sides* and *downstream*, as shown by the small yellow arrows.

7. Once formed, a curve continues to affect the flow by causing faster flow and increased erosion on the *outside* of the bend. Some secondary currents develop in the bend area and further excavate the pools, speeding flow and enhancing the cutbank.

8. As meanders migrate back and forth across the lowlands, they continuously erode and deposit the loosely bound floodplain sediment. This is the main way in which a *floodplain* forms, and the old meanders remain as *scars* on the floodplain.

9. Meanders migrate until they encounter a resistant riverbank, until the volume and velocity of flow drop too low for erosion to continue, or until two parts of a meander intersect.

10. Meanders sometimes join as they migrate toward each other, in the direction of the yellow arrows. This cuts off the meander.

11. The narrow neck of a looping meander can also get cut off during a flood event, when the river rises above the channel and across the floodplain, connecting two segments of the river. The part of the meander that is abandoned is a *cutoff meander*.

12. Cutoff meanders formed in either way (10 or 11) can become filled with water, forming isolated, curved lakes, called *oxbow lakes*.

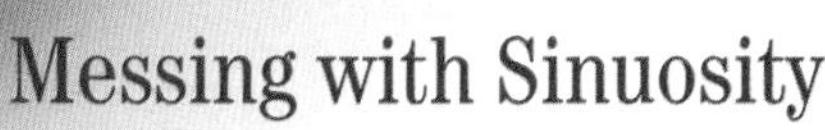

Messing with Sinuosity

Rivers and streams have attained their characteristic sinuosity through natural processes. Their sinuosity represents the interplay between variations in channel depth, water velocity, erosion, deposition, and transport of sediment. In many cases, humans upset this balance by straightening rivers and eliminating their natural variability. These engineering solutions often cause trouble downstream because they upset the dynamics and equilibrium of the system. Rivers that have been channelized may exit the channelized segment with a higher velocity, lower sinuosity, and less sediment than is natural. Areas downstream of the channelized segment, therefore, can experience extreme erosion and destruction of river-bank property. [Alps]

16.05.mtb1

Before You Leave This Page Be Able To

- ✓ Sketch and describe the difference between braided, low-sinuosity, and high-sinuosity rivers.
- ✓ Sketch or describe how flow velocity and channel profile vary in a meandering river, and what features form along different parts of bends.
- ✓ Sketch or describe the evolution of a meander, including how a cutoff meander forms and how it can lead to an oxbow lake.

What Features Characterize Mountain Rivers and Streams?

MOST LARGE RIVER SYSTEMS originate in mountains and are fed by rain, snowmelt, and springs. Mountain streams are steep and actively erode the land with turbulent, fast-moving water. Such erosion produces steep-sided, narrow channels and other landforms that reflect this high-energy environment.

A What Landforms Characterize the Headwaters of Rivers and Streams?

Mountain river systems begin in areas of relatively high relief and, in many cases, high elevation. In such settings, moving water is energetic, wearing rock down and sculpting the bedrock into landscapes with moderate to high relief. Steep streams and rivers are capable of carrying sediment out of the mountains.

Channel Formation

16.06.a1

◁ As water flows over the surface, it accumulates in natural cracks and low spots, such as these small channels, rather than spreading uniformly across the land. [Norway]

16.06.a2

Concentrated flow erodes or dissolves materials, especially those that are weak or loose, eventually carving a small channel or *gully*.

Once formed, a channel captures additional runoff within its small *drainage basin*, and the increased flow leads to further erosion and deepening of the channel.

Channels occur at all scales. Microscopic channels feed into small channels that feed into larger ones, ultimately forming a stream.

Landforms in the Headwaters of Rivers and Streams

16.06.a3

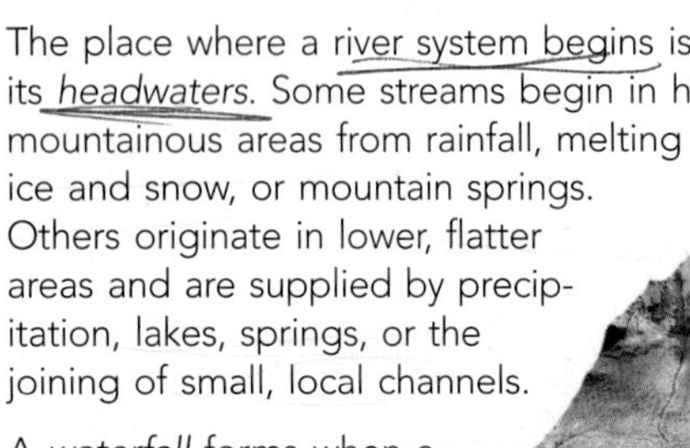

The place where a river system begins is called its *headwaters*. Some streams begin in high mountainous areas from rainfall, melting ice and snow, or mountain springs. Others originate in lower, flatter areas and are supplied by precipitation, lakes, springs, or the joining of small, local channels.

A *waterfall* forms when a stream's gradient is so steep that water cascades over a cliff or ledge. Cliffs and ledges typically develop where a hard, erosion-resistant rock type impedes downcutting by the stream. [Gullfoss, Iceland] ▽

Rapids

Lake

1 Km

16.06.a4

△ *Lakes* are common in mountains where water is impounded by some obstruction, such as a landslide, or water fills a natural low spot. If a lake is created by a constructed dam, it is a *reservoir*.

A *rapid* is a segment of rough, turbulent water along a stream or river. Most rapids develop when the gradient of a river steepens or the channel is constricted by narrow bedrock walls, large rocks, or other debris that partially blocks the channel. Many rapids form where tributaries have deposited fans of debris that crowd or clog the main channel. These obstructions cause water to flow chaotically over and around obstacles, creating extreme turbulence and big rapids. [Grand Canyon, Arizona] ▷

16.06.a5

16.06.a6

B What Landforms Form Along Mountain Rivers and Streams?

As mountain rivers flow toward lower elevations, they interact with tributaries and commonly decrease in gradient as they pass through foothills or mountain fronts. In response, they form other types of landforms.

16.06.b2

1. Many mountain streams and rivers cut down into, or *incise*, bedrock. Early in their history, many rivers incise steep-walled notches and canyons. A canyon is narrow if downward incision is faster than widening of the canyon walls by landslides and other types of slope failure and by erosion along tributaries.

16.06.b1

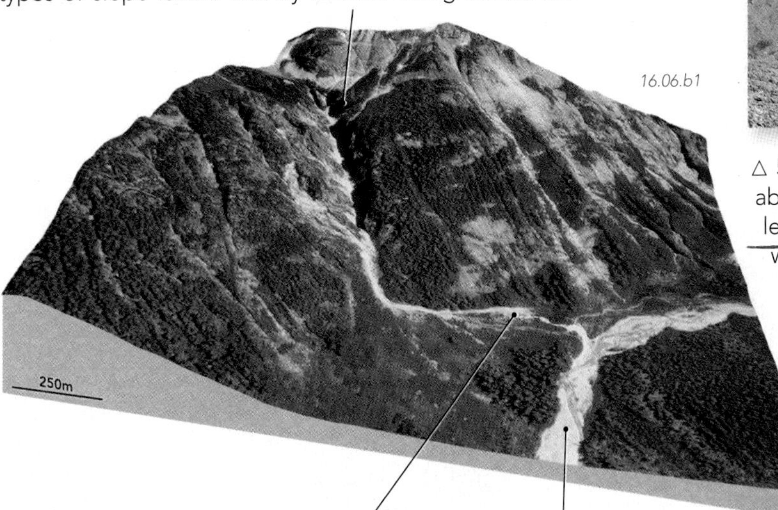

16.06.b3

△ 5. Where a steep, narrow drainage abruptly enters a broader, more gentle valley, coarse sediment carried by running water or by muddy debris flows piles up just below the mouth of the drainage, forming an *alluvial fan*. Deposition occurs here because of the decrease in gradient and the less confined nature of the channel, both of which decrease the velocity of moving water and mud. [Death Valley National Park, California]

△ 2. This narrow canyon is cut into limestone layers that are resistant to erosion in a dry climate. [Buckfarm Canyon, Arizona]

16.06.b4

◁ 3. Side tributaries play a key role in the downstream variations in the gradient, morphology, and turbulence of a mountain river. Tributaries carry sediment and deposit some of it where the tributary and main drainage meet. This sediment can *constrict* the channel, causing a *rapid* at the constriction and backing up and slowing water above the rapid, forming a *pool*. [Grand Canyon, Arizona]

4. When they reach less confined spaces, mountain streams and rivers commonly spread out in a network of sediment-filled *braided channels*. These channels are not strongly incised, so the river spreads out and deposits its sediment along its channel and over a broad plain. [Waiapu River, New Zealand] ▷

16.06.b5

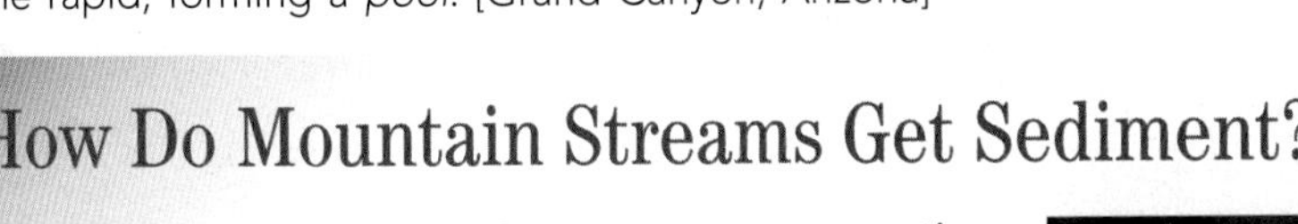

How Do Mountain Streams Get Sediment?

Mountain rivers and streams are energetic primarily because their channels have steep gradients. Erosion dominates over deposition, forming deep *V-shaped valleys* with waterfalls and rapids. Steep valley walls promote landslides and other types of slope failure that widen the canyon and deliver material to the river for removal (▷). Soil on the slopes slides downhill toward the drainage. Tributaries carry debris flows that scour their channels, providing more sediment. Sediment in mountain rivers ranges from car-sized boulders down to silt and clay. Larger clasts start out angular, but begin to round within the turbulent waters. [Tibet–Nepal border region]

16.06.mtb1

Before You Leave This Page Be Able To

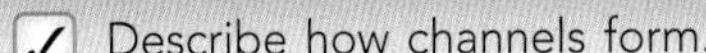

- ✓ Describe how channels form.
- ✓ Describe some of the landforms associated with the headwaters of mountain rivers and streams.
- ✓ Describe what conditions result in a narrow canyon.
- ✓ Describe why sediment is deposited along mountain fronts in alluvial fans.
- ✓ Describe how mountain streams get their sediment.

What Features Characterize Braided Rivers?

MANY RIVERS AND STREAMS ARE BRAIDED SYSTEMS, with a network of channels that split and rejoin, giving an intertwined appearance. Braided rivers generally have a plentiful supply of sediment and steep to moderate gradient, and typically carry and deposit rather coarse sediment. Braided rivers can migrate across broad plains, coating them with a veneer of sediment.

A What Conditions Lead to Braided Streams and Rivers?

Braided streams and rivers are most common in flat-bottomed valleys nestled within mountains and on broad, sloping plains that flank such mountain ranges. They can also form farther from the mountains, in areas where the sediment supply is close to overwhelming the river's capacity to carry it.

Many braided rivers drain from high mountains, such as these, modeled after the South Island of New Zealand.

The Southern Alps of New Zealand are an actively uplifted and steep range. Glaciers, steep slopes, and locally heavy precipitation in the headwaters of the rivers contribute abundant sediment to the streams and rivers. ▷

Braided rivers deposit sediment within and beside their shallow channels and can escape their channels, especially during floods. Sediment in the riverbank is not cemented or otherwise tightly held together, so the material is easy to erode and redistribute, and the river can more easily change position. As the channels migrate back and forth across the broad plain, they cover the broad, low-relief area with a layer of river-deposited sediment. [Denali National Park, Alaska] ▽

Braided streams form where there are steep gradients, a plentiful supply of coarse sediment, and conditions that produce variable flows. In this close-up view, individual channels are braided at various scales, but the overall path of the river is fairly straight. ▽

B What Type of Sediment Does a Braided River Deposit?

Braided rivers are characterized by a wide range of sediments, more so than meandering rivers. Braided rivers are energetic and can carry and deposit coarse gravels and sands in addition to finer materials.

16.07.b1

Braided rivers form when the river has a relatively high sediment load dominated by sand and larger sediment. Sediment is constantly picked up in one place and deposited in another.

16.07.b2

This braided river shows numerous braided channels. The river is clogged with sediment. [Waimakariri River, New Zealand]

16.07.b3

Sand and gravels are the dominant clasts in this part of the river, but braided rivers also carry finer materials, such as mud and silt derived from glaciers and other sources. [Waiapu River, New Zealand]

16.07.b4

This braided-river plain in Tibet contains large, partially rounded boulders in addition to finer sediment. The steep Himalaya range in the distance exceeds 6 km (20,000 ft) in elevation.

Making and Investigating Braided Rivers in the Laboratory

One way geologists and engineers study rivers is to make small-scale versions or models in large water tanks in a laboratory. These tanks can be several meters wide and tens of meters long, and are sloped so that the water flows downhill. The tanks are loaded with sediment, usually sand, silt, and mud, but sometimes glass beads or other materials. Valves are opened to allow water to enter the high side of the tank and flow toward the low end. Geologists then observe the small-scale river that develops, investigating the processes that occur and the features that form. Different variables, such as slope, sediment supply, and consistency of flow, can be specifically varied or controlled to isolate how each factor affects the dynamics of the river system.

The sequence of images here shows successive stages during an experiment in a 2 × 15 meter tank at the National Center for Earth-surface Dynamics in Minneapolis. In this experiment, a braided river developed early on (far left), but became progressively less braided as alfalfa seeds embedded in the sediment sprouted and grew more dense. These experiments indicate that riverside vegetation plays a key role in stabilizing river banks, and can actually influence whether a river remains braided.

16.07.mtb1–6

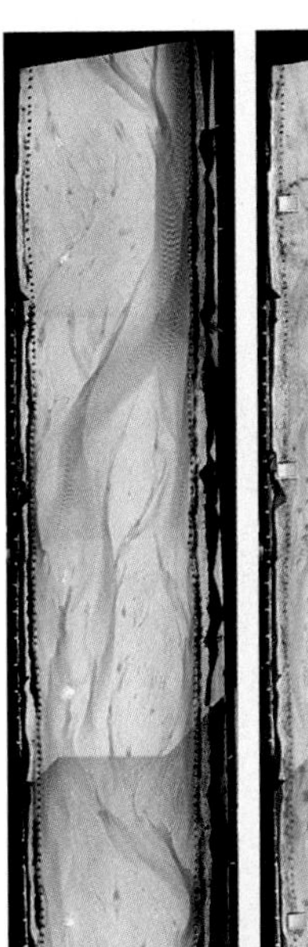

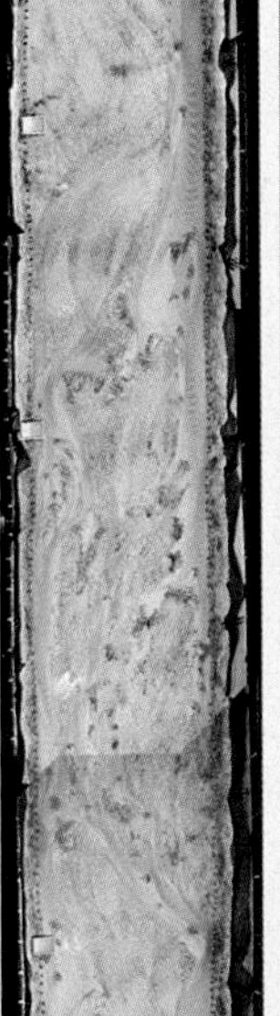

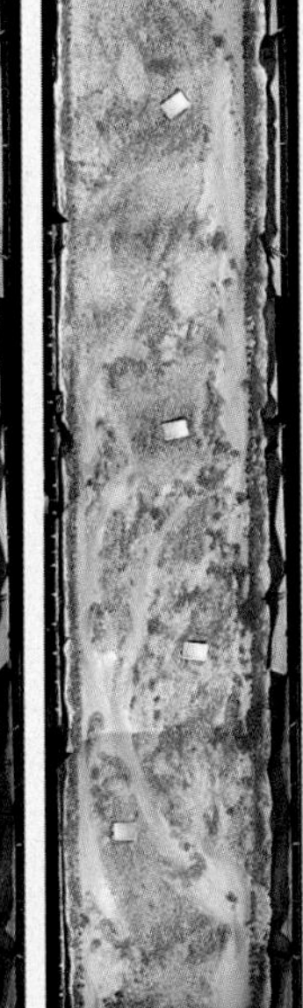

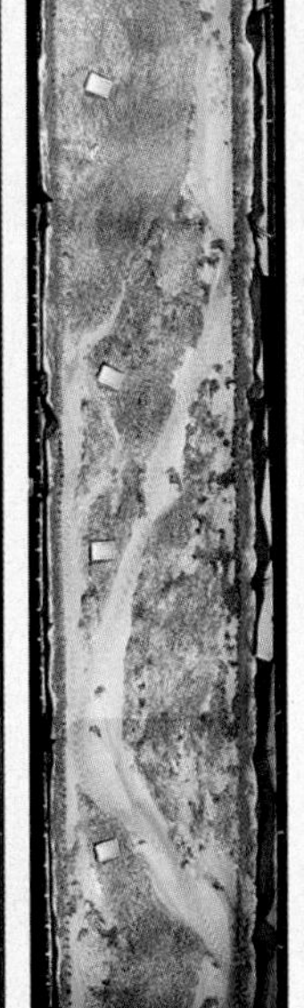

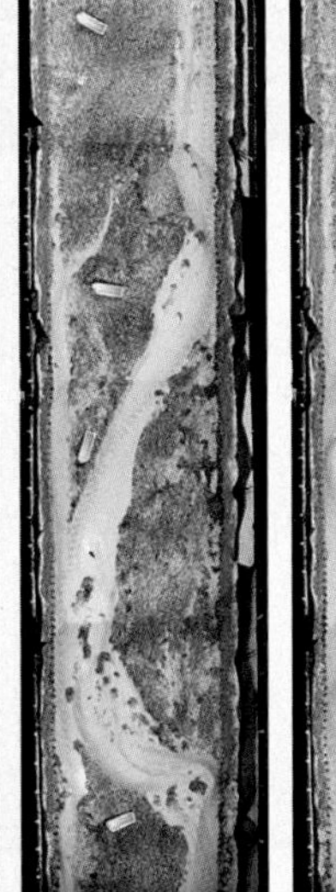

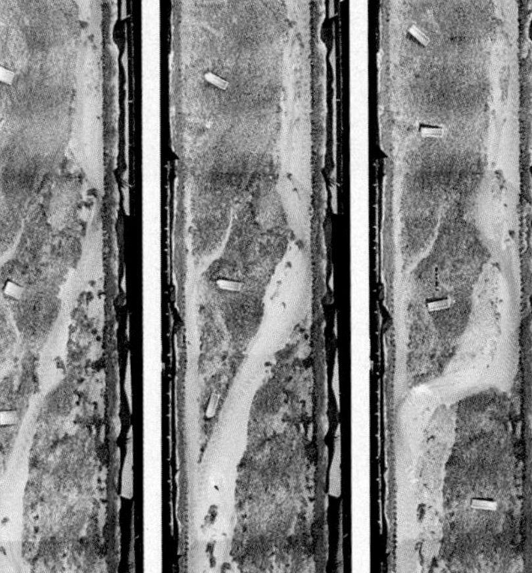

Before You Leave This Page Be Able To

- ✓ Describe the characteristics and setting of braided rivers and streams.
- ✓ Describe the types of sediment that braided rivers carry and deposit.
- ✓ Describe how and why river processes are investigated in laboratory tanks.

What Features Characterize Low-Gradient Rivers?

IF A RIVER SYSTEM CROSSES AREAS of low relief, the gradient of its channel decreases and the river may spread out once it is no longer confined by a narrow valley. Sediments transported and deposited on low-relief plains are mostly clay to sand size, but can include fine gravels. The landforms reflect the interaction of river velocity and sediment size with the more gentle landscape.

A What Landforms Characterize Rivers with Low Gradients?

Many rivers flow across plains that have gentle overall slopes. Such rivers reflect their environs, being dominated by the erosion, transport, and deposition of relatively fine-grained sediment. The features characteristic of these single-channel rivers occur at all scales, from those along small creeks to those along the mighty Mississippi River. Features include meanders, floodplains, and low river terraces.

One Main Channel

Rivers on gentle plains usually occupy a single channel rather than being braided. This single-channel characteristic is linked to the gentle downstream gradient of the river and its floodplain. Notice the low gradient river here occurs on a gentle plain within a mountainous region, so it is important to focus on the characteristics of the river rather than its surrounding environment. Farther upstream, this river is confined to a narrow and deep bedrock canyon. [Animas River, Colorado]

Floodplain

All rivers on gentle plains have floodplains beside the channel. Floodplains represent the area covered with water when the river floods out of its channel.

16.08.a1

Animas River

Meanders

Floodplain

Oxbow Lakes

1 Km

River Terraces

Many rivers have older, stranded floodplains, called *terraces,* perched above and outside the current floodplain. It is common to find matching terrace levels on either side of the existing floodplain. This particular stretch of river lacks obvious terraces.

Point Bars

Meandering rivers often have arcuate deposits of sand and gravel that parallel the inside bend of a meander. Such a deposit is called a *point bar.*

Meanders

Rivers on gentle plains typically flow in dramatically curved paths. The degree to which the single channel is curved varies from rare straight segments to sinuous curves called *meanders.*

Meander Scars and Oxbow Lakes

Meandering rivers leave behind arcuate scars on the landscape, as low curved ridges, lines of vegetation, or curved dry or water-filled depressions. When such depressions contain water, they are *oxbow lakes.*

Scale

River channels, meanders, floodplains and other features can occur at very different scales. Compare the two images to the right. The first is an aerial image of the same Animas River segment shown above. The second is a few meander loops on the Mississippi River. The images are at the same scale! The much smaller scale Animas River has 15 times more meanders than the Mississippi for the same downstream distance.

16.08.a2–3

B How Do Meandering Rivers Traverse Their Floodplains?

Many major and smaller rivers meander across gentle plains, carrying large quantities of water and fine-grained sediment away from foothills or broad, low uplands. Meandering rivers, at some scale, are present in most low-relief regions.

16.08.b1

The meandering Mississippi River begins in a lake in Minnesota and winds its way southward, across the center of the continent. Its length is not constant because of its shifting meanders, but is about 3,700 km (2,300 miles).

From Minneapolis to the sea, a distance of ~2,900 km (1,800 miles), the river drops only 236 m (775 ft), for a very low gradient of less than 0.1 m/km.

At its mouth, the river deposits its load of sediment in a large delta southeast of New Orleans, Louisiana.

16.08.b2

The very broad floodplain of the Mississippi River has countless crescent-shaped scars of ancient meanders, abandoned by the shifting of the river.

Many cutoff meanders are filled with water, forming curved *oxbow lakes*.

Formation of a Levee

Along the edge of many channels is a raised embankment, or *levee*. Natural levees are created by the river, and artificial levees are constructed by humans to try to keep floodwaters from spilling onto the floodplain.

16.08.b3

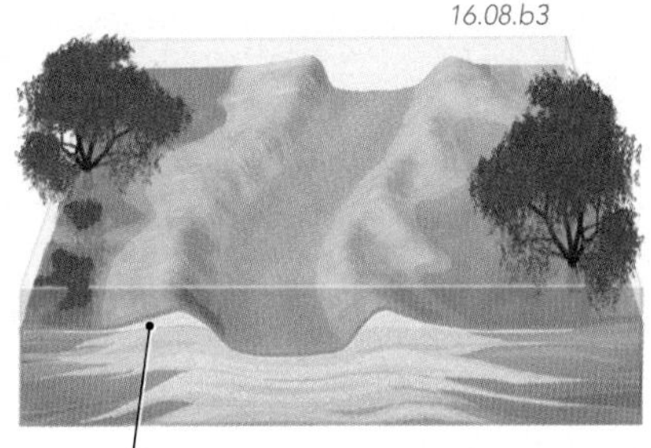

During flooding, sediment-carrying floodwater rises above the channel and begins to spread out. As it does, the current slows and deposits sediment in long mounds next to and paralleling the channel.

16.08.b4

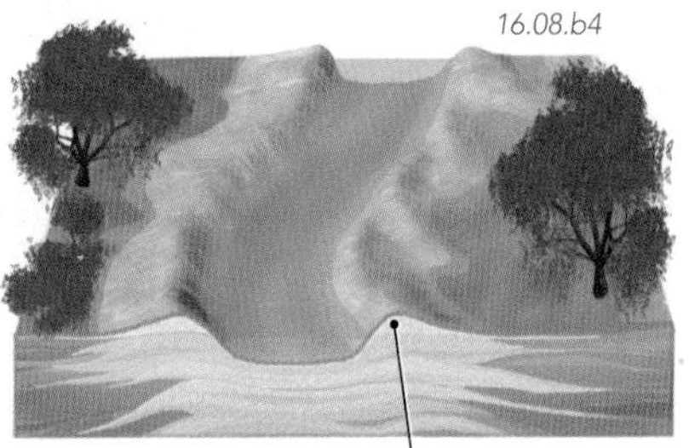

When the flood recedes, sediment that was piled up next to the channel remains as a *levee*. Levees are barriers to water flow from the channel to the floodplain, and from the floodplain back into the channel after a flood.

Levees — Boon or Bust?

While the word *levee* likely leads to thoughts of flooding along the Mississippi, the state of California has 8,000 kilometers (5,000 miles) of human-constructed levees that keep seasonal rainfall from inundating some of the nation's most productive farmlands. Without levees much of this land would be permanently submerged because it has subsided and is now lower than the adjacent rivers. One problem with levee systems is that they invariably fail. It is nearly impossible to engineer an *affordable* levee system that can handle the *largest* flood events. This image shows the 1986 Linda levee failure near Marysville, California. The failure occurred nine days after the floodwaters had crested. The flood caused $400 million in damages.

16.08.mtb1

Before You Leave This Page Be Able To

- ✓ Sketch or describe the features that accompany low-gradient rivers.
- ✓ Describe the character of meander scars and oxbow lakes on the floodplains of meandering rivers.
- ✓ Sketch or describe how natural levees form, and describe the benefits and problems associated with levees.

What Happens When a River Reaches Its Base Level?

BASE LEVEL IS ULTIMATELY THE OCEAN, where rivers slow down and drop their bed load and suspended load. Temporary base levels are established when a river is dammed by a landslide or other natural causes, or by human engineering. The new base level causes changes in the river system both above and below the obstruction. Such changes, however, are temporary—rivers win in the end.

A What Happens as a River Approaches Base Level?

Several landscape-building processes occur when a river enters the ocean, lake, or a temporary base level. Large rivers, like the Amazon and Mississippi rivers, pump freshwater far into the ocean and carry fine sediment out to sea. They deposit coarser sediment as soon as the current slows, forming a *delta* along the shoreline.

16.09.a1

16.09.a2

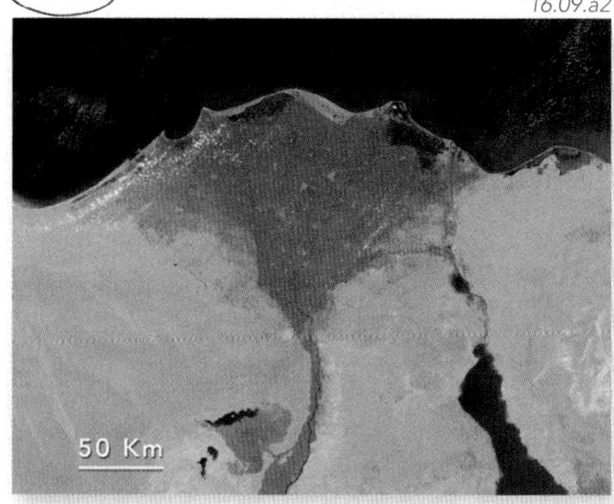

△ **1.** What is a delta? This satellite view above shows the green, triangular-shaped delta formed where sediment from the Nile River is deposited out into the Mediteranean.

2. A delta also forms where the Mississippi River meets the Gulf of Mexico near New Orleans. In this satellite image, the river changes from a meandering river within a broad floodplain to a series of smaller channels that branch apart and spread out in various directions. This branching drainage pattern is a *distributary system*.

3. Dark blue colors on this image indicate clear, deeper waters of the Gulf, whereas lighter blue areas contain suspended sediment and mostly are over shallower water. Sediment from the river accumulates and builds up the delta, which is eroded by waves and by underwater slumps of the steep, unstable delta front.

4. Over the last 7,000 years, the Mississippi has created and then abandoned six huge mounds of sediment, each of which marks a former location of the river mouth and its associated delta; some of these are labeled *Abandoned Delta* on the figure. A new delta (*Active Delta*) is forming where the Mississippi River currently enters the Gulf of Mexico. Eventually, the river will shift and abandon this delta too.

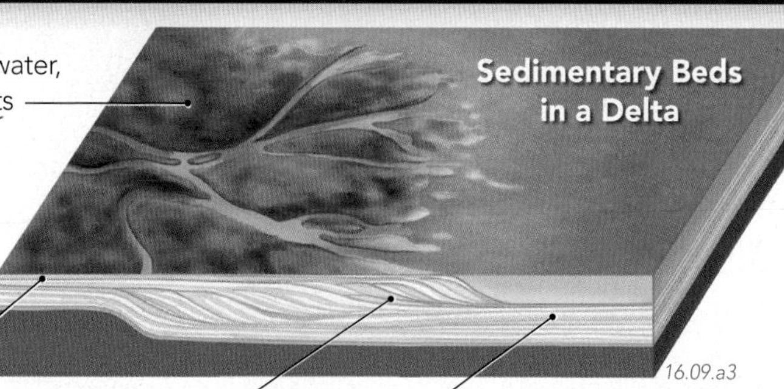

16.09.a3

5. As a delta builds out into water, it forms new land and deposits a characteristic sequence of sedimentary beds. As the river's current slows, sand and larger particles become too heavy to be carried and are deposited in three types of beds. A set of horizontal beds forms on top of the delta.

6. A set of dipping beds forms when sediment is deposited over the edge of the delta, moving the front of the delta seaward.

7. Silt and clay are carried farther out into the ocean (or lake) and are deposited as nearly flat beds in front of the delta.

B What Controls the Deposition of Sediment in a Delta?

Deposition in a delta occurs where a river or stream slows, losing capacity and depositing its load of sediment. The morphology of a delta and the type of sediment deposited reflect the sediment load and discharge of the river, as well as other factors, such as wave activity and the amount of vegetation or ice.

16.09.b1

The Lena Delta of Siberia provides one of the most beautiful satellite images of Earth. This image, taken in the summer, shows a thawed East Siberian Sea and abundant vegetation on the delta. The *distributary pattern* of drainages is obvious. This delta nicely displays the factors that control deposition of sediment in a delta.

Vegetation—The amount of vegetation and seasonal changes in vegetation affect the number and location of delta channels. Generally, deltas that have dense vegetation have fewer channels, whereas deltas with sparse vegetation have more channels. Part of the explanation is that vegetation binds the soil and stabilizes channel positions.

Sediment Load—Coarser sediment, such as sand, is carried in the bed load and deposited first as the velocity drops. Finer material, carried in suspension, can be carried farther. If the river carries more sediment and is closer to its capacity, it will deposit more sediment and drop it sooner.

Discharge—High-discharge flows tend to extend farther out into the ocean. The deposited sediment can then be affected by waves and by currents parallel to the shoreline.

Wave Erosion—Deltas that form along shorelines with strong or continuous wave action tend to be dissected and have somewhat serrated (jagged) edges, or they can be completely truncated.

River and Ocean Ice—Seasonal changes in the amount of ice in the river and along the coast affect discharge and deposition patterns. River ice makes flow more sluggish, and sea ice tends to trap more sediment closer to shore. This satellite image shows the Lena Delta surrounded by sea ice in the winter.

16.09.b2

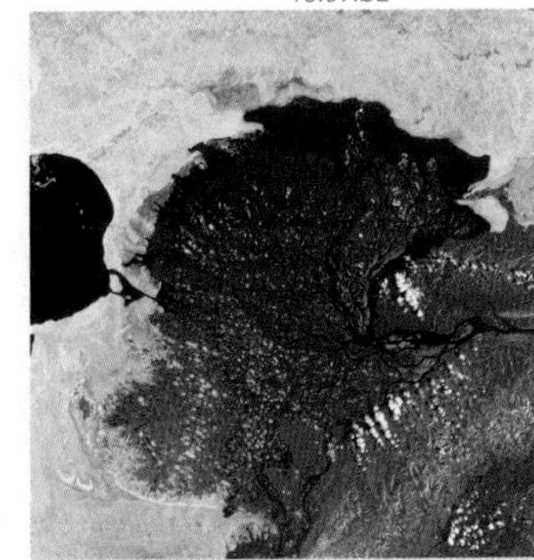

C What Are the Depositional Consequences of Dams?

Human-constructed dams provide hydroelectric power generation, water storage, or flood control, but they stop a river's normal flow and transport of sediment. The reservoir behind the dam represents a *temporary base level*, and so causes the river to deposit sediment behind the dam, limiting the dam's longevity.

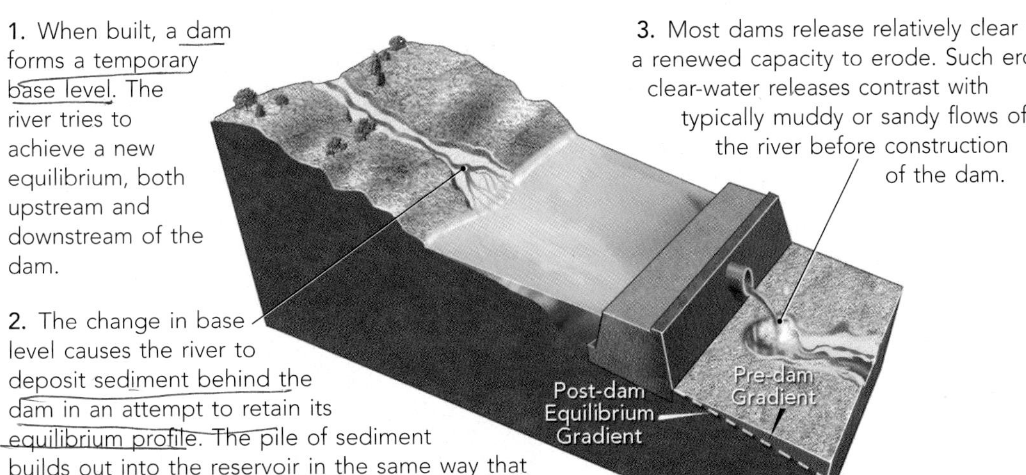

1. When built, a dam forms a temporary base level. The river tries to achieve a new equilibrium, both upstream and downstream of the dam.

2. The change in base level causes the river to deposit sediment behind the dam in an attempt to retain its equilibrium profile. The pile of sediment builds out into the reservoir in the same way that a natural delta builds out into the sea. This sediment can eventually fill up the reservoir, shortening its lifespan.

3. Most dams release relatively clear water that is starved of sediment and that has a renewed capacity to erode. Such erosion occurs below many dams, whose clear-water releases contrast with typically muddy or sandy flows of the river before construction of the dam.

16.09.c1

Before You Leave This Page Be Able To

- ✓ Describe what happens when a river enters an ocean or lake, and what factors control deposition of sediment.
- ✓ Sketch and describe the stratigraphy of delta sediments and the setting in which each type of sediment formed.
- ✓ Describe how construction of a dam affects a river.

How Do Rivers Change over Time?

IN GEOLOGIC TERMS, rivers come and go. Some rivers are old and others are surprisingly young. The age and history of a river are important considerations when evaluating how the river might respond to tectonic, climatic, and sea-level changes. Human activities can also evoke dramatic responses in rivers.

A How Old Are Rivers?

Rivers flow from their source to base level as long as enough water and slope are available to maintain downstream flow. A river's life can begin or end due to changes in water and sediment supply at the source, to changes in the slopes across which the river flows, or changes in the elevation of its base level. Rivers can exist for millions of years, although their characteristics may change due to climatic, glacial, and tectonic events.

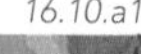

16.10.a1

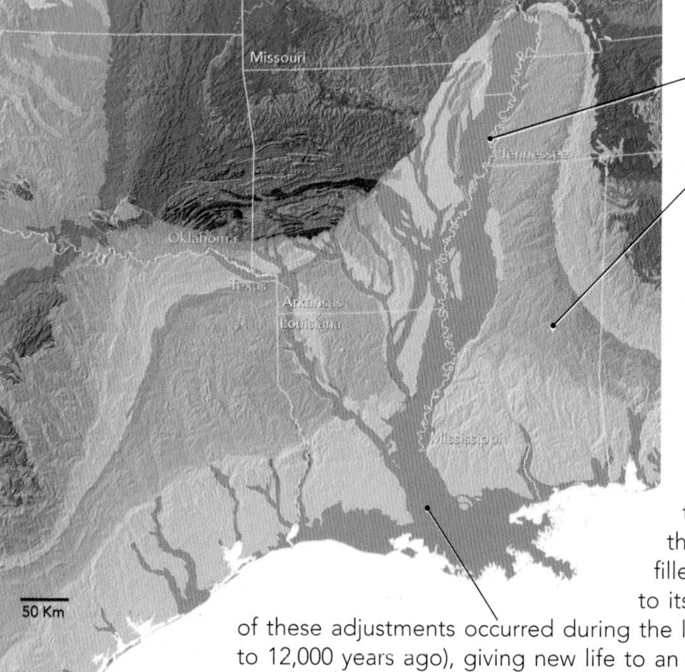

Lower Mississippi River

1. On this geologic map, the river and its tributaries are shown in gray, representing recent sediments. The oldest preserved river sediments indicate that the lower Mississippi began draining the continent during Mesozoic time.

2. The river and its tributaries eroded across a series of Cenozoic sedimentary layers (shown in yellow and orange), with the river incising a valley when sea level was low.

3. Subsequent sea-level rise decreased the river's gradient and the river's sediment filled the excavated valley to its present level. Some of these adjustments occurred during the last ice age (2 million to 12,000 years ago), giving new life to an old river.

4. The river flows along a continent-scale low, the *Mississippi Embayment*, shown in this geologic cross section through Memphis, Tennessee. The embayment originated from Precambrian continental rifting, which thinned the crust and set the stage for the river's formation hundreds of millions of years later. The region has subsided well into the Cenozoic.

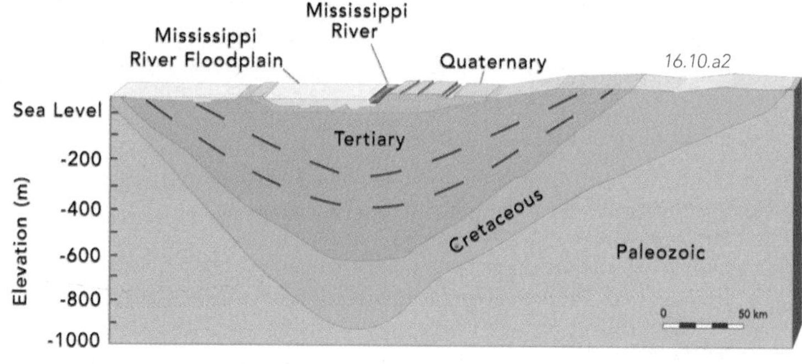

16.10.a2

Upper Mississippi River

16.10.a3

16.10.a4

The upper Mississippi River is young. It formed since the retreat of the last ice sheets, some 10,000 years ago. During the last ice age (◁), ice sheets and glaciers covered the northern half of North America, so northern rivers like the upper Mississippi did not exist. The weight of the ice sheets depressed the crust, causing large regions to slope northward (opposite to today).

◁ Melting of the ice released huge discharges of water that carved completely new river channels, including the upper part of the Mississippi.

The Fall Line

A major boundary, called the *Fall Line*, winds its way between the Appalachian Mountains and the east coast of the United States. The Fall Line, shown here as a red line, is marked by water falls formed along the contact between soft sediment of the coastal plain and harder bedrock in the foothills of the mountains. The Great Falls of the Potomac River, upstream rom Washington, D.C., illustrate how the Fall Line developed. Before the ice age, the Potomac River occupied a broad valley. A drop in sea level during the ice age caused the river to incise deeper. Erosion proceeded upstream, stripping away the soft sediment until it encountered the harder rocks at the Great Falls.

16.10.a5

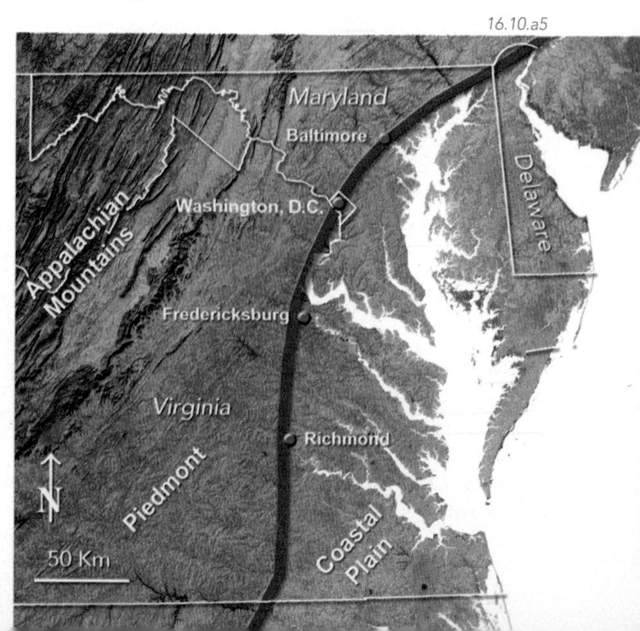

B How Do River Systems Respond to Changing Conditions?

Rivers are sensitive to their environment, including local effects, such as rainfall, and more distant effects, such as changes in sea level. Rivers respond to changes in climate, tectonics, base level, human intervention, and the type of geology they encounter as they deposit sediments or cut deeper into the landscape.

Runoff

1. The amount of flow is the most important factor in how a river develops, and this depends mostly on the amount and timing of precipitation. Direct runoff during rainfall and delayed runoff from snowmelt supply most water to most rivers. The amount of runoff varies dramatically. The flood in the top image did this destruction to condominiums built too close to the bank. [Rillito River, Tucson, Arizona]

16.10.b1

16.10.b2

Glacial and Sea-Level Effects

16.10.b3–4

2. Global cooling and growth of ice sheets and glaciers lowers sea level. It can load and depress the crust, causing drainages to flow toward the ice sheets.

3. Melting of this ice releases huge amounts of meltwater, creating new or larger channels. Isostatic rebound due to ice removal can reverse the regional drainage patterns.

Tectonism

16.10.b5

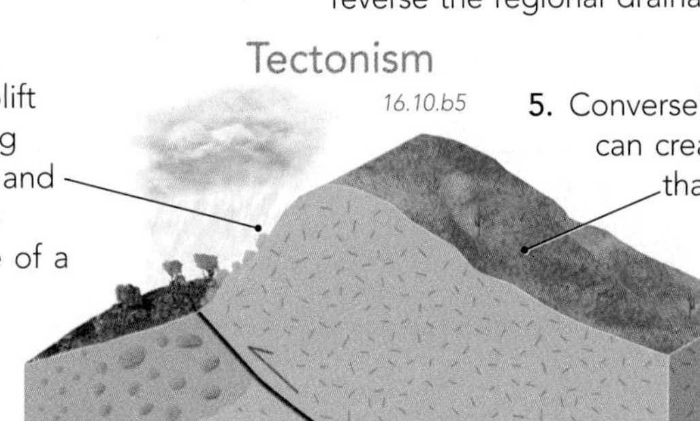

4. Tectonism can uplift mountains, increasing slope, precipitation, and the supply of coarse sediment. The slope of a river and supply of sediment determine whether a river is braided or meandering.

5. Conversely, mountain uplift can create a rain shadow that decreases precipitation on the opposite side of the mountain, reducing the amount of runoff.

Geology

16.10.b6

6. Rivers can more easily erode unconsolidated sediment and soft rock types than harder ones. Rivers that are eroding downward may encounter rocks that have different characteristics, causing a change in the geometry of the river. The impressive Niagara Falls (◁) along the Canadian–U.S. border formed when the post-ice-age Niagara River encountered a more resistant dolostone layer underlain by less resistant shale.

Human Engineering

16.10.b7

7. Dams and other flood-control structures change base level, the amount of discharge, and the supply of sediment, all of which affect the river system upstream and downstream.

Climate

▽ 8. When early settlers came to the American Southwest in the mid-1800s, many alluvial streams were flowing on broad valleys. Settlers built farms on the moisture-rich floodplains.

▷ 9. Climatic effects around 1880 caused streams to *incise (erode down) into* their floodplains. This incision dried up the previous floodplain and many of the farms.

16.10.b8

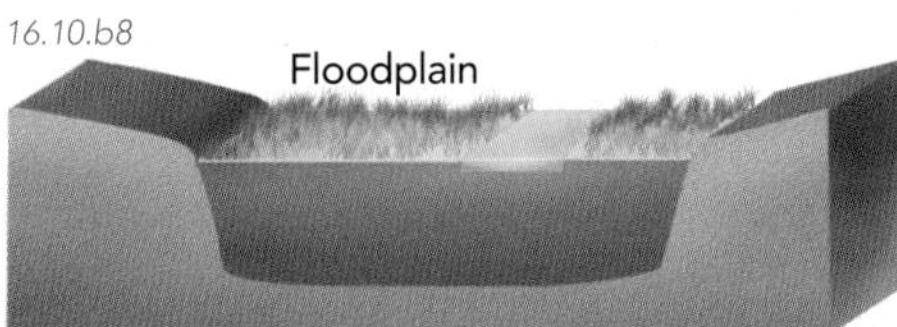

Before 1880 - Streams on Floodplain

16.10.b9

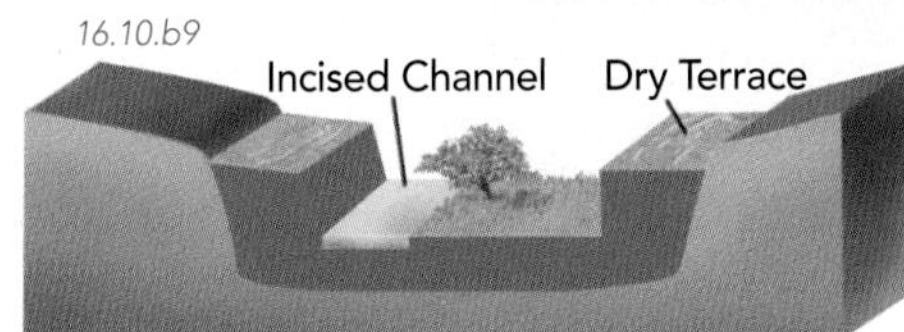

About 1880 - Incision of Channels

▽ 10. Around 1940, the channels began to deposit sediment and build up again.

16.10.b10

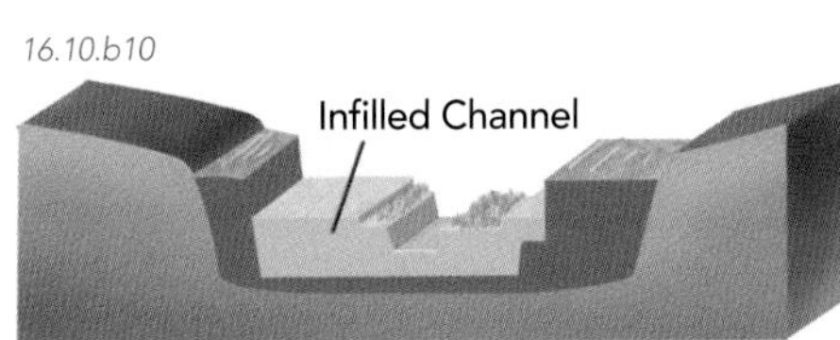

About 1940 - Channel Filling Begins

Before You Leave This Page Be Able To

- ✓ Describe how rivers can be old or young, using the Mississippi River as an example.
- ✓ Describe how river systems respond to changes imposed by climate, tectonism, geology, and human engineering.
- ✓ Summarize the effect that glaciers have on river systems.

What Happens During River Incision?

RIVERS CAN INCISE INTO LANDSCAPES, forming a variety of features, such as multiple levels of terraces. Rivers also carve some unusual canyons, such as those that take odd routes across the landscape, cutting right across mountains that would seem to be insurmountable obstacles. What sequence of events led to the development of these features?

A How Are River Terraces Formed?

River terraces are relatively flat benches that are perched above a river or stream and that stair-step up and outward from the active channel. Most terraces are composed of river-derived sediment and are essentially abandoned floodplains and alluvial plains. Other terraces are cut directly into bedrock and form by erosion. Terraces record different stages in the river's history and indicate that the river or stream has incised into the land.

16.11.a1

Terraces form a series of flat to gently sloping benches or steps, flanked by steeper slopes. Terraces successively step up and away from the channel. [Tibet]

Successively lower terraces step down toward the river, culminating in the lowest terrace, which commonly is only a meter or so above the channel and is often flooded, perhaps nearly every year.

This series of terraces flank the Snake River in Jackson Hole, Wyoming. The terraces are numbered from highest (1) to lowest (3). The modern floodplain also is labeled (F). Which of these terraces formed first and which one formed last?

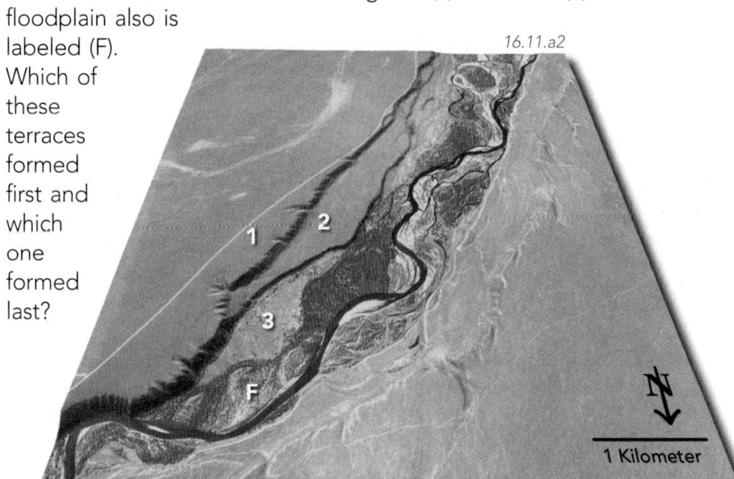

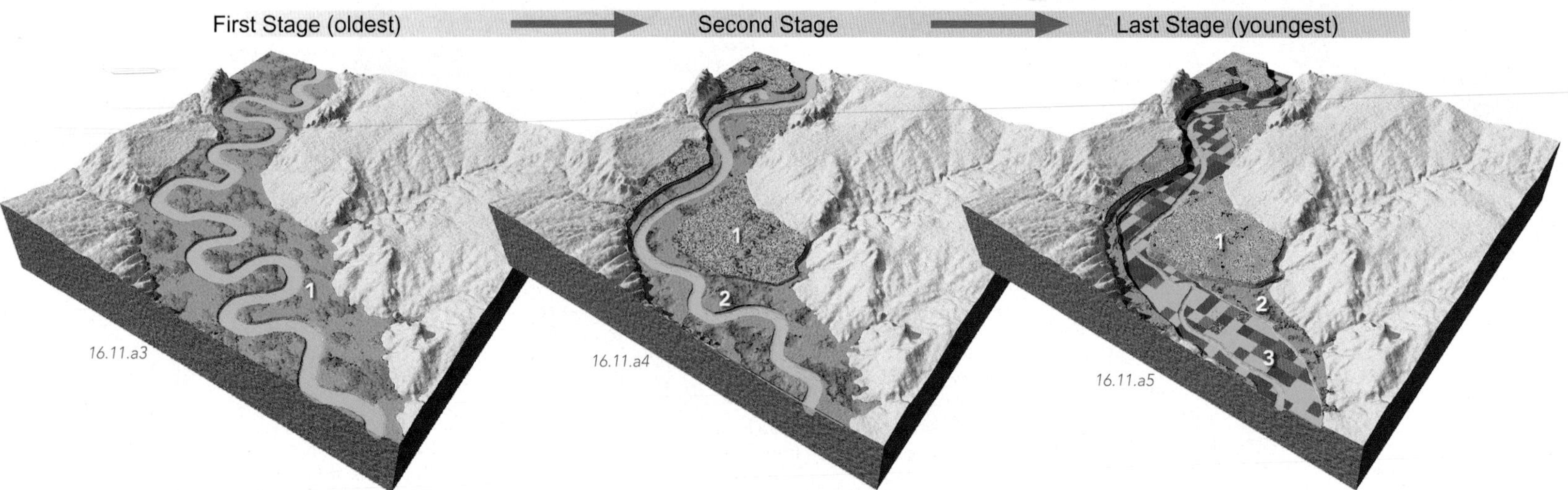

The first stage in terrace formation is deposition of sediment, such as on the floodplain (1) shown above. At this stage in its history, the river is nearly at the same level as the floodplain (i.e., is not incised). The flat surface of the floodplain will later become the flat part of a terrace.

A change in conditions, such as a drop in base level, causes the river to downcut through its floodplain deposits, forming a second, lower floodplain (2). Remnants of the first floodplain are stranded on both sides of the river (1) and, if high enough, are unlikely to be flooded again.

With further downcutting, the river abandons the second floodplain (2), creating a third, even lower one (3). The oldest floodplain (1) is now high and dry. This series of downcutting events creates a stair-step appearance to the land, like those shown in the first two figures on this page.

B How Are Entrenched Meanders Formed?

The landforms we know as meanders form only in loose sediments, such as those on floodplains. However, in the Four Corners region of the American Southwest, and in some other regions, meanders with typical sweeping bends are deeply incised in hard bedrock, forming some puzzling canyons. What do these winding canyons, called *entrenched meanders*, tell us about the history of rivers in this area?

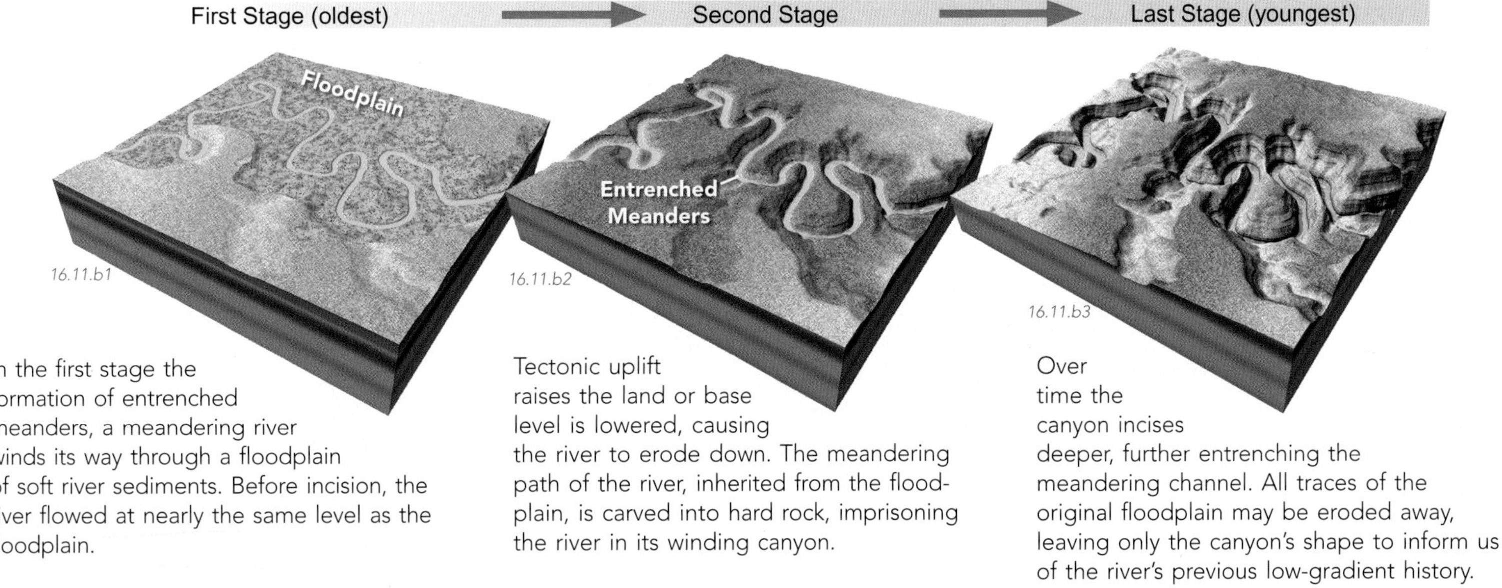

In the first stage the formation of entrenched meanders, a meandering river winds its way through a floodplain of soft river sediments. Before incision, the river flowed at nearly the same level as the floodplain.

Tectonic uplift raises the land or base level is lowered, causing the river to erode down. The meandering path of the river, inherited from the floodplain, is carved into hard rock, imprisoning the river in its winding canyon.

Over time the canyon incises deeper, further entrenching the meandering channel. All traces of the original floodplain may be eroded away, leaving only the canyon's shape to inform us of the river's previous low-gradient history.

Rivers That Cross Geologic Structures

Sometimes rivers appear to perform impossible tasks—cutting a deep canyon directly across a mountain. The Green River (below) flows across a mountain, appropriately called *Split Mountain*, as shown in the photograph to the right. This mountain ridge is an anticline of hard sandstone in Dinosaur National Monument of northern Utah.

These odd canyons can be interpreted in at least two ways. A river may have been flowing over a region that was being actively uplifted and deformed, but the river was able to erode through the structures as fast as they were formed. Such a river is called *antecedent*, meaning it predated formation of the structure.

Alternatively, a river may establish its route when it is flowing on soft, easily eroded rocks, uninfluenced by what lies at depth. As the river begins to incise, it becomes trapped in its own canyon, unable to avoid any geologic structures it encounters as it erodes down through the rocks. Such rivers are *superposed*, meaning they were superimposed on already existing features. The Green River is best interpreted as a superposed river that established a meandering course on soft rocks and then downcut into harder ones.

5 Kilometers

16.11.mtb1

16.11.mtb2

Before You Leave This Page Be Able To

- ✓ Sketch and explain a series of steps showing how river terraces form.
- ✓ Describe one way in which entrenched meanders form.
- ✓ Explain how antecedent and superposed rivers are different.

16.11

What Is and What Is Not a Flood?

THROUGHOUT HISTORY, PEOPLE HAVE LIVED along rivers and streams. Rivers are sources of water for consumption, agriculture, and industry, and provide transportation routes and energy. River valleys offer a relatively flat area for construction, but people who live along rivers are subject to an ever changing flow of water. High amounts of water flowing in rivers and streams often lead to flooding. In many parts of the world, flooding is a very common and costly type of natural and human-caused disaster.

A What Is the Difference Between a Flood and a Normal Flow Event?

Rivers and streams are dynamic systems, and they respond to changes in the amount of water entering the system. When more water enters the system than can be held within the natural confines, the result is a *flood.*

1. Flow in a channel, even when there is not a flood, may cause riverbank erosion. Such erosion can destroy structures built close to the river and make the river change position over time, turning what was floodplain into channel, and what was channel into floodplain.

2. A flood occurs when there is too much water for the channel to hold, and water spills out onto the adjacent land.

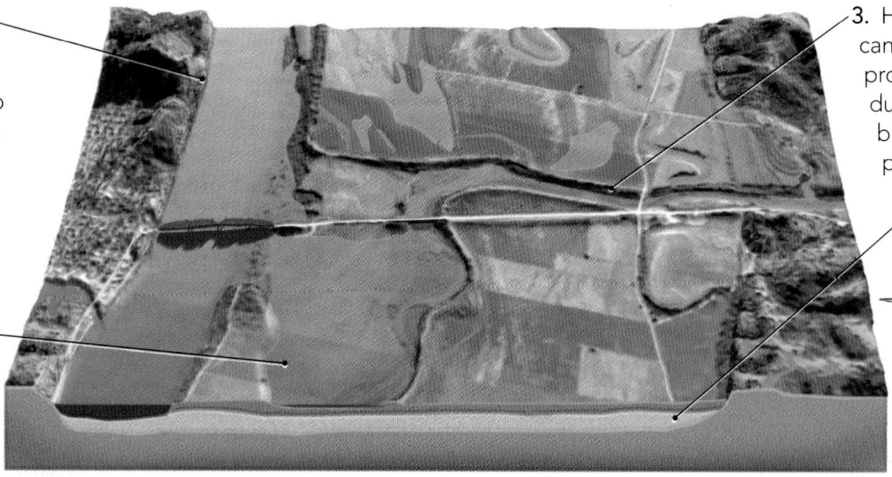

3. Human-constructed levees can sometimes protect property from flooding during large flood events but trap water after the peak flooding ends.

4. Large floods can expand the width of the floodplain, by burying preexisting rocks and material with sediment deposited by the river.

16.12.a1

Normal, Bank-Full Flows

16.12.a2

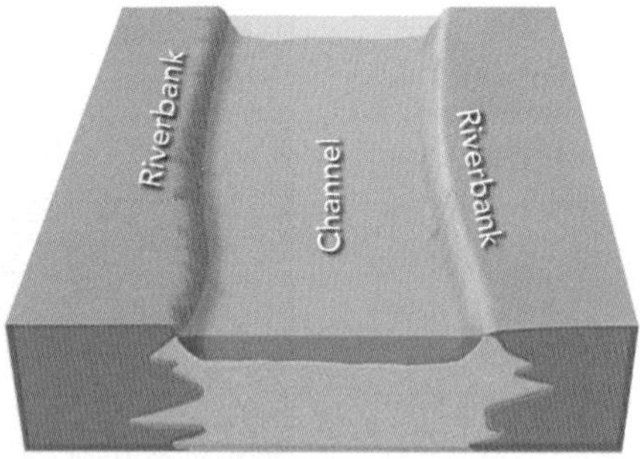

Riverbank
Channel
Riverbank

5. Normal (i.e., non-flooding) flows in rivers and streams can range from nearly dry to bank-full. Although there may be abundant water flowing down the channel, it is generally not considered a flood unless the water overflows the banks.

A river's natural floodplain is an excellent place to contain excess floodwaters—as long as it remains undeveloped by society.

16.12.a3

Flood Stage
Discharge
Stream Flow
Time

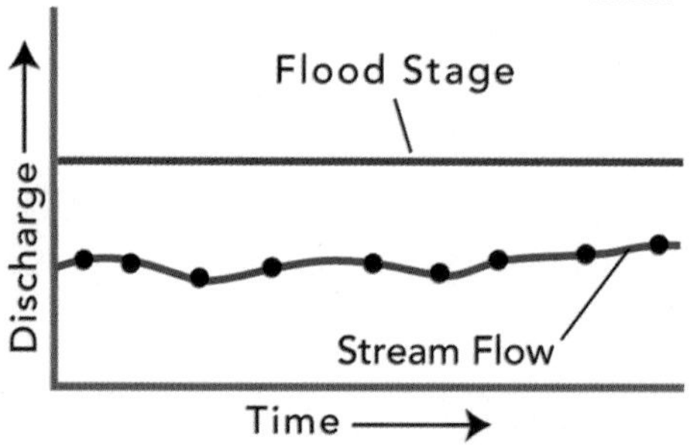

6. This hydrograph shows a typical non-flood flow. The line labeled *Flood Stage* shows the amount of discharge required for the river to overtop its banks and spill out onto the floodplain (i.e., a *flood*). During extended times of dry conditions, or at least weather that is normal for the region, hydrographs may show little change in stream flow over time, as shown here.

Flows During a Flood

7. When the amount of water in a river exceeds the channel capacity, a flood occurs, inundating the floodplain. This hydrograph shows prolonged precipitation or snowmelt upstream that causes a flood event downstream, as represented by discharge greater than flood stage.

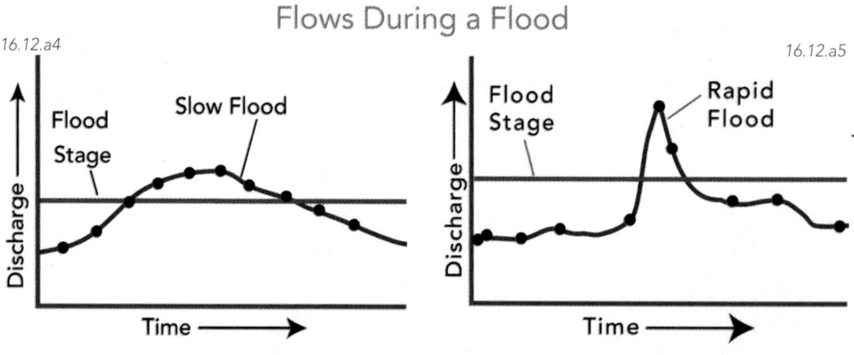

8. Intense rainfall can unleash a brief *flash flood*, with a rapid rise in water levels and an increase in discharge that lasts only a short time. Similarly, rapid *onsets* of flooding result from failure of a natural or constructed dam, but flows last longer.

Step 2: Calculate Discharge for a Profile Across the River

The diagram below on the left is a profile across the river, showing the widths of the *notch* and the *bottomland.* You will calculate discharges along this main profile, which crosses the river near the front of the model on the right. Your instructor may provide you with a second profile (farther back in the model), because the river has different dimensions at different places. This means that the same amount of discharge may reach different heights up and down the river. For your profile(s), complete the following steps:

1. To calculate the discharge needed to fill the notch, first calculate the cross-sectional area of the notch in the profile:

Cross-sectional Area = Width × Depth

2. Next, calculate how much discharge is needed to fill the notch and begin to spill water out onto the bottomland. To calculate discharge, multiply the cross-sectional area of the notch by the average velocity of the river, which is 0.7 m/sec when the notch is filled:

Discharge = Cross-sectional Area × Stream Velocity

3. Repeat the calculations, but this time determine the additional discharge needed to flood the bottomland to a height where floodwater would begin to spill onto the middle bench. The river flows faster when there is more water, so use an average water velocity of 2.0 m/sec. Enter your calculated discharges in the table on the worksheet or on a sheet of paper. You should have two discharge calculations, one to fill and overtop the notch, and another that fills up the notch and bottomland and then begins to spill out onto the middle bench.

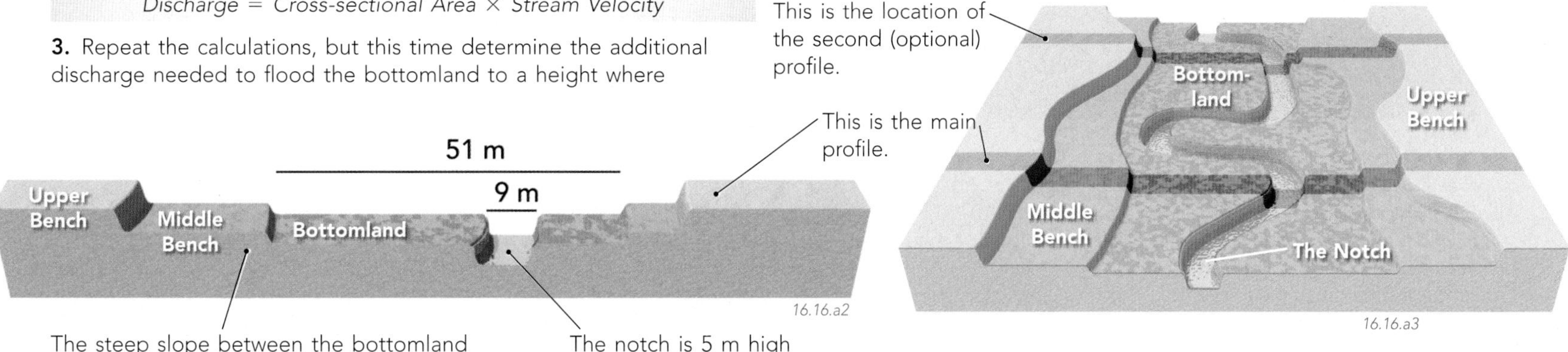

Step 3: Evaluate Flooding Risk Using Exceedance Probability

To determine the probability that each area will be flooded, compare both of your calculated discharges against the following plot, which is an *exceedance probability plot.* Follow the steps below and list in the worksheet the estimated probabilities for overfilling the notch and for overfilling the bottomland on the profile.

1. For each discharge calculation, find the position of that *discharge value* on the *vertical* axis of the plot.

2. Draw a horizontal line from that value to the right until you intersect the probability line (which slopes from lower left to upper right).

3. From the point of intersection, draw a vertical line down to the *horizontal axis* of the plot and read off the corresponding *chance of exceedance* (probability of flooding) on the horizontal axis. The *probability of exceedance* indicates the probability of the calculated amount of discharge being exceeded in any given year.

4. Repeat this procedure for both of your discharge calculations.

5. Consider the implications of each of these probabilities for your choice of site for cropland and a homesite. Use this information to choose final sites for cropland and a house, and explain your reasons on the worksheet.

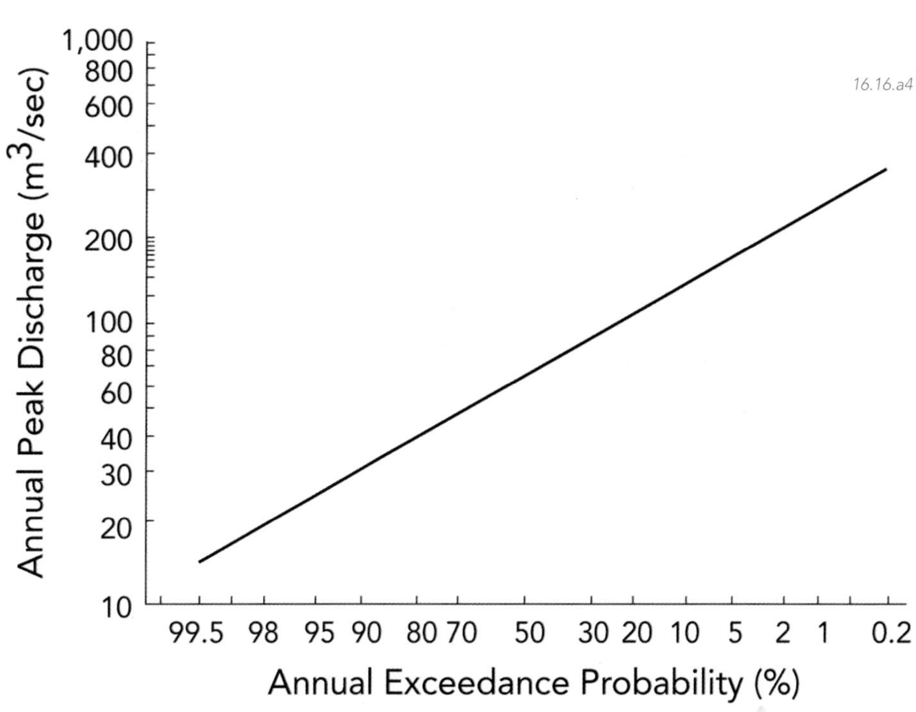

CHAPTER

17 Water Resources

WATER IS OUR MOST IMPORTANT RESOURCE—life on Earth depends on water to live. We are most aware of *surface water*, which occurs in streams, rivers, and lakes, but the amount of freshwater in these settings is much less than the amount of freshwater that is frozen in ice and snow or that occurs in the subsurface as *groundwater*. This chapter is about surface water and groundwater and the important ways in which they interact.

17.00.a2

△ In the center of the Snake River Plain, the Snake River occupies a canyon cut into layers of basalt. The fertile farmlands receive water from rivers, springs, and wells drilled into groundwater. [Shoshone, Idaho]

The Snake River Plain, shown in this large satellite-based image is an arcuate swath of low, basalt-covered land that cuts through the mountains of southern Idaho. It contains a mixture of dry, sage-covered plains, water-filled reservoirs, green agricultural fields, and recent lava flows of dark-colored basalt. Most of Idaho's population lives on the Snake River Plain near the rivers and reservoirs.

The Big Lost River, Little Lost River, and adjacent streams that enter the plain from the north never reach the Snake River. Instead, the water from the rivers and streams seeps into the ground between the grains in the sediment and through narrow fractures in the basalt. For this reason the rivers are called "lost."

Where does water that seeps into the subsurface go?

17.00.a1

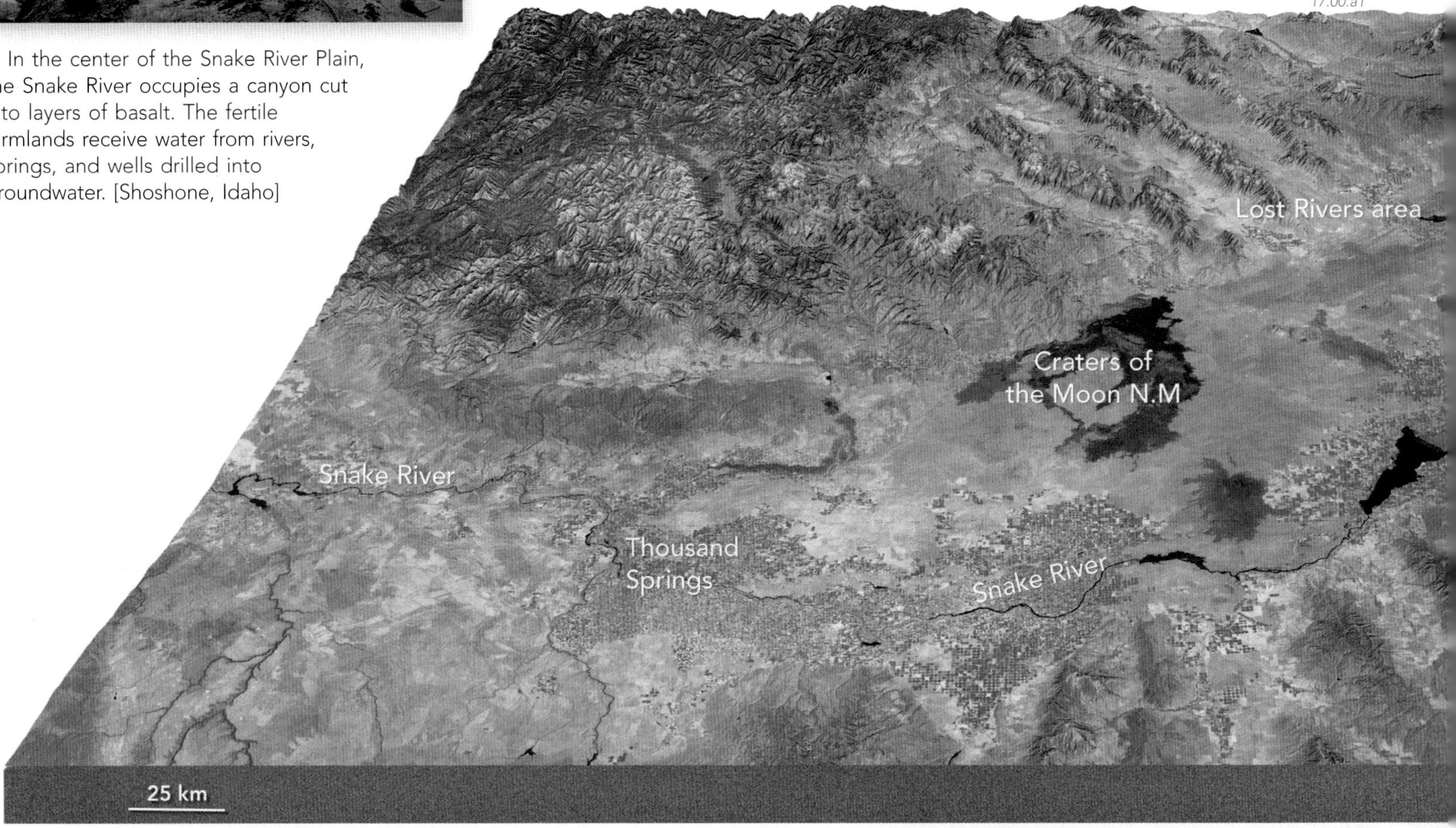

Within the canyon of the Snake River huge springs gush from the steep volcanic walls. The Snake River features 15 of the 65 largest springs in the United States, including those in an area called *Thousand Springs*. The largest commercial trout farms in the United States use ponds fed by these springs.

What causes water to flow from beneath the ground as a spring, and where does the water in a spring come from?

17.00.a3

▷ At Thousand Springs, water pours from fractures and thin layers of sediment within basalt flows that underlie the Snake River Plain.

TOPICS IN THIS CHAPTER

The Snake River is the principal drainage of the region. It winds through the mountains and then flows southwest and west across the Snake River Plain.

Where does this river, flowing across such a dry plain, receive its water?

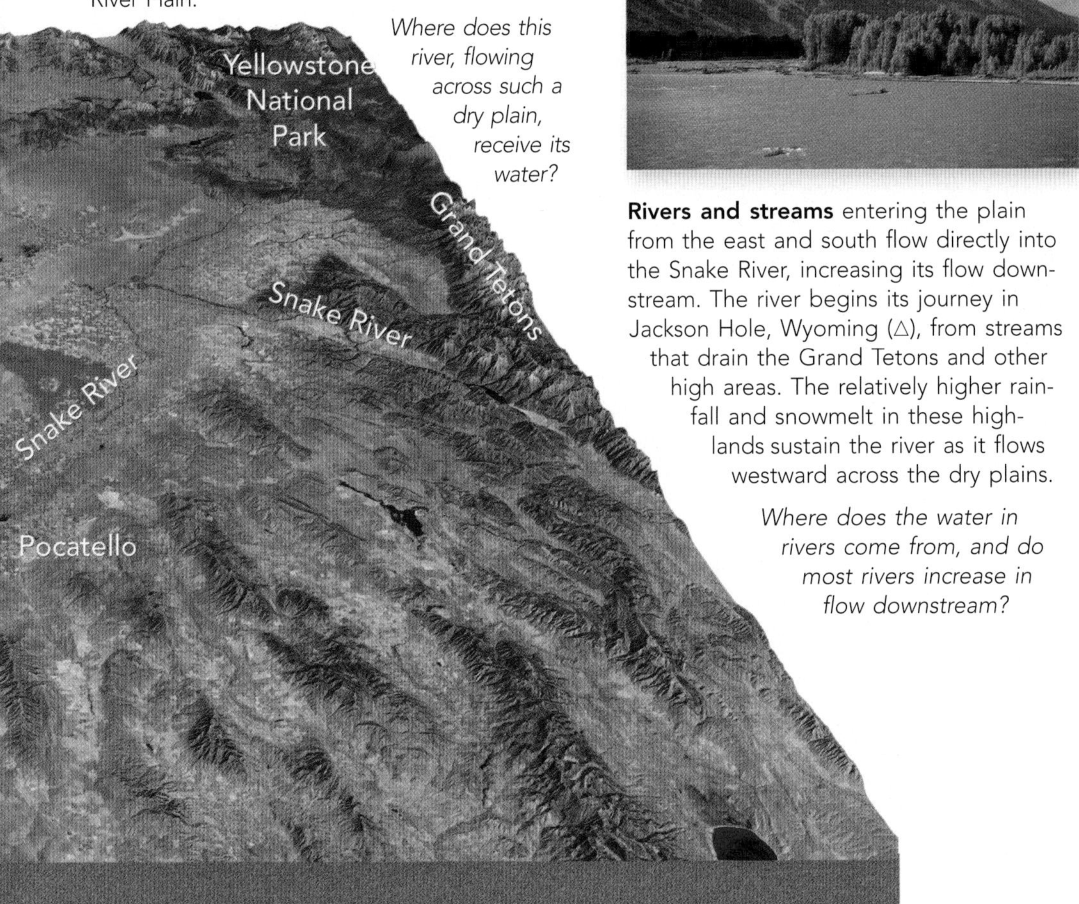

17.00.a4

Rivers and streams entering the plain from the east and south flow directly into the Snake River, increasing its flow downstream. The river begins its journey in Jackson Hole, Wyoming (△), from streams that drain the Grand Tetons and other high areas. The relatively higher rainfall and snowmelt in these highlands sustain the river as it flows westward across the dry plains.

Where does the water in rivers come from, and do most rivers increase in flow downstream?

17.00.a5

A series of lakes and farms occur along the Snake River as it winds through the Snake River Plain, such as near Pocatello (◁). Millions of acres of agriculture are irrigated by surface water derived from the reservoirs, lakes, and rivers, and from groundwater pumped to the surface.

What happens if groundwater is pumped from the subsurface faster than it is replaced by precipitation and other sources?

Disappearing Waters of the Northern Snake River Plain

Groundwater beneath the Snake River Plain is an essential resource for the region, providing most of the drinking water as well as irrigation for farms and ranches away from the actual river. Geologists and other scientists study where this water comes from, how it moves through the subsurface, and the potential limits on using this resource.

Some water enters the subsurface from the Big and Little Lost Rivers, which flow into the basin from the north and then abruptly or gradually disappear as their water sinks between the grains in the porous ground. Other groundwater is contributed directly from the main Snake River and from tributaries that enter the basin from the south and east. Surprisingly, the largest influx of water to the subsurface is seepage from irrigated fields and associated canals.

The surface of the Snake River Plain slopes from northeast to southwest. The flow of groundwater follows this same pattern, flowing southwest and west through the rocks in the subsurface. Groundwater derived from the disappearing rivers flows southwest, along the north side and center of the basin. The groundwater does not flow like an underground river but as water between the grains and within fractures in the rocks. Where the flow of groundwater is intersected by the Snake River Canyon, the water reemerges on the surface at Thousand Springs. This region illustrates a main theme of this chapter—surface water and groundwater are a related and interconnected resource.

Where Does Water Occur on Our Planet?

WATER IS ABUNDANT ON EARTH and occurs in many settings. Most water is in the oceans, with a surprisingly small percentage in lakes, wetlands, and rivers. Water exists in its frozen form as ice and snow, and occurs below the surface as *groundwater* within pore spaces, fractures, and other openings. It is present in the atmosphere as water vapor, clouds, and precipitation, and in plants, animals, and soils.

A Where Did Earth's Water Come from and Where Does It Occur Today?

Most water on Earth was probably incorporated during formation of the planet or from comets and other icy celestial objects that smashed down onto the surface. Over time, much of this water moved toward the surface, such as when magma released water vapor during eruptions.

Oceans—Of Earth's total inventory of surface and near-surface water, an estimated 96.5 percent occurs in the oceans and seas as *saline* (salty) water. The remaining 3.5 percent is *freshwater* held in ice sheets and glaciers, groundwater, and lakes, swamps, and other features on the surface.

Atmosphere—A small, but very important, amount of Earth's water is contained in the atmosphere (0.001 percent). It occurs as invisible water vapor, as water droplets in clouds, and as rain, falling snow, and other types of precipitation.

Glaciers—Nearly 69 percent of Earth's freshwater is tied up in ice and snow in ice caps, glaciers, and permanent snow. A small amount also exists in permafrost and ground ice.

Rivers—Rivers are extremely important to us and are the main source of drinking water for many areas. They contain, however, only a very small amount of Earth's freshwater.

Soil Moisture—Earth's soils contain about as much water as the atmosphere (not much), but like water in the atmosphere, this water is crucial to our existence.

Lakes—Water occurs on the surface in lakes of various sizes. Most are freshwater lakes, but those in dry climates are saline or *brackish* (between fresh and saline). Lakes contain a majority of Earth's liquid freshwater on the surface.

Biological Water—Water is tied up within the cells and structures of plants and animals. It is clearly important to us but represents an exceptionally small percentage of Earth's total water (0.0001 percent)

Groundwater—About 30 percent of Earth's total freshwater occurs as groundwater. Groundwater occurs mostly in the open pores between sediment grains or within fractures that cut rocks.

Swamps and Other Wetlands—These wet places have water lying on the surface and water within the plants and shallow soils. Such places contain about 11 percent of the liquid freshwater on the surface.

Deep-Interior Waters—An unknown, but perhaps very large amount of water is chemically bound in minerals of the crust and mantle. Some scientists think Earth's interior may contain more water than the oceans.

17.01.a1

▷ These bar graphs show USGS estimates of the distribution of water on Earth's surface or in the uppermost levels of the crust. The left bar shows that the oceans contain 96.5 percent of Earth's total free water (i.e., not bound up in minerals), but this water is saline. Only 3.5 percent of Earth's water is fresh. These graphs do not factor in water in Earth's deep interior.

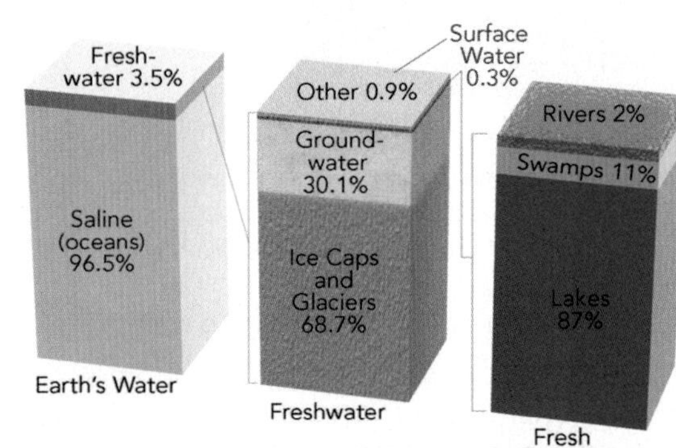

17.01.a2

◁ The middle bar shows that most of Earth's freshwater occurs within ice caps and glaciers followed by groundwater. Less than 1 percent occurs as liquid surface water in lakes and rivers.

◁ The right bar shows where Earth's small percentage of fresh, liquid, surface water resides. Most is in lakes, followed by swamps and rivers.

B How Does Water Move from One Setting to Another?

Water is in constant motion on Earth's surface, moving from ocean to atmosphere, from atmosphere back to the surface, and in and out of the subsurface. The circulation of water from one part of this water system to another is called the *hydrologic cycle*. From the perspective of living things, the hydrologic cycle is the critical system on Earth. It involves a number of important and mostly familiar processes. It is driven by energy from the Sun.

Evaporation—As water is heated by the Sun, its surface molecules become energized enough to break free of the attractive forces binding them together. Once free, they rise into the atmosphere as water vapor.

Condensation—As water vapor cools, such as when it rises, more water molecules join together. Through this process, water vapor becomes a liquid or turns directly into a solid (ice, hail, or snow). These water drops and ice crystals then collect and form clouds.

Precipitation—Rain, hail, snow, and sleet are all forms of precipitation. Clouds are moved by air currents, redistributing moisture across Earth's surface. When clouds cool, like when they rise over a mountain range, the water molecules become less energetic, bond together, and can fall as rain, snow, or hail, depending on the temperature of the air. Precipitation may reach the ground, evaporate as it falls, or be captured by leaves and other vegetation before reaching the ground.

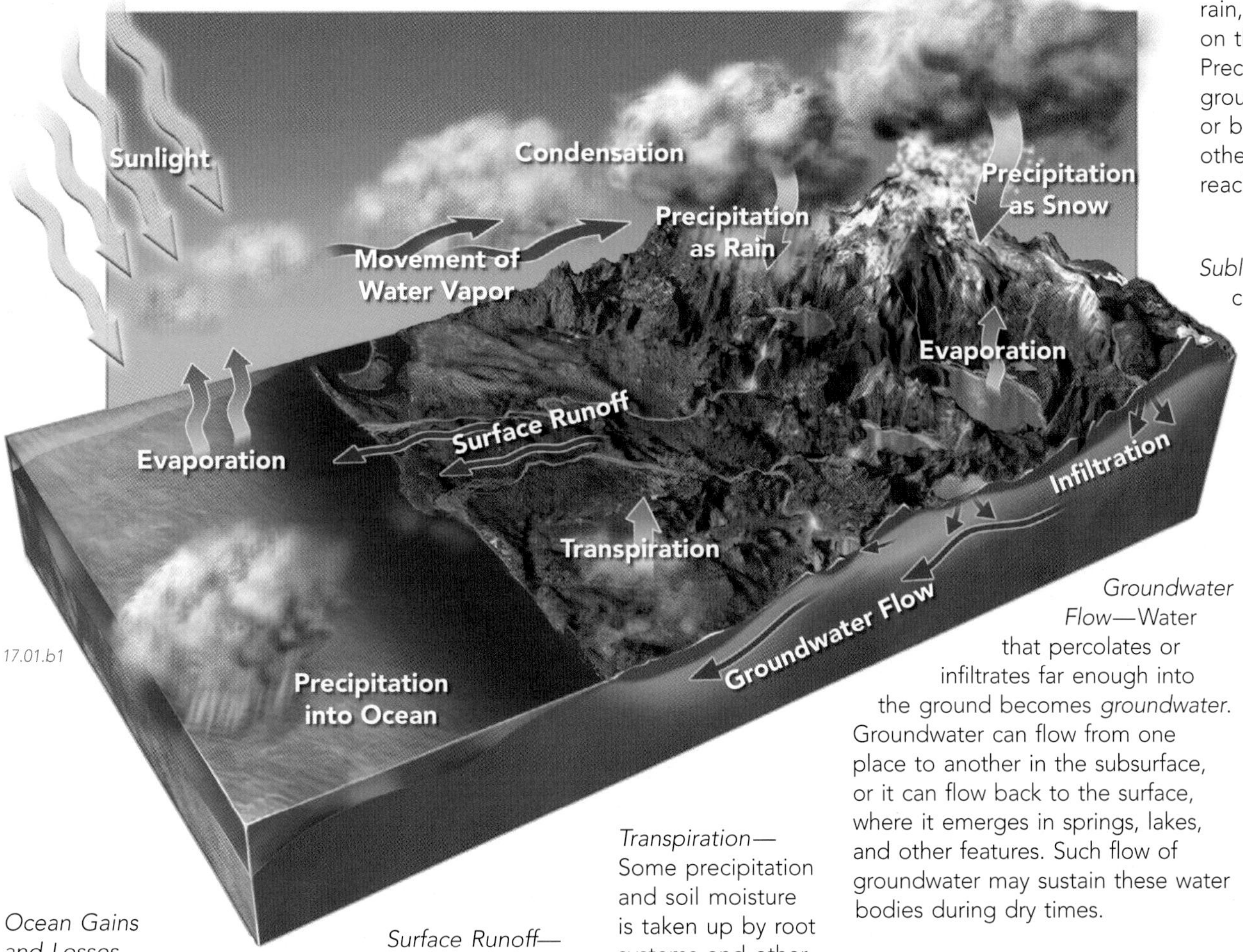

17.01.b1

Sublimation—Water molecules can go directly from a solid (ice) to vapor, a process called *sublimation* (not shown here).

Infiltration—Some precipitation moves downward into the ground, *infiltrating* through fractures and pores in soil and rocks. Some of this water becomes groundwater, some remains within the soil, and some rises back up to the surface. Water can also infiltrate into the ground from lakes, rivers, streams, canals, or any body of water.

Groundwater Flow—Water that percolates or infiltrates far enough into the ground becomes *groundwater*. Groundwater can flow from one place to another in the subsurface, or it can flow back to the surface, where it emerges in springs, lakes, and other features. Such flow of groundwater may sustain these water bodies during dry times.

Transpiration—Some precipitation and soil moisture is taken up by root systems and other water-collecting mechanisms of plants. Through their leaves, plants emit water vapor into the atmosphere by the process of *transpiration*.

Surface Runoff—Rainfall or snowmelt can produce water that flows across the surface as *runoff*. Runoff from direct precipitation can be joined by runoff from melting snow and ice and by the flow of groundwater onto the surface. The various types of runoff collect in streams, rivers, and lakes. Most is eventually carried to the ocean by rivers where it can be evaporated, completing the hydrologic cycle.

Ocean Gains and Losses—Most precipitation falls directly into the ocean, but the ocean loses much more water to *evaporation* than it gains from *precipitation*. The difference is made up by runoff from land.

Before You Leave This Page Be Able To

- ✓ Summarize where most of Earth's total water resides.
- ✓ Describe the different settings where freshwater occurs, identifying which settings contain the most water.
- ✓ Describe or sketch the hydrologic cycle, summarizing the processes that shift water from one part to another.

How Do We Use Freshwater Supplies?

MORE THAN 408 BILLION GALLONS OF FRESHWATER are used in the United States every day. We use freshwater for a variety of purposes, especially power generation and irrigation of farms. Some uses, such as irrigation, actually consume much of the water, whereas others, like recreation, simply use the water but let it continue on its way. How much water does each of our activities consume, and where does the water come from?

A What Are the Main Ways in Which We Consume Freshwater?

The U.S. Geological Survey studied water usage in the United States for the year 2000. The USGS discovered that we consume freshwater in six or seven main ways, depending on how you classify the data.

17.02.a1

◁ 1. *Irrigation*—Farms and ranches are one of the two largest users of freshwater, using nearly 40 percent of freshwater. Farms use water from groundwater, rivers, streams, lakes, and constructed reservoirs to irrigate grain, fruit, vegetables, cotton, animal feed, and other crops.

▷ 2. *Thermoelectric Power*—Electrical-generating power plants are the other large user of freshwater in the United States, also using slightly less than 40 percent. Such plants drive their turbines by converting water into steam and also use large amounts of water to cool hot components. Power plants are also the largest user of saline (salty) water.

17.02.a2

▽ 3. *Public and Domestic Water Uses*—The third largest use of freshwater is by public water suppliers and other domestic uses. We consume water by drinking, bathing, watering lawns and other landscaping, and washing clothes, dishes, and cars. Much water from public water suppliers also goes to businesses. Most of the water for public and domestic use comes from rivers and groundwater.

17.02.a3

Freshwater Usage

Percent	Use
39.6%	Irrigation
39.3%	Thermoelectric Power
13.5%	Public and Domestic Use
5.3%	Industrial Use
1.1%	Aquaculture
0.6%	Mining
0.5%	Livestock

17.02.a4

17.02.a5

△ 4. *Industrial and Mining Uses*—Freshwater and saline water are extensively used by industries, including factories, mills, and refineries. Water is integral to many manufacturing operations, such as making paper, steel, plastics, and concrete. Mining and related activities also use water in the extraction of metals and minerals from crushed rock.

17.02.a6

◁ 5. *Aquaculture*—According to the USGS study, approximately 1 percent of freshwater was used to raise fish and aquatic plants. Most of this usage occurs in Idaho, near the Thousand Springs area. Such water is not totally consumed—much is released back into the Snake River.

▷ 6. *Livestock*—Watering of cows, sheep, horses, and other livestock accounts for only 0.5 percent of freshwater use, but much water is used for irrigation to raise hay, alfalfa, and other animal feed. Many ranches use small constructed reservoirs and ponds as the main water source for animals.

17.02.a7

B How Do We Use and Store Water?

17.02.b1

Electrical Generation—The movement of surface water can be used to generate electricity. For example, a dam helps channel water through turbines in a hydroelectric power plant.

17.02.b2

Transportation—Many large waterways, such as the Mississippi River, are used as energy-efficient transportation systems to transport agricultural products, chemicals, and other industrial products.

17.02.b3

Recreation—Surface water in lakes and rivers is used for many types of recreation, including swimming, tubing, rafting, boating, and fishing. Freshwater is also used to fill swimming pools.

17.02.b4

◁ Surface waters are commonly stored in natural lakes and in constructed *reservoirs* behind concrete and earthen dams. Water for drinking and other municipal uses can be stored in underground or above-ground *storage tanks*.

17.02.b5

◁ Freshwater can be moved from one place to another in canals. Large amounts of freshwater are also moved through raised aqueducts and large and small pipelines. Water is also moved by pumping groundwater to the surface.

C How Do We Refer to Volumes of Water?

Water-resource studies typically report volumes of water in one of three units: *gallons*, *liters*, or *acre-feet*. Gallons and liters are familiar terms, but the concept of an *acre-foot* of water requires some explanation.

How big is an acre? An acre covers an area of 4,047 m² (43,560 ft²). If it was a perfect square, an acre would be 64 m (210 ft) on a side. An acre is equivalent to 91 yards of an American football field.

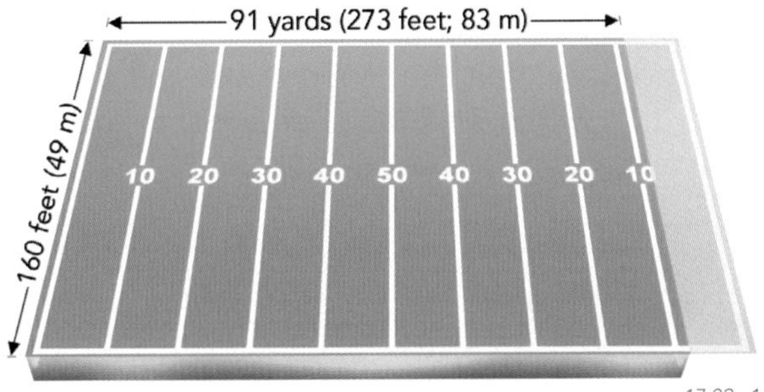

17.02.c1

An acre-foot of water is defined as the volume of water required to cover an acre of land to a height of one foot. Imagine covering 91 percent of a football field with a foot of water. An acre-foot is equivalent to about 326,000 gallons or more than 1.2 million liters.

Drinking-Water Standards in the United States

The U.S. Environmental Protection Agency (EPA) sets standards for safe drinking water. Nearly all *public* water supplies in the United States meet these standards, which can be found at *www.epa.gov*. These standards set a limit on the concentration of a particular contaminant in water. Small municipalities commonly have more trouble meeting these standards than large cities because of limited budgets for building and running facilities to remove contaminants. EPA standards do not apply to private wells.

Many people prefer the taste of bottled water to public tap water, but there are no generally accepted health reasons to buy bottled water so long as the public water provider meets all the federal, state, and local drinking water regulations. Commercially bottled water is monitored by the Food and Drug Administration (FDA) but is not as closely monitored as public water systems. The FDA requires a bottler to test their water source only once a year. Also, bottled water can cost as much as 1,000 times more than municipal drinking water.

Before You Leave This Page Be Able To

- ✓ Describe ways we use freshwater, and which four uses consume the most.
- ✓ Describe how we use and store freshwater.
- ✓ Describe in familiar terms how much water is in an acre-foot.
- ✓ Describe what a drinking water standard is, who sets the limits, and to whom they do and do not apply.

17.2

What Is the Setting of Groundwater?

BENEATH EARTH'S SURFACE is a huge supply of water—*groundwater*. Groundwater occurs beneath all areas but can be far below the surface in some areas and very near the surface in others. Where does this water come from and where does it find room in the solid Earth beneath us?

A What Is Groundwater?

Groundwater is water that is beneath Earth's surface and that exists as a liquid rather than being chemically bonded in minerals. The upper part of rocks and soil can be relatively dry, while lower parts are saturated with water. Groundwater has three different settings that reflect the type of rocks and features that host the water.

17.03.a1

In sediment or sedimentary rocks, adjacent grains do not fit together so there is some space between the grains. These spaces, called *pore spaces*, hold groundwater. Here, the tan objects are grains, the brown represents pore spaces, and the blue indicates pore spaces that are saturated with water.

All types of rocks have fractures that provide openings in which groundwater can accumulate. If fractures connect with one another (are interconnected), then the groundwater can flow. Here, the gray rock is cut by fractures that are unsaturated (brown) or filled with water (blue).

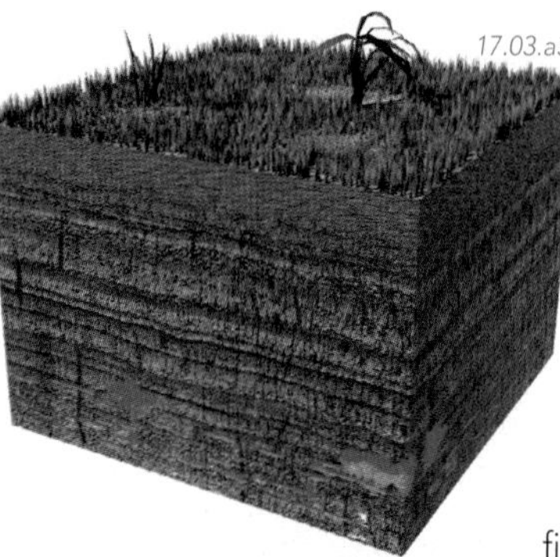

Most groundwater occurs in pore spaces and fractures, but some resides in subterranean cavities and caves. Caves can be filled or partially filled with water, or can be completely dry. The rock shown here is a soluble limestone with wide fractures, bedding planes, and small cavities.

B How Does Groundwater Accumulate?

Groundwater forms from precipitation and snowmelt that seeps from the surface down into the subsurface and accumulates in pores, fractures, and cavities within soil, loose sediment, and rock.

1. When rain falls on the surface or snow melts, the water can either evaporate, be absorbed by plant roots, flow downhill as *runoff*, or seep into the subsurface.

2. Water that soaks into the soil first encounters an upper part of the subsurface where most of the pore spaces are filled with air rather than water. This upper part, called the *unsaturated zone*, can be only centimeters thick or can continue to depths of hundreds of meters. It can become completely dry during long periods without rain.

3. As water penetrates deeper into the subsurface, it eventually enters a zone where all the pore spaces and fractures are filled with water. This zone is called the *saturated zone* and this is where most water occurs in the subsurface.

4. The top of the saturated zone is the *water table*, shown as a dashed red line. *Below* the water table, water fills and can flow through the interconnected pore spaces. Above the water table, some air remains in the pore spaces, and water within the pores can seep downward but is not connected enough to flow.

5. The water table can intersect the surface in lakes, streams, or swamps.

 What Controls How Water Flows Through Rocks?

The rate of groundwater flow is controlled by the steepness of the water table and two important properties of the material—*porosity* and *permeability*. These two properties are related but measure different things. Porosity indicates how much water a rock can hold, but permeability controls whether groundwater can flow.

Porosity

Porosity is the proportion of the volume of rock that is open space (pore space). Porosity varies from less than one percent to more than 50 percent and determines how much water the rock can contain.

1. Well-rounded and well-sorted sediment usually has higher porosity than angular or poorly sorted sediment because the grains do not fit together well. This jar of marbles is analogous to well-rounded cobbles or sand grains and illustrates that a lot of pore space exists in such materials, provided the pore spaces are not filled with a natural cement.

17.03.c1–2

2. Sediment that is poorly sorted, has angular grains, or is held together by a natural cement tends to have *lower* porosity because the finer grained or angular clasts or cement fill in the pore spaces. There is less pore space in this jar of marbles because the marbles are surrounded by fine-grained, angular sediment.

3. Clay consists of small particles shaped like plates or sheets that do not fit tightly together. There is abundant open space (*porosity*) between them, but such pores, like the clay particles, are very small making movement of water difficult. Clay particles can become compacted or can swell when wet, reducing porosity.

17.03.c3 17.03.c4

4. In igneous and metamorphic rocks, porosity is usually low because the minerals are tightly intergrown leaving little free space. Some igneous rocks have less than 1 percent porosity. Fractures cutting the rocks, however, open up narrow spaces and increase the porosity by some amount.

Permeability

Permeability is a measure of the degree to which the pore spaces are interconnected. Low porosity usually results in low permeability, but high porosity does not necessarily indicate high permeability.

5. Loosely cemented gravel and sand commonly contain interconnected pore spaces that allow relatively easy groundwater flow. Such materials have *high permeability* and are the main groundwater host in many areas.

17.03.c5 17.03.c6 17.03.c7 17.03.c8

6. When clay particles are compacted, they tend to become aligned parallel to one another. This decreases the porosity and causes the pore spaces to be very small. Such a rock (e.g., shale) will have very low permeability or may have no permeability.

7. Fractures cut most rock units, but the opened spaces typically represent a small volume of the rock (low porosity). Fractures that are well connected allow water to flow and provide *higher permeability*. Fractures are the only significant permeability in many rocks such as granite.

8. It is possible to have a highly porous rock with little or no permeability. A good example is a vesicular volcanic rock. The bubbles that once contained gas give the rock a high porosity but a *low permeability* since the vesicles are not connected.

Below are examples of high permeability rocks. The cobbles and sand on the left are well rounded, and the fractures on the right are interconnected. Both examples allow water to accumulate in large quantities and move easily through the material. Permeability can be measured in the laboratory or tested in drill holes to express mathematically how groundwater flows using an equation called *Darcy's law*.

17.03.c9

17.03.c10

Before You Leave This Page Be Able To

- ✓ Sketch how groundwater accumulates and occurs in rock and sediment.
- ✓ Sketch and describe what the water table represents.
- ✓ Discuss porosity and permeability, distinguishing between the two and providing examples of materials with high and low values for each attribute.

How and Where Does Groundwater Flow?

GROUNDWATER FLOWS BENEATH THE SURFACE in ways that are controlled by several straight-forward principles. The direction and rate of groundwater flow is largely controlled by the slope of the water table and the geometry and nature of the subsurface rock. Some rock types allow easy groundwater flow, whereas others essentially preclude any significant movement.

A What Is the Geometry of the Water Table?

The water table is usually not a horizontal surface but instead has a three-dimensional shape that mimics the shape of the overlying land surface. There are ridges and hills in a water table as well as valleys. These features control which way groundwater flows.

1. The water table typically has the same shape as the overlying land surface but is more subdued. Where the land surface is high, the water table is also high.

2. The water table slopes from higher to lower areas. It generally is deeper below the surface under mountains than under lowlands, so its slope is less steep than that of the land surface. The shape of the water table is largely independent of the geometry of rock units through which the water table passes.

6. Where the water table intersects the land surface, there may be lakes, wetlands, or a flowing river. Not all rivers coincide with the water table, because some flowing rivers are underlain by unsaturated materials. The river in this figure does represent the water table being at the surface.

17.04.a1

3. Groundwater just below the water table flows *down the slope of the water table*. In this example, it flows from left to right, from areas of higher elevation to areas of lower elevation.

4. Where the water table is horizontal, such as near this stream, groundwater may flow very slowly or not at all. At depth, groundwater may flow in directions other than the near-surface flow.

5. The terminology used to describe features of a water table is derived from topography. A high part of the water table separating parts sloping in opposite directions is called a *groundwater divide*. On either side of a groundwater divide, groundwater flows in opposite directions.

B What Controls the Rate of Groundwater Flow?

The rate of groundwater flow is strongly influenced by the steepness of the water table because flow is driven by the force of gravity. Other factors being equal, water flows faster down a steep water-table slope and slower down a more gentle one. The slope of the water table is called the *hydraulic gradient*.

17.04.b1

Direction of Groundwater Flow

17.04.b2

The rate of groundwater flow is also strongly controlled by the *permeability* of the rock type. In this diagram, flow is fastest in highly permeable cavernous limestone.

Flow is moderately fast in a porous conglomerate or well-sorted sandstone.

Flow is slower in shale, which has small pores, and in a granite with poorly connected fractures.

B How Do Hydrogeologists Depict the Water Table?

Once the appropriate field, drilling, and geophysical data are collected, hydrogeologists produce various types of maps, especially maps on which the elevation of the water table is contoured.

The most important piece of information about groundwater is a map showing variations in the elevation of the water table.

The first step in constructing such a map is to collect and plot elevations of the water table in all available wells. Each number on this map indicates the elevation (in meters above sea level) of the water table at a well in that location.

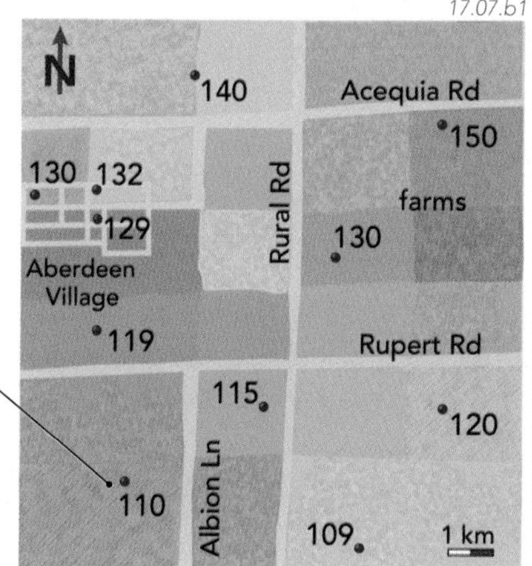

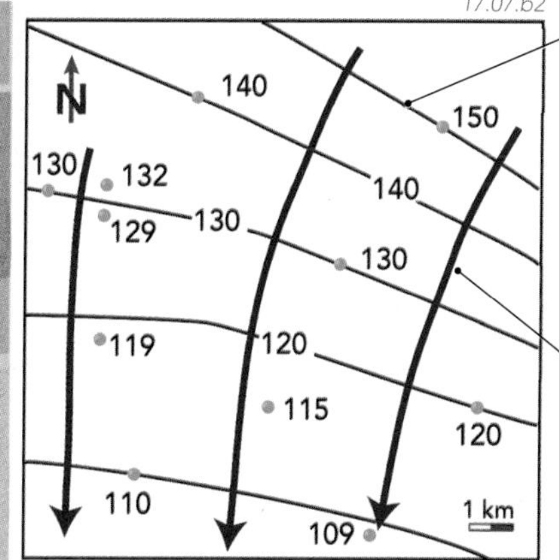

Hydrogeologists then draw contours to show the elevation of the top of the water table. These contours indicate the elevation of the water table in meters. Each contour follows a specific elevation on the water table.

Arrows drawn perpendicular to the contours show the direction of groundwater flow, which is down the slope of the water table.

Other Depictions

Hydrogeologists compare contour maps of water-table elevations to other features, such as the locations of wells, rivers, farms, and other sites that may affect the groundwater, such as by taking water out of the ground.

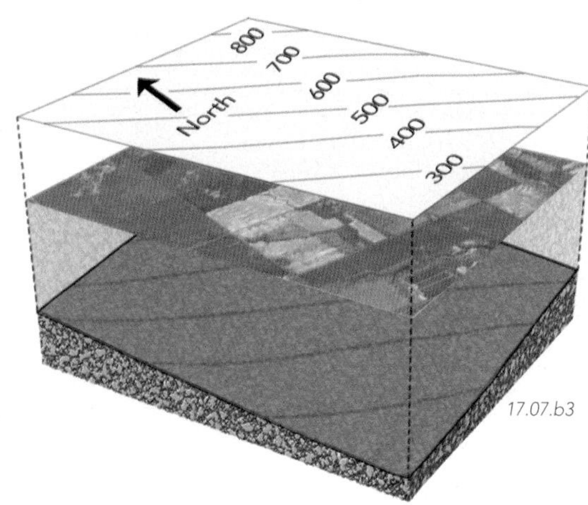

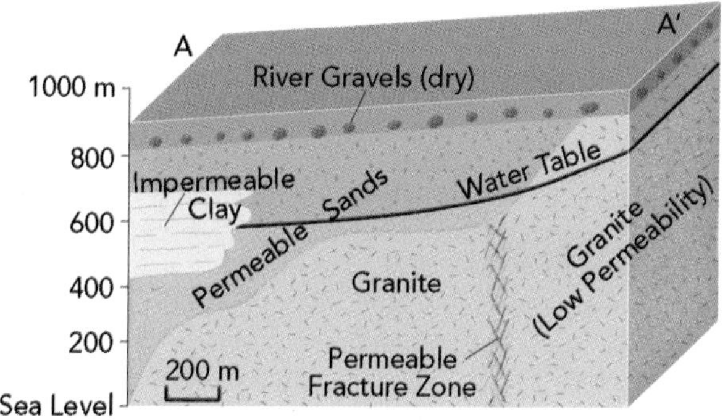

A cross section or block diagram, usually drawn with some vertical exaggeration, helps explore how the water table is related to the subsurface geology. Key considerations include the geometry and distribution of different geologic materials, especially those of different permeability, and how much of each unit is below the water table (i.e., in the saturated zone where it could yield water).

Hydrogeologists incorporate the geologic information and well data into computer programs to produce three-dimensional depictions of the water table, as shown below. They then model the directions and rates of groundwater flow and calculate the volumes of freshwater that will be available for drinking and other uses.

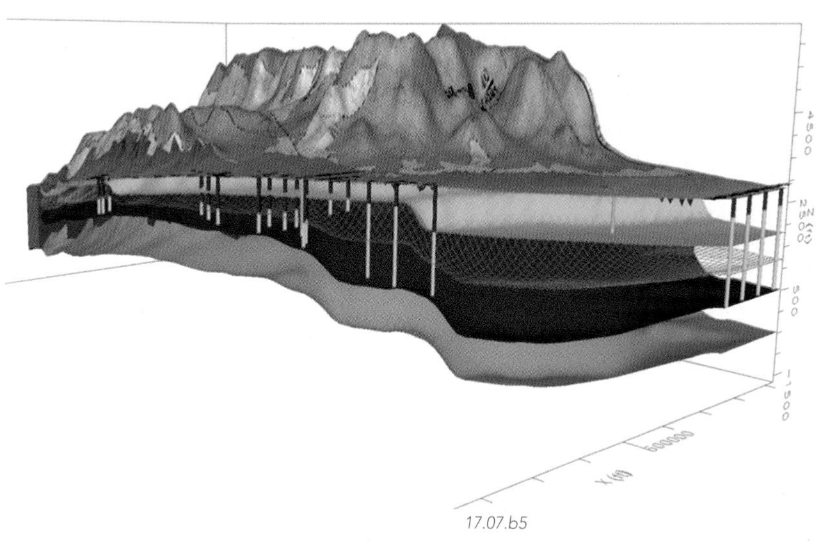

The goal of the various depictions is to understand the three-dimensional geometry of the basin, rock units, water table, and topography. These factors control where and how much water accumulates, where and how it flows, and how it interacts with features we see on the surface.

Before You Leave This Page Be Able To

- ✓ Summarize the types of information that hydrogeologists collect and what each indicates about the subsurface.
- ✓ Describe how a contour map of water-table elevations is constructed and how it would be used to predict the direction of groundwater flow.
- ✓ Describe factors to show in a cross section or block diagram if groundwater is the focus of the study.

What Problems Are Associated with Groundwater Pumping?

THE SUPPLY OF GROUNDWATER IS FINITE so pumping too much groundwater, a practice called *overpumping*, can result in serious problems. Overpumping can cause neighboring wells to dry up, land to subside, and gaping fissures to open across the land surface.

A What Happens to the Water Table If Groundwater Is Overpumped?

Demands on water resources increase if an area's population grows, the amount of land being cultivated increases, or open space is replaced by industry. Groundwater is viewed as a way to acquire additional supplies of freshwater, so new wells are drilled or larger wells replace smaller ones.

How Overpumping Affects the Geometry of the Water Table

1. The problems with overpumping are illustrated by examining a simple case. The topography of this area is fairly flat, there are no bodies of surface water, and a single type of porous and permeable sediment composes the subsurface.

2. As people move into a nearby town, a small well is drilled down to the water table to provide freshwater. The small well pulls out so little groundwater that the water table remains as it has for thousands of years, nearly flat and featureless. The well remains a dependable source of water because its bottom is *below* the water table.

3. Across the entire area, groundwater flows from right to left, down the gentle slope of the water table.

4. As more people move into the surrounding area, a larger well is drilled to extract larger volumes of water to satisfy the growing demand.

5. The new, larger well pumps water so rapidly that groundwater around the well cannot flow in fast enough to replenish what is lost. This causes the local water table to drop and form a funnel-shaped *cone of depression* around the well.

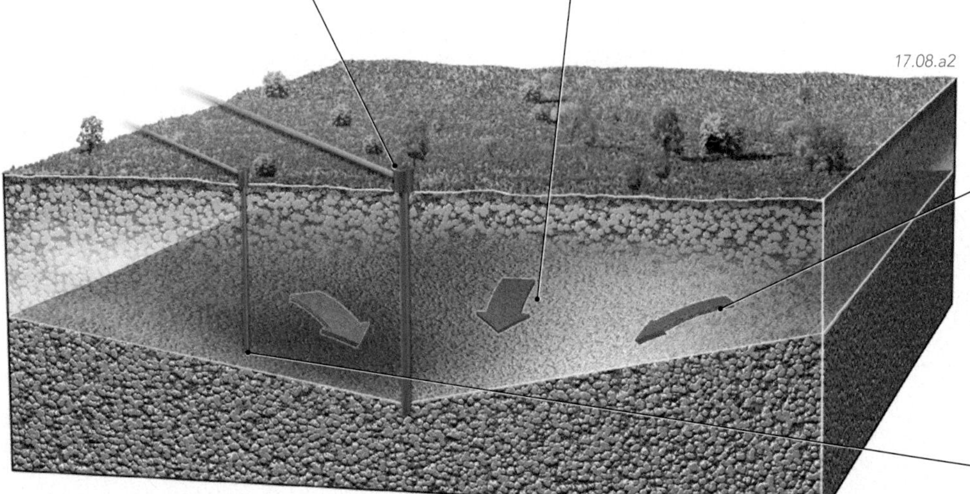

6. The direction of groundwater flow changes dramatically across the entire area. Instead of flowing in one direction, groundwater now flows toward the larger well and into the cone from all directions. The change in flow direction has unintended consequences. It can bring contaminated water into previously fresh wells and may cause serious safety issues, since waste-disposal sites, such as landfills, are generally planned with the groundwater-flow direction in mind.

7. The original well dries up because it no longer reaches the water table, which has been lowered by the larger well's cone of depression.

C How Is Groundwater Contamination Tracked and Remediated?

Once groundwater contamination is identified, what do we do next? Hydrogeologists compile available information to compare the distribution of contamination with all relevant geologic factors. One commonly used option to clean up, or *remediate*, a site of contamination is called "pump-and-treat." Remediation is much more expensive than not causing the problem to begin with.

1. The first step to remediation is to properly understand the situation from all angles—what is the nature of the contamination, where is the contamination now, where did it come from, where is it going, and what are the geologic controls?

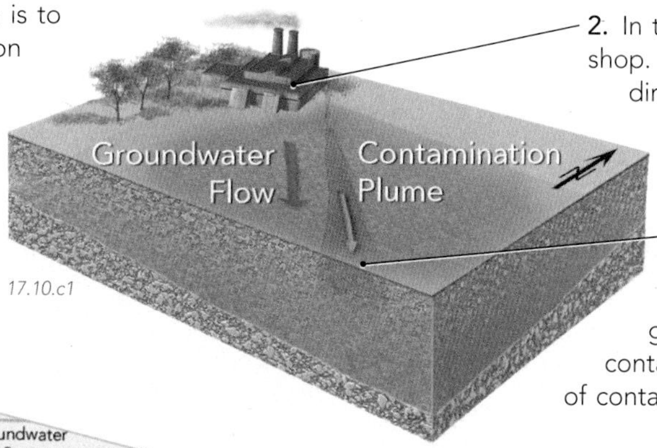

2. In this area, contamination was caused by a chrome-plating shop. The water table slopes to the southeast so this is the direction that the upper levels of groundwater flow and the direction in which contamination is expected to move.

3. Groundwater flow physically carries the chromium ions faster than the chromium ions can chemically diffuse through the water, so the flow of groundwater forms a *plume* of contamination. There is no contamination up-flow (northwest) of the shop, but the plume of contamination will spread to the southeast.

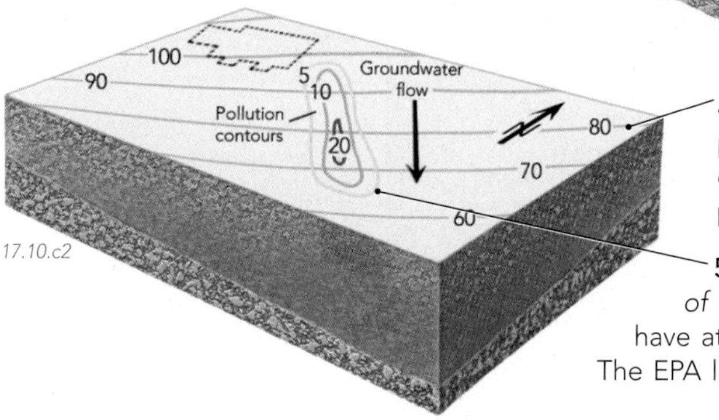

4. To investigate the situation, we contour *elevations of the water table* to more precisely determine which way groundwater is flowing. In this case, the contours decrease in elevation to the southeast. Groundwater flows to the southeast, perpendicular to the contours (and toward lower-elevation contours).

5. We draw a second set of contours based on chemical analyses of the *concentration of contamination*, in this case chromium. For example, areas within the 5 mg/L contour have at least 5 mg/L chromium, and those within the 10 mg/L contour have at least 10 mg/L. The EPA limit for chromium is 0.1 mg/L, so these values are well above EPA standards.

6. From these maps, we can now determine where the contamination is, which way it is moving, and where it will go in the future (down the slope of the water table). If from interviews or historical records we can determine how long ago the contamination occurred, we can use simple calculations (distance/time) to get the rate of flow. We also can use computer simulations to model past and future movement.

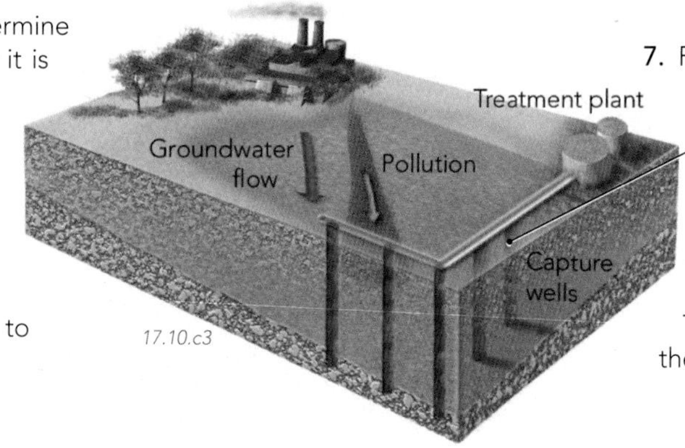

7. Finally, we try to clean up the contamination. One strategy is to drill wells in front of the projected path of the contamination to contain, capture, and extract the contaminated water. The contaminated water is pumped to the surface, where it is processed with carbon filters or other appropriate technology to separate the contaminant from the water. The cleaned water is typically released back into the environment.

A Civil Action

Woburn, Massachusetts, a small town 10 miles north of Boston, was the site of a classic legal case involving groundwater contamination. The case was made famous in the book *A Civil Action* by Jonathan Harr and in a movie of the same name starring John Travolta.

The trouble began in the 1960s when the city drilled two new groundwater wells for municipal water supplies. The wells were drilled into glacial and river sediments that had filled an old valley. After the wells were installed, some residents complained that the water tasted odd and had a chemical odor. Over the next 20 years, residents began to show a high incidence of leukemia and other serious health problems. Chemical analyses showed that the groundwater was contaminated with trichaloroethylene (TCE) and other volatile organic compounds. Local families filed a lawsuit against several chemical companies that were potentially responsible. The verdict remains complex, but the site is a classic example of the interaction of geology, water, health, and environmental law.

Before You Leave This Page Be Able To

- ✓ Sketch a plume of contamination, showing how it relates to the source of contamination and the direction of groundwater flow.
- ✓ Describe some ways that geologists investigate groundwater contamination.
- ✓ Sketch how chemical analyses define a plume of contamination, and one way a plume could be remediated.

What Is Going on with the Ogallala Aquifer?

THE MOST IMPORTANT AQUIFER IN THE UNITED STATES lies beneath the High Plains, stretching from South Dakota to Texas. It provides groundwater for about 20 percent of all cropland in the country, but is severely threatened by overpumping. The setting, characteristics, groundwater flow, and water-use patterns of this aquifer bring together many different aspects of water resources and their relationship to geology.

A What Is the Setting of the Ogallala Aquifer?

1. The *Ogallala aquifer*, also called the *High Plains aquifer*, covers much of the High Plains area. The lightly shaded area on this map shows the outline of the main part of the aquifer. The aquifer forms an irregularly shaped north-south belt from South Dakota and Wyoming through Nebraska, Colorado, Kansas, the panhandles of Oklahoma and Texas, and eastern New Mexico.

2. The Ogallala aquifer covers about 450,000 km^2 (174,000 mi^2) and is currently the largest source of groundwater in the country. It provides 30 percent of all groundwater used for irrigation in the United States. In 1980, near the height of the aquifer's use, 17.6 million acre-feet of water were withdrawn to irrigate 13 million acres of land. The water is used mostly for agriculture and rangeland. The main agricultural products include corn, wheat, soybeans, and feed for livestock.

17.11.a1

3. The aquifer is named for the Ogallala Group, the main geologic formation in the aquifer. The unit was named by a geologist in the early 1900s after the small Nebraskan town of Ogallala.

4. Much of the Ogallala Group consists of sediment deposited by rivers and wind during the later parts of the Cenozoic, mostly between 19 and 5 million years ago. Braided rivers carried abundant sediment eastward from the Rocky Mountains, spreading over the landscape and depositing a relatively continuous layer of sediment. Deposition stopped when regional uplift and tilting caused the rivers to downcut and erode rather than continuing to deposit sediment. Present-day rivers continue to erode into the aquifer.

The Aquifer in Cross Section

5. This vertically exaggerated cross section shows the thickness of the aquifer from west to east. The aquifer is shown in various colors, and rocks below the aquifer are shaded gray. Note that the aquifer is at the surface and is an *unconfined aquifer*.

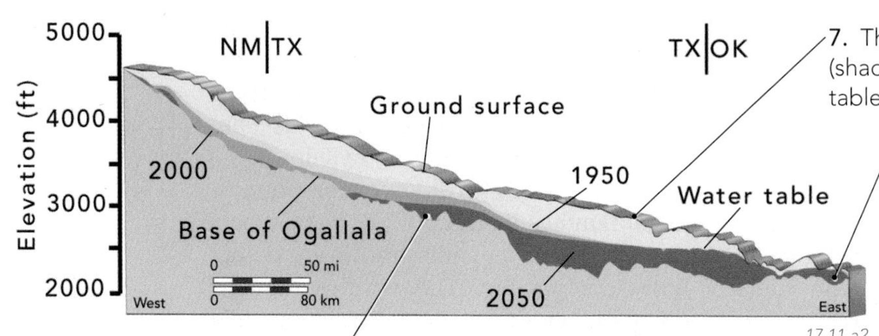

17.11.a2

6. The irregular base of the aquifer is an unconformity that reflects erosion of the land before deposition of the aquifer.

7. The upper part of the aquifer (shaded yellow) is above the water table and in the *unsaturated zone*.

8. Blue shows levels of the water table for 1950 and 2000, and purple shows the predicted levels for 2050. Note that water levels in the aquifer have fallen due to *overpumping*. The western part is predicted to be totally depleted by 2050 (no purple).

B Where Does Groundwater in the Aquifer Come from and How Is It Used?

Most of the water going into the aquifer is from local precipitation. This map, colored according to the amount of precipitation received across the area, with darker shades indicating more precipitation, shows that the western part receives much less precipitation (rain, snow, and hail) than the eastern part.

The parts of the aquifer that receive the least precipitation—the southwestern parts—are also those predicted to go dry by 2050.

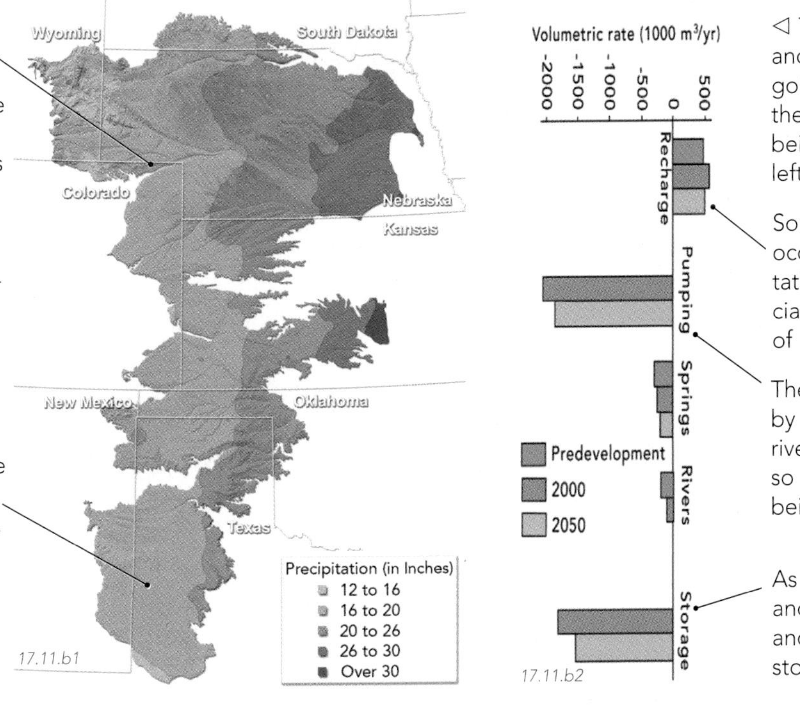

◁ This graph shows the water balance for the Ogallala aquifer. Water going into the aquifer is shown to the right of the axis, whereas water being lost by the aquifer is on the left side of the axis.

Some groundwater recharge is occurring where water from precipitation seeps into the aquifer, especially in areas with higher amounts of precipitation.

The amount taken out of the aquifer by pumping, springs, and inflow into rivers greatly exceeds the recharge so most parts of the aquifer are being dewatered.

As the aquifer dewaters it compacts and causes a decrease in porosity and a loss of pore space in which to store water.

C How Has Overpumping Affected Water Levels in the Ogallala Aquifer?

The USGS estimates that the aquifer contains 3.2 billion acre-feet of water! That is enough to cover the entire lower 48 states with 1.7 feet of water. How much has overpumping affected the aquifer's water levels, and what will happen to the region and the country if much of the aquifer dries up?

This map shows the thickness (in feet) of the *saturated zone* within the aquifer. In some of its northern parts, more than 1,000 feet (300 m) of the aquifer is saturated with water, whereas less than 100 to 200 feet remain saturated in the southern parts.

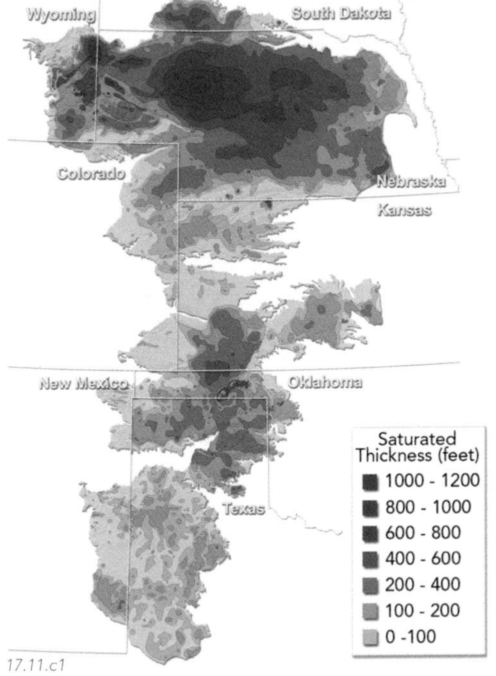

This map shows how many feet the water table dropped in elevation between 1980 and 1995 as a consequence of overpumping. The largest drops, exceeding 40 feet, occurred in southwestern Kansas and the northern part of Texas. Compare this map to the one for precipitation.

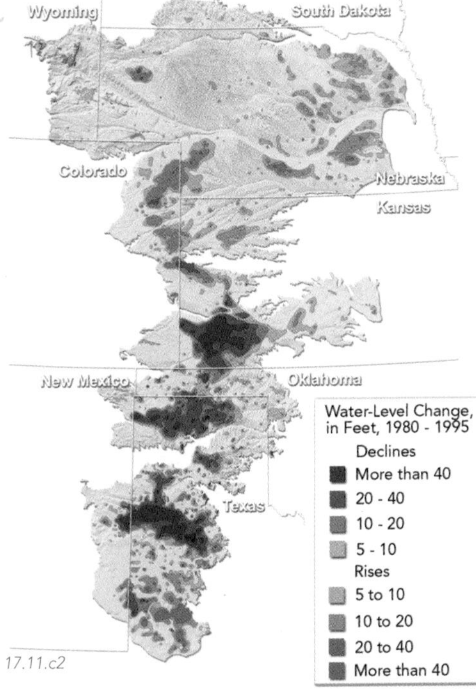

Future Predictions—It is uncertain what will happen, but hydrogeologists have done detailed studies of key areas to try to predict what will happen in the next decades. Projections of current water use and numerical models of the water balance predict that some parts of the aquifer will go dry by 2050. This will have catastrophic consequences for the local farmers, ranchers, and businesses, and for people across the country who depend on the aquifer for much of their food. Subsidence related to groundwater withdrawal and compaction of the aquifer will be an increasing concern. What do you think would happen to the region if this aquifer is partly pumped dry?

Before You Leave This Page Be Able To

- ✓ Summarize the location, characteristics, origin, and importance of the Ogallala aquifer.
- ✓ Summarize the water balance for the aquifer and how water levels have changed in the last several decades.

Who Polluted Surface and Groundwater in This Place?

SURFACE WATER AND GROUNDWATER IN THIS AREA are contaminated. You will use the geology of the area, along with elevations of the water table and chemical analyses of the contaminated water, to determine where the contamination is, where it came from, and where it is going. From your conclusions, you will decide where to drill new wells for uncontaminated groundwater.

Goals of This Exercise:

- Observe the landscape to interpret the area's geologic setting.
- Read descriptions of various natural and constructed features.
- Use well data and water chemistry to draw a map showing where contamination is and which way groundwater is flowing.
- Use the map and other information to interpret where contamination originated, which facilities might be responsible, and where the contamination is headed.
- Determine a well location that is unlikely to be contaminated.
- Suggest a way to remediate some of the contamination.

1. The region contains a series of ridges to the east and a broad, gentle valley to the west. Small towns are scattered across the ridges and valleys. There are also several farms, a dairy, and a number of industrial sites, each of which is labeled with a unique name.

2. A main river, called the *Black River* for its unusual dark, cloudy color, flows westward (right to left) through the center of the valley. The river contains water all year, even when it has not rained in quite a while. Both sides of the valley slope inward, north and south, toward the river.

Procedures

Use the available information to complete the following steps, entering your answers in appropriate places on the worksheet.

1. This figure shows geologic features, rivers, springs, and human-constructed features, including a series of wells (lettered A through P). Observe the distribution of rock units, sediment, rivers, springs, and other features on the landscape. Compare these observations with the cross sections on the sides of the terrain to interpret how the geology is expressed in different areas.
2. Read the descriptions of key features and consider how this information relates to the geologic setting, to the flow of surface water and groundwater, and to the contamination.
3. The data table on the next page shows elevation of the water table in each lettered well. Use these data and the base map on the worksheet to construct a groundwater map with contours of the water table at the following elevations: 100, 110, 120, 130, and 140 meters. On the contoured map, draw arrows pointing down the slope of the water table to show the direction of groundwater flow.
4. Use the data table showing concentrations of a contaminant, purposely unnamed here, in groundwater to shade in areas where there is contamination. Use darker shades for higher levels of contamination.
5. Use the groundwater map to interpret where the contamination most likely originated and which facilities were probably responsible. Mark a large X over these facilities on the map and explain your reasons in the worksheet.
6. Determine which of the lettered well sites will most likely remain free of contamination, and draw circles around two such wells.
7. Devise a plan to remediate the groundwater contamination by drilling wells in front of the plume of contamination; mark these on the map with the letter R.

17.12.a1

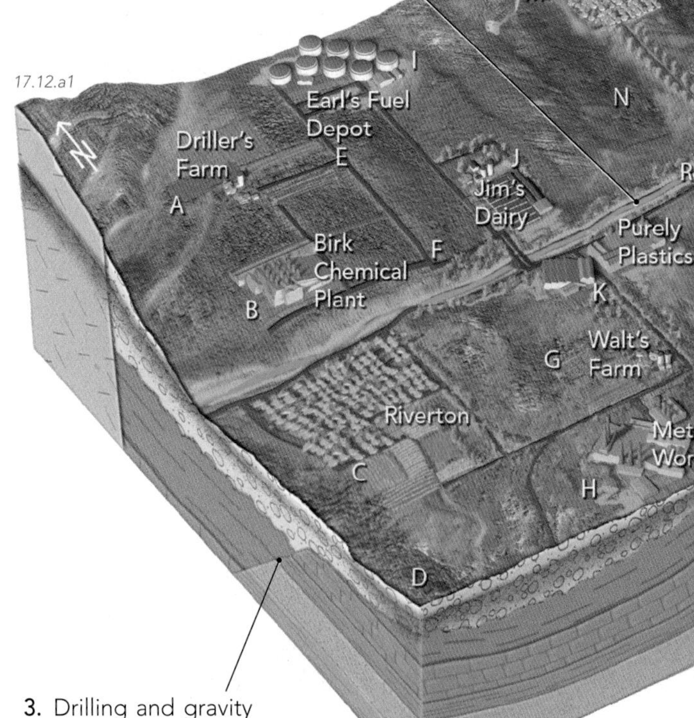

3. Drilling and gravity surveys have shown that the valley is underlain by a thick sequence of relatively unconsolidated and weakly cemented sand and gravel. The deepest part of the basin has been downdropped by normal faults, one of which is buried beneath the gravel.

4. Based on shallow drilling, the water table (the top of the blue shading) mimics the topography, being higher beneath the ridges than beneath the valleys. Overall, the water table slopes from east to west (right to left), parallel to the regional slope of the land.

5. From mapping and other studies on the surface, geologists have determined the sequence of rock units, as summarized in the stratigraphic section in the upper right corner of this page. These studies also document a broad anticline and syncline beneath the eastern part of the region.

6. Bedrock units cross the landscape in a series of north-south stripes, parallel to the strike of the rock layers. One of the north-south valleys is named *Coal Mine Valley* because it contains several large coal mines and a coal-burning, electrical-generating plant. An unsubstantiated rumor says that one of the mines had some sort of chemical spill that was never reported. Activity at the mines and power plant has caused fine coal dust to be blown around by the wind and washed into the smaller rivers that flow along the valley.

7. A north-south ridge is composed of sandstone, which geologists call the *lower sandstone*. A few nice-tasting, freshwater springs issue from the sandstone where it is cut by small streams.

17.12.a2

Stratigraphic Section

Gravel–Unconsolidated sand and gravel in the lower parts of the valley

Upper Sandstone–Well-sorted, permeable sandstone

Upper Shale–Impermeable, with coal

Sinkerton Limestone–Porous, cavernous limestone

Middle Shale–Impermeable shale

Lower Sandstone–Permeable sandstone

Lower Shale–Impermeable shale

Basal Conglomerate–Poorly sorted with salty water

Granite–Sparsely fractured; oldest rock in area

8. The highest part of the region is a ridge of granite and sedimentary rocks along the east edge of the area. This ridge receives quite a bit of rain during the summer and snow in the winter. Several clear streams begin in the ridge and flow westward toward the lowlands.

9. A coal-burning power plant was built over tilted beds of a unit named the *Sinkerton Limestone,* so-called because it is associated with many sinkholes, caves, and *karst topography*. The limestone is so permeable that the power plant has had difficulty keeping water in ponds built to dispose waste waters that are rich in the chemical substances that are naturally present in coal.

10. The tables below list water-table elevations in meters and concentrations of contamination in milligrams per liter (mg/L) for each of the lettered wells (A–P), and the concentration of contamination in samples from four springs (S1–S4) and eight river segments (R1–R8). The location of each sample site is marked on the figure.

Well	Elev. WT	mg/L
A	110	0
B	100	0
C	105	0
D	110	20
E	120	10
F	115	0
G	120	0
H	120	50

Well	Elev. WT	mg/L
I	130	30
J	125	0
K	120	0
L	130	0
M	140	50
N	140	0
O	140	0
P	140	0

Spring	mg/L
S1	50
S2	0
S3	0
S4	0

River	mg/L
R1	0
R2	20
R3	0
R4	0

River	mg/L
R5	0
R6	0
R7	5
R8	5

17.12

CHAPTER 18

Energy and Mineral Resources

NATURAL RESOURCES ARE THE FOUNDATIONS OF SOCIETY. They provide us with the necessary material to sustain our way of life. After water and soils, the most important natural resources are *energy resources* for electricity, transportation, and factories, and *mineral resources*, the starting materials for metals, concrete, bricks, and many other things.

The Arabian Peninsula, most of which is encompassed by the country of Saudi Arabia, is a dry, desert land, bounded on the west by the *Red Sea* and on the east by the *Persian Gulf*. The peninsula is asymmetrical: Its western flank along the Red Sea is a series of steep escarpments, whereas on the east it gradually decreases in elevation until it slips beneath the shallow water of the Persian Gulf.

How did the peninsula form, why is it asymmetric, and why is the Red Sea much deeper than the Persian Gulf?

The Persian Gulf region produces a quarter of the world's oil—about 23 million barrels a day! The established oil reserves are more than 700 billion barrels of oil, or 57 percent of the world's reserves of crude oil. The region also has 45 percent of the world's known gas reserves, and the USGS estimates that the region has the greatest potential for undiscovered oil of any part of the world.

Where do oil and gas come from, and why does this region have such a large share of these critical resources?

18.00.a2

The Zagros Mountains run parallel to the Persian Gulf through the western part of Iran. As shown in this satellite image (△), these mountains contain large folds formed as the Arabian plate is thrust beneath the southwestern edge of the Eurasian plate. This region also is rich in oil.

How are oil and gas related to folded mountain belts?

18.00.a1

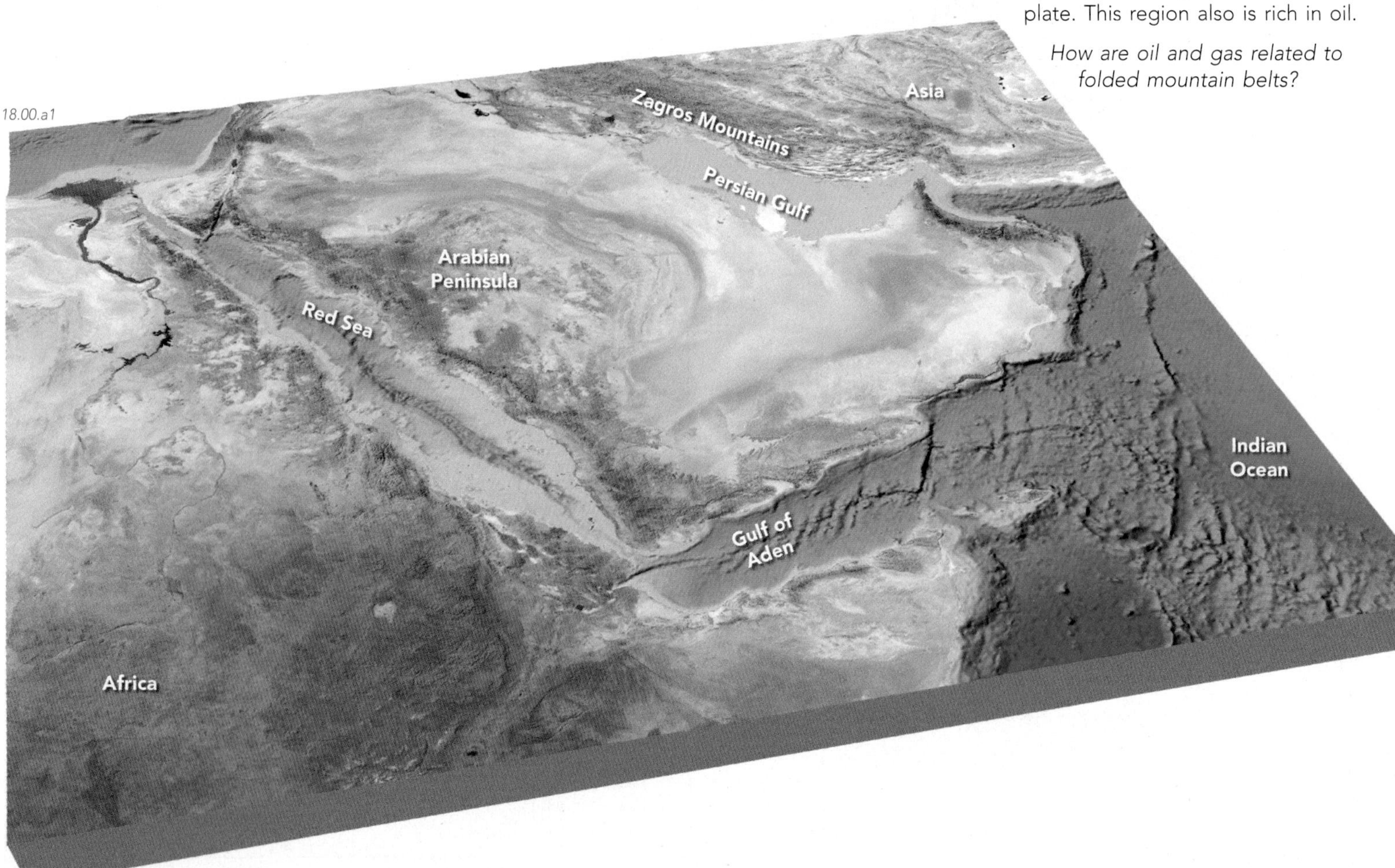

Asia
Zagros Mountains
Persian Gulf
Arabian Peninsula
Red Sea
Indian Ocean
Gulf of Aden
Africa

TOPICS IN THIS CHAPTER

18.00.a3

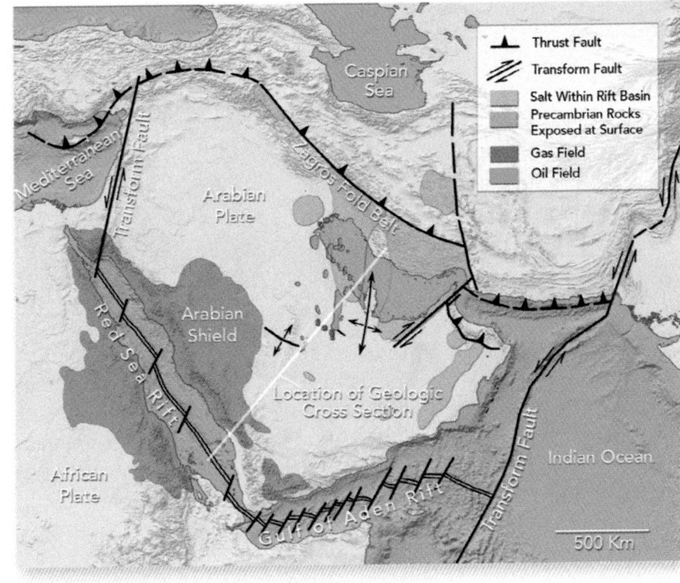

▷ This map shows the main tectonic features of the Arabian Peninsula, the oil and gas fields, and areas where Precambrian rocks are exposed at the surface. The Persian Gulf and Zagros Mountains are forming along a convergent boundary where the Arabian plate is pushing beneath Iran. Most oil is found west of the collision zone.

The Red Sea is a divergent boundary, where the Arabian Peninsula is spreading away from Africa. The flanks of the rift have been uplifted, exposing deeper Precambrian rocks. The Arabian Peninsula lies on the Arabian plate and contains folds, faults, and the world's largest oil and gas fields.

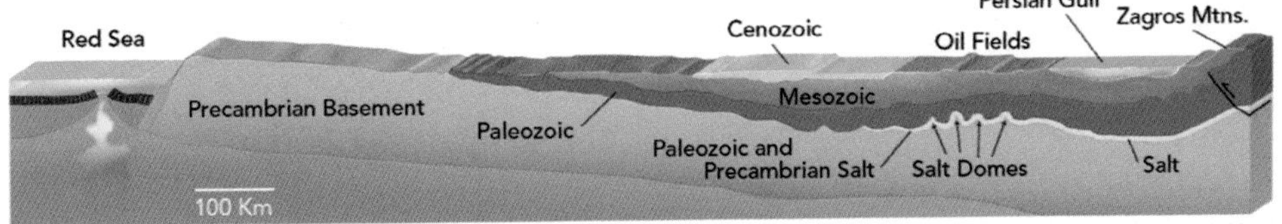

18.00.a4

△ In cross section, the Arabian Peninsula is tilted—the western part was rifted and uplifted, while the eastern part subsided beneath the weight of the Zagros thrusts. Upper sedimentary layers thicken toward the Persian Gulf and contain folds, salt domes, and oil fields.

18.00.a5

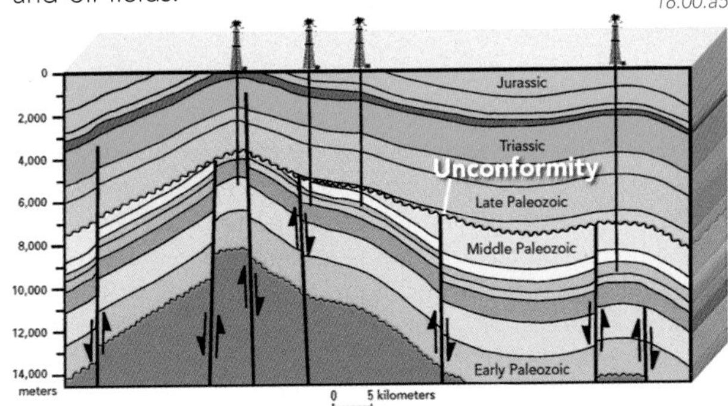

◁ This more detailed cross section of a small part of eastern Saudi Arabia shows the Ghawar oil field, the largest in the world. The oil wells pump oil from within the sedimentary layers. Some wells are centered over anticlines, whereas others were drilled into layers beneath an angular unconformity.

Natural Resources

We rely on a variety of natural resources, some of which are obvious to us, like oil and natural gas, and others that work behind the scenes. It is often said—and it is true—that if a material we use is not grown, it probably came from geology and was found by a geologist. This is especially true of *energy and mineral resources.*

Energy resources include oil and gas, as well as coal, nuclear power, and energy derived from dams, wind, and the Sun. Such resources are not equally distributed in every part of the world. Some areas, like Saudi Arabia and Wyoming, are rich in energy resources, whereas others, like South Africa and the Upper Peninsula of Michigan are rich in mineral resources. Some areas have neither. Why is this so? What factors cause some areas to be rich in resources and others to have so few?

The answer, of course, is geology. Each region has its own unique geologic history, which means that some areas have thick sequences of sedimentary rocks, whereas others have granite and metamorphic rocks. Some areas have folds and faults, whereas others have horizontal layers.

With these variations in geologic history come differences in the abundance and kinds of energy and mineral resources. Certain geologic processes are required to form oil or a copper deposit. For such resources, we are at the mercy of geologic events, most of which happened millions of years ago. The political and economic systems of the world must function around the geologic reality.

How Do Oil and Natural Gas Form?

OIL AND NATURAL GAS SEEM LIKE ODD SUBSTANCES to find in solid rock. Where do they come from? Oil and natural gas, together called *petroleum*, are produced naturally when sediment rich in organic material is deposited, buried, and heated to slightly elevated temperatures. Once formed, petroleum can escape to the surface or be trapped at depth.

A What Is Petroleum and Where Does It Come From?

Naturally occurring petroleum is an organic substance, largely composed of carbon atoms chemically bonded with hydrogen atoms and smaller amounts of other elements. The dominance of hydrogen and carbon atoms is the reason we use the term *hydrocarbons* to refer to oil and natural gas, as well as to their refinery-produced derivative products, such as gasoline and diesel fuel. The organic material comes from several sources.

18.01.a1

Reefs teem with life, including fish and marine organisms that are themselves microscopic but build visible structures, like this coral. Other creatures live in deeper and colder water, and contribute organic material to deep-ocean sediment.

18.01.a2

Plants contribute organic material to sediment, and can grow on land or can be aquatic (i.e., be in water), such as kelp and other aquatic plants. Land plants can make coal and gas when buried, but generally do not decompose to oil.

18.01.a3

Most petroleum comes from *microorganisms*, which can occur in amazing variety and abundance in seas and lakes. Such microorganisms include algae, bacteria, and tiny creatures that live near the surface or in deeper water.

B What Processes Turn Organic Material into Oil and Gas?

Natural organic material begins as relatively unordered substances that are similar to wax and fat. When progressively heated, these organic materials are converted to a succession of other hydrocarbons, including oil. The conversion of organic material to oil, like most geologic processes, generally takes millions of years.

18.01.b1

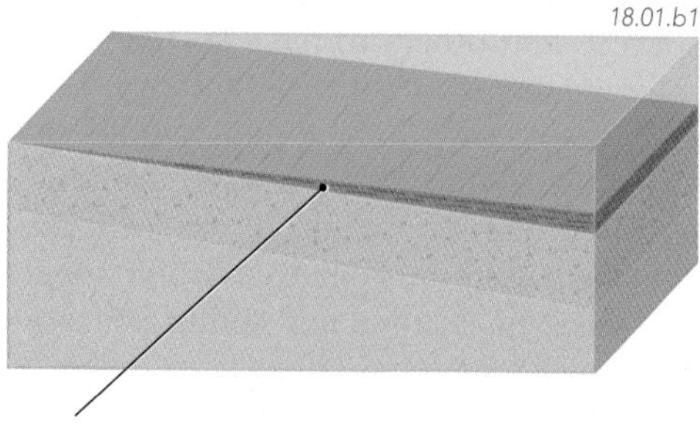

The first stage in formation of oil and gas is *accumulation* of organic material, such as this layer of dark, organic-rich mud. A rock that contains enough organic material to produce petroleum is called a *source rock*. At low temperatures near the surface, the organic material is relatively unordered or still retains some of the structure of the animals or plants from which the material was derived.

18.01.b2

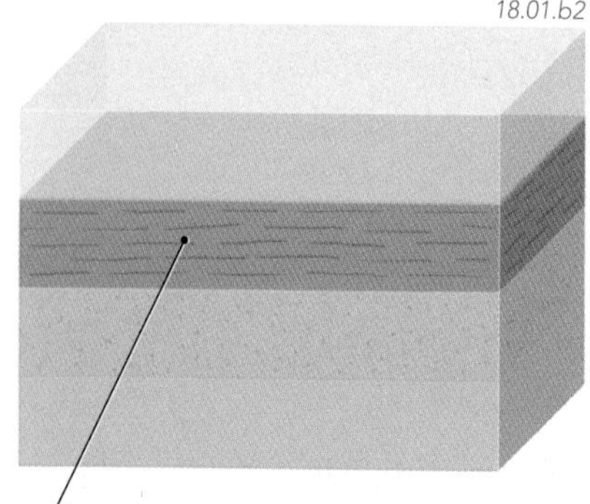

To end up as oil, the organic material must be *preserved*, which means being deposited in oxygen-poor conditions and buried under other layers. When buried to shallow depths and heated to less than ~60°C, the organic starting material is converted into *kerogen*, a thick substance composed of long chains of hydrocarbons.

18.01.b3

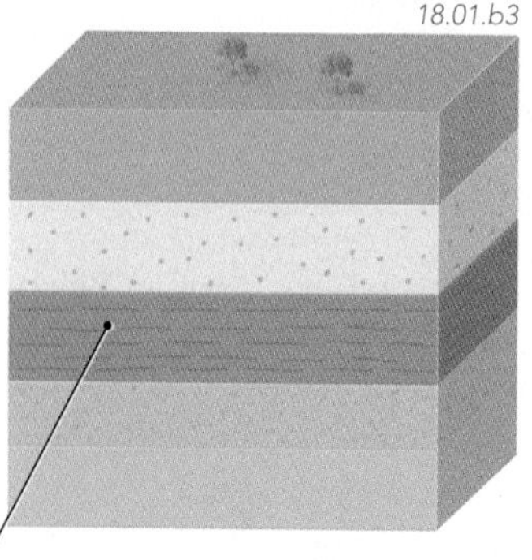

Over time, the source rocks can be progressively *buried* by younger layers and heated because temperature increases with depth. When heated to 60 to 120°C, the long hydrocarbon chains in kerogen break down into heavy and light *oils*. *Natural gas* is also produced, but especially at higher temperatures (120°C to about 200°C).

C Where Do Oil and Gas Go?

Once oil and gas are formed, what happens? Both are mobile, fluid materials and can travel along fractures and through pore spaces between grains. In many situations, they remain within, or fairly close to, the source rock in which they originated. In other cases, they migrate far from where they were generated.

1. As you can observe for yourself by placing several drops of any oil in a bowl of water, oil and gas are lighter (i.e., less dense) than water. They float on the surface of water and will buoyantly rise through groundwater toward the surface. Water under pressure can force oil and gas upward or laterally (sideways) through the rock.

2. Oil and gas, like groundwater, can flow through rocks that are *permeable* enough to allow them to pass through. Some rocks, like many sandstones, have open spaces between the grains and along fractures and so are relatively permeable. Oil and gas can flow through layers, such as the inclined sandstone layer shown here.

18.01.c1

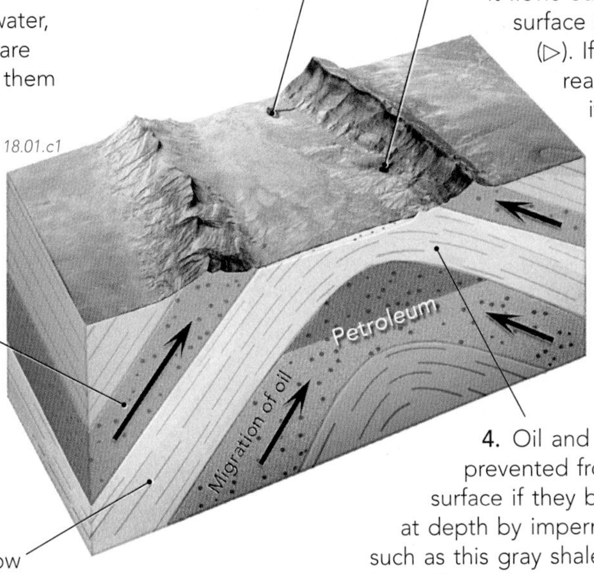

Petroleum

Migration of oil

3. Other rocks are less permeable and block the flow of oil, gas, and groundwater. A rock unit can be relatively *impermeable* if it lacks interconnected pore spaces and through-going fractures. Rocks that are typically impermeable include (1) shale, which has very small pore spaces, (2) unfractured granite, which has crystals that generally fit tightly together, and (3) salt, which flows easily and closes up any open spaces.

7. If oil flows into sandy sediments, it can form *oil sands* or *tar sands*. Large deposits in Alberta, Canada, are mined in large pits to extract the hydrocarbons from the sandy host.

6. Oil can migrate upward until it reaches the surface, where it flows out onto the surface as an *oil seep* (▷). If natural gas reaches the surface, it just blows away in the wind.

18.01.c2

5. To trap oil and gas at depth, a rock unit must have no through-going fractures or faults to provide an easy pathway to the surface. Severely deformed rocks, therefore, generally are less able to trap hydrocarbons. Some faults, however, effectively block the flow of fluids because the faulting has produced finely crushed rock fragments that filled open pore spaces.

4. Oil and gas can be prevented from reaching the surface if they become trapped at depth by impermeable rocks, such as this gray shale. Oil and gas rise as far as they can, floating on top of water within the rock (▷). Gas is lighter than oil, so it floats on top of the oil, which floats on top of the water. The groundwater associated with petroleum is generally very saline (salty), commonly with a higher salt content than sea water.

18.01.c3

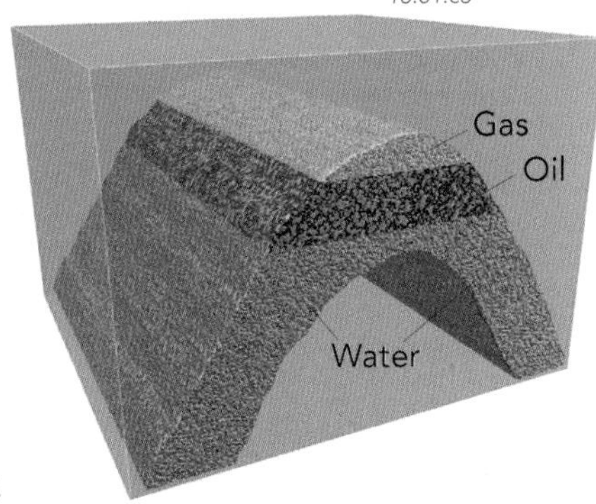

The La Brea Tar Pits

Los Angeles, California, contains one of the world's best known fossil sites, at the *Rancho La Brea Tar Pits*. The tar formed—and is still forming—as oil seeps onto the surface, where it loses its lighter, more easily evaporated components and leaves behind a sticky, dense material called *tar*. The oil was formed at depth by the same processes that formed the numerous oil fields near Los Angeles, but in the case of the tar pits, the subsurface geology was unable to trap the oil at depth.

Recovered from the tar pits are more than *one million bones* of modern and ice-age animals unlucky enough to have been stuck in the tar. Among the animals represented are now-extinct wolves, saber-tooth cats, ground sloths, and many smaller mammals, all of which roamed the area in the last 30,000 years.

18.01.mtb1

Before You Leave This Page Be Able To

- ✓ Summarize where the organic material in petroleum comes from.
- ✓ Summarize how oil and gas are produced by burial and heating.
- ✓ Sketch or describe how oil and gas move through rocks and how they can end up on the surface or be trapped at depth.
- ✓ Briefly describe the La Brea Tar Pits.

In What Settings Are Oil and Gas Trapped?

OIL AND NATURAL GAS CAN BE TRAPPED in the subsurface by various arrangements of rock types and geologic structures. To trap oil or gas, there must be a rock in which the hydrocarbons can accumulate. Such a rock unit is known as a *reservoir*, and the area of the reservoir rock at depth that actually contains hydrocarbons is called an oil or gas *field*. In addition to the reservoir, the oil or gas must be overlain by one or more impermeable rock units to prevent them from rising all the way to the surface.

A How Do Folded Layers Trap Oil and Gas?

The classic view of an oil and gas field is of an upward-bent fold (anticline) that traps petroleum near its crest. Many of the world's oil and gas fields, including some of the largest, are indeed in anticlines.

1. Rocks can be folded by compression into a series of upward-bending folds called *anticlines* and downward-bending folds called *synclines*. In an anticline (shown here), the layers on the flanks of the fold rise upward toward the center or *crest* of the fold.

2. Oil and gas migrate (as shown by the arrows) up the flanks of the folds until they reach the crest. If there is no impermeable cap, the petroleum can escape to the surface, as in this brown, unconfined sandstone layer.

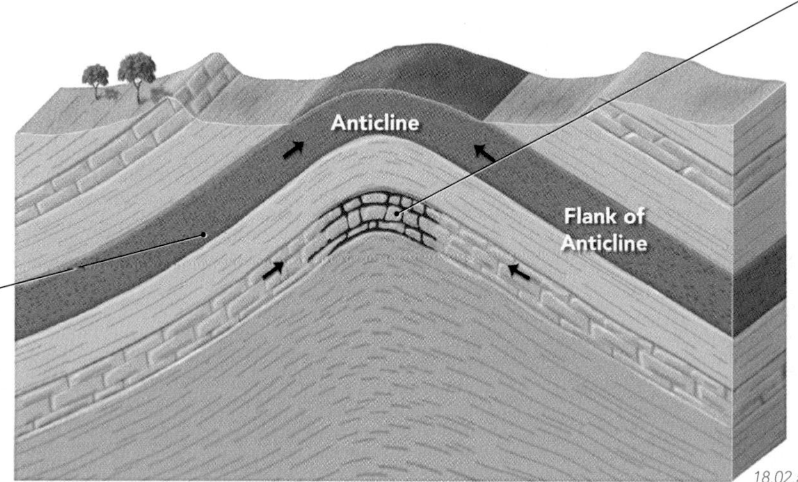

3. If a reservoir rock is capped by an impermeable unit, then oil and gas (shown in black) can accumulate in the crest of the fold. In this case, a permeable limestone is the *reservoir* and an overlying shale is the impermeable cap. An impermeable layer that traps oil is called a *cap rock*.

4. Note that the petroleum does not form an "open pool" of hydrocarbons; instead it fills the pore spaces between grains and the narrow open spaces along fractures and beds.

B How Do Salt Domes Help Trap Oil and Gas?

Salt is a geologically weak material that flows relatively easily when subjected to forces. Salt masses buried at depth can be mobilized by forces imposed by the weight of overlying rocks. The salt flows to try to escape the high-stress situation, and in doing so can rise toward the surface in a domelike structure called a *salt dome*. Arching of rocks over and adjacent to a salt dome can trap petroleum, as along the Gulf Coast and Persian Gulf regions.

1. The weight of a sequence of sedimentary rocks presses downward on a layer of salt at the base of the diagram. The salt responds by flowing as a weak, but solid mass. Salt rises upward, piercing through the overlying rocks as a salt dome.

2. At depth, salt flows along the layer to replenish and perpetuate the rising salt mass.

Oil

Salt

Oil

18.02.b1

3. Rocks over the salt dome are bowed upward and can be eroded into circular or oval features on the surface. In a region where oil is known to be present, such features are targets for oil and gas exploration.

4. Petroleum can accumulate in the crest of folded layers directly above the salt dome, such as in this dome-shaped fold of a limestone reservoir rock.

5. Petroleum is also commonly trapped on the flanks of a salt dome, where the petroleum migrated upward along an uptilted layer, until it encountered the central mass of impermeable salt. Many oil wells are drilled to explore the uptilted rocks that encircle a salt dome.

C What Are Other Ways That Petroleum Can Be Trapped?

Faults can trap oil and gas by juxtaposing permeable against impermeable rock, or by causing folding as rock layers are moved over bends in the fault. Two other common traps are an *unconformity* and a trap formed by a sedimentary layer thinning or changing in character from one rock type to another.

1. In this diagram, a normal fault has displaced sedimentary layers downward against a granite. Petroleum migrated up the tilted layers until it encountered the fault and granite, which in this case are both impermeable enough to stop further upward flow.

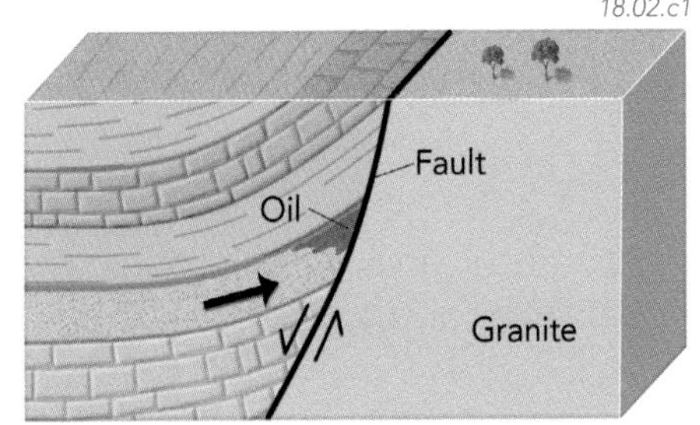

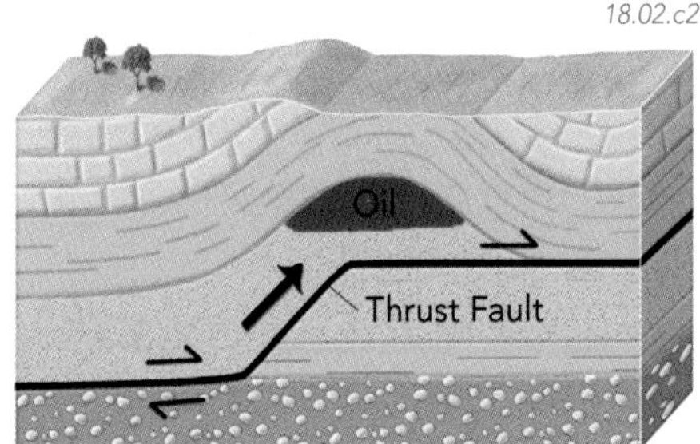

2. These rock layers in a thrust belt have moved up and over a bend or step in a *thrust* fault. As the layers above the fault move over the bend, they fold upward into an anticline. Petroleum migrates up the layers until trapped in the crest of the anticline.

3. An unconformity is an old erosion surface separating two different sequences of rocks. Rocks below an unconformity were tilted and eroded before the layers above the unconformity were deposited.

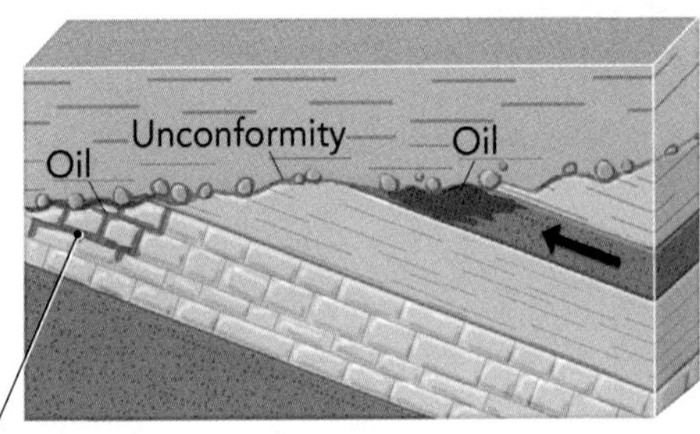

5. All sedimentary rock units eventually end when traced laterally, either because they decrease in thickness or because they change character into another type of rock in a *facies change*.

4. Petroleum can migrate up the tilted layers below the unconformity and be trapped by impermeable sedimentary layers along or above the unconformity. In this case, petroleum accumulated along the unconformity and in the underlying rocks.

6. This *permeable* sandstone bed is encased within a thick, *impermeable* shale. Petroleum migrating up the sandstone layer was trapped where the sandstone thinned and changed laterally into shale. Ancient reefs can form lenses and trap oil in similar ways.

Petroleum Basins of the United States

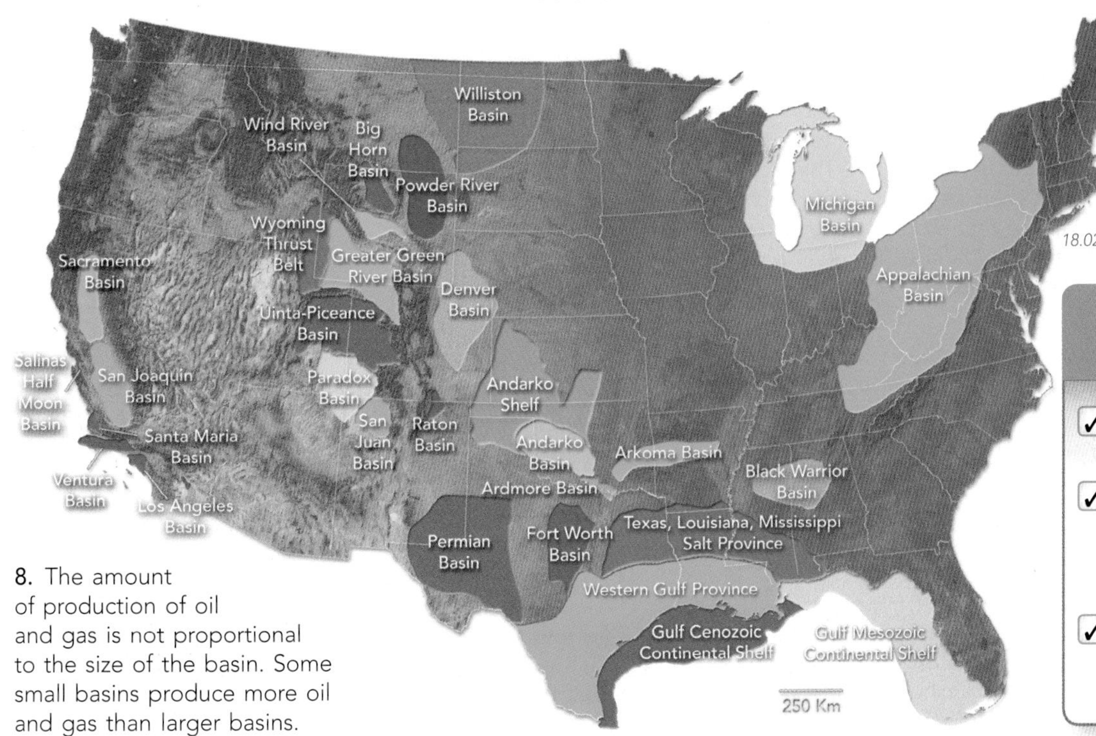

7. This map shows the distribution of the main petroleum basins in the lower 48 states. Note which parts of the country have petroleum, and which do not. The top-producing oil field in the United States is Prudhoe Bay in Alaska (not shown).

8. The amount of production of oil and gas is not proportional to the size of the basin. Some small basins produce more oil and gas than larger basins.

Before You Leave This Page Be Able To

- ✓ Describe the role of a reservoir rock and impermeable cap.
- ✓ Sketch and describe how petroleum is trapped by an anticline, salt dome, fault, unconformity, and facies change.
- ✓ Briefly summarize where petroleum basins are and are not located in the lower 48 states.

How Do Coal and Coal-Bed Methane Form?

COAL IS ANOTHER CARBON-BASED RESOURCE that provides energy and the raw materials to make other products. Coal forms from buried and compacted plants. There are different types of coal and different ways in which we mine, transport, and process coal, generally with the aim of producing electrical energy. Coal beds can also release a type of natural gas called *coal-bed methane.*

A How Does Coal Form?

The development of coal begins with accumulation of plant matter and other organic materials on the surface. Progressive burial, compaction, and heating change the coal from one type to another, improving its quality.

Processes of Coal Formation

1. Formation of coal requires that plants accumulate on the surface in sufficient amounts that the plant matter overwhelms the input of other sediment, such as sand and clay. The most common setting for this is in swamps and other wetlands.

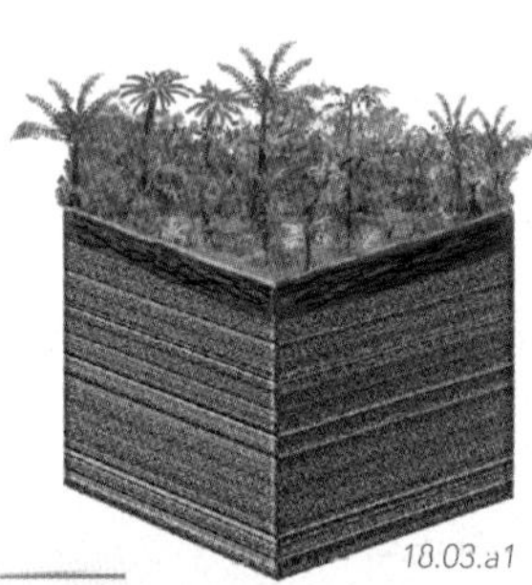

18.03.a1

Types of Coal

18.03.a2

2. The organic material in coal starts as compressed and partially decomposed plant matter, such as *peat*, a water-soaked mass of relatively unconsolidated plant remains found in bogs. Peat was an important energy-fuel resource in the past and is still used in many regions.

3. The plant matter must then be rapidly buried in a way that it is not oxidized or otherwise totally destroyed. Burial can occur in various ways, such as rising sea level that covers the land with water and sediment. With the pressures that accompany burial, water and other impurities begin to be squeezed out, converting the decomposing plant material to an impure variety of coal called *lignite*. Lignite has less carbon than other coals.

18.03.a3

18.03.a4

4. Lignite is a brown, not-very-dense coal.

5. With further burial, the lignite is compressed by the increasing weight of the overlying rock layers, becoming more dense and compact. The thickness of the coal layer will decrease as the material is compressed into less and less space. With burial comes an increase in ambient temperature with depth. The higher temperature begins to cook the coal, driving off chemical constituents, such as sulfur, that are relatively *volatile* (i.e., any substance that evaporates, or turns to vapor, easily). Compaction and increased temperature convert lignite into *subbituminous* coal and then *bituminous* coal. The processes by which coal changes as it is buried and heated is called *maturation*.

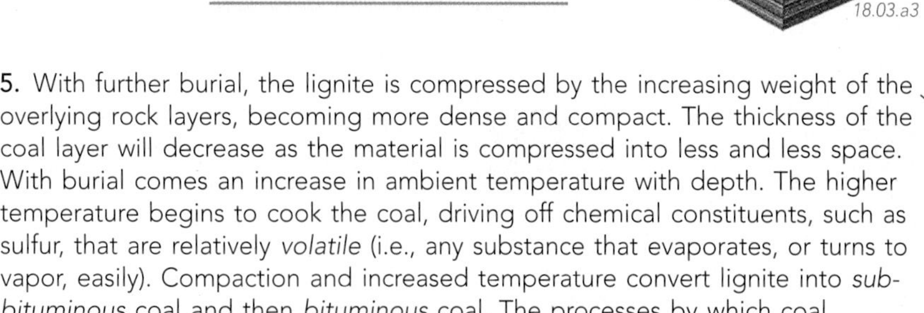

18.03.a5

6. Bituminous coal is a black, fairly dense coal.

18.03.a6

7. As burial continues, the coal becomes even more compacted and, therefore, thinner and more dense. The further increase in temperature drives off even more impurities, resulting in coal with a higher concentration of free carbon (92 percent to 98 percent). Such coal, called *anthracite*, is the most highly prized variety because it burns cleaner and has a higher energy content for a given volume of coal. The energy content of coal is given in terms of *calories*: the amount of heat produced by combustion. Anthracite can have more than twice the calorie content of lignite. The conversion of bituminous coal into anthracite commonly involves a change from a sedimentary host rock to a metamorphic one.

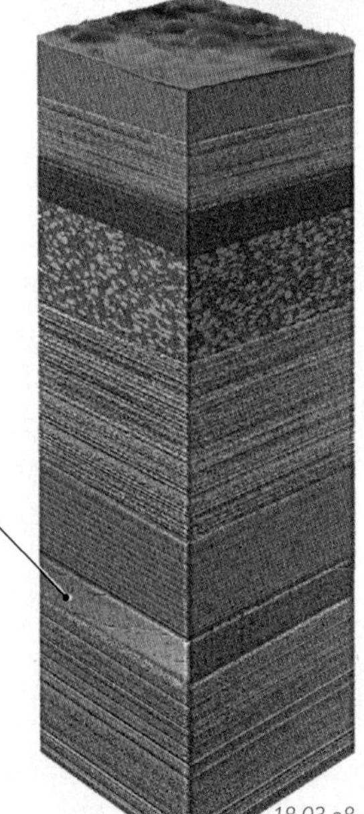

18.03.a8

8. Anthracite is black, dense, and shiny.

18.03.a7

9. During the progression from plant matter to high-quality coal, hydrocarbons released from the coal can include methane (CH_4), a colorless and odorless gas. Methane generated during the maturation of coal is called *coal-bed methane*. Coal that is subbituminous and bituminous is most favorable for coal-bed methane. Compared to ideal conditions, lignite has not matured enough but anthracite is too mature (has been heated too much).

C How Do We Determine What Is in the Subsurface?

It is one thing to document the characteristics, distribution, and attitudes of units on the surface, but exploring for geologic resources also generally requires understanding what is going on in the subsurface.

An excellent way to start understanding what rock units lie at depth is to find places where nature has exposed the sequence, such as in deep canyons or on the sides of mountains.

Observations made at the surface are then extrapolated to the subsurface. Layers can be projected downward using the thicknesses and structural attitudes measured on the surface. Confidence in such extrapolations, however, decreases with increasing depth, as the interpretation gets farther from the actual data. In this example, layers are horizontal at the surface and should continue that way for some distance downward.

Exploration for fossil fuels is challenging because of uncertainties in how the geology may change with depth and because there may be geologic structures that are not expressed on the surface. To limit the number of possibilities, exploration companies invest large sums in geophysical investigations, such as gravity and seismic-reflection surveys.

The ultimate test of a geologist's interpretation of the subsurface is to drill an exploration hole through the rock sequence. The drill core, cuttings, and various physical measurements obtained from a drill hole provide a clear test of the subsurface interpretation. Such data may support the interpretation or require going back, literally, to the drawing board.

18.05.c1

D What Tools Do Geologists Use to Visualize Subsurface Geology?

The main tool used to interpret and visualize the subsurface geometry of rock units and geologic structures is a *geologic cross section.* Various techniques help geologists infer the geometry of folds and faults at depth, based on the faults and orientation of layers exposed at the surface. This cross section shows an interpretation of part of the Arctic National Wildlife Refuge (ANWR), an area in Alaska that may contain large oil reserves but that is ecologically and politically sensitive. The area has rocks and structures that are favorable for oil.

Projection of Faults into Air
Syncline
Late Mesozoic Rocks
Anticline
Thrust Faults
Late Paleozoic Rocks
Precambrian and Early Paleozoic Rocks
0
-5
-10 km
5 Km
18.05.d1

18.05.d2

△ Computer-based visualization programs and expensive visualization rooms are key components of modern exploration efforts. They help geologists integrate information from surface studies, drill holes, and geophysical surveys, especially seismic-reflection profiles.

The Costs of Exploration

The stakes in exploring for fossil fuels are high, both in terms of the nearly prohibitive cost and the possibility, however remote, of lucrative return on investment. Some companies employ hundreds of geologists, along with all the necessary support staff, including business people hired to obtain exploration leases from federal, state, tribal, and private landowners. Doing field work is commonly one of the least expensive parts of the operation, mostly requiring money for four-wheel-drive vehicles, fuel, accommodations, and a team of well-paid geologists.

One of the most expensive aspects involves conducting *geophysical surveys* which may use dozens of large trucks and expensive sensors and computer gear. Drilling exploration holes is astoundingly expensive: An individual drill hole can cost tens of millions of dollars. Drilling costs, quoted in dollars per foot of depth, range from less than \$100/ft for shallow wells to more than \$500/ft for very deep ones—and some wells reach depths of 20,000 feet! An offshore drilling platform can cost a company more than \$100,000 a day!

Before You Leave This Page Be Able To

- ✓ Describe aspects to consider regarding an area's potential for fossil fuels.
- ✓ Summarize the types of field studies geologists conduct in exploring for fossil fuels.
- ✓ Summarize or sketch how geologists infer what is in the subsurface, and describe the tools they use to visualize these data.
- ✓ Describe why exploration is so costly.

How Is Nuclear Energy Produced?

NUCLEAR REACTIONS PRODUCE ENORMOUS ENERGY that can be harnessed to power electrical generators. Currently, nuclear power is the second largest source of electricity in the United States (after coal-fired plants), supplying 20 percent of the nation's electricity. Nuclear power provides nearly 80 percent of France's electricity and 50 percent of Switzerland's. How do nuclear reactions supply so much electricity, and in what geologic settings do we find deposits of uranium, the key component in the process?

A How Does Nuclear Fission Produce Energy?

Present-day nuclear power plants are based on the process of *fission*, in which an unstable isotope of a radioactive element splits into two parts. There is a significant amount of research being done to develop a reactor based on a sustained *fusion* reaction, in which two atoms collide and combine into a larger and heavier element. Earth's radioactive heat arises from *fission*, whereas the Sun's energy comes from *fusion*. Here, we'll discuss *fission*, which is used in power plants today.

1. Uranium is a large and heavy element, with an average atomic mass of 238. The main isotopes of uranium, ^{238}U and ^{235}U, both experience radioactive decay by fission. ^{235}U decays more rapidly than ^{238}U, and so produces more energy, but is much less abundant (more than 99 percent of uranium is ^{238}U).

2. When a uranium atom splits apart by fission, it releases relatively large amounts of energy, partly in the form of heat. In a reactor, ^{235}U atoms are bombarded by neutrons, and this induces fission. The heat produced by fission is used to convert water to steam, which is then used to turn the turbines in electrical generators.

18.06.a1

3. Most reactors are designed to keep the uranium and its decay products isolated from the water that is converted to steam. Only heat is exchanged between the two parts of the system, and so the steam being released does not contain radioactive materials.

B In What Settings Do Uranium Deposits Form?

Uranium atoms are large compared to other elements and so have difficulty fitting into the structure of common minerals. Partly for this reason, uranium is mobile and is transported by groundwater and other fluids.

Some deposits form where uranium, carried by groundwater, encounters water with a different chemistry. Uranium accumulates along the boundary between the two waters, forming an arcuate deposit called a *roll-front*.

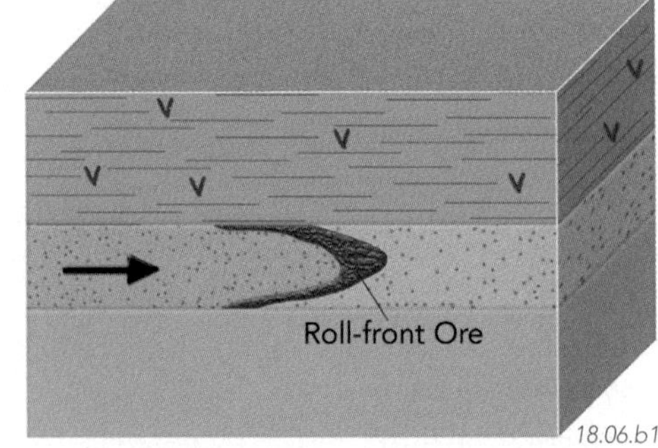

18.06.b1

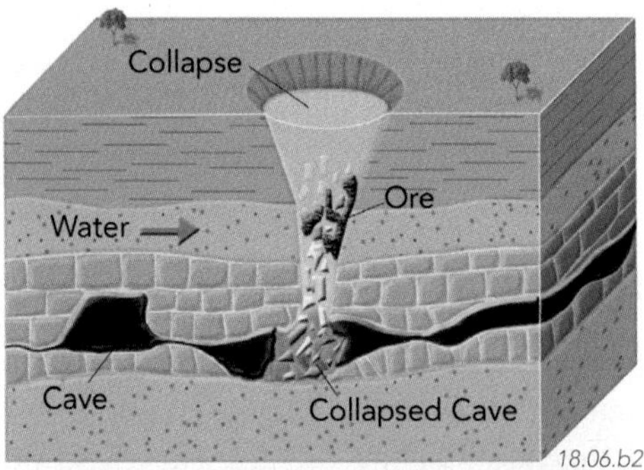

18.06.b2

Some uranium occurs in pipe- or cone-shaped structures filled with angular fragments (breccia). Such *breccia pipes* are formed by collapse of limestone caves, with the uranium introduced later by groundwater.

Some large deposits of uranium ore were deposited by groundwater along *unconformities* during Precambrian time, when conditions on Earth were different (such as less oxidizing) than they are now.

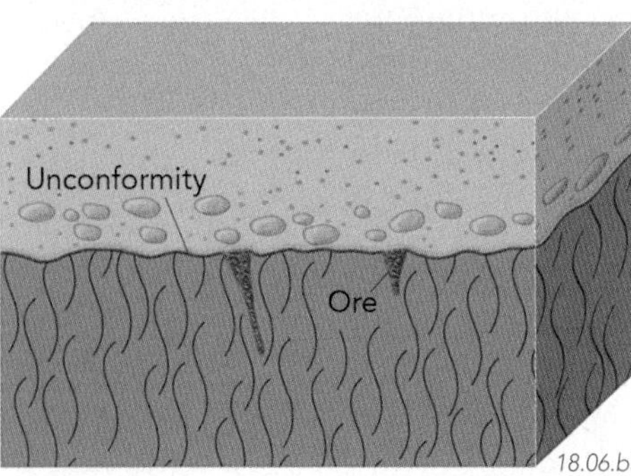

18.06.b3

18.06.b4

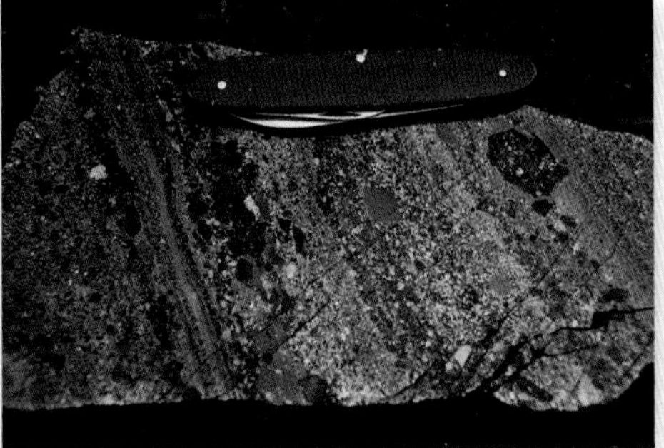

There are many other types of uranium deposits, including those associated with granite pegmatite, with lake beds, and with unusual iron-rich rocks, like the one shown here. [Olympic Dam deposit, southern Australia]

C How Is Uranium Used to Generate Electricity?

A number of steps, some of them technically difficult, are required to obtain material that is rich enough in uranium to sustain a controlled fission reaction.

18.06.c1

Uranium ore is mined by open-pit or underground methods (△) and processed in a mill that physically or chemically concentrates the uranium minerals.

18.06.c2

The most technically challenging part of the operation is using devices called *gas centrifuges* (◁) to preferentially *enrich* the uranium in ^{235}U, the more energetic isotope. Since ^{235}U and ^{238}U have nearly identical properties, uranium enrichment is the major obstacle for countries trying to produce nuclear materials for power-generation—or for nuclear weapons. The *enriched uranium,* now with enough ^{235}U to sustain a fission reaction, is processed into pellets or rods for the reactor.

18.06.c3

The rods of *enriched uranium* are used in one of several different types of nuclear reactors. In each type, some type of material, called the *moderator*, is placed between the rods to slow the neutrons so that they cause fission upon colliding with atoms of ^{235}U. Most commercial reactors in the United States use water as a moderator and are called *light-water reactors*. The Chernobyl disaster in the Soviet Union used a totally different design, where graphite was the moderator. In most reactors, boron-containing fluid is mixed into the water around the rods to control the rate of the reaction. Heat from nuclear fission converts water into steam to drive electrical generators.

D What Are Some Environmental Issues Associated with Nuclear Energy?

Nuclear power holds the promise of generating large amounts of electrical energy without adding more greenhouse gases to the atmosphere, but environmental and political issues have limited its wider use in the United States.

18.06.d1

Uranium mining and milling involve materials that are relatively rich in uranium, which can contaminate the water, land, and air around facilities if materials are handled poorly. Also, some miners and mill operators have become gravely ill from overexposure to radioactive materials. These uranium-contaminated materials in Moab, Utah, lie right next to the Colorado River.

18.06.d2

After fission has depleted some percentage of their uranium, the fuel rods are removed from the reactor and mostly stored on-site. Such *spent fuel rods* contain high concentrations of uranium, as well as other radioactive elements, so pose a large risk of radioactive contamination.

18.06.d3

A steam explosion, followed by a number of tragic mistakes, led to other explosions and a reactor meltdown at the Chernobyl nuclear reactor in the Ukraine in 1986. The disaster killed at least 56 people and contaminated a very large area, requiring resettlement of more than 200,000 people.

Yucca Mountain

Yucca Mountain is a mesa capped by volcanic rocks in an isolated part of the Mojave Desert west of Las Vegas, Nevada. It is currently designated as the official future repository for spent fuel rods and other high-level (uranium-rich) radioactive waste from commercial reactors. More than $6 billion have been spent at the site on scientific studies, initial construction, and related activities. The scientific studies have focused on the surface and subsurface geology of the site, especially whether the region is stable and whether the deep water table will keep radioactive materials out of the environment for thousands to hundreds of thousands of years. The scientific jury is still out, but politics will decide the site's fate.

18.06.mtb1

Before You Leave This Page Be Able To

- ✓ Summarize how nuclear fission releases energy and is used to generate electricity.
- ✓ Summarize or sketch some settings in which uranium deposits form.
- ✓ Describe positive and negative aspects of the use of nuclear energy.
- ✓ Briefly describe what material will be stored at Yucca Mountain.

How Is Water Used to Generate Energy?

WATER IS PLENTIFUL ON OUR PLANET, and so we have found ways to use this abundant resource to generate energy. Using the movement of water to generate electrical energy is called *hydroelectric power*, which commonly employs some type of dam across a river valley, but can also involve tapping the energy of moving water in ocean tides and currents.

A How Is Electricity Generated by Hydroelectric Dams?

Nearly all hydroelectric power comes from dams and provides about 10 percent of electrical energy for the United States and about 20 percent of the world's electricity. It is the main source of electricity for some topographically rugged western states, like Idaho, Washington, and Oregon, as well as nations like Norway and Iceland.

How Electricity Is Generated in Hydroelectric Dams

To generate hydroelectric power, dams capture water from rivers or streams in a *reservoir* behind the dam. The dam is generally constructed out of concrete, but some dams consist of compacted clay, rock, and other material. Large steel pipes or concrete tubes within the dam guide water to flow from higher elevations to lower ones, under the constant—and free—force of gravity.

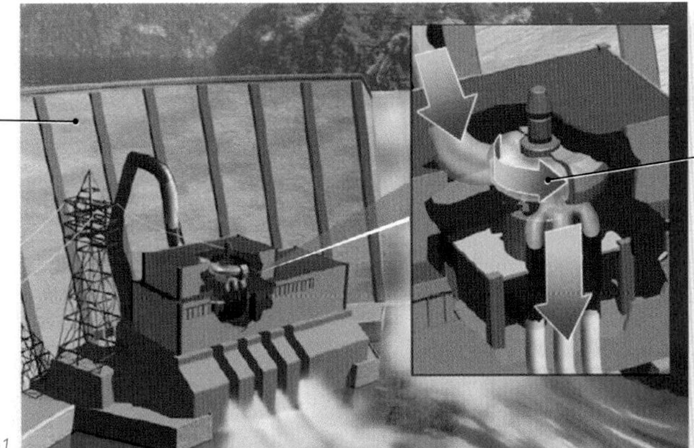

18.07.a1

When power is needed, the water is allowed to flow through the pipes and tubes within the dam and then down into a powerhouse, where the moving water turns turbines or blades in electrical generators. The amount of energy produced is related to the velocity and volume of the water passing through the turbines. An ideal dam, from an energy perspective, is high and has a high-volume reservoir. An advantage of hydroelectric energy is that the *potential energy* of the water can be stored until the electricity (which cannot be stored) is needed.

Geologic Factors Important to Hydroelectric Dams

1. Geologic factors control the suitability of a site for a dam, in many cases, making a site unsuitable for a dam of any type. These factors are summarized here.

2. A main geologic factor is whether there is a deep enough canyon or valley in which to build a dam. Depth of a canyon or valley is in turn controlled by the types and sequence of rocks and by the area's history of erosion and deposition.

3. Most dams are built where the two walls of the canyon or valley are relatively close together. This keeps the width of the dam to a minimum, both for structural stability and to keep construction costs down.

4. Rocks beside and beneath the dam should be relatively impermeable to limit seepage of water around and under the dam. In addition to the loss of valuable water, such seepage can dissolve or loosen material in the rocks and weaken the foundation of the dam over time.

5. Rocks below and adjacent to the dam should also be *chemically nonreactive*. Most limestone is unsuitable because it is soluble, and volcanic tuff can convert to clays and other weak materials if exposed to water over sufficient time. Some dams built on such altered rocks have been abandoned or required a total redesign of the dam structure because the foundation was simply too weak to support a dam.

6. Faults, fractures, and other zones of weakness and high permeability can be a potentially fatal flaw, both in terms of site suitability and for public safety downstream of a dam.

7. Dams cannot safely be built in sites where the adjacent and underlying rock types, like this layer of shale, are too weak to anchor or support the dam.

8. A concrete dam is anchored to the rocks on either side, which are called the *abutments* of the dam. Rocks in the abutments need to be relatively strong and unfractured, and be composed of insoluble rock types.

9. The canyon shape determines the width of the reservoir, as well as the depth. Deep, narrow lakes lose less water to evaporation than shallow, wide ones.

18.07.a2

B How Is Electricity Generated from Ocean Tides and Currents?

Hydroelectric power can be generated by using the change in local sea level that accompanies rising and falling tides. Power can also be generated by submerging propellers or turbines in the shallow ocean, to be spun by moving water.

1. One way to generate electricity from tidal changes is to construct a dam-like structure across a narrow, shallow inlet, such as an estuary. The dam-like barrier impedes the flow of water, causing water to pile up on one side of the barrier during high tide and on the other side during low tide. An ideal location is one with a large difference between high and low tide levels. One favorable site for tidal power is the Bay of Fundy, which can have a tidal range of 16 meters (52 feet).

2. When the tide is rising, tidal forces pull ocean water toward the land and piles it up on the *seaward* side of the barrier. When the water level on the seaward side is sufficiently higher than the water level on the landward side, the gate is opened and the inward-rushing water turns the turbines.

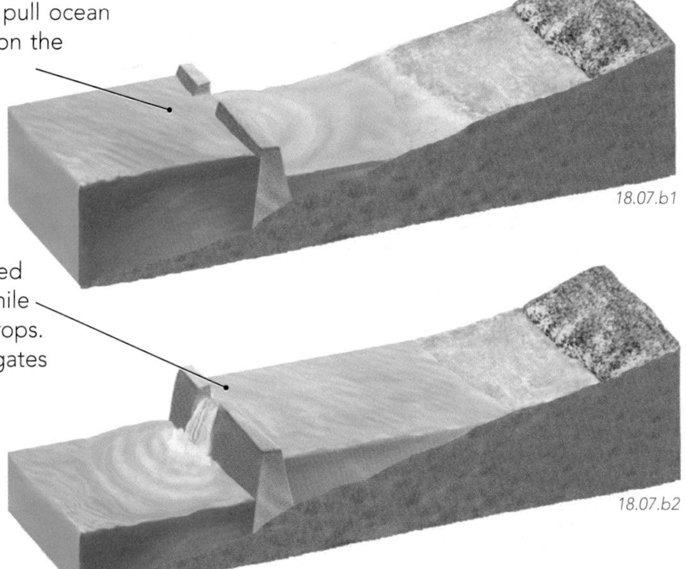

3. During a falling tide, water is trapped on the *landward* side of the barrier, while the water level on the seaward side drops. The trapped water flows through the gates toward the sea, turning the turbines. Electricity cannot be generated at all times, only when the water levels on the two sides of the barrier are sufficiently different. Tidal power generation occurs only during tidal *changes*, not at high or low tides, when water does not move.

4. Alternatively, large turbines or propellers can be anchored to the shallow seafloor in an area where water flows past because of tides or prevailing ocean currents. The moving water spins the turbines or propeller blades, which turn the shaft of an electrical generator. Ocean currents can generate electricity most of the time.

Environmental Issues Associated with Dams

Dams hold the promise of a nearly constant supply of electrical power with no associated emissions of greenhouse gases or toxic contaminants. They have a huge advantage over many other ways of generating electricity—the amount of electricity being generated can be changed rapidly just by increasing or decreasing the amount of water released through the turbines. This is important because the demand for electricity varies greatly between daytime and nighttime, and from hot to cold seasons, and large quantities of electricity are not easily stored. With other systems, if the amount of electricity being generated is greater than the amount being used, the excess is simply lost.

Dams, like *Glen Canyon Dam* in southern Utah (shown here), and those along rivers of the Tennessee Valley, have some important negative environmental aspects to consider. When dams are constructed and reservoirs filled, the canyon or valley behind the dam is inundated by the rising waters. This will destroy any farmlands, houses, or even cities that are located where the reservoir will be. Flooding will destroy all the existing vegetation and animal habitat, as well as any special natural places and archeological sites.

Dams also interfere with the natural river dynamics, changing the natural flow patterns, such as from large spring floods to a more consistent and managed flow. The dam traps sediment carried by the rivers and streams flowing into the reservoir, and the reservoir will eventually fill up with sediment (called *silting up*). This blockage deprives the downstream river of sediment and associated nutrients. This can drastically change the downstream habitat, as clear, cold water from the depths of the reservoir replaces the warmer, muddy water that flowed down the river before construction of the dam.

Before You Leave This Page Be Able To

- ✓ Sketch or describe how electricity is generated by hydroelectric dams and from tides and ocean currents.
- ✓ Summarize how geology affects the location of a dam.
- ✓ Summarize some advantages and disadvantages of hydroelectric dams.

What Are Some Other Sources of Energy?

FOSSIL FUELS, NUCLEAR ENERGY, AND DAMS HAVE DRAWBACKS, so considerable research and development have gone into exploring other ways to produce energy. The goal is to find energy sources that are friendly to the environment in terms of not producing greenhouse gases and that are *renewable resources*, meaning that their supply is essentially limitless and using them doesn't remove something irreplaceable from Earth. Such approaches include using wind, solar energy, and heat from within Earth.

A What Is Geothermal Energy and What Sites Are Most Favorable for Its Use?

Geothermal energy uses Earth's natural heat as an energy source. Natural *hot* water is converted to steam to power electrical generators, and *warm* water can be piped from the ground to places where it can be used to heat buildings and greenhouses, or to keep streets and sidewalks free of ice and snow.

18.08.a2

1. In some regions, very hot water reaches the surface, such as in hot springs, steaming or smoldering pools of water, and an intermittent rising fountain of hot water and steam called a *geyser*. This geyser is at Geyser, Iceland, where geysers derive their name.

2. Temperature increases with depth, so water circulating through the crust can become heated at depth and then rise to the surface. This is how hot springs form.

3. The ideal combination of high temperatures at relatively shallow depths is most likely in areas of recent volcanic activity, such as within a collapsed caldera formed by eruption of volcanic ash.

18.08.a1

5. To generate electricity, hot water is piped to the surface and into power plants. In the plant, the confining pressure on the overheated water is released, and it flashes into steam to drive the turbines in the electrical generators, like these in Iceland. ▽

18.08.a3

6. Regions with recent faulting can also be promising sites for geothermal energy. Faults disrupt the continuity of aquifers and provide a conduit for heated water to rise to the surface and issue from hot springs. The Iceland site shown above is within an active rift zone.

4. Shallow magma or solidified magma chambers that are still hot can heat water to high temperatures, exceeding 200 to 300°C (~500°F). Although water of these temperatures would boil on the surface, it generally does not boil at depth because the confining pressures work against the great expansion in volume required to convert liquid water into gaseous steam.

7. If rocks are hot at a shallow depth but dry, water derived from some other source can be pumped down drill holes to be heated by the rocks and then pumped back to the surface and used for heating or power.

B How Is Electricity Produced from the Wind?

Wind is another clean and renewable energy source. It currently provides very little of the world's and North America's power requirements, but is one of the fastest growing sources of energy, both here and abroad.

Large-scale generation of electricity from wind requires a site that has strong winds much of the time. Important geologic factors to consider are how the surface topography interacts with or controls the winds, and whether a site is suitable for building the necessary facilities. Each wind turbine has its own small electrical generator.

18.08.b1

An *advantage* of wind power is that it is renewable, is nonpolluting, and can be used in remote locations and in areas that have little other infrastructure.

One *disadvantage* of wind power is that winds, and the resulting power, are variable. Wind turbines affect the *aesthetics* of the site, being large and conspicuous, even if they are painted to help blend in with the environment. They can be noisy, are relatively expensive to maintain, and kill birds. The downsides are weighed against the obvious benefits of clean, renewable power.

C How Is Solar Energy Used?

Solar energy involves using the Sun's free electromagnetic energy to heat buildings and generate electricity. There are many strategies for using solar energy, including *passive solar*, *active solar*, and *photovoltaic panels*. All solar-energy approaches work best in sunny climates and in sites with unrestricted views of the Sun.

▷ In *passive solar* uses, light and infrared energy from the Sun enter a space through glass windows, naturally heating the inside air. Passive solar does not use any moving parts (hence the name *passive)* and is as easy as designing a house with large windows facing south (in the Northern Hemisphere) to collect the winter sun.

Active solar implies that there are moving parts and some use of electrical energy, such as a fan for moving heated air or an electric pump for circulating heated fluids from the solar panel to the interior of the building. Active solar allows heating of rooms farther inside the building.

18.08.c1

18.08.c2

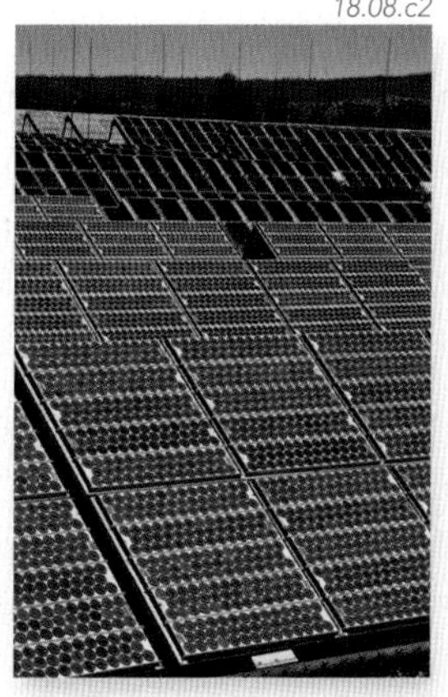

Photovoltaic panels convert sunlight directly into electricity. Such panels, although expensive to produce and install, provide nonpolluting renewable energy, even to remote locations and small sites.

D What Are Some Other Alternative Sources of Energy?

18.08.d1

Biomass—Energy can be produced by burning scrap wood, by burning methane released from decaying organic material in landfills, and in other ways.

18.08.d2

Ethanol—Ethanol is a type of alcohol that can be used to fuel cars. To produce ethanol corn, sugar cane, and other plant material is soaked in ammonia, fermented, and distilled.

Fuel Cells—In this potentially important technology, electricity is used to break water molecules into hydrogen and oxygen, which then can be used as fuel. Some other source of energy must be used to generate the electricity.

18.08.d3

Trade-offs Between Different Ways of Producing Energy

In an ideal world, our energy sources would be renewable, nonpolluting, ubiquitous, portable, cheap, and easily extracted without affecting the ecology or aesthetics of the site. All current energy sources have *trade-offs* of one sort or another—cheap and portable but polluting; renewable and nonpolluting, but expensive and requiring huge facilities. No single energy source does it all.

The decision of which type of energy to use involves identifying the *advantages* and *disadvantages* of each type, and then *carefully weighing* them against each other to see if the good outweighs the bad, or vice versa. This decision process is similar to the scientific method—we pose a question and collect observations and data to better understand the variables. We then make predictions or models for several scenarios, and then test the predictions by observing or numerically modeling the system.

The final decision, however, also involves *values* of individuals, companies, and governments. Such values can involve such nonscientific aspects as aesthetics, ethics, emotions, and politics. An excellent example is the *Cape Wind Project*, where people living around Cape Cod, Massachusetts, known for their strong environmental sentiments, are resisting, for aesthetic reasons, the installation of wind turbines to generate electricity using environment-friendly wind energy.

Before You Leave This Page Be Able To

- ✓ Describe or sketch the surface and subsurface geologic factors favorable for geothermal energy.
- ✓ Summarize how wind can produce electricity.
- ✓ Describe how energy is produced from solar power, biomass, ethanol, and fuel cells.
- ✓ Discuss some of the trade-offs involved in each of the various energy sources.

What Are Mineral Deposits and How Do They Form?

PHYSICAL AND CHEMICAL PROCESSES concentrate and disseminate elements and minerals, causing rocks to have a higher content of some elements and minerals, and a lower content of others. If a volume of rock is enriched enough in an element or mineral to potentially be valuable, we call it a *mineral deposit*. Materials extracted from mineral deposits provide the very foundation for our modern world.

A What Is a Mineral Deposit and What Is an Ore?

Most rocks are not considered to be mineral deposits, even though they are indeed composed of minerals. Instead, the term *mineral deposit* means the rock is especially rich in some commodity that might be valuable, and such rocks are said to be *mineralized*. If a mineral deposit contains enough of a commodity to be mined at a profit, then it is called an *ore deposit*, and the valuable rocks in that deposit are called *ore*.

18.09.a1

An outcrop of plain white quartz is not a mineral deposit, but one that contains flecks of gold is. If rich enough in gold, like the fist-sized sample in the inset photograph, the piece of quartz may also be *ore*.

18.09.a2

Ore can be conspicuous, like this rock that contains shiny, brass-colored sulfide minerals. Some ore is much more subtle, being enriched in some element but otherwise looking like a typical igneous, metamorphic, or sedimentary rock.

B What Determines Whether a Mineral or Rock Is an Ore?

Many factors, some of them nongeologic, determine whether a rock is considered an ore. These include concentration of the commodity in the rock, how easy it is to extract the commodity from the rock, the proximity to markets, and the economics (supply and demand), which controls prices.

18.09.b1

Grade of Ore—The concentration of the valuable commodity in a rock is called the *grade*. A rock that is very rich in the commodity, like this sulfide–rich copper ore, is called *high grade*; the opposite is *low grade*.

18.09.b2

Type of Ore—A commodity can occur in different types of minerals, such as copper in this blue-green copper-oxide mineral. It is cheaper to extract copper from oxide minerals than from the copper sulfide minerals shown in the left-most photograph.

18.09.b3

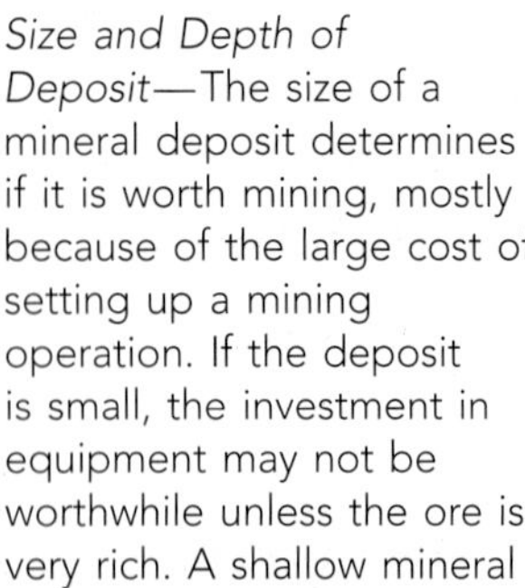

Size and Depth of Deposit—The size of a mineral deposit determines if it is worth mining, mostly because of the large cost of setting up a mining operation. If the deposit is small, the investment in equipment may not be worthwhile unless the ore is very rich. A shallow mineral deposit will be cheaper to mine than a deeper one. A large, open-pit mine is more economic to operate than a small mine or a deep, underground one. [Morenci Copper Mine, Arizona]

18.09.b4

Location of Deposit—A deposit that is close to markets and to infrastructure, such as railroads and electrical lines, will be more economic than one that is far from civilization or in an environmentally sensitive place. Economic factors, such as the price for which the commodity can be sold, and political factors, such as whether the area has a stable government, can determine whether a deposit can be mined. [Bingham Mine, Utah]

C What Processes Can Form a Mineral Deposit?

Many geologic processes can concentrate minerals or chemical elements, but it takes special circumstances to form a mineral deposit, especially one rich enough to become ore. Some ore-forming environments involve hot or deep processes and others involve low-temperature processes typical of near-surface environments.

Hot or Deep Processes

Igneous Crystallization—A crystallizing magma can form a mineral deposit by having one or two minerals crystallize at one time. Heavy crystals can sink to the bottom of a magma chamber, forming enriched parts, such as this dark layer of iron-oxide minerals.

18.09.c1

Hydrothermal Deposition—Hot water, called a *hydrothermal fluid*, can precipitate minerals in fractures, on the surface in hot springs, or from hydrothermal vents on the seafloor. Mineralization within or along a fracture generally involves a hydrothermal fluid and is called a *vein*.

18.09.c3

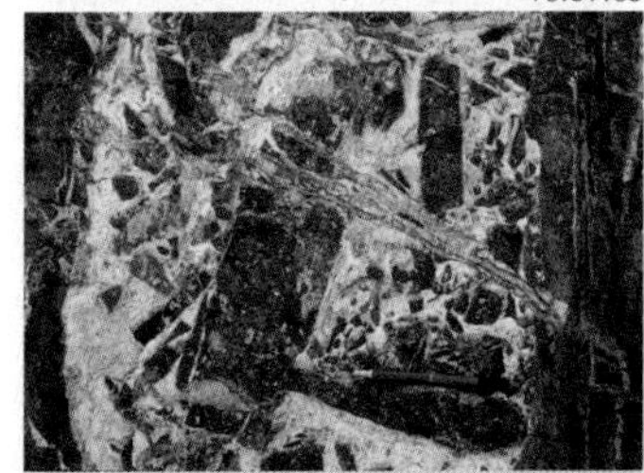

Hydrothermal Replacement—A hydrothermal fluid can permeate through a rock, replacing the materials with new minerals containing chemical components that were carried by the fluid. This dark, tin-rich ore formed by such replacement.

18.09.c5

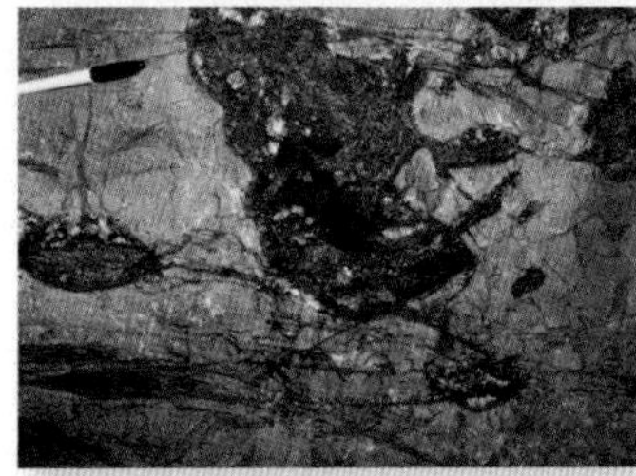

Metamorphism—Metamorphism, due to increased temperatures and pressures, can convert existing minerals in a rock into new minerals that may be valuable. This blue kyanite grew during metamorphism of an initially clay-rich rock.

18.09.c7

Surficial Processes

18.09.c2

Weathering Enrichment—As mineralized rocks are exposed at the surface, valuable elements can be *leached* from the rocks by weathering, carried by groundwater, and *deposited* elsewhere. Leaching removed copper from this rock and enriched rocks below.

18.09.c4

Formation by Weathering—Some ore deposits represent *residual* materials left behind as other chemical components are leached away. Deposits enriched in aluminum (bauxite) and in nickel deposits can form as residual soils from weathering.

18.09.c6

Mechanical Concentration—As materials are transported by rivers and washed by waves, minerals can be sorted and concentrated, usually on the basis of size and density. These river gravels in South Africa were mined for diamonds, which are dense.

18.09.c8

Low-Temperature Precipitation—Valuable minerals are deposited by evaporating sea or lake water, or are deposited by groundwater flowing through permeable rocks. These gypsum beds were deposited by evaporation of sea-water.

The Geologic Setting of Diamonds

Diamonds are a classic ore mineral; a little bit of diamond can be worth a lot, and it is worthwhile to mine a lot of rock just to get a little bit of diamond. Natural diamonds have an unusual origin in that they form at depths of more than 100 kilometers (60 miles) within the mantle and then are violently carried toward the surface in pipe-like volcanic conduits.

18.09.mtb1

South Africa, the largest producer of diamonds, has two main types of diamond deposits. Most diamonds are mined from vertical volcanic conduits called *diamond pipes*, such as the mine shown here. Some diamonds are mined from gravels deposited by streams, whose flowing water carried diamonds eroded from diamond pipes.

Before You Leave This Page Be Able To

- ✓ Explain the meaning of mineral deposit, mineralization, and ore.
- ✓ Summarize geologic and non-geologic factors that determine whether a mineralized body can be mined.
- ✓ Summarize the processes that can form a mineral deposit.
- ✓ Describe how diamond-bearing deposits form.

How Do Precious Metal Deposits Form?

GOLD, SILVER, AND PLATINUM are three members of a family of valuable metallic elements called *precious metals*. These metals are widely used for industrial and monetary purposes in addition to jewelry, but occur only in relatively minor concentrations in Earth's crust and so are high-cost materials. Where do such familiar, but precious, metals come from, and how do we find new deposits?

 In What Settings Do Gold- and Silver-Rich Mineral Deposits Form?

Gold and silver occur together in many geologic environments because these two elements behave similarly under many geologic conditions. Mines that produce gold usually get some silver from the ore as well. Since gold is much more valuable than silver, the discussion below focuses on gold, but mostly also applies to silver.

Gold and Silver Veins

Gold and silver often occur in narrow, steep, tabular bodies, called *veins*, which represent fractures through which hydrothermal fluids passed and deposited minerals. Veins are typically not extensive, but ore can be very high grade. Many veins are associated with volcanic areas or igneous intrusions.

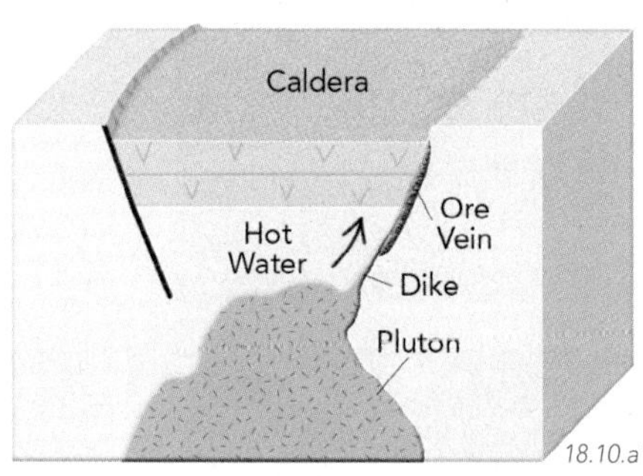

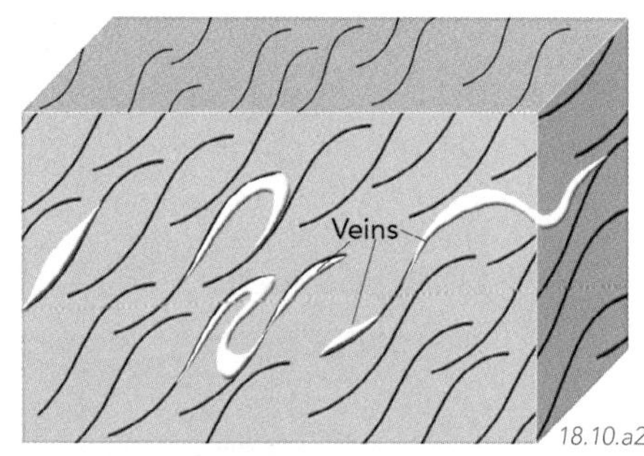

Veins are common in metamorphic rocks, and may contain enough gold to be mined. In some cases, the gold in the veins was already present in the rocks before metamorphism, but was redistributed and concentrated by the metamorphic processes. In other cases, gold and silver were introduced into the rocks by hydrothermal fluids.

Modern and Ancient Placer Deposits

Pieces of gold, liberated from bedrock by weathering and erosion, can be carried away by mountain streams and rivers. Gold is more dense than any other grains being carried by the running water so will be deposited at the base of gravels. Such deposits are called *placer gold deposits*, or simply *placers*.

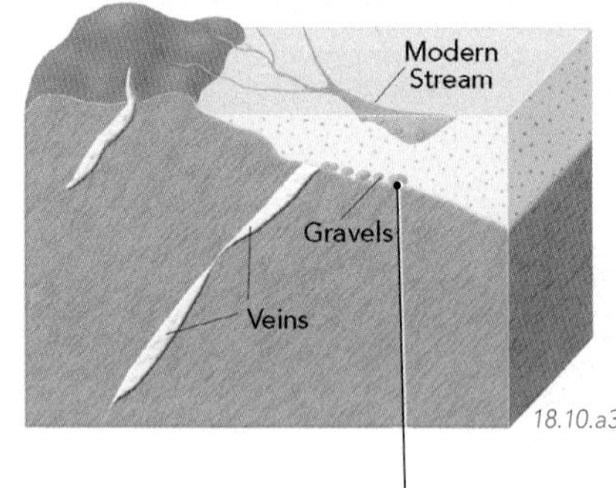

If gold-bearing gravels are buried by later rocks, the unconsolidated gravels will be compacted and lithified into *conglomerate*. The famous gold mines of South Africa are in ancient conglomerate beds in a region called the *Witwatersrand*. Miners have followed such gold-bearing conglomerates to depths of nearly 4,000 meters (13,000 feet)!

Large, Low-Grade Gold Deposits

If gold prices are high enough, it becomes profitable to mine rocks that contain low concentrations of gold. These deposits, typically hosted by volcanic or sedimentary rocks, are low grade but fairly large, and are mined by open-pit methods. The gold typically is not visible, but can be extracted.

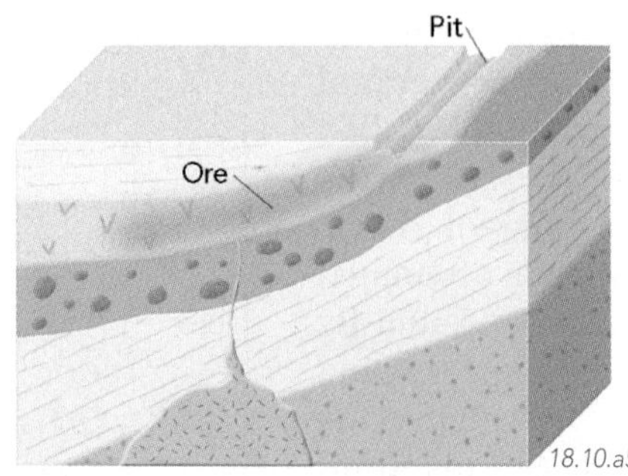

By-Product Gold

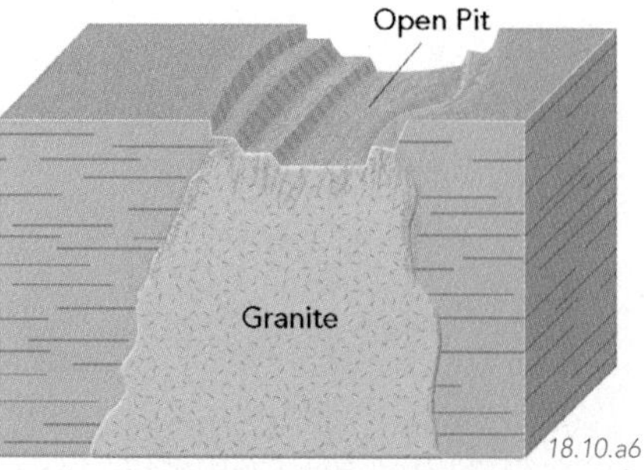

A substantial amount of gold is recovered from other types of mineral deposits, especially from large, open-pit copper mines. The ore is mined for the copper content, but the net worth of the gold recovered as a *by-product* of the copper mining can make the mine profitable during times of low copper prices.

B What Parts of the United States Have Large Gold Deposits?

In the United States, most gold has been mined from bedrock and rivers of the mountainous west and Alaska. In Canada, gold deposits are common in the western mountains and in Precambrian rocks of the Canadian Shield.

1. This map shows larger gold deposits of the lower 48 states. Gold rushes in Alaska and along the Yukon River of northwestern Canada (not shown on this map) were largely in river gravels and so are modern placer deposits.

18.10.b1

2. The famous California Gold Rush was touched off by flakes of gold found in river gravels at Sutter's Mill, but the gold originated within veins in metamorphic rocks of the Sierra Nevada foothills, where there are large gold specimens. [Murphys, California]

3. The deep-underground Homestake Mine of the Black Hills is the largest historic producer of gold in the Western Hemisphere. The gold is in layers and veins within Precambrian metamorphic rocks, and probably originated from hydrothermal vents on the seafloor and subsequent redistribution of the gold into veins during metamorphism.

18.10.b2

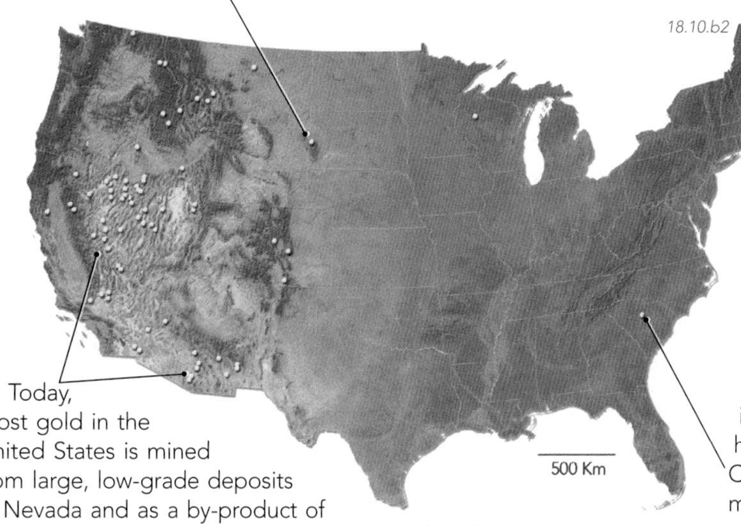

4. The first discovery of gold in the United States was in North Carolina, which was the main source of gold in coins minted during the early 1800s. Gold here and in adjacent South Carolina is present as veins in metamorphic rocks and as placers.

5. Today, most gold in the United States is mined from large, low-grade deposits in Nevada and as a by-product of copper mining in Arizona, New Mexico, and Utah.

C Where Does Platinum Occur?

Platinum is chemically very different from gold and silver and occurs in very different settings. Most platinum deposits are associated with mafic or ultramafic igneous rocks, which rose from the mantle and into the crust.

South Africa is the largest supplier of platinum, typically producing 70 percent of the world's platinum. Most mines are within a large mafic to ultramafic intrusion that developed layers as different minerals crystallized. This intrusion, the Bushveld complex, is the largest layered intrusion on Earth!

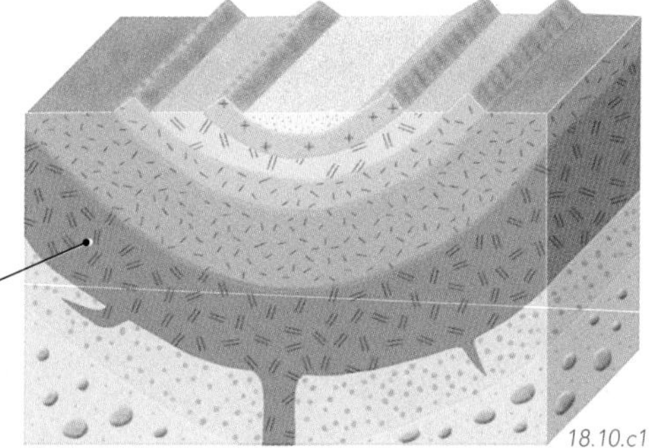

18.10.c1

18.10.c2

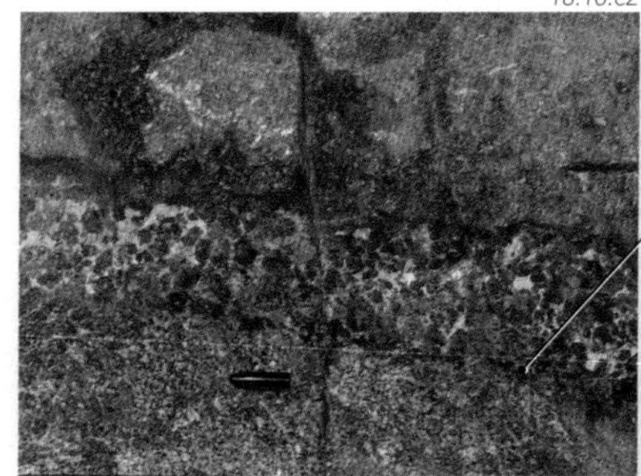

Much of the platinum occurs in a specific zone that is less than 1 meter thick. The layer has centimeter-sized igneous crystals and a high chromium and platinum content. Note the black pen cap for scale.

Precious Metals and Deep Mines

Precious metals are so precious that they are priced by the ounce, specifically by the *troy ounce*, which is about 1.1 normal ounces. What do you think is the going rate for an ounce of gold, silver, or platinum?

Gold is more precious than silver, typically costing 50 times as much per ounce. The price of gold fluctuates tremendously, but when this book was written, gold was more than \$650 per troy ounce. Silver is nearly \$12 per troy ounce, and platinum is more than \$1,300 per troy ounce.

Such high prices make it worthwhile going deep to mine gold. The Homestake Gold Mine is nearly 2.4 km (8,000 ft) deep. Some gold mines in South Africa are nearly 4 km (>13,000 ft) deep to mine a gold-rich pebbly layer less than 30 cm thick. [Ventersdorp, South Africa] ▽

18.10.mtb1

Before You Leave This Page Be Able To

- ✓ Describe or sketch the main geologic settings for gold and silver deposits.
- ✓ Identify where gold is mined in the United States and describe the type of gold deposits in each region.
- ✓ Describe the geologic setting of the world's largest platinum deposits.
- ✓ Explain why precious metal mines can afford to be so deep.

How Do Base Metal Deposits Form?

SOME METALS ARE MUCH MORE COMMON than precious metals and, unlike gold, tarnish fairly easily in air. Such metals are called *base metals*, and include iron, nickel, copper, lead, and zinc, which are fundamental to our daily lives. Where and how do deposits of these metallic elements form?

A How Do Iron Deposits Form?

Iron is arguably our most important metal because it is the main ingredient in steel. Iron deposits generally contain two iron oxide minerals: *magnetite*, which is magnetic, and *hematite*, which has a characteristic red streak.

Most iron ore is extracted from sedimentary sequences called *banded iron formations*, which contain many thin layers of iron-rich and quartz-rich rocks. Nearly all large banded iron formations are Precambrian (about 2 billion years old) and are interpreted to mark a time when increasing oxygen in Earth's atmosphere caused dissolved iron in the ocean to precipitate as iron minerals.

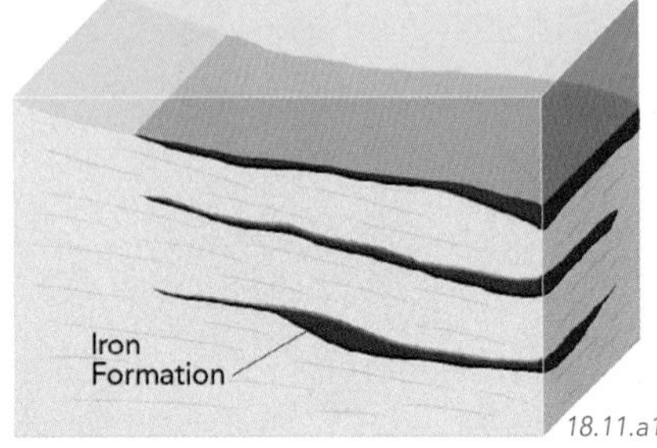

18.11.a1

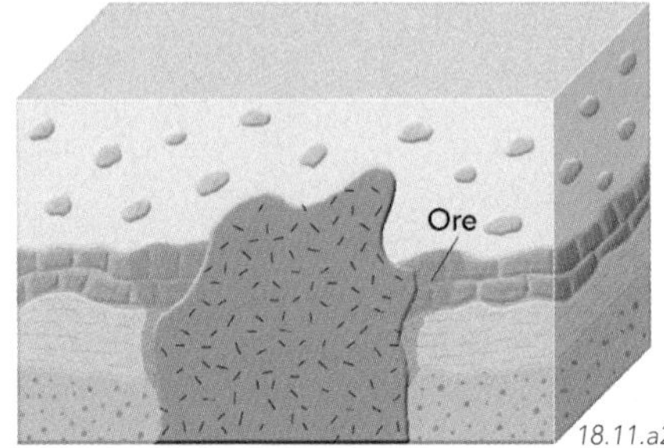

18.11.a2

Other iron mines occur along the flanks of igneous intrusions. When these intrusions were emplaced into the crust, they released metal-rich fluids that permeated adjacent rocks. Limestone and other rocks that are chemically reactive were partly or completely replaced by iron-rich minerals, producing a dark, heavy, magnetic rock, which many people mistake for a meteorite.

B How Do Most Copper Deposits Form?

Copper is a relatively mobile element in water and so occurs in a variety of different types of mineral deposits. The three discussed here illustrate the wide spectrum of copper deposits that exist around the world.

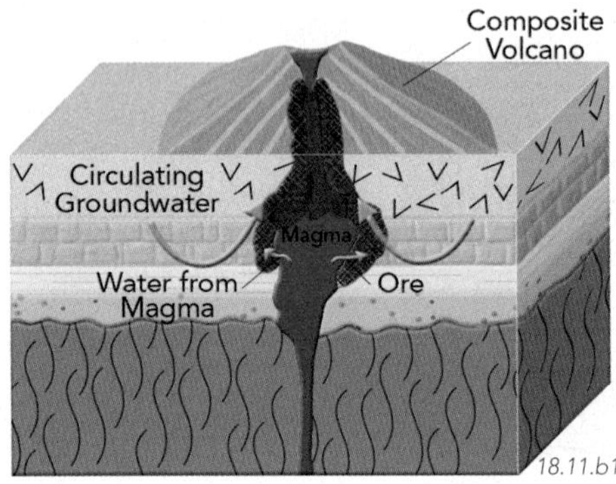

18.11.b1

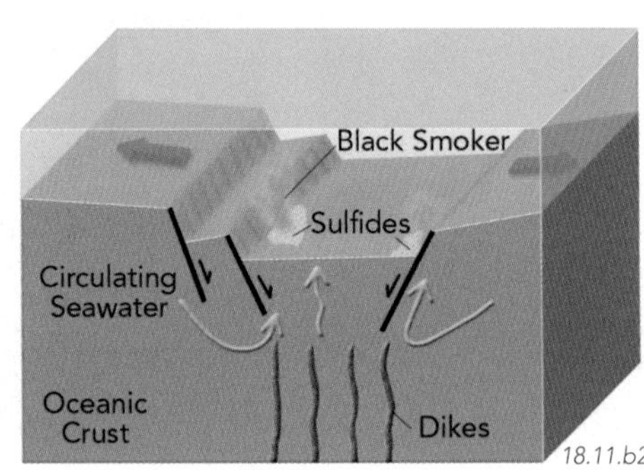

18.11.b2

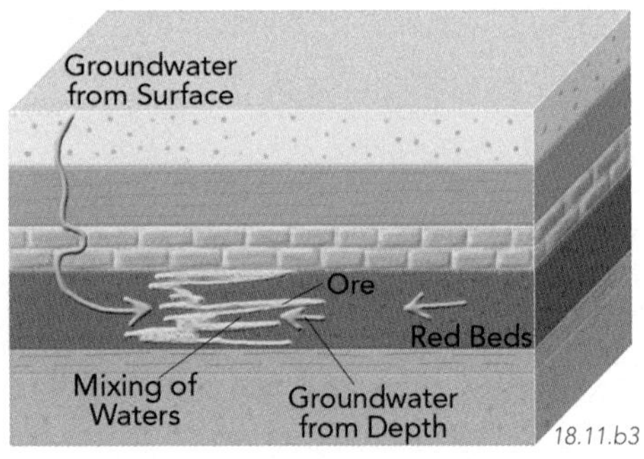

18.11.b3

Most of the world's copper ore is mined from large open pits within or adjacent to intermediate to felsic *intrusions*. Many ore-related intrusions have a *porphyritic* texture (larger crystals in a finer matrix) and the deposits are called *porphyry copper deposits*. They are fairly low grade (<1% copper sulfides and oxides). The photograph below shows samples of weathered ore and porphyritic rocks.

Other copper deposits are much higher grade, but smaller. They contain lenses and pods mainly of sulfide minerals, and are referred to as *massive sulfide deposits*. The bronze-colored rocks below consist almost entirely of sulfide minerals. Many massive sulfides formed in association with volcanic rocks that were erupted in seawater, and represent submarine hydrothermal vents called *black smokers*.

A different type of copper deposit forms in sedimentary rocks. These *sedimentary copper deposits* are common in central Europe and in the copper belt of west-central Africa. They probably formed when copper-rich groundwater mixed with chemically different groundwater, depositing copper. The photograph below shows blue-green copper minerals replacing plant fossils in sandstone.

18.11.b4

18.11.b5

18.11.b6

C What Happens When a Copper Deposit Is Weathered?

Porphyry copper deposits are relatively low grade, and some are barely economic, especially when world markets drive the price of copper down. Weathering can increase the ore grade, making a deposit profitable.

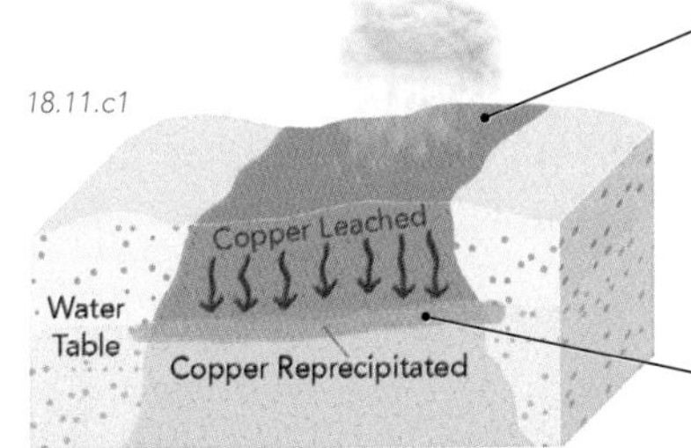

When a copper deposit is exposed at the surface, the upper part becomes weathered and oxidized. Weathering leaches copper from the top of the deposit and leaves these rocks a reddish color.

The leached copper is carried down by percolating groundwater, until it reaches the water table, where the copper is redeposited, making this part of the deposit higher grade (richer in copper).

18.11.c2

D How Do Some Other Base Metal Deposits Form?

There is a wide spectrum of base-metal deposits, with diverse modes of formation, as illustrated by two end members: one formed at the bottom of a large magma chamber and another formed in sedimentary layers.

Famous lead-zinc deposits of the mid-continental United States are hosted by limestone and other carbonate rocks that were deposited over hills of an older, eroded granite. Metal-rich groundwater flowing through the sedimentary layers encountered the buried hills and deposited the lead and zinc minerals as replacements.

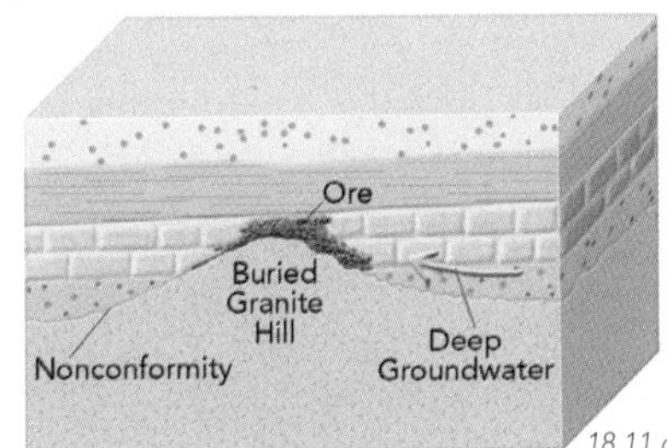

18.11.d1

This type of deposit, called a *Mississippi-Valley deposit,* contains showy crystals of galena (lead sulfide), calcite, and green and purple fluorite (a calcium fluoride mineral). Specimens of minerals from this region can be found in museums and rock shops around the world. ▷

18.11.d2

Near Sudbury, Ontario, Canada, large nickel-sulfide deposits formed at the base of a mafic to ultramafic magma chamber. Most of the sulfide ore minerals crystallized from the large magma chamber.

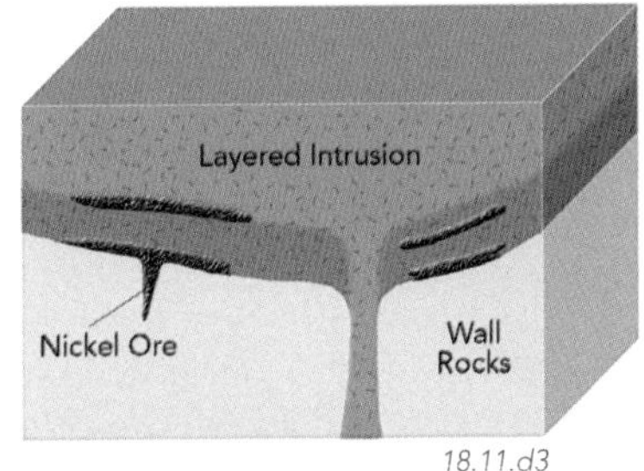

18.11.d3

▷ Rocks around Sudbury contain unusual features, such as these unusual cone-shaped joints. This type of joint, called a *shatter cone,* is common around craters interpreted to have formed from a meteoroid impact. From this and other evidence, many geologists conclude that an ancient meteoroid-impact event fractured the crust and triggered the magmatism that led to the formation of the Sudbury mineral deposits.

18.11.d4

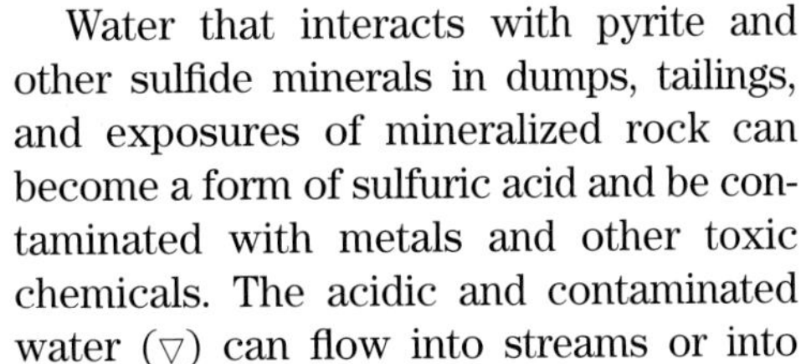

Some Environmental Issues Associated with Mining

Extracting and processing minerals has consequences to the environment. Mines disrupt the land by leaving pits in the ground, along with large piles or *dumps* of rock and other materials were moved to get to the ore. In some cases, ore is crushed and pulverized to extract the precious commodity, and this produces heaps of light-colored powder called *mill tailings.* Dumps and tailings are unsightly, cover whatever used to be there, and can collapse downhill, sending these loose materials into streams and even houses.

18.11.mtb1

Water that interacts with pyrite and other sulfide minerals in dumps, tailings, and exposures of mineralized rock can become a form of sulfuric acid and be contaminated with metals and other toxic chemicals. The acidic and contaminated water (▽) can flow into streams or into groundwater. Such waters kill many types of life and make the water unusable to humans. Finally, some sulfide ores must be roasted in mineral-extraction facilities called *smelters,* which release sulfur-dioxide gas to the atmosphere, a major contributor to acid rain.

Before You Leave This Page Be Able To

- ✓ Describe or sketch two ways that iron deposits form.
- ✓ Describe or sketch ways that copper deposits form, and what happens when a deposit is weathered.
- ✓ Summarize how the Mississippi Valley lead-zinc and Sudbury nickel deposits formed.
- ✓ Discuss a few of the environmental issues involved with mining and processing ore.

How Do We Explore for Mineral Deposits?

MINERAL DEPOSITS ARE ESSENTIAL TO SOCIETY, and much time, effort, and money are spent exploring for them. If you wanted to find a new mineral deposit, where would you start looking? Geologists explore for new mineral deposits by first studying ones they know about—to become familiar with any diagnostic attributes that would help the geologist recognize a new deposit. Geologists then use various strategies and tools to find deposits that are partially exposed on the surface and those that are completely buried.

A How Do Field Geologists Explore for Mineral Deposits?

Much of the search for new mineral deposits occurs in the field. Mineral-exploration geologists conduct various investigations, including geologic mapping, structural studies, and collecting samples for chemical analyses. Mineral exploration takes geologists to many far-off places, such as Peru, Indonesia, and Mongolia.

1. Exploration geologists use existing geologic maps or construct new geologic maps to document the geometry and distribution of rock units. Such geologists pay special attention to geologic structures that provide conduits for hydrothermal fluids, because many mineral deposits are in mineral-filled *veins*.

2. In addition to mapping the rock units, exploration geologists also map mineral changes caused by the passage of hydrothermal fluids through fractures and other pathways. The fluids change the adjacent rocks by altering existing minerals and adding new minerals, such as quartz and pyrite. Such changes are called *alteration*, and can be recognized in the field and on satellite images that are processed to emphasize alteration.

18.12.a2

◁ 3. These rocks show reddish alteration with iron-oxide minerals and a later set of crosscutting fractures that have gray alteration. Such alteration effects can extend far beyond the limits of the actual ore body and thus can help us find a new ore body that is nearby.

4. Mineralizing fluids leave chemical traces in rocks through which they pass. Geologists collect samples of rocks and soils to later analyze the *geochemistry* of the samples for the element of interest (e.g., copper) and for other elements (such as arsenic) that are associated with this type of mineralization.

5. From the various types of geologic information, exploration geologists reconstruct the geologic history of the region. They determine the sequence of rocks and structures and also when the mineralization and alteration occurred. Events that could have destroyed, preserved, enriched, or moved a mineral deposit are carefully considered. Such events include weathering and erosion that could remove or enrich a deposit, faulting that moves a deposit, and sediment deposition, which can bury and hide a deposit. Exploring for mineral deposits can involve most aspects of geology.

B How Do Geologists Use Plate Tectonics to Explore for Ore Deposits?

Plate tectonics is a critical consideration in mineral exploration because many mineral deposits are associated with rocks formed in specific plate-tectonic settings.

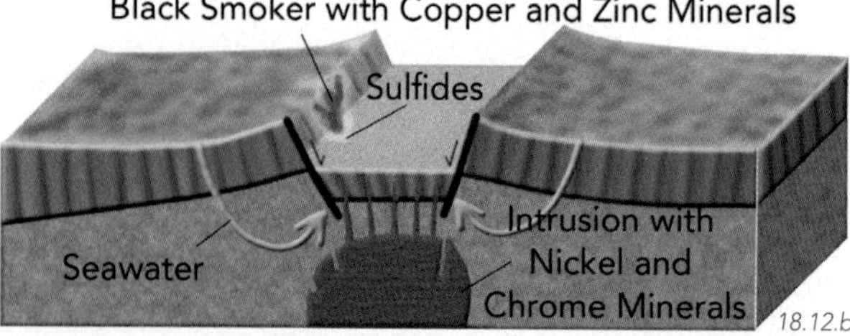

Along a divergent oceanic boundary, plates spread apart, magma invades toward the surface, and heated seawater forms submarine hot springs that deposit sulfide minerals in black smokers.

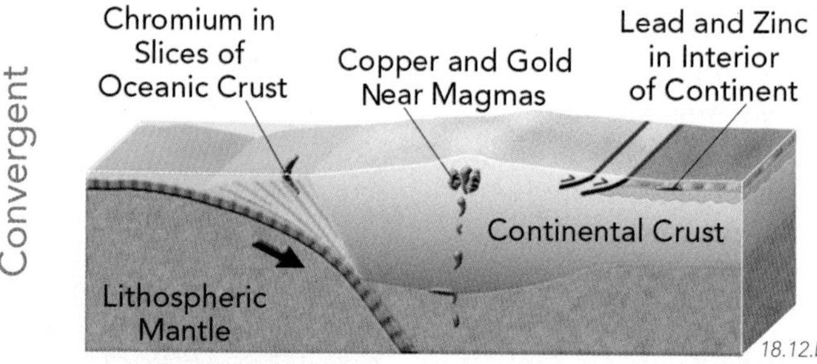

In convergent boundaries, slices of oceanic crust are scraped off the oceanic plate, magma invades the overlying plate, and metal-rich fluids escape from thrust zones.

C How Do We Find Buried Mineral Deposits?

Many mineral deposits are exposed on the surface, and most of these have already been found. Undiscovered mineral deposits are most likely partially or completely buried, requiring different exploration strategies.

1. To locate mineral deposits that are buried beneath younger deposits or obscured by thick soils and vegetation, geologists conduct surveys across potential exploration targets using several *geophysical techniques*.

2. This porphyry copper deposit is buried beneath younger sediment (shown in yellow), but is flanked by a zone of high-grade replacement ore that contains abundant magnetic iron minerals (magnetite) in addition to copper minerals.

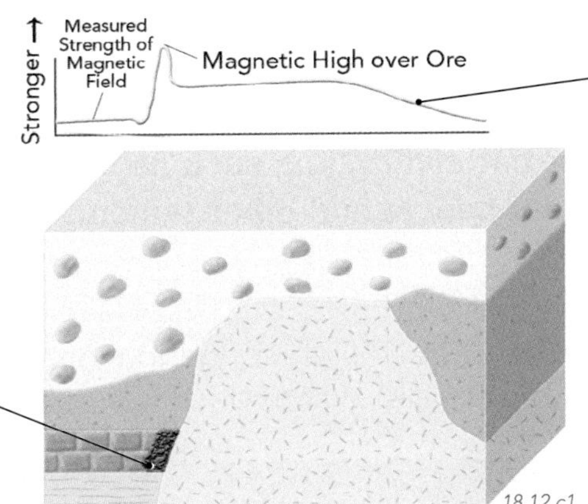

18.12.c1

3. This graph shows the strength of Earth's magnetic field across the area, as determined by a magnetic survey. The survey recorded a strong magnetic signal, called a *magnetic high*, over the ore. Exploration geologists, applying the knowledge that deposits of replacement ore commonly contain magnetic minerals, might hypothesize that such ore lies at depth. This hypothesis can be tested by drilling an exploration drill hole into the location of the magnetic high. Other types of information could be obtained by measuring variations in the strength of gravity or by running electrical current through the ground.

D How Are Mineral Deposits Extracted and Processed?

Once a new mineral deposit has been discovered, it is extensively explored with drill holes so that samples can be chemically analyzed to determine the size, shape, depth, and grade of the deposit. If all these factors are favorable for development, geologists work with engineers to design plans to mine and process the ore.

1. *Open-Pit Mine*—If an ore deposit is shallow enough, it is cost effective to mine the ore in an open pit (▷). Ore can be blasted loose with explosives and loaded into huge ore trucks.

18.12.d1

18.12.d2

3. *Mill*—From the mine, the ore goes to the mill, where it is crushed and run through various processes to separate the ore minerals from the rest of the rock. These large, rotating cylinders contain hard metal spheres that crush the pieces of ore.

2. *Underground Mine*—Deeper ore deposits must be mined by more expensive underground methods. Ore is blasted loose at depth and hauled to the surface by train (▷), ore carts, elevators, or large trucks that drive through even larger tunnels.

18.12.d3

18.12.d4

4. *Smelter or Leach Pads*—Some ore is roasted in furnaces in a *smelter*. Other ore is crushed and placed on pads (◁) that are sprinkled with chemical solutions that dissolve (leach) soluble minerals so they can be recovered.

Why We Explore

Finding mineral resources is essential to support our modern society. Most people are unaware of the amounts of mineral resources consumed in the United States per person. These amounts include the materials used to construct roads, gypsum in wallboard, and copper in wiring. The National Mining Association estimates that average consumption in the United States in 2005 was 47,000 pounds of minerals per person! Nearly half of this amount was sand, gravel, and stone, but includes large amounts of coal, natural gas, and petroleum used for fuels and to make plastics and many other items.

Material	Per Capita Consumption (pounds)
Sand, Gravel, Stone	22,060
Petroleum Products	7,667
Coal	7,589
Natural Gas	6,866
Cement	940
Iron Ore	425
Salt	400
Phosphate Rock	302
Clays	276
Aluminum (Bauxite)	77

Before You Leave This Page Be Able To

- ✓ Describe how we explore for mineral deposits, including buried ones.
- ✓ Describe why the plate-tectonic setting of a region is an important consideration for mineral exploration.
- ✓ Briefly summarize how minerals are extracted and processed.
- ✓ Discuss how much minerals we use.

Why Are Industrial Rocks and Minerals So Important to Society?

GOLD, PLATINUM, AND DIAMONDS get all the attention, but modern society really relies on everyday rocks and minerals, like limestone and gravel. These common materials, called *industrial rocks and minerals*, are used to build much of the infrastructure of our civilization, such as highways, bridges, water pipes, sewer lines, and much of the material used to build houses and other buildings.

A How Are Cement and Concrete Produced?

Cement and concrete are everywhere in our cities. They form the foundations of our buildings, overpasses across our highways, and sidewalks, curbs, and block walls. Where does this material come from?

18.13.a1

1. The starting material for cement and concrete is *limestone*, especially one that is free of impurities, such as certain detrimental clay minerals. Limestone is extracted (mined) mostly in open pits called *quarries.* The price of limestone rarely justifies expensive underground mining. [Marana, Arizona]

18.13.a2

2. Limestone is crushed and processed in a cement plant, where the crushed limestone is mixed with clays and other materials and then roasted in large kilns. The material is cooled and mixed with gypsum to make *cement.* A different product, called *lime*, is also produced by roasting limestone and is used to produce paper, to help make steel, and to treat soils.

Cement Versus Concrete

3. *Cement* is a whitish powder produced at the plant. When the powder is mixed with water (and usually other materials), complex reactions take place and the mixture sets to a solid mass.

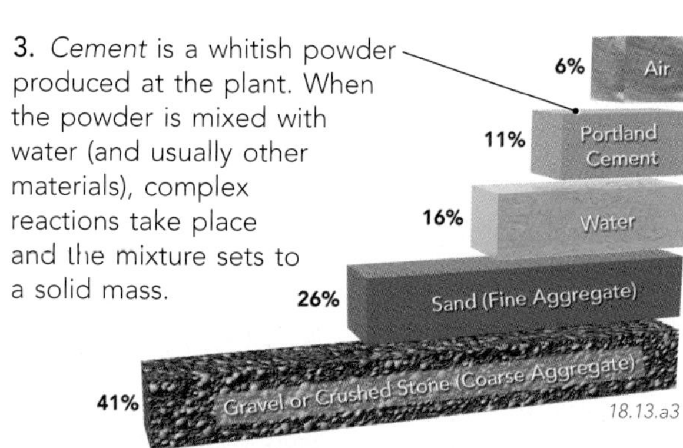

18.13.a3

4. *Concrete* is a mixture (△) of sand, gravel, or crushed stone, which together are called *aggregate*, and cement. Concrete is a thick, wet slurry when it is being transported and laid down. It can be mixed at the plant in truckloads or on site in small batches. Cement makes up only 10 percent to 15 percent of concrete, but it is what holds concrete together.

B Where Do Sheetrock and Plaster Come From?

Sheetrock, also called wallboard, and plaster are the main materials covering inside walls of houses and buildings. These are both produced using *gypsum*, a calcium-sulfate mineral.

18.13.b1

Gypsum is an evaporite mineral, meaning that it can be precipitated by evaporation of water. Most gypsum forms when water in a sea or lake evaporates, leaving behind gypsum, salt, and other substances that were dissolved in the water.

18.13.b2

Most gypsum forms in specific sedimentary layers in the sequence, like these shown here. Mining operations strip off any overlying rocks, mine away the gypsum, and stop at the base of the gypsum-rich layer. [Cuba, New Mexico]

Most interior walls in buildings constructed in the United States in the last 40 years are made from wallboard containing gypsum. Since gypsum is not a very expensive material, the distance from mine to market is critical in determining if the mining operation is profitable. Most gypsum mines are shallow open pits, which are cheaper to operate than an underground mine. Gypsum is common enough that many communities have fairly local sources. In 2006, the leading producers of gypsum in the United States were Oklahoma, Iowa, Nevada, New York, California, and Arkansas. The total amount of gypsum mined in the United States in 2006 was 21.2 million tons, worth $159 million dollars—a lot of gypsum and a lot of wallboard.

C How Are Sand, Gravel, and Crushed Rock Used?

The Earth materials we use the most are *sand and gravel*. Our society uses enormous quantities of sand and gravel, along with crushed rock, to mix into concrete. We use sand to produce brick, roof tile, and glass, as well as to build up or fill in spaces, such as beneath highways and the foundations of buildings. An average of 10 metric tons (22,000 pounds) of sand, gravel, and crushed stone is mined *each year* for each person in the United States.

The largest source of sand and gravel is rivers and streams. Some is quarried from within the active river channel or adjacent low terraces, because these sediments are unconsolidated and individual clasts are generally not highly weathered. If a modern-day channel is not available, sand and gravel can be quarried from older stream and river deposits.

18.13.c1

Various types of hard rocks, such as quartzite, are crushed for use as aggregate, decorative rock, road and railroad beds, and various other uses.

18.13.c2

D What Are Some Other Important Industrial Mineral Deposits?

Many other industrial rocks and minerals are mined and used by society. You will recognize some of these, like salt and clay, but others will be unfamiliar. Certain minerals are used as chemical filters, whereas others are used for fertilizer. Here, we present a small selection of these mineral commodities.

Silica sand is like a typical sand, but consists of nearly all quartz grains, with few other impurities. Quartz sand this pure comes from windblown deposits and some beach deposits, and is used to make glass, ceramics, and microchips.

18.13.d1

18.13.d2

Clay refers to a size of particle and to a family of platy minerals. Clay minerals are a main ingredient in tiles and other ceramics, bricks, and expensive paper. There are several varieties of clay, each with a different value and origin.

Salt is a necessary commodity for nutrition and for many industrial and chemical uses. Salt is an *evaporite* and mostly comes from salt deposits that formed when a sea or lake evaporates enough to precipitate salt. Common salt is the mineral *halite*.

18.13.d3

18.13.d4

Phosphate rock is a general term for rocks that are rich in the element phosphorous, which is essential for plant growth. It is mined from recent marine deposits in Florida and elsewhere and from older (Paleozoic) marine rocks. It is used primarily for fertilizer.

Ways to Obtain Salt

Rock salt, which is the mineral *halite* (NaCl), can be obtained in several ways, depending on the geologic setting in which the salt occurs. Salt can be harvested by trapping salty water and allowing it to evaporate in the sunlight. Salt is harvested this way from Great Salt Lake, Utah (below). Salt can be mined from underground layers of salt or from salt domes, such as along the Gulf Coast. It can also be extracted by pumping freshwater into underground salt bodies and pumping the resulting salty water back to the surface, as in the facility shown below. [Glendale, Arizona]

18.13.mtb1

18.13.mtb2

Before You Leave This Page Be Able To

- ✓ Summarize how limestone is used to make cement, concrete, and lime.
- ✓ Summarize where the gypsum in wallboard and plaster comes from.
- ✓ Discuss why sand, gravel, and other aggregate are our most-used mineral resource.
- ✓ Briefly describe the origins and uses of silica sand, salt, clay, and phosphate rock.

Why Is Wyoming So Rich in Energy Resources?

WYOMING IS INCREDIBLY RICH IN ENERGY RESOURCES. It contains as many large and diverse energy resources as any part of the world. It has the most prolific coal region in the world, and the three largest coal mines in the United States. It ranks first in the country in uranium production and second in natural gas, including coal-bed methane. It shares with adjacent states the largest resource of oil shale in the world. What happened in the geologic history of Wyoming to make the state so energy rich?

A What Is the Geology of Wyoming?

Wyoming has a long and interesting geologic history, spanning more than 2.5 billion years. Nearly all of the state's energy resources, however, resulted from geologic events during the last 60 million years.

The topography of Wyoming reflects its geology, as expressed in this geologic map on a shaded-relief base. Colors reflect types and ages of the exposed rock units. On this map, large uplifted mountain ranges are surrounded by broad valleys (called basins).

Heavy lines on the map represent faults, many of which are along the edges of mountains, such as the Grand Tetons.

The details of the geologic units on this map are not critical here, but the colors represent the following ages of rock, from youngest to oldest:

Cenozoic sedimentary (yellow) and volcanic (red) rocks

Mesozoic sedimentary rocks (green)

Paleozoic sedimentary rocks (blue)

Precambrian metamorphic and granitic rocks (brown and purple)

The mountain ranges contain older (Precambrian and Paleozoic) rocks, whereas the basins are covered by Mesozoic and Cenozoic sedimentary rocks.

18.14.a1

▽ On this east-west cross section (A–A') across the state, the western edge of the state is a fold and thrust belt, called the Overthrust Belt, which has shuffled the Paleozoic and Mesozoic sedimentary rocks.

The cores of the mountain ranges expose ancient Precambrian rocks (brown), which include 2.5 billion-year-old metamorphic and granitic rocks, such as in the Wind River Range.

Dipping off or faulted beneath the Precambrian rocks is a thin veneer of Paleozoic and Mesozoic sedimentary rocks, such as those in the Casper Arch.

The largest basin is the Powder River Basin in the eastern part of the state (the yellow area beneath the "G" in Wyoming on the map). It is a broad downward-bending fold (syncline) that preserves the youngest Cenozoic rocks in the center of the basin. The rock layers rise up toward both flanks of the basin, climbing westward toward the Casper Arch and eastward toward the Black Hills uplift of Wyoming and South Dakota.

18.14.a2

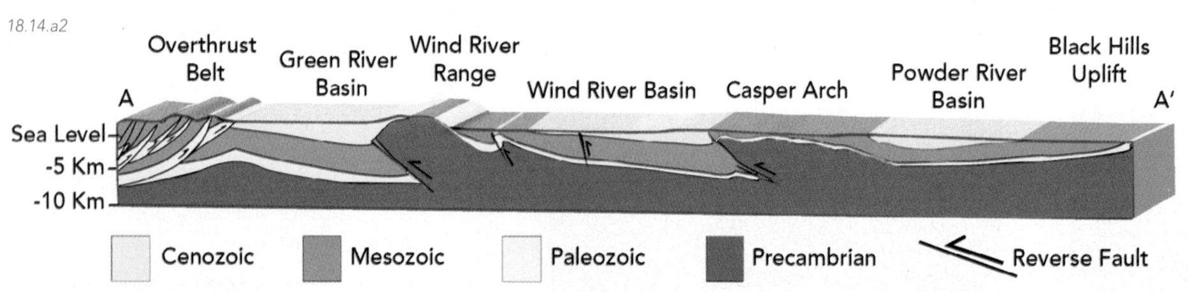

B What Is the Geologic Setting of Wyoming's Natural Resources?

The large-scale geologic features of Wyoming, such as uplifts and basins, control the distribution of oil and gas fields, coal, oil shale, uranium deposits, and hot springs (potential for geothermal energy).

Oil and Gas, and Oil Shale

18.14.b1

1. The oil fields (blue-green) and the gas fields (orange) are in the basins, where sedimentary rocks are preserved and relatively thick. There are almost no oil and gas fields in the uplifted mountain ranges, where the sedimentary layers are eroded away, exposing the underlying Precambrian crystalline rocks.

2. The southwestern part of the state, including the Overthrust Belt, has mostly gas fields instead of oil because rocks in this area were heated more than rocks farther east.

3. The world's largest oil shale deposits (light purple) are in southwest Wyoming and nearby states, where large lakes formed in early Cenozoic time within the basins.

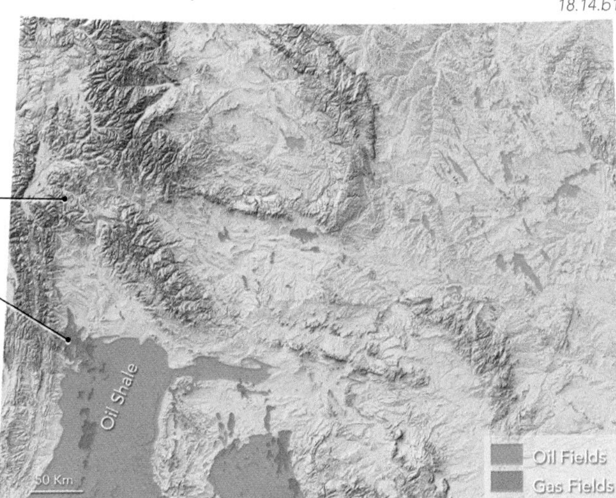
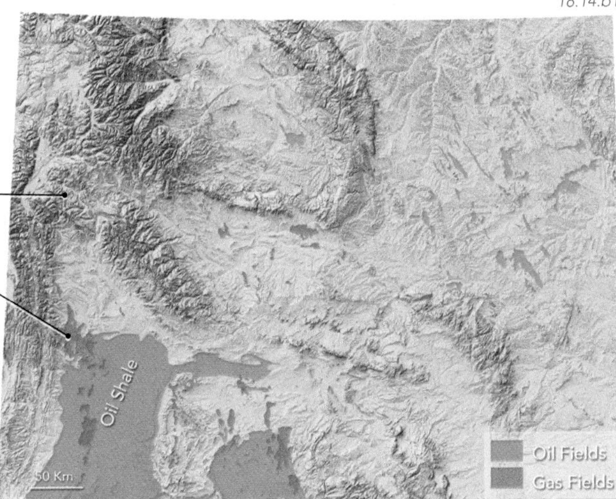

4. The oil and gas are mostly in the Paleozoic and Mesozoic sedimentary layers, because these rocks are young enough to have been organic-rich and were buried deep enough to turn the organic material into oil and gas. The reservoir rocks for the oil and gas include a Paleozoic limestone and a Mesozoic sandstone, both of which have high permeability. The main oil and gas traps are anticlines and facies changes.

Coal and Coal-Bed Methane

5. Coal fields (dark gray) are also in the basins, especially in those parts of the basins that have the youngest Mesozoic and Cenozoic sedimentary rocks. Many of the coal-bearing rocks formed in swampy deltas.

6. There are no coal fields in the large mountain ranges because some coal-bearing layers were eroded off the uplifts and others were only deposited in the basins.

7. Areas of coal-bed methane (light gray) are more widely distributed than the coal mines because coal can be mined only where the coal-bearing units are near the surface, whereas the gas wells can be deeper.

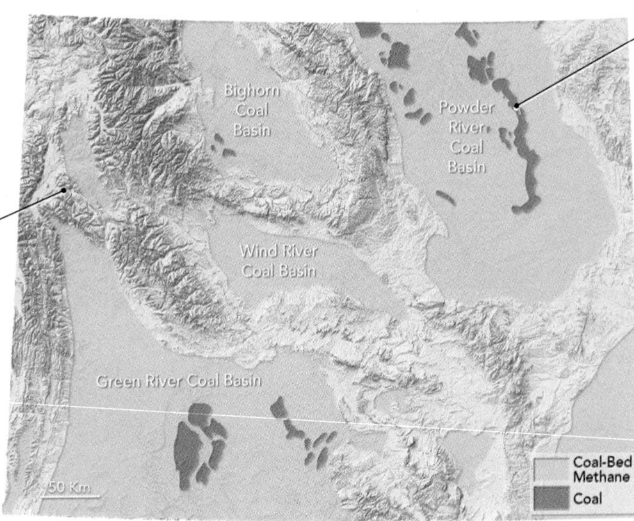

18.14.b2

8. The Powder River Basin produces more coal than any place in the world and also has the largest remaining coal reserves. In 2004, it produced 382 million tons of coal! Most of the mined coal goes to coal-fired electrical-generating plants, some of which are located close to the coal fields to avoid transporting the coal great distances. Large quantities of coal are shipped to power plants in other parts of the country.

Uranium and Potential for Geothermal Energy

9. Areas with potential for geothermal energy commonly have hot springs (shown in red) and geysers. Most hot springs (>100°F) are near areas of recent volcanic activity and faulting near Yellowstone National Park.

10. The largest uranium deposits (yellow) are in Cenozoic sandstone and were deposited by groundwater. The uranium has been mined by traditional methods and has been extracted by pumping water through subsurface ores to dissolve the uranium.

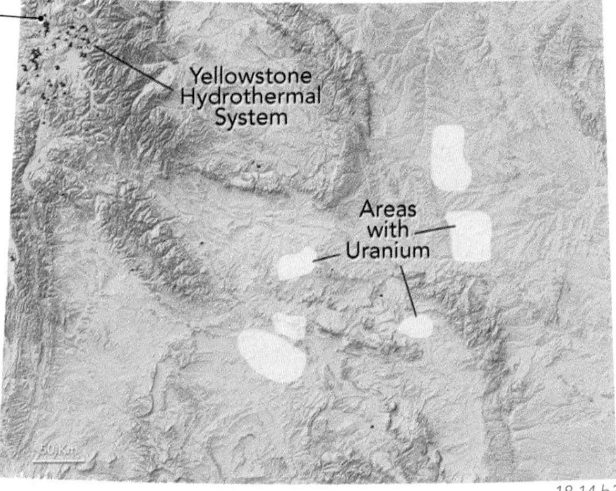

18.14.b3

Before You Leave This Page Be Able To

✓ Summarize or sketch in cross section the main types of geologic features present in Wyoming.

✓ Summarize how large-scale geologic features control the distribution of oil, gas, oil shale, coal, coal-bed methane, uranium, and geothermal energy.

Where Would You Explore for Fossil Fuels in This Place?

THIS REGION EXHIBITS CLUES that it contains oil, gas, coal, or other hydrocarbon-based sources of energy. There are oil seeps, exposed coal seams, and other features that may be related to hydrocarbons. You will use the character and distribution of different rock types on the surface, along with some subsurface information, to identify places to explore for hydrocarbon-based energy sources. This investigation involves many aspects of geology, including sedimentary environments, structural geology, groundwater, and energy resources.

Goals of This Exercise:

- Observe the landscape to understand the geologic setting of different areas, and read the descriptions of each location in order to interpret the significance for exploring for fossil fuels.
- Use a stratigraphic section and descriptions of the rocks to interpret the environment in which each rock layer formed.
- Use surface observations of geologic structures, along with the geologic section shown on the side of the diagram, to interpret the subsurface geology.
- Determine the best locations to explore for hydrocarbon-based fossil fuels.

Procedures

Use the available information to complete the following steps, entering your answers in the appropriate places on the worksheet.

1. Observe the features shown on the landscape to the right. Read the text box associated with each area and consider what that statement implies about the rock types and geologic structure in that area.
2. On the stratigraphic section, read the description of each rock unit and interpret in what environment each unit formed. Next, consider what implications each rock's character has for that rock's potential role in the generation, preservation, or trapping of fossil fuels.
3. Use the various types of structural information to characterize the main geologic structures that cross the area.
4. Integrate your understanding of the rock sequence and the structural geometries to predict what rocks would lie at depth beneath any area, and whether any particular rock layer will be at a shallow, medium, or great depth below the surface.
5. Draw the letters O, G, C, and S any place on the map that you think has potential for oil (O), natural gas (G) including coal-bed methane, coal (C), and oil shale (S). Note that not all of these types of hydrocarbons may be present. You may decide to write a letter (such as C for coal) in more than one location. If you do, label them C1, C2, C3, etc., in order of highest to lowest potential.
6. Write a justification for each of your proposed sites, including what you think would be present in the subsurface and how you intend to extract the resource.

1. The highest feature in the area is a large ridge known as *Tan Mountain*, which is largely composed of tan-colored sandstone. The sandstone is well sorted and is reportedly a good source of groundwater (i.e., is an aquifer). The sandstone unit is named after the mountain, and is called the *Tan Mountain Sandstone.*

2. On both flanks of the mountain, the sandstone dips away from the ridge crest, defining an *anticline* that has nearly the same shape as the mountain.

18.15.a1

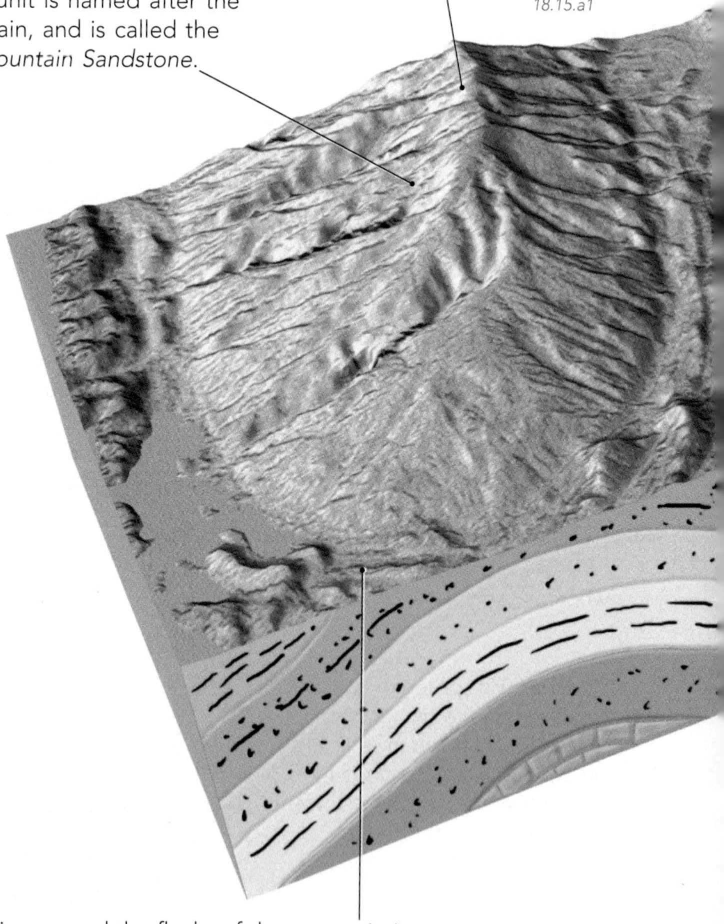

3. Wrapping around the flanks of the mountain is a sequence of reddish and gray sedimentary rocks, which are shown on the cross section with a pale reddish color. Local people informally call this unit the *carbon beds* because the bottom part of the unit contains a layer of coal up to 5 meters (16 feet) thick (too thin to show on the cross section). The coal is exposed only here and there, but digging beneath the surface has shown that the coal layer is fairly continuous. Miners drilling into the unit while exploring for coal experienced some minor explosions due to some type of flammable gas.

4. The two valleys in the center of the area are underlain by tan, dark gray, and brown shale, which geologists call the *upper shale*. This unit, colored tan on the cross section, is the highest unit exposed in the area, except for some thin gravels along the rivers and mountain fronts. The dark-gray and brown parts of the shale emit an oily smell. Some long-time residents claim that the shale actually burns, but this has not been verified.

5. In one valley, the rock layers do not quite match up when geologists compare the rocks exposed on the mountains to either side. A zone of crushed rocks marks a fault zone along one of the mountain fronts. There are some springs and small oil seeps along the fault, but overall the fault does not seem to be very permeable because some type of natural cement has filled in the pore spaces between the broken pieces of rock.

6. The eastern ridge, like the other two, is an anticline. It is more eroded than the other two anticlines, and so exposes deeper rock layers. The Tan Mountain Sandstone caps the highest peaks and overlies a thick sequence of yellowish-tan shale. This shale is called the *muddy shale,* because water will not sink into the unit when it rains and remains on the surface, making a muddy mess.

7. Some recently deposited gravels cover the older rocks in a few places. These gravels are loose and unconsolidated.

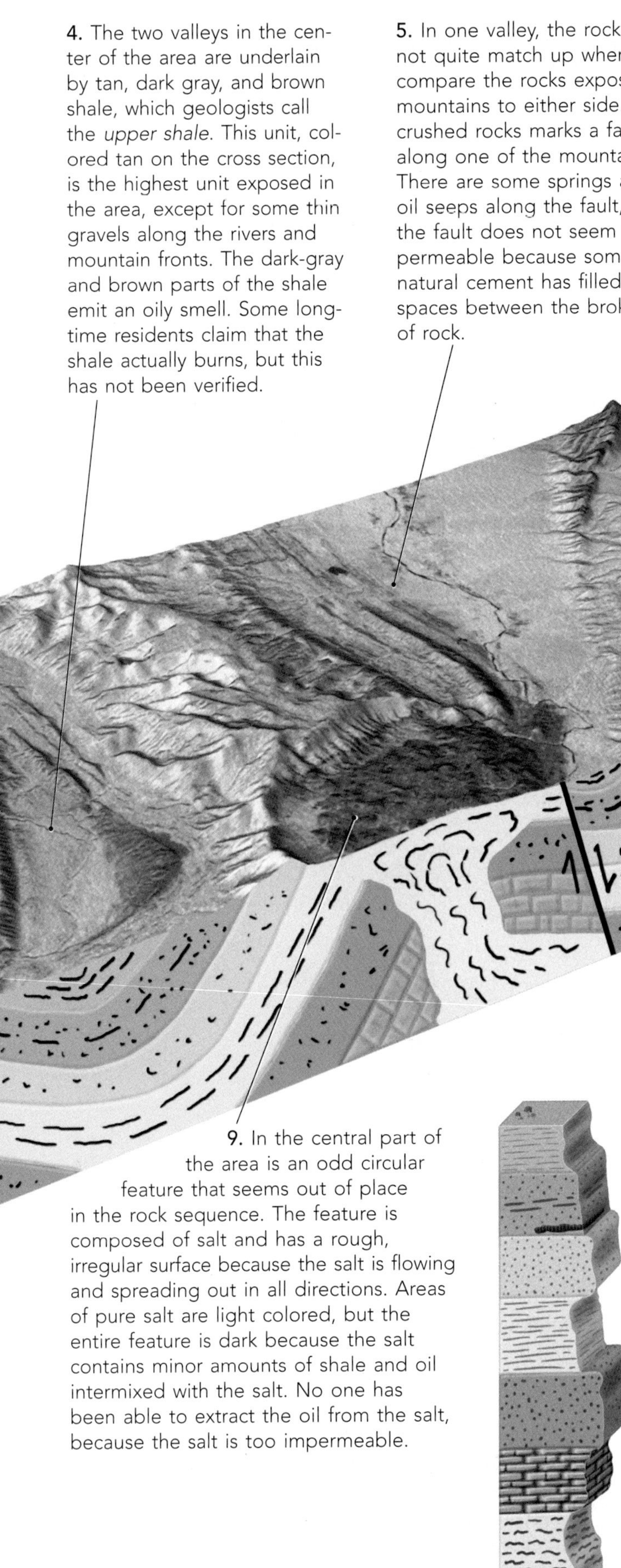

8. Following up on the presence of the oil seeps and an encouraging seismic-reflection survey, oil companies drilled an exploration drill hole in a nearby area, outside the area shown here. The drill hole started in the upper shale and encountered the sequence shown in the stratigraphic section below. The drilling discovered units not exposed at the surface, including a brown, permeable, and oil-stained sandstone, called the *lower sandstone,* which underlies the muddy shale. Farther down is an organic-rich limestone with favorable traces of oil. The lowest unit encountered in the drilling is a layer of salt. The salt has contorted layers formed by flow.

9. In the central part of the area is an odd circular feature that seems out of place in the rock sequence. The feature is composed of salt and has a rough, irregular surface because the salt is flowing and spreading out in all directions. Areas of pure salt are light colored, but the entire feature is dark because the salt contains minor amounts of shale and oil intermixed with the salt. No one has been able to extract the oil from the salt, because the salt is too impermeable.

Stratigraphic Section

18.15.a2

Upper Shale: Tan, dark gray, and dark brown shale with nonmarine fish fossils; dark layers emit an oily smell.

Carbon Beds: Reddish and gray sedimentary beds with coal near the base.

Tan Mountain Sandstone: Tan, locally cross-bedded sandstone with both marine fossils and land plants.

Muddy Shale: Light-gray shale that contains few beds or fractures.

Lower Sandstone: Porous and permeable, well-sorted sandstone with very little natural cement between the grains; locally oil stained.

Lower Limestone: Dark-gray limestone with abundant marine fossils; contains many open fractures and bedding planes, some of which are locally oil stained.

Salt: A mostly cream-colored to light gray salt with some thin organic-rich shales. Tests show that the salt, shale, and overlying limestone have all been heated to 80 to 100°C.

CHAPTER 19

Geology of the Solar System

GEOLOGIC PROCESSES ARE NOT RESTRICTED TO EARTH. Geologic features are exposed on the surfaces of our solar system's four innermost planets, on our own Moon, and on moons of other planets, such as those of Saturn and Jupiter. The planets and moons highlighted in this chapter provide a brief portrait of the most important or interesting bodies in our solar system. Compared to illustrations throughout the chapter, the planets and moons are vastly farther apart and are more different in size than can be shown.

The four inner planets are called terrestrial planets because they have solid rocky surfaces (terra means earth).

Mercury is a small, heavily cratered planet with almost no atmosphere. It has a 650°C difference between night and day temperatures!

Why is the planet so heavily cratered?

Venus has a thick atmosphere of carbon dioxide that captures much of the solar radiation that reaches the planet. This extreme greenhouse effect causes high surface temperatures of 450°C.

What is the land like beneath the clouds?

Earth has plate tectonics and also a strong magnetic field caused by its rotating molten-iron outer core. Abundant surface water sustains a diversity of life.

Why is Earth so different than the other inner planets?

Mars has been recently explored, yielding data that supports the idea that water once flowed on the martian surface. But Mars lost most of its atmosphere sometime in the past and now is so cold that liquid water cannot exist in large quantities on its surface.

What is the evidence for water on Mars?

Asteroids are rocky fragments concentrated in an orbit between Mars and Jupiter. They are similar in composition to certain meteorites and are interpreted to be fragments left over from the formation of the solar system, probably with some pieces of small planetary objects that broke apart.

Mercury

Venus

Earth

Mars

19.00.a1

How the Solar System Formed

The Sun formed about 5 billion years ago from the remnants of previous stars and cosmic dust, all of which had a beginning in what is called the "big bang." According to modern theories the entire universe arose from the *big bang* 10 to 15 billion years ago, so the universe was 5 to 10 billion years old before our solar system began to form. Current theories for formation of our solar system suggest that the Sun and planets condensed from a shapeless cloud, or *nebula*, of gas and dust. Particles of dust clung together to form small chunks and then larger and larger pieces, eventually ending up as planets. The Sun, meantime, became massive enough through its collection of particles to begin atomic fusion and emit protons and electrons in a *solar wind*. The solar wind scrubbed the inner planets of hydrogen, helium, and other light elements near the surface, leaving only heavy materials. Later, Earth gradually re-acquired its supply of hydrogen and other light elements. The outer planets were less affected by solar wind and had enough gravity to retain hydrogen and helium atoms.

During the early stages of its formation, Earth likely collided with another large object that was not quite yet a planet. This catastrophic collision ripped away part of Earth, forming our Moon. It also likely knocked Earth off its original axis of spin, giving the planet the present 23.5 degree tilt of its spin axis. An interesting implication is that we have a moon and seasons because of this collision nearly 4.5 billion years ago.

Jupiter, the largest planet in the solar system, consists of hydrogen and helium with a small rocky core, making it compositionally more similar to the Sun than to Earth. It has a banded, swirling atmosphere and many interesting moons.

Do Jupiter's moons look like ours and are they all the same?

TOPICS IN THIS CHAPTER

Jupiter

Saturn is similar to Jupiter, but has a beautiful system of delicate, icy rings around the planet. Our spacecraft are actively exploring Saturn and its moons.

What have our spacecraft found so far?

Saturn

Uranus

Uranus is large, but smaller than Jupiter and Saturn. It is much farther out in the solar system, being as far from Saturn as Saturn is from the Sun. Uranus has arguably the oddest moon in the solar system.

Neptune

Neptune is very similar in size and composition to Uranus. Both planets are gaseous and have a blue color.

Why are Neptune and Uranus blue?

Pluto is a tiny body with an icy surface and an unusual orbit. Once considered to be the ninth planet, Pluto has been reclassified to *dwarf planet* status by astronomers, leaving our solar system with only eight true planets.

Pluto

The four large outer planets—Jupiter, Saturn, Uranus, and Neptune—are known as the *gas giants* because of their gas-rich character, which is quite different from the terrestrial planets. All four planets have their own moons and some type of rings. One way to think of the solar system is as an inner zone of rocky planets, followed by gaseous ones, and finally by smaller objects, like Pluto, dominated by ice. This outward progression is related to how the solar system formed and evolved.

Outside Our Solar System

Our galaxy, the *Milky Way Galaxy*, is only one of countless huge galaxies in the universe. Each galaxy is composed of millions of stars, like the Sun, and many of these stars are probably orbited by planets. Astronomers have captured some amazing images of other galaxies and of nebulae, which are large accumulations of space dust and stars. Below are the Whirlpool Galaxy (top) and Eagle Nebula (bottom).

19.00.a2

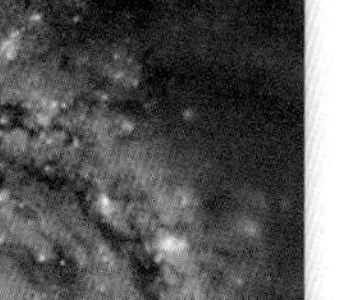

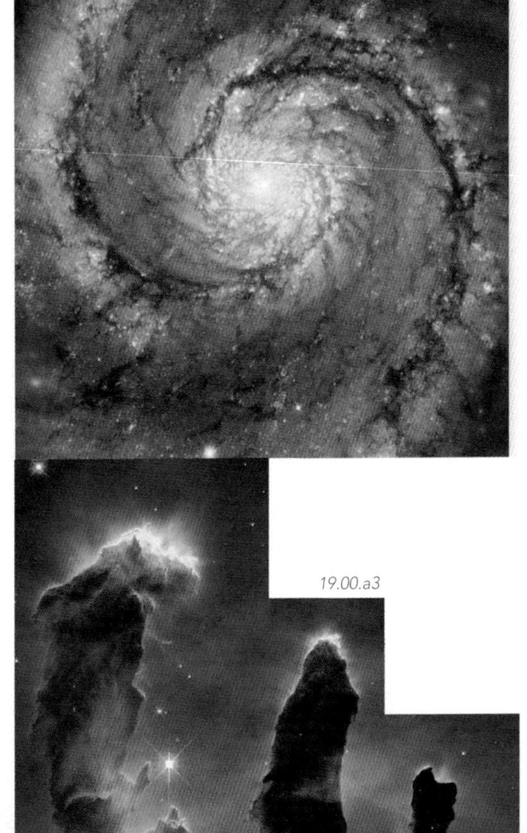

19.00.a3

How Do We Explore Other Planets and Moons?

EXPLORING THE GEOLOGY of other planets and moons is not as easy as studying geology on Earth, even compared to the more remote parts of our planet, such as Antarctica. Nearly all exploration of other planets and their moons has to be done remotely by examining them with telescopes and instruments either on Earth and in orbit around Earth, or by sending spacecraft to visit these distant objects. The only place, other than Earth, where geologists and other humans have walked is the Moon.

A What Can We Observe with Telescopes on Earth and in Earth Orbit?

Historically, most investigations of other planets and moons in the solar system have been made using Earth-based telescopes. Astronomers still rely heavily on Earth-based telescopes but also use telescopes in orbit around Earth or launched into space to avoid the distorting effect of Earth's unpredictable atmosphere.

▽ These images of Mars were taken several months apart from the *Hubble Space Telescope*, which orbits Earth.

19.01.a1

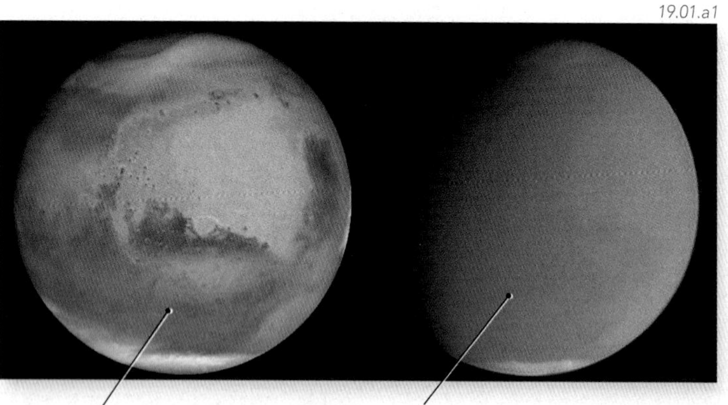

In this view, Mars has an overall orange-red color, but contains dark rocky areas, as well as ice caps at the north and south poles.

Several months later, a planet-wide dust storm had covered most of Mars, obscuring it from view.

19.01.a2

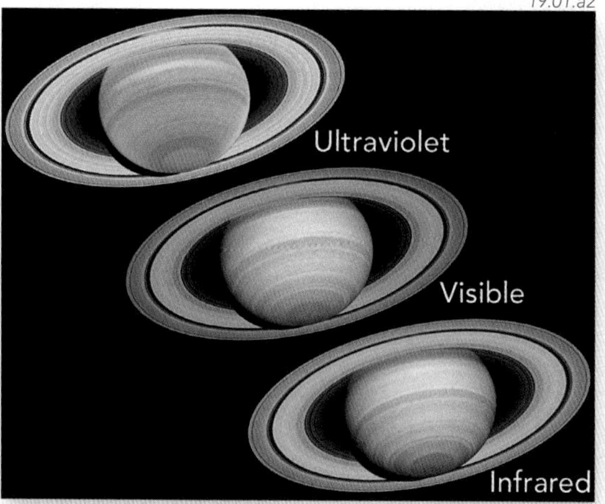

Common telescopes view the night sky using *visible light*, but astronomers use telescopes that observe other parts of the *electromagnetic spectrum*. These three views of Saturn from the *Hubble Space Telescope* show how the planet appears when imaged in visible light (center), short-wavelength ultraviolet (top), and long-wavelength infrared (bottom). Each type of image provides data about different aspects of an object.

B What Do Radar Observations Indicate About a Planetary Surface?

We can observe the surfaces of other planets and moons remotely (from a distance) by using other methods, such as radar, which can penetrate the obscuring atmosphere of a planet and reveal topography of the surface.

1. In this technique, radio waves are transmitted down and sideways toward a planetary surface from a satellite. The instrument measures the amount of the radar signal reflected back from the surface.

2. If the planetary surface is *rough* or slopes *toward* the satellite, the surface will reflect more radar waves back toward the satellite. The area will appear *bright* on the radar image (lots of returned energy).

3. If the planetary surface is *smooth* or slopes away from the satellite, most of the radar energy will bounce away rather than return to the instrument, and the area will appear *dark* on the radar image.

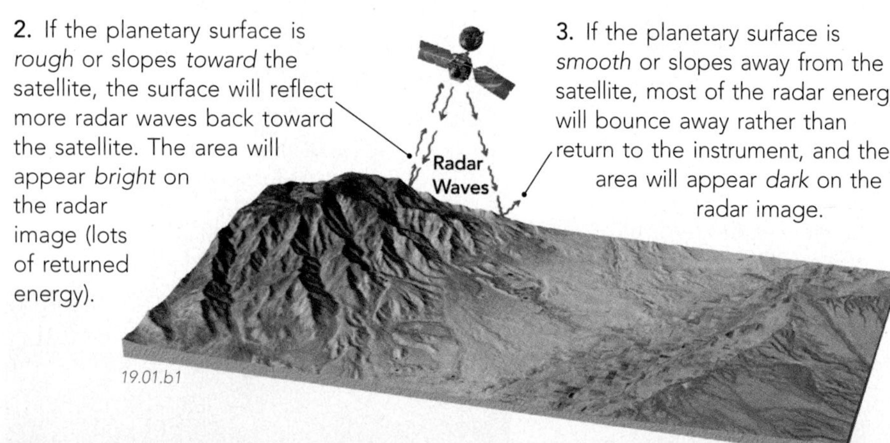

19.01.b1

19.01.b2

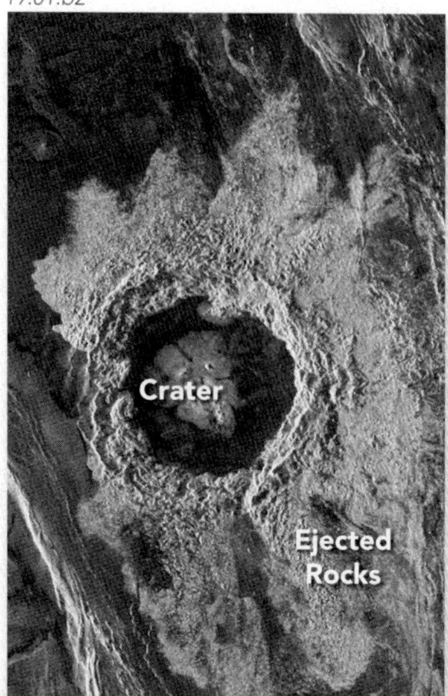

This radar image from Venus shows a crater produced by a meteoroid impact. Much of the interior of the crater is dark and appears to be smooth, whereas the broken rocks ejected from the crater form an *apron* around the crater and appear bright because they are rough.

C What Makes Earth So Special?

Earth is very different from its neighbors because it is just the right distance from the Sun to contain abundant water and to allow a thick, but not too thick, atmosphere to develop, along with oceans, rivers, and life.

19.03.c1

1. In this computer-rendered view of Earth, the planet is dominated by its blue oceans, green and brown continents, and white clouds and ice.

2. Earth's surface conditions allow water to exist as vapor, liquid, and solid phases in the atmosphere, oceans, and ice sheets, respectively. Moving water and ice erode the land and deposit sediment, in the process modifying or even remaking the land surface (resurfacing).

3. The planet is big enough and generates enough heat to allow plate tectonics to operate and form the large-scale features observed today.

4. On Earth, weathering produces abundant sediment, which can be transported by water and atmospheric winds. Most of the land and seafloor are covered by at least a thin veneer of sediments and sedimentary rocks. This image shows a large dust storm.

19.03.c2

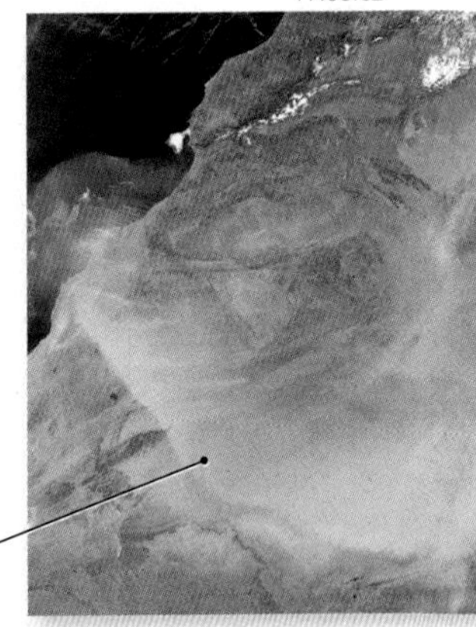

D What Is on the Surface of Mars, the Red Planet?

Compared to Earth, Mars is smaller and has a thinner atmosphere. The color of the planet is due to an abundance of reddish, windblown dust. Mars has been orbited by spacecraft with sophisticated cameras and instruments, and has been visited by landers and rovers. Mars is the focus of the application pages near the end of this chapter.

19.03.d1

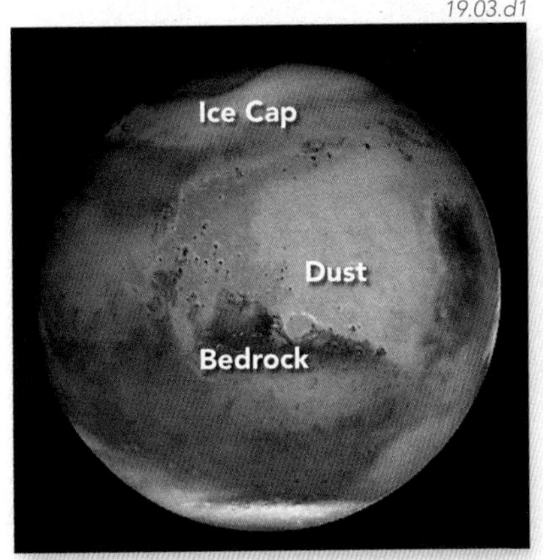

Mars has enough atmosphere to maintain ice caps on the south and north poles, as well as patches of ice in some shaded areas.

Much of the surface consists of reddish dust that has been weathered from rocks and blown around the surface by strong martian winds. The reddish color is from iron-oxide minerals, like hematite.

Martian bedrock is interpreted to consist of basaltic lavas, but some areas have layered rocks that many geologists think are sedimentary in origin.

Mars contains craters, volcanoes, canyons, and channels (▷), which scientists interpret as being formed by water that flowed on the surface more than 2 billion years ago. There are also smooth areas where the moving water deposited its load of sediment, burying preexisting features.

19.03.d2

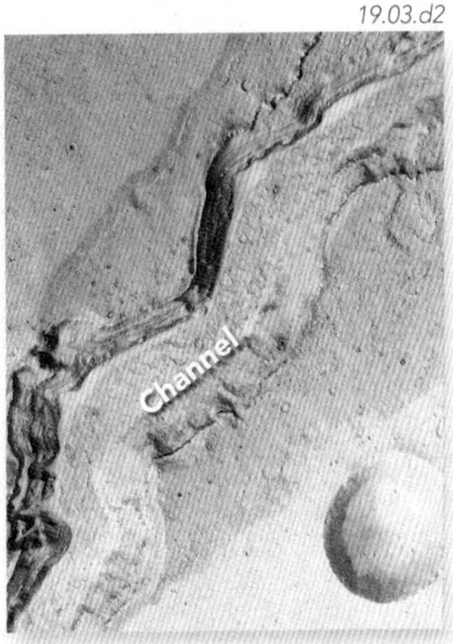

Encounters with Asteroids

Farther from the Sun than Mars, but this side of Jupiter, is a belt of more than 90,000 large and small rocky chunks drifting in space. Some are more than 500 kilometers (300 miles) across, others are less than 1 km. Most asteroids are thought to be debris left over from formation of the solar system 4.5 billion years ago, but some are probably parts of objects that broke up. Several space missions have passed close enough to asteroids to take detailed photographs or to even land on the surface. NASA's *Galileo* spacecraft, on its way to Jupiter, passed two asteroids within the main belt. Photographs of the two asteroids (*Ida* and *Graspa*), taken two years apart, are superimposed here. ▷

19.03.mtb1

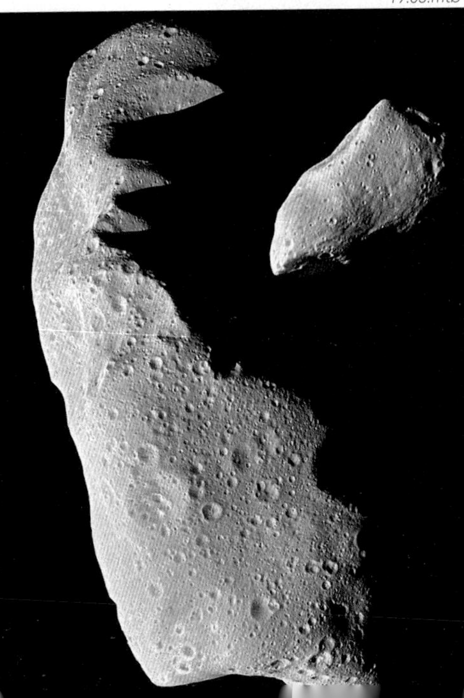

Before You Leave This Page Be Able To

- ✓ Explain why the surface of Mercury is so heavily cratered.
- ✓ Describe why radar was required to investigate the surface of Venus, and what type of features were found.
- ✓ Discuss the factors that make the surface of Earth so different from its neighbors.
- ✓ Summarize the materials and features present on the surface of Mars.
- ✓ Describe what asteroids are and where most are located.

What Is the Geology of Our Moon?

OUR NEAREST NEIGHBOR IN SPACE is the Moon. It is much closer than any other object and can be observed in detail with the simplest of telescopes or binoculars. It was the first object in the solar system to be systematically studied by geologists, who observed the Moon with telescopes, mapped the topography, geology, and composition by sending spacecraft to orbit the Moon, and walked and drove on the Moon's surface, observing the features and collecting rock samples to bring back to Earth for detailed study.

A What Are the Main Geologic Features of the Moon?

Observe this large image of the Moon. The surface of the Moon is not all the same. It has lighter colored areas, with some dark, somewhat-circular patches. With binoculars, we can observe individual craters.

1. This view of the Moon shows the features on the side of the Moon that always faces Earth, called the *near side*. The other, or *far side*, cannot be seen from Earth but has been photographed by spacecraft.

19.04.a1

Mare

Highlands

Mare

Highlands

Rayed Crater

2. Much of the near side and nearly all of the far side consists of light-colored material that is heavily cratered. This material mostly occurs at higher elevations and is called the *lunar highlands* or the *cratered highlands*. ▷

19.04.a2

3. From samples collected by astronauts and other information, we know that the cratered highlands contain igneous rocks that are light colored because they consist almost entirely of feldspar. Rocks from the highlands have been dated to more than 4 billion years. ▷

19.04.a3

4. The dark patches are lower, flatter, and less cratered; they are called *maria* (plural of *mare*). The maria have far fewer craters than the highland and so are much younger. The *maria* consist of dark basalt erupted onto the surface as lava flows, burying and filling craters that existed in the highland material. ▽

19.04.a6

5. Samples from maria consist of basalt lava (▽), mostly dated at 3.8 to 2.5 billion years old. At these distant times, the Moon retained enough heat to allow volcanism.

19.04.a5

6. The other obvious features on the Moon are *impact* craters, some of which have bright rays of material radiating outward. The rays overlie and cut across the top of the maria. Such *rayed craters* are some of the youngest features on the Moon (locally less than 100 million years old). Samples from lunar craters are breccia containing angular rock fragments generated during impacts. ▷

19.04.a4

B What Other Features Are Observed on the Moon?

The Moon is tectonically and volcanically inactive, and has been so for more than two billion years. The vast majority of craters on the Moon are due to impact of meteoroids and other objects onto the Moon's surface, but some are volcanic in origin. These impacts and volcanoes led to several types of features similar to those on Earth.

19.04.b1

After a crater is excavated by an impact, the shattered walls of the crater are commonly too steep to withstand the Moon's gravity, even though it is only 1/6 that of Earth's. Slope failures are common along the walls, like the rotated slump blocks and loose debris shown here.

19.04.b2

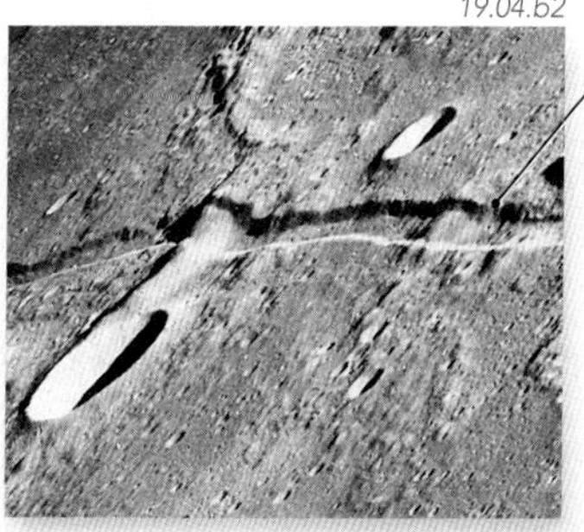

Several types of troughs cut across the Moon's surface. One type, shown here, is fairly straight and is thought to represent a down-faulted block (graben), such as those observed on Earth. The tensional forces that pulled the surface apart were probably related to adjustments after a nearby impact.

C What Causes the Phases of the Moon?

Every month the Moon appears to change its illumination on a regular schedule, going from completely dark to fully lit and back to dark again. Why is this cycle, called *phases of the Moon*, happening?

19.04.c1

1. At times, the side of the Moon facing Earth (*near* side) appears fully illuminated by sunlight. This is called a *full moon.*

19.04.c2

2. Seven days later, only half of the near side is illuminated as viewed from Earth. This is a *quarter moon.*

19.04.c3

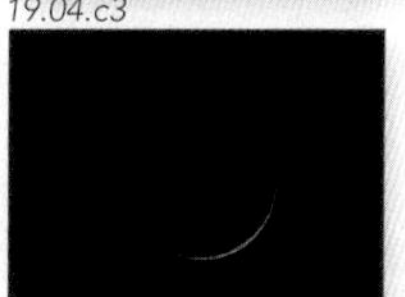

3. Six days later, a thin sliver of the near side is illuminated. The next day, none is illuminated, and this is a *new moon.*

4. Half of the Moon is always illuminated by the Sun, but from Earth we may not be able to see the entire lighted half because of the Moon's position. Because the Moon orbits Earth in 28 days, we see different amounts of the Moon's lighted side.

5. During a *new moon*, the side being illuminated by sunlight is *away* from our view. The Moon appears dark to us, but the other side of the Moon is still completely lighted.

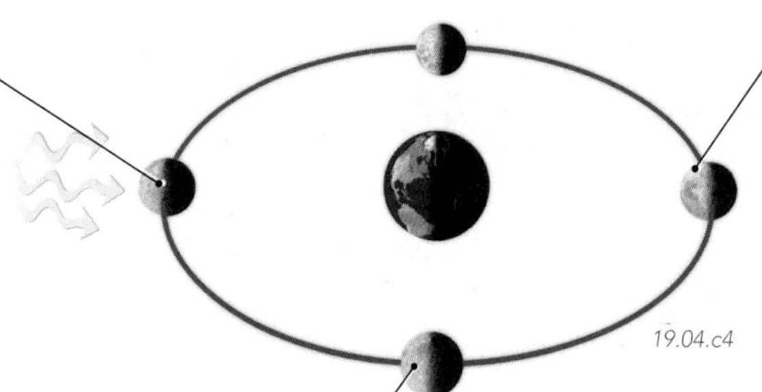

19.04.c4

7. During a *full moon*, the side being illuminated by sunlight is *facing* Earth, so we see all of it. The other side is dark.

6. At other times, we see half of the lighted half of the Moon, so it is a *quarter moon*. When the moon is in this position, we can tell that the phases of the Moon are *not* in any way related to Earth's shadow.

A Model for the Formation of the Moon

Where did the Moon come from, and how did it form? Geologists and other scientists have investigated this question by examining several types of data. The age of the Moon has been estimated by dating actual lunar samples. The chemical composition of these samples, including isotopic analyses, showed some unexpected similarities to rocks on Earth. This led to a theory, currently favored by many scientists, that the Moon was formed when a Mars-sized object collided with Earth early in the history of the solar system. The collision ejected a huge part of Earth's mantle into space, where it later aggregated under the force of gravity, and formed the Moon.

As the Moon was forming and soon thereafter, it became hot enough for large parts to melt. As the magma began to solidify, heavier crystals sank downward, while less dense crystals, such as feldspar, floated upward. The floating crystals accumulated near the surface, forming the light-colored igneous rocks of the highlands. The highlands were cratered by early, intense impacts. Later, basaltic magmas from depth erupted, forming the dark-colored maria. Rayed craters formed even later.

19.04.mtb1

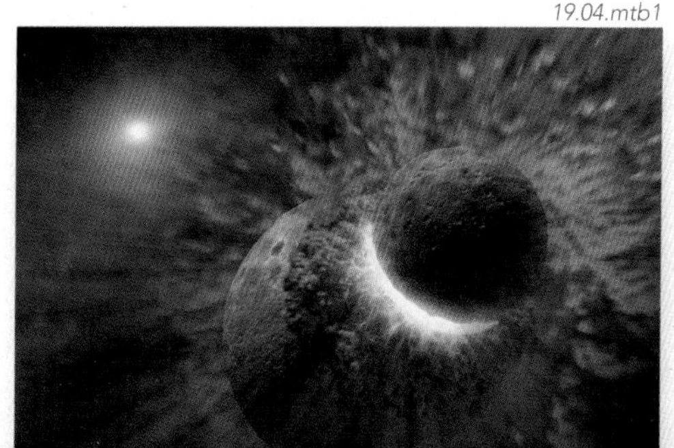

Before You Leave This Page Be Able To

- ✓ Summarize the physical characteristics and rock compositions of the lunar highlands, maria, and craters, and explain how each feature formed.
- ✓ Sketch and describe what causes the phases of the Moon.
- ✓ Summarize one model for how the Moon and its different parts formed.

What Is Observed on Jupiter and Its Moons?

JUPITER, THE LARGEST PLANET IN THE SOLAR SYSTEM, is a gas giant more than three times farther from the Sun than Mars. Jupiter is orbited by, at present count, more than 60 moons, including the largest moon in the solar system. To geologists, the icy and rocky moons are of greater interest than the gas-dominated planet itself because of their spectacular geologic features and their wide diversity.

Jupiter is nearly 780,000,000 km from the Sun, but is so large that we can see it on most clear nights. It is about 2.5 times more massive than all the other planets combined, and contains more than 300 times the mass of Earth. Examine the large image, which is computer-generated, produced by wrapping actual images of Jupiter around a sphere. What do you observe on the surface of the planet?

19.05.a2

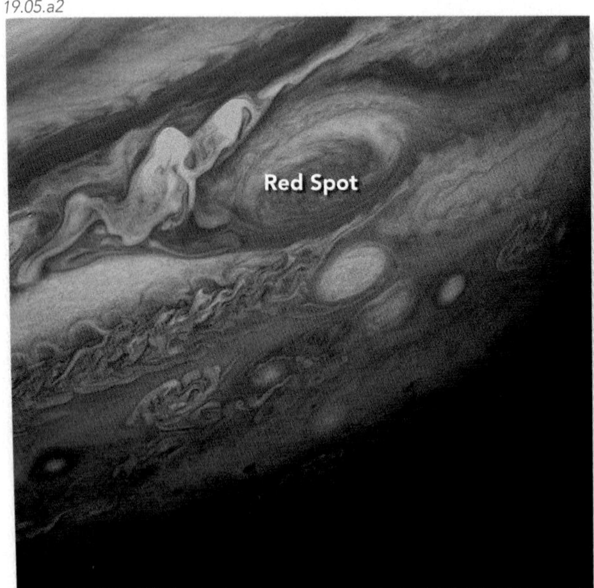

19.05.a1

1. Jupiter is so far from the Sun that it takes nearly 12 Earth-years to orbit the Sun—a Jupiter year is more than 4,300 Earth-days long. As viewed from Jupiter, the Sun appears much smaller and dimmer than it does from Earth.

2. The dominant features of Jupiter are the colorful bands and swirls of the planet's atmosphere. The atmosphere is mostly hydrogen with lesser amounts of helium, and trace amounts of methane, ammonia, and other gases. The interior of the planet consists of hydrogen in liquid and liquid-metallic forms, surrounding a solid core of iron and silicate minerals. Most of the planet is gas, so its overall density is less than that of Earth.

◁ 3. One of the most distinctive features in Jupiter's atmosphere is the Great Red Spot, which is a storm that has existed for at least a hundred years. It is three times wider than the diameter of Earth.

19.05.a3

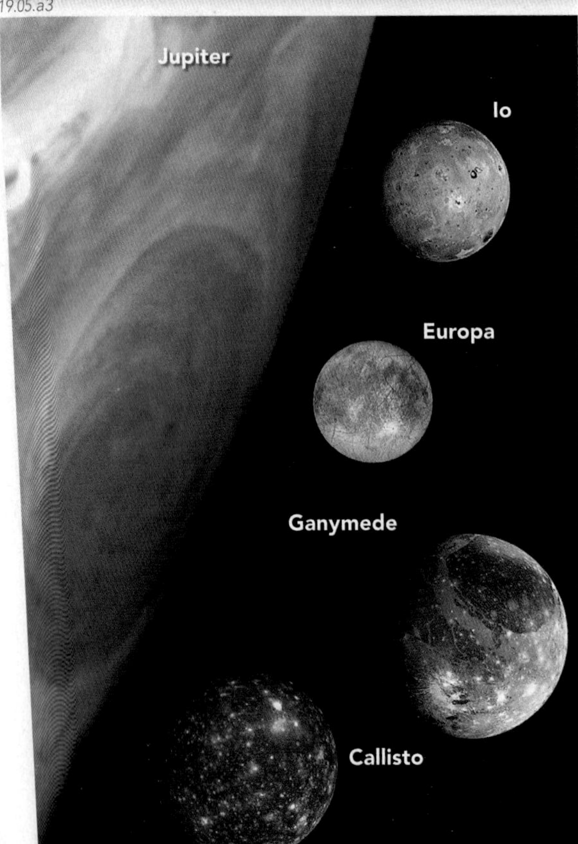

◁ 4. Jupiter's four largest satellites were discovered by Galileo Galilei in 1610 when he observed the planet with a telescope. These four moons are called the *Galilean moons*, and are named *Io, Europa, Ganymede,* and *Callisto.* The dramatic differences between the moons are largely due to differences in their proximity to the massive gas giant (Jupiter) around which they revolve. The moons are not as close to one another nor to Jupiter as shown here.

19.05.a4

◁ 5. This image shows Jupiter's moon Io and the shadow of Io on Jupiter's surface. The image was taken by the Hubble Space Telescope, which orbits Earth.

19.05.a5

6. Of the four Galilean moons, Io is closest to Jupiter. It is slightly larger than Earth's moon. Because it is so close to massive Jupiter, it is subjected to extreme tidal forces that deform its *land surface* up and down by as much as 100 meters, in the same way that our Moon pulls Earth's *water* up a few meters, causing tides.

7. The pulling and squeezing of rocks by tidal forces generates heat, making Io the most volcanically active object in the solar system. Its surface is covered with sulfur-rich lava. A volcanic eruption of such lava was photographed by NASA's *Galileo* spacecraft. ▷

19.05.a6

19.05.a7

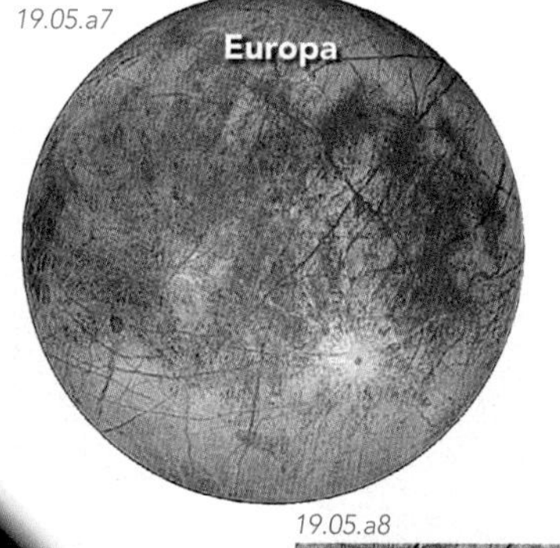

◁ **8.** Europa is farther away but still heated by the tidal forces of Jupiter and the other Galilean moons. These forces have allowed Europa to remain volcanically and tectonically active longer than would be merited by its size. These processes have extensively reworked the surface, accounting for the nearly complete lack of craters. Beneath the icy crust is probably an ocean of liquid water.

19.05.a8

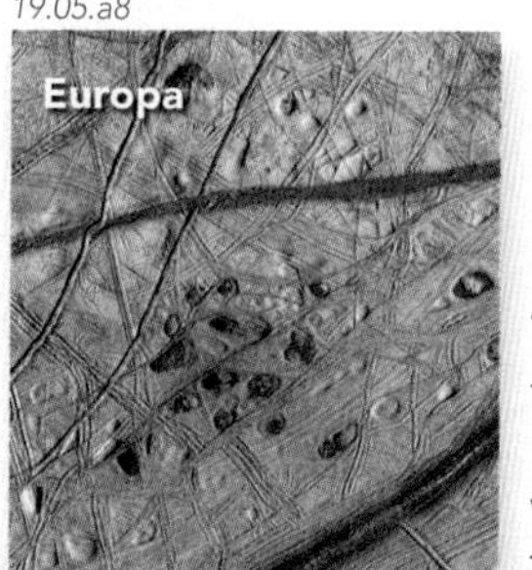

◁ **9.** The surface of Europa is a crust of ice (mostly frozen water) marked by intersecting lines. These lines appear to be fissures that have allowed liquid water to erupt onto the surface.

19.05.a9

△ **10.** Parts of Europa's surface are covered by huge blocks of ice that broke apart and then froze in place.

19.05.a10

△ **11.** Ganymede, the largest moon in the solar system, is thought to consist of a rocky core with a water-ice mantle, and a crust of water-ice and rocks.

19.05.a11

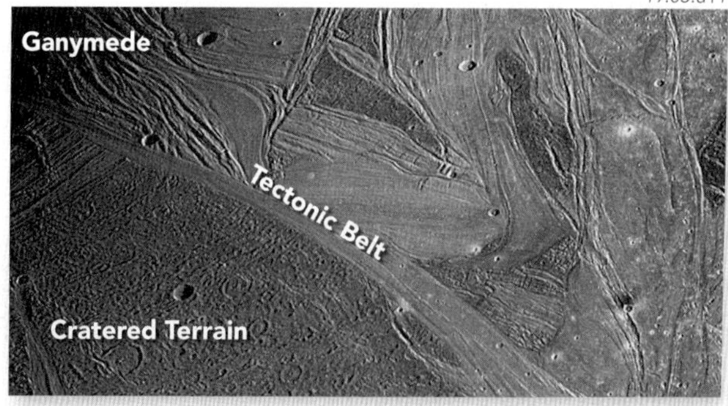

△ **12.** Ganymede's surface consists of dark, cratered patches that are older, and younger patches and belts that crosscut the older surfaces and contain tectonic features similar to those seen on Europa, including fissures.

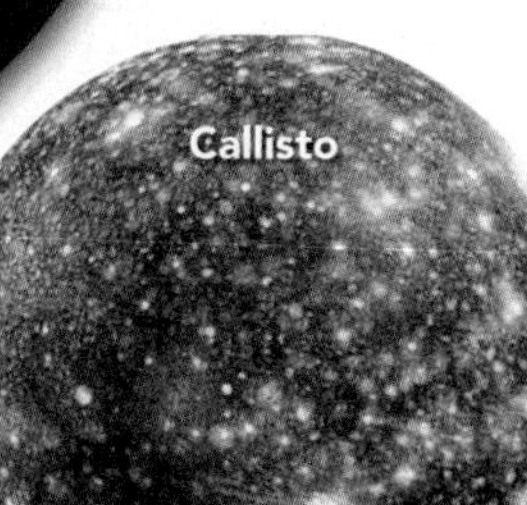

19.05.a12

◁ **13.** Callisto, the third largest moon in the solar system, is the most heavily cratered object in the solar system. It is far enough from Jupiter's tidal forces that its surface has remained largely intact for the last 4 billion years.

Before You Leave This Page Be Able To

- ✓ Summarize the key characteristics of Jupiter, such as its size, internal composition, and atmospheric composition.
- ✓ Briefly summarize the main characteristics of each Galilean moon.

19.5

What Is Observed on Saturn and Its Moons?

SATURN IS A BEAUTIFUL PLANET, a gas giant encircled by a set of spectacular rings. Saturn is the second largest planet in the solar system and is orbited by 47 known moons. Saturn's moons are quite diverse, reflecting differences in the materials and in the role of different geologic processes.

Saturn is farther from the Sun than Jupiter, and the distance between the two planets is greater than the distance between the Sun and Mars. Saturn is nearly twice as far away from the Sun as Jupiter. The large photograph was taken by the *Voyager* spacecraft, and the colors have been enhanced to accentuate the bands.

1. Saturn, like Jupiter, consists mostly of hydrogen and helium, which make up the gaseous atmosphere. The gases become liquid as they are compressed farther down in the planet. The center of the planet is thought to have a solid core of rock and metal. Saturn is more than 1.4 billion kilometers from the Sun, and takes 29.5 Earth-years to orbit the Sun.

2. Like Jupiter, Saturn is like a mini solar system, orbited by a collection of large and small moons. Our knowledge of the geology of these moons has increased dramatically in the last several years due to the arrival of the *Cassini-Huyguns* spacecraft in 2004. The image to the right shows Saturn, its rings, and some of its moons (small light-brown spots), four of which are discussed here: Titan, Iapetus, Enceladus, and Mimas. Titan is the largest moon, and the other three are included, not because they are the next largest, but because they are geologically interesting and nicely illustrate the geologic diversity of Saturn's moons.

▽ 3. Saturn is best known for its rings, which extend outward from the planet a distance nearly equal to the distance from Earth to the Sun. The rings consist of widely separated icy chunks floating in space. Most of the icy chunks are the size of sand, pebbles, and boulders, with some larger pieces. Close-ups of the rings display intricate details of concentric thick and thin rings separated by dark-colored, more-empty space, as viewed in the image below which is colored by particle size. Purple color indicates regions where particles are larger than 5 cm, green and blue indicate particles smaller than 5 centimeters (2 inches), and white bands mark where particles blocked the radio signals used to determine size.

19.06.a2

19.06.a1

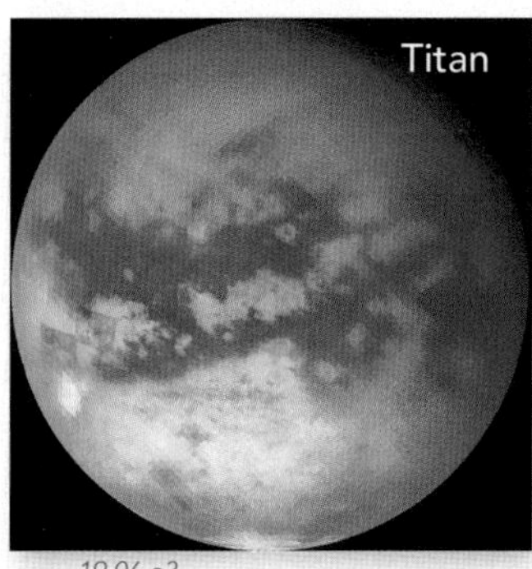

19.06.a3

◁ **4.** Titan is the largest of Saturn's moons and the second largest moon in the solar system, even larger than the planet Mercury. Its surface is obscured by a thick, cloudy atmosphere of mostly nitrogen and methane, but this image generated from various types of data shows some of Titan's features. The surface is inferred to contain solid materials (ices) and liquids, including liquid methane and other hydrocarbons.

5. The *Cassini* spacecraft released the *Huygens* probe, which parachuted through Titan's atmosphere and softly landed on the surface. On the way down, it captured images (▷) of drainage networks and a lake or ocean, confirming that liquids are widespread on Titan's surface. Once on the surface, the Huygens probe sent back an image of well-rounded icy boulders, presumably rounded by transport in flowing liquid.

19.06.a4

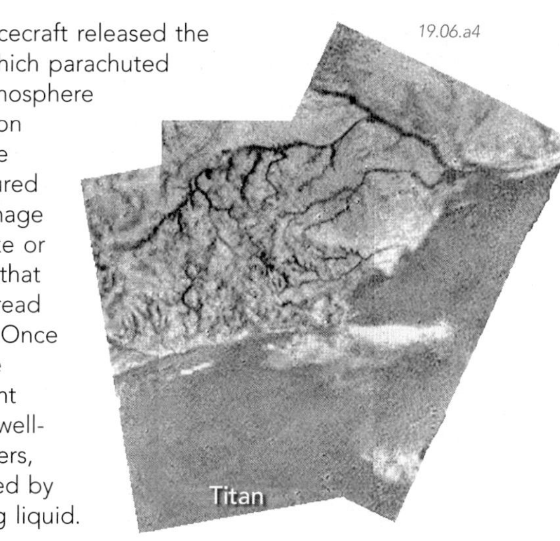

19.06.a5

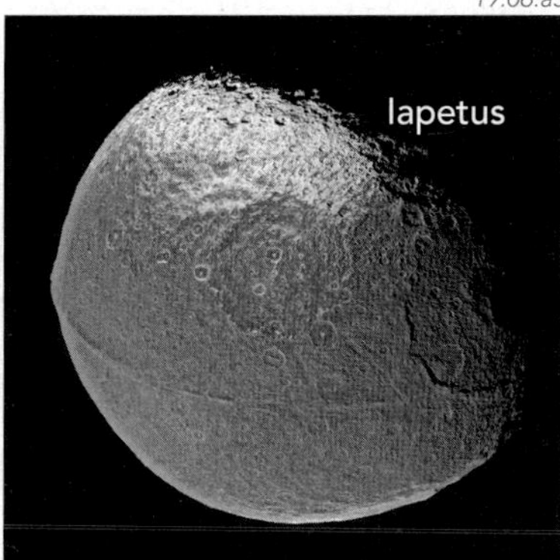

◁ **6.** Iapetus is distinctive in that most of its icy, cratered surface is light colored, but part is quite dark. The light-colored side is water ice; the darker side is thought to be a coating of dust that perhaps escaped from an adjacent moon and has been plastered on the leading edge of Iapetus as it orbits Saturn.

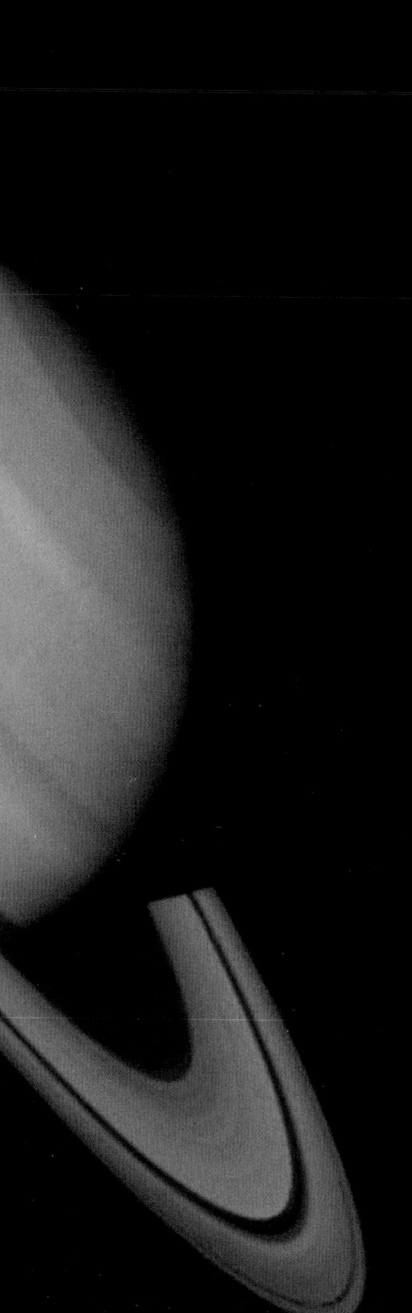

▽ **7.** Enceladus is one of the lightest-colored objects in the solar system, possibly because an icy frost is continuously forming on the surface.

19.06.a6

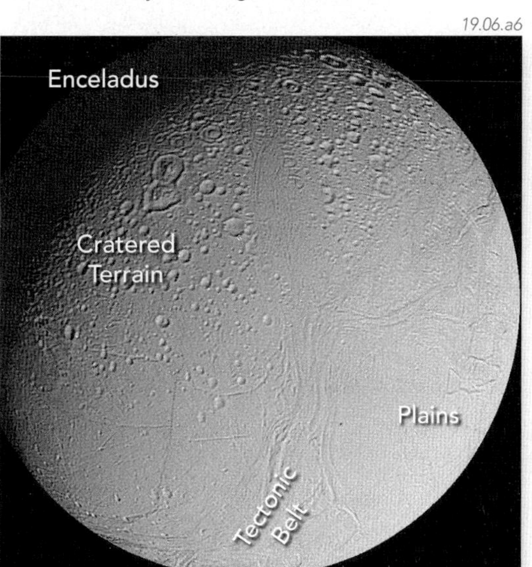

8. Its surface consists of at least three distinct types of terrain, the oldest of which is more heavily cratered.

9. Broad plains lie adjacent to the cratered terrain. The plains are much less cratered and are therefore younger. They are interpreted to have been resurfaced by the eruption of water onto the surface.

10. The third type of terrain consists of belts that slice through the heavily cratered material and through the plains. These belts have linear ridges and troughs probably formed as fissures through which water erupted to the surface. The *Cassini* team discovered active geysers of liquid water.

▽ **11.** Mimas is a relatively small moon whose pock-marked surface contains a large crater 130 km in diameter, with walls nearly 10 kilometers (~33,000 feet) high. This crater was formed by an impact that scientists have calculated nearly blasted the moon apart.

19.06.a7

Before You Leave This Page Be Able To

- ✓ Summarize the key characteristics of Saturn, such as its size and composition.
- ✓ Describe what materials compose the rings of Saturn.
- ✓ Summarize the main characteristics of the four moons of Saturn described here and the geologic processes expressed on the surface of each.

What Is the Geology of the Outer Planets and Their Moons?

THE OUTER PLANETS OF THE SOLAR SYSTEM and their moons are less well known than those from Saturn inward toward the Sun. Many of the observations are based on images taken by the *Voyager 1* and *2* spacecraft that flew through the outer reaches of the solar system in the late 1980s. These images provided a wealth of new information about the planets Uranus and Neptune, and some of their moons.

A What Features Characterize Uranus and Its Unusual Moons?

The next large planet out from Saturn is another large gaseous world called *Uranus*. This planet and some of its moons have some unusual characteristics and features.

1. The planet Uranus is nearly 2.9 billion kilometers from the Sun. The distance from Uranus to Saturn, the next planet in, is comparable to the distance from Saturn to the Sun. It takes Uranus 84 Earth-years to orbit the Sun.

2. Uranus consists largely of liquid and icy materials, including water, methane, and ammonia. The atmosphere is a mixture of hydrogen, helium, and methane. The blue-green color is caused by methane, which absorbs red light and reflects blue light. Uranus does not appear to have a solid surface.

3. Uranus has rings and at least 27 moons, named after characters from the works of William Shakespeare and Alexander Pope, including Oberon, Titania, Juliette, Puck, Ariel, and Miranda.

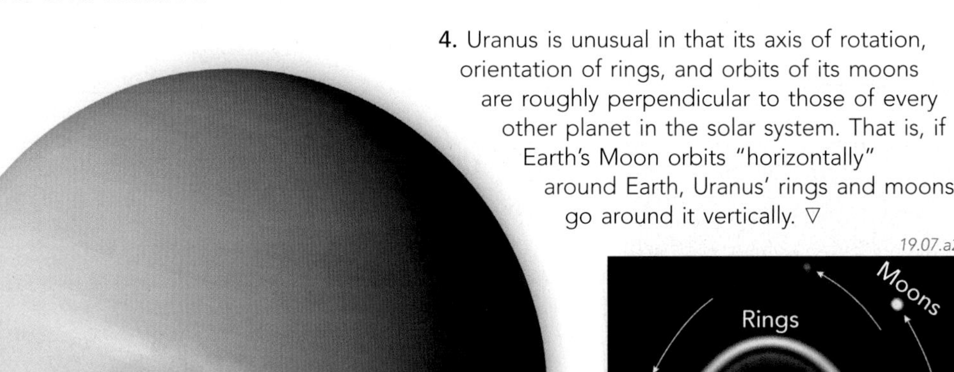

19.07.a1

4. Uranus is unusual in that its axis of rotation, orientation of rings, and orbits of its moons are roughly perpendicular to those of every other planet in the solar system. That is, if Earth's Moon orbits "horizontally" around Earth, Uranus' rings and moons go around it vertically. ▽

19.07.a2

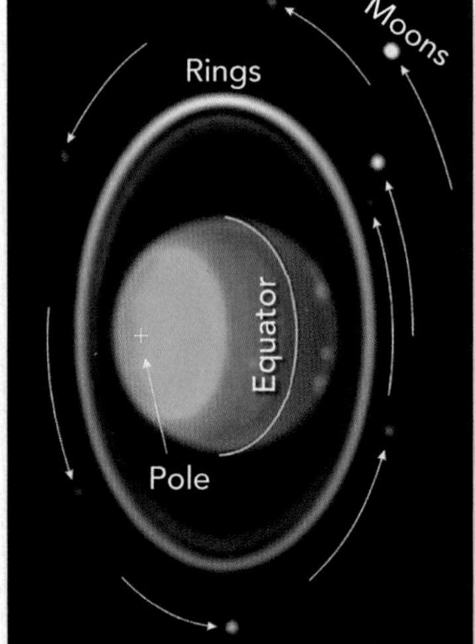

Ariel

▽ 5. Ariel is a moderate-sized moon, approximately 580 kilometers (360 miles) in diameter, and covered by ice. The surface is cut by long fractures, some of which have been filled by upwelling liquid water. This moon is thought to be mostly inactive.

19.07.a3

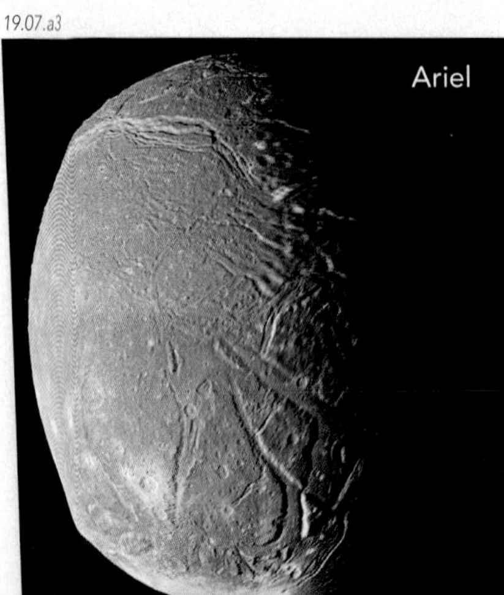

Miranda

▷ 6. To many geologists, Miranda, a small (236-km-diameter) moon of Uranus, is the most bizarre world in our solar system. The surface of Miranda is covered with ice and displays several distinct types of terrain.

7. There is highly cratered terrain that is lighter colored and relatively old.

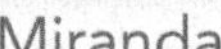

8. Disrupting the heavily cratered terrain are huge oval to angular features, each of which is a *corona*. The origin of these features is unresolved among planetary geologists, but they involve some normal faulting and probably upwelling of deeper materials.

19.07.a4

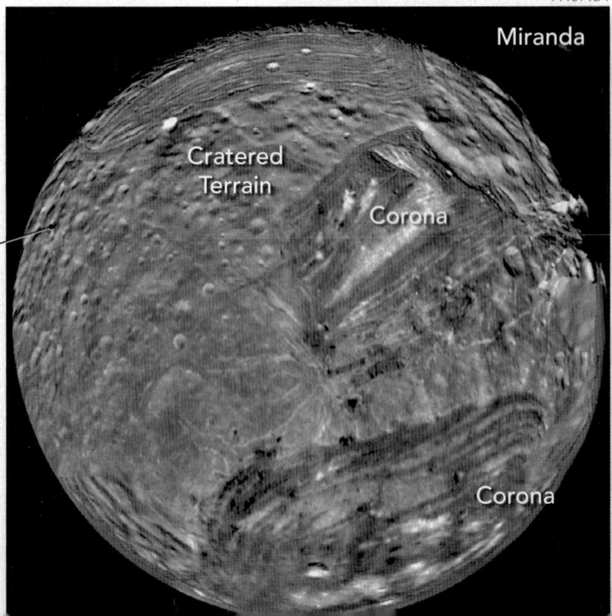

B What Is the Geology of Neptune and Its Moon, Triton?

Neptune is the eighth planet from the Sun and is another gas giant similar in many ways to Uranus. The existence of Neptune was predicted mathematically before the planet was discovered by telescope.

19.07.b1

Neptune is nearly 4.5 billion km from the Sun, and the distance from Uranus to Neptune is more than that between Uranus and Saturn. In other words, the distance between adjacent planets increases as one moves outward through the solar system. Neptune is so far from the Sun that it has not completed even one orbit since it was discovered. It takes 165 Earth-years to go around the Sun!

Neptune is about the same size as Uranus (~50,000 km in diameter) and has a similar composition, with ices and liquids inside and an atmosphere of hydrogen, helium, and methane. Its blue color is due to methane.

19.07.b2

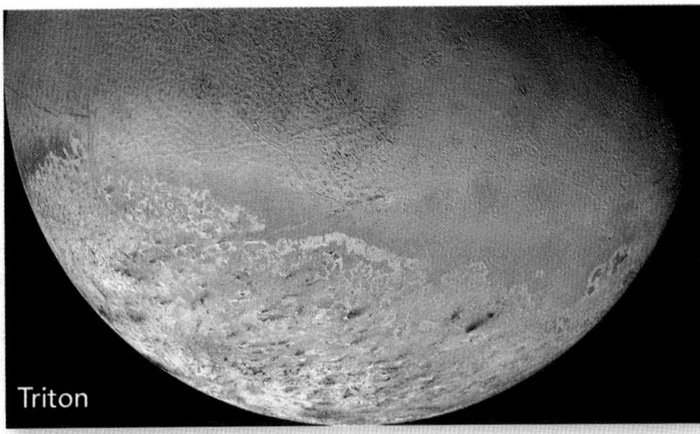

△ The surface of Triton, Neptune's only large satellite, consists of ices of nitrogen and carbon substances. It has two distinct halves, one part is like the surface of a cantaloup and is interpreted to represent activity from eruptions and active geysers.

C What Do We Know About Pluto and Its Companions?

Pluto was once considered to be the ninth and outermost planet, but an international organization of astronomers recently reclassified Pluto to *dwarf planet* status. So our solar system has eight planets, not nine.

Pluto was always an oddity compared to the eight planets. It is a relatively small, icy object, even smaller than Earth's moon. Pluto orbits the Sun in a very elliptical orbit that sometimes brings it closer to the Sun than Neptune. A circuit around the Sun takes 248 Earth-years. Pluto has a large companion called *Charon*, which is half Pluto's size, plus several very small moons.

19.07.c1

No spacecraft has yet visited Pluto, but NASA's *New Horizons* spacecraft is on the way. This spacecraft will also visit the *Kuiper Belt*, a disk-like zone of objects that lies beyond the orbit of Neptune. This belt has a number of objects that are similar to and far beyond Pluto. Some of these Kuiper Belt objects are shown in this artist's conception. ▷

19.07.c2

Comets

Comets are one of the more interesting spectacles in the night sky. Comets are small, icy and rocky objects with very elliptical orbits around the Sun. Some comets, such as Halley's Comet (▷), visit the inner solar system regularly, whereas others visit at very long intervals. Comets are thought to come from a very outer part of the solar system, well beyond the orbit of Neptune. As a comet nears the Sun, gas and dust are stripped off the comet by the solar wind and carried outward, forming a tail that always points away from the Sun.

19.07.mtb1

Before You Leave This Page Be Able To

- ✓ Describe some key features of Uranus and Neptune, and explain how they are similar.
- ✓ Describe unusual features on Ariel, Miranda, and Triton, and identify the materials that comprise the surfaces of these moons.
- ✓ Describe what is known about Pluto and its companions.
- ✓ Describe what comets are and why they have a tail.

What Have We Learned About the Geology of Mars?

THE MOST EXCITING DEVELOPMENTS in planetary geology involve the planet Mars, the *Red Planet*. Recently, Mars has been explored by a number of orbiting spacecraft that carried sophisticated cameras and other instruments, many designed and controlled by planetary geologists. Spacecraft have landed on the planet and unleashed small, robotic rovers that travel across the surface, exploring and collecting data.

A What Have We Learned from Instruments Orbiting Mars?

Several spacecraft, including *Mars Odyssey* and *Mars Express*, have orbited the planet. As they pass over the planet's surface, they record images and take measurements designed to detect water and determine the composition of rocks, sediment, and ice. Using these spacecraft, we recently have made incredible discoveries.

19.08.a1

◁ **1.** Mars contains a huge canyon system called *Valles Marineris* that is 4,000-km long—a length equivalent to the width of the United States. The canyon began as a large rift and has been widened by inward collapse of the steep canyon walls and by other types of erosion.

19.08.a3

◁ **2.** Mars also has the solar system's largest volcano, *Olympus Mons*. It is 600 km across and 27 km high, nearly three times the height of Mount Everest. It is a large shield volcano, like those on Hawaii, but is inactive. Huge landslides and debris avalanches came off its flanks and formed areas of hummocky topography.

19.08.a2

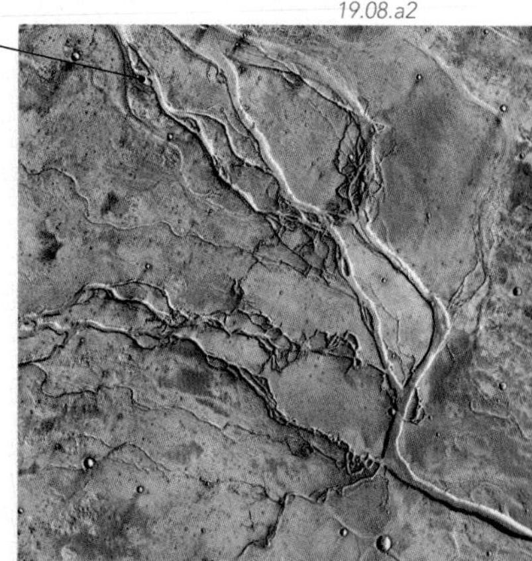

▷ **4.** Some parts of Mars have spectacular channels, interpreted to have been formed by torrents of running water flowing on the surface, at some unknown, but presumably long time ago. Where the channels encountered the gentle plains, they deposited piles of sediments, like in deltas and alluvial fans on Earth.

19.08.a4

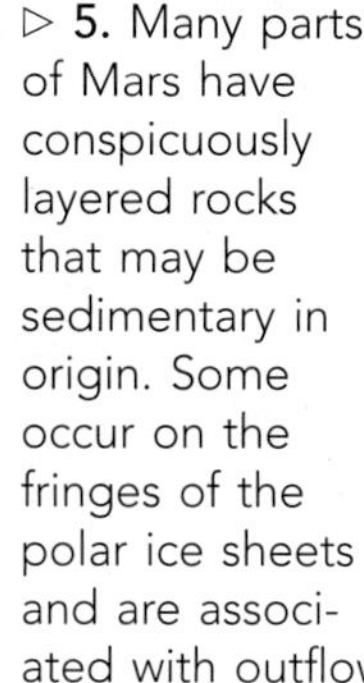

19.08.a5

▷ **5.** Many parts of Mars have conspicuously layered rocks that may be sedimentary in origin. Some occur on the fringes of the polar ice sheets and are associated with outflow channels.

3. The image below shows the *Candor Chasm*, which is part of the Valles Marineris system. The steep walls of the chasm have collapsed downslope, providing some of the most spectacular examples of slope failure in the solar system.

19.08.a6

19.08.a7

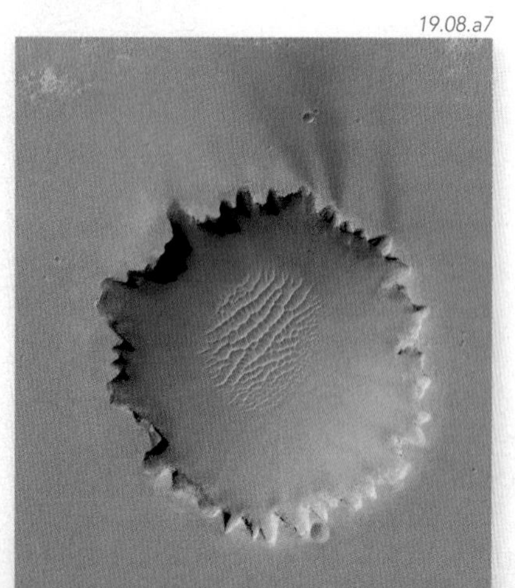

◁ **6.** The atmosphere of Mars is less dense than Earth's, but the winds are strong. Lace-like sand dunes occur in the center of the beautiful Victoria Crater. The crater was visited by one of the Mars Exploration Rovers.

B What Have We Learned from Landers and Rovers on Mars?

NASA has had three successful landings of spacecraft that carried small motorized vehicles with wheels, called *rovers*, to navigate and photograph the Martian landscape. The two most recent rovers, *Spirit* and *Opportunity*, provided a wealth of new geologic information, including a few surprises.

19.08.b1

▷ **1.** Each rover-bearing spacecraft "landed" on the surface by cushioning itself with large air bags that inflated just before the spacecraft bounced onto the surface. Then the rover inside, shown here in an artist's conception, rolled off to explore nearby parts of the planet.

19.08.b2

◁ **2.** The rovers rolled across the surface on wheels, stopping to inspect any outcrops or interesting rocks. They can spin their wheels to dig up sediment on the surface or use a tool to scratch at the rocks. They carry cameras and scientific instruments to measure composition, temperature, and other data.

19.08.b3

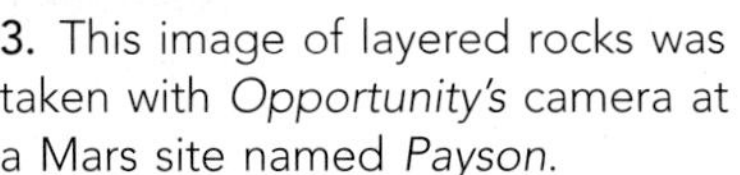

3. This image of layered rocks was taken with *Opportunity's* camera at a Mars site named *Payson*.

19.08.b4

4. *Layers*—One of the first features that caught geologists' attention was the layers exposed in the walls of several craters. The layers are interpreted as being formed by running water or by fast-moving debris ejected from nearby impacts.

19.08.b5

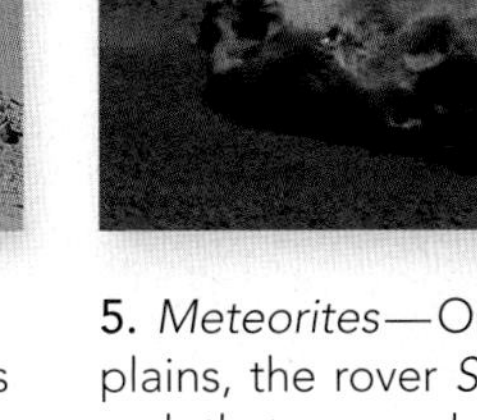

5. *Meteorites*—On its travels across the plains, the rover *Spirit* encountered a lone rock that seemed out of place to a geologist on the team. When the rover was redirected to investigate the rock, it discovered the first meteorite found on Mars.

19.08.b6

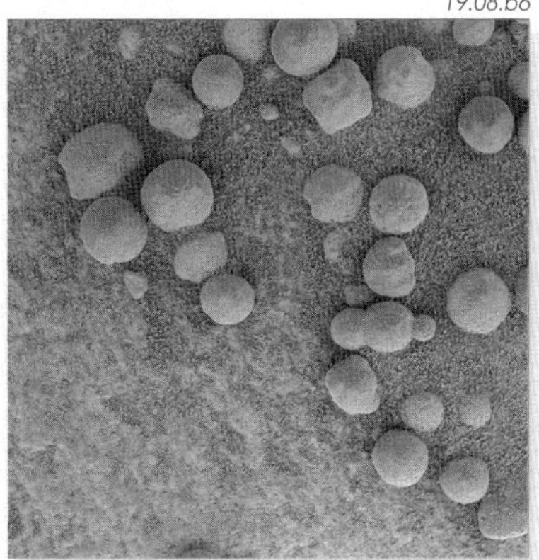

6. *Blueberries*—Within the layered rocks and weathering out onto the surface are millimeter-size, spherical objects, nicknamed *blueberries*. Measurements document that these contain the mineral *hematite*, which on Earth needs water to form.

Choosing a Landing Site

How do researchers choose where to land? For the *Opportunity* rover, they chose the site on the basis of *infrared measurements* that, to geologists, indicated the presence of abundant hematite in the area. On Earth, hematite most commonly forms under wet, oxidizing conditions. The geologists therefore concluded that if you were looking for water on Mars, this would be a good place to start. When the *Opportunity* rover rolled off its platform and started exploring, it was a nice confirmation of the method to find blueberries, composed of hematite, lying on the ground and weathering out of the rocks.

19.08.mtb1

Before You Leave This Page Be Able To

- ✓ Summarize two of the ways that geologists have explored Mars.
- ✓ Describe some features found by orbiting spacecraft and what they imply about processes that have occurred on the planet's surface.
- ✓ Describe some features discovered by the rovers *Spirit* and *Opportunity*.
- ✓ Explain how *Opportunity's* discoveries were made possible by prior spacecraft measurements.

How and When Did Geologic Features on This Alien World Form?

TRAVELING THROUGH SPACE YOU ENCOUNTER AN UNKNOWN WORLD. Your spacecraft orbits the planet and takes images and measurements of the different geologic areas and of the most interesting geologic features. You will use these images, and some initial observations about the features, to interpret how each feature formed and in what chronological order the features formed.

Goals of This Exercise:

- Observe the planet to identify large regions that have a similar geologic appearance.
- Examine close-up images of key features and read descriptions for each, in order to interpret how each feature might have formed.
- Use several strategies to reconstruct the sequence in which the different features formed.
- Summarize the geologic features and history of this planet.

Procedures

Use the available information to complete the following steps, entering your answers in the appropriate places on the worksheet.

1. Observe the image of the entire planet on the next page. Interpret where boundaries are located between areas that have different geologic characteristics.
2. Observe each of the close-up images and read the description that accompanies each, looking for further clues about the types of geologic feature present and how each might have formed.
3. Determine the relative ages of the different geologic regions and features using crosscutting relationships and density of impact craters.
4. Your instructor may have you draw a simple geologic map of the planet, on which each map unit is a different type of geologic area or geologic feature. Draw a legend to accompany your map that has (1) a small box with the color or pattern you have chosen to depict that geologic terrain, (2) the name of the geologic terrain or feature, and (3) a brief description, less than 30 words, that conveys the key characteristics of this terrain and your interpretation of the terrain's origin.
5. Write a short report or list summarizing the geology of the planet and its geologic history. Your instructor will guide you about the length and detail expected. This report should demonstrate the breadth of knowledge you have gained in this course, not just the concepts from this last chapter. In other words, use this final investigation to bring together concepts you have learned throughout the course.

The large image shows one side of the planet as illuminated by the local sun. The surface contains different types of geologic terrain as well as several obvious large geologic features. North is up in this view, south is down, west is to the left, and east is to the right. Some observations about this place are listed below, labeled with letters corresponding to the name or character of the place. Corresponding letters mark the place on the large view of the planet.

Western Terrain (W)—The western side of the planet consists of a heavily cratered terrain with many large and small craters. Samples of the rocks are very shattered and contain many angular fragments of highly weathered basalt.

Dark Terrain (D)—A dark, wide strip curves across the planet from south to northwest. Radar measurements indicate that it has a rough upper surface. A few normal faults cut across the dark material. As shown in the image below, the dark material locally protrudes into the adjacent cratered terrain, covering it and filling some craters. The dark material is partly weathered basalt. The dark terrain has some small impact craters, but fewer than the western terrain. ▽

19.09.a2

Chasm (C)—Cutting across the highly cratered terrain is a deep chasm that narrows progressively toward the south. Toward the north, the chasm has some important relationships with the dark terrain and a reddish-brown sedimentary area (S). The close-up below shows one wall of the chasm. ▽

19.09.a3

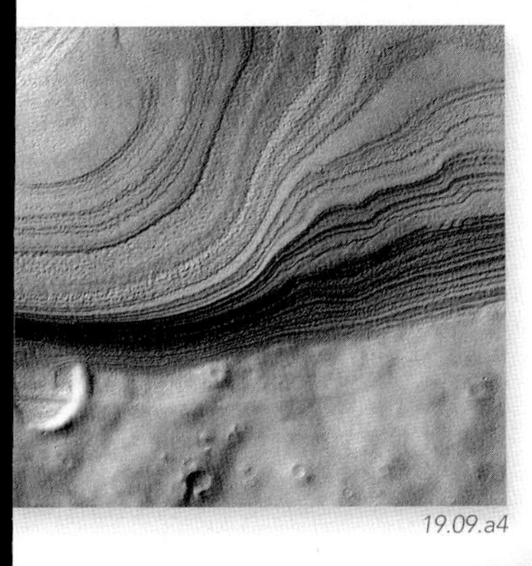

19.09.a4

Polar Ice (P)—The north and south poles are covered with water ice year round. The close-up image to the left shows the edge of the layered ice overlapping a crater. The ice has almost no craters.

Sedimentary Terrain (S)—Adjacent to the north pole and a northern ice cap is a distinctive reddish-brown area. The unit has layers and appears to be sedimentary in origin. Along the southern edge of the terrain, the soft-looking, loose sediment is in contact with terrain that is more heavily cratered, as shown in the detailed image to the right. The sedimentary area has very few craters. Similar material may be present near the south pole but is not visible in this particular view.

19.09.a5

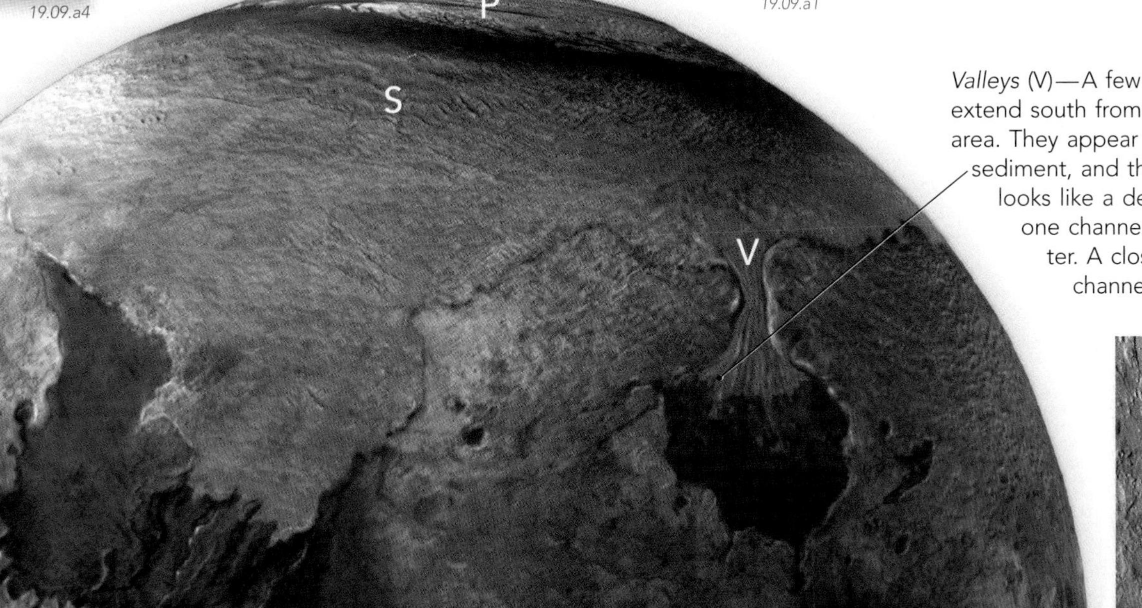

19.09.a1

Valleys (V)—A few valleys or channels extend south from the sedimentary area. They appear to be filled with sediment, and there is a feature that looks like a delta or fan where one channel empties into a crater. A close-up view of one channel is shown below. ▽

19.09.a6

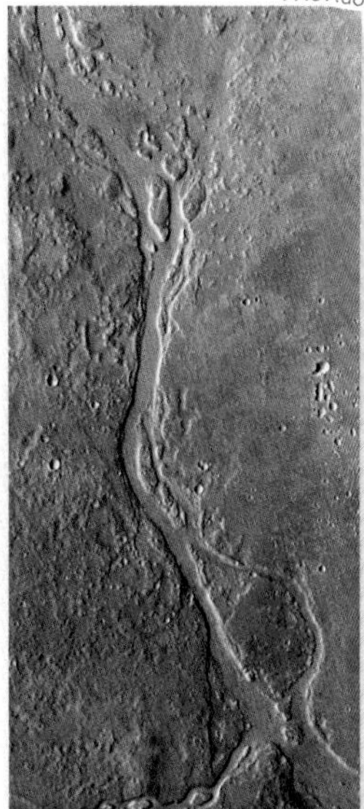

Mountains (M)—In the southeastern part of this view, three large mountains rise above the plain. The close-up below shows one of the mountains. ▽

19.09.a7

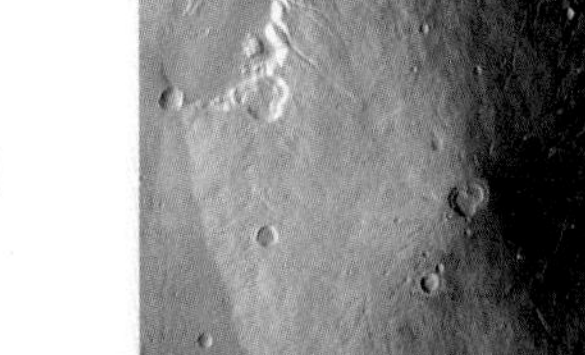

Note that the mountain is cone shaped with a central crater. The flank is indented by small craters, and the lower part of the mountain appears to be missing on one side (upper left side in this view).

Credits

Photographs

Unless otherwise credited: © Stephen J. Reynolds.

Chapter 1

Image Number: 01.00.a2: © David Boyle/Animals Animals; 01.00.a3: Mike Doukas/U.S. Geological Survey; 01.00.a5: © Getty Royalty Free; 01.01.mtb1: U.S. Geological Survey; 01.03.a5: © Dr. Parvinder/Sethi; 01.05.b2: C.G. Newhall/U.S. Geological Survey; 01.05.b3: J.D. Griggs/U.S. Geological Survey; 01.07.a2: Photo by Susanne Gillatt; 01.07.a4: Photo by Michael M. Kelly; 01.09.a2: Woodbridge Williams/National Park Service; 01.09.b2: Photo by Perry H. Rahn.

Chapter 2

Image Number: 02.00.a2: NASA; 02.03.d1: David E. Wieprecht/U.S. Geological Survey; 02.04.mtb1: Daniel Ball/Arizona State University; 02.06.a5: Wendell Duffield/U.S. Geological Survey; 02.07.a1: Cyrus Read/U.S. Geological Survey; 02.07.a2: Kate Bull/U.S. Geological Survey; 02.07.a3: T.A. Plucinski/U.S. Geological Survey; 02.07.b3: Karl String/U.S. Geological Survey; 02.10.b1: Ventura County Star/Associated Press; 02.11.a1: Photo by Paul Fitzgerald; 02.11.a4: U.S. Geological Survey; 02.11.b1: P.R. Christensen/NASA/JPL/Arizona State University; 02.11.b2: Jason Leigh/Electronic Visualization Lab, University of Illinois, Chicago; 02.12.a1: U.S. Geological Survey.

Chapter 3

Image Number: 03.06.mtb1: Robert E. Wallace/U.S. Geological Survey; 03.07.d3: Peter J. Haeussler/U.S. Geological Survey.

Chapter 4

Image Number: 04.00.a4: © Phototake/Alamy; 04.01.a1: © Dr. Parvinder/Sethi; 04.01.a2: © Paul Collis/ALAMY; 04.01.a4: Photograph by N.K. Huber, USGS, Photo Library, Denver, CO; 04.01.a5: © Herve Conge/ISM/Phototake; 04.01.a6: © SS38/PhotoDisc/Getty Images; 04.01.b1: © Vol. 30/PhotoDisc/Getty Images; 04.01.b2: © Corbis Royalty Free; 04.01.b3: © The McGraw-Hill Companies, Inc./Doug Sherman, photographer; 04.01.b5: © Doug Sherman/Geofile; 04.01.b6: Photo by Susanne Gillatt; 04.01.b7: © Dr. Parvinder/Sethi; 04.01.b8: © The McGraw-Hill Companies, Inc./Bob Coyle, photographer; 04.01.b9: © The McGraw-Hill Companies, Inc./Stephen Frisch, photographer; 04.01.b10: © The McGraw-Hill Companies, Inc./John Karachewski, photographer; 04.02.mtb1 (Obsidian, Basalt): © The McGraw-Hill Companies, Inc./Bob Coyle, photographer; 04.02.mtb1: © Parvinder Sethi; 04.03.a4: © The McGraw-Hill Companies, Inc./Ken Cavanagh, photographer; 04.03.a5: © Robert Rutford/James Carter, photographer; 04.03.a6: © The McGraw-Hill Companies, Inc./Jacques Cornell, photographer; 04.03.a7–04.03.a8: © The McGraw-Hill Companies Inc./Ken Cavanagh, photographer; 04.03.a9: © The McGraw-Hill Companies, Inc./Jacques Cornell, photographer; 04.03.a10: © Parvinder Sethi; 04.03.a11: © The McGraw-Hill Companies Inc./Ken Cavanagh, photographer; 04.03.a12, 04.03.b1, 04.03.b2, 04.03.b3: © Doug Sherman/Geofile; 04.03.b4: © Dr. Parvinder/Sethi; 04.04.a1: © The McGraw-Hill Companies, Inc./Bob Coyle, photographer; 04.04.a3 © The McGraw-Hill Companies, Inc./Bob Coyle, photographer; 04.05.a2, 04.05.b3: © Robert Rutford/James Carter, photographer; 04.05.c2, 04.05.c6, 04.05.c8, 04.05.c10: © Doug Sherman/Geofile; 04.05.c4: © Charles E. Jones; 04.05.c11: © The McGraw-Hill Companies, Inc./Doug Sherman, photographer; 04.05.mtb1: © The McGraw-Hill Companies, Inc./Bob Coyle, photographer; 04.06.b1: © Gary Dyson/ALAMY; 04.06.b3: © The McGraw-Hill Companies, Inc./Jacques Cornell, photographer; 04.06.b5: Thomas Sharp/Arizona State University; 04.06.b6: © The McGraw-Hill Companies, Inc./Doug Sherman, photographer; 04.06.b7: © Doug Sherman/Geofile; 04.08.a2: © Robert Rutford/James Carter, photographer; 04.08.a3: © Doug Sherman/Geofile; 04.08.a4: © Charles E. Jones; 04.08.b1, 04.08.b2: Thomas Sharp/Arizona State University; 04.08.b3: © Doug Sherman/Geofile; 04.08.b4, 04.08.b5, 04.09.a1, 04.09.a2: Thomas Sharp/Arizona State University; 04.08.mtb1: © Wally Eberhart/Visuals Unlimited; 04.09.a3: © The McGraw-Hill Companies, Inc./John Karachewski, photographer; 04.09.a6: Thomas Sharp/Arizona State University; 04.09.a7, 04.09.a8, 04.09.a10, 04.09.a12: © Doug Sherman/Geofile; 04.09.a13: Thomas Sharp/Arizona State University; 04.10.a2, 04.10.a3: © Dr. Parvinder Sethi; 04.14.a4: © Vol. 31/PhotoDisc/Getty Images; 04.14.a7: © Brand X/Images RF; 04.14.a8: © EP100/PhotoDisc/Getty Images; 04.14.b5: © Brand X/FotoSearch; 04.14.b6 (gold bars): © Vol. 1/PhotoDisc/Getty Images; 04.15.a1, 04.15.a2: Thomas Sharp/Arizona State University; 04.15.a3: © Doug Sherman/Geofile; 04.15.a5, 04.15.a6, 04.15.a7: Thomas Sharp/Arizona State University.

Chapter 5

Image Number: 05.01.mtb1: © Herve Conge/ISM/Phototake; 05.09.a3: © Arctic Images/ALAMY; 05.10.c2: Thomas Sharp/Arizona State University; 05.10.c3, 05.10.c4: © Robert Rutford/James Carter, photographer; 05.10.c5: Thomas Sharp/Arizona State University; 05.10.c6: © Charles E. Jones; 05.12.b2: Photo by Steven Semken; 05.12.mtb1: Anonymous/National Park Service; 05.12.mtb2: U.S. Geological Survey; 05.13.a2: Photo by Allen Glazner; 05.13.a3, 05.13.a4: Photo by Michael Ort; 05.13.a5: Photo by Scott Chandler; 05.13.a6: Photo by Michael Ort; 05.13.mtb1: Photo by Allen Glazner.

Chapter 6

Image Number: 06.00.a1: Donald A. Swanson/U.S. Geological Survey; 06.00.a3: John Pallister/U.S. Geological Survey; 06.00.a4: Austin Post/U.S. Geological Survey; 06.00.a5: Lyn Topinka/U.S. Geological Survey; 06.00.a6: Dan Dzurisin/U.S. Geological Survey; 06.01.a1: J.D. Griggs/U.S. Geological Survey; 06.01.a2: E. Klett/US Fish and Wildlife Service; 06.01.c8: U.S. Geological Survey; 06.02.a1: J. Judd/U.S. Geological Survey; 06.02.a2: J.D. Griggs/U.S. Geological Survey; 06.02.a3: Don Swanson/U.S. Geological Survey; 06.02.a4: R. Clucas/U.S. Geological Survey; 06.02.a5: U.S. Geological Survey; 06.02.a6: M.E. Yount/Alaska Volcano Observatory, U.S. Geological Survey; 06.02.c1: John Pallister/U.S. Geological Survey; 06.02.c2: Hawaii Volcano Observatory/U.S. Geological Survey; 06.03.a1: J.D. Griggs/U.S. Geological Survey; 06.03.a3: U.S. Geological Survey; 06.03.a4: Photo by Michael M. Kelly; 06.03.a5, J.D. Griggs/U.S. Geological Surrvey; 06.03.a6: B. Runk/S. Schoenberger/Grant Heilman Photography; 06.03.a7: J.D. Griggs/U.S. Geological Survey; 06.03.c2: Photo by Michael M. Kelly; 06.03.c4: © Chris Hellier/Corbis; 06.03.mtb1: Photo by Michael M. Kelly; 06.04.a2: Hawaii Volcano Observatory/U.S. Geological Survey; 06.04.a3: C.C. Heliker/U.S. Geological Survey; 06.04.a4: Hawaii Volcano Observatory/U.S. Geological Survey; 06.04.b1, 06.04.b2, 06.04.c1: J.D. Griggs/U.S. Geological Survey; 06.05.a2: Photo by D.P. Schwert; 06.05.a3: Hawaii Volcano Observatory/U.S. Geological Survey; 06.06.a2: © Jacques Collet/EPA/Corbis; 06.06.b1, 06.06.c1, 06.06.c2: J.D. Griggs/U.S. Geological Survey; 06.06.d2: © AP Images; 06.07.a1: U.S. Geological Survey; 06.07.a3: C. Newhall/U.S. Geological Survey; 06.07.a4: Game McGimsey/Alaska Volcano Observatory, U.S. Geological Survey; 06.07.a5: Rick Wessels/Alaska Volcano Observatory, U.S. Geological Survey; 06.07.b1: R. Christensen/U.S. Geological Survey; 06.07.b4: C.A. Neal/Alaska Volcano Observatory, U.S. Geological Survey; 06.07.b6: R.G. McGimsey/Alaska Volcano Observatory, U.S. Geological Survey; 06.07.mtb1: © Vol. 23/PhotoDisc/Getty Images; 06.07.mtb2: International Space Station/NASA; 06.07.mtb3: Image by Jim Williams, NASA, GSFC Scientific Visualization Studio, and the Landsat 7 Science Team; 06.07.mtb4: © Roger Ressmeyer/Corbis; 06.08.a2: © Bettmann/CORBIS; 06.08.b2: © Popperfoto/ALAMY; 06.08.b3: © Corbis; 06.08.c3: Harry Glicken/U.S. Geological Survey; 06.08.c4-c6: © Copyright Gary Rosenquist, 1980; 06.08.c7: Peter W. Lipman/U.S. Geological Survey; 06.09.a1: T. Miller/U.S. Geological Survey; 06.09.a2: Steve Schilling/U.S. Geological Survey; 06.09.c1: Janet Schaefer/U.S. Geological Survey; 06.09.c3: © Brand X/Punchstock; 06.09.mtb1, 06.09.mtb2: U.S. Geological Survey; 06.10.b1: S.R. Brantley/U.S. Geological Survey; 06.10.b2: R.A. Bailey/U.S. Geological Survey; 06.10.mtb1: © Dr. Parvinder Sethi; 06.11.a1: NASA; 06.13.b1: © Roger Ressmeyer/Corbis; 06.13.b2: U.S. Geological Survey; 06.13.c1: Maxim Sorokin/U.S. Geological Survey; 06.13.c2: Steve J. Smith, Alaska Volcano Observatory/U.S. Geological Survey/Geophysical Institute, University of Alaska, Fairbanks; 06.13.d1: M. Sako/U.S. Geological Survey; 06.13.d2: J.D. Griggs/U.S. Geological Survey; 06.13.d3: S.R. Brantley/U.S. Geological Survey; 06.13.e1: U.S. Geological Survey; 06.13.mtb1: Dan Dzurisin/U.S. Geological Survey; 06.14.a2: Lyn Topinka/U.S. Geological Survey; 06.14.c1: David Wieprecht/U.S. Geological Survey.

Chapter 7

Image Number: 07.00.a2: © Digital Vision/PunchStock; 07.01.a3: © The McGraw-Hill Companies, Inc./John Karachewski, photographer; 07.01.a4: © Digital Vision; 07.01.a6: © Peter Bowater/ALAMY; 07.01.a8: © Charles Palek/Animals Animals; 07.01.a9: © Getty Royalty Free; 07.02.a2: © PhotoDisc/Getty Images; 07.02.a3: © Getty Royalty Free; 07.02.a4: © Digital Vision/Getty Images; 07.02.a5: Tim McCabe/NRCS; 07.02.a7: NASA; 07.04.a2–4: © The McGraw-Hill Companies, Inc./Bob Coyle; photographer; 07.04.a8: U.S. Geological Survey; 07.04.b3: Photo by Susanne Gillatt; 07.04.mtb1: Rickly Hydrological Company, Inc.; 07.05.b2: © The McGraw-Hill Companies, Inc./John A. Karachewski, photographer; 07.06.a1, 07.06.a5: © Digital Vision/Getty Images; 07.06.a7: © Creatas/PunchStock; 07.06.a8: © Beth Davidow/Visuals Unlimited; 07.06.b2: © Bruce Molina, Terra Photographics/Earth Science World Image Bank; http://www.earthscienceworld.org/imagebank; 07.06.b5: © The McGraw-Hill Companies, Inc./Jacques Cornell, photographer; 07.06.b10: © Dorling Kindersley/Courtesy of the Natural History Museum, London; 07.07.b6: © Sheila Terry/Photo Researchers, Inc.; 07.07.b9: © Bruce Molina, Terra Photographics/Earth Science World Image Bank; http://www.earthscienceworld.org/imagebank; 07.08.a6: Danita Delimont/ALAMY; 07.08.b8: © Gerald and Buff Corsi/Visuals Unlimited; 07.09.b1: Photo by Susanne Gillatt; 07.09.d1: © The McGraw-Hill Companies, Inc./John Karachewski, photographer; 07.09.d3: © Jerome Nuefeld/The Experimental Nonlinear Physics Group/The University of Toronto; 07.09.d4: © Dr. Marli Miller/Visuals Unlimited; 07.10.b2: © Julia Waterlow/Eye Ubiquitous/Corbis; 07.10.c1: © The McGraw-Hill Companies, Inc./John A. Karachewski, photographer; 07.10.c4: © John Buitenkant/Photo Researchers, Inc.; 07.11.b2: © Digital Vision; 07.11.d4: U.S. Geological Survey; 07.13.a4: © Dr. Roger Slatt, University of Oklahoma/Earth Science World Image Bank; http://www.earthscienceworld.org/imagebank; 07.14.b1: Guy Gelfenbaum/U.S. Geological Survey; 07.15.b1–07.15.b3, 07.15.c2–07.15.c4: Vincent Matthews/Colorado Geological Survey; 07.15.c5–07.15.c7: © Ron Blakey/Geology-Northern Arizona University. http://jan.ucc.nau.edu/~rcb7/.

Chapter 8

Image Number: 08.00.a3: © Bill Beatty/Visuals Unlimited; 08.00.a4: © Duncan Heron; 08.00.a5: © Abi Howe, American Geological Institute/Earth Science World Image Bank; http://www.earthscienceworld.org/imagebank; 08.01.b6: Vincent Matthews/Colorado Geological Survey; 08.01.mtb1: Spokane Research Lab/NIOSH/CDC; 08.02.mtb1: Courtesy of J.M. Logan and F.M. Chester, Center for Tectonophysics, Texas A & M University; 08.03.c6: © Dean Conger/Corbis; 08.10.a9: © Dr. Marli Miller/Visuals Unlimited; 08.11.c5: NASA; 08.14.a1: Photo by Michael M. Kelly; 08.14.c4: © Phil Degginger/Animals Animals; 08.16.a2: © Charles Ver Straeten/New York State Museum; 08.16.a3: © Gary Braash/Peter Arnold, Inc.

Chapter 9

Image Number: 09.00.a4: Photo by Julia K. Johnson; 09.02.c6: © Dr. Mark Schurr, University of Notre Dame, Collier Lodge Site-2005; 09.03.c3: © The McGraw-Hill Companies, Inc./John A. Karachewski, photographer; 09.04.a5: Photo by Michael M. Kelly; 09.05.a1: © John Cancalosi/Peter Arnold, Inc.; 09.05.a6: © Phil Degginger/ Carnegie Museum/ALAMY Royalty Free; 09.05.a7: © Breck P. Kent/Animals Animals; 09.05.a8: © Francois Gohier/Photo Researchers, Inc.; 09.06.a1: © Getty Royalty Free; 09.06.a4: © Michael Freeman/Corbis; 09.06.b1: © Digital Vision/PunchStock; 09.06.b3-09.06.b4: Photo by Susanne Gillatt; 09.06.b6: © Digital Vision; 09.06.mtb1: Photo by Susanne Gillatt; 09.09.a2: © Yann Arthus-Bertrand/Corbis; 09.09.a3: © Dr. Parvinder Sethi; 09.09. b1: © George Bernard/Animals Animals/Earth Scenes; 09.09.b2: U.S. Geological Survey; 09.09.c1: With permission of the Royal Ontario Museum © ROM; 09.09. d1: Daniel Ball/Arizona State University; 09.10a1: © Breck P. Kent/Animals Animals; 09.10.a2: © Georgette Douwma/ Naturepl.com; 09.10.a3: Wetzel and Company/Janice McDonald; 09.10.b1: © Getty Royalty Free; 09.10.b2: © Sinclair Stammers/Photo Researchers, Inc.; 09.10.b3: © Albert Copley/Visuals Unlimited; 09.10.c1: "Ordovician Marine Environment" © Karen Carr and courtesy Indiana State Museum Foundation; 09.10.c2: "Devonian Marine Environment" © Karen Carr and courtesy Indiana State Museum Foundation; 09.10.c3: "Permian Riverside" © Karen Carr and courtesy Indiana State Museum Foundation; 09.11.a1: "Triassic Landscape," © Karen Carr; 09.11.a2: "Jurassic Landscape," © Karen Carr; 09.11.a3: "Cretaceous Coastal Landscape," © Karen Carr; 09.11.a4: "Cretaceous Marine Environment," © Karen Carr; 09.11.b1: "Alaskan Dinosaurs," © Karen Carr; 09.11. c1: "Miocene River Landscape," © Karen Carr; 09.11.c2: "North American Pleistocene Landscape," © Karen Carr; 09.13.d1: © David W. Benn, Bear Creek Archeology, Inc.; 09.13.d2: Photo by Ramon Arrowsmith.

Chapter 10

Image Number: 10.01.a2: © Ralph White/Corbis; 10.01.a3, 10.01.b1: William Crawford, Integrated Ocean Drilling Program/Texas A&M University; 10.01.b2: Weiss/U.S. Geological Survey; 10.01.mtb1: NOAA PMEL Vents Program; 10.01.mtb2: © Tom McHugh/Photo Researchers, Inc.; 10.02.b1: Gary Wilson-Paleomagnetic Research Facility, University of Otago; 10.03.c1: © Dr. Ken MacDonald/ Photo Researchers, Inc.; 10.03.mtb1, 10.03.mtb2: Photo by Woods Hole Oceanographic Institution and Richard Lutz/Rutgers University; 10.05.c3: NASA; 10.09.a2: © Vol. 89/PhotoDisc/Getty Images; 10.09.a3 NASA; 10.09.a4: © Paul Souders/Corbis; 10.09.a5: International Space Station/NASA; 10.10.c1: National Center for Earth-surface Dynamics; 10.10.c2: James Samples/Northern Arizona University; 10.11.a1: Photograph by Martin P. A. Jackson; 10.11.a3: © The McGraw-Hill Companies, Inc./ John A. Karachewski, photographer; 10.11.c3: Photograph by Martin P.A. Jackson; 10.12.a1-10.12.a6: © Ron Blakey/ Geology-Northern Arizona University. http://jan.ucc.nau. edu/~rcb7/.

Chapter 11

Image Number: 11.00.a3: Photo by Susanne Gillatt; 11.03.a2: © DAJ/ALAMY; 11.03.a3: © Corbis; 11.03.b3: © George D. Lepps/Corbis; 11.03.d4: © JRC, Inc./ALAMY; 11.03.d5: © SCPhotos/ALAMY; 11.08.b2: © Michael Gadomski/Animals Animals; 11.08.b3: © Dr. Marli Miller/ Visuals Unlimited; 11.13.a2, 11.13.a4, 11.13.a5, 11.13.a7, 11.13.a9, 11.13.a13: © Ron Blakey/Geology-Northern Arizona University. http://jan.ucc.nau.edu/~rcb7/.

Chapter 12

Image Number: 12.00.a2: U.S. Geological Survey; 12.00.a3-12.00.a4: NASA; 12.01.d2: Jet Propulsion Laboratory/ NASA; 12.02.b6: J.R. Stacy/U.S. Geological Survey; 12.06.a2: H.W. Wilshire/U.S. Geological Survey; 12.06.a3: J.D. Nakata/U.S. Geological Survey; 12.06.b1: © Visions of America, LLC/ALAMY; 12.06.b2: U.S. Geological Survey; 12.06.b3: © Karl V. Steinbrugge Collection, Earthquake Engineering Research Center, University of California, Berkeley; 12.06c2: U.S. Geological Survey; 12.07.a2: U.S. Army; 12.07.a3: W.C. Mendenhall/U.S. Geological Survey; 12.07.a4: Mehmet Celebi/U.S. Geological Survey; 12.07. a5: M. Celebi/U.S. Geological Survey; 12.07.a6: J.R. Stacy/ U.S. Geological Survey; 12.07.a8: J.K. Hillers/U.S. Geological Survey; 12.08.a2: R.D. Brown, Jr/U.S. Geological Survey; 12.08.a3: Pierre St. Amand/U.S. Geological Survey; 12.08.a4: © North Wind Picture Archives; 12.08.a5: U.S. Geological Survey; 12.08.a6: C.J. Langer/U.S. Geological Survey; 12.08.a7: Dr. Roger Hutchison/NOAA; 12.08.a8: Fengyuan/ U.S. Geological Survey; 12.09.a3: © Lynette Cook/SPL/ Photo Researchers, Inc.; 12.09.b2: National Geophysical Data Center; 12.09.b3: U.S. Navy; 12.09.b4: Commander Dennis J. Sigrist/International Tsunami Information Center; 12.09.b5: Hugh Davies/University of Papua New Guinea; 12.09.mtb1: © AP Images/Harbor Branch Oceanographic Institution and NOAA; 12.11.c1: © Roger Ressmeyer/Corbis; 12.12.a2: W.C. Mendenhall/U.S. Geological Survey; 12.12.a3: C.E. Meyer/U.S. Geological Survey; 12.12.a5: Photo by Ramon Arrowsmith; 12.12.a6: Robert E. Wallace/U.S. Geological Survey; 12.13.a11: U.S. Geological Survey; 12.15.a2: Courtesy of J.M. Logan and F.M. Chester, Center for Tectonophysics, Texas A & M University; 12.16.a2: W.R. Hansen/U.S. Geological Survey; 12.16.a3: A. Post/U.S. Geological Survey; 12.16.a4: U.S. Army; 12.16.b1: G. Plafker/U.S. Geological Survey; 12.16.b2: U.S. Navy.

Chapter 13

Image Number: 13.00.a2, 13.00.a3: NASA; 13.00.a4: © Ed Degginger/Animals Animals; 13.00.a7: NASA; 13.02.b1: Norman Kuring/NASA; 13.02.b3: © Robert Preston/ ALAMY; 13.02.c2: NASA; 13.02.c3: © Daniel Borzynski/ ALAMY; 13.02.c5: Photo Courtesy of Illinois Geological Survey, Joel Dexter; 13.02.d1: Photo by N.H. Daron, USGS; 13.02.d2: © Colin Harris/LighTouch Images/ALAMY; 13.04.b2: Greg Lundeen/NOAA; 13.04.c1: © Eric Nguyen/ Corbis; 13.05.d3: NASA; 13.06.a2: © George F. Mobley/ National Geographic/Getty Images; 13.06.a3-13.06.a4: NASA; 13.07.a1, 13.07.a2, 13.07.a3, 13.07.a4: Photo by Michael M. Kelly; 13.07.c2: © Digital Vision/Punchstock; 13.07.c3: © Vol. 44/PhotoDisc/Getty Images; 13.07.d2: © Corbis Royalty Free; 13.08.d2: © Photolink/Getty Images; 13.09.a2: © Marc Garanger/Corbis; 13.09.a3: © Blaine Harrington III/Corbis; 13.09.a4: © Tony Waltham/Robert Harding World Imagery/Corbis; 13.10.a5: Photo by Mark Twickler, University of New Hampshire, NOAA; 13.10.mtb1: National Polar-orbiting Satellite System/NOAA; 13.11.a3: Wikimedia; 13.12.a1: Photo by Susanne Gillatt; 13.12.a2: U.S. Geological Survey; 13.12.a3: Game McGimsey/U.S. Geological Survey; 13.12.b1, b2, b3: © Ron Blakey/Geology-Northern Arizona University. http://jan.ucc.nau.edu/~rcb7/; 13.12.c1: © Corbis Royalty Free; 13.13.a3: Photo by Susanne Gillatt; 13.13.c1: Corbis Royalty Free; 13.13.c2, c3: © Digital Vision/PunchStock; 13.14.a2: Goddard Space Flight Center/NOAA; 13.15.a2: Photo by Susanne Gillatt; 13.15.a3: © Vol. 36/PhotoDisc/Getty Images; 13.15.a6: © Vol. 31/PhotoDisc/Getty Images; 13.15.a7: © Rab Harling/ ALAMY; 3.15.a10, 13.15.a12: Photo by Susanne Gillatt; 13.15.a13: NASA.

Chapter 14

Image Number: 14.01.a2: © The McGraw-Hill Companies, Inc./John A. Karachewski, photographer; 14.01.a3: © Vol. 44/PhotoDisc/Getty Images; 14.01.a4: © Miles Ertman/ Masterfile; 14.01.a5: Corbis Royalty Free; 14.01.b2: © Dr. Parvinder Sethi; 14.01.b3: NASA; 14.02.mtb1, mtb2: © Everett C. Johnson/Science Faction/Getty Images; 14.03.a1: © Corbis Royalty Free; 14.05.a1: © Vol. 174/ PhotoDisc/Getty Images; 14.05.a2: © Getty Royalty Free; 14.05.b1: © Mark Karrass/Corbis; 14.05.b2: © Dr. Marli Miller/Visuals Unlimited; 14.05.b3: Dr. Frank M. Hanna/ Visuals Unlimited; 14.05b4, 14.06.a1: © Corbis Royalty Free; 14.06.a2: © Stock Connection Distribution/ALAMY; 14.06.a3: © Vol. 31/PhotoDisc/Getty Images; 14.06.a4: 14.06.a5, 14.06.a6, 14.06.a7: U.S. Geological Survey; 14.06.b1: Photographed by Robert Morton; 14.06.b2: U.S. Army Corps of Engineers, Portland District; 14.06.b3: U.S. Army Corps of Engineers; 14.06.b4: © Florida Images/ALAMY; 14.07.a2: © Adam Woolfitt/Corbis; 14.07mtb1-mtb5: U.S. Geological Survey; 14.08.c1: NASA; 14.08.c3: U.S. Geological Survey; 14.08.d2: © The McGraw-Hill Companies, Inc./John A. Karachewski, photographer; 14.10.a3: Photo by Susanne Gillatt; 14.10.a4: National Park Service; 14.10.a6: © Dr. Parvinder Sethi; 14.10.b1: Photo by Gary Fleeger, PaGS, 14.10.c1: © Ron Blakey/ Geology-Northern Arizona University. http://jan.ucc.nau. edu/~rcb7/; 14.12.a6: © Dr. Marli Miller/Visuals Unlimited; 14.12.a9: © Tom Bean Photography; 14.12.a10: Austin Post, U.S. Geological Survey; 14.13.a7: © Gustav Verderber/ Visuals Unlimited; 14.13.a8: Lilja Run Bjarnadottir/ University of Iceland; 14.13.a9 © Steve McCutcheon/Visuals Unlimited; 14.14c2: P. Weis/U.S. Geological Survey; 14.14. c3: © Dean Conger/Corbis; 14.14.c4: © Digital Visions; 14.14.mtb2: Photo by P. Weiss, U.S. Geological Survey; 14.14.mtb3: E. Soldo, National Park Service; 14.16.a2: Photo by Susanne Gillatt; 14.16.b1-14.16.b2: University of Colorado/NASA.

Chapter 15

Image Number: 15.00.a2, 15.00.a3, 15.00.a4: Matthew Larsen/U.S. Geological Survey; 15.03.b3: © Creatas/ PunchStock; 15.03.d1: © Doug Sherman/Geofile; 15.03.d2: Thomas Sharp/Arizona State University; 15.05.a1: © Kenneth Fink/Photo Researchers, Inc.; 15.05.c1: © W.K. Fletcher/Photo Researchers, Inc.; 15.05.c4, 15.05.c5: USDA/Department of Soil Conservation Service; 15.06.a1: © Brand X/Punchstock; 15.06.a2: Photo by Jeff Vaunga, courtesy of USDA Natural Resources Conservation Services; 15.06.b1: © Loetscher Chlaus/ALAMY; 15.06.b2: Photo by Susanne Gillatt; 15.06.b3: © Getty Royalty Free; 15.06.b4: Photo by Tim McCabe, USDA Natural Resources Conservation Center; 15.06.c7: C.E. Meyer/U.S. Geological Survey; 15.06.c8: Photograph courtesy U.S. Geological Survey Photo Library, Denver, CO; 15.06.c9: © Victor de Schwanberg/Photo Researchers, Inc.; 15.06.mtb1: © James

L. Amos/Corbis; 15.07.b3: Photo by Susanne Gillatt; 15.07.mtb1: P. Carrara/U.S. Geological Survey; 15.08.a2: Photograph courtesy of USGS Earthquake Hazards Program; 15.08.a4: Edwin L. Harp/U.S. Geological Survey; 15.08.a5: Photograph by J.T. McGill, USGS Photo Library, Denver, CO; 15.08.b2: © AP Images/Rene Macura; 15.08.b5: Matthew Larsen/U.S. Geological Survey; 15.09.a2: © AP Images; 15.09.a6: © J. David Rogers, Missouri University of Science and Technology; 15.10.a2: © Ralph Lee Hopkins/Photo Researchers, Inc.; 15.10.a4: Reproduced with the permission of the Minister of Public Works and Government Services Canada, 2007 and Courtesy of Natural Resources Canada, Geological Survey of Canada; 15.10.a8: Matthew Larson/U.S. Geological Survey; 15.10.a10: © Lloyd Cluff/Corbis; 15.10.mtb1: © AP Images/Kevork Djansezian; 15.10.mtb2: R.L. Schuster/U.S. Geological Survey; 15.11.a1: Peter Haeussler/U.S. Geological Survey; 15.11.a2: Photograph by J.T. McGill, courtesy U.S. Geological Survey; 15.11.a3: Gerald Wieczorek/U.S. Geological Survey; 15.11.a5: © The McGraw-Hill Companies, Inc./ John A. Karachewski, photographer; 15.12.a1: Matthew Larsen/U.S. Geological Survey; 15.12.a2: Lyn Topinka/U.S. Geological Survey; 15.12.b1: Lyn Topinka/U.S. Geological Survey; 15.12.b2: Matthew Larsen/U.S. Geological Survey; 15.12b3: © Tom Bean/Corbis; 15.12.b4: U.S. Geological Survey; 15.12.c2: Jim Peaco/National Park Service; 15.12.c6: Jennifer Adleman, Alaska Volcano Observatory/ U.S. Geological Survey; 15.13.a1: D.J. Varnes/U.S. Geological Survey; 15.14.a2: © Yvette Cardozo/Corbis.

Chapter 16

Image Number: 16.00.a4: Joel Schmutz/U.S. Geological Survey; 16.00.a5: NASA; 16.01.a1: NASA; 16.02.b2: © Larry Lefever/Grant Heilman Photography; 16.02.d1, 16.02.d3, 16.03.c2: Photo by Michael M. Kelly; 16.06.a4: © PhotoDisc/ Getty Images; 16.06.b5, 16.07.b1: Michal Tal/National Center for Earth-surface Dynamics; 16.07.b2: © Dick Roberts/Visuals Unlimited; 16.07.b3, 16.07.mtb1: Michal Tal/National Center for Earth-surface Dynamics; 16.08.mtb1: Contra Costa Historical Society; 16.09.a2: NASA; 16.09.b1: U.S. Geological Survey EROS Data Center/NASA; 16.09.b2: NASA; 16.12.b2: U.S. Geological Survey; 16.12.b4: Lyn Topinka/U.S. Geological Survey; 16.12.b5: © AP Images/U.S. Bureau of Reclamation; 16.13.b3: W.R. Hansen/U.S. Geological Survey; 16.13.b4: Photo by W.R. Hansen, U.S. Geological Survey.

Chapter 17

Image Number: 17.00.a3: Peg Owens/Department of Tourism; 17.02.a1: Photo by Tim McCabe, USDA Natural Resources Conservation Center; 17.02.a2: © Corbis Royalty Free; 17.02.a3: Photo by Jeff Vanuga, USDA Natural Resources Conservation Center; 17.02.a6: © Vol. 19/ PhotoDisc/Getty Images; 17.02.a7: © Vol. 31/PhotoDisc/ Getty Images; 17.02.b1: © Vol. 39/PhotoDisc/Getty Images; 17.02.b2: © Vol. 16/PhotoDisc/Getty Images; 17.02.b3: © PhotoDisc/Getty Images; 17.02.b4: © The McGraw-Hill Companies, Inc./John A. Karachewski, photographer; 17.02.b5: © Vol. 22/PhotoDisc/Getty Images; 17.05.a2: © The McGraw-Hill Companies, Inc./John A. Karachewski, photographer; 17.06.a2: U.S. Geological Survey; 17.06.b1: © Corbis Royalty Free; 17.06.b2: Photo by William K. Jones; 17.06.b3: © Corbis Royalty Free; 17.06.c2: © Dr. Parvinder Sethi; 17.06.mtb1: D Luchsinger/National Park Service; 17.07.a2: © Terry Whittaker/Photo Researchers, Inc.; 17.07.a5: Dustin Reed/USGS Ohio Water Science Center/U.S. Geological Survey; 17.08.mtb1: Photo by Richard O. Ireland, USGS; 17.09.a2: © Gyori Antoine/ Corbis/Sygma; 7.09.mtb1, mtb2: U.S. Geological Survey International; 17.10.b1: © Colin Cuthbert/Newcastle University/Photo Researchers, Inc.; 17.10.b2: Doug Bartlett/Clear Creek Associates.

Chapter 18

Image Number: 18.00.a2: NASA; 18.01.a1: Photo by Cindy Shaw; 18.01.a2: © Steven P. Lynch; 18.01.a3: Photograph by Belinda Rain, Courtesy of EPA/National Archives; 18.01.mtb1: © David Peevers/Lonely Planet Images; 18.03.a2: © Mark Schneider/Visuals Unlimited; 18.03.a4: © Scientifica/Visuals Unlimited; 18.03.a6, 18.03.a7: © Breck P. Kent/Animals Animals; 18.03.b4: © AP Image; 18.04.a2: Photo by Julia K. Johnson; 18.04.b1: © Science VU/ NOAA/Visuals Unlimited; 18.04.b2: © S. Scribner and J.C. Santamarina, Georgia Institute of Technology; 18.04.d1: © Igor Kostin/Sygma/Corbis; 18.05.b3: © Lowell Georgia/ Corbis; 18.05.d2: © Greg Smith/Corbis; 18.06.c2: Department of Energy; 18.06.c3: © Tim Wright/Corbis; 18.06.d1: © Dewitt Jones/Corbis; 18.06.d2: Courtesy of Department of Energy; 18.06.d3: © Igor Kostin/Sygma/ Corbis; 18.06.mtb1: Department of Energy; 18.07.b3: © Marine Current Turbines TM Ltd.; 18.08.b1: © The McGraw-Hill Companies, Inc./John Flournoy, photographer; 18.08.c1: Photo by Susanne Gillatt; 18.08.c2: © Corbis Royalty Free; 18.08.d1: Photo by K.J. Kolb; 18.08.d2: Photo by Georgi Banchev; 18.08.d3: Jet Propulsion Laboratory/NASA; 18.09.c4: © Indiapicture/ALAMY; 18.13.a2: © Vol. 39/PhotoDisc/Getty Images; 18.13.c1 © The McGraw-Hill Companies, Inc./John A. Karachewski, Photographer; 18.13.d1: © Robert Garvey/Corbis; 18.13.d2: © Andrew Brown/Ecoscen/Corbis; 18.13.d3: © The McGraw-Hill Companies, Inc./John A. Karachewski, photographer; 18.13.d4: © Vol. 31/PhotoDisc/Getty Images; 18.13.mtb1: © Vol. 39/PhotoDisc/Getty Images.

Chapter 19

Image Number: 19.00.a2: NASA and The Hubble Heritage Team (STScI/AURA) Acknowledgement: N. Scoville (Caltech) and T. Rector (NOAO); 19.00.a3: J. Hester, P. Scowen (ASU), HST, NASA; 19.01.a1: NASA, James Bell (Cornell Univ.), Michael Wolff (Space Science Inst.), and the Hubble Heritage Team (STcI/AURA); 19.01.b2: NASA/ JPL; 19.01.c1: P.R. Christensen/NASA/JPL/Arizona State University; 19.01.d1: NASA; 19.01.d2, 19.01.d3: NASA/ JPL; 19.01.mtb1: NASA/JPL-Caltech/UMD; 19.02.a1: JPL/ NASA/University of Arizona; 19.02.a2: NASA; 19.02.a3, 19.02.a4: P.R. Christensen/NASA/JPL/Arizona State University; 19.02.a5: NASA/JPL; 19.02.a6: NASA/JPL/ Malin Space Science Systems; 19.02.b2: NASA/JPL/MSSS; 19.02.d1, 19.02.d2: P.R. Christensen/NASA/JPL/Arizona State University; 19.02.d3: NASA/JPL/MSSS; 19.02.mtb1: NASA; 19.03.a1: NASA/U.S. Geological Survey; 19.03.a2: NASA/JPL/Northwestern University; 19.03.b1: NASA/JPL; 19.03.b2: NASA/U.S. Geological Survey; 19.03.b3, 19.03.b4: NASA/JPL; 19.03.c2: NASA; 19.03.d1: NASA, James Ben (Cornell Univ.), Michael Wolff (Space Science Inst.), and the Hubble Heritage Team (STScI/AURA); 19.03.d2: NASA/JPL/MSSS; 19.03.mtb1: NASA/U.S. Geological Survey; 19.04.a1: U.S. Geological Survey; 19.04.a2: NASA/ U.S. Geological Survey; 19.04.a3, 19.04.a4, 19.04.a5, 19.04.a6, 19.04.b1, 19.04.b2: NASA; 19.04.c1, 19.04.c2, 19.04.c3: Photo by Donald Burt; 19.04.mtb1: NASA/JPS-Caltech; 19.05.a2: NASA; 19.05.a4: John Spencer (Lowell Observatory) and NASA; 19.05.a5: PIRL/University of Arizona; 19.05.a6: NASA/University of Arizona; 19.05.a7: NASA; 19.05.a8: NASA/JPL/R. Pappalardo (University of Colorado); 19.05.a9, 19.05.a10: NASA/JPL; 19.05.a11: NASA/Brown University; 19.05.a12: PIRL/University of Arizona; 19.06.a1: JPL; 19.06.a2: NASA/JPL; 19.06.a3: NASA/University of Arizona/LPL; 19.06.a4: NASA/JPL/ ESA/University of Arizona; 19.06.a4: ESA/NASA/JPL/ University of Arizona; 19.06.a5: CICLOPS/Space Science Institute; 19.06.a6: NASA; 19.06.a7: NASA/Cassini Imaging Team; 19.07.a3: Calvin J. Hamilton/NASA; 19.07.a4, 19.07.b2, 19.07.mtb1: NASA; 19.08.a1: NASA/ JPL/ASU; 19.08.a2: P.R. Christensen/NASA/JPL/Arizona State University; 19.08.a3: NASA/U.S. Geological Survey; 19.08.a5: NASA/JPL/Arizona State University; 19.08.a6: NASA/STScI; 19.08.a7: NASA/JPL-Caltech/Univ. of Arizona; 19.08.b2: NASA/JPL/Cornell; 19.08.b3: NASA/JPL-Caltech/ USGS/Cornell; 19.08b4, 19.08.b5, 19.08.b6: NASA/JPL/ Cornell; 19.09.a2: P.R. Christensen/NASA/JPL/Arizona State University; 19.09.a3: NASA/JPL; 19.09.a4, 19.09.a5, 19.09.a6, 19.09.a7: P.R. Christensen/NASA/JPL/Arizona State University.

Text and Line Art

Chapter 1

01.02.d1, 01.02.d2: Martin Jakobsson/Stockholm Geo Visualisation Laboratory; 01.02.d2: 01.09.a3 p. 20 After U.S. Geological Survey HA-743.

Chapter 2

02.03.d2: After U.S. Geological Survey Fact Sheet 100-03; 02.05.a1, 02.05.b1: Institute for Geophysics, University of Texas; 02.07.d1: S. Gusiakov/NTL, Russia; 02.09.c1: Kentucky Geological Survey; 02.10.b2: National Atlas, U.S. Geological Survey; 02.11.a2: G.N. Meyer and L. Swanson, eds./Minnesota Geological Survey; 02.11.b3: After Luc Ikelle/Texas A&M University; 02.11.b4: D. Setterholm/Minnesota Geological Survey; 02.13.b2: After T. Kenkmann and D. Scherler, *Lunar Planetary Science*, 2002.

Chapter 3

03.02.a1, 03.02.b1, 03.03.a1, 03.03.c1, 03.05.c1, 03.07.b1: Data from Michelle K. Hall-Wallace; 03.07.c1: Jet Propulsion Laboratory/NASA; 03.08.b1: Geologic Survey of Canada/ Digital Data Cornell University.

Chapter 5

05.08.mtb1: Ocean Drilling Project, Site 1201, Texas A&M University; 05.10.mtb1: Data from GEOROC/Max-Planck Society; 05.11.mtb1: P. Modified from King and H. Beikman/ U.S. Geological Survey; 05.13.a1: Modified from U.S. Geological Survey Digital Data Series 11.

Chapter 6

06.04.a1: Inset from Schmincke, 2004, Springer-Verlag; 06.05.a1: After D. Swanson, *American Journal of Science*, 1975; 06.05.c4: After Chris Jenkins, Institute of Arctic & Alpine Research, University of Colorado at Boulder; 06.08.a1: After L. Gurioli, *Geology*, 2005; 06.08.c1: After J. Verhoogen/California University, *Department of Geological Sciences Bulletin;* 06.08.c2: U.S. Geological Survey; 06.10.a2: After F. Goff, *Geo-Heat Center Bulletin*, 2002; 06.11.c1: U.S. Geological Survey Fact Sheet 100-03; 06.11.c2: U.S. Geological Survey Fact Sheet 2005-3024; 06.11c3: U.S. Geological Survey Fact Sheet 100-03; 06.11.c4: U.S. Geological Survey Fact Sheet 100-03; 06.11.mtb1: U.S. Geological Survey Open-File Report 95-59; 06.13.a1: After E. Wolfe/ U.S. Geological Survey, 1996; 06.13.a2: U.S. Geological Survey; 06.13.b3: After A. Daag/ U.S. Geological Survey 1996; 06.13.d4: After C. Wicks/ U.S. Geological Survey; 06.14.a3: U.S. *Geological Survey Bulletin 1292;* 06.14.c2: U.S. Geological Survey Open-File Report 98-428; 06.14.c3: U.S. Geological Survey Open-File Report 98-428; 06.14.mtb1: U.S. Geological Survey Open-File Report 94-585.

Chapter 7

07.15.a1: Colorado Geological Survey; 07.15.c1: Colorado Geological Survey.

Chapter 8

08.00.a2: After P. King, Princeton University Press, 1977; 08.16.a1: P. King and H. Beikman/U.S. Geological Survey, 1974; 08.16.b1: R. Stanley and N. Ratcliffe, *Geological Society of America Bulletin,* 1985; 08.16.c1: Geology Department, Union College; 08.16.c2: After F. Spear, *Journal of Petrology,* 2002.

Chapter 9

09.02.b1: After McGraw-Hill; 09.13.c1: P. King and H. Beikman/U.S. Geological Survey.

Chapter 10

10.00.a1: After G. Hatcher, Monterey Bay Aquarium Research Institute; 10.00.a2: After D. Wagner and others/ California Geological Survey CD 2002-04; 10.01.d1: After M. Fisher and others, U.S. Geological Survey Professional Paper 1687; 10.02.b2: After J. Kious/U.S. Geological Survey, 1996; 10.02.c1: U.S. Geological Survey; 10.02.d1: U.S. Geological Survey; 10.03.b1: After K. Macdonald, Academic Press, 2001; 10.04.a2: After Gabi Laske, University of California, San Diego; 10.04.a3: After David Sandwell/Scripps Institution of Oceanography; 10.04.a4: After R. Dietmar Muller/School of Geosciences University of Sidney; 10.05.c1: Don Anderson/Seismological Laboratory, California Institute of Technology; 10.06.d1: Smithsonian Global Volcanism Program; 10.07.a5-10.07.a6: After P. Gans, *Tectonics,* 1997; 10.08.a2: After N. White, Oxford University Press, 1990; 10.08.a3: After Z. Beydoun, *Episodes,* 1998; 10.09.c1: Chris Jenkins/Institute of Arctic & Alpine Research, University of Colorado at Boulder; 10.10.a1: D.L. Divins/National Geophysical Data Center; 10.10.c3: Gerry Hatcher/Monterey Bay Aquarium Research Institute; 10.10.c4: D.L. Divins/National Geophysical Data Center; 10.11.c2: After T. Affolter and J-P Gratier, Journal of Geophysical Research, 2004; 10.11.d1: After F. Diegel, American Association of Petroleum Geologists, Memoir 65; 10.13.a2-10.13.a7: After J. Pindell and L. Kennan, GCSSCPM Conference, 2001.

Chapter 11

11.03.c1: J. Shaw, *Science,* 1999; 11.04.b1: *U.S. Geological Survey Bulletin 2146-D,* 11.04.mtb1: King and H. Beikman/U.S. Geological Survey; 11.04.mtb2: A. Bally, Geologic Society of America DNAG, 1989; 11.05.mtb1: After B. Sageman, *Chemical Geology,* 2003; 11.08.a2: P. King, Princeton University Press, 1977; 11.09.mtb1: W. Nokleberg and others, U.S. Geological Survey Open-File Report 97-161; 11.10.a1: U.S. Geological Survey Tapestry of Time; 11.10.b1: Geologic Survey of Canada/ Digital Data Cornell University; 11.10.mtb1: After E. Moores and others, Geological Society of America Special Paper 338, 1999; 11.12.a6-11.12.mtb1: After L. Fichter, 1993/James Madison University.

Chapter 12

12.00.a1: Travel time: Kenji Satake/Geological Survey of Japan, Rupture and epicenters: Atul Nayak/Scripps Institution of Oceanography; 12.01.d3: After J. Zucca/ Lawrence Livermore National Laboratory; 12.03.a1: Paula Dunbar/National Oceanographic and Atmospheric Administration; 12.05.a1: U.S. Geological Survey; 12.05.a2: Data from National Earthquake Information Center/U.S. Geological Survey; 12.05.a4: Data from National Earthquake Information Center/U.S. Geological Survey; 12.05.b1: Data from National Earthquake Information Center/U.S. Geological Survey; 12.05.b2: After N. Short, 2006/National Aeronautics and Space Administration; 12.05.c1: After C. Stover, 1993, U.S. Geological Survey Professional Paper 1527; I. Wong/Utah Geological Survey—Public Information Series 76; 12.07.a1: Paula Dunbar/National Oceanographic and Atmospheric Administration; 12.07.a7: Kathleen M. Haller/U.S. Geological Survey; 12.09.b1: National Oceanic and Atmospheric Agency; 12.10.b1: After J. Calzia, 2005, EOS Transactions AGU (86)52; 12.10.b3: After C. Wentworth/U.S. Geological Survey; 12.10.b4: After C. Wentworth/U.S. Geological Survey; 12.10.c3-12.10.c4: Y. Klinger and others, *Bulletin of the Seismological Society of America,* 2003; 12.10.mtb1: After Earthscope/U.S. Geological Survey; 12.11.a1: D. Giardani, Global Seismic Hazard Assessment Program; 12.11.a2: U.S. Geological Survey; 12.11.a3: After R. Wesson and others, 1999, U.S. Geological Survey Map I-2679; 12.11.a4: After F. Klein and others, 2000, U.S. Geological Survey Map I-2724; 12.11.b1: G. Plafker eds., U.S. Geological Survey Circular 1045, 1989; 12.11.b2: U.S. Geological Survey; 12.11.c2: U.S. Geological Survey; 12.11.c3: U.S. Geological Survey; 12.12.a1: Kathleen M. Haller/U.S. Geological Survey; 12.14.c1: Incorporated Research Institutions for Seismology; 12.14.c2: Joint Earth Science Education Initiative—Royal Society of Chemistry; 12.15.a3: Edward Garnero/Arizona State University; 12.15.c1-12.15.c3: Edward Garnero/Arizona State University; 12.16.a1: E. Eckel, U.S. Geological Society Professional Paper 546, 1970; 12.16.c1-12.16.c3: After G. Plafker, U.S. Geological Society Professional Paper 543, 1969.

Chapter 13

13.01.a1: After M. Hackworth/Idaho State University; 13.04.a1: Earth Observatory/National Aeronautics and Space Administration; 13.04.a3: Goddard Space Flight Center/National Aeronautics and Space Administration; 13.04.b1: National Severe Storms Laboratory—National Oceanic and Atmospheric Agency; 13.04.c2: After N. Short, 2006, National Aeronautics and Space Administration; 13.05.a1: After United Nations Atlas of the Oceans; 13.05.b1: International Arctic Science Committee; 13.06.c1-13.06.c2: National Weather Center National Oceanic and Atmospheric Agency; 13.07.b1: National Center for Atmospheric Research; 13.08.b1: Image Science and Analysis Laboratory—NASA-Johnson Space Center; 13.08.d1: After P. Reich—World Soil Resources/United States Department of Agriculture; 13.10.a1-13.10.a4 National Academy of Sciences; 13.10.a6-13.10.a8 National Academy of Sciences; 13.11.a4: Carbon Dioxide Information Analysis Center Oak Ridge National Laboratory; 13.11.a5: National Research Council, 2006; 13.11.a6: National Research Council, 2006; 13.12.c3-13.12.c4: After C. Duncan, 2003, Geology, 31, 75-78; 13.13.b1: After W. Hargrove and R. Luxmoore, 1998 Oak Ridge National Laboratory; 13.14.a1: After National Hurricane Center National Oceanic and Atmospheric Agency, 2005; 13.14.b1-13.14.c1: After W. Gray and P. Klotzbach, 2004 Colorado State University.

Chapter 14

14.08.a1: Coastal Services Center, NOAA; 14.09.mtb1: After S. Dutch/University of Wisconsin, Green Bay; 14.10.c2: University of British Columbia; 14.14.b2: After PACE21 Network, European Science Foundation; 14.14.c1: After J. Feth U.S. Geological Survey Professional Paper 424B, 1961; 14.14.mtb1: After L. Topinka, Cascade Volcano Observatory/U.S. Geological Survey; 14.16.d1: U.S. Geological Survey Fact Sheet 076-00.

Chapter 15

15.09.mtb1: After G. Kiersch, *Civil Engineering,* 1964; 15.11.b1: U.S. Geological Survey Open-File Report 97-0289; 15.11.b2: E. Brabb and others, U.S. Geological Survey Map 2329, 1999; 15.13.b1: U.S. Geological Survey; 15.13.c1-15.13.c2: D. Varnes and W. Savage, *U.S. Geological Survey Bulletin 2130,* 1996; 15.13.d1-15.13.d2: D. Varnes and W. Savage, *U.S. Geological Survey Bulletin 2130,* 1996.

Chapter 16

16.03.b1: After J. Mount, 1995, *California Rivers and Streams: The Conflict Between Fluvial Process and Land Use.* University of California Press; 16.04.a2: C. Bailey/William and Mary College, 1998; 16.10.a1: U.S. Geological Survey Tapestry of Time; 16.10.a2: U.S. Geological Survey; 16.10.a3–a4: Martin Jakobsson/ Stockholm Geo Visualisation Laboratory; 16.13.a1: K. Wahl, U.S. Geological Survey Circular 1120B; 16.13.a2: C. Parett, U.S. Geological Survey Circular 1120-A; 16.13.a5: U.S. Geological Survey; 16.13.b2: R. Maddox and others, *Monthly Weather Review,* 1977; 16.14.b1: Data from National Water Information System U.S. Geological Survey; 16.14.b2: Data from National Water Information System U.S. Geological Survey; 16.15.a11: After D. Topping, 2003, U.S. Geological Survey Professional Paper 1677.

Chapter 17

17.01.a2: After U.S. Geological Survey; 17.02.a4: U.S. Geological Survey; 17.07.b5: D. Bartlett/Clear Creek Associates; 17.11.a1: High Plains Regional Ground Water Study/U.S. Geological Survey; 17.11.a2: Texas Water Development Board; 17.11.b1: U.S. Geological Survey; 17.11.b2: Texas Water Development Board; 17.11.b4: U.S. Geological Survey; 17.11.c2: U.S. Geological Survey.

Chapter 18

18.00.a3: R. Pollastro, *U.S. Geological Survey Bulletin 2202-H;* 18.00.a4: Various sources; 18.00.a5: R. Pollastro, *U.S. Geological Survey Bulletin 2202-H;* 18.02.c5: W. Perry, *U.S. Geological Survey Bulletin 2146-D,* 1997; 18.04.a1: Energy Information Administration, DOE; 18.04.b3: U.S. Geological Survey Fact Sheet 021-01, 2001; 18.04.c2: Energy Information Administration, DOE; 18.04.d2: Energy Information Administration, DOE; 18.05.d1: T. Moore, U.S. Geological Survey Open-File Report 98-34, 1999; 18.10.b2: U.S. Geological Survey Mineral Resources; 18.12.mtb1: National Mining Association; 18.14.a1: P. King and H. Beikman/U.S. Geological Survey; 18.14.a2: Wyoming Geological Survey; 18.14.b1, 18.14.b2, 18.14.b3: Wyoming Geological Survey.

Chapter 19

19.01.a2: NASA and E. Karkoschka/University of Arizona; 19.02.b1: R. Evans/NASA/HST Comet Science Team; 19.03.c1: NASA/Johns Hopkins University; 19.05.a3: NASA; 19.07.a2: Erich Karkoschka/University of Arizona/ NASA; 19.07.c2: A. Feild/NASA/ESA; 19.08.b1: NASA/ JPL; 19.08.mtb1: P.R. Christensen/NASA/JPL/Arizona State University.

Index

D

H

I

J

K

L

M